EARTH DYNAMICS

Deformations and Oscillations of the Rotating Earth

The Earth is a dynamic system. It has a fluid, mobile atmosphere, a continually changing global distribution of ice, snow and water, a fluid core, a thermally convecting mantle and mobile tectonic plates. Internal dynamic processes, together with external gravitational forces of the Sun, Moon and planets, exert torques on the solid Earth or displace its mass, affecting the Earth's shape, rotation and gravitational field.

D. E. Smylie provides a rigorous overview of the dynamical behaviour of the solid Earth, explaining the theory and presenting methods for numerical implementation. Topics include advanced digital analysis, earthquake displacement fields, free core nutations observed by the very long baseline interferometric technique, translational modes of the solid inner core observed by the superconducting gravimeters and dynamics of the fluid outer core.

This book is fully supported by open source computer code, available online for students to explore and test the theory. Also online are a suite of graphics generated from the numerical analysis, which combine with 100 graphical worked examples in the book to make this an ideal tool for researchers and graduate students in the fields of geodesy, seismology and solid Earth geophysics.

DOUGLAS SMYLIE is a Professor of Geophysics at York University, Toronto. He has conducted research on earthquake displacement fields, the rotation of the Earth and the dynamics of the deep interior while lecturing in geophysics at the University of Toronto, the University of Western Ontario, the University of British Columbia and York University in Toronto. In 2002 he was awarded the John Tuzo Wilson Medal by the Canadian Geophysical Union for his achievements. Professor Smylie is a Fellow of the Royal Astronomical Society, a member of the American Geophysical Union and has served as Founder and President of the Canadian Geophysical Union.

EARTH DYNAMICS

Deformations and Oscillations of the Rotating Earth

D. E. SMYLIE

York University, Toronto

CAMBRIDGE
UNIVERSITY PRESS

CAMBRIDGE
UNIVERSITY PRESS

University Printing House, Cambridge CB2 8BS, United Kingdom

One Liberty Plaza, 20th Floor, New York, NY 10006, USA

477 Williamstown Road, Port Melbourne, VIC 3207, Australia

4843/24, 2nd Floor, Ansari Road, Daryaganj, Delhi - 110002, India

79 Anson Road, #06-04/06, Singapore 079906

Cambridge University Press is part of the University of Cambridge.

It furthers the University's mission by disseminating knowledge in the pursuit of education, learning and research at the highest international levels of excellence.

www.cambridge.org
Information on this title: www.cambridge.org/9781108445825

First published 2013
First paperback edition 2017

A catalogue record for this publication is available from the British Library

Library of Congress Cataloging in Publication data
Smylie, D. E.
Earth dynamics : deformations and oscillations of the rotating Earth / D.E. Smylie, York University, Toronto.
pages cm
ISBN 978-0-521-87503-5 (Hardback)
1. Geodynamics. 2. Earth–Rotation. I. Title.
QE517.5.S678 2013
551.1–dc23

2012040769

ISBN 978-0-521-87503-5 Hardback
ISBN 978-1-108-44582-5 Paperback

Dedication

This book is dedicated to my wife Susan, my children Diane, Janet, Hugh and Andrea and my grandchildren Grace, Gavin, Ben, Ella, Jay, Quinn, Andreas, Alan, Lance, Bob and Ashley.

Contents

Preface and acknowledgments

The study of Earth's dynamics, from near surface earthquake displacement fields to the translational modes of the solid inner core, has long been a fascination for me. In the present work I have been influenced by *Numerical Recipes*, published by Cambridge University Press, to include computer code so often omitted in scientific publications. I have gone one step further to include, on the website www.cambridge.org/smylie, open source downloadable software through the Oracle Virtual Machine, allowing a full Fedora Linux operating system to be installed on users' machines along with the TRIUMF graphics system, giving full access to Fortran, LaTeX and TeX as well as codes from the book itself and possible updates.

Throughout the writing of this book I have been very ably assisted by Dr Gary Henderson in every aspect. While I take full responsibility for any remaining errors, his high skills in English, theory and Fortran have been very much appreciated.

I have been fortunate to have benefited from many teachers and professors in my studies. Reg Daniels kindled my interest in mathematics in high school, Fraser Grant introduced me to mathematical geophysics, while Tuzo Wilson hosted the geophysics laboratory at 49 St. George St. in the University of Toronto.

I have also been fortunate to have worked with many colleagues and students. My doctoral thesis supervisor, Michael Rochester, continued in research with me for many years. Dr Xianhua Jiang through his variational calculations discovered the prograde free core nutation and his thesis won the Canadian Association of Graduate Schools Doctoral Dissertation Prize for all fields in Canada in 1994 and the 1995 Annual Dissertation Award of the Northeastern U.S. Association of Graduate Schools. While he was unable to detect the prograde free core nutation definitively in the spectra of the VLBI nutation record at the time, this task was left to Dr Andrew Palmer. Dr Palmer, using singular value decomposition, found both the prograde and retrograde nutations in the unequally spaced VLBI observations and from the ring down of the two modes was able to measure the viscosity at the top of the fluid outer core. Another brilliant doctoral student, Dr Hong Ma, won the Governor General's Gold Medal as the best graduating student at York University in all fields in 1996. Other brilliant doctoral students I should mention are Dr Ian Johnson and Dr James B. Merriam.

Finally, I thank my wife Susan and all my family for their support in all my endeavours.

Doug Smylie
June 25, 2012

The book website www.cambridge.org/smylie

Readers can access the website www.cambridge.org/smylie to find open source downloadable software through the Oracle Virtual Machine, allowing a full Fedora Linux operating system to be installed on their own machines along with the TRIUMF graphics system, giving full use of Fortran, LaTeX and TeX. Also included are Fortran source codes from the book, data input files such as Earth models, sample output files and results, as well as instructions for their use, including installation and compilation directions. The website will also include addenda for the book. Those interested in an e-book form of the book can find a link for purchase on the website www.cambridge.org/smylie.

1
Introduction and theoretical background

The Earth's dynamical behaviour is a complex and fascinating subject with many practical ramifications. Its description requires the language of mathematics and computation. In this book, we attempt to make the theoretical foundations of the description of Earth's dynamics as complete as possible, and we accompany the theoretical descriptions with computer code and graphics for the implementation of the theory.

1.1 Scalar, vector and tensor analysis

We will make extensive use of scalars, vectors and tensors throughout the book. In this section, we will summarise the properties most often used. It is assumed that the reader is familiar with the elementary results of vector analysis summarised in Appendix A.

1.1.1 Scalars

Physical quantities determined by a single number such as mass, temperature and energy are *scalars*. Scalars are invariants under a change of co-ordinates, they remain the same in all co-ordinate systems. They are sometimes simply referred to as *invariants*. A scalar field is a function of space and time.

1.1.2 Vectors

Vectors require both magnitude and direction for their specification. They may be described by their components, their projections on the co-ordinate axes. An arbitrary vector then associates a scalar with each direction in space through an expression that is linear and homogeneous in the direction cosines.

In general, a vector can be defined in a space of arbitrary dimensions numbering two or greater. Our applications will be confined to a space of three dimensions and we will adopt this limitation. Let $\boldsymbol{r}$ be the radius vector to a point P specified by the three curvilinear co-ordinates u^1, u^2, u^3. Then $\boldsymbol{r}$ is described by the function

$$\boldsymbol{r} = \boldsymbol{r}(u^1, u^2, u^3). \tag{1.1}$$

The change in $\boldsymbol{r}$ due to differential displacements along the co-ordinate curves is then

$$d\boldsymbol{r} = \frac{\partial \boldsymbol{r}}{\partial u^1} du^1 + \frac{\partial \boldsymbol{r}}{\partial u^2} du^2 + \frac{\partial \boldsymbol{r}}{\partial u^3} du^3. \tag{1.2}$$

If two of the curvilinear co-ordinates are held fixed, the third describes a curve in space. Moving along u^j a unit distance from the point P, the change in $\boldsymbol{r}$ is equal to $\boldsymbol{b}_j = \partial \boldsymbol{r} / \partial u^j$. The *unitary vectors*

$$\boldsymbol{b}_1 = \frac{\partial \boldsymbol{r}}{\partial u^1}, \quad \boldsymbol{b}_2 = \frac{\partial \boldsymbol{r}}{\partial u^2}, \quad \boldsymbol{b}_3 = \frac{\partial \boldsymbol{r}}{\partial u^3}, \tag{1.3}$$

associated with the point P, form a base system for all vectors there. Any vector at that point can be expressed as a linear, homogeneous combination of the unitary base vectors. In particular,

$$d\boldsymbol{r} = \boldsymbol{b}_1 du^1 + \boldsymbol{b}_2 du^2 + \boldsymbol{b}_3 du^3. \tag{1.4}$$

The vector space, thus defined, may have different units measured along the co-ordinate directions. For instance, in thermodynamics the co-ordinates may represent pressure, volume and temperature, all in different units. This is an example of *affine geometry*. More commonly, we will be concerned with vectors in *metric geometry*, where the unitary vectors can be referred to a common unit of length. This allows measurement of the absolute value of a vector of arbitrary orientation and the distance between neighbouring points.

The three unitary base vectors $\boldsymbol{b}_1, \boldsymbol{b}_2, \boldsymbol{b}_3$ define a parallelepiped with volume

$$V = \boldsymbol{b}_1 \cdot (\boldsymbol{b}_2 \times \boldsymbol{b}_3) = \boldsymbol{b}_2 \cdot (\boldsymbol{b}_3 \times \boldsymbol{b}_1) = \boldsymbol{b}_3 \cdot (\boldsymbol{b}_1 \times \boldsymbol{b}_2), \tag{1.5}$$

using the properties (A.1) of the triple scalar product. A new triplet of base vectors, defined as

$$\boldsymbol{b}^1 = \frac{1}{V}(\boldsymbol{b}_2 \times \boldsymbol{b}_3), \quad \boldsymbol{b}^2 = \frac{1}{V}(\boldsymbol{b}_3 \times \boldsymbol{b}_1), \quad \boldsymbol{b}^3 = \frac{1}{V}(\boldsymbol{b}_1 \times \boldsymbol{b}_2), \tag{1.6}$$

are, in turn, orthogonal to the planes formed by the pairs $(\boldsymbol{b}_2 \times \boldsymbol{b}_3)$, $(\boldsymbol{b}_3 \times \boldsymbol{b}_1)$ and $(\boldsymbol{b}_1 \times \boldsymbol{b}_2)$. Adopting the *range convention*, whereby superscripts and subscripts are implied to range over the values 1, 2, 3, the two triplets of base vectors are found to obey

$$\boldsymbol{b}^i \cdot \boldsymbol{b}_j = \delta^i_j, \tag{1.7}$$

where δ^i_j is the *Kronecker delta*, which is equal to unity for $i = j$, zero otherwise. The original three unitary base vectors can be recovered from the new triplet of base vectors, using the properties (A.4) of the quadruple vector product, giving

$$\boldsymbol{b}_1 = V\left(\boldsymbol{b}^2 \times \boldsymbol{b}^3\right), \quad \boldsymbol{b}_2 = V\left(\boldsymbol{b}^3 \times \boldsymbol{b}^1\right), \quad \boldsymbol{b}_3 = V\left(\boldsymbol{b}^1 \times \boldsymbol{b}^2\right). \tag{1.8}$$

The triplet, $\boldsymbol{b}^1$, $\boldsymbol{b}^2$ and $\boldsymbol{b}^3$, are called *reciprocal unitary vectors*. They may be used as a base system of the vector space as an alternative to the original three unitary base vectors. In this base system, the differential $d\boldsymbol{r}$ becomes expressible as

$$d\boldsymbol{r} = \boldsymbol{b}^1 du_1 + \boldsymbol{b}^2 du_2 + \boldsymbol{b}^3 du_3. \tag{1.9}$$

Adopting the *summation convention*, whereby a repeated superscript or subscript implies summation over that index, equating the expressions (1.4) and (1.9) for the differential $d\boldsymbol{r}$ gives

$$d\boldsymbol{r} = \boldsymbol{b}_i\, du^i = \boldsymbol{b}^j du_j. \tag{1.10}$$

Taking the scalar product of this equation, first with $\boldsymbol{b}^k$, then with $\boldsymbol{b}_k$, and using the orthogonality relation (1.7), produces

$$du^k = \boldsymbol{b}^k \cdot \boldsymbol{b}^j du_j, \quad du_k = \boldsymbol{b}_k \cdot \boldsymbol{b}_i du^i. \tag{1.11}$$

Replacing the superscript k by i in the first relation, and replacing the subscript k by j in the second, relates the components of $d\boldsymbol{r}$ in the unitary and reciprocal unitary base systems by

$$du^i = g^{ij} du_j, \quad du_j = g_{ji} du^i, \tag{1.12}$$

with

$$g^{ij} = \boldsymbol{b}^i \cdot \boldsymbol{b}^j = g^{ji}, \quad g_{ji} = \boldsymbol{b}_j \cdot \boldsymbol{b}_i = g_{ij}. \tag{1.13}$$

An arbitrary vector $\boldsymbol{V}$ may be expressed as a linear combination of its components in the unitary base system $\boldsymbol{b}_1$, $\boldsymbol{b}_2$, $\boldsymbol{b}_3$, or in the reciprocal unitary base system $\boldsymbol{b}^1$, $\boldsymbol{b}^2$, $\boldsymbol{b}^3$, as

$$\boldsymbol{V} = v^i \boldsymbol{b}_i = v_j \boldsymbol{b}^j, \tag{1.14}$$

where, by (1.7), the components in each system are

$$v^i = \boldsymbol{V} \cdot \boldsymbol{b}^i, \quad v_j = \boldsymbol{V} \cdot \boldsymbol{b}_j. \tag{1.15}$$

Again, with scalar multiplication and using the orthogonality relation (1.7), the components in the unitary base system and in the reciprocal unitary base system are found to be related by

$$v^i = g^{ij} v_j, \quad v_j = g_{ji} v^i. \tag{1.16}$$

Replacing v^i and v_j in (1.14) by these expressions gives

$$\boldsymbol{V} = g^{ij} v_j \boldsymbol{b}_i = \boldsymbol{b}^i \cdot \boldsymbol{b}^j v_j \boldsymbol{b}_i = (\boldsymbol{b}^i \cdot \boldsymbol{V})\, \boldsymbol{b}_i \tag{1.17}$$

and

$$\boldsymbol{V} = g_{ji} v^i \boldsymbol{b}^j = \boldsymbol{b}_j \cdot \boldsymbol{b}_i v^i \boldsymbol{b}^j = (\boldsymbol{b}_j \cdot \boldsymbol{V})\, \boldsymbol{b}^j. \tag{1.18}$$

The components v^i are called the *contravariant components* of the vector $\boldsymbol{V}$ while the components v_j are called the *covariant components* of $\boldsymbol{V}$. By convention, contravariant components are indicated by a superscripted index and covariant components are indicated by a subscripted index.

Physical components of $\boldsymbol{V}$ can be defined by resolving the vector along a system of vectors of *unit* length. These may be defined as parallel to the unitary base system, with each reduced to unit length, by

$$\hat{e}_1 = \frac{\boldsymbol{b}_1}{\sqrt{\boldsymbol{b}_1 \cdot \boldsymbol{b}_1}} = \frac{1}{\sqrt{g_{11}}} \boldsymbol{b}_1, \quad \hat{e}_2 = \frac{1}{\sqrt{g_{22}}} \boldsymbol{b}_2, \quad \hat{e}_3 = \frac{1}{\sqrt{g_{33}}} \boldsymbol{b}_3, \tag{1.19}$$

thus,

$$\boldsymbol{V} = V_1 \hat{e}_1 + V_2 \hat{e}_2 + V_3 \hat{e}_3, \tag{1.20}$$

with physical components

$$V_1 = \sqrt{g_{11}} v^1, \quad V_2 = \sqrt{g_{22}} v^2, \quad V_3 = \sqrt{g_{33}} v^3. \tag{1.21}$$

Hence, the physical components, V_i, are of the same dimensions as the vector $\boldsymbol{V}$ itself.

We can now give dimensions to the vector space. The differential vector $d\boldsymbol{r}$ represents a displacement from the point P with co-ordinates (u^1, u^2, u^3) to the point with co-ordinates $(u^1 + du^1, u^2 + du^2, u^3 + du^3)$. Denoting the magnitude of this displacement by ds,

$$ds^2 = d\boldsymbol{r} \cdot d\boldsymbol{r} = \boldsymbol{b}_i \cdot \boldsymbol{b}_j \, du^i \, du^j = \boldsymbol{b}^i \cdot \boldsymbol{b}^j \, du_i \, du_j, \tag{1.22}$$

or

$$ds^2 = g_{ij} \, du^i \, du^j = g^{ij} \, du_i \, du_j. \tag{1.23}$$

The coefficients g_{ij} and g^{ij} appear in bilinear forms expressing the square of the incremental displacement in terms of the increments in the co-ordinates u^i, or in terms of the increments in the reciprocal co-ordinates u_i. They are called *metrical coefficients*.

1.1.3 Vectors and co-ordinate transformations

Consider how the components of the vector $\boldsymbol{V}$ transform under a transformation of co-ordinates. Suppose we adopt a new set of curvilinear co-ordinates (u'^1, u'^2, u'^3) in place of (u^1, u^2, u^3). The unitary base vectors $\boldsymbol{b}_i$ are expressible as

$$\boldsymbol{b}_i = \frac{\partial \boldsymbol{r}}{\partial u^i} = \frac{\partial \boldsymbol{r}}{\partial u'^j}\frac{\partial u'^j}{\partial u^i} = \boldsymbol{b}'_j\frac{\partial u'^j}{\partial u^i}, \tag{1.24}$$

using the chain rule for partial derivatives. The vectors

$$\boldsymbol{b}'_j = \frac{\partial \boldsymbol{r}}{\partial u'^j} \tag{1.25}$$

form a new triplet of unitary base vectors in the new co-ordinate system. The vector $\boldsymbol{V}$ may be expressed by its contravariant components in either co-ordinate system as

$$\boldsymbol{V} = v^i\boldsymbol{b}_i = v'^j\boldsymbol{b}'_j. \tag{1.26}$$

Substituting for $\boldsymbol{b}_i$ from (1.24), we find that

$$\boldsymbol{V} = v^i\boldsymbol{b}_i = \frac{\partial u'^j}{\partial u^i}v^i\boldsymbol{b}'_j = v'^j\boldsymbol{b}'_j. \tag{1.27}$$

Thus, the transformation law for the contravariant components of the vector $\boldsymbol{V}$ is

$$v'^j = \frac{\partial u'^j}{\partial u^i}v^i \tag{1.28}$$

The new covariant components of the vector $\boldsymbol{V}$ may be found from its new contravariant components, using the second of relations (1.16) and the metrical coefficient g'_{kj} in the new co-ordinate system, where

$$g'_{kj} = \boldsymbol{b}'_k \cdot \boldsymbol{b}'_j, \tag{1.29}$$

with $\boldsymbol{b}'_i$ being a unitary base vector in the new co-ordinate system. Using the chain rule for partial derivatives, the base vectors in the new co-ordinate system can be related to those in the original co-ordinate system by

$$\boldsymbol{b}'_k = \frac{\partial \boldsymbol{r}}{\partial u'^k} = \frac{\partial \boldsymbol{r}}{\partial u^l}\frac{\partial u^l}{\partial u'^k} = \boldsymbol{b}_l\frac{\partial u^l}{\partial u'^k}, \tag{1.30}$$

$$\boldsymbol{b}'_j = \frac{\partial \boldsymbol{r}}{\partial u'^j} = \frac{\partial \boldsymbol{r}}{\partial u^m}\frac{\partial u^m}{\partial u'^j} = \boldsymbol{b}_m\frac{\partial u^m}{\partial u'^j}. \tag{1.31}$$

Then,

$$g'_{kj} = \boldsymbol{b}'_k \cdot \boldsymbol{b}'_j = \boldsymbol{b}_l \cdot \boldsymbol{b}_m\frac{\partial u^l}{\partial u'^k}\frac{\partial u^m}{\partial u'^j} = \frac{\partial u^l}{\partial u'^k}\frac{\partial u^m}{\partial u'^j}g_{lm}. \tag{1.32}$$

Multiplying (1.28) through by g'_{kj} and summing over j, the new covariant components of $\boldsymbol{V}$ are found to be given by

$$v'_k = \frac{\partial u^l}{\partial u'^k}\frac{\partial u^m}{\partial u'^j}\frac{\partial u'^j}{\partial u^i} g_{lm} v^i, \tag{1.33}$$

where

$$\frac{\partial u^m}{\partial u'^j}\frac{\partial u'^j}{\partial u^i} = \frac{\partial u^m}{\partial u^i} = \delta^m_i. \tag{1.34}$$

Hence, (1.33) reduces to

$$v'_k = \frac{\partial u^l}{\partial u'^k}\delta^m_i g_{lm} v^i = \frac{\partial u^l}{\partial u'^k} v_l. \tag{1.35}$$

Thus, the transformation law for the covariant components of the vector $\boldsymbol{V}$ is

$$v'_j = \frac{\partial u^i}{\partial u'^j} v_i. \tag{1.36}$$

1.1.4 Tensors

In the analysis of the state of stress in a solid, it was realised that physical quantities more complicated than vectors were required for the description of the state of stress. Considering an imaginary surface within the stressed solid, with orientation described by its outward normal vector, the force on the surface is different for each orientation of the surface. Thus, the stress associates a force vector with each spatial direction. This has led to the definition of a *second-order tensor* as a physical quantity that associates a vector (*a first-order tensor*) with each spatial direction, and the generalisation that a tensor of order n associates a tensor of order $n-1$ with each spatial direction.

The transformation laws, for contravariant vectors (1.28) and for covariant vectors (1.36), are easily generalised to those for tensors of second and higher order. In fact, we have already met a *second-order, twice covariant tensor*, the metrical coefficient g_{ij}, whose transformation law is given by (1.32) as

$$g'_{kl} = \frac{\partial u^i}{\partial u'^k}\frac{\partial u^j}{\partial u'^l} g_{ij}. \tag{1.37}$$

Replacing the index j by k in both of the relations (1.12), and i by j in the second, gives

$$du^i = g^{ik} du_k, \quad du_k = g_{kj} du^j. \tag{1.38}$$

Then,

$$du^i = g^{ik} du_k = g^{ik} g_{kj} du^j. \tag{1.39}$$

Comparing the left and right sides of this equation, we find that

$$g^{ik} g_{kj} = \delta^i_{\ j}. \tag{1.40}$$

Thus, the 3×3 matrix of components of g^{ik} multiplied by the 3×3 matrix of components of g_{kj} yields the unit matrix. Hence, the matrix representing g^{ik} is the inverse of the matrix representing g_{kj}. Since the inverse of a matrix is equal to the matrix of cofactors divided by the determinant, we have

$$g^{ik} = \frac{\Delta^{ik}}{g}, \tag{1.41}$$

where Δ^{ik} is the cofactor of the element g_{ik} of the 3×3 matrix of components of g_{kj} and g is the determinant given by

$$g = \begin{vmatrix} g_{11} & g_{12} & g_{13} \\ g_{21} & g_{22} & g_{23} \\ g_{31} & g_{32} & g_{33} \end{vmatrix}. \tag{1.42}$$

Symmetry of g^{ik} follows directly from the symmetry of g_{ik}. Now, suppose $\bar{g}^{ik}$ is a *second-order, twice contravariant tensor* with components equal to those of g^{ik} in one particular co-ordinate system. Then, by (1.40), in this co-ordinate system,

$$\bar{g}^{ik} g_{kj} = \delta^i_{\ j}. \tag{1.43}$$

This is now a tensor equation valid in all co-ordinate systems. In another system of co-ordinates it becomes

$$\bar{g}'^{ik} g'_{kj} = \delta^i_{\ j}. \tag{1.44}$$

Given g'_{kj}, we can take the inverse of the 3×3 matrix of its components as g'^{ik}, obeying

$$g'^{ik} g'_{kj} = \delta^i_{\ j}. \tag{1.45}$$

Comparing (1.44) and (1.45) it is found that

$$g'^{ik} = \bar{g}'^{ik} \tag{1.46}$$

in the new co-ordinate system as well, and thus they are identical in all co-ordinate systems. The metrical coefficient of the reciprocal unitary base system is then a *second-order, twice contravariant tensor* obeying the transformation law

$$g'^{kl} = \frac{\partial u'^k}{\partial u^i} \frac{\partial u'^l}{\partial u^j} g^{ij}. \tag{1.47}$$

The transformation laws (1.28), for first-order contravariant tensor components, and (1.36), for first-order covariant tensor components, apply to physical quantities more commonly known as vectors. The transformation laws (1.37) and (1.47) for the *metric tensors* are representative of those defining second-order, twice covariant and twice contravariant tensors. These transformation laws are easily generalised to define tensors of arbitrary order and even tensors of mixed contravariance and covariance.

When two tensors are multiplied and summed over one or more indices, such as in (1.43), the result is called the *contracted product*. The contracted product can be used to generate a *test for tensor character*. Suppose we have a quantity $t(i, j, k)$ with three indices i, j, k. Further suppose we have two contravariant vectors, $\boldsymbol{u}$ and $\boldsymbol{w}$, and a covariant vector, $\boldsymbol{v}$, and we form the triple contracted product

$$t(i, j, k)\, u^i v_j w^k = t^j_{ik}, \tag{1.48}$$

and find that it is invariant in all co-ordinate systems. Then t^j_{ik} is a third-order tensor, once contravariant in the index j, twice covariant in the indices i and k.

Such tests for tensor character can take many forms. For example, when we introduced the Kronecker delta in relation (1.7), we wrote it as though it were a second-order mixed tensor. If this were so, it would transform as

$$\delta'^{\,k}_{l} = \frac{\partial u'^k}{\partial u^i}\frac{\partial u^j}{\partial u'^l}\delta^i_j, \tag{1.49}$$

where $\delta'^{\,k}_{l}$ is the Kronecker delta in the new co-ordinate system. Taking the implied summation over j, the right side reduces to

$$\frac{\partial u'^k}{\partial u^i}\frac{\partial u^i}{\partial u'^l} = \frac{\partial u'^k}{\partial u'^l} = \delta'^{\,k}_{l}, \tag{1.50}$$

as required by the assumed transformation law (1.49). This identifies the Kronecker delta as a second-order mixed tensor, as assumed by our adopted notation.

1.1.5 Metric tensors and elements of arc, surface and volume

An infinitesimal displacement $d\boldsymbol{s}_1$, along the u^1 direction from a point P with co-ordinates (u^1, u^2, u^3), from (1.4), is

$$d\boldsymbol{s}_1 = \boldsymbol{b}_1\, du^1. \tag{1.51}$$

The magnitude of this infinitesimal displacement is

$$ds_1 = |\boldsymbol{b}_1|\, du^1 = \sqrt{\boldsymbol{b}_1 \cdot \boldsymbol{b}_1}\, du^1 = \sqrt{g_{11}}\, du^1, \tag{1.52}$$

using the second of relations (1.13). Similarly,

$$d\boldsymbol{s}_2 = \boldsymbol{b}_2\, du^2, \quad d\boldsymbol{s}_3 = \boldsymbol{b}_3\, du^3, \tag{1.53}$$

and

$$ds_2 = \sqrt{g_{22}}\, du^2, \quad ds_3 = \sqrt{g_{33}}\, du^3, \tag{1.54}$$

represent infinitesimal displacements in the u^2 and u^3 directions, respectively.

Next, we consider an element of surface area da_1, on the surface defined by u^1 constant, contained by the parallelogram formed by the infinitesimal displacements ds_2, ds_3 along the directions u^2, u^3. The area of the parallelogram is

$$da_1 = |ds_2 \times ds_3| = |b_2 \times b_3|\, du^2 du^3, \tag{1.55}$$

with

$$|b_2 \times b_3| = \sqrt{(b_2 \times b_3) \cdot (b_2 \times b_3)}. \tag{1.56}$$

Using the vector identity (A.3),

$$(b_2 \times b_3) \cdot (b_2 \times b_3) = (b_2 \cdot b_2)(b_3 \cdot b_3) - (b_2 \cdot b_3)(b_3 \cdot b_2), \tag{1.57}$$

we find that

$$da_1 = \sqrt{g_{22}\, g_{33} - g_{23}^2}\, du^2 du^3. \tag{1.58}$$

Similarly, for elements on the u^2 and u^3 surfaces their areas are

$$da_2 = \sqrt{g_{33}\, g_{11} - g_{31}^2}\, du^3 du^1 \tag{1.59}$$

and

$$da_3 = \sqrt{g_{11}\, g_{22} - g_{12}^2}\, du^1 du^2. \tag{1.60}$$

An element of volume, bounded by the three co-ordinate surfaces, is given by

$$dv = ds_1 \cdot (ds_2 \times ds_3) = b_1 \cdot (b_2 \times b_3)\, du^1 du^2 du^3. \tag{1.61}$$

If we set $V = b_2 \times b_3$ in the expression (1.17), we get

$$b_2 \times b_3 = \left[b^i \cdot (b_2 \times b_3) \right] b_i. \tag{1.62}$$

Replacing b^i by its expressions (1.6) and taking the scalar product with b_1, we have

$$b_1 \cdot (b_2 \times b_3) = \frac{b_1}{b_1 \cdot (b_2 \times b_3)} \cdot \Big[(b_2 \times b_3) \cdot (b_2 \times b_3)\, b_1 + (b_3 \times b_1) \cdot (b_2 \times b_3)\, b_2 + (b_1 \times b_2) \cdot (b_2 \times b_3)\, b_3 \Big]. \tag{1.63}$$

Permuting subscripts in the first factor on the left of the vector identity (1.57), two additional vector identities emerge,

$$(\mathbf{b}_3 \times \mathbf{b}_1) \cdot (\mathbf{b}_2 \times \mathbf{b}_3) = (\mathbf{b}_3 \cdot \mathbf{b}_2)(\mathbf{b}_1 \cdot \mathbf{b}_3) - (\mathbf{b}_3 \cdot \mathbf{b}_3)(\mathbf{b}_1 \cdot \mathbf{b}_2) \tag{1.64}$$

and

$$(\mathbf{b}_1 \times \mathbf{b}_2) \cdot (\mathbf{b}_2 \times \mathbf{b}_3) = (\mathbf{b}_1 \cdot \mathbf{b}_2)(\mathbf{b}_2 \cdot \mathbf{b}_3) - (\mathbf{b}_1 \cdot \mathbf{b}_3)(\mathbf{b}_2 \cdot \mathbf{b}_2). \tag{1.65}$$

Using the three vector identities, expression (1.63) can be reduced to

$$\begin{aligned}[\mathbf{b}_1 \cdot (\mathbf{b}_2 \times \mathbf{b}_3)]^2 &= (\mathbf{b}_1 \cdot \mathbf{b}_1)[(\mathbf{b}_2 \cdot \mathbf{b}_2)(\mathbf{b}_3 \cdot \mathbf{b}_3) - (\mathbf{b}_2 \cdot \mathbf{b}_3)(\mathbf{b}_3 \cdot \mathbf{b}_2)] \\ &\quad + (\mathbf{b}_1 \cdot \mathbf{b}_2)[(\mathbf{b}_2 \cdot \mathbf{b}_3)(\mathbf{b}_3 \cdot \mathbf{b}_1) - (\mathbf{b}_2 \cdot \mathbf{b}_1)(\mathbf{b}_3 \cdot \mathbf{b}_3)] \\ &\quad + (\mathbf{b}_1 \cdot \mathbf{b}_3)[(\mathbf{b}_2 \cdot \mathbf{b}_1)(\mathbf{b}_3 \cdot \mathbf{b}_2) - (\mathbf{b}_2 \cdot \mathbf{b}_2)(\mathbf{b}_3 \cdot \mathbf{b}_1)] \\ &= \begin{vmatrix} \mathbf{b}_1 \cdot \mathbf{b}_1 & \mathbf{b}_1 \cdot \mathbf{b}_2 & \mathbf{b}_1 \cdot \mathbf{b}_3 \\ \mathbf{b}_2 \cdot \mathbf{b}_1 & \mathbf{b}_2 \cdot \mathbf{b}_2 & \mathbf{b}_2 \cdot \mathbf{b}_3 \\ \mathbf{b}_3 \cdot \mathbf{b}_1 & \mathbf{b}_3 \cdot \mathbf{b}_2 & \mathbf{b}_3 \cdot \mathbf{b}_3 \end{vmatrix}. \end{aligned} \tag{1.66}$$

Replacing the scalar products with components of the covariant metric tensor defined in the second of relations (1.13), the determinant in expression (1.66) is identical to that defined by (1.42). Hence,

$$[\mathbf{b}_1 \cdot (\mathbf{b}_2 \times \mathbf{b}_3)]^2 = V^2 = g. \tag{1.67}$$

Then, the expression (1.61) for the volume element becomes

$$dv = \sqrt{g}\, du^1 du^2 du^3. \tag{1.68}$$

1.1.6 The cross product and differential operators

The three unitary base vectors define a parallelepiped with volume V given by any of the three scalar triple products as expressed in (1.5). On taking the square root of (1.67) this volume becomes

$$V = \sqrt{g}. \tag{1.69}$$

Suppose that, in addition to the arbitrary vector $\boldsymbol{V}$, expressed by (1.14) as a linear combination of its contravariant components and the unitary base vectors, we have a second arbitrary vector $\boldsymbol{W}$ similarly expressed. Then, the cross product of the two vectors is

$$\begin{aligned} \boldsymbol{V} \times \boldsymbol{W} &= \left(v^1 \boldsymbol{b}_1 + v^2 \boldsymbol{b}_2 + v^3 \boldsymbol{b}_3\right) \times \left(w^1 \boldsymbol{b}_1 + w^2 \boldsymbol{b}_2 + w^3 \boldsymbol{b}_3\right) \\ &= \left(v^2 w^3 - v^3 w^2\right)(\boldsymbol{b}_2 \times \boldsymbol{b}_3) + \left(v^3 w^1 - v^1 w^3\right)(\boldsymbol{b}_3 \times \boldsymbol{b}_1) \\ &\quad + \left(v^1 w^2 - v^2 w^1\right)(\boldsymbol{b}_1 \times \boldsymbol{b}_2) \\ &= V\left[\left(v^2 w^3 - v^3 w^2\right)\boldsymbol{b}^1 + \left(v^3 w^1 - v^1 w^3\right)\boldsymbol{b}^2 + \left(v^1 w^2 - v^2 w^1\right)\boldsymbol{b}^3\right] \\ &= \sqrt{g}\begin{vmatrix} \boldsymbol{b}^1 & \boldsymbol{b}^2 & \boldsymbol{b}^3 \\ v^1 & v^2 & v^3 \\ w^1 & w^2 & w^3 \end{vmatrix}, \end{aligned} \tag{1.70}$$

making use of the definitions (1.6) of the reciprocal base vectors.

The *curl* of an arbitrary vector $\boldsymbol{V}$ is written as the cross product of the vector differential operator ∇ and $\boldsymbol{V}$. The integral theorem of Stokes (A.25) gives,

$$\iint_S (\nabla \times \boldsymbol{V}) \cdot \hat{\boldsymbol{\nu}} dS = \int_C \boldsymbol{V} \cdot d\boldsymbol{s}, \tag{1.71}$$

where the surface S is bounded by a closed curve C, with $d\boldsymbol{s}$ an infinitesimal tangential vector displacement along the bounding curve, in a direction such that the surface is on the left. $\hat{\boldsymbol{\nu}}$ is the unit outward normal to S. If the bounding curve is allowed to shrink to a point, in the limit the component of the curl in the direction of $\hat{\boldsymbol{\nu}}$ is given by Stokes' theorem as

$$(\nabla \times \boldsymbol{V}) \cdot \hat{\boldsymbol{\nu}} = \lim_{C \to 0} \frac{1}{S} \int_C \boldsymbol{V} \cdot d\boldsymbol{s}, \tag{1.72}$$

with S the infinitesimal enclosed area. If we take the bounding curve to be a parallelogram in the u^1 surface formed by sides $\boldsymbol{b}_2\, du^2$ and $\boldsymbol{b}_3\, du^3$ in the u^2 and u^3 directions, the sides parallel to the u^3 direction contribute

$$\left(\boldsymbol{V} \cdot \boldsymbol{b}_3\, du^3\right)_{u^2+du^2} - \left(\boldsymbol{V} \cdot \boldsymbol{b}_3\, du^3\right)_{u^2}, \tag{1.73}$$

while the sides parallel to the u^2 direction contribute

$$-\left(\boldsymbol{V} \cdot \boldsymbol{b}_2\, du^2\right)_{u^3+du^3} + \left(\boldsymbol{V} \cdot \boldsymbol{b}_2\, du^2\right)_{u^3}, \tag{1.74}$$

to the integral on the right side of (1.72). In the limit as the bounding curve shrinks to a point, the first contribution is replaced by the partial derivative with respect to u^2, times du^2, times du^3, and the second is replaced by the negative of the partial derivative with respect to u^3, times du^3, times du^2. Thus, the integral on the right side of (1.72) becomes

$$\left[\frac{\partial}{\partial u^2}(\boldsymbol{V} \cdot \boldsymbol{b}_3) - \frac{\partial}{\partial u^3}(\boldsymbol{V} \cdot \boldsymbol{b}_2)\right] du^2 du^3. \tag{1.75}$$

This is to be divided by the infinitesimal enclosed area, S, of the parallelogram,

$$S = \sqrt{(\boldsymbol{b}_2 \times \boldsymbol{b}_3) \cdot (\boldsymbol{b}_2 \times \boldsymbol{b}_3)}\, du^2 du^3, \tag{1.76}$$

to give the right side of (1.72). The unit outward normal vector $\hat{\boldsymbol{\nu}}$ is in the direction of the reciprocal base vector $\boldsymbol{b}^1$. Hence,

$$\hat{\boldsymbol{\nu}} = \frac{\boldsymbol{b}^1}{\sqrt{\boldsymbol{b}^1 \cdot \boldsymbol{b}^1}} = V \frac{\boldsymbol{b}^1}{\sqrt{(\boldsymbol{b}_2 \times \boldsymbol{b}_3) \cdot (\boldsymbol{b}_2 \times \boldsymbol{b}_3)}}, \tag{1.77}$$

using the first expression in (1.6) for the reciprocal base vector $\boldsymbol{b}^1$. Replacing V with its expression in (1.69) as $\sqrt{g}$, and collecting terms, the component of the curl in the direction of $\boldsymbol{b}^1$ is given by

$$\begin{aligned}(\nabla \times \boldsymbol{V}) \cdot \boldsymbol{b}^1 &= \frac{1}{\sqrt{g}} \left[\frac{\partial}{\partial u^2} (\boldsymbol{V} \cdot \boldsymbol{b}_3) - \frac{\partial}{\partial u^3} (\boldsymbol{V} \cdot \boldsymbol{b}_2) \right] \\ &= \frac{1}{\sqrt{g}} \left(\frac{\partial v_3}{\partial u^2} - \frac{\partial v_2}{\partial u^3} \right),\end{aligned} \tag{1.78}$$

where we have used the definition of the covariant components of $\boldsymbol{V}$ given by the second of expressions (1.15). The remaining two components of the curl are found by simply permuting indices. Expanding $\nabla \times \boldsymbol{V}$ in the unitary base system, as in (1.17), it then becomes

$$\begin{aligned}\nabla \times \boldsymbol{V} &= \frac{1}{\sqrt{g}} \left[\left(\frac{\partial v_3}{\partial u^2} - \frac{\partial v_2}{\partial u^3} \right) \boldsymbol{b}_1 + \left(\frac{\partial v_1}{\partial u^3} - \frac{\partial v_3}{\partial u^1} \right) \boldsymbol{b}_2 + \left(\frac{\partial v_2}{\partial u^1} - \frac{\partial v_1}{\partial u^2} \right) \boldsymbol{b}_3 \right] \\ &= \frac{1}{\sqrt{g}} \begin{vmatrix} \boldsymbol{b}_1 & \boldsymbol{b}_2 & \boldsymbol{b}_3 \\ \partial/\partial u^1 & \partial/\partial u^2 & \partial/\partial u^3 \\ v_1 & v_2 & v_3 \end{vmatrix}.\end{aligned} \tag{1.79}$$

The expressions for the cross product (1.70) and the curl (1.79) can be condensed by introduction of the *permutation symbol* ξ_{ijk}. This is defined as

$$\xi_{ijk} = \begin{cases} 1, & i, j, k \text{ cyclic}, \\ -1, & i, j, k \text{ anticyclic}, \\ 0, & \text{otherwise}. \end{cases} \tag{1.80}$$

Cyclic permutations of the subscripts are $(1, 2, 3)$, $(2, 3, 1)$, $(3, 1, 2)$, while anticyclic permutations are $(1, 3, 2)$, $(3, 2, 1)$, $(2, 1, 3)$. In a right-handed Cartesian system of co-ordinates, we will later find that the permutation symbol is a third-order tensor, called the *alternating tensor*. Using the permutation symbol, the ith covariant component of the cross product (1.70) becomes

$$(\boldsymbol{V} \times \boldsymbol{W})_i = \sqrt{g}\, \xi_{ijk} v^j w^k. \tag{1.81}$$

The ith contravariant component of the curl of a vector (1.79) becomes

$$(\nabla \times \boldsymbol{V})^i = \frac{1}{\sqrt{g}} \xi_{ijk} \frac{\partial v_k}{\partial u^j}. \tag{1.82}$$

The *divergence* of an arbitrary vector $\boldsymbol{V}$ is written as the scalar product of the vector differential operator ∇ and $\boldsymbol{V}$. The divergence theorem of Gauss (A.17) gives

$$\iiint_{\mathcal{V}} \nabla \cdot \boldsymbol{V} d\mathcal{V} = \iint_{\mathcal{S}} \boldsymbol{V} \cdot \hat{\boldsymbol{\nu}} d\mathcal{S}, \tag{1.83}$$

where $\mathcal{V}$ is a volume enclosed by a surface $\mathcal{S}$. Thus, the integral of the divergence of $\boldsymbol{V}$ throughout the volume is equal to the integral of the outward normal component of $\boldsymbol{V}$ over the surface. If we now let the surface shrink, in the limit we find that

$$\nabla \cdot \boldsymbol{V} = \lim_{\mathcal{S} \to 0} \frac{1}{V} \iint_{\mathcal{S}} \boldsymbol{V} \cdot \hat{\boldsymbol{\nu}} d\mathcal{S}, \tag{1.84}$$

with V representing the infinitesimal volume. Now take the infinitesimal volume to be the parallelepiped bounded by the surfaces u^1 and $u^1 + du^1$, u^2 and $u^2 + du^2$, u^3 and $u^3 + du^3$. First, consider the integral of the outward normal component of $\boldsymbol{V}$ over the surfaces u^1 and $u^1 + du^1$. It is closely

$$[\boldsymbol{V} \cdot (\boldsymbol{b}_2 \times \boldsymbol{b}_3)]_{u^1+du^1} du^2 du^3 - [\boldsymbol{V} \cdot (\boldsymbol{b}_2 \times \boldsymbol{b}_3)]_{u^1} du^2 du^3. \tag{1.85}$$

In the limit, as the bounding surface shrinks to zero, it becomes the partial derivative with respect to u^1, times du^1, times $du^2 du^3$, or

$$\frac{\partial}{\partial u^1} \left[\boldsymbol{V} \cdot (\boldsymbol{b}_2 \times \boldsymbol{b}_3) \right] du^1 du^2 du^3. \tag{1.86}$$

From (1.6) and (1.5),

$$\boldsymbol{b}_2 \times \boldsymbol{b}_3 = V \boldsymbol{b}^1 = \boldsymbol{b}_1 \cdot (\boldsymbol{b}_2 \times \boldsymbol{b}_3) \, \boldsymbol{b}^1 = \sqrt{g} \, \boldsymbol{b}^1, \tag{1.87}$$

on taking the square root of (1.67) to replace the scalar triple product. Hence, (1.86) becomes

$$\frac{\partial}{\partial u^1} \left(\boldsymbol{V} \cdot \boldsymbol{b}^1 \sqrt{g} \right) du^1 du^2 du^3 = \frac{\partial}{\partial u^1} \left(v^1 \sqrt{g} \right) du^1 du^2 du^3, \tag{1.88}$$

recognising $\boldsymbol{V} \cdot \boldsymbol{b}^1$ as the contravariant component v^1 of $\boldsymbol{V}$, defined by (1.15). Adding in the contributions of the other two pairs of sides of the parallelepiped, the total integral of the outward normal component of $\boldsymbol{V}$ over the surface is

$$\frac{\partial}{\partial u^i} \left(v^i \sqrt{g} \right) du^1 du^2 du^3. \tag{1.89}$$

To obtain the divergence as expressed by (1.84), this must be divided by the volume of the parallelepiped given by (1.68), leading to

$$\nabla \cdot \boldsymbol{V} = \frac{1}{\sqrt{g}} \frac{\partial}{\partial u^i} \left(v^i \sqrt{g} \right). \tag{1.90}$$

The *gradient* of a scalar function ϕ, written $\nabla\phi$, is a vector representing the maximum rate of change of ϕ in direction and magnitude. A differential displacement $d\boldsymbol{r}$ results in a differential change in ϕ given by

$$d\phi = \nabla\phi \cdot d\boldsymbol{r} = \frac{\partial \phi}{\partial u^i} du^i. \tag{1.91}$$

The contravariant components of $d\boldsymbol{r}$, by the first of relations (1.15), are

$$du^i = \boldsymbol{b}^i \cdot d\boldsymbol{r}, \tag{1.92}$$

leading to

$$\left(\nabla\phi - \boldsymbol{b}^i \frac{\partial \phi}{\partial u^i} \right) \cdot d\boldsymbol{r} = 0. \tag{1.93}$$

Since the displacement $d\boldsymbol{r}$ is arbitrary, the gradient of an arbitrary scalar function ϕ is

$$\nabla\phi = \boldsymbol{b}^i \frac{\partial \phi}{\partial u^i}, \tag{1.94}$$

expressed in terms of the reciprocal base vectors. Substituting $\boldsymbol{b}^i$ for the vector $\boldsymbol{V}$ in (1.17) gives

$$\boldsymbol{b}^i = \boldsymbol{b}^i \cdot \boldsymbol{b}^j \boldsymbol{b}_j = g^{ij} \boldsymbol{b}_j, \tag{1.95}$$

so that, in terms of the base vectors,

$$\nabla\phi = \boldsymbol{b}_j g^{ij} \frac{\partial \phi}{\partial u^i}. \tag{1.96}$$

Finally, the *Laplacian* of a scalar field ϕ is $\nabla^2\phi$, and is considered to be the divergence of the gradient. The contravariant components of the gradient from (1.96) are

$$g^{ji} \frac{\partial \phi}{\partial u^j} = g^{ij} \frac{\partial \phi}{\partial u^j}, \tag{1.97}$$

and direct substitution in (1.90) yields

$$\nabla \cdot \nabla\phi = \nabla^2\phi = \frac{1}{\sqrt{g}} \frac{\partial}{\partial u^i} \left(\sqrt{g}\, g^{ij} \frac{\partial \phi}{\partial u^j} \right). \tag{1.98}$$

1.1.7 Orthogonal co-ordinates

Up to this point, we have not imposed any requirement on the unitary base vectors other than that they must not be coplanar. Our applications of scalar, vector and tensor analysis will be, without exception, carried out in orthogonal co-ordinate systems. Thus, from now onwards, we will assume that we are dealing with orthogonal co-ordinate systems.

The squared magnitudes of the reciprocal unitary base vectors, defined in terms of the unitary base vectors by (1.6), are given by

$$\begin{aligned} \boldsymbol{b}^1 \cdot \boldsymbol{b}^1 &= \frac{1}{V^2}(\boldsymbol{b}_2 \times \boldsymbol{b}_3) \cdot (\boldsymbol{b}_2 \times \boldsymbol{b}_3) \\ &= \frac{1}{V^2}\Big[(\boldsymbol{b}_2 \cdot \boldsymbol{b}_2)(\boldsymbol{b}_3 \cdot \boldsymbol{b}_3) - (\boldsymbol{b}_2 \cdot \boldsymbol{b}_3)(\boldsymbol{b}_3 \cdot \boldsymbol{b}_2)\Big], \end{aligned} \tag{1.99}$$

using the vector identity (1.57). Permuting subscripts in the vector identity yields,

$$\begin{aligned} \boldsymbol{b}^2 \cdot \boldsymbol{b}^2 &= \frac{1}{V^2}(\boldsymbol{b}_3 \times \boldsymbol{b}_1) \cdot (\boldsymbol{b}_3 \times \boldsymbol{b}_1) \\ &= \frac{1}{V^2}\Big[(\boldsymbol{b}_3 \cdot \boldsymbol{b}_3)(\boldsymbol{b}_1 \cdot \boldsymbol{b}_1) - (\boldsymbol{b}_3 \cdot \boldsymbol{b}_1)(\boldsymbol{b}_1 \cdot \boldsymbol{b}_3)\Big], \end{aligned} \tag{1.100}$$

and

$$\begin{aligned} \boldsymbol{b}^3 \cdot \boldsymbol{b}^3 &= \frac{1}{V^2}(\boldsymbol{b}_1 \times \boldsymbol{b}_2) \cdot (\boldsymbol{b}_1 \times \boldsymbol{b}_2) \\ &= \frac{1}{V^2}\Big[(\boldsymbol{b}_1 \cdot \boldsymbol{b}_1)(\boldsymbol{b}_2 \cdot \boldsymbol{b}_2) - (\boldsymbol{b}_1 \cdot \boldsymbol{b}_2)(\boldsymbol{b}_2 \cdot \boldsymbol{b}_1)\Big]. \end{aligned} \tag{1.101}$$

In an orthogonal co-ordinate system, the metric tensor has only three non-vanishing components,

$$g_{11} = \boldsymbol{b}_1 \cdot \boldsymbol{b}_1, \quad g_{22} = \boldsymbol{b}_2 \cdot \boldsymbol{b}_2, \quad g_{33} = \boldsymbol{b}_3 \cdot \boldsymbol{b}_3. \tag{1.102}$$

Then, from (1.66) and (1.67), the three non-vanishing components of the metric tensor yield

$$V^2 = (\boldsymbol{b}_1 \cdot \boldsymbol{b}_1)(\boldsymbol{b}_2 \cdot \boldsymbol{b}_2)(\boldsymbol{b}_3 \cdot \boldsymbol{b}_3). \tag{1.103}$$

Thus, from (1.99), (1.100) and (1.101), the squared magnitudes of the reciprocal unitary base vectors become

$$\boldsymbol{b}^1 \cdot \boldsymbol{b}^1 = \frac{1}{\boldsymbol{b}_1 \cdot \boldsymbol{b}_1}, \quad \boldsymbol{b}^2 \cdot \boldsymbol{b}^2 = \frac{1}{\boldsymbol{b}_2 \cdot \boldsymbol{b}_2}, \quad \boldsymbol{b}^3 \cdot \boldsymbol{b}^3 = \frac{1}{\boldsymbol{b}_3 \cdot \boldsymbol{b}_3}, \tag{1.104}$$

the reciprocals of the squared magnitudes of the original unitary base vectors. In an orthogonal co-ordinate system, the reciprocal unitary base vectors defined by (1.6) are parallel to the unitary base vectors but are subject to the scalings

$$\begin{aligned} \boldsymbol{b}^1 &= \frac{1}{\boldsymbol{b}_1 \cdot \boldsymbol{b}_1}\boldsymbol{b}_1 = \frac{1}{g_{11}}\boldsymbol{b}_1, \\ \boldsymbol{b}^2 &= \frac{1}{\boldsymbol{b}_2 \cdot \boldsymbol{b}_2}\boldsymbol{b}_2 = \frac{1}{g_{22}}\boldsymbol{b}_2, \\ \boldsymbol{b}^3 &= \frac{1}{\boldsymbol{b}_3 \cdot \boldsymbol{b}_3}\boldsymbol{b}_3 = \frac{1}{g_{33}}\boldsymbol{b}_3. \end{aligned} \tag{1.105}$$

Taking the scalar product of each of these equations with its corresponding reciprocal unitary base vector, and using the orthogonality relation (1.7), we recover the required squared magnitude ratios (1.104).

In an orthogonal co-ordinate system, with the abbreviations

$$h_1 = \sqrt{g_{11}}, \quad h_2 = \sqrt{g_{22}}, \quad h_3 = \sqrt{g_{33}}, \tag{1.106}$$

the infinitesimal displacements, (1.52) and (1.54), along the co-ordinate directions become

$$ds_1 = h_1 du^1, \quad ds_2 = h_2 du^2, \quad ds_3 = h_3 du^3, \tag{1.107}$$

while the surface elements, (1.58), (1.59) and (1.60), are

$$da_1 = h_2 h_3 du^2 du^3, \quad da_2 = h_3 h_1 du^3 du^1, \quad da_3 = h_1 h_2 du^1 du^2. \tag{1.108}$$

The volume element (1.68) is

$$dv = h_1 h_2 h_3 \, du^1 du^2 du^3 = \sqrt{g} \, du^1 du^2 du^3. \tag{1.109}$$

The off-diagonal elements of the determinant (1.42) defining g vanish, so it is simply $g_{11}g_{22}g_{33}$, giving

$$g = h_1^2 h_2^2 h_3^2. \tag{1.110}$$

The physical components of an arbitrary vector, $\boldsymbol{V}$, in an orthogonal system, are expressed by its projections on the three unit vectors, as in (1.20), except that now the unit vectors are orthogonal, giving $\hat{e}_i \cdot \hat{e}_j = \delta^i_{\ j}$. From (1.19) and (1.106), the unitary base system of vectors is $\boldsymbol{b}_i = h_i \hat{e}_i$, while from (1.105), the reciprocal base system vectors are $\boldsymbol{b}^i = \hat{e}_i / h_i$. From (1.21) and (1.106), the contravariant components of $\boldsymbol{V}$ are $v^i = V_i / h_i$, while from (1.16), the covariant components of $\boldsymbol{V}$ are $v_i = h_i V_i$. In these last four expressions, no summation is implied by the repeated subscript.

By substitution in the formula (1.70) for the cross product of two vectors, $\boldsymbol{V}$ and $\boldsymbol{W}$, it is found that

$$\boldsymbol{V} \times \boldsymbol{W} = h_1 h_2 h_3 \begin{vmatrix} \hat{\mathbf{e}}_1/h_1 & \hat{\mathbf{e}}_2/h_2 & \hat{\mathbf{e}}_3/h_3 \\ V_1/h_1 & V_2/h_2 & V_3/h_3 \\ W_1/h_1 & W_2/h_2 & W_3/h_3 \end{vmatrix}$$

$$= \begin{vmatrix} \hat{\mathbf{e}}_1 & \hat{\mathbf{e}}_2 & \hat{\mathbf{e}}_3 \\ V_1 & V_2 & V_3 \\ W_1 & W_2 & W_3 \end{vmatrix}. \tag{1.111}$$

The cross product takes its familiar form in all orthogonal co-ordinate systems. From this, it follows that the triple scalar product (A.1) can be written in the form of a determinant as

$$\boldsymbol{a} \cdot (\boldsymbol{b} \times \boldsymbol{c}) = \begin{vmatrix} a_1 & a_2 & a_3 \\ b_1 & b_2 & b_3 \\ c_1 & c_2 & c_3 \end{vmatrix}. \tag{1.112}$$

Substitution in formula (1.79) for the curl gives

$$\nabla \times \boldsymbol{V} = \frac{1}{h_1 h_2 h_3} \begin{vmatrix} h_1 \hat{\mathbf{e}}_1 & h_2 \hat{\mathbf{e}}_2 & h_3 \hat{\mathbf{e}}_3 \\ \partial/\partial u^1 & \partial/\partial u^2 & \partial/\partial u^3 \\ h_1 V_1 & h_2 V_2 & h_3 V_3 \end{vmatrix}, \tag{1.113}$$

while substitution in formula (1.90) for the divergence gives

$$\nabla \cdot \boldsymbol{V} = \frac{1}{h_1 h_2 h_3} \left[\frac{\partial}{\partial u^1}(h_2 h_3 V_1) + \frac{\partial}{\partial u^2}(h_3 h_1 V_2) + \frac{\partial}{\partial u^3}(h_1 h_2 V_3) \right]. \tag{1.114}$$

The gradient (1.96) of a scalar function ϕ becomes

$$\nabla \phi = \frac{1}{h_1} \frac{\partial \phi}{\partial u^1} \hat{\mathbf{e}}_1 + \frac{1}{h_2} \frac{\partial \phi}{\partial u^2} \hat{\mathbf{e}}_2 + \frac{1}{h_3} \frac{\partial \phi}{\partial u^3} \hat{\mathbf{e}}_3, \tag{1.115}$$

on replacing $g^{ij} = \boldsymbol{b}^i \cdot \boldsymbol{b}^j$ by $1/h_i^2$.

The Laplacian (1.98) of ϕ, in an orthogonal system, takes the form

$$\nabla^2 \phi = \frac{1}{h_1 h_2 h_3} \left[\frac{\partial}{\partial u^1} \left(\frac{h_2 h_3}{h_1} \frac{\partial \phi}{\partial u^1} \right) + \frac{\partial}{\partial u^2} \left(\frac{h_3 h_1}{h_2} \frac{\partial \phi}{\partial u^2} \right) + \frac{\partial}{\partial u^3} \left(\frac{h_1 h_2}{h_3} \frac{\partial \phi}{\partial u^3} \right) \right]. \tag{1.116}$$

In Cartesian co-ordinates, the vector identity (A.14),

$$\nabla \times (\nabla \times \boldsymbol{V}) = \nabla (\nabla \cdot \boldsymbol{V}) - \nabla^2 \boldsymbol{V}, \tag{1.117}$$

is easily established. In Cartesian co-ordinates the unit vectors are independent of position, and the operator $\nabla^2 \boldsymbol{V}$ can be interpreted as the Laplacian operating on each of the Cartesian components of $\boldsymbol{V}$, or

$$\nabla^2 \boldsymbol{V} = \hat{\mathbf{e}}_j \nabla^2 V_j. \tag{1.118}$$

In orthogonal curvilinear co-ordinates, the operation $\nabla^2 \boldsymbol{V}$ is not defined directly but may be expressed as the difference

$$\nabla(\nabla \cdot \boldsymbol{V}) - \nabla \times (\nabla \times \boldsymbol{V}). \tag{1.119}$$

We conclude this subsection with the specific orthogonal, curvilinear co-ordinates we will most often use: the spherical polar system (r, θ, ϕ). A point $P(u^1, u^2, u^3)$ is at radius $u^1 = r$, co-latitude $u^2 = \theta$ and east longitude $u^3 = \phi$ and has Cartesian co-ordinates (x, y, z) given by

$$x = r \sin\theta \cos\phi, \quad y = r \sin\theta \sin\phi, \quad z = r\cos\theta. \tag{1.120}$$

The unit vectors $\hat{\boldsymbol{r}}, \hat{\boldsymbol{\theta}}, \hat{\boldsymbol{\phi}}$ in the directions of increasing r, θ and ϕ form a right-handed system, and the metrical coefficients are $h_1 = 1$, $h_2 = r$ and $h_3 = r\sin\theta$. Substitution in expression (1.113) for the curl yields

$$\nabla \times \boldsymbol{V} = \begin{vmatrix} \hat{\boldsymbol{r}}/r^2 \sin\theta & \hat{\boldsymbol{\theta}}/r\sin\theta & \hat{\boldsymbol{\phi}}/r \\ \partial/\partial r & \partial/\partial\theta & \partial/\partial\phi \\ V_r & rV_\theta & r\sin\theta\, V_\phi \end{vmatrix}, \tag{1.121}$$

with (V_r, V_θ, V_ϕ) the spherical polar components of the vector $\boldsymbol{V}$. Substitution in expression (1.114) for the divergence gives

$$\nabla \cdot \boldsymbol{V} = \frac{1}{r^2}\frac{\partial}{\partial r}\left(r^2 V_r\right) + \frac{1}{r\sin\theta}\frac{\partial}{\partial\theta}\left(\sin\theta\, V_\theta\right) + \frac{1}{r\sin\theta}\frac{\partial V_\phi}{\partial\phi}. \tag{1.122}$$

From expression (1.115), the gradient of an arbitrary scalar field ψ becomes

$$\nabla\psi = \hat{\boldsymbol{r}}\frac{\partial\psi}{\partial r} + \frac{\hat{\boldsymbol{\theta}}}{r}\frac{\partial\psi}{\partial\theta} + \frac{\hat{\boldsymbol{\phi}}}{r\sin\theta}\frac{\partial\psi}{\partial\phi}, \tag{1.123}$$

while, from expression (1.116), its Laplacian becomes

$$\begin{aligned}\nabla^2\psi &= \frac{1}{r^2}\frac{\partial}{\partial r}\left(r^2\frac{\partial\psi}{\partial r}\right) + \frac{1}{r^2\sin\theta}\frac{\partial}{\partial\theta}\left(\sin\theta\frac{\partial\psi}{\partial\theta}\right) + \frac{1}{r^2\sin^2\theta}\frac{\partial^2\psi}{\partial\phi^2} \\ &= \frac{\partial^2\psi}{\partial r^2} + \frac{2}{r}\frac{\partial\psi}{\partial r} + \frac{1}{r^2}\frac{\partial^2\psi}{\partial\theta^2} + \frac{\cot\theta}{r^2}\frac{\partial\psi}{\partial\theta} + \frac{1}{r^2\sin^2\theta}\frac{\partial^2\psi}{\partial\phi^2}. \end{aligned} \tag{1.124}$$

1.1.8 Pseudo-tensors

There is a wide class of physical quantities that have many of the properties of true tensors but are not invariant in all co-ordinate systems. They are called *pseudo-tensors*.

Vectors are first-order tensors whose properties under co-ordinate transformations we have already examined in Subsection 1.1.3. Vectors following the regular transformation laws for all co-ordinate systems are called *polar vectors*. Thus far, we have been careful to use right-handed co-ordinate systems. In a right-handed co-ordinate system, rotating the first co-ordinate axis towards the second would cause a right-threaded screw to advance in the direction of the third co-ordinate axis. In a left-handed co-ordinate system, it would advance in a direction opposite to that of the third co-ordinate axis. The components of an *axial vector* change sign in going from a right-handed co-ordinate system to a left-handed system.

One of the most common examples of an axial vector is the vector that is the cross product (1.111) of two other vectors. The reference co-ordinate system is transformed from right-handed to left-handed by interchanging the (1, 2) components and unit vectors. The cross product then becomes

$$\begin{vmatrix} \hat{e}_2 & \hat{e}_1 & \hat{e}_3 \\ V_2 & V_1 & V_3 \\ W_2 & W_1 & W_3 \end{vmatrix}. \tag{1.125}$$

Since the first and second columns in this determinant have been switched compared with (1.111), the sign of the cross product has been reversed compared with that in a right-handed co-ordinate system. Similarly, interchanging first and second co-ordinates, components and unit vectors, it can be shown that the vector produced by the curl (1.113) of a vector reverses sign in going from a right-handed co-ordinate system to a left-handed system. These axial vectors can be shown to be special cases of second-order tensors.

Another class of pseudo-tensors is *relative tensors*, which transform as tensors, except that they may be multiplied, or divided, by an integral power of the Jacobian, in which case they are called *tensor capacities*, or *tensor densities*, respectively.

One of the simplest examples of a tensor capacity is the volume element considered in Subsection 1.1.5. In transforming to a new system of co-ordinates (u'^1, u'^2, u'^3) from the old system (u^1, u^2, u^3), we assume the relations

$$\begin{aligned} u'^1 &= u'^1(u^1, u^2, u^3) \\ u'^2 &= u'^2(u^1, u^2, u^3) \\ u'^3 &= u'^3(u^1, u^2, u^3). \end{aligned} \tag{1.126}$$

The volume element in the old co-ordinate system is the parallelepiped bounded by the surfaces u^1 and $u^1 + du^1$, u^2 and $u^2 + du^2$, u^3 and $u^3 + du^3$. The volume of the parallelepiped is given by the triple scalar product of its bounding vectors. In the new system of co-ordinates, the bounding vector in the u^1 direction has components

$$h_1 \left(\frac{\partial u'^1}{\partial u^1}, \frac{\partial u'^2}{\partial u^1}, \frac{\partial u'^3}{\partial u^1} \right) du^1, \tag{1.127}$$

while that in the u^2 direction has components

$$h_2\left(\frac{\partial u'^1}{\partial u^2}, \frac{\partial u'^2}{\partial u^2}, \frac{\partial u'^3}{\partial u^2}\right) du^2, \tag{1.128}$$

and that in the u^3 direction has components

$$h_3\left(\frac{\partial u'^1}{\partial u^3}, \frac{\partial u'^2}{\partial u^3}, \frac{\partial u'^3}{\partial u^3}\right) du^3. \tag{1.129}$$

Writing the triple scalar product in its determinant form (1.112), the volume element becomes

$$\begin{vmatrix} \partial u'^1/\partial u^1 & \partial u'^2/\partial u^1 & \partial u'^3/\partial u^1 \\ \partial u'^1/\partial u^2 & \partial u'^2/\partial u^2 & \partial u'^3/\partial u^2 \\ \partial u'^1/\partial u^3 & \partial u'^2/\partial u^3 & \partial u'^3/\partial u^3 \end{vmatrix} h_1h_2h_3\, du^1du^2du^3. \tag{1.130}$$

The determinant is recognised as the familiar Jacobian and, from (1.109), the volume element in the new co-ordinate system becomes

$$dv' = J\left(\frac{u'^1, u'^2, u'^3}{u^1, u^2, u^3}\right) dv. \tag{1.131}$$

Thus, the volume element is not a true scalar but is multiplied by the Jacobian in transforming to a new co-ordinate system. Quantities that transform in such a fashion are called *tensor capacities*. The volume element is a *scalar capacity*.

As an example of a tensor density, consider a quantity resembling a vector associated with a twice covariant, antisymmetric, second order tensor,

$$\begin{pmatrix} t_{11} & t_{12} & t_{13} \\ t_{21} & t_{22} & t_{23} \\ t_{31} & t_{32} & t_{33} \end{pmatrix} = \begin{pmatrix} 0 & t_{12} & -t_{31} \\ -t_{12} & 0 & t_{23} \\ t_{31} & -t_{23} & 0 \end{pmatrix}. \tag{1.132}$$

Because of its antisymmetry, there are only three independent components, and these may be taken as the components $\tau^1 = t_{23}$, $\tau^2 = t_{31}$, $\tau^3 = t_{12}$, of a quantity resembling a contravariant vector, $\boldsymbol{\tau}$. In Cartesian co-ordinates, the double contraction $\xi_{ijk}t_{jk}$ is called the *dual vector* of the second order tensor. It has components $t_{23} - t_{32}$, $t_{31} - t_{13}$, $t_{12} - t_{21}$, and, hence, it is equal to $2\boldsymbol{\tau}$. In transforming from an old system of co-ordinates (u^1, u^2, u^3) to a new system (u'^1, u'^2, u'^3), the twice covariant, second-order tensor, t_{ij}, transforms as (1.37),

$$t'_{kl} = \frac{\partial u^i}{\partial u'^k}\frac{\partial u^j}{\partial u'^l}t_{ij}. \tag{1.133}$$

Since t_{ij} is antisymmetric,

$$t'_{kl} = -\frac{\partial u^i}{\partial u'^k}\frac{\partial u^j}{\partial u'^l}t_{ji}. \tag{1.134}$$

On interchanging the indices k and l, as well as i and j, we find

$$t'_{lk} = -\frac{\partial u^j}{\partial u'^l}\frac{\partial u^i}{\partial u'^k}t_{ij} = -t'_{kl}, \qquad (1.135)$$

using relation (1.133). The antisymmetry of t_{ij} is preserved in the transformation to the new co-ordinates, giving $t'_{kl} = -t'_{lk}$. Then, $\tau'^1 = t'_{23}, \tau'^2 = t'_{31}, \tau'^3 = t'_{12}$. Replacing the antisymmetric components of t_{ij} by τ^1, τ^2, τ^3 and the antisymmetric components of t'_{ij} by $\tau'^1, \tau'^2, \tau'^3$, we find that

$$\begin{aligned}
\tau'^1 &= \left(\frac{\partial u^2}{\partial u'^2}\frac{\partial u^3}{\partial u'^3} - \frac{\partial u^3}{\partial u'^2}\frac{\partial u^2}{\partial u'^3}\right)\tau^1 + \left(\frac{\partial u^3}{\partial u'^2}\frac{\partial u^1}{\partial u'^3} - \frac{\partial u^1}{\partial u'^2}\frac{\partial u^3}{\partial u'^3}\right)\tau^2 \\
&\quad + \left(\frac{\partial u^1}{\partial u'^2}\frac{\partial u^2}{\partial u'^3} - \frac{\partial u^2}{\partial u'^2}\frac{\partial u^1}{\partial u'^3}\right)\tau^3, \\
\tau'^2 &= \left(\frac{\partial u^2}{\partial u'^3}\frac{\partial u^3}{\partial u'^1} - \frac{\partial u^3}{\partial u'^3}\frac{\partial u^2}{\partial u'^1}\right)\tau^1 + \left(\frac{\partial u^3}{\partial u'^3}\frac{\partial u^1}{\partial u'^1} - \frac{\partial u^1}{\partial u'^3}\frac{\partial u^3}{\partial u'^1}\right)\tau^2 \qquad (1.136) \\
&\quad + \left(\frac{\partial u^1}{\partial u'^3}\frac{\partial u^2}{\partial u'^1} - \frac{\partial u^2}{\partial u'^3}\frac{\partial u^1}{\partial u'^1}\right)\tau^3, \\
\tau'^3 &= \left(\frac{\partial u^2}{\partial u'^1}\frac{\partial u^3}{\partial u'^2} - \frac{\partial u^3}{\partial u'^1}\frac{\partial u^2}{\partial u'^2}\right)\tau^1 + \left(\frac{\partial u^3}{\partial u'^1}\frac{\partial u^1}{\partial u'^2} - \frac{\partial u^1}{\partial u'^1}\frac{\partial u^3}{\partial u'^2}\right)\tau^2 \\
&\quad + \left(\frac{\partial u^1}{\partial u'^1}\frac{\partial u^2}{\partial u'^2} - \frac{\partial u^2}{\partial u'^1}\frac{\partial u^1}{\partial u'^2}\right)\tau^3.
\end{aligned}$$

The coefficients of the transformation law (1.36) for the covariant components of a vector may be abbreviated as

$$a_{ij} = \frac{\partial u^j}{\partial u'^i}. \qquad (1.137)$$

These may be regarded as the elements of a 3×3 matrix,

$$\mathbf{A} = \begin{pmatrix} \partial u^1/\partial u'^1 & \partial u^2/\partial u'^1 & \partial u^3/\partial u'^1 \\ \partial u^1/\partial u'^2 & \partial u^2/\partial u'^2 & \partial u^3/\partial u'^2 \\ \partial u^1/\partial u'^3 & \partial u^2/\partial u'^3 & \partial u^3/\partial u'^3 \end{pmatrix}. \qquad (1.138)$$

Let **C** be the matrix of cofactors of the determinant, $|\mathbf{A}|$, of the matrix **A** (1.138), with elements c_{ij}. Then, the transformation (1.136) of the quantities τ^j to the quantities τ'^i may be written

$$\tau'^i = c_{ij}\tau^j. \qquad (1.139)$$

The inverse of the matrix **A** is given by

$$\mathbf{A}^{-1} = \frac{\mathbf{C}}{|\mathbf{A}|} = \mathbf{B}^T, \qquad (1.140)$$

where $\mathbf{B}^T$ is the inverse of $\mathbf{A}$ with elements b_{ij}. Thus, the transformation (1.139) is expressible as

$$\tau'^i = |\mathbf{A}|\, b_{ij}\tau^j. \tag{1.141}$$

Writing out the implied summation in expression (1.34), we have, by the familiar chain rule for partial derivatives,

$$\frac{\partial u^i}{\partial u^j} = \frac{\partial u^i}{\partial u'^1}\frac{\partial u'^1}{\partial u^j} + \frac{\partial u^i}{\partial u'^2}\frac{\partial u'^2}{\partial u^j} + \frac{\partial u^i}{\partial u'^3}\frac{\partial u'^3}{\partial u^j} = \delta^i_{\ j}. \tag{1.142}$$

Choosing

$$b_{kj} = \frac{\partial u'^k}{\partial u^j}, \tag{1.143}$$

relation (1.142) can be written as

$$a_{ki} b_{kj} = \delta^i_{\ j}, \tag{1.144}$$

or, in matrix notation, as

$$\mathbf{A}^T\mathbf{B} = \mathbf{B}^T\mathbf{A} = \mathbf{I}, \tag{1.145}$$

where $\mathbf{I}$ is the unit matrix, confirming the choice (1.143) of the elements of $\mathbf{B}^T = \mathbf{A}^{-1}$. Taking the determinant of both sides of (1.145),

$$|\mathbf{A}|\,|\mathbf{B}| = 1, \tag{1.146}$$

since $|\mathbf{A}^T| = |\mathbf{A}|$ and $|\mathbf{B}^T| = |\mathbf{B}|$. The transformation (1.141) then takes the form

$$\tau'^i = \frac{1}{|\mathbf{B}|} b_{ij}\tau^j = \frac{\partial u'^i}{\partial u^j}\tau^j \Big/ J\left(\frac{u'^1, u'^2, u'^3}{u^1, u^2, u^3}\right). \tag{1.147}$$

The quantity τ^j transforms as a contravariant vector, except that in the transformation it is divided by the Jacobian. Thus, it is not a true tensor but is a *tensor density*.

1.1.9 Cartesian tensors

Tensor analysis is considerably simplified when carried out in Cartesian co-ordinate systems. Continuing to confine our attention to right-handed systems, a point $P(u^1, u^2, u^3)$ is at Cartesian co-ordinates $u^1 = x = x_1$, $u^2 = y = x_2$, $u^3 = z = x_3$. The unit vectors $\hat{e}_1$, $\hat{e}_2$, $\hat{e}_3$ in the directions of increasing x_1, x_2, x_3 form a right-handed system with metrical coefficients $h_1 = h_2 = h_3 = 1$. Then, the distinction between covariant and contravariant tensors vanishes as does the distinction between tensor and physical components. In addition, both the unitary base vectors and the reciprocal unitary base vectors become identical to the unit vectors $\hat{e}_1$, $\hat{e}_2$, $\hat{e}_3$.

In the transformation from co-ordinates (x_1, x_2, x_3) to a new system of right-handed Cartesian co-ordinates (x'_1, x'_2, x'_3), the partial derivatives

$$\frac{\partial u'^j}{\partial u^i} \quad \text{and} \quad \frac{\partial u^i}{\partial u'^j} \tag{1.148}$$

both become $\cos(x'_j, x_i) = \cos(x_i, x'_j) = c_{ji}$, the direction cosine between x'_j and x_i, which is equal to the direction cosine between x_i and x'_j. Thus, in Cartesian co-ordinate systems, the transformation laws (1.28) and (1.36) for the components (V_1, V_2, V_3) of an arbitrary vector $\boldsymbol{V}$ become

$$V'_j = c_{ji} V_i. \tag{1.149}$$

In contrast to the partial derivatives (1.148) in curvilinear co-ordinates, the direction cosines c_{ji} are not functions of position. The transformation of co-ordinates follows the same law as that for vectors,

$$x'_j = c_{jk} x_k. \tag{1.150}$$

The inverse transformation from the new co-ordinates back to the original co-ordinates follows

$$x_k = c_{lk} x'_l. \tag{1.151}$$

Then,

$$x'_j = c_{jk} c_{lk} x'_l. \tag{1.152}$$

The co-ordinates x'_j and x'_l are independent if $j \neq l$, and if $j = l$ they are identical. Hence,

$$c_{jk} c_{lk} = \delta_{jl}, \tag{1.153}$$

analogous to (1.34), where δ_{jl} is the Kronecker delta, originally defined following (1.7) and shown to follow the transformation law (1.49) for a second-order mixed tensor. In Cartesian co-ordinates, it is a simple second-order tensor as our notation implies.

If $\boldsymbol{a}$, $\boldsymbol{b}$, $\boldsymbol{c}$ are three arbitrary vectors, the volume of the parallelepiped they bound in a right-handed co-ordinate system is the triple scalar product (1.112). In a Cartesian co-ordinate system, the ith component of the cross product $\boldsymbol{b} \times \boldsymbol{c}$, from (1.81), becomes

$$(\boldsymbol{b} \times \boldsymbol{c})_i = \xi_{ijk} b_j c_k, \tag{1.154}$$

using the permutation symbol defined by (1.80). Then the triple scalar product is given by the triple contraction $\xi_{ijk} a_i b_j c_k$. Since this leads to the scalar volume of the bounding parallelepiped, the permutation symbol in right-handed Cartesian co-ordinates is a third-order tensor, called the *alternating tensor*.

The transformation (1.149) for a vector may be described as associating a scalar with each spatial direction by a linear, homogeneous relation in the direction cosines. Then a second-order tensor associates a vector with each spatial direction by a linear, homogeneous relation in the direction cosines. This leads to a generalisation that a tensor of order n associates a tensor of order $n-1$ with each spatial direction by a linear, homogeneous relation in the direction cosines. For a second-order tensor T_{ij}, the transformation law (1.149) generalises to

$$T'_{kl} = c_{ki}c_{lj}T_{ij}, \tag{1.155}$$

giving the tensor T'_{kl} in the new co-ordinate system. For a tensor of arbitrary order, the transformation generalises to

$$T'_{kl\ldots} = c_{ki}c_{lj}\cdots T_{ij\ldots}. \tag{1.156}$$

The double contraction

$$v_i = \xi_{ijk}T_{jk} \tag{1.157}$$

of the second-order tensor T_{jk} leads to the vector v_i, called the *dual vector* of T_{jk}, with components

$$v_1 = T_{23} - T_{32}, \quad v_2 = T_{31} - T_{13}, \quad v_3 = T_{12} - T_{21}, \tag{1.158}$$

using the properties of the alternating tensor. The dual vector therefore depends only on the antisymmetric part of T_{jk}. Conversely, if the dual vector vanishes, the tensor T_{jk} is symmetric. The contraction

$$\xi_{ijk}v_i = \xi_{ijk}\xi_{ilm}T_{lm} \tag{1.159}$$

of the dual vector leads to a second-order tensor determined by evaluating the product $\xi_{ijk}\xi_{ilm}$ of alternating tensors. If (i, j, k) is a cyclic permutation of indices, the permutation (i, l, m) will be cyclic if $j = l$ and $k = m$, giving the product the value +1. If (i, j, k) and (i, l, m) have opposite cyclicity, the product has the value −1. Otherwise, the product vanishes. The product can then be written in terms of Kronecker deltas as $\xi_{ijk}\xi_{ilm} = \delta_{lj}\delta_{mk} - \delta_{lk}\delta_{mj}$ and the contraction of the dual vector becomes

$$\xi_{ijk}v_i = \left(\delta_{lj}\delta_{mk} - \delta_{lk}\delta_{mj}\right)T_{lm} = T_{jk} - T_{kj}. \tag{1.160}$$

For the antisymmetric part of T_{jk} we have $T_{jk} = -T_{kj}$, giving $T_{jk} - T_{kj} = 2T_{jk}$. Hence, the antisymmetric part of T_{jk} can be recovered from the dual vector as

$$T_{jk} = \frac{1}{2}\xi_{ijk}v_i. \tag{1.161}$$

This tensor is called the *dual antisymmetric tensor* of the vector v_i.

1.2 Separation of vector fields

A vector field $\boldsymbol{u}$ may be represented in terms of its curl and divergence. If we write $\boldsymbol{u}$ as the sum of the gradient of a scalar potential ϕ and the curl of a vector potential $\boldsymbol{\psi}$, as in the classical Helmholtz separation, we have

$$\boldsymbol{u} = \nabla\phi + \nabla \times \boldsymbol{\psi}. \tag{1.162}$$

Since the divergence of a curl vanishes identically, as does the curl of a gradient, we find Poisson equations for the scalar and vector potentials,

$$\nabla^2\phi = \nabla \cdot \boldsymbol{u}, \tag{1.163}$$

and, using the identity $-\nabla \times (\nabla \times \boldsymbol{\psi}) + \nabla(\nabla \cdot \boldsymbol{\psi}) = \nabla^2\boldsymbol{\psi}$, valid in Cartesian co-ordinates,

$$\nabla^2\boldsymbol{\psi} = -\nabla \times \boldsymbol{u} + \nabla(\nabla \cdot \boldsymbol{\psi}). \tag{1.164}$$

The potentials in the Helmholtz separation are not unique, since the potentials

$$\phi' = \phi + C \quad \text{and} \quad \boldsymbol{\psi}' = \boldsymbol{\psi} + \nabla\chi, \tag{1.165}$$

where C is any constant and χ is an arbitrary scalar function of position, will yield the identical vector field $\boldsymbol{u}$. Removing the uncertainty $\nabla\chi$ in $\boldsymbol{\psi}$ is called *choosing the gauge*. If we choose the gauge so that $\nabla \cdot \boldsymbol{\psi} = 0$, then for another choice of vector potential, $\boldsymbol{\psi}'$, we have $\nabla \cdot \boldsymbol{\psi}' = \nabla \cdot \boldsymbol{\psi} + \nabla^2\chi = \nabla^2\chi$, and we have another Poisson equation for χ,

$$\nabla^2\chi = \nabla \cdot \boldsymbol{\psi}'. \tag{1.166}$$

In seismology, it is usual to choose the gauge so that $\nabla \cdot \boldsymbol{\psi} = 0$. The vector potential $\boldsymbol{\psi}$ then obeys the vector Poisson equation

$$\nabla^2\boldsymbol{\psi} = -\nabla \times \boldsymbol{u}. \tag{1.167}$$

It is easy to verify that the solution of the Poisson equation for the scalar potential is

$$\phi(\boldsymbol{r}) = -\frac{1}{4\pi} \iiint \frac{\nabla' \cdot \boldsymbol{u}(\boldsymbol{r}')}{|\boldsymbol{r} - \boldsymbol{r}'|} dV', \tag{1.168}$$

where the potential at the field point at $\boldsymbol{r}$ is found by integrating over all source points at $\boldsymbol{r}'$. The distance from the source point with co-ordinates (x_1', x_2', x_3') to the field point with co-ordinates (x_1, x_2, x_3) is

$$R = \sqrt{(x_1 - x_1')^2 + (x_2 - x_2')^2 + (x_3 - x_3')^2} = |\boldsymbol{r} - \boldsymbol{r}'|. \tag{1.169}$$

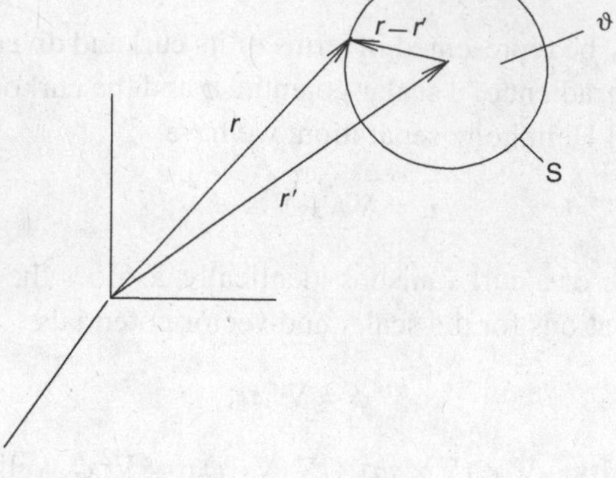

Figure 1.1 Small sphere of radius $\epsilon = R = |\boldsymbol{r} - \boldsymbol{r}'|$, surrounding the source point $\boldsymbol{r}'$, as the source point and field point, $\boldsymbol{r}$, approach coincidence.

For $\boldsymbol{r} \neq \boldsymbol{r}'$, the gradient of $1/R$ is

$$\nabla\left(\frac{1}{R}\right) = \nabla\left(\frac{1}{|\boldsymbol{r} - \boldsymbol{r}'|}\right) = -\frac{\boldsymbol{r} - \boldsymbol{r}'}{R^3}. \tag{1.170}$$

Taking the divergence yields

$$\begin{aligned}\nabla^2\left(\frac{1}{R}\right) &= \nabla^2\left(\frac{1}{|\boldsymbol{r} - \boldsymbol{r}'|}\right) \\ &= \frac{3}{2}\frac{\left[2\left(x_1 - x_1'\right)^2 + 2\left(x_2 - x_2'\right)^2 + 2\left(x_3 - x_3'\right)^2\right]}{R^5} - \frac{3}{R^3} = 0.\end{aligned} \tag{1.171}$$

Now suppose the source point is close to the field point so that R is the very small quantity ϵ. Surround the source point with a sphere of radius ϵ, as illustrated in Figure 1.1.

Now take the Laplacian of expression (1.168). As the source point and field point approach coincidence, the right side approaches

$$-\frac{1}{4\pi}\nabla \cdot \boldsymbol{u}(\boldsymbol{r}) \iiint \nabla^2\left(\frac{1}{R}\right) d\mathcal{V}'. \tag{1.172}$$

By the divergence theorem of Gauss, the volume integral throughout the small sphere is the integral over its surface of the outward normal component of $\nabla(1/R)$, equal to $-\epsilon/\epsilon^3$. The surface integral is then $-\epsilon/\epsilon^3 \times 4\pi\epsilon^2 = -4\pi$. Thus, the volume integral (1.168) is the general solution of the Poisson equation (1.163).

An arbitrary vector field $\boldsymbol{u}$ can then be broken into a lamellar part, determined within a constant by the scalar potential ϕ given by (1.168), and a solenoidal part uniquely determined, once the gauge has been chosen, by the vector potential $\boldsymbol{\psi}$, given by the solution of the vector Poisson equation (1.167).

Equation (1.167) is a vector Poisson equation, and solving it, component by component, yields

$$\boldsymbol{\psi}(\boldsymbol{r}) = \frac{1}{4\pi} \iiint \frac{\nabla' \times \boldsymbol{u}(\boldsymbol{r}')}{|\boldsymbol{r} - \boldsymbol{r}'|} d\mathcal{V}'. \tag{1.173}$$

We see from expression (1.173) that the solenoidal part of the vector field $\boldsymbol{u}$ depends on its curl.

A further breakdown is common in geomagnetism, where the solenoidal magnetic field is decomposed into poloidal and toroidal or torsional parts. This decomposition was first used by Lamb (1881) and formally demonstrated by Backus (1958). The Lamb–Backus decomposition allows $\boldsymbol{u}$ to be expressed as the sum of the lamellar part $\boldsymbol{L}$, the poloidal part $\boldsymbol{P}$ and the toroidal or torsional part $\boldsymbol{T}$. Then,

$$\boldsymbol{u} = \boldsymbol{L} + \boldsymbol{P} + \boldsymbol{T} = \nabla L + \nabla \times (\nabla \times \boldsymbol{r} P) + \nabla \times \boldsymbol{r} T. \tag{1.174}$$

The scalar L is, to within a constant, identical to the scalar potential ϕ, already considered, and the scalar P generates the poloidal field, while the scalar T generates the toroidal or torsional field.

We shall see in the next section that the curl of a poloidal field is toroidal or torsional, and that the curl of a toroidal or torsional field is poloidal. The negative of the curl, taken twice, of a poloidal field is a new poloidal field with scalar P', which is the Laplacian of P, while the negative of the curl, taken twice, of a toroidal or torsional field is a new toroidal or torsional field with scalar T', which is the Laplacian of T. Thus

$$-\nabla \times (\nabla \times \boldsymbol{u}) = \nabla \times (\nabla \times \boldsymbol{r} P') + \nabla \times \boldsymbol{r} T' \tag{1.175}$$

with

$$\nabla^2 P = P', \quad \nabla^2 T = T' \tag{1.176}$$

While the solenoidal vector potential, $\boldsymbol{\psi}$, is dependent on $-\nabla \times \boldsymbol{u}$, and determined by solution of the vector Poisson equation (1.167), the poloidal and toroidal or torsional fields are dependent on the negative of the curl, taken twice, $-\nabla \times (\nabla \times \boldsymbol{u})$, through its scalars P', T', and determined by the solutions of the two scalar Poisson equations (1.176).

1.3 Vector spherical harmonics

The Earth departs from spherical shape by only about one part in three hundred. For this reason spherical harmonics are widely used in geophysics. Often, the fields being described are vector fields in spherical co-ordinate systems and the basis functions are Legendre functions of the first kind.

We will make extensive use of the associated Legendre functions $P_n^m(x)$ defined in terms of the Legendre polynomials $P_n(x)$ by

$$P_n^m(x) = (-1)^m \left(1 - x^2\right)^{m/2} \frac{d^m P_n(x)}{dx^m}. \tag{1.177}$$

This is Hobson's definition (Copson, 1955). Sometimes the factor $(-1)^m$ is omitted, as is commonly done in the USA (Dahlen and Tromp, 1998). In the classical literature, Ferrers used the notation $T_n^m(x)$ to distinguish this case. In a spherical polar co-ordinate system (r, θ, ϕ), the angular dependence is conveniently described by series of the spherical harmonic functions

$$P_n^m(\cos\theta)\, e^{im\phi} \tag{1.178}$$

for $m \geq 0$. For all m, we take

$$P_n^{-m} = (-1)^m \frac{(n-m)!}{(n+m)!} P_n^m. \tag{1.179}$$

Spherical harmonics are orthogonal under integration over a sphere. Thus,

$$\int_0^{2\pi} \int_0^{\pi} P_n^m(\cos\theta)\, P_l^k(\cos\theta) \sin\theta\, e^{i(m+k)\phi}\, d\theta\, d\phi = (-1)^m \frac{4\pi}{2n+1} \delta_l^n \delta_{-k}^m, \tag{1.180}$$

where $\delta_l^n \delta_{-k}^m$ is a product of Kronecker deltas. We can then expand an arbitrary, time dependent, function on a sphere $f(r, \theta, \phi; t)$, as

$$f(r, \theta, \phi; t) = \sum_{n=0}^{\infty} \sum_{m=-n}^{n} q_n^m(r, t)\, P_n^m(\cos\theta)\, e^{im\phi}. \tag{1.181}$$

Multiplying both sides of (1.181) by $P_l^k(\cos\theta)\, e^{ik\phi}$, integrating over the unit sphere and using the orthogonality relation (1.180), yields

$$q_n^m(r, t) = (-1)^m \frac{2n+1}{4\pi} \int_0^{2\pi} \int_0^{\pi} f(r, \theta, \phi; t)\, P_n^{-m}(\cos\theta)\, e^{-im\phi} \sin\theta\, d\theta\, d\phi \tag{1.182}$$

as the expression for the radial coefficients $q_n^m(r, t)$. If the function $f(r, \theta, \phi; t)$ is real, then

$$q_n^{-m} = (-1)^m \frac{(n+m)!}{(n-m)!} q_n^{m*}, \tag{1.183}$$

where the asterisk denotes the complex conjugate.

The lamellar, poloidal and toroidal or torsional scalars can be expanded in the form (1.181) to give

$$\begin{Bmatrix} L \\ P \\ T \end{Bmatrix} = \sum_{n=0}^{\infty} \sum_{m=-n}^{n} \begin{Bmatrix} l_n^m(r,t) \\ p_n^m(r,t) \\ t_n^m(r,t) \end{Bmatrix} P_n^m(\cos\theta)\, e^{im\phi}. \tag{1.184}$$

The radial coefficients l_n^m, p_n^m, t_n^m, in general, can be functions of time, so we need to use partial derivatives when differentiating them.

The resulting vector spherical harmonics can be found from expression (1.174). In what follows, where convenient, we take the summations over m and n to be implied.

The components, in the spherical polar co-ordinate directions, of the lamellar vector are

$$\boldsymbol{L}_n^m = \left(\frac{\partial l_n^m}{\partial r} P_n^m,\ \frac{l_n^m}{r} \frac{dP_n^m}{d\theta},\ \frac{im}{r\sin\theta} l_n^m P_n^m \right) e^{im\phi}, \tag{1.185}$$

while those of the poloidal vector are

$$\boldsymbol{P}_n^m = \left(\frac{n(n+1)}{r} p_n^m P_n^m,\ \frac{1}{r}\frac{\partial}{\partial r}(r p_n^m) \frac{dP_n^m}{d\theta},\ \frac{im}{r\sin\theta}\frac{\partial}{\partial r}(r p_n^m) P_n^m \right) e^{im\phi}, \tag{1.186}$$

where Legendre's equation (B.1) has been used to simplify the expression for the radial component. The components of the toroidal or torsional vector spherical harmonic are

$$\boldsymbol{T}_n^m = \left(0,\ \frac{im}{\sin\theta} t_n^m P_n^m,\ -t_n^m \frac{dP_n^m}{d\theta} \right) e^{im\phi}. \tag{1.187}$$

The divergence of the lamellar vector $\boldsymbol{L}$ is a scalar L'' that, with the use of Legendre's equation (B.1), can be shown to be given by

$$L'' = \nabla^2 L. \tag{1.188}$$

The negative of the curl of the poloidal vector $\boldsymbol{P}$ is a toroidal or torsional vector with scalar P'' given by

$$P'' = \nabla^2 P. \tag{1.189}$$

The negative of the curl, taken twice, of the poloidal vector $\boldsymbol{P}$ is a poloidal vector $\boldsymbol{P}''$ given by

$$\boldsymbol{P}'' = \nabla \times (\nabla \times \boldsymbol{r} P''). \tag{1.190}$$

The curl of the toroidal or torsional vector $\boldsymbol{T}$ is a poloidal vector with scalar T. The negative of the curl, taken twice, of the toroidal or torsional vector $\boldsymbol{T}$ is a toroidal or torsional vector $\boldsymbol{T}''$ given by

$$\boldsymbol{T}'' = \nabla \times \boldsymbol{r} T'', \tag{1.191}$$

with T'' given by

$$T'' = \nabla^2 T. \tag{1.192}$$

The scalars L'', P'', T'' may, in turn, be expanded in spherical harmonics as

$$\left\{\begin{array}{c} L'' \\ P'' \\ T'' \end{array}\right\} = \sum_{n=0}^{\infty} \sum_{m=-n}^{n} \left\{\begin{array}{c} l_n''^m(r,t) \\ p_n''^m(r,t) \\ t_n''^m(r,t) \end{array}\right\} P_n^m(\cos\theta)\, e^{im\phi}, \tag{1.193}$$

with

$$\left\{\begin{array}{c} l'' \\ p'' \\ t'' \end{array}\right\}_n^m = \frac{1}{r}\frac{\partial^2}{\partial r^2}\left(r \left\{\begin{array}{c} l \\ p \\ t \end{array}\right\}_n^m \right) - \frac{n(n+1)}{r^2} \left\{\begin{array}{c} l \\ p \\ t \end{array}\right\}_n^m . \tag{1.194}$$

In seismology, it is usual to combine the lamellar and poloidal fields into the single spheroidal field, $\boldsymbol{S}_n^m$, with components

$$\boldsymbol{S}_n^m = \boldsymbol{L}_n^m + \boldsymbol{P}_n^m = \left(u_n^m P_n^m,\ v_n^m \frac{dP_n^m}{d\theta},\ \frac{im}{\sin\theta} v_n^m P_n^m \right) e^{im\phi}, \tag{1.195}$$

where $u_n^m(r,t)$ is the radial spheroidal coefficient and $v_n^m(r,t)$ is the transverse spheroidal coefficient. Comparison with expressions (1.185) and (1.186) show these to be related to the lamellar and poloidal coefficients by

$$u_n^m = \frac{\partial l_n^m}{\partial r} + \frac{n(n+1)}{r} p_n^m, \quad v_n^m = \frac{1}{r} l_n^m + \frac{1}{r}\frac{\partial}{\partial r}(r p_n^m). \tag{1.196}$$

In turn, inversion of these relations allows expression of the lamellar and poloidal coefficients, in terms of the radial spheroidal and transverse spheroidal coefficients, as

$$l_n''^m = \frac{1}{r}\frac{\partial^2}{\partial r^2}(r l_n^m) - \frac{n(n+1)}{r^2} l_n^m = \frac{\partial u_n^m}{\partial r} + \frac{(2u_n^m - n(n+1) v_n^m)}{r}, \tag{1.197}$$

and

$$p_n''^m = \frac{1}{r}\frac{\partial^2}{\partial r^2}(r p_n^m) - \frac{n(n+1)}{r^2} p_n^m = \frac{\partial v_n^m}{\partial r} + \frac{(v_n^m - u_n^m)}{r}. \tag{1.198}$$

As is the case in the application of Legendre functions to the expansion of scalar fields in the form (1.181), the usefulness of vector spherical harmonics derives from their orthogonality under integration over a sphere. Suppose we have two spheroidal vector fields with radial coefficients u_n^m, v_n^m and $u_l'^k$, $v_l'^k$, respectively.

If we take their scalar product, multiply by $\sin\theta$, and integrate over a sphere, we find that

$$\begin{aligned}\int_0^{2\pi}\int_0^{\pi} & \boldsymbol{S}_n^m \cdot \boldsymbol{S}_l'^{\,k} \sin\theta\, d\theta\, d\phi \\ &= \int_0^{2\pi}\int_0^{\pi} u_n^m u_l'^{\,k} P_n^m P_l^k \sin\theta\, e^{i(m+k)\phi}\, d\theta\, d\phi \\ &+ \int_0^{2\pi}\int_0^{\pi} v_n^m v_l'^{\,k} \left[\frac{dP_n^m}{d\theta}\frac{dP_l^k}{d\theta} - \frac{mk}{\sin^2\theta} P_n^m P_l^k\right] \sin\theta\, e^{i(m+k)\phi}\, d\theta\, d\phi. \end{aligned} \tag{1.199}$$

From the orthogonality of Legendre functions (1.180), the first integral on the right is

$$(-1)^m \frac{4\pi}{2n+1} \delta_l^n \delta_{-k}^m u_n^m u_l'^{\,k}. \tag{1.200}$$

After integrating over ϕ, the second integral on the right gives

$$2\pi\delta_{-k}^m v_n^m v_l'^{\,k} \int_0^{\pi} \left[\frac{dP_n^m}{d\theta}\frac{dP_l^k}{d\theta} + \frac{m^2}{\sin^2\theta} P_n^m P_l^k\right] \sin\theta\, d\theta. \tag{1.201}$$

Further integration by parts of the first term yields

$$\begin{aligned}2\pi\delta_{-k}^m v_n^m v_l'^{\,k} \Bigg[& P_l^k \frac{dP_n^m}{d\theta} \sin\theta \bigg|_0^{\pi} - \int_0^{\pi} \frac{1}{\sin\theta}\frac{d}{d\theta}\left(\sin\theta \frac{dP_n^m}{d\theta}\right) P_l^k \sin\theta\, d\theta \\ & + \int_0^{\pi} \frac{m^2}{\sin^2\theta} P_n^m P_l^k \sin\theta\, d\theta \Bigg]. \end{aligned} \tag{1.202}$$

Since $\sin\theta$ vanishes at both limits, and on replacing

$$\frac{1}{\sin\theta}\frac{d}{d\theta}\left(\sin\theta \frac{dP_n^m}{d\theta}\right) \quad \text{by} \quad -\left\{n(n+1) - \frac{m^2}{\sin^2\theta}\right\} P_n^m, \tag{1.203}$$

in accordance with Legendre's equation (B.1), we are left with

$$\begin{aligned}2\pi\delta_{-k}^m v_n^m v_l'^{\,k} n(n+1) & \int_0^{\pi} P_n^m P_l^k \sin\theta\, d\theta \\ &= (-1)^m \frac{4\pi}{2n+1} \delta_l^n \delta_{-k}^m\, n(n+1)\, v_n^m v_l'^{\,k}. \end{aligned} \tag{1.204}$$

The spheroidal vector fields $\boldsymbol{S}_n^m$ and $\boldsymbol{S}_l'^{\,k}$ then obey the orthogonality relation

$$\begin{aligned}\int_0^{2\pi}\int_0^{\pi} & \boldsymbol{S}_n^m \cdot \boldsymbol{S}_l'^{\,k} \sin\theta\, d\theta\, d\phi \\ &= (-1)^m \frac{4\pi}{2n+1} \left[u_n^m(r) u_l'^{\,k}(r) + n(n+1)\, v_n^m v_l'^{\,k}(r)\right] \delta_l^n \delta_{-k}^m. \end{aligned} \tag{1.205}$$

From relations (1.196), the orthogonality relations for two purely lamellar vectors, $\boldsymbol{L}_n^m$ and $\boldsymbol{L}_l'^k$, or two purely poloidal vectors, $\boldsymbol{P}_n^m$ and $\boldsymbol{P}_l'^k$, follow directly,

$$\int_0^{2\pi}\int_0^{\pi} \boldsymbol{L}_n^m \cdot \boldsymbol{L}_l'^k \sin\theta\, d\theta\, d\phi$$
$$= (-1)^m \frac{4\pi}{2n+1}\left[\frac{\partial l_n^m}{\partial r}\frac{\partial l_l'^k}{\partial r} + \frac{n(n+1)}{r} l_n^m l_l'^k\right]\delta_l^n \delta_{-k}^m, \qquad (1.206)$$

$$\int_0^{2\pi}\int_0^{\pi} \boldsymbol{P}_n^m \cdot \boldsymbol{P}_l'^k \sin\theta\, d\theta\, d\phi$$
$$= (-1)^m \frac{4\pi}{2n+1}\left[\frac{n(n+1)}{r^2} p_n^m p_l'^k + \frac{1}{r^2}\frac{\partial}{\partial r}(r p_n^m)\frac{\partial}{\partial r}\left(r p_l'^k\right)\right]\delta_l^n \delta_{-k}^m. \qquad (1.207)$$

The integral over a unit sphere, of the scalar product of two toroidal or torsional vector spherical harmonics, multiplied by $\sin\theta$, gives

$$\int_0^{2\pi}\int_0^{\pi} \boldsymbol{T}_n^m \cdot \boldsymbol{T}_l'^k \sin\theta\, d\theta\, d\phi$$
$$= 2\pi \delta_{-k}^m t_n^m t_l'^k \int_0^{\pi}\left[\frac{dP_n^m}{d\theta}\frac{dP_l^k}{d\theta} + \frac{m^2}{\sin^2\theta} P_n^m P_l^k\right]\sin\theta\, d\theta. \qquad (1.208)$$

The same integral was encountered before in expression (1.201). By the previous result, the orthogonality relation for two toroidal or torsional vectors $\boldsymbol{T}_n^m$ and $\boldsymbol{T}_l'^k$ becomes

$$\int_0^{2\pi}\int_0^{\pi} \boldsymbol{T}_n^m \cdot \boldsymbol{T}_l'^k \sin\theta\, d\theta\, d\phi = (-1)^m \frac{4\pi}{2n+1} n(n+1)\, t_n^m t_l'^k \delta_l^n \delta_{-k}^m. \qquad (1.209)$$

Toroidal or torsional vector spherical harmonics do not interact with lamellar, poloidal or spheroidal vector spherical harmonics. If we take their scalar product with any of these, multiply by $\sin\theta$ and integrate over a unit sphere, the result is found, in all cases, to depend on the integral

$$\int_0^{\pi} \frac{d}{d\theta}\left(P_n^m P_n^{-m}\right) d\theta = P_n^m(\cos\theta)\, P^{-m}(\cos\theta)\,\Big|_0^{\pi}$$
$$= P_n^m(-1)\, P_n^{-m}(-1) - P_n^m(1)\, P_n^{-m}(1). \qquad (1.210)$$

For $m \neq 0$, we have $P_n^{\pm m}(\pm 1) = 0$. For $m = 0$, we have $P_n(-1) = (-1)^n$ and $P_n(1) = 1$. Thus, in all cases, the integral (1.210) vanishes.

In the four orthogonality relations (1.205), (1.206), (1.207) and (1.209), the radial coefficients $u_l'^k$, $v_l'^k$, $l_l'^k$, $p_l'^k$ and $t_l'^k$ are at our disposal. By choosing $v_l'^k = 0$ and $u_l'^k = 1$ in (1.205), we find the radial spheroidal coefficient. By choosing $u_l'^k = 0$ and $v_l'^k = 1$, we find the transverse spheroidal coefficient. By choosing

$\partial l_l'^k/\partial r = 0$ and $l_l'^k = 1$ in (1.206), we find the lamellar radial coefficient. By choosing $\partial(rp_l'^k)/\partial r = 0$ and $p_l'^k = 1$ in (1.207), we find the poloidal radial coefficient. Finally, by choosing $t_l'^k = 1$ in (1.209), we find the toroidal or torsional radial coefficient.

It follows that we can expand an arbitrary vector field in a series of lamellar, poloidal and toroidal vector spherical harmonics, verifying the Lamb–Backus decomposition, frequently used in geomagnetism. Similarly, an arbitrary vector field may be expanded in a series of spheroidal and torsional vector spherical harmonics, as is commonly done in seismology.

1.4 Elasticity theory

Although the Earth is a rotating, self-gravitating body with significant radial variation in its elastic properties, and large pre-existing stress, in some cases it may be treated, locally, as uniform, not self-gravitating, and free of pre-stress. Much of classical elasticity theory then applies. We delay a fuller description of realistic Earth deformations to Chapter 3.

1.4.1 *Analysis of stress*

The state of stress in a continuum is studied by examining the forces acting on an imaginary surface element in the medium, as illustrated in Figure 1.2.

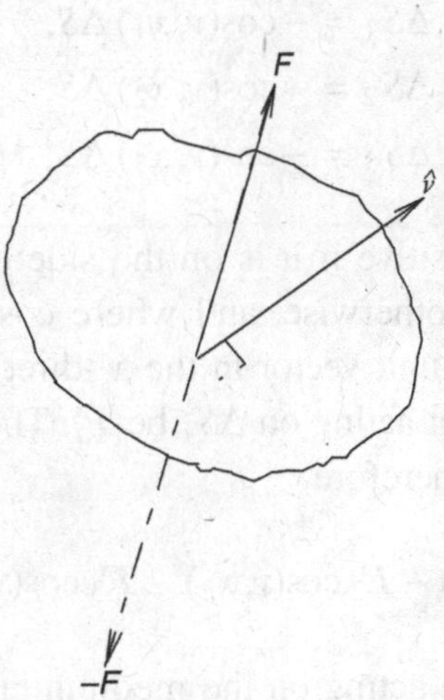

Figure 1.2 Small surface element in a stressed medium. Vector $\hat{\boldsymbol{\nu}}$ is the unit outward normal vector of the surface element. The force $\boldsymbol{F}$ acts on the outside of the element, while an equal and opposite force $-\boldsymbol{F}$ acts on the inside.

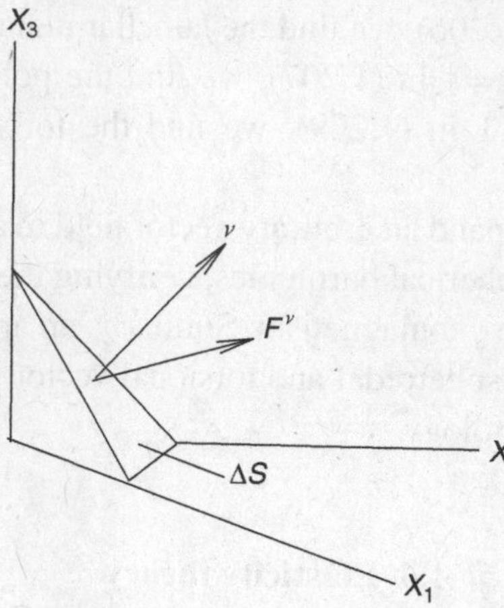

Figure 1.3 A small tetrahedron, in a stressed medium, cut out by the Cartesian co-ordinate planes.

A force $\boldsymbol{F}$ acts on the outside of the small surface element, while an equal and opposite force $-\boldsymbol{F}$ acts on the inside. The orientation of the surface element is specified by its unit outward normal vector $\hat{\boldsymbol{\nu}}$. To specify the state of stress in the medium we must, in general, specify both $\boldsymbol{F}$ and $\hat{\boldsymbol{\nu}}$. To do this, consider the tetrahedron cut out by the Cartesian co-ordinate planes shown in Figure 1.3.

The faces of the tetrahedron orthogonal to the three co-ordinate directions have areas ΔS_1, ΔS_2 and ΔS_3, respectively, while the fourth face has area ΔS. Then,

$$\begin{aligned} \Delta S_1 &= -\cos(\nu, x_1)\,\Delta S, \\ \Delta S_2 &= -\cos(\nu, x_2)\,\Delta S, \\ \Delta S_3 &= -\cos(\nu, x_3)\,\Delta S, \end{aligned} \tag{1.211}$$

where an area is counted positive if it is on the side of the normal vector or co-ordinate direction, negative otherwise, and where $\cos(\nu, x_i)$ is the cosine of the angle between $\hat{\boldsymbol{\nu}}$ and $\hat{e}_i$, the unit vector in the x_i-direction. Let the force per unit area acting on ΔS be $\boldsymbol{F}^\nu$, that acting on ΔS_i be $\boldsymbol{F}_i$. The sum of the surface forces acting on the tetrahedron is therefore

$$\left[\boldsymbol{F}^\nu - \boldsymbol{F}_1\cos(\nu, x_1) - \boldsymbol{F}_2\cos(\nu, x_2) - \boldsymbol{F}_3\cos(\nu, x_3)\right]\Delta S. \tag{1.212}$$

Body forces could also be acting on the medium enclosed by the tetrahedron. Body forces are proportional to volume. Suppose L is a characteristic length of the tetrahedron. Then, the surface forces are proportional to L^2, the body forces are proportional to L^3. If we let $\Delta S \to 0$ with $\hat{\boldsymbol{\nu}}$ fixed, it is evident that the body

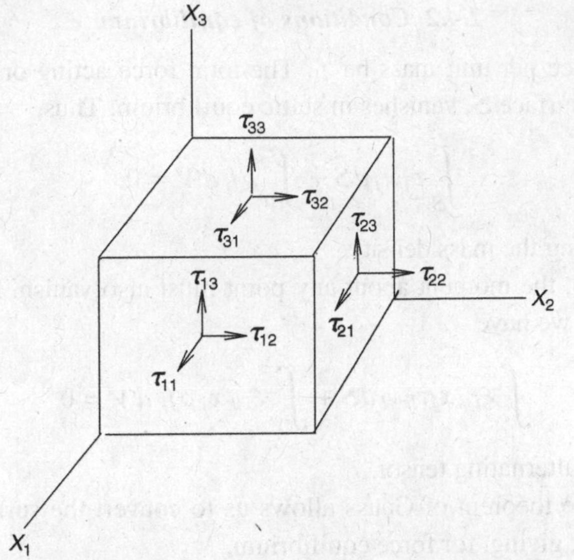

Figure 1.4 A rectangular parallelepiped, with faces parallel to the co-ordinate axes, in the stressed medium.

forces vanish more strongly than the surface forces, and that, in the limit, force equilibrium demands that

$$\boldsymbol{F}^{\nu} = \boldsymbol{F}_1 \cos(\nu, x_1) + \boldsymbol{F}_2 \cos(\nu, x_2) + \boldsymbol{F}_3 \cos(\nu, x_3). \qquad (1.213)$$

The state of stress in a medium can therefore be described by a quantity, which associates a vector with each spatial direction, by means of an expression that is linear and homogeneous in the direction cosines. In other words, *stress is a second-order tensor*.

If we write the components of $\boldsymbol{F}_1$ as $(\tau_{11}, \tau_{12}, \tau_{13})$, those of $\boldsymbol{F}_2$ as $(\tau_{21}, \tau_{22}, \tau_{23})$ and those of $\boldsymbol{F}_3$ as $(\tau_{31}, \tau_{32}, \tau_{33})$, we have

$$F_i^{\nu} = \tau_{ji}\nu_j, \qquad (1.214)$$

where the range and summation conventions are taken to apply. The τ_{ji} are the Cartesian components of the *second-order stress tensor*. They represent the components of the forces per unit area on the faces of the rectangular parallelepiped shown cut from the medium in Figure 1.4. The components τ_{11}, τ_{22}, τ_{33} are referred to as *normal stresses*, while τ_{12}, τ_{13}, τ_{21}, τ_{23}, τ_{31}, τ_{32} are referred to as *shear stresses*.

1.4.2 Conditions of equilibrium

Let the body force per unit mass be f_i. The total force acting on the volume $\mathcal{V}$, bounded by the surface $\mathcal{S}$, vanishes in static equilibrium. Thus,

$$\int_{\mathcal{S}} \tau_{ji}\nu_j \, d\mathcal{S} + \int_{\mathcal{V}} \rho f_i \, d\mathcal{V} = 0, \tag{1.215}$$

with ρ representing the mass density.

In equilibrium, the moment about any point must also vanish. For the moment about the origin, we have

$$\int_{\mathcal{S}} \xi_{ijk} x_j \tau_{lk} \nu_l \, d\mathcal{S} + \int_{\mathcal{V}} \xi_{ijk} x_j \rho f_k \, d\mathcal{V} = 0. \tag{1.216}$$

where ξ_{ijk} is the alternating tensor.

The divergence theorem of Gauss allows us to convert the surface integrals to volume integrals, giving, for force equilibrium,

$$\int_{\mathcal{V}} \left[\frac{\partial \tau_{ji}}{\partial x_j} + \rho f_i \right] d\mathcal{V} = 0, \tag{1.217}$$

and, for moment equilibrium,

$$\begin{aligned} &\int_{\mathcal{V}} \xi_{ijk} \left[\frac{\partial}{\partial x_l} \left(x_j \tau_{lk} \right) + \rho x_j f_k \right] d\mathcal{V} \\ &= \int_{\mathcal{V}} \xi_{ijk} \left[\tau_{jk} + x_j \left(\frac{\partial \tau_{lk}}{\partial x_l} + \rho f_k \right) \right] d\mathcal{V} = 0. \end{aligned} \tag{1.218}$$

Since $\mathcal{V}$ is an arbitrary volume,

$$\frac{\partial \tau_{ji}}{\partial x_j} = -\rho f_i \tag{1.219}$$

and

$$\xi_{ijk} \left[\tau_{jk} + x_j \left(\frac{\partial \tau_{lk}}{\partial x_l} + \rho f_k \right) \right] = \xi_{ijk} \tau_{jk} = 0. \tag{1.220}$$

The latter condition for moment equilibrium expands to three equations: $\tau_{23} - \tau_{32} = 0, -\tau_{13} + \tau_{31} = 0, \tau_{12} - \tau_{21} = 0$. Hence, moment equilibrium requires that $\tau_{ij} = \tau_{ji}$, or that *the stress tensor is symmetric*. The condition for force equilibrium can then alternatively be written as

$$\frac{\partial \tau_{ij}}{\partial x_j} = -\rho f_i, \tag{1.221}$$

referred to as the *equation of equilibrium*.

The stress tensor has only six independent Cartesian components: three normal stresses, $\tau_{11}, \tau_{22}, \tau_{33}$ and three shear stresses, $\tau_{12}, \tau_{23}, \tau_{13}$.

1.4.3 Analysis of deformation

The deformation of a body is described by the *vector displacement field* $u_i(x_1, x_2, x_3)$, which is the vector displacement the material particle originally at (x_1, x_2, x_3) experienced during deformation. The displacement of a material particle P', originally at $x_k + dx_k$, relative to another material particle P, originally at x_k, is then

$$du_i = \frac{\partial u_i}{\partial x_j} dx_j. \tag{1.222}$$

The *vector gradient* $\partial u_i/\partial x_j$ is a second-order tensor. Since du_i and dx_j are vectors, in a new co-ordinate system they become

$$du'_i = c_{ij} du_j, \quad dx'_j = c_{jk} dx_k, \tag{1.223}$$

and in reverse transformation,

$$dx_k = c_{lk} dx'_l. \tag{1.224}$$

Then,

$$du'_i = \frac{\partial u'_i}{\partial x'_j} dx'_j = c_{ij} du_j = c_{ij} \frac{\partial u_j}{\partial x_k} dx_k = c_{ij} c_{lk} \frac{\partial u_j}{\partial x_k} dx'_l, \tag{1.225}$$

or

$$\left(\frac{\partial u'_i}{\partial x'_l} - c_{ij} c_{lk} \frac{\partial u_j}{\partial x_k} \right) dx'_l = 0. \tag{1.226}$$

Thus,

$$\frac{\partial u'_i}{\partial x'_l} = c_{ij} c_{lk} \frac{\partial u_j}{\partial x_k}, \tag{1.227}$$

in accordance with the transformation law for second-order tensors.

The vector gradient can be separated into symmetric and antisymmetric parts by writing

$$\frac{\partial u_i}{\partial x_j} = \frac{1}{2} \left(\frac{\partial u_i}{\partial x_j} + \frac{\partial u_j}{\partial x_i} \right) + \frac{1}{2} \left(\frac{\partial u_i}{\partial x_j} - \frac{\partial u_j}{\partial x_i} \right) = e_{ji} + \omega_{ji}, \tag{1.228}$$

where

$$e_{ij} = \frac{1}{2} \left(\frac{\partial u_j}{\partial x_i} + \frac{\partial u_i}{\partial x_j} \right), \quad \omega_{ij} = \frac{1}{2} \left(\frac{\partial u_j}{\partial x_i} - \frac{\partial u_i}{\partial x_j} \right) \tag{1.229}$$

are both second-order tensors.

The *dual vector* of a second-order tensor T_{jk} is given by the double contraction $v_i = \xi_{ijk} T_{jk}$. Thus $v_1 = T_{23} - T_{32}$, $v_2 = T_{31} - T_{13}$, $v_3 = T_{12} - T_{21}$. Only the antisymmetric part of T_{jk} contributes to the dual vector.

One-half of the dual vector of the vector gradient $\partial u_k/\partial x_j$ is therefore

$$\Omega_i = \frac{1}{2}\xi_{ijk}\omega_{jk} = \frac{1}{2}\xi_{ijk}\left(\omega_{jk} + e_{jk}\right) = \frac{1}{2}\xi_{ijk}\frac{\partial u_k}{\partial x_j} = \frac{1}{2}(\nabla \times \boldsymbol{u})_i \,. \tag{1.230}$$

If T_{jk} is an antisymmetric tensor, it may be recovered from its dual vector v_i since

$$\xi_{ijk} v_i = \xi_{ijk}\xi_{ilm} T_{lm} = T_{jk} - T_{kj} = 2T_{jk}. \tag{1.231}$$

Thus,

$$T_{jk} = \frac{1}{2}\xi_{ijk} v_i. \tag{1.232}$$

Applied to ω_{ji}, we find $\omega_{ji} = \xi_{kji}\Omega_k$. Then

$$\omega_{ji}\, dx_j = \xi_{kji}\Omega_k\, dx_j = (\boldsymbol{\Omega} \times d\boldsymbol{r})_i \,, \tag{1.233}$$

where $(d\boldsymbol{r})_i = dx_i$.

The displacement of the material particle P', originally at $x_k + dx_k$, relative to the material particle P, originally at x_k, is

$$du_i = e_{ij}\, dx_j + (\boldsymbol{\Omega} \times d\boldsymbol{r})_i \,. \tag{1.234}$$

Hence, ω_{ij} represents a rigid-body rotation of P' with respect to P through the angle $\boldsymbol{\Omega}$. It does not, therefore, represent a true deformation of the medium.

If, on the other hand, the displacement field is entirely rigid-body in the neighbourhood of P and P', and if P'' is a third neighbouring material particle, originally at $x_k + \delta x_k$, the angle $P'PP'' = \theta$ must be preserved and the lengths PP', PP'' must remain unchanged in the deformation. The scalar product $d\boldsymbol{r} \cdot \delta\boldsymbol{r}$ before and after deformation is invariant. Thus,

$$dx_k\, \delta x_k = \left(dx_k + \frac{\partial u_k}{\partial x_j} dx_j\right)\left(\delta x_k + \frac{\partial u_k}{\partial x_l}\delta x_l\right). \tag{1.235}$$

To lowest order in small quantities, we have

$$dx_k\, \delta x_k \approx dx_k\, \delta x_k + \frac{\partial u_k}{\partial x_j} dx_j\, \delta x_k + \frac{\partial u_k}{\partial x_l}\delta x_l\, dx_k, \tag{1.236}$$

or

$$\left(\frac{\partial u_k}{\partial x_j} + \frac{\partial u_j}{\partial x_k}\right) dx_j\,\delta x_k = 0. \tag{1.237}$$

Hence, if the displacement is locally rigid-body, e_{ij} must vanish. We have already shown that if e_{ij} vanishes, the displacement is rigid-body. Thus, a necessary and sufficient condition for a rigid-body displacement field is that $e_{ij} = 0$. Therefore,

$$e_{ij} = \frac{1}{2}\left(\frac{\partial u_j}{\partial x_i} + \frac{\partial u_i}{\partial x_j}\right) \tag{1.238}$$

measures the true deformation of the medium. It is called the *strain tensor*.

Now, let $ds = |d\boldsymbol{r}|$, $\delta s = |\delta \boldsymbol{r}|$, $d\boldsymbol{r} = \hat{\boldsymbol{\mu}}\,ds$, $\delta\boldsymbol{r} = \hat{\boldsymbol{\nu}}\,\delta s$. Then,

$$d\boldsymbol{r} \cdot \delta\boldsymbol{r} = dx_k\,\delta x_k = ds\,\delta s\,\mu_k\nu_k = ds\,\delta s\,(\hat{\boldsymbol{\mu}} \cdot \hat{\boldsymbol{\nu}}) = ds\,\delta s \cos\theta \tag{1.239}$$

before deformation. Since the scalar product is not affected by rigid-body rotation, after deformation it is

$$\begin{aligned}\left(dx_k + e_{jk}dx_j\right)\left(\delta x_k + e_{lk}\delta x_l\right) &\approx dx_k\,\delta x_k + \left(e_{jk} + e_{kj}\right) dx_j\,\delta x_k \\ &= \left(\cos\theta + 2e_{jk}\mu_j\nu_k\right) ds\,\delta s.\end{aligned} \tag{1.240}$$

If P' and P'' coincide, before deformation $dx_k\,\delta x_k = (ds)^2$. After deformation, it is

$$\left[1 + 2e_{jk}\mu_j\mu_k\right](ds)^2. \tag{1.241}$$

The relative extension, or change in length per unit length, in the PP' direction is then $e_{jk}\mu_j\mu_k$. In the x_i-direction, the relative extension is e_{ii}. This is the *normal component of the strain tensor* in the x_i-direction. The diagonal elements of the strain tensor represent the relative extensions in the three co-ordinate directions arising from the deformation.

If PP' and PP'' are orthogonal before deformation, $\theta = \pi/2$ and $\cos\theta = \cos\pi/2 = 0$. After deformation, they are no longer orthogonal and $\cos(\pi/2 + \Delta\theta) = \cos\pi/2 + 2e_{jk}\mu_j\nu_k$. Then,

$$\cos(\pi/2 + \Delta\theta) = -\sin\Delta\theta \approx -\Delta\theta = 2e_{jk}\mu_j\nu_k. \tag{1.242}$$

If, before deformation, PP' is aligned with the x_1-axis and PP'' is aligned with the x_2-axis, the situation is as illustrated in Figure 1.5. The change in the angle $P'PP''$ is

$$\Delta\theta = -2e_{jk}\mu_j\nu_k. \tag{1.243}$$

The unit vectors have components $\mu_j = \delta_j^1$ and $\nu_k = \delta_k^2$. Thus,

$$\Delta\theta = -2e_{12} = -\frac{\partial u_1}{\partial x_2} \tag{1.244}$$

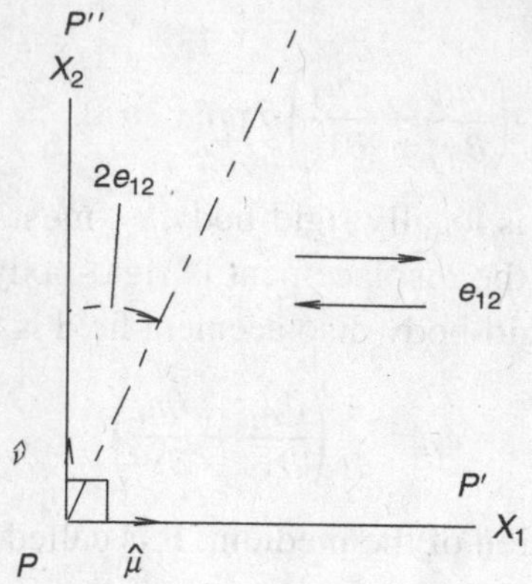

Figure 1.5 Deformation of two lines, PP' and PP'', joining material points in the medium, orthogonal before deformation by the shearing strain e_{12}.

represents the total change in angle, deriving half from e_{12}, the shear strain, and half from $\Omega_3 = -\frac{1}{2}\partial u_1/\partial x_2$, the rigid-body rotation of P'' around P. The off-diagonal elements of the strain tensor represent the shearing strains in planes orthogonal to the three co-ordinate directions.

1.4.4 Hooke's law and the Navier equation

In infinitesimal elasticity theory, the strains are regarded as sufficiently small that their products can be neglected. By one of the oldest laws of physics, enunciated by Hooke in 1678, stress and strain are linearly related. Hooke's law has been found to be broadly applicable for small strains.

Since stress and strain are both second-order tensors, the most general linear relation between stress and strain would theoretically involve as many as 81 coefficients. Fortunately, for an isotropic medium, there are only two independent coefficients and the generalized Hooke's law takes the form,

$$\tau_{ij} = \lambda e_{kk}\delta^i_j + 2\mu e_{ij}, \tag{1.245}$$

where λ and μ are the Lamé elastic coefficients.

The normal strains e_{11}, e_{22}, e_{33} are the relative extensions in the three co-ordinate directions. Their sum $e_{kk} = e_{11} + e_{22} + e_{33} = \Theta$ is called the *cubical dilatation* or simply the *dilatation*. If δx_1, δx_2, δx_3 are the sides of a small parallelepiped cut out by surfaces parallel to the co-ordinate planes, its volume before deformation is $V = \delta x_1\, \delta x_2\, \delta x_3$. After deformation, its volume is

$$\begin{aligned}&(\delta x_1 + e_{11}\,\delta x_1)(\delta x_2 + e_{22}\,\delta x_2)(\delta x_3 + e_{33}\,\delta x_3)\\ &\quad\approx (1 + e_{11} + e_{22} + e_{33})\,\delta x_1\,\delta x_2\,\delta x_3,\end{aligned} \tag{1.246}$$

correct to first order in small quantities. Its volume after deformation is then

$$V + (e_{11} + e_{22} + e_{33})\,V = V + \Delta V. \tag{1.247}$$

The increase in volume, per unit volume, or the cubical dilatation, is therefore

$$\frac{\Delta V}{V} = e_{11} + e_{22} + e_{33} = \frac{\partial u_1}{\partial x_1} + \frac{\partial u_2}{\partial x_2} + \frac{\partial u_3}{\partial x_3} = \nabla \cdot \boldsymbol{u} = \Theta. \tag{1.248}$$

In the case of a pure shear, as illustrated in Figure 1.5, Hooke's law gives the stress as $\tau_{12} = 2\mu e_{12}$. The ratio of the applied shear stress to the deformation angle, $\tau_{12}/2e_{12} = \mu$, is called the *shear modulus* or the *modulus of rigidity*.

Hooke's law (1.245) relating stress to strain can be inverted to give a linear relation between strain and stress. Contracting the stress tensor, we find

$$\tau_{kk} = \lambda\Theta\delta^k_k + 2\mu e_{kk} = 3\lambda\Theta + 2\mu\Theta. \tag{1.249}$$

Thus, the dilatation is given by $\Theta = \tau_{kk}/(3\lambda+2\mu)$. Then Hooke's law can be written

$$\tau_{ij} = \frac{\lambda}{3\lambda + 2\mu}\tau_{kk}\delta^i_{\;j} + 2\mu e_{ij}. \tag{1.250}$$

Solving for e_{ij} gives the desired linear relation,

$$e_{ij} = -\frac{\lambda}{2\mu\,(3\lambda + 2\mu)}\tau_{kk}\delta^i_{\;j} + \frac{1}{2\mu}\tau_{ij}, \tag{1.251}$$

between strain and stress.

Now, suppose the state of stress is entirely uniaxial, and that τ_{11} is the only non-vanishing component of the stress tensor. The off-diagonal components of strain all vanish and

$$e_{11} = \frac{\lambda + \mu}{\mu\,(3\lambda + 2\mu)}\tau_{11}, \quad e_{22} = e_{33} = -\frac{\lambda}{2\mu\,(3\lambda + 2\mu)}\tau_{11}. \tag{1.252}$$

For uniaxial stress,

$$\tau_{11} = \frac{\mu\,(3\lambda + 2\mu)}{\lambda + \mu}e_{11} \tag{1.253}$$

relates stress to strain in the direction of application of the stress. The coefficient of proportionality,

$$E = \frac{\mu\,(3\lambda + 2\mu)}{\lambda + \mu}, \tag{1.254}$$

called *Young's modulus*, arises frequently in engineering applications.

In spite of the applied stress being uniaxial, the material is also strained in the two directions orthogonal to the stress axis. The ratio of the magnitudes of these strains to the axial strain is given by the dimensionless *Poisson's ratio*,

$$\sigma = \frac{\lambda}{2(\lambda + \mu)}. \tag{1.255}$$

If the state of stress is hydrostatic, there are no shear stresses, since by definition a fluid in equilibrium cannot support shear stress, and $\tau_{ij} = -p\delta^i_j$, where p is the hydrostatic pressure. Then, from the relation (1.251) between strain and stress,

$$e_{ij} = -\frac{\lambda}{2\mu(3\lambda + 2\mu)}(-3p)\,\delta^i_j - \frac{p}{2\mu}\delta^i_j = -\frac{p}{3\lambda + 2\mu}\delta^i_j. \tag{1.256}$$

The dilatation is $\Theta = e_{kk} = -3p/(3\lambda + 2\mu)$. The ratio of the pressure to the negative of the dilatation measures the resistance of the fluid to compression and is called the *bulk modulus*, or, as is common in geophysical usage, the *incompressibility*, K. It is

$$\frac{p}{-\Theta} = K = \lambda + \frac{2}{3}\mu. \tag{1.257}$$

The dilatation and the strains are, in turn, connected to the displacement field by

$$\Theta = e_{kk} = \nabla \cdot \boldsymbol{u} \quad \text{and} \quad e_{ij} = \frac{1}{2}\left(\frac{\partial u_j}{\partial x_i} + \frac{\partial u_i}{\partial x_j}\right). \tag{1.258}$$

In terms of the displacement field, the expression for the stress, given by Hooke's law (1.245), becomes

$$\tau_{ij} = \lambda \nabla \cdot \boldsymbol{u}\delta^i_j + \mu\left(\frac{\partial u_j}{\partial x_i} + \frac{\partial u_i}{\partial x_j}\right). \tag{1.259}$$

Betti showed in 1872 (see Love (1927), pp. 173–174) that for two systems of surface tractions and body forces, the work done by the first system, acting through the displacements caused by the second system, is equal to the work done by the second system, acting through the displacements caused by the first system. Let the surface tractions per unit area be t_i and t'_i, the body forces per unit volume be F_i and F'_i and the displacement fields they cause be u_i and u'_i, respectively. Then the work done by the first system of surface tractions and body forces, acting through the displacements caused by the second system is

$$\int_S t_i u'_i \, dS + \int_V F_i u'_i \, dV = \int_S \tau_{ji} \nu_j u'_i \, dS + \int_V F_i u'_i \, dV, \tag{1.260}$$

where we have used (1.214) to express the surface traction t_i in terms of the stress τ_{ji}. Transforming the surface integral to a volume integral by Gauss's theorem, the work done becomes

$$\int_V \left[\frac{\partial}{\partial x_j} \left(\tau_{ji} u_i' \right) + F_i u_i' \right] dV = \int_V \left[\tau_{ji} \frac{\partial u_i'}{\partial x_j} + \frac{\partial \tau_{ji}}{\partial x_j} u_i' + F_i u_i' \right] dV. \tag{1.261}$$

From the equilibrium equation (1.221),

$$\frac{\partial \tau_{ji}}{\partial x_j} = -\rho f_i = -F_i \quad \text{and} \quad \frac{\partial \tau_{ji}}{\partial x_j} u_i' = -F_i u_i'. \tag{1.262}$$

The pair of terms,

$$\frac{\partial \tau_{ji}}{\partial x_j} u_i' + F_i u_i', \tag{1.263}$$

then combine to

$$-F_i u_i' + F_i u_i' = 0. \tag{1.264}$$

Substitution for the stress τ_{ji} from Hooke's law (1.259), gives the work done as

$$\begin{aligned} \int_V & \left[\lambda \Theta \delta_i^j \frac{\partial u_i'}{\partial x_j} + \mu \left(\frac{\partial u_i}{\partial x_j} + \frac{\partial u_j}{\partial x_i} \right) \frac{\partial u_i'}{\partial x_j} \right] dV \\ &= \int_V \left[\lambda \Theta \Theta' + \mu \frac{\partial u_i}{\partial x_j} \frac{\partial u_i'}{\partial x_j} + \mu \frac{\partial u_j}{\partial x_i} \frac{\partial u_i'}{\partial x_j} \right] dV \\ &= \int_V \left[\lambda \Theta \Theta' + \mu \frac{\partial u_i}{\partial x_j} \frac{\partial u_i'}{\partial x_j} + \mu \frac{\partial u_i}{\partial x_j} \frac{\partial u_j'}{\partial x_i} \right] dV. \end{aligned} \tag{1.265}$$

On interchange of primed and unprimed quantities, this expression is unchanged, and thus we have *Betti's reciprocal theorem* as stated,

$$\int_S t_i u_i' \, dS + \int_V F_i u_i' \, dV = \int_S t_i' u_i \, dS + \int_V F_i' u_i \, dV. \tag{1.266}$$

Substituting for the stress, as given by Hooke's law (1.259), in the equation of equilibrium (1.221), yields an equation for the displacement field,

$$\frac{\partial}{\partial x_j} \left[\lambda \nabla \cdot \boldsymbol{u} \delta_j^i + \mu \left(\frac{\partial u_i}{\partial x_j} + \frac{\partial u_j}{\partial x_i} \right) \right] = -F_i. \tag{1.267}$$

In uniform media, λ and μ are not functions of position, and carrying out the differentiation gives,

$$\lambda \frac{\partial}{\partial x_i} (\nabla \cdot \boldsymbol{u}) + \mu \frac{\partial^2 u_i}{\partial x_j \partial x_j} + \mu \frac{\partial^2 u_j}{\partial x_i \partial x_j} = -F_i. \tag{1.268}$$

In symbolic notation this is

$$\lambda\nabla(\nabla\cdot\boldsymbol{u}) + \mu\nabla^2\boldsymbol{u} + \mu\nabla(\nabla\cdot\boldsymbol{u}) = -\boldsymbol{F}. \tag{1.269}$$

Using the vector identity (A.14),

$$\nabla^2\boldsymbol{u} = -\nabla\times(\nabla\times\boldsymbol{u}) + \nabla(\nabla\cdot\boldsymbol{u}), \tag{1.270}$$

we obtain the *Navier equation* for the displacement field,

$$(\lambda+2\mu)\nabla(\nabla\cdot\boldsymbol{u}) - \mu\nabla\times(\nabla\times\boldsymbol{u}) = -\boldsymbol{F}. \tag{1.271}$$

1.4.5 Solutions of the Navier equation

By the Helmholtz separation considered in Section 1.2, the vector displacement field $\boldsymbol{u}$ can be broken into lamellar and solenoidal parts written as

$$\boldsymbol{u} = \nabla L + \nabla\times\boldsymbol{A}, \tag{1.272}$$

where L is the scalar lamellar potential, and $\boldsymbol{A}$ is the vector potential, giving rise to the solenoidal part of the vector displacement field, with the gauge $\nabla\cdot\boldsymbol{A} = 0$. Similarly, the vector body force per unit volume can be separated as

$$\boldsymbol{F} = \nabla L_F + \nabla\times\boldsymbol{A}_F, \tag{1.273}$$

with the gauge $\nabla\cdot\boldsymbol{A}_F = 0$. A further separation derives from the Lamb–Backus decomposition (1.174) of a vector field into lamellar, poloidal and toroidal or torsional parts. Applied to both $\boldsymbol{u}$ and $\boldsymbol{F}$, we have

$$\boldsymbol{u} = \boldsymbol{L} + \boldsymbol{P} + \boldsymbol{T} = \nabla L + \nabla\times(\nabla\times\boldsymbol{r}P) + \nabla\times\boldsymbol{r}T, \tag{1.274}$$

and

$$\boldsymbol{F} = \boldsymbol{L}_F + \boldsymbol{P}_F + \boldsymbol{T}_F = \nabla L_F + \nabla\times(\nabla\times\boldsymbol{r}P_F) + \nabla\times\boldsymbol{r}T_F, \tag{1.275}$$

where L, P, T are the scalars for $\boldsymbol{u}$, and L_F, P_F, T_F are the scalars for $\boldsymbol{F}$. Applying the Helmholtz separation to the Navier equation of equilibrium (1.271) itself yields

$$(\lambda+2\mu)\nabla\left(\nabla^2 L\right) = -\nabla L_F, \tag{1.276}$$

$$-\mu\nabla\times(\nabla\times(\nabla\times\boldsymbol{A})) = -\nabla\times\boldsymbol{A}_F. \tag{1.277}$$

Taking the divergence of (1.276) gives

$$\nabla^4 L = -\frac{1}{\lambda+2\mu}\nabla^2 L_F, \tag{1.278}$$

although the scalar potentials $(\lambda+2\mu)\nabla^2 L$ and $-L_F$ can differ, at most, by a constant C, so that

$$\nabla^2 L = -\frac{L_F + C}{\lambda + 2\mu}. \tag{1.279}$$

Comparison of expressions (1.272) and (1.274) allows the vector potential $\boldsymbol{A}$ to be written

$$\boldsymbol{A} = \nabla \times \boldsymbol{r}P + \boldsymbol{r}T, \tag{1.280}$$

while comparison of expressions (1.273) and (1.275) allows the vector potential $\boldsymbol{A}_F$ to be written

$$\boldsymbol{A}_F = \nabla \times \boldsymbol{r}P_F + \boldsymbol{r}T_F. \tag{1.281}$$

Substitution in equation (1.277) yields, for the poloidal part,

$$\mu\nabla \times \left(\nabla \times \boldsymbol{r}\nabla^2 P\right) = -\nabla \times (\nabla \times \boldsymbol{r}P_F), \tag{1.282}$$

while the toroidal or torsional part yields

$$\mu\nabla \times \boldsymbol{r}\nabla^2 T = -\nabla \times \boldsymbol{r}T_F, \tag{1.283}$$

where expressions (1.189) and (1.190) for the negative of the curl, taken twice, for the poloidal part, and expressions (1.191) and (1.192) for the negative of the curl, taken twice, for the toroidal or torsional part, have been used. Equating the poloidal scalars on each side of expression (1.282) gives

$$\nabla^2 P = -\frac{1}{\mu}P_F, \tag{1.284}$$

and equating the toroidal or torsional scalars on each side of expression (1.283) gives

$$\nabla^2 T = -\frac{1}{\mu}T_F. \tag{1.285}$$

Particular solutions of the equations of equilibrium (1.276) and (1.277) are

$$(\lambda + 2\mu)\nabla^2 L = -L_F, \tag{1.286}$$

$$-\mu\nabla \times (\nabla \times \boldsymbol{A}) = -\boldsymbol{A}_F. \tag{1.287}$$

The potentials for the body force density are given by

$$L_F = -\frac{1}{4\pi}\iiint \frac{\nabla' \cdot \boldsymbol{F}(\boldsymbol{r}')}{R}\, d\mathcal{V}', \tag{1.288}$$

from expression (1.168), and

$$\boldsymbol{A}_F = \frac{1}{4\pi}\iiint \frac{\nabla' \times \boldsymbol{F}(\boldsymbol{r}')}{R}\, d\mathcal{V}', \tag{1.289}$$

from expression (1.173), where $R = |\boldsymbol{r} - \boldsymbol{r}'|$. The volume integrals can be simplified using the theorem of Gauss, generalized to a tensor field $T_{ijk\cdots}$,

$$\iiint \frac{\partial T_{ijk\cdots}}{\partial x_r} d\mathcal{V} = \iint T_{ijk\cdots}\nu_r \, d\mathcal{S}, \tag{1.290}$$

the surface integral being taken over the surface $\mathcal{S}$ with unit outward normal vector ν_i enclosing the volume $\mathcal{V}$. If $T_{ijk\cdots}$ is the vector field u_j,

$$\iiint \frac{\partial u_j}{\partial x_i} d\mathcal{V} = \iint u_j \nu_i \, d\mathcal{S}. \tag{1.291}$$

Contracting both sides, we find

$$\iiint \frac{\partial u_i}{\partial x_i} d\mathcal{V} = \iint u_i \nu_i \, d\mathcal{S}, \tag{1.292}$$

or

$$\iiint \nabla \cdot \boldsymbol{u} \, d\mathcal{V} = \iint \boldsymbol{u} \cdot \hat{\boldsymbol{\nu}} \, d\mathcal{S}, \tag{1.293}$$

the familiar, elementary form of Gauss's theorem. Taking the dual vector of the vector gradient as the tensor field, we get

$$\iiint \xi_{ijk} \frac{\partial u_k}{\partial x_j} d\mathcal{V} = \iint \xi_{ijk} u_k \nu_j \, d\mathcal{S}, \tag{1.294}$$

or

$$\iiint \nabla \times \boldsymbol{u} \, d\mathcal{V} = \iint (\hat{\boldsymbol{\nu}} \times \boldsymbol{u}) \, d\mathcal{S}. \tag{1.295}$$

Using the two vector calculus identities,

$$\nabla' \cdot \left(\frac{\boldsymbol{F}(\boldsymbol{r}')}{R} \right) = \frac{\nabla' \cdot \boldsymbol{F}(\boldsymbol{r}')}{R} + \boldsymbol{F}(\boldsymbol{r}') \cdot \nabla' \left(\frac{1}{R} \right), \tag{1.296}$$

and

$$\nabla' \times \left(\frac{\boldsymbol{F}(\boldsymbol{r}')}{R} \right) = \frac{\nabla' \times \boldsymbol{F}(\boldsymbol{r}')}{R} + \nabla' \left(\frac{1}{R} \right) \times \boldsymbol{F}(\boldsymbol{r}'), \tag{1.297}$$

together with the forms (1.293) and (1.295) of Gauss's theorem, the potentials for the body force density can be converted to

$$L_F = \frac{1}{4\pi} \iiint \boldsymbol{F}(\boldsymbol{r}') \cdot \nabla' \left(\frac{1}{R} \right) d\mathcal{V}' - \frac{1}{4\pi} \iint \frac{\boldsymbol{F}(\boldsymbol{r}')}{R} \cdot \hat{\boldsymbol{\nu}} d\mathcal{S}', \tag{1.298}$$

and

$$\boldsymbol{A}_F = \frac{1}{4\pi} \iiint \boldsymbol{F}(\boldsymbol{r}') \times \nabla' \left(\frac{1}{R} \right) d\mathcal{V}' - \frac{1}{4\pi} \iint \frac{\boldsymbol{F}(\boldsymbol{r}')}{R} \times \hat{\boldsymbol{\nu}} d\mathcal{S}'. \tag{1.299}$$

If $\boldsymbol{F}(\boldsymbol{r}') \to 0$ faster than $1/r'^2$ as $r' \to \infty$, the surface integrals may be neglected, thus

$$L_F = -\frac{1}{4\pi} \iiint \nabla\left(\frac{1}{R}\right) \cdot \boldsymbol{F}(\boldsymbol{r}')\, dV', \tag{1.300}$$

and

$$\boldsymbol{A}_F = \frac{1}{4\pi} \iiint \nabla\left(\frac{1}{R}\right) \times \boldsymbol{F}(\boldsymbol{r}')\, dV', \tag{1.301}$$

where we have used the identity

$$\nabla'\left(\frac{1}{R}\right) = -\nabla\left(\frac{1}{R}\right). \tag{1.302}$$

1.4.6 Kelvin's problem

If the body force density is concentrated at a point, it may be represented as

$$\boldsymbol{F}(\boldsymbol{r}) = \boldsymbol{F}_0\, \delta(\boldsymbol{r} - \boldsymbol{r}_0), \tag{1.303}$$

where $\delta(\boldsymbol{r} - \boldsymbol{r}_0)$ is the Dirac delta distribution. In this case, the body force is a point force $\boldsymbol{F}_0$ at $\boldsymbol{r} = \boldsymbol{r}_0$. If (ξ_1, ξ_2, ξ_3) are the components of $\boldsymbol{r}_0$, we have $R^2 = (x_1 - \xi_1)^2 + (x_2 - \xi_2)^2 + (x_3 - \xi_3)^2$ for $\boldsymbol{R} = \boldsymbol{r} - \boldsymbol{r}_0$. The integrals (1.300) and (1.301) then give,

$$L_F = -\frac{1}{4\pi} \nabla\left(\frac{1}{R}\right) \cdot \boldsymbol{F}_0 = \frac{1}{4\pi} \frac{\boldsymbol{R} \cdot \boldsymbol{F}_0}{R^3}, \tag{1.304}$$

and

$$\boldsymbol{A}_F = \frac{1}{4\pi} \nabla\left(\frac{1}{R}\right) \times \boldsymbol{F}_0 = \frac{1}{4\pi} \frac{\boldsymbol{F}_0 \times \boldsymbol{R}}{R^3}, \tag{1.305}$$

since

$$\nabla\left(\frac{1}{R}\right) = -\frac{\boldsymbol{R}}{R^3}. \tag{1.306}$$

Taking the divergence,

$$\nabla \cdot \left(\frac{\boldsymbol{F}_0 \times \boldsymbol{R}}{R^3}\right) = -3\frac{(\boldsymbol{F}_0 \times \boldsymbol{R}) \cdot \boldsymbol{R}}{R^5} = 0, \tag{1.307}$$

verifies that $\boldsymbol{A}_F$ has the correct gauge, $\nabla \cdot \boldsymbol{A}_F = 0$.

If $\boldsymbol{C}$ is an arbitrary constant vector, we have the gradient

$$\nabla\left(\frac{\boldsymbol{C} \cdot \boldsymbol{R}}{R}\right) = \frac{\boldsymbol{C}}{R} - \frac{(\boldsymbol{C} \cdot \boldsymbol{R})\, \boldsymbol{R}}{R^3}. \tag{1.308}$$

Taking the divergence,

$$\nabla^2\left(\frac{\boldsymbol{C}\cdot\boldsymbol{R}}{R}\right) = -2\frac{\boldsymbol{C}\cdot\boldsymbol{R}}{R^3}. \tag{1.309}$$

Then, taking the constant vector to be

$$\frac{\boldsymbol{F}_0}{8\pi(\lambda+2\mu)}, \tag{1.310}$$

we see that

$$L = \frac{1}{8\pi(\lambda+2\mu)}\frac{\boldsymbol{F}_0\cdot\boldsymbol{R}}{R} \tag{1.311}$$

is a particular solution of (1.286) for the scalar displacement potential. For gauge $\nabla\cdot\boldsymbol{A}=0$, equation (1.287) gives, for the ith component of the vector displacement potential,

$$\mu\nabla^2 A_i = -(\boldsymbol{A}_F)_i = -\frac{1}{4\pi}\frac{\xi_{ijk}F_{0j}(x_k-\xi_k)}{R^3}, \tag{1.312}$$

using expression (1.305) for $\boldsymbol{A}_F$. Next, taking the constant vector $\boldsymbol{C}$ to be the vectors $(\hat{e}_3F_{02}-\hat{e}_2F_{03})$, $(\hat{e}_1F_{03}-\hat{e}_3F_{01})$ and $(\hat{e}_2F_{01}-\hat{e}_1F_{02})$, in turn, we find that

$$\begin{aligned}
\nabla^2\left[\frac{(\hat{e}_3F_{02}-\hat{e}_2F_{03})\cdot\boldsymbol{R}}{R}\right] &= -2\frac{(\hat{e}_3F_{02}-\hat{e}_2F_{03})\cdot\boldsymbol{R}}{R^3},\\
\nabla^2\left[\frac{(\hat{e}_1F_{03}-\hat{e}_3F_{01})\cdot\boldsymbol{R}}{R}\right] &= -2\frac{(\hat{e}_1F_{03}-\hat{e}_3F_{01})\cdot\boldsymbol{R}}{R^3},\\
\nabla^2\left[\frac{(\hat{e}_2F_{01}-\hat{e}_1F_{02})\cdot\boldsymbol{R}}{R}\right] &= -2\frac{(\hat{e}_2F_{01}-\hat{e}_1F_{02})\cdot\boldsymbol{R}}{R^3},
\end{aligned} \tag{1.313}$$

with $(\hat{e}_1, \hat{e}_2, \hat{e}_3)$ representing the Cartesian unit vectors. Combining these results, we obtain

$$\nabla^2\left(\frac{\boldsymbol{F}_0\times\boldsymbol{R}}{R}\right) = -2\frac{\boldsymbol{F}_0\times\boldsymbol{R}}{R^3}. \tag{1.314}$$

We are then led to

$$\boldsymbol{A} = \frac{1}{8\pi\mu}\frac{\boldsymbol{F}_0\times\boldsymbol{R}}{R} \tag{1.315}$$

as a particular solution of (1.287). Since

$$\nabla\cdot\left(\frac{\boldsymbol{F}_0\times\boldsymbol{R}}{R}\right) = -\frac{(\boldsymbol{F}_0\times\boldsymbol{R})\cdot\boldsymbol{R}}{R^3} = 0, \tag{1.316}$$

the assumed gauge, $\nabla\cdot\boldsymbol{A}=0$, is verified.

The vector displacement field is found from (1.272) using expressions (1.311) and (1.315) for L and $\boldsymbol{A}$, respectively. The vector identities,

$$\nabla\left(\frac{\boldsymbol{F}_0 \cdot \boldsymbol{R}}{R}\right) = \frac{\boldsymbol{F}_0}{R} - \frac{(\boldsymbol{F}_0 \cdot \boldsymbol{R})}{R^3}\boldsymbol{R}, \tag{1.317}$$

and

$$\nabla \times \left(\frac{\boldsymbol{F}_0 \times \boldsymbol{R}}{R}\right) = \frac{\boldsymbol{F}_0}{R} + \frac{(\boldsymbol{F}_0 \cdot \boldsymbol{R})}{R^3}\boldsymbol{R}, \tag{1.318}$$

allow simplification of the final expression for $\boldsymbol{u}$ to

$$\boldsymbol{u} = \frac{\lambda + 3\mu}{8\pi\mu(\lambda + 2\mu)}\frac{\boldsymbol{F}_0}{R} + \frac{\lambda + \mu}{8\pi\mu(\lambda + 2\mu)}\frac{(\boldsymbol{F}_0 \cdot \boldsymbol{R})}{R^3}\boldsymbol{R}. \tag{1.319}$$

This result, for the displacement field arising from a concentrated point force in an infinite, uniform medium, was first obtained by Lord Kelvin in 1848, and is generally known as the solution to *Kelvin's problem.*

With the constant

$$H = \frac{\lambda + \mu}{8\pi\mu(\lambda + 2\mu)}, \tag{1.320}$$

the associated normal stresses are

$$\begin{aligned}
\tau_{11} &= -2\mu\frac{H}{R^3}\left[3\frac{(x_1 - \xi_1)^2 \boldsymbol{F}_0 \cdot \boldsymbol{R}}{R^2} + \frac{\mu}{\lambda + \mu}\Big(2F_{01}(x_1 - \xi_1) - \boldsymbol{F}_0 \cdot \boldsymbol{R}\Big)\right],\\
\tau_{22} &= -2\mu\frac{H}{R^3}\left[3\frac{(x_2 - \xi_2)^2 \boldsymbol{F}_0 \cdot \boldsymbol{R}}{R^2} + \frac{\mu}{\lambda + \mu}\Big(2F_{02}(x_2 - \xi_2) - \boldsymbol{F}_0 \cdot \boldsymbol{R}\Big)\right],\\
\tau_{33} &= -2\mu\frac{H}{R^3}\left[3\frac{(x_3 - \xi_3)^2 \boldsymbol{F}_0 \cdot \boldsymbol{R}}{R^2} + \frac{\mu}{\lambda + \mu}\Big(2F_{03}(x_3 - \xi_3) - \boldsymbol{F}_0 \cdot \boldsymbol{R}\Big)\right],
\end{aligned} \tag{1.321}$$

and the associated shear stresses are

$$\begin{aligned}
\tau_{13} = -2\mu\frac{H}{R^3}\Bigg[&3\frac{(x_1 - \xi_1)(x_3 - \xi_3)\,\boldsymbol{F}_0 \cdot \boldsymbol{R}}{R^2}\\
&+ \frac{\mu}{\lambda + \mu}\Big(F_{01}(x_3 - \xi_3) + F_{03}(x_1 - \xi_1)\Big)\Bigg],\\
\tau_{23} = -2\mu\frac{H}{R^3}\Bigg[&3\frac{(x_2 - \xi_2)(x_3 - \xi_3)\,\boldsymbol{F}_0 \cdot \boldsymbol{R}}{R^2}\\
&+ \frac{\mu}{\lambda + \mu}\Big(F_{02}(x_3 - \xi_3) + F_{03}(x_2 - \xi_2)\Big)\Bigg],\\
\tau_{12} = -2\mu\frac{H}{R^3}\Bigg[&3\frac{(x_1 - \xi_1)(x_2 - \xi_2)\,\boldsymbol{F}_0 \cdot \boldsymbol{R}}{R^2}\\
&+ \frac{\mu}{\lambda + \mu}\Big(F_{01}(x_2 - \xi_2) + F_{02}(x_1 - \xi_1)\Big)\Bigg].
\end{aligned} \tag{1.322}$$

1.4.7 The Papkovich–Neuber solution

Other forms of solutions to the Navier equation can be developed. Using the dimensionless Poisson's ratio σ (1.255), it can be recast in the form

$$\nabla^2 \boldsymbol{u} + \frac{1}{1-2\sigma}\nabla(\nabla\cdot\boldsymbol{u}) = -\frac{\boldsymbol{F}}{\mu}. \tag{1.323}$$

Since $\nabla\cdot\boldsymbol{u} = \nabla^2 L$, it follows that $\nabla(\nabla\cdot\boldsymbol{u}) = \nabla^2(\nabla L)$, and

$$\nabla^2\left[\boldsymbol{u} + \frac{1}{1-2\sigma}\nabla L\right] = -\frac{\boldsymbol{F}}{\mu}. \tag{1.324}$$

Writing the vector quantity in square brackets as $\boldsymbol{\Phi}/2\mu$, we have

$$\nabla^2\boldsymbol{\Phi} = -2\boldsymbol{F}, \tag{1.325}$$

with

$$\frac{1}{2\mu}\boldsymbol{\Phi} = \boldsymbol{u} + \frac{1}{1-2\sigma}\nabla L. \tag{1.326}$$

Taking the divergence of both sides of the latter relation gives

$$\frac{1}{2\mu}\nabla\cdot\boldsymbol{\Phi} = \frac{2(1-\sigma)}{1-2\sigma}\nabla^2 L. \tag{1.327}$$

Using the vector identity

$$\nabla^2(\boldsymbol{R}\cdot\boldsymbol{\Phi}) = \boldsymbol{R}\cdot\nabla^2\boldsymbol{\Phi} + 2(\nabla\cdot\boldsymbol{\Phi}), \tag{1.328}$$

and relation (1.325), we obtain

$$\nabla\cdot\boldsymbol{\Phi} = \frac{1}{2}\left(\nabla^2(\boldsymbol{R}\cdot\boldsymbol{\Phi}) + 2\boldsymbol{R}\cdot\boldsymbol{F}\right) = 4\mu\frac{1-\sigma}{1-2\sigma}\nabla^2 L. \tag{1.329}$$

Thus,

$$\nabla^2\left[\frac{2\mu}{1-2\sigma}L - \frac{1}{4(1-\sigma)}\boldsymbol{R}\cdot\boldsymbol{\Phi}\right] = \frac{1}{2(1-\sigma)}\boldsymbol{R}\cdot\boldsymbol{F}. \tag{1.330}$$

Denoting the scalar in square brackets by B,

$$\nabla^2 B = \frac{1}{2(1-\sigma)}\boldsymbol{R}\cdot\boldsymbol{F}, \tag{1.331}$$

and

$$L = \frac{1-2\sigma}{2\mu}\left[B + \frac{1}{4(1-\sigma)}\boldsymbol{R}\cdot\boldsymbol{\Phi}\right]. \tag{1.332}$$

Then, using (1.326), the vector displacement field is found to be represented by

$$2\mu \boldsymbol{u} = \boldsymbol{\Phi} - \nabla \left[B + \frac{1}{4(1-\sigma)} \boldsymbol{R} \cdot \boldsymbol{\Phi} \right]. \tag{1.333}$$

This representation of the displacement field was developed by Papkovich (1932) and Neuber (1934) and is referred to as the *Papkovich–Neuber* solution of the Navier equation. The vector potential $\boldsymbol{\Phi}$ obeys the vector Poisson equation (1.325), while the scalar potential B obeys the scalar Poisson equation (1.331). Outside regions containing body forces, the displacement field is represented by the four scalar, harmonic functions Φ_1, Φ_2, Φ_3 and B. There has been much debate as to whether or not all four are independent, or whether one can be set to zero without loss of generality (see Sokolnikoff (1956), p. 331). From equations (1.279), (1.284) and (1.285), the displacement field appears to depend on the lamellar, poloidal and toroidal or torsional scalars. In the absence of body forces, the lamellar scalar is determined to within a constant as a harmonic function, and the poloidal and toroidal or torsional scalars are harmonic. Indeed, in the case of Kelvin's problem of a force $\boldsymbol{F}_0$ concentrated at (ξ_1, ξ_2, ξ_3) in an infinite medium, only the vector potential,

$$\boldsymbol{\Phi} = \frac{1}{2\pi} \frac{\boldsymbol{F}_0}{R}, \tag{1.334}$$

with $R^2 = (x_1 - \xi_1)^2 + (x_2 - \xi_2)^2 + (x_3 - \xi_3)^2$, is required, and B may be set to zero.

1.4.8 The Galerkin vector

Another representation of the displacement field related to the Papkovich–Neuber solution of the Navier equation is that given by the *Galerkin vector* $\boldsymbol{G}$,

$$2\mu \boldsymbol{u} = 2(1-\sigma) \nabla^2 \boldsymbol{G} - \nabla (\nabla \cdot \boldsymbol{G}). \tag{1.335}$$

If we set

$$\nabla^2 \boldsymbol{G} = \frac{1}{2(1-\sigma)} \boldsymbol{\Phi} \tag{1.336}$$

and

$$\nabla \cdot \boldsymbol{G} = \frac{2\mu}{1-2\sigma} L, \tag{1.337}$$

the Galerkin vector representation reduces to equation (1.326) defining the vector potential $\boldsymbol{\Phi}$,

$$2\mu \boldsymbol{u} = \boldsymbol{\Phi} - \frac{2\mu}{1-2\sigma} \nabla L. \tag{1.338}$$

Taking the Laplacian of both sides of equation (1.336), and using relation (1.325), the Galerkin vector is found to obey

$$\nabla^4 \boldsymbol{G} = -\frac{\boldsymbol{F}}{1-\sigma}. \tag{1.339}$$

This equation can also be obtained by direct substitution of the representation of the displacement field (1.335) by the Galerkin vector into the Navier equation (1.323). From (1.335), we find that

$$\nabla^2 \boldsymbol{u} = \frac{1-\sigma}{\mu}\nabla^4 \boldsymbol{G} - \frac{1}{2\mu}\nabla^2\nabla(\nabla\cdot \boldsymbol{G}) = \frac{1-\sigma}{\mu}\nabla^4 \boldsymbol{G} - \frac{1}{2\mu}\nabla\left(\nabla^2(\nabla\cdot \boldsymbol{G})\right), \tag{1.340}$$

and

$$\begin{aligned}\nabla(\nabla\cdot \boldsymbol{u}) &= \frac{1-\sigma}{\mu}\nabla\left(\nabla\cdot(\nabla^2 \boldsymbol{G})\right) - \frac{1}{2\mu}\nabla\left(\nabla^2(\nabla\cdot \boldsymbol{G})\right)\\ &= \frac{1-2\sigma}{2\mu}\nabla\left(\nabla^2(\nabla\cdot \boldsymbol{G})\right).\end{aligned} \tag{1.341}$$

Substitution of these expressions into (1.323) gives

$$\frac{1-\sigma}{\mu}\nabla^4 \boldsymbol{G} = -\frac{\boldsymbol{F}}{\mu}, \tag{1.342}$$

confirming equation (1.339). The Galerkin vector for Kelvin's problem of a force $\boldsymbol{F}_0$ concentrated at (ξ_1, ξ_2, ξ_3) in an infinite medium is

$$\boldsymbol{G} = \frac{\lambda+\mu}{4\pi(\lambda+2\mu)} R\boldsymbol{F}_0. \tag{1.343}$$

1.4.9 The Boussinesq problem

Kelvin's problem may be specialised to the case where the concentrated force is applied at $(x_1', x_2', 0)$, on the surface $x_3 = 0$, and entirely in the x_3-direction. The Galerkin vector then has the form $\boldsymbol{G} = AR_0\hat{\boldsymbol{e}}_3$, with A an arbitrary constant, and where R_0 is the radius to the field point,

$$R_0 = \sqrt{(x_1 - x_1')^2 + (x_2 - x_2')^2 + x_3^2}. \tag{1.344}$$

The displacement components are then

$$u_1 = \frac{A}{2\mu}\frac{x_3}{R_0^3}\left(x_1 - x_1'\right),$$
$$u_2 = \frac{A}{2\mu}\frac{x_3}{R_0^3}\left(x_2 - x_2'\right), \tag{1.345}$$
$$u_3 = \frac{A}{2\mu}\frac{1}{R_0}\left[\frac{\lambda + 3\mu}{\lambda + \mu} + \frac{x_3^2}{R_0^2}\right].$$

The associated normal stresses are

$$\tau_{11} = -A\frac{x_3}{R_0^3}\left[3\frac{(x_1 - x_1')^2}{R_0^2} - \frac{\mu}{\lambda + \mu}\right], \tag{1.346}$$
$$\tau_{22} = -A\frac{x_3}{R_0^3}\left[3\frac{(x_2 - x_2')^2}{R_0^2} - \frac{\mu}{\lambda + \mu}\right], \tag{1.347}$$
$$\tau_{33} = -A\frac{x_3}{R_0^3}\left[3\frac{x_3^2}{R_0^2} + \frac{\mu}{\lambda + \mu}\right], \tag{1.348}$$

while the shear stresses are

$$\tau_{13} = -A\frac{x_1 - x_1'}{R_0^3}\left[3\frac{x_3^2}{R_0^2} + \frac{\mu}{\lambda + \mu}\right], \tag{1.349}$$
$$\tau_{23} = -A\frac{x_2 - x_2'}{R_0^3}\left[3\frac{x_3^2}{R_0^2} + \frac{\mu}{\lambda + \mu}\right], \tag{1.350}$$
$$\tau_{12} = -3A\frac{(x_1 - x_1')(x_2 - x_2')x_3}{R_0^5}. \tag{1.351}$$

On the plane $x_3 = 0$, the normal stresses vanish but the two shear stresses, τ_{13} given by (1.349) and τ_{23} given by (1.350), do not.

For the plane $x_3 = 0$ to be a free surface, we need an additional, lamellar solution of the Navier equation, represented by the scalar potential in the Papkovich–Neuber formulation, or by the scalar potential, L, in the Helmholtz separation (1.272). Take $L = B\log(R_0 + x_3)$, with B a constant to be determined. The extra displacement field, given by the gradient of L, is

$$u_1 = B\frac{x_1 - x_1'}{R_0\,(R_0 + x_3)}, \quad u_2 = B\frac{x_2 - x_2'}{R_0\,(R_0 + x_3)}, \quad u_3 = \frac{B}{R_0}. \tag{1.352}$$

This displacement field is both lamellar and solenoidal, and it has associated normal stresses,

$$\tau_{11} = \frac{2\mu B}{R_0^2(R_0 + x_3)}\left[\frac{(x_2 - x_2')^2 + x_3^2}{R_0} - \frac{(x_1 - x_1')^2}{R_0 + x_3}\right], \tag{1.353}$$

$$\tau_{22} = \frac{2\mu B}{R_0^2(R_0 + x_3)}\left[\frac{(x_1 - x_1')^2 + x_3^2}{R_0} - \frac{(x_2 - x_2')^2}{R_0 + x_3}\right], \tag{1.354}$$

$$\tau_{33} = -2\mu B\frac{x_3}{R_0^3}, \tag{1.355}$$

with shear stresses,

$$\tau_{13} = -2\mu B\frac{x_1 - x_1'}{R_0^3}, \tag{1.356}$$

$$\tau_{23} = -2\mu B\frac{x_2 - x_2'}{R_0^3}, \tag{1.357}$$

$$\tau_{12} = -2\mu B\frac{(x_1 - x_1')(x_2 - x_2')(2R_0 + x_3)}{R_0^3(R_0 + x_3)^2}. \tag{1.358}$$

The normal stress (1.355) on the surface $x_3 = 0$ again vanishes, and the stresses τ_{13}, expressed by (1.356), and τ_{23}, expressed by (1.357), have the same form as their counterparts, expressed by (1.349) and (1.350). This suggests that a linear combination of the displacement fields (1.345) and (1.352) would leave the surface $x_3 = 0$ stress free.

The problem of a concentrated force acting normal to the free surface of an elastic half-space was first solved by J. Boussinesq in 1878 and is known as the *Boussinesq problem.* As suggested, we form a linear combination of the displacement fields (1.345) and (1.352) to get

$$u_1 = \frac{A}{2\mu}\frac{x_3}{R_0^3}\left(x_1 - x_1'\right) + B\frac{x_1 - x_1'}{R_0(R_0 + x_3)}, \tag{1.359}$$

$$u_2 = \frac{A}{2\mu}\frac{x_3}{R_0^3}\left(x_2 - x_2'\right) + B\frac{x_2 - x_2'}{R_0(R_0 + x_3)}, \tag{1.360}$$

$$u_3 = \frac{A}{2\mu}\left[\frac{\lambda + 3\mu}{\lambda + \mu}\frac{1}{R_0} + \frac{x_3^2}{R_0^3}\right] + \frac{B}{R_0}. \tag{1.361}$$

For determination of the constants, A and B, two conditions are required. One is provided by the condition that the surface $x_3 = 0$ is stress free, except at the point of application of the normal force. A second is provided by the condition that the total traction on a hemisphere in the medium, centred on the point of application of the normal force, must balance the applied point force. If the hemisphere has radius a, the traction per unit area on it is given by relation (1.214), where the components of the normal vector are $\nu_1 = (x_1 - x_1')/a$, $\nu_2 = (x_2 - x_2')/a$, and $\nu_3 = x_3/a$ while the stresses τ_{ij} are the sum of those given by relations (1.346) through (1.351) and those given by relations (1.353) through (1.358). The components of the tractions per unit area are

$$F_1^\nu = -3A\frac{(x_1 - x_1')x_3}{a^4} - 2\mu B\frac{(x_1 - x_1')}{a^2\,(a + x_3)}, \tag{1.362}$$

$$F_2^\nu = -3A\frac{(x_2 - x_2')x_3}{a^4} - 2\mu B\frac{(x_2 - x_2')}{a^2\,(a + x_3)}, \tag{1.363}$$

$$F_3^\nu = -A\frac{1}{a^2}\left(3\frac{x_3^2}{a^2} + \frac{\mu}{\lambda + \mu}\right) - 2\mu B\frac{1}{a^2}. \tag{1.364}$$

With (a, θ, ϕ) as the spherical polar co-ordinates on the surface of the hemisphere, the components of the total traction on the hemisphere are

$$T_i = a^2 \int_0^{2\pi}\int_0^{\pi/2} F_i^\nu \sin\theta\, d\theta\, d\phi. \tag{1.365}$$

The Cartesian co-ordinates on the surface of the hemisphere, referred to its centre, are $x_1 - x_1' = a\sin\theta\cos\phi$, $x_2 - x_2' = a\sin\theta\sin\phi$, and $x_3 = a\cos\theta$. Carrying out the integrations in (1.365), we find that

$$T_1 = T_2 = 0, \quad T_3 = -2\pi\left(\frac{\lambda + 2\mu}{\lambda + \mu}A + 2\mu B\right). \tag{1.366}$$

While the normal stresses, (1.348) and (1.355), vanish on the surface $x_3 = 0$, the shear stresses τ_{13} and τ_{23} do not vanish, but add, to become

$$\tau_{13} = -\frac{x_1 - x_1'}{R_0^3}\left(\frac{\mu}{\lambda + \mu}A + 2\mu B\right), \tag{1.367}$$

$$\tau_{23} = -\frac{x_2 - x_2'}{R_0^3}\left(\frac{\mu}{\lambda + \mu}A + 2\mu B\right). \tag{1.368}$$

The condition for the surface $x_3 = 0$ to be stress free is then

$$\frac{\mu}{\lambda + \mu}A + 2\mu B = 0. \tag{1.369}$$

The second condition, that the total traction on the surface of the hemisphere, given by (1.366), balances the applied point force, becomes

$$\frac{\lambda + 2\mu}{\lambda + \mu}A + 2\mu B = \frac{P}{2\pi}. \tag{1.370}$$

Solving (1.369) and (1.370) gives

$$A = \frac{P}{2\pi}, \quad B = -\frac{P}{4\pi(\lambda+\mu)}. \tag{1.371}$$

Substituting these values in (1.345) and (1.352) gives the total displacement field for the Boussinesq problem as

$$u_1 = \frac{P}{4\pi\mu}\frac{x_1 - x_1'}{R_0}\left[\frac{x_3}{R_0^2} - \frac{\mu}{\lambda+\mu}\frac{1}{(R_0+x_3)}\right], \tag{1.372}$$

$$u_2 = \frac{P}{4\pi\mu}\frac{x_2 - x_2'}{R_0}\left[\frac{x_3}{R_0^2} - \frac{\mu}{\lambda+\mu}\frac{1}{(R_0+x_3)}\right], \tag{1.373}$$

$$u_3 = \frac{P}{4\pi\mu}\frac{1}{R_0}\left[\frac{x_3^2}{R_0^2} + \frac{\lambda+2\mu}{\lambda+\mu}\right]. \tag{1.374}$$

The Galerkin vector for the Boussinesq problem can be found by adding the Galerkin vector $\boldsymbol{G} = AR_0\hat{\mathbf{e}}_3$ to the Galerkin vector corresponding to the extra displacement field (1.352). The former amounts to

$$\frac{P}{2\pi}R_0\hat{\mathbf{e}}_3, \tag{1.375}$$

while the latter is

$$\begin{aligned} &-2\mu B\frac{\lambda+\mu}{\lambda-\mu}\left[\boldsymbol{R}_0\log(R_0+x_3) - x_3\hat{\mathbf{e}}_3\right] \\ &= \frac{P}{2\pi}\frac{\mu}{\lambda-\mu}\left[\boldsymbol{R}_0\log(R_0+x_3) - x_3\hat{\mathbf{e}}_3\right], \end{aligned} \tag{1.376}$$

as can be shown with the aid of the vector identities

$$\nabla\cdot\left[\boldsymbol{R}_0\log(R_0+x_3) - x_3\hat{\mathbf{e}}_3\right] = 3\log(R_0+x_3), \tag{1.377}$$

and

$$\nabla^2\left[\boldsymbol{R}_0\log(R_0+x_3) - x_3\hat{\mathbf{e}}_3\right] = 2\frac{\boldsymbol{R}_0 + R_0\hat{\mathbf{e}}_3}{R_0(R_0+x_3)}. \tag{1.378}$$

The total Galerkin vector for the Boussinesq problem is then

$$\boldsymbol{G} = \frac{P}{2\pi}\left[R_0\hat{\mathbf{e}}_3 + \frac{\mu}{\lambda-\mu}\left(\boldsymbol{R}_0\log(R_0+x_3) - x_3\hat{\mathbf{e}}_3\right)\right]. \tag{1.379}$$

1.4.10 Cerruti's problem

When the concentrated force is applied tangentially at $(x_1', x_2', 0)$ on the free surface, $x_3 = 0$, of an elastic half-space, the problem is associated with the name of Cerruti, who first solved it in 1882.

As in the Boussinesq problem, we begin by specialising Kelvin's problem to the case where the concentrated force is applied at $(x_1', x_2', 0)$, on the surface $x_3 = 0$, and now entirely in the x_1-direction. Then, the Galerkin vector is $\boldsymbol{G} = AR_0\hat{\mathbf{e}}_1$, with A again an arbitrary constant. The displacement components are then

$$u_1 = \frac{A}{2\mu}\frac{1}{R_0}\left[\frac{\lambda + 3\mu}{\lambda + \mu} + \frac{(x_1 - x_1')^2}{R_0^2}\right], \tag{1.380}$$

$$u_2 = \frac{A}{2\mu}\frac{(x_1 - x_1')(x_2 - x_2')}{R_0^3}, \tag{1.381}$$

$$u_3 = \frac{A}{2\mu}\frac{x_3}{R_0^3}(x_1 - x_1'). \tag{1.382}$$

The associated normal stresses are

$$\tau_{11} = -A\frac{x_1 - x_1'}{R_0^3}\left[3\frac{(x_1 - x_1')^2}{R_0^2} + \frac{\mu}{\lambda + \mu}\right], \tag{1.383}$$

$$\tau_{22} = -A\frac{x_1 - x_1'}{R_0^3}\left[3\frac{(x_2 - x_2')^2}{R_0^2} - \frac{\mu}{\lambda + \mu}\right], \tag{1.384}$$

$$\tau_{33} = -A\frac{x_1 - x_1'}{R_0^3}\left[3\frac{x_3^2}{R_0^2} - \frac{\mu}{\lambda + \mu}\right], \tag{1.385}$$

while the shear stresses are

$$\tau_{13} = -A\frac{x_3}{R_0^3}\left[3\frac{(x_1 - x_1')^2}{R_0^2} + \frac{\mu}{\lambda + \mu}\right], \tag{1.386}$$

$$\tau_{23} = -3A\frac{(x_1 - x_1')(x_2 - x_2')x_3}{R_0^5}, \tag{1.387}$$

$$\tau_{12} = -A\frac{x_2 - x_2'}{R_0^3}\left[3\frac{(x_1 - x_1')^2}{R_0^2} + \frac{\mu}{\lambda + \mu}\right]. \tag{1.388}$$

On the plane $x_3 = 0$, the two shear stresses, τ_{13} and τ_{23}, vanish but the normal stress, τ_{33}, does not.

In contrast to the Boussinesq problem, we need two additional solutions for the plane $x_3 = 0$ to be a free surface.

The first extra displacement field is given by the gradient of the scalar potential,

$$L = B\frac{x_1 - x_1'}{R_0 + x_3}, \tag{1.389}$$

with B a free constant, in the Helmholtz separation (1.272), and is

$$u_1 = -\frac{B}{R_0 + x_3}\left[\frac{(x_1 - x_1')^2}{R_0\,(R_0 + x_3)} - 1\right], \tag{1.390}$$

$$u_2 = -B\frac{(x_1 - x_1')(x_2 - x_2')}{R_0\,(R_0 + x_3)^2}, \tag{1.391}$$

$$u_3 = -B\frac{x_1 - x_1'}{R_0\,(R_0 + x_3)}. \tag{1.392}$$

This displacement field is both lamellar and solenoidal, and it has associated normal stresses

$$\tau_{11} = -2\mu B\frac{x_1 - x_1'}{R_0\,(R_0 + x_3)^2}\left[3 - \left(x_1 - x_1'\right)^2 \frac{3R_0 + x_3}{R_0^2\,(R_0 + x_3)}\right], \tag{1.393}$$

$$\tau_{22} = -2\mu B\frac{x_1 - x_1'}{R_0\,(R_0 + x_3)^2}\left[1 - \left(x_2 - x_2'\right)^2 \frac{3R_0 + x_3}{R_0^2\,(R_0 + x_3)}\right], \tag{1.394}$$

$$\tau_{33} = 2\mu B\frac{x_1 - x_1'}{R_0^3}, \tag{1.395}$$

with shear stresses

$$\tau_{13} = -\frac{2\mu B}{R_0\,(R_0 + x_3)}\left[1 - \left(x_1 - x_1'\right)^2 \frac{2R_0 + x_3}{R_0^2\,(R_0 + x_3)}\right], \tag{1.396}$$

$$\tau_{23} = 2\mu B\frac{(x_1 - x_1')(x_2 - x_2')}{R_0^3}\left[\frac{2R_0 + x_3}{(R_0 + x_3)^2}\right], \tag{1.397}$$

$$\tau_{12} = -2\mu B\frac{x_2 - x_2'}{R_0\,(R_0 + x_3)^2}\left[1 - \left(x_1 - x_1'\right)^2 \frac{3R_0 + x_3}{R_0^2\,(R_0 + x_3)}\right]. \tag{1.398}$$

The second extra displacement field derives from the Galerkin vector,

$$C(x_1 - x_1')\log(R_0 + x_3)\,\hat{e}_3, \tag{1.399}$$

with C a free constant. From (1.335), the corresponding displacement field is found to be

$$u_1 = \frac{C}{2\mu}\frac{1}{R_0}\left[\frac{(x_1 - x_1')^2}{R_0^2} - 1\right], \tag{1.400}$$

$$u_2 = \frac{C}{2\mu}\frac{(x_1 - x_1')(x_2 - x_2')}{R_0^3}, \tag{1.401}$$

$$u_3 = \frac{C}{2\mu}\frac{x_1 - x_1'}{R_0}\left[\frac{x_3}{R_0^2} + \frac{\lambda + 2\mu}{\lambda + \mu}\frac{2}{R_0 + x_3}\right]. \tag{1.402}$$

This displacement field has associated normal stresses

$$\tau_{11} = -C\frac{x_1 - x_1'}{R_0^3}\left[3\frac{(x_1 - x_1')^2}{R_0^2} - \frac{2\lambda + 3\mu}{\lambda + \mu}\right], \tag{1.403}$$

$$\tau_{22} = -C\frac{x_1 - x_1'}{R_0^3}\left[3\frac{(x_2 - x_2')^2}{R_0^2} - \frac{\mu}{\lambda + \mu}\right], \tag{1.404}$$

$$\tau_{33} = -C\frac{x_1 - x_1'}{R_0^3}\left[3\frac{x_3^2}{R_0^2} + \frac{2\lambda + 3\mu}{\lambda + \mu}\right], \tag{1.405}$$

with shear stresses

$$\tau_{13} = -\frac{C}{R_0^3}\left[x_3\left(3\frac{(x_1 - x_1')^2}{R_0^2} - 1\right)\right.$$
$$\left. - \frac{\lambda + 2\mu}{\lambda + \mu}\frac{R_0^2}{R_0 + x_3}\left(1 - (x_1 - x_1')^2\frac{2R_0 + x_3}{R_0^2(R_0 + x_3)}\right)\right], \tag{1.406}$$

$$\tau_{23} = -C\frac{(x_1 - x_1')(x_2 - x_2')}{R_0^3}\left[3\frac{x_3}{R_0^2} + \frac{\lambda + 2\mu}{\lambda + \mu}\frac{2R_0 + x_3}{(R_0 + x_3)^2}\right], \tag{1.407}$$

$$\tau_{12} = -C\frac{x_2 - x_2'}{R_0^3}\left[3\frac{(x_1 - x_1')^2}{R_0^2} - 1\right]. \tag{1.408}$$

In order for the two shear stresses, τ_{13} and τ_{23}, to vanish on the plane $x_3 = 0$, the two free constants of the extra displacement fields must be related by

$$2\mu B = \frac{\lambda + 2\mu}{\lambda + \mu}C. \tag{1.409}$$

Combining the extra displacement fields, they are found to have the associated normal stresses

$$\tau_{11} = -C(x_1 - x_1')\left[\frac{\lambda + 2\mu}{\lambda + \mu}\frac{1}{R_0(R_0 + x_3)^2}\left(3 - (x_1 - x_1')^2\frac{3R_0 + x_3}{R_0^2(R_0 + x_3)}\right)\right.$$
$$\left. + \frac{1}{R_0^3}\left(3\frac{(x_1 - x_1')^2}{R_0^2} - \frac{2\lambda + 3\mu}{\lambda + \mu}\right)\right], \tag{1.410}$$

$$\tau_{22} = -C(x_1 - x_1')\left[\frac{\lambda + 2\mu}{\lambda + \mu}\frac{1}{R_0(R_0 + x_3)^2}\left(1 - (x_2 - x_2')^2\frac{3R_0 + x_3}{R_0^2(R_0 + x_3)}\right)\right.$$
$$\left. + \frac{1}{R_0^3}\left(3\frac{(x_2 - x_2')^2}{R_0^2} - \frac{\mu}{\lambda + \mu}\right)\right], \tag{1.411}$$

$$\tau_{33} = -C\frac{x_1 - x_1'}{R_0^3}\left[3\frac{x_3^2}{R_0^2} + 1\right], \tag{1.412}$$

with shear stresses

$$\tau_{13} = -C\frac{x_3}{R_0^3}\left[3\frac{(x_1 - x_1')^2}{R_0^2} - 1\right], \tag{1.413}$$

$$\tau_{23} = -3C\frac{(x_1 - x_1')(x_2 - x_2')x_3}{R_0^5}, \tag{1.414}$$

$$\tau_{12} = -C\frac{x_2 - x_2'}{R_0^3}\left[3\frac{(x_1 - x_1')^2}{R_0^2} - 1\right.$$
$$\left. + \frac{\lambda + 2\mu}{\lambda + \mu}\frac{R_0^2}{(R_0 + x_3)^2}\left(1 - (x_1 - x_1')^2\frac{3R_0 + x_3}{R_0^2(R_0 + x_3)}\right)\right]. \tag{1.415}$$

From expressions (1.383) and (1.388), for the total normal stress to vanish on the free surface $x_3 = 0$, the remaining free constants must be related by

$$\frac{\mu}{\lambda + \mu}A = C. \tag{1.416}$$

The combined extra displacement fields produce a traction per unit area on the hemisphere, centred on the point of application of the tangential force, with components,

$$F_1^\gamma = \frac{C}{a^2} - C\frac{2\lambda + \mu}{\lambda + \mu}\frac{(x_1 - x_1')^2}{a^4}$$
$$+ C\frac{\lambda + 2\mu}{\lambda + \mu}\frac{a^2 - x_3^2}{a^2(a + x_3)^2}\left[\frac{(x_1 - x_1')^2}{a^2} - 2\frac{(x_1 - x_1')^2 x_3}{a(a^2 - x_3^2)} - 1\right], \tag{1.417}$$

$$F_2^\gamma = -C\frac{(x_1 - x_1')(x_2 - x_2')}{a^2}\left[\frac{3}{a^2} - \frac{\lambda + 2\mu}{\lambda + \mu}\frac{2}{(a + x_3)^2}\right], \tag{1.418}$$

$$F_3^\gamma = -3C\frac{(x_1 - x_1')x_3}{a^4}. \tag{1.419}$$

The components of the total traction on the hemisphere are

$$T_i = a^2\int_0^{2\pi}\int_0^{\pi/2} F_i^\gamma \sin\theta \, d\theta \, d\phi = a^2\int_0^{2\pi}\int_0^1 F_i^\gamma \, du \, d\phi, \tag{1.420}$$

where the integration variable in the last expression has been changed to $u = \cos\theta$. Thus,

$$T_1 = C\int_0^{2\pi}\int_0^1 du\,d\phi - C\frac{2\lambda+\mu}{\lambda+\mu}\int_0^{2\pi}\int_0^1 \left(1-u^2\right) du\,\cos^2\phi\,d\phi$$
$$-2C\frac{\lambda+2\mu}{\lambda+\mu}\int_0^{2\pi}\int_0^1 u\frac{1-u}{1+u}\,du\,\cos^2\phi\,d\phi$$
$$+C\frac{\lambda+2\mu}{\lambda+\mu}\int_0^{2\pi}\int_0^1 (1-u)^2\,du\,\cos^2\phi\,d\phi$$
$$-C\frac{\lambda+2\mu}{\lambda+\mu}\int_0^{2\pi}\int_0^1 \frac{1-u}{1+u}\,du\,d\phi. \tag{1.421}$$

The evaluation of T_1 then requires the evaluation of the five elementary integrals,

$$I_1 = \int_0^{2\pi}\cos^2\phi\,d\phi = \int_0^{2\pi}\frac{1}{2}(1+\cos 2\phi)\,d\phi$$
$$= \frac{1}{2}\phi + \frac{1}{4}\sin 2\phi\,\Big|_0^{2\pi} = \pi, \tag{1.422}$$

$$I_2 = \int_0^1 \left(1-u^2\right) du = u - \frac{1}{3}u^3\,\Big|_0^1 = \frac{2}{3}, \tag{1.423}$$

$$I_3 = \int_0^1 \frac{u(1-u)}{1+u}\,du = \int_0^1 (1-u)\,du - 2\int_0^1 \frac{1}{1+u}\,du + \int_0^1 du$$
$$= u - \frac{1}{2}u^2 - 2\log(1+u) + u\,\Big|_0^1 = \frac{3}{2} - 2\log 2, \tag{1.424}$$

$$I_4 = \int_0^1 (1-u)^2\,du = -\frac{1}{3}(1-u)^3\,\Big|_0^1 = \frac{1}{3}, \tag{1.425}$$

$$I_5 = \int_0^1 \frac{1-u}{1+u}\,du = 2\int_0^1 \frac{1}{1+u}\,du - \int_0^1 du$$
$$= 2\log(1+u) - u\,\Big|_0^1 = 2\log 2 - 1. \tag{1.426}$$

Then,

$$T_1 = C\cdot 2\pi - C\frac{2\lambda+\mu}{\lambda+\mu}\cdot\pi\cdot\frac{2}{3} - 2C\frac{\lambda+2\mu}{\lambda+\mu}\cdot\pi\cdot\left(\frac{3}{2} - 2\log 2\right)$$
$$+C\frac{\lambda+2\mu}{\lambda+\mu}\cdot\pi\cdot\frac{1}{3} - C\frac{\lambda+2\mu}{\lambda+\mu}\cdot 2\pi\cdot(2\log 2 - 1) = 0. \tag{1.427}$$

Consideration of the integrals over ϕ shows that $T_2 = T_3 = 0$. The extra displacement fields thus make no contribution to the traction on the hemisphere centred on the point of application of the tangential force.

The components of the traction per unit area on the hemisphere, coming from the stresses, (1.383) through (1.388), generated directly by the application of the tangential force, are

$$F_1^\gamma = -\frac{A}{a^2}\left[3\frac{(x_1 - x_1')^2}{a^2} + \frac{\mu}{\lambda + \mu}\right], \tag{1.428}$$

$$F_2^\gamma = -3A\frac{(x_1 - x_1')(x_2 - x_2')}{a^4}, \tag{1.429}$$

$$F_3^\gamma = -3A\frac{(x_1 - x_1')x_3}{a^4}. \tag{1.430}$$

Substituting these in (1.420), we find that they contribute the components of the total traction on the hemisphere,

$$T_1 = -2A\pi\frac{\lambda + 2\mu}{\lambda + \mu}, \quad T_2 = 0, \quad T_3 = 0. \tag{1.431}$$

For equilibrium, the component T_1 must balance the applied force, giving $T_1 = -P$ and

$$A = \frac{P}{2\pi}\frac{\lambda + \mu}{\lambda + 2\mu}. \tag{1.432}$$

From (1.409) and (1.416), it follows that

$$B = \frac{P}{4\pi}\frac{1}{\lambda + \mu}, \quad C = \frac{P}{2\pi}\frac{\mu}{\lambda + 2\mu}. \tag{1.433}$$

Replacing the three constants in their respective displacement field expressions by these values, we find the total displacement field for the Cerruti problem to be

$$u_1 = \frac{P}{4\pi\mu}\frac{1}{R_0}\left[1 + \frac{(x_1 - x_1')^2}{R_0^2} + \frac{\mu}{\lambda + \mu}\left(\frac{R_0}{R_0 + x_3} - \frac{(x_1 - x_1')^2}{(R_0 + x_3)^2}\right)\right], \tag{1.434}$$

$$u_2 = \frac{P}{4\pi\mu}\frac{(x_1 - x_1')(x_2 - x_2')}{R_0}\left[\frac{1}{R_0^2} - \frac{\mu}{\lambda + \mu}\frac{1}{(R_0 + x_3)^2}\right], \tag{1.435}$$

$$u_3 = \frac{P}{4\pi\mu}\frac{x_1 - x_1'}{R_0}\left[\frac{x_3}{R_0^2} + \frac{\mu}{\lambda + \mu}\frac{1}{R_0 + x_3}\right]. \tag{1.436}$$

1.4.11 The Mindlin problems

When the downward point force is below the surface of the elastic half-space, the Boussinesq problem becomes Mindlin problem I. Similarly, when the horizontal point force is below the surface, the Cerruti problem becomes Mindlin problem II. The *Mindlin problems* were first solved by Mindlin (1936) after whom they are named. The original solutions were generalised and expanded to consideration of nuclei of strain by Mindlin and Cheng (1950).

In contrast to the Boussinesq and Cerruti problems, where the conditions of equilibrium were established by integrating tractions over a hemisphere centred on the point of application of the force, for the Mindlin problems we use the *method of images* and combine additional solutions, which are singular at the image point in the upper half-space, to ensure equilibrium. The additional solutions are also chosen to make the common bounding surface of the two half-spaces stress free, allowing removal of the upper half-space.

A point force of strength P is located at (ξ_1, ξ_2, ξ_3) in the lower half-space. The radius to the field point at (x_1, x_2, x_3) is once again

$$R = \sqrt{(x_1 - \xi_1)^2 + (x_2 - \xi_2)^2 + (x_3 - \xi_3)^2}. \tag{1.437}$$

The additional solutions combine at the image point to produce a point force of strength $-P$, producing overall force equilibrium. The image point is located at $(\xi_1, \xi_2, -\xi_3)$, in the upper half-space. The radius from there to the field point is then

$$Q = \sqrt{(x_1 - \xi_1)^2 + (x_2 - \xi_2)^2 + (x_3 + \xi_3)^2}. \tag{1.438}$$

For Mindlin problem I, the Galerkin vector, $2\mu HPR\hat{e}_3$, for Kelvin's problem with a downward point force in the lower half-space is the starting point. For a point force with magnitude $8\pi\mu(\lambda + 2\mu) / (\lambda + \mu)$ in the x_3 direction, Mindlin and Cheng (1950) give the Galerkin vector for Mindlin problem I as

$$2\mu\hat{e}_3\Big\{R + \big[8\sigma(1-\sigma) - 1\big]Q + 4(1-2\sigma)\big[(1-\sigma)x_3 - \sigma\xi_3\big]\log(Q + x_3 + \xi_3) - \frac{2\xi_3 x_3}{Q}\Big\}, \tag{1.439}$$

with the elastic constants (excepting μ) expressed in terms of the dimensionless Poisson's ratio (1.255). It is thus seen to be a linear combination of the Galerkin vector for the Kelvin problem of a downward point force in the lower half-space, proportional to $\hat{e}_3 R$ (omitting the factor $2\mu HP$), with the Galerkin vectors,

$$G_1 = A\hat{e}_3 \log(Q + x_3 + \xi_3), \tag{1.440}$$

$$G_2 = B\hat{e}_3 (x_3 + \xi_3) \log(Q + x_3 + \xi_3), \tag{1.441}$$

$$G_3 = C\hat{e}_3 \frac{x_3}{Q}, \tag{1.442}$$

$$G_4 = D\hat{e}_3 Q. \tag{1.443}$$

The last Galerkin vector of the group, G_4, represents a Kelvin problem generated by rotating a lower half-space solution by 180° into the upper half-space, as an image Kelvin solution. All four of the Galerkin vectors represent solutions singular at the image point. The arbitrary constants A, B, C, D of the linear combination remain to be determined.

For their determination, three conditions arise from the vanishing of the normal stress τ_{33} and the shear stresses τ_{13}, τ_{23} on the surface $x_3 = 0$ for it to be a free surface, and a fourth condition arises to ensure that the total force on a sphere surrounding the image point balances the point force in the lower half-space.

The displacement field corresponding to each of the four Galerkin vectors is given by relation (1.335).

For the first Galerkin vector G_1, we have

$$\nabla(\nabla \cdot \boldsymbol{G}_1) = -\hat{e}_1 \frac{A(x_1 - \xi_1)}{Q^3} - \hat{e}_2 \frac{A(x_2 - \xi_2)}{Q^3} - \hat{e}_3 \frac{A(x_3 + \xi_3)}{Q^3}, \tag{1.444}$$

and

$$\nabla^2 \boldsymbol{G}_1 = A\hat{e}_3 \frac{\partial^2}{\partial x_j \partial x_j}\big(\log(Q + x_3 + \xi_3)\big) = 0. \tag{1.445}$$

Then, from (1.335)

$$2\mu \boldsymbol{u} = -\nabla(\nabla \cdot \boldsymbol{G}_1), \tag{1.446}$$

and the displacement field is found to have the components

$$u_1 = \frac{A}{2\mu}\frac{x_1 - \xi_1}{Q^3}, \quad u_2 = \frac{A}{2\mu}\frac{x_2 - \xi_2}{Q^3}, \quad u_3 = \frac{A}{2\mu}\frac{x_3 + \xi_3}{Q^3}. \tag{1.447}$$

Taking derivatives, we find that

$$\frac{\partial u_1}{\partial x_1} = \frac{A}{2\mu}\left[\frac{1}{Q^3} - 3\frac{(x_1 - \xi_1)^2}{Q^5}\right], \tag{1.448}$$

$$\frac{\partial u_2}{\partial x_2} = \frac{A}{2\mu}\left[\frac{1}{Q^3} - 3\frac{(x_2 - \xi_2)^2}{Q^5}\right], \tag{1.449}$$

$$\frac{\partial u_3}{\partial x_3} = \frac{A}{2\mu}\left[\frac{1}{Q^3} - 3\frac{(x_3 + \xi_3)^2}{Q^5}\right]. \tag{1.450}$$

Thus, the cubical dilatation, the divergence of the displacement field, is

$$\nabla \cdot \boldsymbol{u} = \frac{A}{2\mu}\left[\frac{3}{Q^3} - \frac{3Q^2}{Q^5}\right] = 0. \tag{1.451}$$

This solenoidal displacement field has the associated normal stresses

$$\tau_{11} = \frac{A}{Q^3}\left[1 - 3\frac{(x_1 - \xi_1)^2}{Q^2}\right], \tag{1.452}$$

$$\tau_{22} = \frac{A}{Q^3}\left[1 - 3\frac{(x_2 - \xi_2)^2}{Q^2}\right], \tag{1.453}$$

$$\tau_{33} = \frac{A}{Q^3}\left[1 - 3\frac{(x_3 + \xi_3)^2}{Q^2}\right], \tag{1.454}$$

and the associated shear stresses

$$\tau_{13} = -3A\frac{(x_1 - \xi_1)(x_3 + \xi_3)}{Q^5}, \tag{1.455}$$

$$\tau_{23} = -3A\frac{(x_2 - \xi_2)(x_3 + \xi_3)}{Q^5}, \tag{1.456}$$

$$\tau_{12} = -3A\frac{(x_1 - \xi_1)(x_2 - \xi_2)}{Q^5}. \tag{1.457}$$

This stress field generated by the Galerkin vector $\boldsymbol{G}_1$ produces a traction per unit area, on a small sphere of radius a surrounding the image point, with components

$$F_1 = \tau_{11}\frac{(x_1 - \xi_1)}{a} + \tau_{12}\frac{(x_2 - \xi_2)}{a} + \tau_{13}\frac{(x_3 + \xi_3)}{a} = -2A\frac{(x_1 - \xi_1)}{a^4}, \tag{1.458}$$

$$F_2 = \tau_{12}\frac{(x_1 - \xi_1)}{a} + \tau_{22}\frac{(x_2 - \xi_2)}{a} + \tau_{23}\frac{(x_3 + \xi_3)}{a} = -2A\frac{(x_2 - \xi_2)}{a^4}, \tag{1.459}$$

$$F_3 = \tau_{13}\frac{(x_1 - \xi_1)}{a} + \tau_{23}\frac{(x_2 - \xi_2)}{a} + \tau_{33}\frac{(x_3 + \xi_3)}{a} = -2A\frac{(x_3 + \xi_3)}{a^4}. \tag{1.460}$$

The components of the total traction on the small sphere are then

$$T_i = a^2 \int_0^{2\pi}\int_0^{\pi} F_i \sin\theta \, d\theta \, d\phi. \tag{1.461}$$

Thus, $T_1 = T_2 = T_3 = 0$.

For the second Galerkin vector $\boldsymbol{G}_2$, we have

$$\begin{aligned} \nabla(\nabla \cdot \boldsymbol{G}_2) = {} & B\hat{\mathbf{e}}_1 (x_1 - \xi_1)\left[\frac{1}{Q(Q + x_3 + \xi_3)} - \frac{x_3 + \xi_3}{Q^3}\right] \\ & + B\hat{\mathbf{e}}_2 (x_2 - \xi_2)\left[\frac{1}{Q(Q + x_3 + \xi_3)} - \frac{x_3 + \xi_3}{Q^3}\right] \\ & + B\hat{\mathbf{e}}_3\left[\frac{2}{Q} - \frac{(x_3 + \xi_3)^2}{Q^3}\right], \end{aligned} \tag{1.462}$$

and

$$\nabla^2 \boldsymbol{G}_2 = B\hat{\mathbf{e}}_3 \frac{2}{Q}. \tag{1.463}$$

From relation (1.335),

$$2\mu \boldsymbol{u} = \frac{\lambda + 2\mu}{\lambda + \mu} \nabla^2 \boldsymbol{G}_2 - \nabla (\nabla \cdot \boldsymbol{G}_2), \tag{1.464}$$

giving a displacement field with components

$$u_1 = \frac{B}{2\mu} (x_1 - \xi_1) \left(\frac{x_3 + \xi_3}{Q^3} - \frac{1}{Q(Q + x_3 + \xi_3)} \right), \tag{1.465}$$

$$u_2 = \frac{B}{2\mu} (x_2 - \xi_2) \left(\frac{x_3 + \xi_3}{Q^3} - \frac{1}{Q(Q + x_3 + \xi_3)} \right), \tag{1.466}$$

$$u_3 = \frac{B}{2\mu} \frac{(x_3 + \xi_3)^2}{Q^3} + \frac{B}{\lambda + \mu} \frac{1}{Q}, \tag{1.467}$$

and derivatives

$$\begin{aligned} \frac{\partial u_1}{\partial x_1} &= \frac{B}{2\mu} \left[\frac{x_3 + \xi_3}{Q^3} - \frac{1}{Q(Q + x_3 + \xi_3)} \right] \\ &+ \frac{B}{2\mu} (x_1 - \xi_1)^2 \left[-3 \frac{x_3 + \xi_3}{Q^5} + \frac{1}{Q^3 (Q + x_3 + \xi_3)} + \frac{1}{Q^2 (Q + x_3 + \xi_3)^2} \right], \end{aligned} \tag{1.468}$$

$$\begin{aligned} \frac{\partial u_2}{\partial x_2} &= \frac{B}{2\mu} \left[\frac{x_3 + \xi_3}{Q^3} - \frac{1}{Q(Q + x_3 + \xi_3)} \right] \\ &+ \frac{B}{2\mu} (x_2 - \xi_2)^2 \left[-3 \frac{x_3 + \xi_3}{Q^5} + \frac{1}{Q^3 (Q + x_3 + \xi_3)} + \frac{1}{Q^2 (Q + x_3 + \xi_3)^2} \right], \end{aligned} \tag{1.469}$$

$$\frac{\partial u_3}{\partial x_3} = \frac{B}{2\mu} \frac{x_3 + \xi_3}{Q^3} + \frac{B}{2\mu} (x_3 + \xi_3) \left[\frac{1}{Q^3} - 3 \frac{(x_3 + \xi_3)^2}{Q^5} \right] - \frac{B}{\lambda + \mu} \frac{x_3 + \xi_3}{Q^3}. \tag{1.470}$$

Summing the derivatives, the cubical dilatation is found to be

$$\nabla \cdot \boldsymbol{u} = -\frac{B}{\lambda + \mu} \frac{x_3 + \xi_3}{Q^3}. \tag{1.471}$$

This displacement field has the associated normal stresses

$$\begin{aligned} \tau_{11} &= B \frac{x_3 + \xi_3}{Q^3} \left[\frac{\mu}{\lambda + \mu} - 3 \frac{(x_1 - \xi_1)^2}{Q^2} \right] \\ &+ \frac{B}{Q(Q + x_3 + \xi_3)} \left[\frac{(x_1 - \xi_1)^2 (2Q + x_3 + \xi_3)}{Q^2 (Q + x_3 + \xi_3)} - 1 \right], \end{aligned} \tag{1.472}$$

$$\begin{aligned} \tau_{22} &= B \frac{x_3 + \xi_3}{Q^3} \left[\frac{\mu}{\lambda + \mu} - 3 \frac{(x_2 - \xi_2)^2}{Q^2} \right] \\ &+ \frac{B}{Q(Q + x_3 + \xi_3)} \left[\frac{(x_2 - \xi_2)^2 (2Q + x_3 + \xi_3)}{Q^2 (Q + x_3 + \xi_3)} - 1 \right], \end{aligned} \tag{1.473}$$

$$\tau_{33} = B \frac{x_3 + \xi_3}{Q^3} \left[\frac{\lambda}{\lambda + \mu} - 3 \frac{(x_3 + \xi_3)^2}{Q^2} \right], \tag{1.474}$$

and the associated shear stresses

$$\tau_{13} = B\frac{x_1 - \xi_1}{Q^3}\left[\frac{\lambda}{\lambda + \mu} - 3\frac{(x_3 + \xi_3)^2}{Q^2}\right], \tag{1.475}$$

$$\tau_{23} = B\frac{x_2 - \xi_2}{Q^3}\left[\frac{\lambda}{\lambda + \mu} - 3\frac{(x_3 + \xi_3)^2}{Q^2}\right], \tag{1.476}$$

$$\tau_{12} = B\frac{(x_1 - \xi_1)(x_2 - \xi_2)}{Q^3}\left[\frac{2Q + x_3 + \xi_3}{(Q + x_3 + \xi_3)^2} - 3\frac{x_3 + \xi_3}{Q^2}\right]. \tag{1.477}$$

The stress field generated by the Galerkin vector G_2 produces the traction per unit area, on a small sphere of radius a surrounding the image point, with components

$$F_1 = B\frac{x_1 - \xi_1}{a^2}\left[\frac{1}{a + x_3 + \xi_3} - 3\frac{x_3 + \xi_3}{a^2}\right], \tag{1.478}$$

$$F_2 = B\frac{x_2 - \xi_2}{a^2}\left[\frac{1}{a + x_3 + \xi_3} - 3\frac{x_3 + \xi_3}{a^2}\right], \tag{1.479}$$

$$F_3 = \frac{B}{a^2}\left[\frac{\lambda}{\lambda + \mu} - 3\frac{(x_3 + \xi_3)^2}{a^2}\right]. \tag{1.480}$$

The total traction on the small sphere is found to have components $T_1 = T_2 = 0$ and

$$\begin{aligned} T_3 &= B\int_0^{2\pi}\int_0^{\pi}\left[\frac{\lambda}{\lambda + \mu} - 3\cos^2\theta\right]\sin\theta \, d\theta \, d\phi \\ &= 2\pi B\left[-\frac{\lambda}{\lambda + \mu}\cos\theta + \cos^3\theta\right]_0^{\pi} \\ &= 2\pi B\left[\frac{2\lambda}{\lambda + \mu} - 2\right] \\ &= -4\pi B\frac{\mu}{\lambda + \mu}. \end{aligned} \tag{1.481}$$

For the third Galerkin vector G_3, we have

$$\begin{aligned} \nabla(\nabla \cdot G_3) = &-C\hat{e}_1\frac{x_1 - \xi_1}{Q^3}\left[1 - 3\frac{x_3(x_3 + \xi_3)}{Q^2}\right] - C\hat{e}_2\frac{x_2 - \xi_2}{Q^3}\left[1 - 3\frac{x_3(x_3 + \xi_3)}{Q^2}\right] \\ &- C\hat{e}_3\left[2\frac{x_3 + \xi_3}{Q^3} + \frac{x_3}{Q^3}\left(1 - 3\frac{(x_3 + \xi_3)^2}{Q^2}\right)\right], \end{aligned} \tag{1.482}$$

$$\nabla^2 G_3 = -2C\hat{e}_3\frac{x_3 + \xi_3}{Q^3}. \tag{1.483}$$

From relation (1.335) the displacement field is found to have the components

$$u_1 = \frac{C}{2\mu}\frac{x_1-\xi_1}{Q^3}\left[1-3\frac{x_3\,(x_3+\xi_3)}{Q^2}\right], \tag{1.484}$$

$$u_2 = \frac{C}{2\mu}\frac{x_2-\xi_2}{Q^3}\left[1-3\frac{x_3\,(x_3+\xi_3)}{Q^2}\right], \tag{1.485}$$

$$u_3 = \frac{C}{2\mu}\frac{x_3}{Q^3}\left[1-3\frac{(x_3+\xi_3)^2}{Q^2}\right]-\frac{C}{\lambda+\mu}\frac{x_3+\xi_3}{Q^3}, \tag{1.486}$$

with derivatives

$$\frac{\partial u_1}{\partial x_1} = \frac{C}{2\mu Q^3}\left[1-3\frac{(x_1-\xi_1)^2}{Q^2}\right]-\frac{3C}{2\mu}\frac{x_3\,(x_3+\xi_3)}{Q^5}\left[1-5\frac{(x_1-\xi_1)^2}{Q^2}\right], \tag{1.487}$$

$$\frac{\partial u_2}{\partial x_2} = \frac{C}{2\mu Q^3}\left[1-3\frac{(x_2-\xi_2)^2}{Q^2}\right]-\frac{3C}{2\mu}\frac{x_3\,(x_3+\xi_3)}{Q^5}\left[1-5\frac{(x_2-\xi_2)^2}{Q^2}\right], \tag{1.488}$$

$$\frac{\partial u_3}{\partial x_3} = \frac{C}{2\mu Q^3}\left[1-3\frac{(x_3+\xi_3)^2}{Q^2}\right]-\frac{3C}{2\mu}\frac{x_3\,(x_3+\xi_3)}{Q^5}\left[1-5\frac{(x_3+\xi_3)^2}{Q^2}\right]$$
$$-\frac{6C}{2\mu}\frac{x_3\,(x_3+\xi_3)}{Q^5}-\frac{C}{\lambda+\mu}\frac{1}{Q^3}\left[1-3\frac{(x_3+\xi_3)^2}{Q^2}\right]. \tag{1.489}$$

Summing the derivatives, the cubical dilatation is found to be

$$\nabla\cdot\boldsymbol{u} = -\frac{C}{\lambda+\mu}\frac{1}{Q^3}\left[1-3\frac{(x_3+\xi_3)^2}{Q^2}\right]. \tag{1.490}$$

This displacement field has the associated normal stresses

$$\tau_{11} = C\frac{\mu}{\lambda+\mu}\frac{1}{Q^3}+3C\frac{\lambda}{\lambda+\mu}\frac{(x_3+\xi_3)^2}{Q^5}-3C\frac{x_3\,(x_3+\xi_3)}{Q^5}$$
$$-3C\frac{(x_1-\xi_1)^2}{Q^5}\left[1-5\frac{x_3\,(x_3+\xi_3)}{Q^2}\right], \tag{1.491}$$

$$\tau_{22} = C\frac{\mu}{\lambda+\mu}\frac{1}{Q^3}+3C\frac{\lambda}{\lambda+\mu}\frac{(x_3+\xi_3)^2}{Q^5}-3C\frac{x_3\,(x_3+\xi_3)}{Q^5}$$
$$-3C\frac{(x_2-\xi_2)^2}{Q^5}\left[1-5\frac{x_3\,(x_3+\xi_3)}{Q^2}\right], \tag{1.492}$$

$$\tau_{33} = -\frac{\mu}{\lambda+\mu}\frac{1}{Q^3}\left[1-3\frac{(x_3+\xi_3)^2}{Q^2}\right]-\frac{x_3\,(x_3+\xi_3)}{Q^5}\left[3-5\frac{(x_3+\xi_3)^2}{Q^2}\right], \tag{1.493}$$

while the associated shear stresses of this displacement field are

$$\tau_{13} = -3C\frac{(x_1-\xi_1)\,x_3}{Q^5}\left[1-5\frac{(x_3+\xi_3)^2}{Q^2}\right]-3C\frac{\lambda}{\lambda+\mu}\frac{(x_1-\xi_1)(x_3+\xi_3)}{Q^5}, \quad (1.494)$$

$$\tau_{23} = -3C\frac{(x_2-\xi_2)\,x_3}{Q^5}\left[1-5\frac{(x_3+\xi_3)^2}{Q^2}\right]-3C\frac{\lambda}{\lambda+\mu}\frac{(x_2-\xi_2)(x_3+\xi_3)}{Q^5}, \quad (1.495)$$

$$\tau_{12} = -3C\frac{(x_1-\xi_1)(x_2-\xi_2)}{Q^5}\left[1-5\frac{x_3\,(x_3+\xi_3)}{Q^2}\right]. \quad (1.496)$$

The stress field generated by the Galerkin vector G_3 produces the traction per unit area, on a small sphere of radius a surrounding the image point, with components

$$F_1 = 3C\frac{x_1-\xi_1}{a^4}\left[\frac{(x_3+\xi_3)^2}{a^2}+3\frac{x_3\,(x_3+\xi_3)}{a^2}-\frac{1}{3}\left(\frac{3\lambda+2\mu}{\lambda+2\mu}\right)\right], \quad (1.497)$$

$$F_2 = 3C\frac{x_2-\xi_2}{a^4}\left[\frac{(x_3+\xi_3)^2}{a^2}+3\frac{x_3\,(x_3+\xi_3)}{a^2}-\frac{1}{3}\left(\frac{3\lambda+2\mu}{\lambda+2\mu}\right)\right], \quad (1.498)$$

$$F_3 = 3C\left\{\frac{x_3+\xi_3}{a^4}\left[\frac{(x_3+\xi_3)^2}{a^2}+3\frac{x_3\,(x_3+\xi_3)}{a^2}-\frac{2}{3}\left(\frac{3\lambda+2\mu}{\lambda+\mu}\right)\right]+\frac{\xi_3}{a^4}\right\}. \quad (1.499)$$

The components of the total traction on the small sphere are found to be $T_1 = T_2 = 0$ and

$$\begin{aligned}
T_3 &= a^2\int_0^{2\pi}\int_0^{\pi} F_3\sin\theta\,d\theta\,d\phi \\
&= 2\pi a^2\int_0^{\pi} F_3\sin\theta\,d\theta \\
&= 6\pi C a^2\int_0^{\pi}\left\{\frac{\cos\theta}{a^3}\left[\cos^2\theta+3\frac{(a\cos\theta-\xi_3)}{a}\cos\theta\right.\right. \\
&\qquad\left.\left.-\frac{2}{3}\left(\frac{3\lambda+2\mu}{\lambda+\mu}\right)\right]+\frac{\xi_3}{a^4}\right\}\sin\theta\,d\theta \\
&= \frac{6\pi C}{a}\left[-\frac{1}{4}\cos^4\theta+\frac{1}{3}\left(\frac{3\lambda+2\mu}{\lambda+\mu}\right)\cos^2\theta+\frac{\xi_3}{a}\left(\cos^3\theta-\cos\theta\right)\right]_0^{\pi} \\
&= 0.
\end{aligned} \quad (1.500)$$

For Kelvin's problem at the image point, the stresses are found from expressions (1.321) and (1.322) for Kelvin's problem at the source point, by replacing the co-ordinate ξ_3 by $-\xi_3$. The Galerkin vector $G_4 = \hat{e}_3 DQ$ then yields the normal stresses

$$\tau_{11} = D\frac{x_3+\xi_3}{Q^3}\left[\frac{\mu}{\lambda+\mu} - 3\frac{(x_1-\xi_1)^2}{Q^2}\right], \tag{1.501}$$

$$\tau_{22} = D\frac{x_3+\xi_3}{Q^3}\left[\frac{\mu}{\lambda+\mu} - 3\frac{(x_2-\xi_2)^2}{Q^2}\right], \tag{1.502}$$

$$\tau_{33} = -D\frac{x_3+\xi_3}{Q^3}\left[\frac{\mu}{\lambda+\mu} + 3\frac{(x_3+\xi_3)^2}{Q^2}\right], \tag{1.503}$$

and the shear stresses

$$\tau_{13} = -D\frac{x_1-\xi_1}{Q^3}\left[\frac{\mu}{\lambda+\mu} + 3\frac{(x_3+\xi_3)^2}{Q^2}\right], \tag{1.504}$$

$$\tau_{23} = -D\frac{x_2-\xi_2}{Q^3}\left[\frac{\mu}{\lambda+\mu} + 3\frac{(x_3+\xi_3)^2}{Q^2}\right], \tag{1.505}$$

$$\tau_{12} = -3D\frac{(x_1-\xi_1)(x_2-\xi_2)(x_3+\xi_3)}{Q^5}. \tag{1.506}$$

The stress field generated by the Galerkin vector G_4 produces the traction per unit area, on a small sphere of radius a surrounding the image point, with components

$$F_1 = -3D\frac{(x_1-\xi_1)(x_3+\xi_3)}{a^4}, \tag{1.507}$$

$$F_2 = -3D\frac{(x_2-\xi_2)(x_3+\xi_3)}{a^4}, \tag{1.508}$$

$$F_3 = -D\frac{1}{a^2}\left[\frac{\mu}{\lambda+\mu} + 3\frac{(x_3+\xi_3)^2}{a^2}\right]. \tag{1.509}$$

Thus, $T_1 = T_2 = 0$ and

$$\begin{aligned} T_3 &= -a^2\int_0^{2\pi}\int_0^{\pi}\frac{D}{a^2}\left[\frac{\mu}{\lambda+\mu} + 3\cos^2\theta\right]\sin\theta\, d\theta\, d\phi \\ &= 2\pi D\left[\frac{\mu}{\lambda+\mu}\cos\theta + \cos^3\theta\right]_0^{\pi} = -4\pi D\frac{\lambda+2\mu}{\lambda+\mu}. \end{aligned} \tag{1.510}$$

The arbitrary constants A, B, C, D are to be determined by a balance of the point force at the source point (ξ_1, ξ_2, ξ_3) and traction at the image point $(\xi_1, \xi_2, -\xi_3)$, and by the further condition that the surface $x_3 = 0$ is stress free ($\tau_{13} = \tau_{23} = \tau_{33} = 0$).

Omitting the factor $2\mu HP$ is equivalent to setting it to unity, or taking the magnitude of the point force at the source point to be

$$P = 4\pi\frac{\lambda+2\mu}{\lambda+\mu}. \tag{1.511}$$

Force equilibrium for the infinite elastic medium then requires that

$$-4\pi\frac{\lambda+2\mu}{\lambda+\mu}D-4\pi\frac{\mu}{\lambda+\mu}B=-4\pi\frac{\lambda+2\mu}{\lambda+\mu}, \tag{1.512}$$

or

$$B\frac{\mu}{\lambda+2\mu}+D=1. \tag{1.513}$$

On the surface $x_3=0$, we have $Q=R$. The condition that the normal stress vanishes on this surface is then

$$\begin{aligned}\frac{\xi_3}{R^3}\left[\frac{\mu}{\lambda+\mu}+3\frac{\xi_3^2}{R^2}\right]-D\frac{\xi_3}{R^3}\left[\frac{\mu}{\lambda+\mu}+3\frac{\xi_3^2}{R^2}\right]+A\frac{1}{R^3}\left[1-3\frac{\xi_3^2}{R^2}\right]&\\ +B\frac{\xi_3}{R^3}\left[\frac{\lambda}{\lambda+\mu}-3\frac{\xi_3^2}{R^2}\right]-C\frac{\mu}{\lambda+\mu}\frac{1}{R^3}\left[1-3\frac{\xi_3^2}{R^2}\right]&=0.\end{aligned} \tag{1.514}$$

The condition that the shear stresses vanish on the surface $x_3=0$ is

$$\begin{aligned}-\frac{1}{R^3}\left[\frac{\mu}{\lambda+\mu}+3\frac{\xi_3^2}{R^2}\right]-D\frac{1}{R^3}\left[\frac{\mu}{\lambda+\mu}+3\frac{\xi_3^2}{R^2}\right]-3A\frac{\xi_3}{R^5}&\\ +B\frac{1}{R^3}\left[\frac{\lambda}{\lambda+\mu}-3\frac{\xi_3^2}{R^2}\right]-3C\frac{\lambda}{\lambda+\mu}\frac{\xi_3}{R^5}&=0.\end{aligned} \tag{1.515}$$

In both cases, the coefficients of terms in $1/R^3$ and of terms in $1/R^5$ must vanish. For the normal stress this yields

$$A+B\xi_3\frac{\lambda}{\lambda+\mu}-C\frac{\mu}{\lambda+\mu}-D\xi_3\frac{\mu}{\lambda+\mu}=-\xi_3\frac{\mu}{\lambda+\mu}, \tag{1.516}$$

$$A+B\xi_3-C\frac{\mu}{\lambda+\mu}+D\xi_3=\xi_3, \tag{1.517}$$

while for the shear stresses we have

$$B\lambda-D\mu=\mu, \tag{1.518}$$

$$A+B\xi_3+C\frac{\lambda}{\lambda+\mu}+D\xi_3=-\xi_3. \tag{1.519}$$

Equations (1.513) and (1.518) can be solved to give

$$B=2\mu\frac{\lambda+2\mu}{(\lambda+\mu)^2}=4(1-\sigma)(1-2\sigma), \tag{1.520}$$

$$D=2\lambda\frac{\lambda+2\mu}{(\lambda+\mu)^2}-1=8\sigma(1-\sigma)-1. \tag{1.521}$$

Subtracting equation (1.519) from equation (1.517) gives

$$C = -2\xi_3, \tag{1.522}$$

and then equation (1.517) yields

$$A = -4\xi_3 \frac{\mu}{\lambda + \mu} = -4\xi_3 (1 - 2\sigma). \tag{1.523}$$

These values of the linear combination coefficients A, B, C, D are found to satisfy equation (1.516). Once again including the factor 2μ, they also produce the Galerkin vector (1.439), as quoted for Mindlin problem I.

The linear combination of the Galerkin vector $R\hat{e}_3$, for Kelvin's problem of a downward point force in the lower half-space, with the preceding four Galerkin vectors produces a displacement field with components

$$\begin{aligned} u_1 = {} & \frac{1}{2\mu} \frac{(x_1 - \xi_1)(x_3 - \xi_3)}{R^3} + \frac{A}{2\mu} \frac{x_1 - \xi_1}{Q^3} \\ & + \frac{B}{2\mu} (x_1 - \xi_1) \left[\frac{x_3 + \xi_3}{Q^3} - \frac{1}{Q(Q + x_3 + \xi_3)} \right] \\ & + \frac{C}{2\mu} \frac{x_1 - \xi_1}{Q^3} \left[1 - 3 \frac{x_3 (x_3 + \xi_3)}{Q^2} \right] + \frac{D}{2\mu} \frac{(x_1 - \xi_1)(x_3 + \xi_3)}{Q^3}, \end{aligned} \tag{1.524}$$

$$\begin{aligned} u_2 = {} & \frac{1}{2\mu} \frac{(x_2 - \xi_2)(x_3 - \xi_3)}{R^3} + \frac{A}{2\mu} \frac{x_2 - \xi_2}{Q^3} \\ & + \frac{B}{2\mu} (x_2 - \xi_2) \left[\frac{x_3 + \xi_3}{Q^3} - \frac{1}{Q(Q + x_3 + \xi_3)} \right] \\ & + \frac{C}{2\mu} \frac{x_2 - \xi_2}{Q^3} \left[1 - 3 \frac{x_3 (x_3 + \xi_3)}{Q^2} \right] + \frac{D}{2\mu} \frac{(x_2 - \xi_2)(x_3 + \xi_3)}{Q^3}, \end{aligned} \tag{1.525}$$

$$\begin{aligned} u_3 = {} & \frac{1}{2\mu} \frac{1}{R} \left[3 - 4\sigma + \frac{(x_3 - \xi_3)^2}{R^2} \right] + \frac{A}{2\mu} \frac{x_3 + \xi_3}{Q^3} \\ & + \frac{B}{2\mu} \frac{1}{Q} \left[2 - 4\sigma + \frac{(x_3 + \xi_3)^2}{Q^2} \right] - \frac{C}{2\mu} (2 - 4\sigma) \frac{x_3 + \xi_3}{Q^3} \\ & + \frac{C}{2\mu} \frac{x_3}{Q^3} \left[1 - 3 \frac{(x_3 + \xi_3)^2}{Q^2} \right] + \frac{D}{2\mu} \frac{1}{Q} \left[3 - 4\sigma + \frac{(x_3 + \xi_3)^2}{Q^2} \right]. \end{aligned} \tag{1.526}$$

Substituting the values of the constants A, B, C, D, and multiplying by 2μ, we find the components of the displacement field, given by the Galerkin vector (1.439) for Mindlin problem I for a point force of magnitude $8\pi\mu(\lambda+2\mu)/(\lambda+\mu)$ to be

$$
\begin{aligned}
u_1 &= (x_1-\xi_1)\left[\frac{x_3-\xi_3}{R^3}+\frac{(3-4\sigma)(x_3-\xi_3)}{Q^3}\right.\\
&\qquad\left.+\frac{6\xi_3 x_3(x_3+\xi_3)}{Q^5}-\frac{4(1-\sigma)(1-2\sigma)}{Q(Q+x_3+\xi_3)}\right],\\
u_2 &= (x_2-\xi_2)\left[\frac{x_3-\xi_3}{R^3}+\frac{(3-4\sigma)(x_3-\xi_3)}{Q^3}\right.\\
&\qquad\left.+\frac{6\xi_3 x_3(x_3+\xi_3)}{Q^5}-\frac{4(1-\sigma)(1-2\sigma)}{Q(Q+x_3+\xi_3)}\right],\\
u_3 &= \frac{3-4\sigma}{R}+\frac{8(1-\sigma)^2-(3-4\sigma)}{Q}+\frac{(x_3-\xi_3)^2}{R^3}\\
&\qquad+\frac{(3-4\sigma)(x_3+\xi_3)^2-2\xi_3 x_3}{Q^3}+\frac{6\xi_3 x_3(x_3+\xi_3)^2}{Q^5}.
\end{aligned}
\tag{1.527}
$$

In this form of the displacement field, the elastic constants appear exclusively through the dimensionless Poisson's ratio (1.255).

For Mindlin problem II, the Galerkin vector, $2\mu HPR\hat{e}_1$, for Kelvin's problem with a horizontal point force in the lower half-space is the starting point. For a point force with magnitude $8\pi\mu(\lambda+2\mu)/(\lambda+\mu)$ in the x_1 direction, Mindlin and Cheng (1950) give the Galerkin vector for Mindlin problem II as

$$
\begin{aligned}
2\mu\Big\{\hat{e}_1&\left[R+Q-\frac{2\xi_3^2}{Q}+4(1-\sigma)(1-2\sigma)\big((x_3+\xi_3)\log(Q+x_3+\xi_3)-Q\big)\right]\\
&+\hat{e}_3\left[\frac{2\xi_3(x_1-\xi_1)}{Q}+2(1-2\sigma)(x_1-\xi_1)\log(Q+x_3+\xi_3)\right]\Big\}.
\end{aligned}
\tag{1.528}
$$

This Galerkin vector is a linear combination of the Galerkin vector for the Kelvin problem of a horizontal point force in the lower half-space with the additional Galerkin vectors, $\hat{e}_1 Q$, $\hat{e}_1(x_3+\xi_3)\log(Q+x_3+\xi_3)$, $\hat{e}_3(x_1-\xi_1)\log(Q+x_3+\xi_3)$, $\hat{e}_1/Q$ and $\hat{e}_3(x_1-\xi_1)/Q$. All of the additional Galerkin vectors represent solutions singular at the image point. Once again, the solution can be found, as in Mindlin problem I, through the determination of the constants of the linear combination by imposing the conditions of total force balance and that the surface, $x_3=0$, be stress free. The details of the solution are left to the reader. The components of the resulting displacement are,

$$
\begin{aligned}
u_1 = {} & \frac{3-4\sigma}{R} + \frac{1}{Q} + \frac{2\xi_3 x_3}{Q^3} + \frac{4(1-\sigma)(1-2\sigma)}{Q + x_3 + \xi_3} \\
& + (x_1 - \xi_1)^2 \left[\frac{1}{R^3} + \frac{3-4\sigma}{Q^3} - \frac{6\xi_3 x_3}{Q^5} - \frac{4(1-\sigma)(1-2\sigma)}{Q(Q + x_3 + \xi_3)^2} \right], \\
u_2 = {} & (x_1 - \xi_1)(x_2 - \xi_2) \left[\frac{1}{R^3} + \frac{3-4\sigma}{Q^3} - \frac{6\xi_3 x_3}{Q^5} - \frac{4(1-\sigma)(1-2\sigma)}{Q(Q + x_3 + \xi_3)^2} \right], \\
u_3 = {} & (x_1 - \xi_1) \left[\frac{x_3 - \xi_3}{R^3} + \frac{(3-4\sigma)(x_3 - \xi_3)}{Q^3} - \frac{6\xi_3 x_3 (x_3 + \xi_3)}{Q^5} \right. \\
& \left. + \frac{4(1-\sigma)(1-2\sigma)}{Q(Q + x_3 + \xi_3)} \right].
\end{aligned}
\tag{1.529}
$$

The displacement fields for the two Mindlin problems were used by Press (1965) as the basis of the first demonstration, using dislocation theory, of the very extensive nature of earthquake displacement fields.

1.5 Linear algebraic systems

Linear algebraic systems of equations are perhaps the most common systems of equations arising in the study of Earth's dynamics. They are also extensively described in the literature of mathematics and computation (see, for example, Press *et al.* (1992), Wilkinson (1965)). The intention here is to provide a basic description, rather than repeat the exhaustive treatises available elsewhere.

In general, we will write the linear algebraic system of equations as

$$
\mathbf{A}\boldsymbol{x} = \boldsymbol{b}, \tag{1.530}
$$

where $\mathbf{A}$ is the coefficient matrix, $\boldsymbol{x}$ is the unknown vector, and $\boldsymbol{b}$ is the constant vector. In the most common case, $\mathbf{A}$ is taken to be an $n \times n$ square matrix, so the vectors $\boldsymbol{x}$ and $\boldsymbol{b}$ are n-length. In addition to solving for the unknown vector $\boldsymbol{x}$, we may wish to calculate the inverse, $\mathbf{A}^{-1}$, of the coefficient matrix, as well as its determinant, $|\mathbf{A}|$.

For simplicity and clarity in expounding methods of solution for the three foregoing problems, we will consider a 3×3 system. We will append the constant vector and the unit matrix to the coefficient matrix in (1.530) to give

$$
\begin{pmatrix}
a_{11} & a_{12} & a_{13} & b_1 & 1 & 0 & 0 \\
a_{21} & a_{22} & a_{23} & b_2 & 0 & 1 & 0 \\
a_{31} & a_{32} & a_{33} & b_3 & 0 & 0 & 1
\end{pmatrix}, \tag{1.531}
$$

called the *augmented matrix*.

We may regard this as a system of equations with four successive constant vectors. The usual elimination procedure for solving the linear system (1.530) may be formalised by row operations on (1.531). Any row can be divided or multiplied by a constant, and linear combinations of two or more rows can be used to replace any row, without changing the solution of the system (1.530).

If the first row is divided by the first element of the coefficient matrix (the pivot), appropriate multiples of the resulting row can be subtracted from the following rows to make the rest of the elements of the first column vanish. If this procedure is followed sequentially, reducing elements below the diagonal to zero, the last row will have only its last element not zero and the coefficient matrix will have upper triangular form. The last row can then easily be solved for the last unknown, which can be substituted into the second to last row to find the second to last unknown, and so on, in a procedure called *back-substitution*. This method of solving the linear system is called *Gaussian elimination*. When it is modified to reduce elements both above and below the diagonal to zero, with the resulting reduction of the coefficient matrix to the unit matrix, it is called *Gauss–Jordan elimination*.

Following the completion of Gauss–Jordan elimination on the augmented matrix (1.531), we obtain the equivalent of the system (1.530) with the unit matrix added to its right-hand side multiplied by $\mathbf{A}^{-1}$, the inverse of the coefficient matrix. Thus, in the reduced system, the coefficient matrix is replaced by the unit matrix, the constant vector is replaced by the solution vector and the appended unit matrix is replaced by the inverse of the coefficient matrix.

The numerical stability of the Gauss–Jordan procedure is enhanced by selection of the available pivot of greatest magnitude at each step (Carnahan *et al.*, 1969). If the selection of the pivot is confined to row operations, this is called *partial pivoting*.

We illustrate the procedure by a numerical example. In each case, the equivalent matrix multiplication for operations on the system is given first. Suppose we have the system of equations

$$\begin{aligned} 2x_1 \quad -7x_2 \quad +4x_3 \quad &= 9 \\ -3x_1 \quad +8x_2 \quad +5x_3 \quad &= 6 \\ x_1 \quad +9x_2 \quad -6x_3 \quad &= 1. \end{aligned} \tag{1.532}$$

For this system, the array (1.531) becomes

$$\mathbf{C} = \begin{pmatrix} 2 & -7 & 4 & 9 & 1 & 0 & 0 \\ -3 & 8 & 5 & 6 & 0 & 1 & 0 \\ 1 & 9 & -6 & 1 & 0 & 0 & 1 \end{pmatrix}. \tag{1.533}$$

The element of the coefficient matrix with largest magnitude is $a_{32} = 9$. Selecting this element as the pivot, the array takes the form

$$\begin{pmatrix} 1 & 0 & 0 \\ 0 & 1 & 0 \\ 0 & 0 & 1/9 \end{pmatrix} \mathbf{C} = \mathbf{P}_1\mathbf{C}$$

$$= \mathbf{D} = \begin{pmatrix} 2 & -7 & 4 & 9 & 1 & 0 & 0 \\ -3 & 8 & 5 & 6 & 0 & 1 & 0 \\ 1/9 & 1 & -2/3 & 1/9 & 0 & 0 & 1/9 \end{pmatrix}. \tag{1.534}$$

Multiplying the last row by 7 and adding it to the first, then multiplying the last row by 8 and subtracting it from the second, yields the array

$$\begin{pmatrix} 1 & 0 & 7 \\ 0 & 1 & -8 \\ 0 & 0 & 1 \end{pmatrix} \mathbf{D} = \mathbf{Q}_1\mathbf{D}$$

$$= \mathbf{E} = \begin{pmatrix} 25/9 & 0 & -2/3 & 88/9 & 1 & 0 & 7/9 \\ -35/9 & 0 & 31/3 & 46/9 & 0 & 1 & -8/9 \\ 1/9 & 1 & -2/3 & 1/9 & 0 & 0 & 1/9 \end{pmatrix}. \tag{1.535}$$

Elements on the row and column of the initial pivot are no longer available as pivots. Of the available elements, $a_{23} = 31/3$ has the largest magnitude. Selecting this element as the next pivot, and using the resulting second row to reduce the other elements in the third column to zero, gives the array

$$\begin{pmatrix} 1 & 2/3 & 0 \\ 0 & 1 & 0 \\ 0 & 2/3 & 1 \end{pmatrix} \begin{pmatrix} 1 & 0 & 0 \\ 0 & 3/31 & 0 \\ 0 & 0 & 1 \end{pmatrix} \mathbf{E} = \mathbf{Q}_2\mathbf{P}_2\mathbf{E}$$

$$= \mathbf{F} = \begin{pmatrix} 235/93 & 0 & 0 & 940/93 & 1 & 2/31 & 67/93 \\ -35/93 & 0 & 1 & 46/93 & 0 & 3/31 & -8/93 \\ -13/93 & 1 & 0 & 41/93 & 0 & 2/31 & 5/93 \end{pmatrix}. \tag{1.536}$$

The only available pivot is now $a_{11} = 235/93$. Accordingly, the resulting first row is used to reduce the remaining elements in the first column to zero, producing the array

$$\begin{pmatrix} 1 & 0 & 0 \\ 35/93 & 1 & 0 \\ 13/93 & 0 & 1 \end{pmatrix} \begin{pmatrix} 93/235 & 0 & 0 \\ 0 & 1 & 0 \\ 0 & 0 & 1 \end{pmatrix} \mathbf{F} = \mathbf{Q}_3\mathbf{P}_3\mathbf{F}$$

$$= \mathbf{G} = \begin{pmatrix} 1 & 0 & 0 & 4 & 93/235 & 6/235 & 67/235 \\ 0 & 0 & 1 & 2 & 7/47 & 5/47 & 1/47 \\ 0 & 1 & 0 & 1 & 13/235 & 16/235 & 22/235 \end{pmatrix}. \tag{1.537}$$

Finally, with the interchange of the second and third rows, the completed reduction of the array (1.531) takes the form

$$\begin{pmatrix} 1 & 0 & 0 \\ 0 & 0 & 1 \\ 0 & 1 & 0 \end{pmatrix} \mathbf{G} = \mathbf{RG}$$

$$= \mathbf{H} = \begin{pmatrix} 1 & 0 & 0 & 4 & 93/235 & 6/235 & 67/235 \\ 0 & 1 & 0 & 1 & 13/235 & 16/235 & 22/235 \\ 0 & 0 & 1 & 2 & 7/47 & 5/47 & 1/47 \end{pmatrix}. \qquad (1.538)$$

The unit matrix is on the left of the array, followed by the solution vector in the next column, and the inverse of the original coefficient matrix is on the right.

The reduction of the coefficient matrix $\mathbf{A}$, in the system (1.530), to the unit matrix $\mathbf{I}$ is seen to be achieved by the sequence of matrix multiplications,

$$\mathbf{R Q_3 P_3 Q_2 P_2 Q_1 P_1 A} = \mathbf{I}. \qquad (1.539)$$

Taking the determinant of both sides of this equation gives

$$\det \mathbf{R} \cdot \det \mathbf{P_3} \cdot \det \mathbf{P_2} \cdot \det \mathbf{P_1} \cdot \det \mathbf{Q_3} \cdot \det \mathbf{Q_2} \cdot \det \mathbf{Q_1} \cdot \det \mathbf{A} = 1. \qquad (1.540)$$

Now, $\det \mathbf{R} = -1$, showing that row interchange reverses the sign of the determinant, while $\det \mathbf{P_3} = 93/235$, $\det \mathbf{P_2} = 3/31$, $\det \mathbf{P_1} = 1/9$, and $\det \mathbf{Q_3} = \det \mathbf{Q_2} = \det \mathbf{Q_1} = 1$. Hence,

$$\det \mathbf{A} = -1 \times 9 \times \frac{31}{3} \times \frac{235}{93} = -235. \qquad (1.541)$$

The subroutine LINSOL solves for the inverse of the coefficient matrix, returned in A, the solution vector, returned in B and the determinant of the coefficient matrix, returned in DET. The integer vectors, IR and IC, record the row and column of each successive pivot, allowing the rows to be unscrambled, and the integer vector IDET records the rearranged sequence of the rows, allowing the sign of the determinant to be determined.

```
      SUBROUTINE LINSOL(A,B,N,C,DET,M1,M2)
C
C LINSOL solves the linear system Ax=b by Gauss-Jordan elimination
C with partial pivoting. A is the N x N coefficient matrix
C and B is the N-length constant vector b. The inverse of A
C is returned in A, the solution vector x is returned in B,
C and the determinant of A is returned in DET. C is the N x 2N+1
C augmented matrix used internally. M1 is the maximum dimension
C for N specified in the main programme and M2=2M1+1 is
C the maximum dimension for 2N+1 used in the main programme.
C
      IMPLICIT DOUBLE PRECISION(A-H,O-Z)
      DIMENSION A(M1,M1),B(M1),C(M1,M2),HOLD(M1,M2),
     1  IR(M1),IC(M1),IDET(M1)
C Construct the augmented matrix C.
```

```
      N2=2*N+1
      NP1=N+1
      NM1=N-1
C Insert the matrix A.
      DO 10 I=1,N
        DO 10 J=1,N
          C(I,J)=A(I,J)
 10   CONTINUE
C Insert the unknown vector.
      DO 11 J=1,N
        C(J,NP1)=B(J)
 11   CONTINUE
C Augment with the identity matrix.
      DO 12 I=1,N
        DO 12 J=1,N
          JJ=J+NP1
          C(I,JJ)=0.D0
          IF(I.EQ.J)C(I,JJ)=1.D0
 12   CONTINUE
C Begin Gauss-Jordan elimination.
      DET=1.D0
      DO 13 I=1,N
        IM1=I-1
C Search for largest pivot.
        PIVOT=0.D0
        DO 14 J=1,N
          DO 14 K=1,N
C Scan IR and IC vectors for invalid pivot subscripts.
            IF(I.EQ.1)GO TO 15
            DO 16 ISC=1,IM1
              DO 16 JSC=1,IM1
                IF(J.EQ.IR(ISC))GO TO 14
                IF(K.EQ.IC(JSC))GO TO 14
 16         CONTINUE
 15         IF(DABS(C(J,K)).LE.DABS(PIVOT))GO TO 14
            PIVOT=C(J,K)
            IR(I)=J
            IC(I)=K
 14     CONTINUE
        DET=DET*PIVOT
C Label next row.
        IRP1=IR(I)+1
C Label previous row.
        IRM1=IR(I)-1
C Divide row by pivot.
        DO 17 J=1,N2
          C(IR(I),J)=C(IR(I),J)/PIVOT
 17     CONTINUE
C Reduce elements below pivot to zero.
        IF(IRP1.GT.N)GO TO 18
        DO 19 K=IRP1,N
          ELL=C(K,IC(I))
          DO 20 J=1,N2
            C(K,J)=C(K,J)-C(IR(I),J)*ELL
 20       CONTINUE
 19     CONTINUE
 18     CONTINUE
C Reduce elements above pivot to zero.
        IF(IRM1.LT.1)GO TO 21
        DO 22 K=1,IRM1
          ELL=C(K,IC(I))
```

```
      DO 23 J=1,N2
        C(K,J)=C(K,J)-C(IR(I),J)*ELL
23      CONTINUE
22     CONTINUE
21     CONTINUE
13   CONTINUE
C Unscramble equations and record row interchanges.
     DO 24 I=1,N
       IDET(IR(I))=IC(I)
       DO 24 J=1,N2
         HOLD(IC(I),J)=C(IR(I),J)
24   CONTINUE
C Put equations back in original matrix.
     DO 25 I=1,N
       DO 25 J=1,N2
         C(I,J)=HOLD(I,J)
25   CONTINUE
C Determine sign of determinant.
     ICH=0
     DO 26 I=1,NM1
       IP1=I+1
       DO 26 J=IP1,N
         IF(IDET(J).GE.IDET(I))GO TO 26
         IHOLD=IDET(J)
         IDET(J)=IDET(I)
         IDET(I)=IHOLD
         ICH=ICH+1
26   CONTINUE
     IF((ICH/2)*2.NE.ICH)DET=-DET
C Put inverse in A and solution in B.
     DO 27 I=1,N
       B(I)=C(I,NP1)
       DO 27 J=1,N
         JJ=J+NP1
         A(I,J)=C(I,JJ)
27   CONTINUE
     RETURN
     END
```

1.6 Interpolation and approximation

In the study of Earth's dynamics, we are more often concerned with functions represented numerically than by simple analytical expressions. Thus, interpolation and approximation are of central importance.

Commonly, we have a numerical representation of a function, $f(x)$, as a table of function values $(f_1, \ldots, f_N)$ at N discrete locations, or nodes, $(x_1, \ldots, x_N)$ along the x-axis. The question is then how best to represent the function between its tabulated values. Let us consider its representation $s(x)$ on the interval (x_i, x_{i+1}).

The crudest approximation is to take it to be a constant equal to one of the values at the end points, either f_i or f_{i+1}, or the mean value, giving

$$s(x) = \frac{f_i + f_{i+1}}{2}. \tag{1.542}$$

The approximating function then appears as a histogram.

The next approximation would be to represent it as a linear interpolation between the end point values,

$$s(x) = f_i + \frac{x - x_i}{x_{i+1} - x_i}(f_{i+1} - f_i). \tag{1.543}$$

The approximating function then appears as a series of trapeziums. While this approximating function is now continuous, it has discontinuous first derivatives at the nodes. This leads us to consider higher order approximation. As with all higher order methods, a caveat is to compare the interpolation with the function values, especially in poorly constrained intervals, as they can lead to divergences from expected behaviours.

If we wish to match the function values and first derivatives at the end points of the interval, we require four free parameters and $s(x)$ becomes a cubic. If we allow discontinuities in the second derivatives at the nodes, the cubic approximating function is called a *local* or *Hermite* spline. The term *spline* comes from the name of a draughting tool used to draw curves that are smooth and have continuous curvature. If the approximating function $s(x)$ is continuous at the nodes, and has there not only continuous first derivatives, but also continuous second derivatives, it is called a *natural spline*. We will first consider interpolation and approximation by natural splines.

1.6.1 Natural splines

By analogy to (1.543), the second derivatives at the end points of the interval (x_i, x_{i+1}) will be correctly represented if

$$s''(x) = f_i'' + \frac{x - x_i}{x_{i+1} - x_i}\left(f_{i+1}'' - f_i''\right), \tag{1.544}$$

where each superscript prime indicates differentiation. This leads us to postulate the form

$$\begin{aligned} s(x) = {} & f_i + \frac{x - x_i}{x_{i+1} - x_i}(f_{i+1} - f_i) \\ & + \frac{(x - x_i)(x - x_{i+1})}{6}\left[K + f_i'' + \frac{x - x_i}{x_{i+1} - x_i}\left(f_{i+1}'' - f_i''\right)\right], \end{aligned} \tag{1.545}$$

for an approximating function that matches both function values and second derivatives at the end points of the interval (x_i, x_{i+1}). Note that, while the nodes x_j and function values f_j are known, the constant K and second derivatives f_j'' are as yet unknown. Then,

$$s'(x) = \frac{f_{i+1} - f_i}{x_{i+1} - x_i} + \frac{K + f_i''}{6}\left[(x - x_{i+1}) + (x - x_i)\right]$$
$$+ \frac{1}{6}\frac{f_{i+1}'' - f_i''}{x_{i+1} - x_i}\left[2\,(x - x_{i+1})\,(x - x_i) + (x - x_i)^2\right], \tag{1.546}$$

and

$$s''(x) = \frac{K + f_i''}{3} + \frac{1}{3}\frac{f_{i+1}'' - f_i''}{x_{i+1} - x_i}\left[x - x_{i+1} + 2\,(x - x_i)\right]. \tag{1.547}$$

Thus, to match second derivatives at the end points, the constant is $K = f_{i+1}'' + f_i''$. On the interval (x_i, x_{i+1}), the first derivative of the approximating function is

$$s'(x) = \frac{f_{i+1} - f_i}{x_{i+1} - x_i} + \frac{f_{i+1}'' + 2f_i''}{6}\left[(x - x_{i+1}) + (x - x_i)\right]$$
$$+ \frac{1}{6}\frac{f_{i+1}'' - f_i''}{x_{i+1} - x_i}\left[2\,(x - x_{i+1})\,(x - x_i) + (x - x_i)^2\right], \tag{1.548}$$

while on the interval (x_{i-1}, x_i), it is

$$s'(x) = \frac{f_i - f_{i-1}}{x_i - x_{i-1}} + \frac{f_i'' + 2f_{i-1}''}{6}\left[(x - x_i) + (x - x_{i-1})\right]$$
$$+ \frac{1}{6}\frac{f_i'' - f_{i-1}''}{x_i - x_{i-1}}\left[2\,(x - x_i)\,(x - x_{i-1}) + (x - x_{i-1})^2\right]. \tag{1.549}$$

To ensure continuity of first derivatives, expressions (1.548) and (1.549) are equated at the common node, $x = x_i$, to give

$$(x_i - x_{i-1})\,f_{i-1}'' + 2\,(x_{i+1} - x_{i-1})\,f_i'' + (x_{i+1} - x_i)\,f_{i+1}''$$
$$= \frac{6}{x_i - x_{i-1}}f_{i-1} - \frac{6\,(x_{i+1} - x_{i-1})}{(x_{i+1} - x_i)\,(x_i - x_{i-1})}f_i + \frac{6}{x_{i+1} - x_i}f_{i+1}. \tag{1.550}$$

The condition (1.550) applies at all internal nodes ($i = 2, \ldots, N-1$), giving $N-2$ equations in the N second derivatives.

Two further conditions are required to express the second derivatives at the nodes entirely in terms of the function values there. In the conventional description of *natural splines*, the second derivatives at the end nodes are taken to vanish. This corresponds to the spline draughting tool being clamped at the end points. Instead, we choose to make the highest derivatives as smooth as possible, taking the third derivatives to be the same in each of the first two intervals and in each of the last two intervals. Differentiating expression (1.547), the third derivative on the interval (x_i, x_{i+1}) is found to be

$$s'''(x) = \frac{f_{i+1}'' - f_i''}{x_{i+1} - x_i}. \tag{1.551}$$

Thus, our two further conditions are

$$\frac{f_2'' - f_1''}{x_2 - x_1} = \frac{f_3'' - f_2''}{x_3 - x_2}, \tag{1.552}$$

and

$$\frac{f_N'' - f_{N-1}''}{x_N - x_{N-1}} = \frac{f_{N-1}'' - f_{N-2}''}{x_{N-1} - x_{N-2}}. \tag{1.553}$$

The second derivatives at the end points are then expressed as

$$f_1'' = \frac{(x_3 - x_1)\, f_2'' - (x_2 - x_1)\, f_3''}{x_3 - x_2}, \tag{1.554}$$

and

$$f_N'' = \frac{(x_N - x_{N-2})\, f_{N-1}'' - (x_N - x_{N-1})\, f_{N-2}''}{x_{N-1} - x_{N-2}}. \tag{1.555}$$

We may write the $N - 2$ equations in the N second derivatives implied by the condition (1.550) as,

$$a_{i-1,i-2} f_{i-1}'' + a_{i-1,i-1} f_i'' + a_{i-1,i} f_{i+1}'' = b_{i-1,i-1} f_{i-1} + b_{i-1,i} f_i + b_{i-1,i+1} f_{i+1}, \tag{1.556}$$

for $i = 2, \ldots, N - 1$. For $i = 2$, with substitution for f_1'' from expression (1.554), equation (1.550) yields

$$\left[2\,(x_3 - x_1) + \frac{(x_3 - x_1)\,(x_2 - x_1)}{x_3 - x_2}\right] f_2'' + \left[(x_3 - x_2) - \frac{(x_2 - x_1)^2}{x_3 - x_2}\right] f_3''$$
$$= \frac{6}{x_2 - x_1} f_1 - \frac{6\,(x_3 - x_1)}{(x_3 - x_2)\,(x_2 - x_1)} f_2 + \frac{6}{x_3 - x_2} f_3, \tag{1.557}$$

or

$$a_{11} f_2'' + a_{12} f_3'' = b_{11} f_1 + b_{12} f_2 + b_{13} f_3, \tag{1.558}$$

while for $i = N - 1$, with substitution for f_N'' from expression (1.555), it yields

$$\left[(x_{N-1} - x_{N-2}) - \frac{(x_N - x_{N-1})^2}{x_{N-1} - x_{N-2}}\right] f_{N-2}''$$
$$+ \left[2\,(x_N - x_{N-2}) + \frac{(x_N - x_{N-1})(x_N - x_{N-2})}{x_{N-1} - x_{N-2}}\right] f_{N-1}''$$
$$= \frac{6}{x_{N-1} - x_{N-2}} f_{N-2} - \frac{6\,(x_N - x_{N-2})}{(x_N - x_{N-1})(x_{N-1} - x_{N-2})} f_{N-1} + \frac{6}{x_N - x_{N-1}} f_N, \tag{1.559}$$

or

$$a_{N-2,N-3} f_{N-2}'' + a_{N-2,N-2} f_{N-1}''$$
$$= b_{N-2,N-2} f_{N-2} + b_{N-2,N-1} f_{N-1} + b_{N-2,N} f_N. \tag{1.560}$$

The $(N-2)$-length vector of second derivatives,

$$\boldsymbol{f}''_{N-2} = \left(f''_2, f''_3, \ldots, f''_{N-2}, f''_{N-1}\right)^T, \tag{1.561}$$

is related to the N-length vector of function values,

$$\boldsymbol{f} = (f_1, f_2, \ldots, f_{N-1}, f_N)^T, \tag{1.562}$$

by the matrix equation

$$\mathbf{A}\boldsymbol{f}''_{N-2} = \mathbf{B}\boldsymbol{f}, \tag{1.563}$$

where $\mathbf{A}$ is the $(N-2)\times(N-2)$ tridiagonal matrix

$$\mathbf{A} = \begin{pmatrix} a_{11} & a_{12} & & & \\ 1 & a_{22} & a_{23} & & \\ & & \ddots & & \\ & & 1 & a_{N-3,N-3} & a_{N-3,N-2} \\ & & & a_{N-2,N-3} & a_{N-2,N-2} \end{pmatrix}, \tag{1.564}$$

and $\mathbf{B}$ is the $(N-2)\times N$ matrix

$$\mathbf{B} = \begin{pmatrix} b_{11} & b_{12} & b_{13} & & & & \\ & b_{22} & b_{23} & b_{24} & & & \\ & & \ddots & & & & \\ & & & b_{N-3,N-3} & b_{N-3,N-2} & b_{N-3,N-1} & \\ & & & & b_{N-2,N-2} & b_{N-2,N-1} & b_{N-2,N} \end{pmatrix}. \tag{1.565}$$

Matrix $\mathbf{A}$ can be augmented by the identity matrix and inverted entirely by row operations alone, as in the Gauss–Jordan technique. A forward pass eliminates elements on the sub-diagonal, while a backward pass eliminates elements on the super-diagonal. Then, $\boldsymbol{f}''_{N-2} = (\mathbf{A}^{-1}\mathbf{B})\boldsymbol{f}$. The matrix $\mathbf{A}^{-1}\mathbf{B}$ has dimensions $(N-2)\times N$. A new first row can be added by a combination of its first and second rows, as indicated by (1.554), then a new last row can be added as the combination of its last and second last rows, as indicated by (1.555). The full vector of second derivatives, $\boldsymbol{f}'' = (f''_1, f''_2, \ldots, f''_{N-1}, f''_N)^T$, is then related to the vector of function values by

$$\boldsymbol{f}'' = \mathbf{C}\boldsymbol{f}, \tag{1.566}$$

with $\mathbf{C}$ a full $N\times N$ square matrix. The subroutine SPMAT finds the matrix C connecting the second derivatives to the function values at the nodes:

```
      SUBROUTINE SPMAT(N1,N2,N3,C,R,B,M1,M2,M3)
C
C SPMAT computes the N1 by N1 matrix C connecting
C second derivatives to function values at N1 nodal points R(I).
C B is the N3=N1-2 by N2=2*N1-2 augmented matrix used internally
C in the Gaussian elimination procedure.
C M1, M2, M3 are the respective maximum dimensions for N1, N2, N3
C specified in the main programme.
C
      IMPLICIT DOUBLE PRECISION(A-H,O-Z)
      DIMENSION C(M1,M1),R(M1),B(M3,M2)
      N=N3+1
C Construct augmented matrix A.
      DO 11 I=1,N3
        DO 10 J=1,N2
          B(I,J)=0.D0
 10     CONTINUE
        B(I,I)=2.D0*(R(I+2)-R(I))
        B(I,I+N3)=1.D0
        IF(I.EQ.1.OR.I.EQ.N3)GO TO 11
        B(I,I-1)=R(I+1)-R(I)
        B(I,I+1)=R(I+2)-R(I+1)
 11   CONTINUE
C Apply conditions that third derivatives are constant
C in first two and last two intervals.
      B(1,2)=R(3)-R(2)-((R(2)-R(1))**2.D0)/(R(3)-R(2))
      B(N3,N3-1)=R(N3+1)-R(N3)-((R(N1)-R(N1-1))**2.D0)/(R(N1-1)-R(N1-2))
      B(1,1)=B(1,1)+((R(3)-R(1))*(R(2)-R(1)))/(R(3)-R(2))
      B(N3,N3)=B(N3,N3)
     1  +((R(N1)-R(N1-1))*(R(N1)-R(N1-2)))/(R(N1-1)-R(N1-2))
      NM1=N3-1
C Eliminate sub-diagonal.
      DO 13 I=1,NM1
        DIAG=B(I,I)
        ELL=B(I+1,I)
        DO 12 J=1,N2
          B(I,J)=B(I,J)/DIAG
          B(I+1,J)=B(I+1,J)-B(I,J)*ELL
 12     CONTINUE
 13   CONTINUE
C Eliminate super-diagonal.
      DO 15 I=1,NM1
        NO=N-I
        DIAG=B(NO,NO)
        ELL=B(NO-1,NO)
        DO 14 J=1,N2
          B(NO,J)=B(NO,J)/DIAG
          B(NO-1,J)=B(NO-1,J)-B(NO,J)*ELL
 14     CONTINUE
 15   CONTINUE
C Move inverse to position formerly occupied by A.
      DO 17 I=1,N3
        DO 16 J=1,N3
          B(I,J)=B(I,J+N3)
 16     CONTINUE
 17   CONTINUE
C Construct matrix B in N-2 by N augmentation space.
      DO 19 I=1,N3
        DO 18 J=1,N1
          B(I,J+N3)=0.0D0
 18     CONTINUE
```

```
      IN=I+N3
      FACTOR=6.D0/((R(I+2)-R(I+1))*(R(I+1)-R(I)))
      B(I,IN)=FACTOR*(R(I+2)-R(I+1))
      B(I,IN+1)=-FACTOR*(R(I+2)-R(I))
      B(I,IN+2)=FACTOR*(R(I+1)-R(I))
19    CONTINUE
C Multiply inverse of A into B to get N-2 by N version of C.
      DO 22 I=1,N3
        DO 21 J=1,N1
          SUM=0.D0
          DO 20 K=1,N3
            SUM=SUM+B(I,K)*B(K,J+N3)
20        CONTINUE
          C(I+1,J)=SUM
21      CONTINUE
22    CONTINUE
C Add first and last rows to C.
      DO 23 J=1,N1
        C(1,J)=C(2,J)*(R(3)-R(1))/(R(3)-R(2))
     1 -C(3,J)*(R(2)-R(1))/(R(3)-R(2))
        C(N1,J)=C(N1-1,J)*(R(N1)-R(N1-2))/(R(N1-1)-R(N1-2))
     1 -C(N1-2,J)*(R(N1)-R(N1-1))/(R(N1-1)-R(N1-2))
23    CONTINUE
      RETURN
      END
```

After the subroutine SPMAT has been used to find matrix **C**, we can use expression (1.545) to calculate the interpolate at x, through

$$s(x) = f_i + \frac{x - x_i}{x_{i+1} - x_i}(f_{i+1} - f_i) + \frac{(x - x_i)(x - x_{i+1})}{6}\left[\left(1 + \frac{x - x_i}{x_{i+1} - x_i}\right) f''_{i+1} + \left(2 - \frac{x - x_i}{x_{i+1} - x_i}\right) f''_i\right], \quad (1.567)$$

once the subinterval containing x is identified. The subroutine INTPL carries out the interpolation after SPMAT has been called:

```
      SUBROUTINE INTPL(XV,YV,N1,C,R,S,M1)
C
C INTPL interpolates to find YV at XV, using a cubic spline,
C given the function S(I) at N1 nodal points R(I). C is
C the coefficient matrix, from subroutine SPMAT, relating
C the second derivatives to the function values at the nodal points.
C M1 is the maximum dimension for N1 specified in the main programme.
C
      IMPLICIT DOUBLE PRECISION(A-H,O-Z)
      DIMENSION C(M1,M1),R(M1),S(M1)
      NPM1=N1-1
      J=NPM1
C Locate subinterval.
      DO 10 L=1,NPM1
        IF(XV.GE.R(L).AND.XV.LT.R(L+1)) J=L
10    CONTINUE
C Find second derivative SUM1 at beginning of subinterval
C and second derivative SUM2 at end of subinterval.
      SUM1=0.D0
      SUM2=0.D0
```

```
      DO 11 K=1,N1
        SUM1=SUM1+C(J,K)*S(K)
        SUM2=SUM2+C(J+1,K)*S(K)
 11   CONTINUE
C Find interpolate YV at XV.
      YV=S(J)+(S(J+1)-S(J))*(XV-R(J))/(R(J+1)-R(J))
     1 +(XV-R(J))*(XV-R(J+1))*(SUM1*(2.D0-(XV-R(J))/(R(J+1)-R(J)))
     2 +SUM2*(1.D0+(XV-R(J))/(R(J+1)-R(J))))/6.D0
      RETURN
      END
```

As well as interpolation, differentiation of a tabulated function can easily be accomplished with *spline* approximation. With some rearrangement, expression (1.548) gives the derivative of the approximating function as

$$s'(x) = \frac{f_{i+1} - f_i}{x_{i+1} - x_i} + \frac{(x - x_{i+1})(x - x_i)}{6(x_{i+1} - x_i)}\left(f''_{i+1} - f''_i\right)$$
$$+ \frac{(2x - x_{i+1} - x_i)}{6}\left[\left(1 + \frac{x - x_i}{x_{i+1} - x_i}\right)f''_{i+1} + \left(2 - \frac{x - x_i}{x_{i+1} - x_i}\right)f''_i\right], \qquad (1.568)$$

for the subinterval containing x. The subroutine DERIV calculates the derivative at x after SPMAT has been called:

```
      SUBROUTINE DERIV(XV,YV,N1,C,R,S,M1)
C
C DERIV calculates the derivative YV at XV, using a cubic spline,
C given the function S(I) at N1 nodal points R(I). C is
C the coefficient matrix, from subroutine SPMAT, relating
C the second derivatives to the function values at the nodal points.
C M1 is the maximum dimension for N1 specified in the main programme.
C
      IMPLICIT DOUBLE PRECISION(A-H,O-Z)
      DIMENSION C(M1,M1),R(M1),S(M1)
      NPM1=N1-1
      J=NPM1
C Locate subinterval.
      DO 10 L=1,NPM1
        IF(XV.GE.R(L).AND.XV.LT.R(L+1)) J=L
 10   CONTINUE
C Find second derivative SUM1 at beginning of subinterval
C and second derivative SUM2 at the end of subinterval.
      SUM1=0.D0
      SUM2=0.D0
      DO 11 K=1,N1
        SUM1=SUM1+C(J,K)*S(K)
        SUM2=SUM2+C(J+1,K)*S(K)
 11   CONTINUE
C Find derivative YV at XV.
      YV=(S(J+1)-S(J))/(R(J+1)-R(J))
     1 +(XV-R(J))*(XV-R(J+1))*(SUM2-SUM1)/(6.D0*(R(J+1)-R(J)))
     2 +(2.D0*XV-R(J+1)-R(J))*(SUM1*(2.D0-(XV-R(J))/(R(J+1)-R(J)))
     3 +SUM2*(1.D0+(XV-R(J))/(R(J+1)-R(J))))/6.D0
      RETURN
      END
```

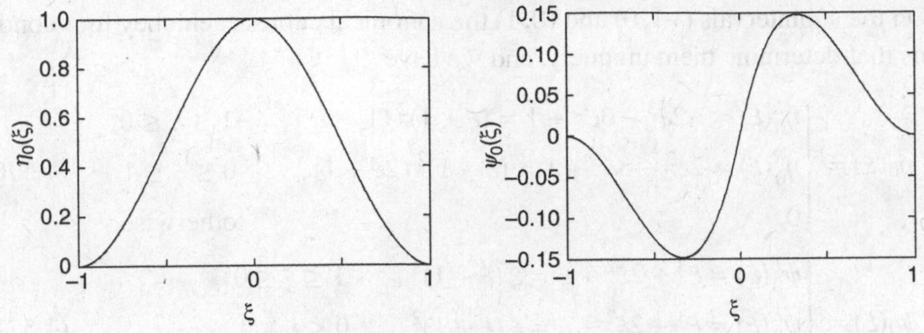

Figure 1.6 The canonical cubics.

1.6.2 Local Hermite splines

In contrast to *natural splines*, for which the approximating cubic depends on all of the function values, $(f_1, \ldots, f_N)$, at the N nodes, $(x_1, \ldots, x_N)$, along the x-axis, *local* or *Hermite spline* interpolation depends only on the local values of the function and its derivatives $(f_1', \ldots, f_N')$. On the interval (x_i, x_{i+1}), the function is represented by the approximating cubic, $s(x)$, with

$$s(x) = f_i\eta_i^R(x) + f_{i+1}\eta_{i+1}^L(x) + f_i'\psi_i^R(x) + f_{i+1}'\psi_{i+1}^L(x). \tag{1.569}$$

The representation depends only on the four quantities $f_i, f_{i+1}, f_i', f_{i+1}'$, the function values and derivatives at the end points. The cubic $\eta_i^R(x)$ has vanishing function value and derivative at $x = x_{i+1}$, but vanishing derivative and unit function value at $x = x_i$, while the cubic $\eta_{i+1}^L(x)$ has vanishing function value and derivative at $x = x_i$, but vanishing derivative and unit function value at $x = x_{i+1}$. The cubic $\psi_i^R(x)$ has vanishing function value and derivative at $x = x_{i+1}$, but vanishing function value and unit derivative at $x = x_i$, while the cubic $\psi_{i+1}^L(x)$ has vanishing function value and derivative at $x = x_i$, but vanishing function value and unit derivative at $x = x_{i+1}$. Thus η_i^R and η_{i+1}^L support the function at the ends of the interval, while ψ_i^R and ψ_{i+1}^L support the derivatives at the ends of the interval.

We may define canonical cubics on the interval $(-1, 1)$ of the ξ-axis with vanishing function values and derivatives at the end points. Support for the function is provided by the canonical cubic $\eta_0(\xi)$, which has unit function value and vanishing derivative at the midpoint, $\xi = 0$, and is shown on the left of Figure 1.6. Support for the derivative is provided by the canonical cubic $\psi_0(\xi)$, which has vanishing function value and unit derivative at the midpoint, $\xi = 0$, and is shown on the right of Figure 1.6.

On the subintervals $(-1, 0)$ and $(0, 1)$ the canonical cubics each obey four conditions that determine them uniquely, and we have

$$\eta_0(\xi) = \begin{cases} \eta_0^L(\xi) = -2\xi^3 - 3\xi^2 + 1 = (\xi + 1)^2\,(1 - 2\xi)\,, & -1 \le \xi \le 0 \\ \eta_0^R(\xi) = 2\xi^3 - 3\xi^2 + 1 = (\xi - 1)^2\,(2\xi + 1)\,, & 0 \le \xi \le 1 \\ 0, & \text{otherwise,} \end{cases} \tag{1.570}$$

$$\psi_0(\xi) = \begin{cases} \psi_0^L(\xi) = \xi^3 + 2\xi^2 + \xi = \xi\,(\xi + 1)^2\,, & -1 \le \xi \le 0 \\ \psi_0^R(\xi) = \xi^3 - 2\xi^2 + \xi = \xi\,(\xi - 1)^2\,, & 0 \le \xi \le 1 \\ 0, & \text{otherwise.} \end{cases} \tag{1.571}$$

The four cubics used in the approximating function (1.569) are then defined by

$$\eta_i^R(x) = \eta_0^R(\xi_i), \qquad \eta_{i+1}^L(x) = \eta_0^L(\xi_{i+1}), \tag{1.572}$$

$$\psi_i^R(x) = \frac{1}{\xi_i'}\psi_0^R(\xi_i), \quad \psi_{i+1}^L(x) = \frac{1}{\xi_{i+1}'}\psi_0^L(\xi_{i+1}), \tag{1.573}$$

with

$$\xi_i = \frac{x - x_i}{x_{i+1} - x_i}, \qquad \xi_{i+1} = \frac{x - x_{i+1}}{x_{i+1} - x_i}, \tag{1.574}$$

$$\frac{1}{\xi_i'} = \frac{1}{\xi_{i+1}'} = x_{i+1} - x_i. \tag{1.575}$$

The term ξ_i maps the interval (x_i, x_{i+1}) onto the interval $(0, 1)$ of the ξ-axis, while ξ_{i+1} maps the interval (x_i, x_{i+1}) onto the interval $(-1, 0)$ of the ξ-axis.

1.6.3 Even or odd local basis functions

Although, traditionally, spherical harmonics have been used to represent scalar and vector fields in geophysics because of Earth's nearly spherical shape, they are global functions that are sometimes inefficient as basis functions. Local basis functions, closely related to *Hermite splines*, can be constructed that are much more efficient, and which can have globally even or odd properties directly incorporated. They are found to be particularly useful in the calculation of long-period core modes, where Coriolis coupling severely limits the efficiency of spherical harmonics (see Chapter 8). In that application, the modes are known to be either purely even or purely odd in the equatorial plane, and these properties can be directly incorporated in the basis functions.

If $z = \cos\theta$, where θ is the co-latitude, we wish to represent a function $f(z)$ on the z-axis. On the interval (z_i, z_{i+1}), if the polynomial approximating function $s(z)$ is to match the function and derivative values, f_i, f_{i+1}, f_i', f_{i+1}', at the end points, we revert to the *local Hermite splines* described in the previous subsection. If, in addition, a purely even function is to be represented, the approximating function takes the form

$$s(z) = c_0 + c_2 z^2 + c_4 z^4 + c_6 z^6, \tag{1.576}$$

where the four free constants, c_0, c_2, c_4, c_6, are used to match the function and its derivatives at the end points. If a purely odd function is to be represented, the approximating function takes the form

$$s(z) = c_1 z + c_3 z^3 + c_5 z^5 + c_7 z^7, \tag{1.577}$$

again leaving the four free constants, c_1, c_3, c_5, c_7, to match the function and its derivatives at the end points. In either case, the approximating function is written

$$s(z) = f_i \eta_i^R(z) + f_{i+1}\eta_{i+1}^L(z) + f_i' \psi_i^R(z) + f_{i+1}' \psi_{i+1}^L(z). \tag{1.578}$$

In the even case, the supporting polynomials are sixth degree and defined in terms of the *canonical Hermite cubics* by

$$\eta_i^R(z) = \eta_0^R(\zeta_i), \qquad \eta_{i+1}^L(z) = \eta_0^L(\zeta_{i+1}), \tag{1.579}$$

$$\psi_i^R(z) = \frac{1}{\zeta_i'(z_i)}\psi_0^R(\zeta_i), \quad \psi_{i+1}^L(z) = \frac{1}{\zeta_{i+1}'(z_{i+1})}\psi_0^L(\zeta_{i+1}), \tag{1.580}$$

with

$$\zeta_i(z) = \frac{z^2 - z_i^2}{z_{i+1}^2 - z_i^2}, \qquad \zeta_{i+1}(z) = \frac{z^2 - z_{i+1}^2}{z_{i+1}^2 - z_i^2}, \tag{1.581}$$

$$\zeta_i'(z) = \zeta_{i+1}'(z) = \frac{2z}{z_{i+1}^2 - z_i^2}. \tag{1.582}$$

The term ζ_i maps the interval (z_i, z_{i+1}) onto the interval $(0, 1)$ of the ζ-axis, while ζ_{i+1} maps the interval (z_i, z_{i+1}) onto the interval $(-1, 0)$ of the ζ-axis. The corresponding derivatives are

$$\eta_i'^R(z) = \zeta_i'(z)\eta_0'^R(\zeta_i), \quad \eta_{i+1}'^L(z) = \zeta_{i+1}'(z)\eta_0'^L(\zeta_{i+1}), \tag{1.583}$$

and

$$\psi_i'^R(z) = \frac{z}{z_i}\psi_0'^R(\zeta_i), \quad \psi_{i+1}'^L(z) = \frac{z}{z_{i+1}}\psi_0'^L(\zeta_{i+1}). \tag{1.584}$$

In the odd case, the supporting polynomials are seventh degree. With some experimentation, they can be expressed in terms of the canonical Hermite cubics by

$$\eta_i^R(z) = \frac{z}{z_i}\left[\eta_0^R(\zeta_i) - \frac{1}{z_i\zeta_i'(z_i)}\psi_0^R(\zeta_i)\right], \tag{1.585}$$

$$\eta_{i+1}^L(z) = \frac{z}{z_{i+1}}\left[\eta_0^L(\zeta_{i+1}) - \frac{1}{z_{i+1}\zeta_{i+1}'(z_{i+1})}\psi_0^L(\zeta_{i+1})\right], \tag{1.586}$$

$$\psi_i^R(z) = \frac{z}{z_i}\frac{1}{\zeta_i'(z_i)}\psi_0^R(\zeta_i), \tag{1.587}$$

$$\psi_{i+1}^L(z) = \frac{z}{z_{i+1}}\frac{1}{\zeta_{i+1}'(z_{i+1})}\psi_0^L(\zeta_{i+1}). \tag{1.588}$$

Their derivatives are

$$\eta_i'^R(z) = \frac{1}{z_i}\eta_0^R(\zeta_i) + \frac{z}{z_i}\zeta_i'(z)\eta_0'^R(\zeta_i)$$
$$- \frac{1}{z_i^2\zeta_i'(z_i)}\psi_0^R(\zeta_i) - \frac{z^2}{z_i^3}\psi_0'^R(\zeta_i), \tag{1.589}$$

$$\eta_{i+1}'^L(z) = \frac{1}{z_{i+1}}\eta_0^L(\zeta_{i+1}) + \frac{z}{z_{i+1}}\zeta_{i+1}'(z)\eta_0'^L(\zeta_{i+1})$$
$$- \frac{1}{z_{i+1}^2\zeta_{i+1}'(z_{i+1})}\psi_0^L(\zeta_{i+1}) - \frac{z^2}{z_{i+1}^3}\psi_0'^L(\zeta_{i+1}), \tag{1.590}$$

and

$$\psi_i'^R(z) = \frac{z^2}{z_i^2}\psi_0'^R(\zeta_i) + \frac{1}{z_i\zeta_i'(z_i)}\psi_0^R(\zeta_i), \tag{1.591}$$

$$\psi_{i+1}'^L(z) = \frac{z^2}{z_{i+1}^2}\psi_0'^L(\zeta_{i+1}) + \frac{1}{z_{i+1}\zeta_{i+1}'(z_{i+1})}\psi_0^L(\zeta_{i+1}). \tag{1.592}$$

The interval (z_1, z_2) beginning at the equator, with $z = z_1 = 0$, is an exception. In the even case, the derivative vanishes at the equator, and in the odd case, the function vanishes there. Thus lower order polynomials can be used for the approximating function. In the even case,

$$s(z) = c_0 + c_2z^2 + c_4z^4 \tag{1.593}$$

can be used to match f_1, f_2, f_2' through the three free constants c_0, c_2, c_4. In the odd case,

$$s(z) = c_1 z + c_3 z^3 + c_5 z^5 \tag{1.594}$$

can be used to match f_1', f_2, f_2' through the three free constants c_1, c_3, c_5. The canonical Hermite cubics are replaced by canonical quadratics defined on the interval $(-1, 1)$ of the ξ-axis by

$$\eta_0(\xi) = \begin{cases} \eta_0^L(\xi) = -3\xi^2 - 2\xi + 1 = -(3\xi - 1)(\xi + 1), & -1 \le \xi \le 0 \\ \eta_0^R(\xi) = \xi^2 - 2\xi + 1 = (\xi - 1)^2 & 0 \le \xi \le 1 \\ 0, & \text{otherwise,} \end{cases} \tag{1.595}$$

$$\psi_0(\xi) = \begin{cases} \psi_0^L(\xi) = \xi^2 + \xi = \xi(\xi + 1), & -1 \le \xi \le 0 \\ \psi_0^R(\xi) = 0, & 0 \le \xi \le 1 \\ 0, & \text{otherwise.} \end{cases} \tag{1.596}$$

In the even case, we can write

$$s(z) = f_1 \eta_1^R(z) + f_2 \eta_2^L(z) + f_2' \psi_2^L(z). \tag{1.597}$$

The supporting polynomials are quartics that can be defined in terms of the canonical quartics by

$$\eta_1^R(z) = \eta_0^R(\zeta_1), \quad \eta_2^L(z) = \eta_0^L(\zeta_2) + 2\psi_0^L(\zeta_2), \tag{1.598}$$

$$\psi_2^L(z) = \frac{z_2}{2}\psi_0^L(\zeta_2), \tag{1.599}$$

with

$$\zeta_1 = \frac{z^2}{z_2^2}, \quad \zeta_2 = \frac{z^2 - z_2^2}{z_2^2}, \tag{1.600}$$

$$\zeta_1'(z) = \zeta_2'(z) = \frac{2z}{z_2^2}. \tag{1.601}$$

The term ζ_1 maps the interval $(0, z_2)$ onto the interval $(0, 1)$ of the ζ-axis, while ζ_2 maps the interval $(0, z_2)$ onto the interval $(-1, 0)$ of the ζ-axis. The corresponding derivatives are

$$\eta_1'^R(z) = 2\frac{z}{z_2^2}\eta_0'^R(\zeta_1), \quad \eta_2'^L(z) = 2\frac{z}{z_2^2}\left(\eta_0'^L(\zeta_2) + 2\psi_0'^L(\zeta_2)\right), \tag{1.602}$$

$$\psi_2'^L(z) = \frac{z}{z_2}\psi_0'^L(\zeta_2). \tag{1.603}$$

In the odd case, we can write

$$s(z) = f_1' \psi_1^R(z) + f_2 \eta_2^L(z) + f_2' \psi_2^L(z). \tag{1.604}$$

The supporting polynomials are quintics that can be defined in terms of the canonical quadratics by

$$\eta_2^L(z) = \frac{z}{z_2}\left(\eta_0^L(\zeta_2) + \frac{3}{2}\psi_0^L(\zeta_2)\right), \tag{1.605}$$

$$\psi_1^R(z) = z\eta_0^R(\zeta_1), \quad \psi_2^L(z) = \frac{z}{2}\psi_0^L(\zeta_2). \tag{1.606}$$

Their derivatives are

$$\eta_2'^L(z) = \frac{z}{z_2}\left(\frac{1}{z}\eta_0^L(\zeta_2) + 2\frac{z}{z_2^2}\eta_0'^L(\zeta_2) + \frac{3}{2z}\psi_0^L(\zeta_2) + 3\frac{z}{z_2^2}\psi_0'^L(\zeta_2)\right), \tag{1.607}$$

and

$$\psi_1'^R(z) = \eta_0^R(\zeta_1) + 2\frac{z^2}{z_2^2}\eta_0'^R(\zeta_1), \quad \psi_2'^L(z) = \frac{1}{2}\psi_0^L(\zeta_2) + \frac{z^2}{z_2^2}\psi_0'^L(\zeta_2). \tag{1.608}$$

2

Time sequence and spectral analysis

New discoveries in Earth dynamics can only be made through the comparison of theory with observations. Often we are looking for signals close to or below the noise level; otherwise, they would have already been observed. Thus, the analysis of observations in both time and frequency domains is of crucial importance.

For several decades now, observations in the time domain have been represented by discrete samples. The samples may be equally spaced along the time axis or unequally spaced. Unequally spaced samples may result from inherent properties of the measurement technique, or from fundamental restrictions such as the visibility of sources at particular times. Unequally spaced samples may also be the result of digitiser failure or other instrument problems, leaving gaps in otherwise equally spaced time sequences. We include the analysis of unequally spaced time sequences and the application of singular value decomposition to their study. Most sequences of interest were originally continuous physical signals. Thus, we examine the effects of the sampling process itself on the results of the analysis.

Often observations are made at several locations and it is desired to bring out common features of the records from different observatories. For this purpose, we describe in detail, the *product spectrum*. This may be regarded as a kind of generalisation of the *cross spectrum* between two records. As is the case in any spectral analysis, the estimation of confidence intervals is of prime importance in establishing the significance of the results. We establish methods of estimating confidence intervals both for the product spectrum and for conventional spectral estimates.

All real observational records are of finite length. It is therefore important to consider the effects of finite record length on any analysis. Often a number of segments of the record are used to estimate spectral densities, reducing the variance of the estimate by averaging over estimates on the individual segments. To make more efficient use of available data, we employ the Welch overlapping segment analysis (WOSA), and obtain an asymptotic formula for the variance inflation arising from the fact that the overlapping segments are not statistically independent.

In conventional spectral density estimation, resolution in the frequency domain is limited to the order of the reciprocal of the record length. For an equally spaced time sequence, the discrete Fourier transform (DFT) representation in the frequency domain is periodic in the record length. In the time domain, the sequence is assumed to vanish outside the finite record. For actual time sequences neither of these properties hold. Information theory suggests that data outside the finite record should not add entropy, as a measure of information, to the result of analysis. In other words, a spectral estimate should maximise the entropy of the available data. This leads to the maximum entropy method (MEM) of spectral analysis, due to Burg, which we describe in detail.

2.1 Time domain analysis

We begin with consideration of the properties and analysis of time sequences in the time domain. Generally, any operation in the time domain has a counterpart in the frequency domain, but we will delay discussion of such relationships until we describe their analysis in the frequency domain.

2.1.1 Classification of time sequences

Perhaps the most common time sequences, and the easiest to treat, are those that arise from equally spaced sampling of a continuous function of time. In general, a time sequence is denoted by the indefinite sequence of complex numbers

$$\ldots, f_{-1}, \underset{\uparrow}{f_0}, f_1, \ldots, f_j, \ldots \tag{2.1}$$

The time index is subscripted, an arrow underneath indicating the origin of the time axis. The whole time sequence, for brevity, may be indicated by its general term f_j. The samples, f_j, will be assumed to be equally spaced, unless otherwise stated.

The average *power* in the finite time sequence

$$f_{-N}, f_{-N+1}, \ldots, \underset{\uparrow}{f_0}, \ldots, f_{N-1}, f_N \tag{2.2}$$

is

$$\frac{1}{2N+1}\sum_{j=-N}^{N}|f_j|^2 = \frac{1}{2N+1}\sum_{j=-N}^{N} f_j f_j^*, \tag{2.3}$$

where the superscript asterisk denotes complex conjugation. Time sequences that obey the restriction

$$\lim_{N\to\infty}\frac{1}{2N+1}\sum_{j=-N}^{N}|f_j|^2 < \infty, \tag{2.4}$$

or equivalently

$$\lim_{N\to\infty} \frac{1}{2N+1} \sum_{j=-N}^{N} f_j f_j^* < \infty, \tag{2.5}$$

are called *power signals*. Power signals are the most intense time sequences we can expect to meet in practice. The Earth tide signal registering on gravimeters is an example of a power signal.

Other signals build up from zero or a very low level and then fade away. Averaged over all time, such signals would have zero power. Instead, they have finite energy and are called *energy signals*. These are such that

$$\sum_{j=-\infty}^{\infty} |f_j| = \sum_{j=-\infty}^{\infty} f_j f_j^* < \infty, \tag{2.6}$$

and, hence, have finite energy. An example of an energy signal would be the barometric pressure signal associated with the passage of a storm front.

A third class of time sequences are zero until a certain time, often when an earthquake or other energy releasing event takes place. They are one-sided and take the form

$$\ldots, 0, 0, 0, \underset{\uparrow}{f_0}, f_1, f_2, \ldots, \tag{2.7}$$

and are called *wavelets*.

We have left unspecified the duration of particular time sequences. Of course, all time sequences resulting from actual records are of *finite duration*. In theoretical discussions they may be taken to be of *unlimited duration*.

2.1.2 Convolution and the z-transform

Perhaps the most common operation that can be performed with two time sequences is their *convolution*. For two time sequences with general terms f_j and g_j or

$$\ldots, f_{-1}, \underset{\uparrow}{f_0}, f_1, \ldots, f_j, \ldots \tag{2.8}$$

and

$$\ldots, g_{-1}, \underset{\uparrow}{g_0}, g_1, \ldots, g_j, \ldots, \tag{2.9}$$

the *convolution* of f_j with g_j is

$$h_j = \sum_{k=-\infty}^{\infty} f_k g_{j-k}. \tag{2.10}$$

Writing $l = j - k$, this becomes

$$h_j = \sum_{l=\infty}^{-\infty} f_{j-l} g_l = \sum_{l=-\infty}^{\infty} g_l f_{j-l} = \sum_{k=-\infty}^{\infty} g_k f_{j-k}. \qquad (2.11)$$

Thus, the convolution of f_j with g_j is the same as the convolution of g_j with f_j.

The convolution of *two energy signals* produces an *energy signal*, and the convolution of an *energy signal* with a *power signal* results in a *power signal*, while the convolution of *two power signals does not exist.*

Convolution, and many other operations with time sequences, are facilitated by the use of the *z-transform*. For the time sequence

$$\ldots, f_{-1}, \underset{\uparrow}{f_0}, \ldots, f_j, \ldots, \qquad (2.12)$$

the z-transform is defined as

$$F(z) = \cdots + \frac{f_{-1}}{z} + f_0 + f_1 z + \cdots + f_j z^j + \cdots \qquad (2.13)$$

If h_j is the sequence resulting from the convolution of the sequence f_j with the sequence g_j, it has the z-transform

$$H(z) = F(z) \cdot G(z). \qquad (2.14)$$

Hence, the z-transform of the convolution of f_j with g_j is the product of the z-transforms of f_j and g_j. The product of the z-transforms is

$$\begin{aligned} F(z) \cdot G(z) &= \left(\cdots + \frac{f_{-1}}{z} + f_0 + f_1 z + \cdots\right)\left(\cdots + \frac{g_{-1}}{z} + g_0 + g_1 z + \cdots\right) \\ &= \left(\cdots + \frac{\cdots f_{-1} g_{-1} \cdots}{z^2} + \frac{\cdots f_{-1} g_0 + f_0 g_{-1} \cdots}{z}\right. \\ &\quad + \cdots f_{-1} g_1 + f_0 g_0 + f_1 g_{-1} \cdots \\ &\quad \left. + (\cdots f_0 g_1 + f_1 g_0 \cdots) z + (\cdots f_1 g_1 \cdots) z^2 + \cdots\right), \end{aligned} \qquad (2.15)$$

where we have included only terms involving time indices −1, 0 and 1 in the sequences f_j and g_j. If all terms are included, the coefficient of the term in z^j takes the form

$$\cdots + f_{-1} g_{j+1} + f_0 g_j + f_1 g_{j-1} + \cdots + f_k g_{j-k} + \cdots, \qquad (2.16)$$

in conformity with the general term in expression (2.10) for the convolution.

The subroutine CONV finds the time sequence h_j that is the convolution of the sequence f_j with the sequence g_j:

```
      SUBROUTINE CONV(H,F,G,L,M,N)
C
C CONV finds the convolution of the M-length sequence F with the N-length
C sequence G as the L-length sequence H. L=M+N-1.
C
      IMPLICIT DOUBLE COMPLEX(A-H,O-Z)
      DIMENSION H(L),F(M),G(N)
      L=M+N-1
      DO 10 I=1,L
        H(I)=(0.D0,0.D0)
10    CONTINUE
      DO 20 I=1,M
        DO 30 J=1,N
          K=I+J-1
          H(K)=H(K)+F(I)*G(J)
30      CONTINUE
20    CONTINUE
      RETURN
      END
```

2.1.3 Expected value, auto- and crosscorrelation

Time sequences may be *deterministic* or *stochastic* (random). In the latter case, a particular time sequence may be regarded as a single *realisation* of the stochastic process. In both cases, for their treatment, we introduce the *expected value* operator $E\{\cdots\}$.

Of course, the expected value of a deterministic time sequence is just the sequence itself. The expected value of a stochastic variable is found by multiplying a given value of the variable by the probability that it can take on this value, and averaging the result over all possible values of the variable.

We can then regard a stochastic time sequence as the realisation of a sequence of random variables generated by a stochastic process. Each trial of the process will result in a new time sequence. The *expected value* of any quantity at a specific time, say t_0, is found by averaging across an infinite *ensemble* of realisations of the stochastic process, as illustrated in Figure 2.1.

Because the expected value operator is an averaging operator, it is linear. Thus, the expected value of a *constant times a stochastic variable* is the *constant times the expected value* of the stochastic variable. Also, the expected value of a *sum of stochastic variables* is equal to the *sum of the expected values* of the stochastic variables.

The *autocorrelation* of the sequence f_j at lag k, time index l, is defined as

$$\phi_{ff}(k,l) = E\left\{f_l f_{l-k}^*\right\}. \tag{2.17}$$

At zero lag, the autocorrelation gives the mean squared amplitude, or power, of the sequence at time index l.

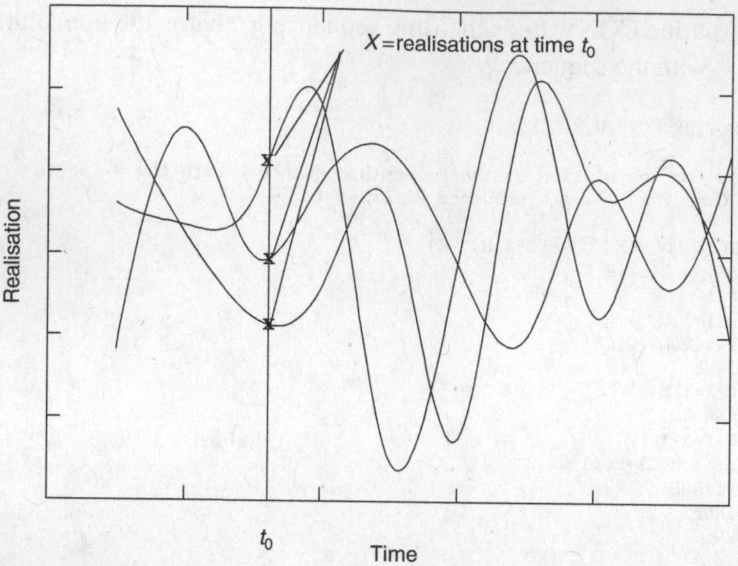

Figure 2.1 Successive realisations of a stochastic process. The expected value of a random variable at a specific time t_0 is found by averaging across an infinite ensemble of such realisations.

Frequently, the stochastic process generating the time sequence will be *stationary*, or assumed to be approximately so. In this case, the statistical properties of the process are independent of translations along the time axis. Then, the autocorrelation is independent of the time index and we write it as

$$\phi_{ff}(k) = E\left\{f_l f_{l-k}^*\right\}. \tag{2.18}$$

The autocorrelation of stationary sequences is a *Hermitian* function of lag, for

$$\phi_{ff}(-k) = E\left\{f_l f_{l+k}^*\right\}, \tag{2.19}$$

and, writing m for $l + k$, we have

$$\phi_{ff}(-k) = E\left\{f_m^* f_{m-k}\right\} = \phi_{ff}^*(k). \tag{2.20}$$

For stationary sequences, averaging across an infinite ensemble of independent realisations is equivalent to averaging along the time axis of a single record. This is known as the *ergodic hypothesis*. For *stationary power signals*, the autocorrelation can then be equivalently defined as

$$\phi_{ff}(k) = \lim_{N\to\infty} \frac{1}{2N+1} \sum_{l=-N}^{N} f_l f_{l-k}^*. \tag{2.21}$$

In general, this is not a satisfactory definition for energy signals since it would give zero for all lags. For energy signals, the autocorrelation is given the relaxed definition

$$\phi_{ff}(k) = \sum_{l=-\infty}^{\infty} f_l f_{l-k}^*. \tag{2.22}$$

While this definition might appear to be inconsistent with the ensemble definition of autocorrelation, we shall see later that if we form a stochastic process in which the deterministic energy signal is taken to begin at random times with random amplitudes, it becomes a consistent definition for such a process.

The *crosscorrelation* of the sequence f_j with the sequence g_j at lag k, time index l, is defined to be

$$\phi_{fg}(k, l) = E\left\{f_l g_{l-k}^*\right\}. \tag{2.23}$$

Once again, for stationary sequences, the dependence on time index disappears and

$$\phi_{fg}(k) = E\left\{f_l g_{l-k}^*\right\} = E\left\{g_m^* f_{m+k}\right\} = \phi_{gf}^*(-k). \tag{2.24}$$

Accepting the ergodic hypothesis for two stationary power signals yields the equivalent definition

$$\phi_{fg}(k) = \lim_{N\to\infty} \frac{1}{2N+1} \sum_{l=-N}^{N} f_l g_{l-k}^*. \tag{2.25}$$

For two energy signals, or one energy signal and one power signal, the definition is relaxed to

$$\phi_{fg}(k) = \sum_{l=-\infty}^{\infty} f_l g_{l-k}^*. \tag{2.26}$$

Again, we shall see later how this definition can be made consistent with the ensemble average definition.

We can make a connection between autocorrelation, crosscorrelation and convolution by introducing the *time reverse* of a sequence. The time reverse of the sequence f_j is simply defined as the sequence f_{-j}^*. If f_j and g_j are two energy signals, the crosscorrelation of f_j with g_j is

$$\phi_{fg}(k) = \sum_{l=-\infty}^{\infty} f_l g_{l-k}^* = \sum_{l=-\infty}^{\infty} f_l h_{k-l}, \tag{2.27}$$

with $h_{k-l} = g_{l-k}^*$ or $h_j = g_{-j}^*$, the time reverse of g_j. Hence, the crosscorrelation of f_j with g_j is the same as the convolution of f_j with the time reverse of g_j. Similarly, the autocorrelation of an energy signal is the same as its convolution with its own time reverse.

2.1.4 White noise and Wold decomposition

Time sequences may have specific statistical properties. Consider a time sequence,

$$\ldots, n_{-1}, \underset{\uparrow}{n_0}, n_1, \ldots, n_j, \ldots, \tag{2.28}$$

which is completely uncorrelated. Its autocorrelation at lag k, time index l, is then

$$\phi_{nn}(k, l) = E\left\{n_l n_{l-k}^*\right\} = \delta_k^0 E\left\{n_l n_l^*\right\}, \tag{2.29}$$

where δ_k^0 is the Kronecker delta, if it is completely uncorrelated. Thus, the autocorrelation vanishes for all lags, except zero lag where it is the power at the particular time index l. Such a time sequence is called a *white noise sequence*.

If it is stationary and of unit power, its autocorrelation is

$$\phi_{nn}(k) = \delta_k^0. \tag{2.30}$$

The sequence

$$\ldots, 0, 0, \underset{\uparrow}{1}, 0, 0, \ldots \tag{2.31}$$

is called the *unit impulse sequence*. Its general term, f_j, is equal to δ_j^0. Hence, in the lag domain, the autocorrelation of a white noise sequence of unit power is the unit impulse sequence.

Wold (1938) proved a very important theorem called the *Wold decomposition theorem* (see Box *et al.* (1994)). The theorem shows that a stationary stochastic sequence may be decomposed into the convolution of a deterministic energy signal with a stationary white noise sequence of unit power.

Suppose we now represent a deterministic energy signal by the stationary stochastic process that results from its convolution with a white noise sequence of unit power. If we convolved the deterministic energy signal f_j with the unit impulse sequence, we would get

$$h_j = \sum_{k=-\infty}^{\infty} f_k \delta_{j-k}^0 = f_j, \tag{2.32}$$

just the deterministic energy signal itself.

If we convolve it with white noise of unit power, we get the stationary sequence

$$h_j = \sum_{k=-\infty}^{\infty} f_k n_{j-k}. \tag{2.33}$$

Thus, this represents the superposition of deterministic energy signals, f_j, starting at random times with random amplitudes. The autocorrelation of the stationary stochastic sequence, h_j, is

$$\begin{aligned}\phi_{hh}(k) &= E\left\{h_l h_{l-k}^*\right\} = E\left\{\sum_{m=-\infty}^{\infty} f_m n_{l-m} \sum_{n=-\infty}^{\infty} f_n^* n_{l-k-n}^*\right\} \\ &= \sum_{m=-\infty}^{\infty}\sum_{n=-\infty}^{\infty} f_m f_n^* E\left\{n_{l-m} n_{l-k-n}^*\right\} \\ &= \sum_{m=-\infty}^{\infty}\sum_{n=-\infty}^{\infty} f_m f_n^* \delta_{k-m+n}^0 = \sum_{m=-\infty}^{\infty} f_m f_{m-k}^* = \phi_{ff}(k),\end{aligned} \tag{2.34}$$

in accordance with the relaxed definition (2.22) of autocorrelation for energy signals. Our relaxed definition of autocorrelation for an energy signal is consistent with the ensemble average for stationary stochastic processes, provided that we use the equivalent stationary stochastic process to represent the energy signal. Similarly, we can construct stationary stochastic processes that lead to (2.26) as a consistent definition of crosscorrelation for two energy signals.

In the case of one energy signal f_j and one stationary power signal g_j, we write

$$h_j = \sum_{k=-\infty}^{\infty} f_k n_{j-k} \tag{2.35}$$

and

$$g_j = \sum_{k=-\infty}^{\infty} a_k n_{j-k}, \tag{2.36}$$

where, by the Wold decomposition theorem, a_k is a deterministic energy signal. Then,

$$\begin{aligned}\phi_{hg}(k) &= E\left\{h_l g_{l-k}^*\right\} \\ &= E\left\{\sum_{m=-\infty}^{\infty} f_m n_{l-m} \sum_{n=-\infty}^{\infty} a_n^* n_{l-k-n}^*\right\} \\ &= \sum_{m=-\infty}^{\infty}\sum_{n=-\infty}^{\infty} f_m a_n^* E\left\{n_{l-m} n_{l-k-n}^*\right\} = \sum_{m=-\infty}^{\infty}\sum_{n=-\infty}^{\infty} f_m a_n^* \delta_{k-m+n}^0 \\ &= \sum_{m=-\infty}^{\infty} f_m a_{m-k}^*.\end{aligned} \tag{2.37}$$

Therefore, in the relaxed definition (2.26) of crosscorrelation, for one energy signal and one power signal, we need to replace the power signal by its equivalent deterministic energy signal found by Wold decomposition.

2.1.5 Properties of wavelets

In general, wavelets have the form

$$\ldots, 0, 0, \underset{\uparrow}{f_0}, f_1, \ldots, f_j, \ldots \tag{2.38}$$

In practice, the wavelets that arise are *finite wavelets*, which we write as

$$(\underset{\uparrow}{f_0}, f_1, f_2, \ldots, f_n). \tag{2.39}$$

The z-transform of a finite wavelet is the polynomial

$$F(z) = f_0 + f_1 z + f_2 z^2 + \cdots + f_n z^n, \tag{2.40}$$

which can be factored to

$$F(z) = f_n (z_1 + z)(z_2 + z) \cdots (z_n + z). \tag{2.41}$$

Since multiplication of z-transforms is equivalent to convolution in the time domain, we see that the finite wavelet is f_n times the successive convolutions of the *dipole wavelets*

$$(\underset{\uparrow}{z_1}, 1), (\underset{\uparrow}{z_2}, 1), \ldots, (\underset{\uparrow}{z_n}, 1). \tag{2.42}$$

The *time reverse* of the wavelet (2.39) is the finite sequence

$$f_n^*, f_{n-1}^*, \ldots, \underset{\uparrow}{f_0^*}. \tag{2.43}$$

If we shift the *time reverse* n units in the positive time direction, we obtain the *reverse wavelet* to (2.39),

$$(\underset{\uparrow}{f_n^*}, f_{n-1}^*, \ldots, f_0^*). \tag{2.44}$$

The reverse wavelet has the z-transform

$$R(z) = f_n^* + f_{n-1}^* z + \cdots + f_0^* z^n. \tag{2.45}$$

The time reverse then has the z-transform

$$\frac{R(z)}{z^n}. \tag{2.46}$$

Dividing the z-transform by z^n gives the z-transform of the wavelet shifted n time units in the negative time direction.

Since the autocorrelation of a wavelet is the same as its convolution with its own time reverse, the z-transform of the autocorrelation of the wavelet (2.39) is

$$\Phi(z) = F(z)\frac{R(z)}{z^n}. \tag{2.47}$$

The reverse wavelet of the reverse wavelet is the original wavelet. The z-transform of the autocorrelation of the reverse wavelet is then

$$R(z)\frac{F(z)}{z^n} = \Phi(z), \tag{2.48}$$

identical to that of the original wavelet.

Now, from (2.45), we have

$$\begin{aligned}\frac{R(z)}{z^n} &= \frac{f_n^*}{z^n} + \frac{f_{n-1}^*}{z^{n-1}} + \cdots + f_0^* \\ &= \left(f_0 + f_1\left(\frac{1}{z^*}\right) + \cdots + f_n\left(\frac{1}{z^*}\right)^n\right)^* \\ &= F^*\left(\frac{1}{z^*}\right) \\ &= f_n^*\left(\frac{1}{z} + z_1^*\right)\left(\frac{1}{z} + z_2^*\right)\cdots\left(\frac{1}{z} + z_n^*\right).\end{aligned} \tag{2.49}$$

On substitution in expression (2.47) from (2.41) and (2.49), we find the z-transform of the autocorrelation takes the form

$$\begin{aligned}\Phi(z) &= f_n f_n^*\,(z_1 + z)\left(\frac{1}{z} + z_1^*\right)\cdots(z_n + z)\left(\frac{1}{z} + z_n^*\right) \\ &= \frac{|f_n|}{z^n}\,(z_1 + z)\left(1 + z_1^* z\right)\cdots(z_n + z)\left(1 + z_n^* z\right).\end{aligned} \tag{2.50}$$

We see that the z-transform of the autocorrelation depends on the product of z-transforms of paired *dipole wavelets*. For example, the jth pair, $(z_j + z)(1 + z_j^* z)$, is the product of the z-transforms of the wavelets

$$(\underset{\uparrow}{z_j}, 1) \quad \text{and} \quad (\underset{\uparrow}{1}, z_j^*),$$

respectively. The first dipole wavelet is the reverse wavelet of the second and vice versa. Thus, any number of the component dipole wavelets can be reversed without changing the autocorrelation. Hence, there are 2^n possible $(n + 1)$-length wavelets with the same autocorrelation. Wavelets have unique autocorrelations but there is no unique wavelet corresponding to a given autocorrelation.

The wavelets corresponding to a given autocorrelation may be classified according to their *delay* properties. If the wavelet has its dipole components arranged so that in magnitude the leading coefficient is greater than, or equal to, the second, it is called *minimum delay*. If the leading coefficient in each dipole component is less than, or equal to, the second, it is called *maximum delay*. Otherwise, it is *mixed delay*.

Wavelets whose coefficients differ at most by a complex constant of unit magnitude have the same autocorrelation and are said to be *equivalent*.

Although there is a unique minimum delay wavelet with the same autocorrelation as a given $(n+1)$-length wavelet, we need to show that for a given autocorrelation there is only one minimum delay wavelet. We begin with the z-transform of the given autocorrelation,

$$\begin{aligned}\Phi(z) &= \phi(-n)\, z^{-n} + \phi(-n+1)\, z^{-n+1} + \cdots \\ &\quad + \phi(0) + \cdots + \phi(n-1)\, z^{n-1} + \phi(n)\, z^{n}. \end{aligned} \tag{2.51}$$

The polynomial

$$\begin{aligned}P(z) &= z^n \Phi(z) \\ &= \phi(-n) + \phi(-n+1)\, z + \cdots \\ &\quad + \phi(0)\, z^n + \cdots + \phi(n-1)\, z^{2n-1} + \phi(n)\, z^{2n} \end{aligned} \tag{2.52}$$

is then of degree $2n$ with $2n$ roots. From the Hermitian property of the autocorrelation, we have

$$\begin{aligned}P(z) &= \phi^*(n) + \phi^*(n-1)\, z + \cdots \\ &\quad + \phi(0)\, z^n + \cdots + \phi^*(-n+1)\, z^{2n-1} + \phi^*(-n)\, z^{2n} \\ &= z^{2n} P^*(1/z^*). \end{aligned} \tag{2.53}$$

In factored form $P(z)$ becomes

$$\begin{aligned}P(z) &= \phi(n)\,(z - z_1)\,(z - z_2)\cdots(z - z_{2n}) \\ &= z^{2n} P^*(1/z^*) = z^{2n} \phi^*(n) \left(1/z - z_1^*\right) \cdots \left(1/z - z_{2n}^*\right) \\ &= \phi^*(n) \left(1 - z_1^* z\right)\left(1 - z_2^* z\right)\cdots\left(1 - z_{2n}^* z\right). \end{aligned} \tag{2.54}$$

Thus, for every root z_j of $P(z)$ there is another root at $1/z_j^*$. Hence, $P(z)$ is the product of two polynomials, one with roots at $z_1, z_2, \ldots, z_n$, the other with roots at $1/z_1^*, 1/z_2^*, \ldots, 1/z_n^*$.

If $G(z)$ is the polynomial of degree n,

$$G(z) = g_n\,(z - z_1)\,(z - z_2)\cdots(z - z_n)\,, \tag{2.55}$$

then

$$G^*(1/z^*) = g_n^* \left(1/z - z_1^*\right)\left(1/z - z_2^*\right)\cdots\left(1/z - z_n^*\right). \tag{2.56}$$

Then, with appropriate choice of $|g_n|^2$, we find that

$$P(z) = G(z)\, z^n G^*(1/z^*) \tag{2.57}$$

and

$$\Phi(z) = G(z)\,G^*(1/z^*). \tag{2.58}$$

The factors of $G(z)$ can be arranged in only one minimum delay way. When in minimum delay form, the factors of $G(z)$ have the general term $-(z_j - z)$ with $|z_j| \geq 1$.

The term $G(z)$ then becomes the z-transform of the only $(n + 1)$-length minimum delay wavelet corresponding to $\Phi(z)$. Shorter wavelets can be minimum delay but do not produce sufficiently long autocorrelations. Longer wavelets with z-transforms of the form

$$G(z)\,z^p, \tag{2.59}$$

where p is a positive integer, give the correct autocorrelation, for then

$$\Phi(z) = G(z)\,z^p \cdot G^*(1/z^*)\,1/z^p = G(z)\,G^*(1/z^*), \tag{2.60}$$

but are not minimum delay since they contain p of the dipole factors $(0 + 1 \cdot z)$ in their z-transforms and, hence, are not minimum delay. Putting them in minimum delay form would yield p of the factors $(1 + 0 \cdot z)$ or $1^p = 1$, and the wavelet would be reduced to the minimum delay wavelet.

It also follows that maximum delay wavelets are not uniquely determined by the autocorrelation. Longer than $(n+1)$-length maximum or mixed delay wavelets will give the correct autocorrelation.

2.2 Linear optimum Wiener filters

In a linear filtering operation, a given member of a time sequence is replaced by a *linear combination* of itself and neighbouring members of the sequence. The time sequence being filtered may be regarded as the input to a linear system characterised by its impulse response. In turn, the filtered sequence may be regarded as the output of the linear system. It is, therefore, the convolution of the input sequence with the impulse response of the system. If f_j is the input sequence, and g_j is a deterministic wavelet representing the impulse response of the linear system, the output sequence h_j is given by the convolution

$$h_j = \sum_{k=0}^{N} g_k f_{j-k}. \tag{2.61}$$

In general, the objective of *optimum* filtering is to make the output sequence h_j of the filtering operation as close as possible to a desired sequence d_j. The departure of the desired sequence from the filtered output sequence is measured by an *error sequence* ϵ_j with

$$\epsilon_j = d_j - h_j. \tag{2.62}$$

Optimum Wiener filters minimise the *error power* or the mean square of the error sequence. The error power is

$$\begin{aligned}
E\left\{\epsilon_j\epsilon_j^*\right\} &= E\left\{(d_j - h_j)(d_j^* - h_j^*)\right\} \\
&= E\left\{\left(d_j - \sum_{k=0}^{N} g_k f_{j-k}\right)\left(d_j^* - \sum_{l=0}^{N} g_l^* f_{j-l}^*\right)\right\} \\
&= E\left\{d_j d_j^* - d_j \sum_{l=0}^{N} g_l^* f_{j-l}^* - d_j^* \sum_{k=0}^{N} g_k f_{j-k}\right\} \\
&\quad + E\left\{\sum_{k=0}^{N} g_k f_{j-k} \sum_{l=0}^{N} g_l^* f_{j-l}^*\right\}.
\end{aligned} \tag{2.63}$$

In the expression for the error power, only the elements $g_0, \ldots, g_N$ of the wavelet representing the impulse response have yet to be specified. When, with respect to the real and imaginary parts of the elements of the impulse response, the partial derivatives of the error power all vanish, the error power will be an extremum. Since there can be no maximum to the error power, or how badly the filter performs, the extremum obtained by setting the partial derivatives to zero must be a minimum. To minimise the error power, we have

$$\begin{aligned}
\frac{\partial}{\partial \mathrm{Re} g_m} E\left\{\epsilon_j\epsilon_j^*\right\} &= E\left\{-d_j \sum_{l=0}^{N} \delta_l^m f_{j-l}^* - d_j^* \sum_{k=0}^{N} \delta_k^m f_{j-k}\right\} \\
&\quad + E\left\{\sum_{k=0}^{N}\sum_{l=0}^{N}\left(\delta_k^m f_{j-k} g_l^* f_{j-l}^* + g_k f_{j-k} \delta_l^m f_{j-l}^*\right)\right\} \\
&= E\left\{-d_j f_{j-m}^* - d_j^* f_{j-m}\right\} \\
&\quad + E\left\{\sum_{k=0}^{N}\left(f_{j-m} g_k^* f_{j-k}^* + g_k f_{j-k} f_{j-m}^*\right)\right\} = 0,
\end{aligned} \tag{2.64}$$

and

$$\begin{aligned}
-i\frac{\partial}{\partial \mathrm{Im} g_m} E\left\{\epsilon_j\epsilon_j^*\right\} &= E\left\{d_j f_{j-m}^* - d_j^* f_{j-m}\right\} \\
&\quad + E\left\{\sum_{k=0}^{N}\left(f_{j-m} g_k^* f_{j-k}^* - g_k f_{j-k} f_{j-m}^*\right)\right\} = 0,
\end{aligned} \tag{2.65}$$

for $m = 0, 1, \ldots, N$. Adding the two equations gives

$$E\left\{-2d_j^* f_{j-m} + 2\sum_{k=0}^{N} f_{j-m} g_k^* f_{j-k}^*\right\} = 0, \tag{2.66}$$

or, on dividing by two and taking complex conjugates,

$$\sum_{k=0}^{N} g_k E\left\{f_{j-k} f_{j-m}^*\right\} = E\left\{d_j f_{j-m}^*\right\}, \quad m = 0, 1, \ldots, N. \tag{2.67}$$

Recognising that the autocorrelation is $\phi_{ff}(m-k) = E\{f_{j-k} f_{j-m}^*\}$ and the crosscorrelation is $\phi_{df}(m) = E\{d_j f_{j-m}^*\}$, the conditional equations for the optimum Wiener filter become

$$\sum_{k=0}^{N} g_k \phi_{ff}(m-k) = \phi_{df}(m), \quad m = 0, 1, \ldots, N. \tag{2.68}$$

In matrix form, they are

$$\begin{pmatrix} \phi_{ff}(0) & \phi_{ff}(-1) & \cdots & \phi_{ff}(-N) \\ \phi_{ff}(1) & \phi_{ff}(0) & \cdots & \phi_{ff}(-N+1) \\ \vdots & \vdots & \ddots & \vdots \\ \phi_{ff}(N) & \phi_{ff}(N-1) & \cdots & \phi_{ff}(0) \end{pmatrix} \begin{pmatrix} g_0 \\ g_1 \\ \vdots \\ g_N \end{pmatrix} = \begin{pmatrix} \phi_{df}(0) \\ \phi_{df}(1) \\ \vdots \\ \phi_{df}(N) \end{pmatrix}. \tag{2.69}$$

The coefficient matrix is then seen to be equidiagonal and Hermitian (the complex conjugate of its transpose is equal to the matrix itself). The latter follows from the properties $\phi_{ff}^*(k) = \phi_{ff}(-k)$ and $\phi_{ff}^*(-k) = \phi_{ff}(k)$. Matrices that are equidiagonal and Hermitian are said to be *Toeplitz* matrices.

It is to be noted that if f_j and d_j are energy signals, the optimum Wiener filter equations retain the same form using the relaxed definitions of autocorrelation (2.22) and crosscorrelation (2.26).

2.2.1 Prediction and prediction error filters

Optimum Wiener filters can be applied to the problem of prediction. For example, we might want to predict the value of a time sequence one time unit after the previous N values. That is, we want to predict f_j from $f_{j-N}, \ldots, f_{j-2}, f_{j-1}$. The prediction is given by the output

$$h_j = \sum_{k=1}^{N} g_k f_{j-k} \tag{2.70}$$

of the linear filter with impulse response g_j. The conditional equations for the optimum Wiener prediction filter are found to be

$$\sum_{k=1}^{N} g_k \phi_{ff}(m-k) = \phi_{df}(m), \quad m = 1, \ldots, N. \tag{2.71}$$

The summation starts at $k = 1$, so there is one less equation, since the member of the time sequence to be predicted cannot be included in the calculation of the prediction. The perfect prediction would be f_j itself. Thus, $d_j = f_j$, and

$$\phi_{df}(j) = E\left\{d_k f_{k-j}^*\right\} = E\left\{f_k f_{k-j}^*\right\} = \phi_{ff}(j). \tag{2.72}$$

The unit prediction equations (2.71) then become

$$\sum_{k=1}^{N} g_k \phi_{ff}(m-k) = \phi_{ff}(m), \quad m = 1, \ldots, N \tag{2.73}$$

with matrix form

$$\begin{pmatrix} \phi_{ff}(0) & \cdots & \phi_{ff}(-N+1) \\ \vdots & \ddots & \vdots \\ \phi_{ff}(N-1) & \cdots & \phi_{ff}(0) \end{pmatrix} \begin{pmatrix} g_1 \\ \vdots \\ g_N \end{pmatrix} = \begin{pmatrix} \phi_{ff}(1) \\ \vdots \\ \phi_{ff}(N) \end{pmatrix}. \tag{2.74}$$

The quality of the unit prediction is measured by the error sequence ϵ_j. It is given by

$$\epsilon_j = d_j - h_j = f_j - \sum_{k=1}^{N} g_k f_{j-k} = \sum_{k=0}^{N} \gamma_k f_{j-k}, \tag{2.75}$$

with $\gamma_0 = 1, \gamma_1 = -g_1, \ldots, \gamma_N = -g_N$. Thus, the filter $1, \gamma_1, \ldots, \gamma_N$ convolved with the sequence f_j gives the prediction error sequence directly. It is called the *prediction error filter*.

Now consider the *prediction error power* of the $(N+1)$-length prediction error filter. It is

$$P_{N+1} = E\left\{\epsilon_j\epsilon_j^*\right\} = E\left\{\left(f_j - \sum_{k=1}^{N} g_k f_{j-k}\right)\left(f_j^* - \sum_{l=1}^{N} g_l^* f_{j-l}^*\right)\right\}$$
$$= E\left\{f_j f_j^* - f_j \sum_{l=1}^{N} g_l^* f_{j-l}^* - f_j^* \sum_{k=1}^{N} g_k f_{j-k}\right\}$$
$$+ E\left\{\sum_{k=1}^{N} g_k f_{j-k} \sum_{l=1}^{N} g_l^* f_{j-l}^*\right\}$$
$$= \phi_{ff}(0) - \sum_{l=1}^{N} g_l^* E\left\{f_j f_{j-l}^*\right\} - \sum_{k=1}^{N} g_k E\left\{f_j^* f_{j-k}\right\}$$
$$+ \sum_{k=1}^{N}\sum_{l=1}^{N} g_k g_l^* E\left\{f_{j-k} f_{j-l}^*\right\}. \tag{2.76}$$

Thus,

$$P_{N+1} = \phi_{ff}(0) - \sum_{l=1}^{N} g_l^* \phi_{ff}(l) - \sum_{k=1}^{N} g_k \phi_{ff}^*(k)$$
$$+ \sum_{k=1}^{N}\sum_{l=1}^{N} g_k g_l^* \phi_{ff}(l-k). \tag{2.77}$$

From the unit prediction equations (2.73), the last term on the right-hand side of expression (2.77) can be transformed to give

$$P_{N+1} = \phi_{ff}(0) - \sum_{l=1}^{N} g_l^* \phi_{ff}(l) - \sum_{k=1}^{N} g_k \phi_{ff}^*(k) + \sum_{l=1}^{N} g_l^* \phi_{ff}(l)$$
$$= \phi_{ff}(0) - \sum_{k=1}^{N} g_k \phi_{ff}^*(k) = \sum_{k=0}^{N} \gamma_k \phi_{ff}^*(k) = \sum_{k=0}^{N} \gamma_k \phi_{ff}(-k). \tag{2.78}$$

Writing the unit prediction equations (2.73) in terms of the prediction error coefficients they become

$$\sum_{k=0}^{N} \gamma_k \phi_{ff}(m-k) = 0, \quad m = 1, \ldots, N. \tag{2.79}$$

Augmenting the system of equations (2.79) with expression (2.78) for the prediction error, the *prediction error equations* take the matrix form

$$\begin{pmatrix} \phi_{ff}(0) & \cdots & \phi_{ff}(-N) \\ \vdots & \ddots & \vdots \\ \phi_{ff}(N) & \cdots & \phi_{ff}(0) \end{pmatrix} \begin{pmatrix} 1 \\ \gamma_1 \\ \vdots \\ \gamma_N \end{pmatrix} = \begin{pmatrix} P_{N+1} \\ 0 \\ \vdots \\ 0 \end{pmatrix}. \tag{2.80}$$

2.2.2 Predictive deconvolution

Frequently, in the study of Earth's dynamics, we have a recorded time sequence y_j that is known to be the response to an excitation x_j. If the impulse response of the particular Earth system is b_j, then the recorded response is the convolution of b_j with x_j, if the system can be modelled as linear. Often, we want to recover the excitation by removing the effect of the linear system. Thus, we want to undo the convolution

$$y_j = \sum_k b_k x_{j-k} \tag{2.81}$$

that produced the observed output sequence y_j. This is the problem of *deconvolution.*

In terms of z-transforms, equation (2.81) becomes

$$Y(z) = B(z)\,X(z). \tag{2.82}$$

The solution to the deconvolution problem then requires the construction of a sequence a_j with z-transform $A(z)$ such that

$$A(z)\,B(z) = 1, \tag{2.83}$$

or such that the convolution of a_j with b_j produces the unit impulse sequence.

On multiplying equation (2.82) through on the left-hand side by $A(z)$, from (2.83) we see that the z-transform of the sequence being sought is given by

$$X(z) = A(z)\,Y(z), \tag{2.84}$$

while, from (2.83), $A(z)$ is given by

$$A(z) = \frac{1}{B(z)}. \tag{2.85}$$

The sequence a_j is called the *inverse* to the impulse response b_j.

In practice, the inverse is modelled as the $(m+1)$-length finite wavelet,

$$a = (\underset{\uparrow}{a_0}, a_1, \ldots, a_m)$$

and the impulse response is modelled as the $(n+1)$-length finite wavelet,

$$b = (\underset{\uparrow}{b_0}, b_1, \ldots, b_n).$$

The convolution c of the wavelets a and b is to approximate the unit impulse,

$$\delta = (\underset{\uparrow}{1}, 0, 0, \ldots, 0).$$

The error sequence is then the $(m+n+1)$-length wavelet,

$$\epsilon = (\underset{\uparrow}{1 - c_0}, -c_1, \ldots, -c_{m+n}), \tag{2.86}$$

and the error energy is

$$\begin{aligned} I &= (1 - c_0)(1 - c_0^*) + c_1 c_1^* + \cdots + c_{m+n} c_{m+n}^* \\ &= 1 - c_0 - c_0^* + \sum_{l=0}^{m+n} c_l c_l^*. \end{aligned} \tag{2.87}$$

Since c is the convolution of a and b,

$$c_l = \sum_{k=0}^{m} a_k b_{l-k}, \quad c_0 = a_0 b_0, \tag{2.88}$$

$$\begin{aligned} I &= 1 - a_0 b_0 - a_0^* b_0^* + \sum_{l=0}^{m+n} \left(\sum_{k=0}^{m} a_k b_{l-k} \sum_{j=0}^{m} a_j^* b_{l-j}^* \right) \\ &= 1 - a_0 b_0 - a_0^* b_0^* + \sum_{j=0}^{m} \sum_{k=0}^{m} \sum_{l=0}^{m+n} a_k b_{l-k} a_j^* b_{l-j}^*. \end{aligned} \tag{2.89}$$

The error energy is minimised by setting to zero its partial derivatives with respect to the real and imaginary parts of the elements of the inverse wavelet.

Then,

$$\frac{\partial I}{\partial \mathrm{Re} a_0} = -b_0 - b_0^* + \sum_{j,k,l} \left(\delta_k^0 b_{l-k} a_j^* b_{l-j}^* + a_k b_{l-k} \delta_j^0 b_{l-j}^* \right) = 0, \tag{2.90}$$

$$i \frac{\partial I}{\partial \mathrm{Im} a_0} = b_0 - b_0^* - \sum_{j,k,l} \left(\delta_k^0 b_{l-k} a_j^* b_{l-j}^* - a_k b_{l-k} \delta_j^0 b_{l-j}^* \right) = 0. \tag{2.91}$$

Adding equations (2.90) and (2.91), we find that

$$\sum_{j,k,l} \left(a_k b_{l-k} \delta_j^0 b_{l-j}^* \right) = \sum_{k=0}^{m} a_k \sum_{l=0}^{m+n} b_{l-k} b_l^* = b_0^*, \tag{2.92}$$

or

$$\sum_{k=0}^{m} a_k\,\phi_{bb}\,(-k) = b_0^*, \tag{2.93}$$

on using the relaxed definition (2.22) of autocorrelation for energy signals.

For $p = 1, \ldots, m$, we have

$$\frac{\partial I}{\partial \mathrm{Re}a_p} = \sum_{j,k,l} \left(\delta_k^p b_{l-k} a_j^* b_{l-j}^* + a_k b_{l-k} \delta_j^p b_{l-j}^*\right) = 0, \tag{2.94}$$

$$i\frac{\partial I}{\partial \mathrm{Im}a_p} = -\sum_{j,k,l} \left(\delta_k^p b_{l-k} a_j^* b_{l-j}^* - a_k b_{l-k} \delta_j^p b_{l-j}^*\right) = 0. \tag{2.95}$$

On adding equations (2.94) and (2.95), we get

$$\sum_{j,k,l} a_k b_{l-k} \delta_j^p b_{l-j}^* = \sum_{k=0}^{m} a_k \sum_{l=0}^{m+n} b_{l-k} b_{l-p}^* = 0. \tag{2.96}$$

Again using the relaxed definition of autocorrelation (2.22), we find that

$$\sum_{k=0}^{m} a_k\,\phi_{bb}(p-k) = 0, \quad p = 1, \ldots, m. \tag{2.97}$$

In matrix form, equations (2.93) and (2.97) give the minimum error energy system for the inverse sequence,

$$\begin{pmatrix} \phi_{bb}\,(0) & \cdots & \phi_{bb}\,(-m) \\ \vdots & \ddots & \vdots \\ \phi_{bb}\,(m) & \cdots & \phi_{bb}\,(0) \end{pmatrix} \begin{pmatrix} a_0 \\ a_1 \\ \vdots \\ a_m \end{pmatrix} = \begin{pmatrix} b_0^* \\ 0 \\ \vdots \\ 0 \end{pmatrix}. \tag{2.98}$$

On scaling the inverse sequence by its first term, this system takes on the same form as the prediction error equations (2.80), giving rise to the description of this method as predictive deconvolution.

Returning to expression (2.89), the error energy can be written

$$I = 1 - a_0 b_0 - a_0^* b_0^* + \sum_{j=0}^{m} a_j^* \sum_{k=0}^{m} a_k\,\phi_{bb}(j-k). \tag{2.99}$$

From equations (2.93) and (2.97), when $I = I_{\min}$,

$$\sum_{k=0}^{m} a_k\,\phi_{bb}(j-k) = b_0^* \delta_j^0, \quad j = 0, \ldots, m. \tag{2.100}$$

Then,

$$\begin{aligned} I_{\min} &= 1 - a_0 b_0 - a_0^* b_0^* + \sum_{j=0}^{m} a_j^* b_0^* \delta_j^0 \\ &= 1 - a_0 b_0 - a_0^* b_0^* + a_0^* b_0^* \\ &= 1 - a_0 b_0 = 1 - c_0. \end{aligned} \tag{2.101}$$

From (2.87), the error energy is

$$I = 1 - c_0 - c_0^* + \sum_{l=0}^{m+n} c_l c_l^*, \tag{2.102}$$

giving

$$\sum_{l=0}^{m+n} c_l c_l^* = c_0^* = c_0 \tag{2.103}$$

for $I = I_{\min}$. Therefore, c_0 must be real and non-negative. From (2.101), since $I_{\min}$ is a sum of squares and must also be real and non-negative definite, we see that $0 \leq c_0 \leq 1$ and $0 \leq I_{\min} \leq 1$. Therefore, $I_{\min}$ provides a quality factor between 0 and 1, which measures how well the deconvolution filter is performing. Here, c_0 is the actual approximation to the desired unit impulse sequence and thus $I_{\min}$ measures the difference, or ϵ_0.

From expression (2.99) for the error energy,

$$I = 1 - a_0 b_0 - a_0^* b_0^* + \sum_{j=0}^{m} \sum_{k=0}^{m} a_k a_j^* \phi_{bb}(j - k). \tag{2.104}$$

Introducing the new index $s = j - k$ and summing first over s and then over j, we get

$$\begin{aligned} I &= 1 - a_0 b_0 - a_0^* b_0^* + \sum_{j=0}^{m} \sum_{s=j-m}^{j} a_j^* a_{j-s} \phi_{bb}(s) \\ &= 1 - a_0 b_0 - a_0^* b_0^* + \sum_{s=-m}^{m} \phi_{aa}^*(s) \phi_{bb}(s). \end{aligned} \tag{2.105}$$

Now suppose that

$$\alpha = (\underset{\uparrow}{\alpha_0}, \alpha_1, \ldots, \alpha_m)$$

is a wavelet with the same autocorrelation as the optimum wavelet a. Since expression (2.105) holds for the error energy, whether or not a is optimum, we have for α

$$I = 1 - \alpha_0 b_0 - \alpha_0^* b_0^* + \sum_{s=-m}^{m} \phi_{aa}^*(s)\, \phi_{bb}(s). \tag{2.106}$$

For the optimum wavelet, I is minimal, so

$$-a_0 b_0 - a_0^* b_0^* \leq -\alpha_0 b_0 - \alpha_0^* b_0^*, \tag{2.107}$$

or

$$2c_0 \geq 2\mathrm{Re}\,(\alpha_0 b_0)\,. \tag{2.108}$$

In terms of their delay properties, wavelets whose coefficients differ at most by a complex constant of unit magnitude are equivalent (see Section 2.1.5), and we may take the phase of α_0 to be the opposite of that of b_0. Then we have $c_0 \geq \alpha_0 b_0$ or $a_0 b_0 \geq \alpha_0 b_0$ or $|a_0| \geq |\alpha_0|$. Thus, the leading coefficient of a is as large, or greater, in magnitude than the leading coefficient of any other wavelet in the suite of wavelets with the same autocorrelation. Hence, the optimum wavelet is the minimum delay wavelet of the suite. The least error energy inverse is minimum delay.

2.2.3 *The Levinson algorithm*

Optimum Wiener filter equations (2.69) and (2.80), and minimum error energy systems of equations (2.98), all have Toeplitz coefficient matrices. The Levinson algorithm takes advantage of the Toeplitz form of the coefficient matrices to construct a rapid, recursive method of solution for such systems of equations. In matrix notation, the system of equations to be solved takes the form,

$$\begin{pmatrix} r_0 & r_{-1} & \cdots & r_{-m} \\ r_1 & r_0 & \cdots & r_{-m+1} \\ \vdots & \vdots & \ddots & \vdots \\ r_m & r_{m-1} & \cdots & r_0 \end{pmatrix} \begin{pmatrix} f_0 \\ f_1 \\ \vdots \\ f_m \end{pmatrix} = \begin{pmatrix} g_0 \\ g_1 \\ \vdots \\ g_m \end{pmatrix}, \tag{2.109}$$

with r_j representing the autocorrelation sequence at lag j in the lag domain. Writing the transpose of the coefficient matrix as R_m, the system of equations in row form becomes

$$(f_0, f_1, \ldots, f_m)\, R_m = (g_0, g_1, \ldots, g_m)\,. \tag{2.110}$$

Since both the autocorrelations and the coefficient matrix itself are Hermitian, $R_m^T = R_m^*$.

The Levinson algorithm is recursive, each solution building on the previous one. Thus, we begin with the simplest case of all, that for $m = 0$. With the first subscript denoting the value of m, the system is $f_{0,0}r_0 = g_0$ with solution $f_{0,0} = g_0/r_0$. Next, for $m = 1$ the system takes the form

$$(f_{1,0}, f_{1,1})\begin{pmatrix} r_0 & r_1 \\ r_1^* & r_0 \end{pmatrix} = (g_0, g_1) = (f_{1,0}, f_{1,1})\,R_1. \tag{2.111}$$

Rather than proceeding to the solution of this system directly, we first solve the auxiliary system

$$(a_{1,0}, a_{1,1})\,R_1 = (\alpha_1, 0)\,. \tag{2.112}$$

The auxiliary system is then of the same form as the prediction error equations (2.80). To solve the auxiliary system, we consider the system

$$(a_{0,0}, 0)\begin{pmatrix} r_0 & r_1 \\ r_1^* & r_0 \end{pmatrix} = (\alpha_0, \beta_0)\,. \tag{2.113}$$

Thus, $\alpha_0 = a_{0,0}r_0$ and $\beta_0 = a_{0,0}r_1$. From the Hermitian property of the coefficient matrix, reversing the order of equations and variables in this system, and taking complex conjugates, leads to the system

$$\left(0, a_{0,0}^*\right)\begin{pmatrix} r_0 & r_1 \\ r_1^* & r_0 \end{pmatrix} = \left(\beta_0^*, \alpha_0^*\right). \tag{2.114}$$

If we multiply this system by k_0 and add it to (2.113), we get

$$\left(a_{0,0}, k_0 a_{0,0}^*\right) R_1 = \left(\alpha_0 + k_0\beta_0^*, \beta_0 + k_0\alpha_0^*\right). \tag{2.115}$$

If this is to be identical to the auxiliary system (2.112),

$$\begin{aligned} a_{1,0} &= a_{0,0} & \alpha_1 &= \alpha_0 + k_0\beta_0^*, \\ a_{1,1} &= k_0 a_{0,0}^* & k_0 &= -\beta_0/\alpha_0^*. \end{aligned} \tag{2.116}$$

This completes the solution of the auxiliary system (2.112) within the scale factor $a_{0,0}$. We set $a_{0,0} = 1$ to complete the solution.

Recursive solutions of the successive auxiliary systems can be appended. For $m = 2$, the extension of (2.113) is

$$(a_{1,0}, a_{1,1}, 0)\begin{pmatrix} r_0 & r_1 & r_2 \\ r_1^* & r_0 & r_1 \\ r_2^* & r_1^* & r_0 \end{pmatrix} = (\alpha_1, 0, \beta_1)\,. \tag{2.117}$$

This system includes the solution of the previous auxiliary system (2.112) as well as the definition

$$\beta_1 = a_{1,0}r_2 + a_{1,1}r_1. \tag{2.118}$$

Reversing the order of equations and variables, and taking complex conjugates, leads to the system

$$\left(0, a_{1,1}^*, a_{1,0}^*\right) R_2 = \left(\beta_1^*, 0, \alpha_1^*\right). \tag{2.119}$$

Multiplying this system by k_1 and adding the result to the original system (2.117) gives

$$\left(a_{1,0},\ a_{1,1} + k_1 a_{1,1}^*,\ k_1 a_{1,0}^*\right) R_2 = \left(\alpha_1 + k_1\beta_1^*,\ 0,\ \beta_1 + k_1\alpha_1^*\right). \tag{2.120}$$

For this to solve the next auxiliary system,

$$(a_{2,0}, a_{2,1}, a_{2,2}) R_2 = (\alpha_2, 0, 0), \tag{2.121}$$

requires

$$\begin{aligned} k_1 &= -\beta_1/\alpha_1^*, \\ a_{2,0} &= a_{1,0}, \\ a_{2,1} &= a_{1,1} + k_1 a_{1,1}^*, \\ a_{2,2} &= k_1 a_{1,0}^*, \\ \alpha_2 &= \alpha_1 + k_1\beta_1^*. \end{aligned} \tag{2.122}$$

The solution of the auxiliary system for $m = n + 1$ is then derived from the solution for $m = n$ through the recurrence relations

$$\begin{aligned} \beta_n &= a_{n,0}r_{n+1} + a_{n,1}r_n + \cdots + a_{n,n}r_1, \\ k_n &= -\beta_n/\alpha_n^*, \\ a_{n+1,0} &= a_{n,0}, \\ a_{n+1,1} &= a_{n,1} + k_n a_{n,n}^*, \\ &\vdots \\ a_{n+1,n} &= a_{n,n} + k_n a_{n,1}^*, \\ a_{n+1,n+1} &= k_n a_{n,0}^*, \\ \alpha_{n+1} &= \alpha_n + k_n\beta_n^*. \end{aligned} \tag{2.123}$$

A similar scheme can be devised to solve the full equations for $m = 1$. We begin with the system

$$(f_{0,0}, 0) \begin{pmatrix} r_0 & r_1 \\ r_1^* & r_0 \end{pmatrix} = (g_0, \gamma_0), \tag{2.124}$$

with $\gamma_0 = f_{0,0} r_1$. Again this includes the equation for $m = 0$ and defines the new parameter γ_0. Reversing the order of equations and variables in the auxiliary system (2.112), and taking complex conjugates, yields the system

$$\left(a_{1,1}^*, a_{1,0}^*\right) R_1 = \left(0, \alpha_1^*\right). \tag{2.125}$$

Multiplying this system by q_0 and adding it to the system (2.124) gives

$$\left(f_{0,0} + q_0 a_{1,1}^*,\ q_0 a_{1,0}^*\right) R_1 = \left(g_0,\ \gamma_0 + q_0 \alpha_1^*\right). \tag{2.126}$$

If this is to be a solution of the full equations (2.111) for $m = 1$, then

$$\begin{aligned} q_0 &= \frac{g_1 - \gamma_0}{\alpha_1^*}, \\ f_{1,0} &= f_{0,0} + q_0 a_{1,1}^*, \\ f_{1,1} &= q_0 a_{1,0}^*. \end{aligned} \tag{2.127}$$

For $m = 2$, we reverse the order of the equations and variables in the next auxiliary system (2.121), and take complex conjugates, to get

$$\left(a_{2,2}^*, a_{2,1}^*, a_{2,0}^*\right) R_2 = \left(0, 0, \alpha_2^*\right). \tag{2.128}$$

If we multiply this system by q_1 and add it to the system

$$\left(f_{1,0}, f_{1,1}, 0\right) R_2 = \left(g_0, g_1, \gamma_1\right), \tag{2.129}$$

with $\gamma_1 = f_{1,0} r_2 + f_{1,1} r_1$, we obtain

$$\left(f_{1,0} + q_1 a_{2,2}^*,\ f_{1,1} + q_1 a_{2,1}^*,\ q_1 a_{2,0}^*\right) R_2 = \left(g_0,\ g_1,\ \gamma_1 + q_1 \alpha_2^*\right). \tag{2.130}$$

For this to be a solution of

$$\left(f_{2,0}, f_{2,1}, f_{2,2}\right) R_2 = \left(g_0, g_1, g_2\right), \tag{2.131}$$

requires

$$\begin{aligned} q_1 &= \frac{g_2 - \gamma_1}{\alpha_2^*}, \\ f_{2,0} &= f_{1,0} + q_1 a_{2,2}^*, \\ f_{2,1} &= f_{1,1} + q_1 a_{2,1}^*, \\ f_{2,2} &= q_1 a_{2,0}^*. \end{aligned} \tag{2.132}$$

The solution of the full equations for $m = n + 1$ is then derived from the solution for $m = n$ through the recurrence relations,

$$
\begin{aligned}
\gamma_n &= f_{n,0}r_{n+1} + f_{n,1}r_n + \cdots + f_{n,n}r_1, \\
q_n &= \frac{g_{n+1} - \gamma_n}{\alpha^*_{n+1}}, \\
f_{n+1,0} &= f_{n,0} + q_n a^*_{n+1,n+1}, \\
f_{n+1,1} &= f_{n,1} + q_n a^*_{n+1,n}, \\
&\vdots \\
f_{n+1,n} &= f_{n,n} + q_n a^*_{n+1,1}, \\
f_{n+1,n+1} &= q_n a^*_{n+1,0}.
\end{aligned}
\tag{2.133}
$$

By making use of the Toeplitz form of the coefficient matrix, the Levinson algorithm reduces the computational effort from the usual N^3, for linear algebraic systems of size $N \times N$, to only N^2.

The subroutine LEVSOL implements the Levinson algorithm through the successive recursions (2.123) and (2.133).

```
      SUBROUTINE LEVSOL(LR,R,G,F,A)
C
C LEVSOL solves a linear system with a Toeplitz coefficient matrix by
C the Levinson recursive method. LR is the dimension of the system and R is
C the LR-length vector of autocorrelations for zero and positive lags,
C or in the general case, the first column vector of the coefficient matrix.
C G is the right-hand side vector, F is the solution vector and A is
C a vector representing the solution of an auxiliary prediction error
C system used in the recursion. Since A is overwritten in the recursion,
C its previous values are stored in the vector AO.
C
      IMPLICIT DOUBLE COMPLEX(A-H,O-Z)
      DIMENSION R(LR),G(LR),F(LR),A(LR),AO(LR)
C Solve 1x1 system.
      F(1)=G(1)/R(1)
      IF(LR.EQ.1) RETURN
C Set first element of the auxiliary vector to unity.
      A(1)=(1.D0,0.D0)
C Initialize alpha, beta and gamma.
      ALPHA=R(1)
      BETA=R(2)
      GAMMA=F(1)*R(2)
C Begin loop over remaining equation systems.
      DO 10 L=2,LR
C Find k for L-1.
        AK=-BETA/DCONJG(ALPHA)
C Find last element of new auxiliary vector.
        LM1=L-1
        A(L)=AK
C 2x2 system is a special case.
        IF(L.EQ.2) GO TO 20
C Store old auxiliary vector elements.
        DO 40 I=2,LM1
```

```
         AO(I)=A(I)
 40     CONTINUE
C Find remaining elements of new auxiliary vector.
        LM2=L-2
        DO 50 J=1,LM2
          JP1=J+1
          K=L-J
          A(JP1)=A(JP1)+AK*DCONJG(AO(K))
 50     CONTINUE
C Update alpha.
 20     ALPHA=ALPHA+AK*DCONJG(BETA)
C Find q for L-1.
        Q=(G(L)-GAMMA)/DCONJG(ALPHA)
C Find last element of solution vector.
        F(L)=Q
C Find remaining elements of the solution vector.
        DO 60 J=1,LM1
          K=L-J+1
          F(J)=F(J)+Q*DCONJG(A(K))
 60     CONTINUE
C Return if solution is complete.
        IF(L.EQ.LR) RETURN
C Find new values of beta and gamma.
        BETA=(0.D0,0.D0)
        GAMMA=(0.D0,0.D0)
        DO 10 I=1,L
          K=L-I+2
          BETA=BETA+A(I)*R(K)
          GAMMA=GAMMA+F(I)*R(K)
 10   CONTINUE
      RETURN
      END
```

2.3 Frequency domain analysis

Time domain sequences can equally well be represented in the frequency domain. The representations in the two domains have a one-to-one correspondence, and any operation on the representation in one domain has an equivalent counterpart in the other. Some operations, such as filtering, are more easily done in the time domain, while the effects of discrete sampling and finite record length, for example, are more clearly understood in the frequency domain. We begin with a description of the *discrete Fourier transform (DFT)*, the main tool for going from the time domain to the frequency domain.

2.3.1 The discrete Fourier transform

The time sequence is taken to be of length $N = 2n + 1$, and we write it as

$$g_{-n}, g_{-n+1}, \ldots, g_{-1}, g_0, g_1, \ldots, g_{n-1}, g_n,$$

consisting of equally spaced samples at intervals Δt of a record of length T, as illustrated in Figure 2.2.

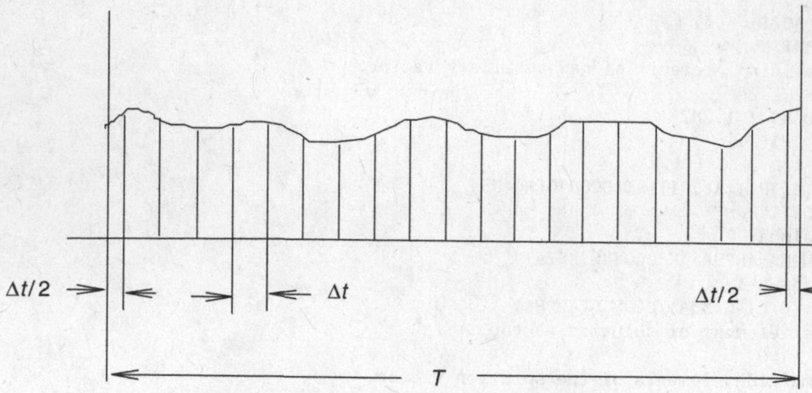

Figure 2.2 Finite time record of length T, sampled at uniform intervals Δt.

The first and last samples are taken to be $\Delta t/2$ from the ends of the record with $T = (2n+1)\,\Delta t$. In the *discrete Fourier transform (DFT)* representation, the time sequence is taken to be a superposition of sinusoids with periods equal to the record length, and overtones, plus a *dc* term (the abbreviation dc, for direct current, is common in electrical engineering usage in referring to a zero frequency or constant term). Using complex sinusoids, the approximation to the time sequence g_j is taken as

$$h_j = \frac{1}{T} \sum_{k=-n}^{n} H_k e^{i2\pi \frac{k}{T} t_j}, \quad j = -n, \ldots, n, \tag{2.134}$$

with the sample points at times $t_j = j\Delta t$. Having $2n+1$ time samples, we expect to be able to estimate $2n+1$ sinusoids at most. If we multiply both sides of (2.134) by

$$e^{-i2\pi \frac{\ell}{T} t_j}$$

and sum over j, we find that

$$\begin{aligned} \sum_{j=-n}^{n} h_j e^{-i2\pi \frac{\ell}{T} t_j} &= \frac{1}{T} \sum_{j=-n}^{n} \sum_{k=-n}^{n} H_k e^{i2\pi \frac{(k-\ell)}{T} t_j} \\ &= \frac{1}{T} \sum_{k=-n}^{n} H_k \sum_{j=-n}^{n} e^{i2\pi \frac{(k-\ell)}{T} t_j}. \end{aligned} \tag{2.135}$$

Now consider the sum

$$\sum_{j=-n}^{n} e^{i2\pi \frac{(k-\ell)}{T} t_j},$$

arising on the right side of (2.135), with $t_j = j\Delta t$. We recognise it as the sum of a finite geometric progression with first term

$$a = e^{-i2\pi\frac{(k-\ell)}{T}n\Delta t} \tag{2.136}$$

and common ratio

$$r = e^{i2\pi\frac{(k-\ell)}{T}\Delta t}. \tag{2.137}$$

Thus, for $T = (2n+1)\Delta t$,

$$\begin{aligned}\sum_{j=-n}^{n} e^{i2\pi\frac{(k-\ell)}{T}t_j} &= \frac{a\left(1 - r^{2n+1}\right)}{1-r} \\ &= \frac{e^{-i2\pi\frac{(k-\ell)}{T}n\Delta t}\left(1 - e^{i2\pi\frac{(k-\ell)}{T}(2n+1)\Delta t}\right)}{1 - e^{i2\pi\frac{(k-\ell)}{T}\Delta t}} \\ &= \frac{e^{-i2\pi\frac{(k-\ell)}{T}n\Delta t}\left(1 - e^{i2\pi(k-\ell)}\right)}{1 - e^{i2\pi\frac{(k-\ell)}{T}\Delta t}}.\end{aligned} \tag{2.138}$$

If $k \neq \ell$, the sum is zero. If $k = \ell$, by l'Hôpital's rule, the sum is $2n+1$. The sinusoids are orthogonal under addition over the sample points, giving

$$\sum_{j=-n}^{n} e^{i2\pi\frac{(k-\ell)}{T}t_j} = (2n+1)\,\delta_\ell^k. \tag{2.139}$$

From (2.135), we then find that

$$\sum_{j=-n}^{n} h_j e^{-i2\pi\frac{\ell}{T}j\Delta t} = \frac{1}{T}\sum_{k=-n}^{n} H_k\,(2n+1)\,\delta_\ell^k = \frac{2n+1}{T}H_\ell. \tag{2.140}$$

Thus, the DFT of the sequence h_j is

$$H_\ell = \Delta t \sum_{j=-n}^{n} h_j e^{-i2\pi\frac{\ell}{T}t_j}, \tag{2.141}$$

for $\ell = -n, -n+1, \ldots, n-1, n$.

As illustrated in Figure 2.3, the frequency domain sequence, H_j, representing the DFT, may be regarded as resulting from a function of frequency being sampled

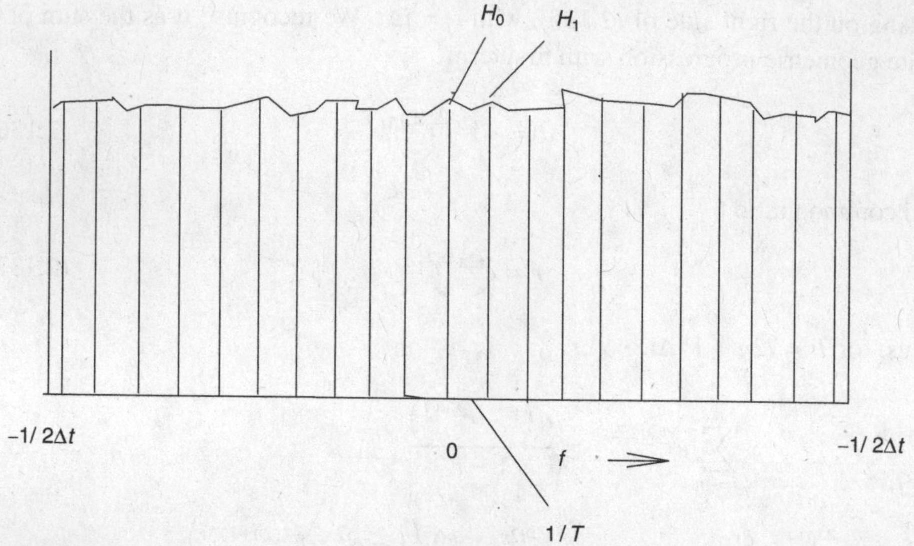

Figure 2.3 Frequency domain representation of the discrete Fourier transform. Frequency samples are spaced at intervals of $1/T$ on the band $-1/2\Delta t$ to $1/2\Delta t$.

at intervals of $\Delta f = 1/T$. Again, the first and last samples are taken at $\Delta f/2$ from the ends of the frequency record. The width of the frequency axis covered is

$$(2n+1)\,\Delta f = \frac{(2n+1)}{T} = \frac{1}{\Delta t}. \tag{2.142}$$

The frequency domain representation of the DFT in Figure 2.3 may be regarded as the counterpart of the time domain representation shown in Figure 2.2.

The question then arises as to how well the representation (2.134), based on the DFT, approximates the original time sequence g_j. The DFT of g_j is

$$G_k = \frac{T}{2n+1} \sum_{j=-n}^{n} g_j e^{-i2\pi \frac{k}{T} t_j}. \tag{2.143}$$

Then, the representation of g_j, based on the DFT, is

$$g_j = \frac{1}{T} \sum_{k=-n}^{n} G_k e^{i2\pi \frac{k}{T} t_j}. \tag{2.144}$$

With back substitution for the DFT from (2.143), the approximation is

$$g_\ell \approx \frac{1}{T} \sum_{k=-n}^{n} \left(\frac{T}{2n+1} \sum_{j=-n}^{n} g_j e^{-i2\pi \frac{k}{T} t_j} \right) e^{i2\pi \frac{k}{T} t_\ell}. \tag{2.145}$$

Thus,

$$g_\ell = \frac{1}{2n+1}\sum_{j=-n}^{n} g_j \sum_{k=-n}^{n} e^{i2\pi\frac{\ell-j}{T}k\Delta t}$$

$$= \frac{1}{2n+1}\sum_{j=-n}^{n} g_j \sum_{k=-n}^{n} e^{i2\pi\frac{(\ell-j)}{T}t_k}$$

$$= \frac{1}{2n+1}\sum_{j=-n}^{n} g_j (2n+1)\,\delta_j^\ell = g_\ell, \tag{2.146}$$

with the use of the orthogonality relation (2.139). The representation is therefore exact, and the DFT pair

$$g_j = \Delta f \sum_{k=-n}^{n} G_k e^{i2\pi f_k t_j} \tag{2.147}$$

and

$$G_k = \Delta t \sum_{j=-n}^{n} g_j e^{-i2\pi f_k t_j}, \tag{2.148}$$

with $f_k = k/T$, provide a one-to-one mapping between the time and frequency domains.

The DFT pair is easily generalised to an arbitrary number of points. For a record of length $T = N\Delta t$, sampled at N equispaced points, the frequency domain representation is given at N points at intervals of $\Delta f = 1/T$ along the positive frequency axis, and

$$\mathsf{g}_j = \Delta f \sum_{k=0}^{N-1} \mathcal{G}_k e^{i\frac{2\pi}{N}kj}, \quad j = 0, 1, \ldots, N-1, \tag{2.149}$$

$$\mathcal{G}_k = \Delta t \sum_{j=0}^{N-1} \mathsf{g}_j e^{-i\frac{2\pi}{N}kj}, \quad k = 0, 1, \ldots, N-1. \tag{2.150}$$

Both of these expressions are periodic in the indices j and k, with period N, as are (2.147) and (2.148), with period $2n+1$. This suggests that we wrap both the time and frequency records into cylinders, as illustrated in Figure 2.4.

For an odd number of points $N = 2n+1$, the generalised DFT (2.150) may be related to the DFT expressed by (2.148). The time sequences (2.147) and (2.149) are related by

$$g_j = \mathsf{g}_{j+n}, \quad j = -n, -n+1, \ldots, n. \tag{2.151}$$

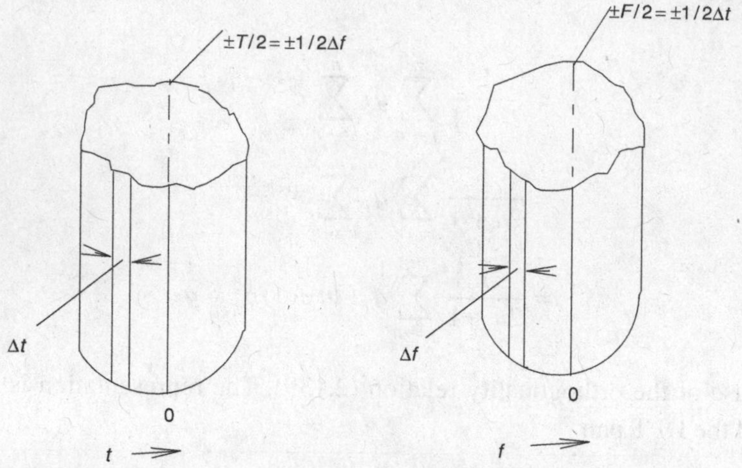

Figure 2.4 The time and frequency records may be imagined to be wrapped in cylindrical form to give the proper periodicity in the N points.

From (2.150),

$$\mathcal{G}_k = \Delta t \sum_{j=0}^{2n} \mathsf{g}_j e^{-i\frac{2\pi}{N}jk} = \Delta t \sum_{j=n+1}^{2n} \mathsf{g}_j e^{-i\frac{2\pi}{N}jk} + \Delta t \sum_{j=0}^{n} \mathsf{g}_j e^{-i\frac{2\pi}{N}jk}. \tag{2.152}$$

Replacing the summation index by $\ell = j - N$, the first term on the right-hand side becomes

$$\Delta t \sum_{\ell=-n}^{-1} \mathsf{g}_{\ell+N} e^{-i\frac{2\pi}{N}k(\ell+N)} = \Delta t \sum_{\ell=-n}^{-1} \mathsf{g}_{\ell+N} e^{-i\frac{2\pi}{N}k\ell}, \tag{2.153}$$

using the period, N, in the indices of the exponential. Using the relation (2.151), we then have that

$$\begin{aligned}\mathcal{G}_k &= \Delta t \sum_{\ell=-n}^{-1} g_{\ell+n+1} e^{-i\frac{2\pi}{N}k\ell} + \Delta t \sum_{j=0}^{n} g_{j-n} e^{-i\frac{2\pi}{N}jk} \\ &= \Delta t \sum_{\ell=-n}^{-1} g_{\ell-n} e^{-i\frac{2\pi}{N}k\ell} + \Delta t \sum_{j=0}^{n} g_{j-n} e^{-i\frac{2\pi}{N}jk},\end{aligned} \tag{2.154}$$

on using the periodicity of g_j expressed by (2.147). On replacing the summation index ℓ by j, we find that

$$\mathcal{G}_k = \Delta t \sum_{j=-n}^{n} g_{j-n} e^{-i\frac{2\pi}{N}jk}. \tag{2.155}$$

This is just the DFT expressed by (2.148) of the time sequence shifted back by n units. Once again, changing the summation index to $\ell = j - n$, we have that

$$\mathcal{G}_k = e^{-i\frac{2\pi}{N}kn}\Delta t \sum_{\ell=-2n}^{0} g_\ell e^{-i\frac{2\pi}{N}k\ell}. \tag{2.156}$$

Because of the periodicity of the input time sequence, the summation range need only cover one complete cycle. This leads to the *shifting theorem* for DFTs,

$$\mathcal{G}_k = e^{-i\frac{2\pi}{N}kn} G_k, \quad k = 0, 1, \ldots, n, \tag{2.157}$$

the DFT of a time sequence shifted back by n units is found by multiplying the DFT of the unshifted time sequence by the factor

$$e^{-i\frac{2\pi}{N}kn}. \tag{2.158}$$

The DFT, G_k, can be found for negative indices using its periodicity, as

$$\mathcal{G}_k = e^{-i\frac{2\pi}{N}kn} G_{-N+k}, \quad k = n+1, n+2, \ldots, 2n. \tag{2.159}$$

Thus, the DFT expressed by (2.148) can be found from the generalised DFT expressed by (2.150) as

$$G_k = e^{i\frac{2\pi}{N}kn}\mathcal{G}_k, \quad k = 0, 1, \ldots, n,$$
$$G_k = e^{i\frac{2\pi}{N}kn}\mathcal{G}_{k+N}, \quad k = -n, -n+1, \ldots, -1. \tag{2.160}$$

2.3.2 The DFT and the z-transform

The discrete Fourier transform (DFT) can be directly related to the z-transform. From the definition of the z-transform in Section 2.1.2, the z-transform of the sequence, $\mathsf{g}_0, \mathsf{g}_1, \ldots, \mathsf{g}_{N-2}, \mathsf{g}_{N-1}$, is

$$\mathcal{G}(z) = \sum_{j=0}^{N-1} \mathsf{g}_j z^j. \tag{2.161}$$

Its DFT, given by (2.150), is

$$\mathcal{G}_k = \Delta t \sum_{j=0}^{N-1} \mathsf{g}_j e^{-i\frac{2\pi}{N}kj} = \Delta t \sum_{j=0}^{N-1} \mathsf{g}_j z^j = \Delta t\, \mathcal{G}(z), \tag{2.162}$$

with $z = e^{-i\frac{2\pi}{N}k}$. If we have a second sequence, $\mathsf{h}_0, \mathsf{h}_1, \ldots, \mathsf{h}_{M-2}, \mathsf{h}_{M-1}$, of length M, its convolution with g_j is

$$\mathsf{e}_j = \sum_{k=0}^{N-1} \mathsf{g}_k \mathsf{h}_{j-k}, \tag{2.163}$$

a sequence of length $L = M + N - 1$. Since the z-transform of the convolution of two sequences is the product of their z-transforms (see equation (2.14)), the z-transform of the sequence e_j is

$$\mathcal{E}(z) = \mathcal{G}(z) \cdot \mathcal{H}(z) = \sum_{j=0}^{M+N-2} \mathsf{e}_j z^j. \tag{2.164}$$

The DFT of the sequence e_j is then

$$\mathcal{E}_k = \Delta t\, \mathcal{E}(z) = \Delta t\, \mathcal{G}(z) \cdot \mathcal{H}(z) = \frac{1}{\Delta t} \mathcal{G}_k \cdot \mathcal{H}_k, \tag{2.165}$$

with $z = e^{-i\frac{2\pi}{L}k}$. The sequence $\Delta t \cdot \mathsf{e}_j$ can then be recovered from the inverse DFT,

$$\Delta t \cdot \mathsf{e}_j = \Delta f \sum_{k=0}^{M+N-2} \mathcal{G}_k \cdot \mathcal{H}_k e^{i\frac{2\pi}{L}kj}. \tag{2.166}$$

Thus, convolution can be accomplished by taking the inverse of the product of the two DFTs.

2.3.3 *The fast Fourier transform*

Direct calculation of the DFT (2.150) and its inverse (2.149) requires N^2 complex multiplications in each case. A much faster method of doing these calculations was originally described by Gauss as far back as 1805 (Heideman *et al.*, 1995), although the algorithm remained relatively unknown. Claerbout (1985, p. 12) cites the first machine computation he knows of as being due to Vern Herbert in 1962 at Chevron Standard in Calgary, Canada. It was rediscovered in 1965 and widely publicized by Cooley and Tukey (1965) and has become known as the Cooley–Tukey algorithm. The explosive growth of fast methods of computing DFTs since 1965 is indicated by the 3400 entries in a recent bibliography of efficient algorithms (Sorenson *et al.*, 1995).

The form (2.166) suggests that we absorb Δt into the time sequence, transforming the DFT (2.150) to the sum

$$\mathcal{G}_k = \sum_{j=0}^{N-1} x_j e^{-i\frac{2\pi}{N}kj}. \tag{2.167}$$

The inverse transform (2.149) then becomes

$$\Delta t \cdot \mathsf{g}_j = x_j = \frac{1}{N} \sum_{k=0}^{N-1} \mathcal{G}_k e^{i\frac{2\pi}{N}kj}. \tag{2.168}$$

In its most elementary form, the fast Fourier transform (FFT) algorithm takes the data length to be an integral power of two, so that $N = 2^m$, where m is an integer. The DFT (2.167) may be expressed as

$$\mathcal{G}_k = \sum_{j=0}^{N-1} x_j W^{kj}, \tag{2.169}$$

with $W = e^{-i\frac{2\pi}{N}}$. The sequence x_j is then divided into two new sequences $y_\ell = x_{2\ell}$ and $z_\ell = x_{2\ell+1}$, with $\ell = 0, 1, \ldots, N/2 - 1$. Each of the new sequences, $y = (x_0, x_2, \ldots, x_{N-2})$ and $z = (x_1, x_3, \ldots, x_{N-1})$, is half as long and may be regarded as having double the sample interval of the original sequence, with DFTs

$$B_k = \sum_{\ell=0}^{N/2-1} y_\ell W^{2k\ell} \quad \text{and} \quad C_k = \sum_{\ell=0}^{N/2-1} z_\ell W^{2k\ell}, \quad k = 0, 1, \ldots, N/2 - 1. \tag{2.170}$$

Expression (2.169) may then be rewritten as

$$\mathcal{G}_k = \sum_{\ell=0}^{N/2-1} \left(y_\ell W^{2k\ell} + z_\ell W^{k(2\ell+1)} \right) \tag{2.171}$$

or

$$\mathcal{G}_k = B_k + W^k C_k, \quad k = 0, 1, \ldots, N/2 - 1. \tag{2.172}$$

This provides the first half of the desired DFT. To obtain the second half, we substitute $k + N/2$ for k in (2.170) to get

$$B_{k+N/2} = \sum_{\ell=0}^{N/2-1} y_\ell W^{2(k+N/2)\ell} = \sum_{\ell=0}^{N/2-1} y_\ell W^{2k\ell} = B_k, \tag{2.173}$$

$$C_{k+N/2} = \sum_{\ell=0}^{N/2-1} z_\ell W^{2(k+N/2)\ell} = \sum_{\ell=0}^{N/2-1} z_\ell W^{2k\ell} = C_k. \tag{2.174}$$

Thus, again substituting $k + N/2$ for k in (2.172), we find that

$$\mathcal{G}_{k+N/2} = B_{k+N/2} + W^{k+N/2} C_{k+N/2} = B_k - W^k C_k, \quad k = 0, 1, \ldots, N/2 - 1, \tag{2.175}$$

giving the second half of the DFT.

The DFT of the full sequence x_j is then found from the DFTs of the two half-length sequences x_j and y_j. The decimation to half-length sequences can be continued until we have only two-length sequences to transform. For $N = 2^m$, $m = \log_2 N$, m successive DFTs are required, each involving $2m = 2\log_2 N$ complex multiplications for each frequency index of which there are N, for a total of $2N \log_2 N$. Direct calculation takes N^2 complex multiplications for a relative speed ratio of $N/2m$. Even for moderately long sequences with $N = 1024$, direct calculation is $1024/20 = 51.2$ times slower.

The successive decimations can be illustrated by the case $N = 16$, beginning with the full input sequence,

$$(x_0, x_1, x_2, x_3, x_4, x_5, x_6, x_7, x_8, x_9, x_{10}, x_{11}, x_{12}, x_{13}, x_{14}, x_{15}),$$

$$\text{then } (x_0, x_2, x_4, x_6, x_8, x_{10}, x_{12}, x_{14})(x_1, x_3, x_5, x_7, x_9, x_{11}, x_{13}, x_{15}),$$

$$\text{then } (x_0, x_4, x_8, x_{12})(x_2, x_6, x_{10}, x_{14})(x_1, x_5, x_9, x_{13})(x_3, x_7, x_{11}, x_{15}),$$

$$\text{then } (x_0, x_8)(x_4, x_{12})(x_2, x_{10})(x_6, x_{14})(x_1, x_9)(x_5, x_{13})(x_3, x_{11})(x_7, x_{15}).$$

The final decimation yields a series of two-length sequences. Their DFTs are easily calculated as sums and differences,

$$(x_0 + x_8,\ x_0 - x_8)(x_4 + x_{12},\ x_4 - x_{12}) \cdots (x_7 + x_{15},\ x_7 - x_{15}),$$

and provide the basis for calculating the series of four-length DFTs through equations (2.172) and (2.175). The doubling process can be continued until the full DFT of the original sequence is found.

The inverse DFT (2.168) can easily be found using the same procedure by simply reversing the sign of the exponent of e and dividing the input DFT sequence by N.

The subroutine FFT2N finds the DFT or the inverse DFT for sequences of length an integral power of two.

```
      SUBROUTINE FFT2N(LG,G,IS)
C
C FFT2N finds the discrete Fourier transform of the sequence G of length
C equal to an integral power, N, of 2, for input IS=-1. For input IS=1,
C it finds the inverse of the DFT sequence G of length LG, again an integral
C power, N, of 2. The input sequences are first decimated into basic pairs,
C which are then successively transformed recursively by doubling to
C the final resulting sequence. On output, the resulting sequence replaces
C the input sequence.
C
      IMPLICIT DOUBLE COMPLEX(A-H,O-Z)
      DOUBLE PRECISION PI,ARG,ARGM,FACT
      DIMENSION G(LG)
C Define pi.
      PI=3.14159265358979D0
C Define common factor in argument of W.
      ARG=PI*DFLOAT(IS)
C Determine if DFT or inverse DFT.
      FACT=1.D0
      IF(IS.EQ.1) FACT=FACT/DFLOAT(LG)
C Begin sequence decimation and scaling.
      J=1
      DO 10 I=1,LG
        IF(I.GT.J) GO TO 20
        HOLD=G(J)*FACT
        G(J)=G(I)*FACT
        G(I)=HOLD
20      M=LG/2
30      IF(J.LE.M) GO TO 40
        J=J-M
        M=M/2
        IF(M.GE.1) GO TO 30
```

```
40      J=J+M
10    CONTINUE
C If input sequence is of unit length, exit.
      IF(LG.EQ.1) GO TO 70
C Start calculation of DFTs by doubling.
C Set initial DFT length as power of 2.
      L=1
50    LT2=L*2
      DO 60 M=1,L
        MM1=M-1
C Find argument of W raised to required power.
        ARGM=DFLOAT(MM1)*ARG/DFLOAT(L)
C Find W to required power.
        W=DCMPLX(DCOS(ARGM),DSIN(ARGM))
C Find double length transforms.
        DO 60 K=M,LG,LT2
          HOLD=W*G(K+L)
          G(K+L)=G(K)-HOLD
          G(K)=G(K)+HOLD
60    CONTINUE
C Update DFT length.
      L=LT2
C Test if transform is complete.
      IF(L.LT.LG) GO TO 50
70    CONTINUE
      RETURN
      END
```

2.3.4 Generalisation of the FFT

The fast Fourier transform presented in the previous section applies only to sequences with lengths equal to an integral power of two. The sequences could be padded with zeros to the required length, but the resultant change in record length would alter the sample interval in frequency for the DFT, and in time for the inverse DFT. Instead, in this section, we present a form of the FFT valid for sequences of arbitrary length.

Our approach is based on the method of Bluestein (1970), which transforms the DFT into a convolution. Beginning with expression (2.169) for the DFT, we have the identities,

$$\mathcal{G}_k = \sum_{j=0}^{N-1} x_j W^{kj} = \sum_{j=0}^{N-1} x_j W^{kj+(j^2-j^2+k^2-k^2)/2}$$

$$= W^{k^2/2} \sum_{j=0}^{N-1} x_j W^{j^2/2} \cdot W^{-(k-j)^2/2}. \tag{2.176}$$

The last sum on the right-hand side may be interpreted as the convolution of the N-length sequence

$$y_j = W^{j^2/2} x_j, \quad j = 0, 1, \ldots, N-1,$$

with the $(2N-1)$-length sequence

$$z_j = W^{-j^2/2}, \quad j = -N+1, \ldots, 0, \ldots, N-1,$$

yielding the $(3N-2)$-length sequence

$$h_k, \quad k = -N+1, \ldots, 0, \ldots, 2N-2.$$

The DFT is then expressible as

$$\mathcal{G}_k = W^{k^2/2} \sum_{j=0}^{N-1} y_j z_{k-j} = W^{k^2/2} h_k, \quad k = 0, 1, \ldots, N-1. \tag{2.177}$$

Thus, only the terms $h_0, h_1, \ldots, h_{N-1}$ of the convolved sequence are required in the calculation of the transform.

The sequences y_j and z_j are then zero padded to length an integral power of two, equal to or larger than $3N-2$. The convolution is found, as outlined in Section 2.3.2, by taking the inverse of the product of the DFTs of the padded sequences y_j and z_j, using calls to the subroutine FFT2N.

The subroutine FFTN finds the DFT or its inverse for sequences of arbitrary length N using the Bluestein convolution formulation; N can be any positive integer. Again, the same subroutine can be used to find the inverse DFT by reversing the sign of the exponent of e and dividing the input DFT sequence by N.

```
      SUBROUTINE FFTN(LG,G,IS,L,WK1,WK2)
C
C FFTN finds the discrete Fourier transform of the sequence G of length LG
C for input IS=-1. For input IS=1, it finds the inverse of the DFT
C sequence G of length LG. LG can be any positive integer. The subroutine
C uses the method of Bluestein in which the transform is converted to
C a convolution. The sequences entering the convolution are padded with
C zeros to length L, an integral power of two, equal to or larger than 3LG-2.
C The convolution is then performed by finding the DFTs of each sequence
C using the subroutine FFT2N, multiplying them, and then finding the inverse
C of the resulting DFT. WK1 and WK2 are work area vectors, defined in
C the main programme, of length L or larger. On output, the transformed
C sequence replaces the input sequence.
C
      IMPLICIT DOUBLE COMPLEX(A-H,O-Z)
      DOUBLE PRECISION PI,ARG,ARGJ2,FACT,DC,DS
      DIMENSION G(LG),WK1(L),WK2(L)
C Define pi.
      PI=3.14159265358979D0
C Define common factor in argument of W.
      ARG=PI*DFLOAT(IS)/DFLOAT(LG)
C Determine if DFT or inverse DFT.
      FACT=1.D0
      IF(IS.EQ.1) FACT=FACT/DFLOAT(LG)
C Find N, the power of two equal to, or next larger than, 3*LG-2,
C the length of the convolution.
      LGT3M2=3*LG-2
      N=0
      I=1
```

```
10   IF(I.GE.LGT3M2) GO TO 20
     I=I*2
     N=N+1
     GO TO 10
20   CONTINUE
     L=2**N
C Zero fill padded sequences being convolved.
     DO 30 I=1,L
       WK1(I)=(0.D0,0.D0)
       WK2(I)=(0.D0,0.D0)
30   CONTINUE
C Find sequences to be convolved.
     DO 40 J=1,LG
       JM1=J-1
       JM1S=JM1*JM1
       ARGJ2=ARG*DFLOAT(JM1S)
       DC=DCOS(ARGJ2)
       DS=DSIN(ARGJ2)
       WK1(J)=FACT*G(J)*DCMPLX(DC,DS)
C Store weights for final DFT calculation.
       G(J)=DCMPLX(DC,DS)
       JPLGM1=J+LG-1
       WK2(JPLGM1)=DCMPLX(DC,-DS)
       IF(J.EQ.1) GO TO 40
       LGMJP1=LG-J+1
       WK2(LGMJP1)=WK2(JPLGM1)
40   CONTINUE
C Find FFTs of sequences entering convolution.
     IS=-1
     CALL FFT2N(L,WK1,IS)
     CALL FFT2N(L,WK2,IS)
C Multiply FFTs to find FFT of convolution.
     DO 50 I=1,L
       WK1(I)=WK1(I)*WK2(I)
50   CONTINUE
C Find convolution by inverse DFT.
     IS=1
     CALL FFT2N(L,WK1,IS)
C Weight convolution to obtain DFT.
     DO 60 I=1,LG
       IPLGM1=I+LG-1
       G(I)=G(I)*WK1(IPLGM1)
60   CONTINUE
     RETURN
     END
```

2.3.5 The DFT for unequally spaced samples

The description of the discrete Fourier transform (DFT), its properties, and methods of calculating it, have thus far assumed that the samples in the time domain are equally spaced. Often, this is not the case. Observations may characteristically be unequally spaced in, for example, a wide range of astronomical and space measurements, which can only be made at certain times, or when sources are visible. Another common type of unequally spaced sequence arises when there are gaps in otherwise equally spaced samples. This may occur when glitches in the signal are removed, or when the digitiser, or instrument, is subject to failure.

In the case of unequally spaced samples, the representation (2.134) of the time sequence g_j may be used if it is understood that the sample points t_j are not at uniform intervals. If the sample points are at equally spaced intervals, we shall see that a least squares adjustment of this representation gives the DFT (2.148).

This suggests the following approach to the analysis of unequally spaced data. If we have a record segment of length M, with $2L + 1$ samples, equally or unequally spaced, we try to represent it in the frequency domain for $N \leq L$ by $2N + 1$ sinusoids, equally spaced in frequency, as

$$g'_j = \frac{1}{M} \sum_{k=-N}^{N} G_k e^{i2\pi(k/M)t_j}. \tag{2.178}$$

The $2N + 1$ values, G_k, of the DFT, once calculated, give a frequency domain representation that is equally spaced in frequency with sample interval $\Delta f = 1/M$. The FFT algorithm can then be used to give an equivalent, equally spaced time sequence.

To find the DFT sequence G_k, we first construct the error sequence ϵ_j as the difference of the sequence g_j and its representation g'_j expressed by the sum of sinusoids (2.178),

$$\epsilon_j = g_j - g'_j. \tag{2.179}$$

If each member of the sequence g_j has an associated standard deviation σ_j, the G_k are chosen to minimise the objective function

$$I = \sum_{j=-L}^{L} \frac{\epsilon_j \epsilon_j^*}{\sigma_j^2}, \tag{2.180}$$

the superscript asterisk again indicating complex conjugation. Substitution for g'_j from (2.178) in expression (2.180) for the objective function gives

$$I = \sum_{j=-L}^{L} \frac{1}{\sigma_j^2} \Bigg\{ g_j g_j^* - \frac{g_j}{M} \sum_{l=-N}^{N} G_l^* e^{-i2\pi(l/M)t_j} - \frac{g_j^*}{M} \sum_{k=-N}^{N} G_k e^{i2\pi(k/M)t_j} + \frac{1}{M^2} \sum_{k=-N}^{N} \sum_{l=-N}^{N} G_k G_l^* e^{i2\pi((k-l)/M)t_j} \Bigg\}. \tag{2.181}$$

The conditions for a minimum of I are that its partial derivatives with respect to the real and imaginary parts of G_m vanish for $m = -N, \ldots, N$. On differentiation of (2.181), we find that

$$\frac{\partial I}{\partial \mathrm{Re} G_m} = \sum_{j=-L}^{L} \frac{1}{\sigma_j^2} \left\{ -\frac{g_j}{M} e^{-i2\pi(m/M)t_j} - \frac{g_j^*}{M} e^{i2\pi(m/M)t_j} + \frac{1}{M^2} \sum_{l=-N}^{N} G_l^* e^{i2\pi((m-l)/M)t_j} + \frac{1}{M^2} \sum_{k=-N}^{N} G_k e^{i2\pi((k-m)/M)t_j} \right\} = 0, \tag{2.182}$$

and

$$i\frac{\partial I}{\partial \mathrm{Im} G_m} = \sum_{j=-L}^{L} \frac{1}{\sigma_j^2} \left\{ -\frac{g_j}{M} e^{-i2\pi(m/M)t_j} + \frac{g_j^*}{M} e^{i2\pi(m/M)t_j} - \frac{1}{M^2} \sum_{l=-N}^{N} G_l^* e^{i2\pi((m-l)/M)t_j} + \frac{1}{M^2} \sum_{k=-N}^{N} G_k e^{i2\pi((k-m)/M)t_j} \right\} = 0. \tag{2.183}$$

On addition of (2.182) and (2.183), we find that

$$\sum_{k=-N}^{N} G_k \sum_{j=-L}^{L} \frac{1}{\sigma_j^2} e^{i2\pi((k-m)/M)t_j} = M \sum_{j=-L}^{L} \frac{g_j}{\sigma_j^2} e^{-i2\pi(m/M)t_j}, \quad m = -N, \ldots, N. \tag{2.184}$$

With

$$C_m = \sum_{j=-L}^{L} \frac{1}{\sigma_j^2} e^{-i2\pi(m/M)t_j}, \qquad d_m = M \sum_{j=-L}^{L} \frac{g_j}{\sigma_j^2} e^{-i2\pi(m/M)t_j}, \tag{2.185}$$

the conditional equations (2.184) become

$$\sum_{k=-N}^{N} G_k C_{m-k} = d_m, \quad m = -N, \ldots, N. \tag{2.186}$$

These conditional equations may be written in array form as (Rochester *et al.*, 1974, Appendix B),

$$\begin{pmatrix} C_0 & C_{-1} & \cdots & C_{-2N} \\ C_1 & C_0 & \cdots & C_{-2N+1} \\ \vdots & \vdots & \ddots & \vdots \\ C_{2N} & C_{2N-1} & \cdots & C_0 \end{pmatrix} \begin{pmatrix} G_{-N} \\ G_{-N+1} \\ \vdots \\ G_N \end{pmatrix} = \begin{pmatrix} d_{-N} \\ d_{-N+1} \\ \vdots \\ d_N \end{pmatrix}. \tag{2.187}$$

First summing over j in expression (2.181), and using the definitions (2.185), the objective function I becomes

$$I = \sum_{j=-L}^{L} \frac{g_j g_j^*}{\sigma_j^2} - \frac{1}{M^2}\sum_{l=-N}^{N} G_l^* d_l - \frac{1}{M^2}\sum_{k=-N}^{N} G_k d_k^* + \frac{1}{M^2}\sum_{l=-N}^{N} G_l^* \sum_{k=-N}^{N} G_k C_{l-k}. \tag{2.188}$$

The objective function will take on its minimum value, $I_{\min}$, when the conditional equations (2.186) hold, and thus,

$$\begin{aligned} I_{\min} &= \sum_{j=-L}^{L} \frac{g_j g_j^*}{\sigma_j^2} - \frac{1}{M^2}\sum_{l=-N}^{N} G_l^* d_l - \frac{1}{M^2}\sum_{k=-N}^{N} G_k d_k^* + \frac{1}{M^2}\sum_{l=-N}^{N} G_l^* d_l \\ &= \sum_{j=-L}^{L} \frac{g_j g_j^*}{\sigma_j^2} - \frac{1}{M^2}\sum_{k=-N}^{N} G_k d_k^* \\ &= \sum_{j=-L}^{L} \frac{g_j g_j^*}{\sigma_j^2} - \sum_{j=-L}^{L} \frac{g_j^*}{\sigma_j^2}\frac{1}{M}\sum_{k=-N}^{N} G_k e^{i2\pi(k/M)t_j}. \end{aligned} \tag{2.189}$$

Finally, with substitution from the representation (2.178), we have that

$$I_{\min} = \sum_{j=-L}^{L} \frac{g_j g_j^*}{\sigma_j^2} - \sum_{j=-L}^{L} \frac{g'_j g_j^*}{\sigma_j^2}. \tag{2.190}$$

If the representation (2.178) is an exact fit to the time sequence, then this minimum value of the objective function will be zero.

In the case when each observation is taken to have equal error, or when no error estimate is available, the objective function (2.180) becomes simply the error energy

$$I = \sum_{j=-L}^{L} \epsilon_j \epsilon_j^*. \tag{2.191}$$

Expressions (2.185) then reduce to

$$C_m = \sum_{j=-L}^{L} e^{-i2\pi(m/M)t_j}, \qquad d_m = M\sum_{j=-L}^{L} g_j e^{-i2\pi(m/M)t_j}. \tag{2.192}$$

In the equally spaced case, t_j can be replaced by $j\Delta t$, and by the orthogonality relation (2.139),

$$C_m = (2L+1)\,\delta_m^0 \tag{2.193}$$

and, as forecast, the coefficient matrix of the conditional equations (2.187) becomes diagonal, and for $N = L$ their solution is easily found to be

$$G_m = \frac{M}{2L+1} \sum_{j=-L}^{L} g_j e^{-i2\pi(m/M)t_j} = \Delta t \sum_{j=-L}^{L} g_j e^{-i2\pi(m/M)t_j}, \quad m = -L, \ldots, L, \tag{2.194}$$

just the DFT for equally spaced samples (2.148).

The mean square of the time domain representation (2.178) can be related to the mean square in the frequency domain as represented by the DFT. Multiplying through by $g_j'^*$ and summing over the sample points, we get

$$\sum_{j=-L}^{L} g_j' g_j'^* = \frac{1}{M^2} \sum_{k=-N}^{N} \sum_{m=-N}^{N} G_k G_m^* \sum_{j=-L}^{L} e^{i2\pi((k-m)/M)j\Delta t}, \tag{2.195}$$

for equally spaced samples. On using the orthogonality relation (2.139), we have

$$\Delta t \sum_{j=-L}^{L} g_j' g_j'^* = \Delta f \sum_{k=-N}^{N} G_k G_k^*, \tag{2.196}$$

for sample interval in time $\Delta t = M/(2L+1)$ and sample interval in frequency $\Delta f = 1/M$. This is the form *Parseval's relation* takes for discrete, equally spaced samples. On multiplying through by Δf, the left side becomes the power in the time sequence,

$$\frac{1}{2L+1} \sum_{j=-L}^{L} g_j' g_j'^* = E\left\{g_j' g_j'^*\right\} = (\Delta f)^2 \sum_{k=-N}^{N} G_k G_k^*, \tag{2.197}$$

with $E\{g_j' g_j'^*\}$ representing the squared magnitude of a particular realisation of the time sequence. For stationary sequences, this expected value is independent of whether the samples are equally spaced or unequally spaced. Under the assumption of stationarity, Parseval's relation holds in either case.

2.3.6 Singular value decomposition

The conditional equations (2.187) for the DFT have a coefficient matrix that is equidiagonal and Hermitian and is thus *Toeplitz*. This suggests that the Levinson algorithm described in Section 2.2.3 might be used in their solution. However, experience shows that for long time sequences, the coefficient matrix is ill-conditioned, and the Levinson algorithm is inadequate for the solution of the conditional equations, even when implemented in double quadruple (octuple) precision (Hida *et al.*, 2000). Instead, we use the *singular value decomposition (SVD)* method. Writing the system (2.187) in matrix notation as

$$\mathbf{C} \cdot G = d, \tag{2.198}$$

the SVD method factors the complex coefficient matrix $\mathbf{C}$ into the triple product

$$\mathbf{C} = \mathbf{U} \cdot \mathbf{W} \cdot \mathbf{V}^{\mathrm{H}}, \tag{2.199}$$

where $\mathbf{U}$ and $\mathbf{V}$ are unitary matrices and $\mathbf{W}$ is a real, diagonal matrix with the positive singular values of $\mathbf{C}$ in descending order down the diagonal. The superscript H indicates the complex conjugate of the transpose, the Hermitian transpose or *adjoint.* A matrix whose Hermitian transpose is equal to the original matrix is itself said to be *Hermitian.* A matrix which multiplied by its Hermitian transpose gives the unit matrix is said to be a *unitary* matrix, the complex analogue of a real orthogonal matrix. Similarly, a *unitary transformation* is the complex analogue of a similarity transformation with real orthogonal matrices, as both preserve the eigenvalues of a matrix. The inverse of a unitary matrix is equal to its Hermitian transpose. The singular value decomposition of a complex rectangular matrix, as well as its properties and methods of calculation, are outlined by Golub and Kahan (1965).

Once the SVD of $\mathbf{C}$, (2.199), has been carried out, the solution vector G of the conditional equations (2.198), giving the DFT, is easily found using the properties of unitary matrices as

$$G = \mathbf{V} \cdot \mathbf{W}^{-1} \cdot \mathbf{U}^{\mathrm{H}} \cdot d. \tag{2.200}$$

Many of the properties of the coefficient matrix $\mathbf{C}$ can be determined from its SVD (Press *et al.* (1992, pp. 53–56)). Its *condition number* is defined as the ratio of the largest singular value to the smallest. If one or more of the singular values vanish, no solution exists. The solution (2.200) will be numerically unstable if the reciprocal of the condition number approaches machine precision. This suggests setting the inverses of the smallest singular values to zero in $\mathbf{W}^{-1}$, giving $\mathbf{W}'^{-1}$, to improve numerical stability, leading to an approximate solution of (2.200) for the DFT,

$$G' = \mathbf{V} \cdot \mathbf{W}'^{-1} \cdot \mathbf{U}^{\mathrm{H}} \cdot d. \tag{2.201}$$

However, the number of singular values to be eliminated remains unspecified.

A basic advance in the application of the SVD technique to the calculation of the DFT was made by Palmer (2005). Any acceptable DFT must obey Parseval's relation (2.197), irrespective of whether the time samples are equally spaced or unequally spaced. Typically, the mean square in the frequency domain, as represented by the DFT, is many orders of magnitude larger than the mean square in the time domain before any singular values are eliminated. As successive singular values are eliminated, starting with the smallest and working upward, the mean square in the frequency domain appears to decrease monotonically until Parseval's relation (2.197) is closely obeyed. Thus, Palmer's procedure is to define the ratio R of the right-hand side to the left-hand side of (2.197),

$$R = \frac{(\Delta f)^2 \sum_{k=-N}^{N} G_k G_k^*}{E\left\{g'_j g'^*_j\right\}}, \tag{2.202}$$

and then to iterate on the number of singular values eliminated (NSVE) until the ratio R is closest to unity.

The first step in the SVD of a Hermitian matrix, as described by Golub and Kahan (1965), is the reduction of the matrix to upper bidiagonal form by a series of Householder transformations. A detailed description of such annihilating transformations as reflections is provided by Maron and Lopez (1991).

The basis of a Householder transformation is the matrix

$$\mathbf{P} = \mathbf{I} - 2\boldsymbol{u} \otimes \boldsymbol{u}^H, \tag{2.203}$$

where $\mathbf{I}$ is the unit matrix and $\boldsymbol{u} \otimes \boldsymbol{u}^H$ is the outer product of the vectors $\boldsymbol{u}$ and $\boldsymbol{u}^H$,

$$\boldsymbol{u} \otimes \boldsymbol{u}^H = \begin{pmatrix} u_1 u_1^* & u_1 u_2^* & \cdots & u_1 u_M^* \\ u_2 u_1^* & u_2 u_2^* & \cdots & u_2 u_M^* \\ \vdots & \vdots & \ddots & \vdots \\ u_M u_1^* & u_M u_2^* & \cdots & u_M u_M^* \end{pmatrix}, \tag{2.204}$$

and the inner product of $\boldsymbol{u}^H$ and $\boldsymbol{u}$ is $\boldsymbol{u}^H \cdot \boldsymbol{u} = |\boldsymbol{u}|^2 = 1$, where $\boldsymbol{u}$ is a yet to be specified unit vector. Thus, the outer product generates a matrix that is itself Hermitian. Then,

$$\begin{aligned} \mathbf{P}\mathbf{P}^{\mathrm{H}} &= \left(\mathbf{I} - 2\boldsymbol{u} \otimes \boldsymbol{u}^H\right) \cdot \left(\mathbf{I} - 2\boldsymbol{u} \otimes \boldsymbol{u}^H\right) \\ &= \mathbf{I} - 4\boldsymbol{u} \otimes \boldsymbol{u}^H + 4\boldsymbol{u} \otimes \boldsymbol{u}^H \cdot \boldsymbol{u} \otimes \boldsymbol{u}^H \\ &= \mathbf{I} - 4\boldsymbol{u} \otimes \boldsymbol{u}^H + 4\boldsymbol{u} \otimes \boldsymbol{u}^H \\ &= \mathbf{I}. \end{aligned} \tag{2.205}$$

The transformation matrix $\mathbf{P}$ is a unitary matrix for an arbitrary unit vector $\boldsymbol{u}$. Now, suppose the vector $\boldsymbol{u}$ is proportional to the difference of two vectors, a vector $\boldsymbol{v}$ with M components and a vector $\boldsymbol{w}$, whose first $n-1$ components are identical to those of $\boldsymbol{v}$, and whose nth component is $-\eta$, with

$$\eta = \sqrt{\sum_{i=n}^{M} v_i v_i^*} \tag{2.206}$$

and complex phase the same as that of the nth component of $\boldsymbol{v}$, and with remaining components having null values. In array form,

$$\boldsymbol{v} - \boldsymbol{w} = \begin{pmatrix} v_1 \\ v_2 \\ \vdots \\ v_{n-1} \\ v_n \\ v_{n+1} \\ \vdots \\ v_M \end{pmatrix} - \begin{pmatrix} v_1 \\ v_2 \\ \vdots \\ v_{n-1} \\ -\eta \exp\left(i \arg(v_n)\right) \\ 0 \\ \vdots \\ 0 \end{pmatrix}. \tag{2.207}$$

The choice of phase in the nth component of $\boldsymbol{w}$ the same as that of v_n avoids the possibility of subtracting nearly equal complex quantities in forming $\boldsymbol{v} - \boldsymbol{w}$. Since $\boldsymbol{u}$ is a unit vector, we write

$$\boldsymbol{u} = \frac{\boldsymbol{v} - \boldsymbol{w}}{|\boldsymbol{v} - \boldsymbol{w}|}. \tag{2.208}$$

Then,

$$\begin{aligned} |\boldsymbol{v} - \boldsymbol{w}|^2 &= (v_n - w_n)(v_n - w_n)^* + \sum_{i=n+1}^{M} v_i v_i^* \\ &= \left(v_n + \eta \exp\left(i \arg(v_n)\right)\right)\left(v_n^* + \eta \exp\left(-i \arg(v_n)\right)\right) + \left(\eta^2 - |v_n|^2\right) \\ &= |v_n|^2 + \eta^2 + 2\eta\, |v_n| + \left(\eta^2 - |v_n|^2\right) \\ &= 2\eta(\eta + |v_n|), \end{aligned} \tag{2.209}$$

and

$$\boldsymbol{u} = \frac{1}{\sqrt{2\eta\,(\eta + |v_n|)}} \begin{pmatrix} 0 \\ \vdots \\ 0 \\ v_n + \eta \exp\left(i \arg(v_n)\right) \\ v_{n+1} \\ \vdots \\ v_M \end{pmatrix}. \tag{2.210}$$

If we then perform the unitary transformation $\mathbf{P}$ on the vector $\boldsymbol{v}$, we find that

$$\begin{aligned}
\mathbf{P} \cdot \boldsymbol{v} &= \boldsymbol{v} - 2\boldsymbol{u} \otimes \boldsymbol{u}^H \cdot \boldsymbol{v} \\
&= \boldsymbol{v} - 2\boldsymbol{u} \frac{\left(\eta^2 + \eta v_n \exp\left(-i \arg\left(v_n\right)\right)\right)}{\sqrt{2\eta\left(\eta + |v_n|\right)}} \\
&= \boldsymbol{v} - \frac{2\eta\left(\eta + |v_n|\right)}{\sqrt{2\eta\left(\eta + |v_n|\right)}} \boldsymbol{u} \\
&= \boldsymbol{v} - (\boldsymbol{v} - \boldsymbol{w}) = \boldsymbol{w}.
\end{aligned} \tag{2.211}$$

Taking the vector $\boldsymbol{v}$ to be a column vector of an $M \times M$ matrix, elements below v_n can be brought to zero by a unitary Householder transformation. Similarly, if $\boldsymbol{v}^T$ is a row vector, the superscript T indicating the transpose of the corresponding column vector, then on taking the transpose of the product $\mathbf{P} \cdot \boldsymbol{v}$, we have

$$\begin{aligned}
\boldsymbol{v}^T \cdot \mathbf{P}^{\mathrm{T}} = \boldsymbol{v}^T \cdot \mathbf{P}^* &= \boldsymbol{v}^T \cdot \left(\mathbf{I} - 2\boldsymbol{u}^* \otimes \boldsymbol{u}^T\right) = \boldsymbol{v}^T - 2\boldsymbol{v}^T \cdot \boldsymbol{u}^* \otimes \boldsymbol{u}^T \\
&= \boldsymbol{v}^T - \left(\boldsymbol{v}^T - \boldsymbol{w}^T\right) = \boldsymbol{w}^T.
\end{aligned} \tag{2.212}$$

Thus, the elements of the row vector beyond v_n can be brought to zero by post-multiplying by the complex conjugate of the Householder matrix. By alternating column and row operations, it is found that previous annihilations are unaffected, and the whole matrix can be reduced to upper bidiagonal form by unitary transformations.

We illustrate the succession of column and row operations required to reduce a complex matrix to upper bidiagonal form, with the following 5×5 matrix:

```
real part of matrix is,
  0.28488D+03 -0.12135D+03 -0.18145D+03  0.27585D+03 -0.53595D+02
 -0.12135D+03  0.28488D+03 -0.12135D+03 -0.18145D+03  0.27585D+03
 -0.18145D+03 -0.12135D+03  0.28488D+03 -0.12135D+03 -0.18145D+03
  0.27585D+03 -0.18145D+03 -0.12135D+03  0.28488D+03 -0.12135D+03
 -0.53595D+02  0.27585D+03 -0.18145D+03 -0.12135D+03  0.28488D+03
imaginary part of matrix is,
  0.00000D+00  0.25772D+03 -0.21954D+03 -0.70624D+02  0.27956D+03
 -0.25772D+03  0.00000D+00  0.25772D+03 -0.21954D+03 -0.70624D+02
  0.21954D+03 -0.25772D+03  0.00000D+00  0.25772D+03 -0.21954D+03
  0.70624D+02  0.21954D+03 -0.25772D+03  0.00000D+00  0.25772D+03
 -0.27956D+03  0.70624D+02  0.21954D+03 -0.25772D+03  0.00000D+00.
```

The first step in the reduction to upper bidiagonal form is to pre-multiply by the unitary Householder matrix constructed to annihilate the elements below the diagonal in the first column, then to post-multiply by the unitary Householder matrix constructed to annihilate the elements above the superdiagonal in the first row. The matrix then takes the form

```
real part of matrix is,
  -0.63681D+03 -0.54262D+03  0.00000D+00  0.00000D+00  0.00000D+00
   0.00000D+00  0.47353D-01 -0.20383D-02  0.77597D-02 -0.24832D-01
   0.00000D+00  0.41176D-01  0.41669D-02 -0.12093D-01 -0.20853D-01
   0.00000D+00  0.14868D+00 -0.19657D-02  0.50075D-01 -0.41175D-01
   0.00000D+00 -0.36038D+00 -0.10301D-02 -0.33330D-01  0.15153D+00
imaginary part of matrix is,
   0.00000D+00  0.11524D+04  0.00000D+00  0.00000D+00  0.00000D+00
   0.00000D+00 -0.70684D-02 -0.44649D-02  0.95123D-02  0.58349D-02
   0.00000D+00  0.82013D-01  0.19453D-02  0.12739D-01 -0.26964D-01
   0.00000D+00 -0.18263D+00  0.28303D-02  0.58515D-03  0.79070D-01
   0.00000D+00 -0.89162D-01 -0.10635D-02 -0.65263D-01  0.63077D-03.
```

These operations are repeated for the second column, second row, and so on, until the final operation is on the second to last column. The matrix is left in the complex upper bidiagonal form

```
real part of matrix is,
  -0.63681D+03 -0.54262D+03  0.00000D+00  0.00000D+00  0.00000D+00
   0.00000D+00 -0.44672D+00 -0.12538D+00  0.00000D+00  0.00000D+00
   0.00000D+00  0.00000D+00 -0.85026D-04 -0.25365D-02  0.00000D+00
   0.00000D+00  0.00000D+00  0.00000D+00 -0.38416D-02  0.51173D-02
   0.00000D+00  0.00000D+00  0.00000D+00  0.00000D+00  0.17430D-02
imaginary part of matrix is,
   0.00000D+00  0.11524D+04  0.00000D+00  0.00000D+00  0.00000D+00
   0.00000D+00  0.66681D-01 -0.15867D+00  0.00000D+00  0.00000D+00
   0.00000D+00  0.00000D+00  0.10496D-01  0.41533D-02  0.00000D+00
   0.00000D+00  0.00000D+00  0.00000D+00 -0.34631D-02  0.50654D-02
   0.00000D+00  0.00000D+00  0.00000D+00  0.00000D+00  0.25834D-02.
```

The basis of the foregoing operations is the subroutine HHOLDER that annihilates the components of the M-length vector $\boldsymbol{v}$ beyond the nth component for $n < M$. It calculates the vector $\boldsymbol{u}$ required to construct the Householder matrix $\mathbf{P}$ and replaces the vector $\boldsymbol{v}$ with the transformed vector $\boldsymbol{w}$.

```
      SUBROUTINE HHOLDER(V,U,N,M,MMAX)
C
C The subroutine HHOLDER annihilates the components of the vector V(M)
C beyond the Nth component for N<M. The Householder vector U is calculated
C and the annihilated vector W replaces V on return.
C
      DOUBLE COMPLEX U(MMAX),V(MMAX),W(MMAX)
      DOUBLE PRECISION ETA,AMAG,VWMAG,VR,VI
C No modification for N=M.
      IF(N.EQ.M) GO TO 10
C Find value of ETA.
      ETA=0.D0
      DO 11 I=N,M
        ETA=ETA+DREAL(V(I)*DCONJG(V(I)))
11    CONTINUE
      ETA=DSQRT(ETA)
C Find magnitude and real and imaginary parts of V(N).
      AMAG=CDABS(V(N))
      VR=DREAL(V(N))
      VI=DIMAG(V(N))
```

```
C Find magnitude of V-W.
      VWMAG=2.D0*ETA*(ETA+AMAG)
      VWMAG=DSQRT(VWMAG)
C Construct vector U.
C Set initial values of components of U to zero.
      DO 12 I=1,M
        U(I)=(0.D0,0.D0)
 12   CONTINUE
C Set value of U(N).
      U(N)=V(N)+ETA*DCMPLX(VR/AMAG,VI/AMAG)
C Set values of remaining components of U.
      NP1=N+1
      DO 13 I=NP1,M
        U(I)=V(I)
 13   CONTINUE
C Normalize U to unit vector.
      DO 14 I=1,M
        U(I)=U(I)/VWMAG
 14   CONTINUE
C Initialize vector W as vector V.
      DO 15 I=1,M
        W(I)=V(I)
 15   CONTINUE
C Construct vector W.
C Set value of W(N).
      W(N)=W(N)-VWMAG*U(N)
C Zero remaining components of W.
      DO 16 I=NP1,M
        W(I)=(0.D0,0.D0)
 16   CONTINUE
C Replace V by W.
      DO 17 I=1,M
        V(I)=W(I)
 17   CONTINUE
 10   CONTINUE
      RETURN
      END
```

Although the sequence of unitary Householder transformations reduces the matrix to upper bidiagonal form, it is still complex valued. G. W. Stewart (Stewart, 2001, p. 217) points out that it can be further reduced to real, positive upper bidiagonal form by a further sequence of unitary transformations, thereby simplifying further numerical operations. We illustrate the procedure with the 3×3 complex, upper bidiagonal matrix

$$\begin{pmatrix} d_1 & e_1 & 0 \\ 0 & d_2 & e_2 \\ 0 & 0 & d_3 \end{pmatrix}. \tag{2.213}$$

We first pre-multiply by a diagonal matrix as

$$\begin{pmatrix} u_{11} & 0 & 0 \\ 0 & 1 & 0 \\ 0 & 0 & 1 \end{pmatrix} \begin{pmatrix} d_1 & e_1 & 0 \\ 0 & d_2 & e_2 \\ 0 & 0 & d_3 \end{pmatrix} = \begin{pmatrix} u_{11}d_1 & u_{11}e_1 & 0 \\ 0 & d_2 & e_2 \\ 0 & 0 & d_3 \end{pmatrix}. \tag{2.214}$$

Choosing $u_{11} = d_1^*/|d_1|$ gives $u_{11}d_1 = |d_1|$, making the first diagonal element real and positive. Notice that the pre-multiplying matrix is unitary, since multiplying by its Hermitian transpose gives the unit matrix. We next post-multiply this result as

$$\begin{pmatrix} |d_1| & u_{11}e_1 & 0 \\ 0 & d_2 & e_2 \\ 0 & 0 & d_3 \end{pmatrix} \begin{pmatrix} 1 & 0 & 0 \\ 0 & v_{22} & 0 \\ 0 & 0 & 1 \end{pmatrix} = \begin{pmatrix} |d_1| & v_{22}u_{11}e_1 & 0 \\ 0 & v_{22}d_2 & e_2 \\ 0 & 0 & d_3 \end{pmatrix}. \tag{2.215}$$

Taking $v_{22} = u_{11}^* \cdot e_1^*/|e_1|$ gives $v_{22}u_{11}e_1 = |e_1|$. Once again, the post-multiplying matrix is unitary, since $v_{22}v_{22}^* = 1$. We then pre-multiply as

$$\begin{pmatrix} 1 & 0 & 0 \\ 0 & u_{22} & 0 \\ 0 & 0 & 1 \end{pmatrix} \begin{pmatrix} |d_1| & |e_1| & 0 \\ 0 & v_{22}d_2 & e_2 \\ 0 & 0 & d_3 \end{pmatrix} = \begin{pmatrix} |d_1| & |e_1| & 0 \\ 0 & u_{22}v_{22}d_2 & u_{22}e_2 \\ 0 & 0 & d_3 \end{pmatrix}. \tag{2.216}$$

Taking $u_{22} = v_{22}^* \cdot d_2^*/|d_2|$ gives $u_{22}v_{22}d_2 = |d_2|$. The pre-multiplying matrix is unitary since $u_{22}u_{22}^* = 1$. Again, we post-multiply as

$$\begin{pmatrix} |d_1| & |e_1| & 0 \\ 0 & |d_2| & u_{22}e_2 \\ 0 & 0 & d_3 \end{pmatrix} \begin{pmatrix} 1 & 0 & 0 \\ 0 & 1 & 0 \\ 0 & 0 & v_{33} \end{pmatrix} = \begin{pmatrix} |d_1| & |e_1| & 0 \\ 0 & |d_2| & v_{33}u_{22}e_2 \\ 0 & 0 & v_{33}d_3 \end{pmatrix}. \tag{2.217}$$

Taking $v_{33} = u_{22}^* \cdot e_2^*/|e_2|$ gives $v_{33}u_{22}e_2 = |e_2|$. The post-multiplying matrix is unitary since $v_{33}v_{33}^* = 1$. Finally, we pre-multiply as

$$\begin{pmatrix} 1 & 0 & 0 \\ 0 & 1 & 0 \\ 0 & 0 & u_{33} \end{pmatrix} \begin{pmatrix} |d_1| & |e_1| & 0 \\ 0 & |d_2| & |e_2| \\ 0 & 0 & v_{33}d_3 \end{pmatrix} = \begin{pmatrix} |d_1| & |e_1| & 0 \\ 0 & |d_2| & |e_2| \\ 0 & 0 & u_{33}v_{33}d_3 \end{pmatrix}. \tag{2.218}$$

Taking $u_{33} = v_{33}^* \cdot d_3^*/|d_3|$ gives $u_{33}v_{33}d_3 = |d_3|$. The pre-multiplying matrix is unitary since $u_{33}u_{33}^* = 1$. The transformed real, positive matrix is then

$$\begin{pmatrix} |d_1| & |e_1| & 0 \\ 0 & |d_2| & |e_2| \\ 0 & 0 & |d_3| \end{pmatrix}. \tag{2.219}$$

The complete transformation is seen to depend on the determination of the unit length phasors $u_{11}, v_{22}, u_{22}, v_{33}, u_{33}$. These can be computed recursively as

$$u_{11} = \frac{d_1^*}{|d_1|}, \quad v_{22} = u_{11}^* \frac{e_1^*}{|e_1|}, \quad u_{22} = v_{22}^* \frac{d_2^*}{|d_2|},$$
$$v_{33} = u_{22}^* \frac{e_2^*}{|e_2|}, \quad u_{33} = v_{33}^* \frac{d_3^*}{|d_3|}. \tag{2.220}$$

The ultimate objective of singular value decomposition is to factor the matrix **C** into the triple product (2.199). Thus far, our reduction of the matrix to upper bidiagonal form by Householder transformations, and our further reduction of the matrix to real, positive, upper bidiagonal form, has been accomplished by pre-multiplication and post-multiplication by unitary matrices. Since the product of two unitary matrices results in a unitary matrix, the products of the pre-multiplying and post-multiplying unitary matrices can be accumulated so that the transformation of **C** takes the form,

$$\mathbf{U}^{\mathrm{H}} \cdot \mathbf{C} \cdot \mathbf{V}. \tag{2.221}$$

We shall see that our further operations on the real, positive, upper bidiagonal matrix also consist of pre-multiplication and post-multiplication by unitary matrices, which again can be accumulated. Thus, our final reduction of **C** to diagonal form is represented by

$$\mathbf{W} = \mathbf{U}^{\mathrm{H}} \cdot \mathbf{C} \cdot \mathbf{V}. \tag{2.222}$$

On multiplying on the left by **U** and on the right by $\mathbf{V}^{\mathrm{H}}$, we arrive at the factorisation (2.199).

In accumulating the unitary matrices resulting from pre-multiplication by the Householder transformation matrix **P**, the product is

$$\mathbf{P}\mathbf{U}^{\mathrm{H}} = \left(\mathbf{I} - 2\boldsymbol{u} \otimes \boldsymbol{u}^{H}\right)\mathbf{U}^{\mathrm{H}} = \mathbf{U}^{\mathrm{H}} - 2u_i u_k^* U_{kj}^H, \tag{2.223}$$

using the range and summation conventions in the last term. Similarly, for post-multiplication by the complex conjugate of the Householder transformation matrix, the product is

$$\mathbf{V}\mathbf{P}^* = \mathbf{V}\left(\mathbf{I} - 2\boldsymbol{u}^* \otimes \boldsymbol{u}^{T}\right) = \mathbf{V} - 2V_{ik}u_k^* u_j. \tag{2.224}$$

In the case of the reduction of the complex bidiagonal matrix to real, positive, upper bidiagonal form, the pre-multiplications amount to column scalings, while the post-multiplications are simple row scalings.

The subroutine BIDIAG reduces the matrix **C** to the real, positive, upper bidiagonal matrix **B**, while accumulating the unitary matrices $\mathbf{U}^{\mathrm{H}}$ and **V**, using calls to the subroutine HHOLDER.

```
      SUBROUTINE BIDIAG(U,C,VH,M,MMAX)
C
C The subroutine BIDIAG converts the complex M x M matrix C into upper
C bidiagonal form by a series of Householder unitary transformations,
C applied alternately, to columns by pre-multiplication and to rows by
C post-multiplication. Finally, rows are scaled by pre-multiplication
C with a unitary matrix and, alternately, columns are scaled by
C post-multiplication with a unitary matrix, to reduce the phase of diagonal
C elements to zero and, alternately, to reduce the phase of above diagonal
C elements to zero, producing an upper bidiagonal matrix with real and
C positive elements. Thus, the final, real and positive upper bidiagonal
C matrix B is found as B=U.C.VH a unitary similarity transformation of C.
C B is returned in C. The unitary matrices U and VH are also returned by
C the subroutine.
```

```
C
      DOUBLE COMPLEX C(MMAX,MMAX),V(MMAX),HU(MMAX),
     1 U(MMAX,MMAX),VH(MMAX,MMAX),H0(MMAX),H1(MMAX),H2(MMAX),CONST
      DOUBLE PRECISION EMAG
C Initialize unitary matrices U and VH.
      DO 10 I=1,M
        DO 11 J=1,M
          U(I,J)=(0.D0,0.D0)
          VH(I,J)=(0.D0,0.D0)
 11     CONTINUE
        U(I,I)=(1.D0,0.D0)
        VH(I,I)=(1.D0,0.D0)
 10   CONTINUE
C Apply Householder annihilation.
      MM1=M-1
      DO 12 N=1,MM1
C Put Nth column in V.
        DO 13 I=1,M
          V(I)=C(I,N)
 13     CONTINUE
C Annihilate Nth column.
        CALL HHOLDER(V,HU,N,M,MMAX)
C Multiply Hermitian transpose of Householder vector HU
C into unitary matrix U and into transformed matrix B.
        DO 14 I=1,M
          H1(I)=(0.D0,0.D0)
          H2(I)=(0.D0,0.D0)
          DO 14 J=N,M
            H0(J)=DCONJG(HU(J))
            H1(I)=H1(I)+H0(J)*U(J,I)
            H2(I)=H2(I)+H0(J)*C(J,I)
 14     CONTINUE
C Update unitary matrix U and transformed matrix B.
        DO 15 I=N,M
          DO 15 J=1,M
            U(I,J)=U(I,J)-2.D0*HU(I)*H1(J)
            C(I,J)=C(I,J)-2.D0*HU(I)*H2(J)
 15     CONTINUE
C Overwrite annihilated Nth column.
        DO 16 I=1,M
          C(I,N)=V(I)
 16     CONTINUE
        IF(N.EQ.MM1) GO TO 12
C Put Nth row in V.
        DO 17 J=1,M
          V(J)=C(N,J)
 17     CONTINUE
        NP1=N+1
C Annihilate Nth row.
        CALL HHOLDER(V,HU,NP1,M,MMAX)
C Multiply unitary matrix VH and transformed matrix C
C into Householder vector HU.
        DO 18 I=1,M
          H1(I)=(0.D0,0.D0)
          H2(I)=(0.D0,0.D0)
          DO 18 J=NP1,M
            H0(J)=DCONJG(HU(J))
            H1(I)=H1(I)+VH(I,J)*H0(J)
            H2(I)=H2(I)+C(I,J)*H0(J)
 18     CONTINUE
C Update unitary matrix VH and transformed matrix B.
```

```
      DO 19 I=1,M
        DO 19 J=NP1,M
          VH(I,J)=VH(I,J)-2.D0*H1(I)*HU(J)
          C(I,J)=C(I,J)-2.D0*H2(I)*HU(J)
 19     CONTINUE
C Overwrite annihilated Nth row.
        DO 20 J=1,M
          C(N,J)=V(J)
 20     CONTINUE
 12   CONTINUE
C Convert upper diagonal matrix B to a real, positive matrix.
      N=1
      CONST=(1.D0,0.D0)
 21   CONTINUE
      NP1=N+1
C Find magnitudes of current row elements.
      EMAG=DREAL(C(N,N)*DCONJG(C(N,N)))
      EMAG=DSQRT(EMAG)
C Update unitary matrix U.
      CONST=DCONJG(CONST)*DCONJG(C(N,N))/EMAG
C Scale current row.
      DO 22 I=1,M
        U(N,I)=CONST*U(N,I)
 22   CONTINUE
C Replace diagonal element.
      C(N,N)=DCMPLX(EMAG,0.D0)
      IF(N.EQ.M) GO TO 23
      EMAG=DREAL(C(N,NP1)*DCONJG(C(N,NP1)))
      EMAG=DSQRT(EMAG)
C Update unitary matrix VH.
      CONST=DCONJG(CONST)*DCONJG(C(N,NP1))/EMAG
C Scale next column.
      DO 24 I=1,M
        VH(I,NP1)=CONST*VH(I,NP1)
 24   CONTINUE
C Replace above diagonal element.
      C(N,NP1)=DCMPLX(EMAG,0.D0)
      N=N+1
      GO TO 21
 23   CONTINUE
      RETURN
      END
```

The subroutine BIDIAG converts the 5×5 complex matrix we are using as an example into the following real, positive, upper bidiagonal matrix:

```
real part of matrix is,
  0.63681D+03 0.12738D+04 0.00000D+00 0.00000D+00 0.00000D+00
  0.00000D+00 0.45167D+00 0.20223D+00 0.00000D+00 0.00000D+00
  0.00000D+00 0.00000D+00 0.10497D-01 0.48666D-02 0.00000D+00
  0.00000D+00 0.00000D+00 0.00000D+00 0.51721D-02 0.72003D-02
  0.00000D+00 0.00000D+00 0.00000D+00 0.00000D+00 0.31164D-02
imaginary part of matrix is,
  0.00000D+00 0.00000D+00 0.00000D+00 0.00000D+00 0.00000D+00
  0.00000D+00 0.00000D+00 0.00000D+00 0.00000D+00 0.00000D+00
  0.00000D+00 0.00000D+00 0.00000D+00 0.00000D+00 0.00000D+00
  0.00000D+00 0.00000D+00 0.00000D+00 0.00000D+00 0.00000D+00
  0.00000D+00 0.00000D+00 0.00000D+00 0.00000D+00 0.00000D+00.
```

In general, after the call to the subroutine BIDIAG, we have the real, positive, upper bidiagonal matrix $\mathbf{B}$ and the accumulated unitary matrices $\mathbf{U}^{\mathrm{H}}$ and $\mathbf{V}$ from the transformation of $\mathbf{C}$,

$$\mathbf{B} = \mathbf{U}^{\mathrm{H}} \cdot \mathbf{C} \cdot \mathbf{V} = \begin{pmatrix} d_1 & e_1 & & & \\ & d_2 & e_2 & & \\ & & \ddots & \ddots & \\ & & & d_{M-1} & e_{M-1} \\ & & & & d_M \end{pmatrix}. \tag{2.225}$$

The product

$$\mathbf{B}^{\mathrm{T}}\mathbf{B} = \begin{pmatrix} d_1^2 & d_1 e_1 & & & \\ d_1 e_1 & d_2^2 + e_1^2 & d_2 e_2 & & \\ & \ddots & \ddots & \ddots & \\ & & d_{M-2}e_{M-2} & d_{M-1}^2 + e_{M-2}^2 & d_{M-1}e_{M-1} \\ & & & d_{M-1}e_{M-1} & d_M^2 + e_{M-1}^2 \end{pmatrix}, \tag{2.226}$$

is then a real, symmetric tridiagonal matrix. The singular value decomposition of $\mathbf{B}$ itself leads to the representation

$$\mathbf{B} = \mathbf{X} \cdot \mathbf{W} \cdot \mathbf{Y}^{\mathrm{T}}, \tag{2.227}$$

where $\mathbf{X}$ and $\mathbf{Y}$ are orthogonal matrices with column vectors $\boldsymbol{x}_1, \boldsymbol{x}_2, \ldots, \boldsymbol{x}_M$ and $\boldsymbol{y}_1, \boldsymbol{y}_2, \ldots, \boldsymbol{y}_M$, respectively. $\mathbf{W}$ is again the diagonal matrix with the singular values $s_1, s_2, \ldots, s_M$ down its diagonal, appearing in the original factorisation (2.199). Multiplying (2.227) on the right by $\mathbf{Y}$ yields the relation

$$\mathbf{B}\boldsymbol{y}_i = s_i \boldsymbol{x}_i. \tag{2.228}$$

Taking the transpose of (2.227) gives

$$\mathbf{B}^{\mathrm{T}} = \mathbf{Y} \cdot \mathbf{W} \cdot \mathbf{X}^{\mathrm{T}}, \tag{2.229}$$

and multiplying this relation on the right by $\mathbf{X}$ yields

$$\mathbf{B}^{\mathrm{T}}\boldsymbol{x}_i = s_i \boldsymbol{y}_i. \tag{2.230}$$

Multiplying (2.228) on the left by $\mathbf{B}^{\mathrm{T}}$ gives

$$\mathbf{B}^{\mathrm{T}}\mathbf{B}\,\boldsymbol{y}_i = s_i \mathbf{B}^{\mathrm{T}}\boldsymbol{x}_i = s_i^2 \boldsymbol{y}_i. \tag{2.231}$$

The eigenvalues of the tridiagonal matrix $\mathbf{B}^{\mathrm{T}}\mathbf{B}$ are then the squares of the singular values of $\mathbf{B}$ and $\mathbf{C}$.

The calculation of the eigenvalues of a symmetric tridiagonal matrix is a classical problem solved by the QR algorithm (Wilkinson, 1968), in which the matrix is factored iteratively into the product of a unitary matrix $\mathbf{Q}$ and an upper triangular matrix $\mathbf{R}$. Beginning with the matrix $\mathbf{A}_s$ at the sth step, it is factored as $\mathbf{Q}_s \cdot \mathbf{R}_s$. Then, in the next step, the matrix $\mathbf{A}_{s+1}$ is formed by the unitary transformation

$$\mathbf{A}_{s+1} = \mathbf{Q}_s^{\mathrm{H}} \cdot \mathbf{A}_s \cdot \mathbf{Q}_s = \mathbf{Q}_s^{\mathrm{H}} \cdot \mathbf{Q}_s \cdot \mathbf{R}_s \cdot \mathbf{Q}_s = \mathbf{R}_s \cdot \mathbf{Q}_s. \tag{2.232}$$

A detailed description of this algorithm applied to the symmetric tridiagonal eigenvalue problem is given by Stewart (2001, pp. 164–167), including the implementation of the *Wilkinson shift* of eigenvalues to produce cubic convergence to upper triangular form, which, in the tridiagonal case, is diagonal. Since unitary transformations preserve the eigenvalues of a matrix, the eigenvalues are then on the diagonal.

While, once the eigenvalues of $\mathbf{B}^{\mathrm{T}}\mathbf{B}$ are found, the singular values can be found as their positive square roots, the QR iteration can be applied directly to the matrix $\mathbf{B}$ (Stewart, 2001, pp. 219–225). The first step mimics the application of the QR algorithm to the bidiagonal matrix $\mathbf{B}^{\mathrm{T}}\mathbf{B}$. With shift σ, a plane rotation of the $(1, 2)$ axes is made to bring to zero the $(1, 2)$ element of the shifted matrix $\mathbf{T} = \mathbf{B}^{\mathrm{T}}\mathbf{B} - \sigma^2\mathbf{I}$. A plane rotation of the (x_1, x_2) axes through an angle θ produces new co-ordinates (x_1', x_2') given by

$$\begin{pmatrix} x_1' \\ x_2' \end{pmatrix} = \begin{pmatrix} \cos\theta & \sin\theta \\ -\sin\theta & \cos\theta \end{pmatrix} \begin{pmatrix} x_1 \\ x_2 \end{pmatrix}. \tag{2.233}$$

The required transformation of the matrix $\mathbf{T}$ is accomplished by post-multiplication by the orthogonal matrix $\mathbf{P}_{12}$, as $\mathbf{TP}_{12}$ equal to

$$\begin{pmatrix} t_{11} & t_{12} & & & & \\ t_{12} & t_{22} & t_{23} & & & \\ & t_{23} & t_{33} & t_{34} & & \\ & & \ddots & \ddots & \ddots & \\ & & & t_{M-2\,M-1} & t_{M-1\,M-1} & t_{M-1\,M} \\ & & & & t_{M-1\,M} & t_{M\,M} \end{pmatrix} \begin{pmatrix} c & s & & & \\ -s & c & & & \\ & & 1 & & \\ & & & \ddots & \\ & & & & 1 & \\ & & & & & 1 \end{pmatrix}$$

$$= \begin{pmatrix} c \cdot t_{11} - s \cdot t_{12} & s \cdot t_{11} + c \cdot t_{12} & & & & \\ c \cdot t_{12} - s \cdot t_{22} & s \cdot t_{12} + c \cdot t_{22} & t_{2,3} & & & \\ -s \cdot t_{23} & c \cdot t_{23} & t_{33} & t_{34} & & \\ & & \ddots & \ddots & \ddots & \\ & & & t_{M-2\,M-1} & t_{M-1\,M-1} & t_{M-1\,M} \\ & & & & t_{M-1\,M} & t_{M\,M} \end{pmatrix}, \tag{2.234}$$

where we have abbreviated $\cos\theta$ as c, and $\sin\theta$ as s. Such an orthogonal transformation is generally referred to as a *Givens rotation*. In order to bring the $(1,2)$ element to zero, we require that

$$s = -\frac{t_{12}}{\sqrt{t_{11}^2 + t_{12}^2}}, \quad c = \frac{t_{11}}{\sqrt{t_{11}^2 + t_{12}^2}}, \tag{2.235}$$

where $t_{11} = d_1^2 - \sigma^2$ and $t_{12} = d_1 e_1$. Instead of applying the Givens rotation to the shifted matrix $\mathbf{T}$, it is applied directly to the positive, upper bidiagonal matrix $\mathbf{B}$ by post-multiplication by $\mathbf{P}_{12}$. This transformation generates a non-zero *rogue element* in the $(2,1)$ position, destroying the upper bidiagonal form of $\mathbf{B}$. By pre-multiplying by an orthogonal matrix with the form of $\mathbf{P}_{12}^{\mathrm{T}}$, the $(2,1)$ rogue element can be brought to zero, generating a new non-zero rogue element in the $(1,3)$ position. In turn, this element can be brought to zero by post-multiplying by an orthogonal rotation matrix for the $(2,3)$ axes, producing a new non-zero rogue element in the $(3,2)$ position. The process of post- and pre-multiplication by orthogonal rotation matrices is continued until the non-zero rogue elements are chased from the matrix and it is returned to its original upper bidiagonal form.

The Wilkinson shift is computed from the trailing 2×2 submatrix of $\mathbf{B}$ written as

$$\mathbf{Z} = \begin{pmatrix} q & r \\ 0 & p \end{pmatrix}. \tag{2.236}$$

The eigenvalues of the matrix

$$\mathbf{Z}^{\mathrm{T}}\mathbf{Z} = \begin{pmatrix} q^2 & qr \\ qr & p^2 + r^2 \end{pmatrix}, \tag{2.237}$$

are the squares of the singular values of $\mathbf{Z}$. The characteristic equation for the eigenvalues of $\mathbf{Z}^{\mathrm{T}}\mathbf{Z}$ is

$$\lambda^2 - \left(p^2 + q^2 + r^2\right)\lambda + p^2 q^2 = 0. \tag{2.238}$$

If $\sigma_{\min}$ is the smallest singular value of $\mathbf{Z}$ and $\sigma_{\max}$ is the largest, then the characteristic equation for the eigenvalues of $\mathbf{Z}^{\mathrm{T}}\mathbf{Z}$ becomes

$$\left(\lambda - \sigma_{\min}^2\right)\left(\lambda - \sigma_{\max}^2\right) = \lambda^2 - \left(\sigma_{\min}^2 + \sigma_{\max}^2\right)\lambda + \sigma_{\min}^2\sigma_{\max}^2 = 0. \tag{2.239}$$

Thus,

$$\sigma_{\min}^2 + \sigma_{\max}^2 = p^2 + q^2 + r^2 \quad \text{and} \quad \sigma_{\min}^2\sigma_{\max}^2 = p^2 q^2. \tag{2.240}$$

These equations are satisfied by

$$\sigma_{\min} = \frac{\sqrt{(p+q)^2 + r^2} - \sqrt{(p-q)^2 + r^2}}{2}, \tag{2.241}$$

$$\sigma_{\max} = \frac{\sqrt{(p+q)^2 + r^2} + \sqrt{(p-q)^2 + r^2}}{2}. \tag{2.242}$$

The Wilkinson shift is chosen as the smallest singular value, $\sigma_{\min}$, of the trailing matrix $\mathbf{Z}$. Its square is the smallest eigenvalue of the matrix $\mathbf{Z}^T\mathbf{Z}$.

The initial QR iteration is carried out on the full $M \times M$ bidiagonal matrix. Thus, the first row is the *leading* row of the initial *working matrix* and the last row, M, is the *trailing* row of the initial working matrix. In a technique called *matrix splitting* (Burden and Faires, 1988, pp. 515–519), after each QR iteration on the bidiagonal matrix $\mathbf{B}$, the upper diagonal element in the leading row is tested for negligibility. If the upper diagonal element in the leading row is negligibly small compared to the absolute value of the smallest of its nearest diagonal element neighbours, the bidiagonal matrix deflates by one dimension. The first row with a non-negligible upper diagonal element becomes the leading row K of the new working matrix. The trailing row of the new working matrix becomes the first row N, below the new leading row, with a negligible upper diagonal element (or, in the case of the last row, a non-existent upper diagonal element). Then, the new working matrix begins at row K and ends at row N. Finally, the last iteration is performed on a 2×2 bidiagonal matrix. For the 5×5 matrix we are using as an example, the final form of $\mathbf{B}$ is

```
upper diagonal of B is,
   0.13105D-72  0.92262D-31 -0.25429D-19 -0.36258D-21  0.00000D+00
diagonal of B is,
   0.14241D+04  0.28591D+00  0.10321D-01  0.76246D-02  0.15188D-02,
```

for an assumed machine precision of 10^{-15}. The unitary matrices in the SVD factorisation (2.199) of the example matrix are $\mathbf{U}$,

```
real part of matrix is,
   0.44717D+00 -0.63165D+00  0.30071D+00 -0.45811D+00 -0.31745D+00
  -0.19052D+00  0.14267D+00 -0.24572D+00 -0.22489D+00 -0.46049D+00
  -0.28493D+00 -0.31439D-02 -0.82148D-01  0.85482D-01 -0.59629D+00
   0.43326D+00  0.30757D+00  0.33106D+00  0.52251D+00 -0.44210D+00
  -0.84196D-01 -0.13213D+00 -0.29110D+00  0.12652D+00 -0.31403D+00
imaginary part of matrix is,
  -0.21841D-16 -0.20379D-13  0.18805D-11 -0.12338D-11  0.17202D-12
  -0.40463D+00  0.28401D+00  0.37157D+00 -0.47903D+00 -0.91814D-01
   0.34475D+00 -0.25425D-02 -0.64460D+00 -0.11350D+00  0.43863D-01
   0.11092D+00  0.80108D-01  0.29806D+00  0.83846D-01  0.15821D+00
  -0.43917D+00 -0.61767D+00  0.75420D-01  0.44029D+00  0.46451D-01,
```

and $\mathbf{V}^H$,

```
real part of matrix is,
   0.44717D+00 -0.19052D+00 -0.28493D+00  0.43326D+00 -0.84196D-01
  -0.63165D+00  0.14267D+00 -0.31439D-02  0.30757D+00 -0.13213D+00
   0.30071D+00 -0.24572D+00 -0.82148D-01  0.33106D+00 -0.29110D+00
  -0.45811D+00 -0.22489D+00  0.85482D-01  0.52251D+00  0.12652D+00
   0.31745D+00  0.46049D+00  0.59629D+00  0.44210D+00  0.31403D+00
imaginary part of matrix is,
   0.00000D+00  0.40463D+00 -0.34475D+00 -0.11092D+00  0.43917D+00
   0.00000D+00 -0.28401D+00  0.25425D-02 -0.80108D-01  0.61767D+00
   0.00000D+00 -0.37157D+00  0.64460D+00 -0.29806D+00 -0.75420D-01
   0.00000D+00  0.47903D+00  0.11350D+00 -0.83846D-01 -0.44029D+00
   0.00000D+00 -0.91814D-01  0.43863D-01  0.15821D+00  0.46451D-01,
```

while the recovered matrix $\mathbf{C} = \mathbf{U} \cdot \mathbf{W} \cdot \mathbf{V}^H$ is

```
real part of matrix is,
   0.28488D+03 -0.12135D+03 -0.18145D+03  0.27585D+03 -0.53595D+02
  -0.12135D+03  0.28488D+03 -0.12135D+03 -0.18145D+03  0.27585D+03
  -0.18145D+03 -0.12135D+03  0.28488D+03 -0.12135D+03 -0.18145D+03
   0.27585D+03 -0.18145D+03 -0.12135D+03  0.28488D+03 -0.12135D+03
  -0.53595D+02  0.27585D+03 -0.18145D+03 -0.12135D+03  0.28488D+03
imaginary part of matrix is,
   0.60215D-31  0.25772D+03 -0.21954D+03 -0.70624D+02  0.27956D+03
  -0.25772D+03  0.15695D-13  0.25772D+03 -0.21954D+03 -0.70624D+02
   0.21954D+03 -0.25772D+03 -0.69882D-13  0.25772D+03 -0.21954D+03
   0.70624D+02  0.21954D+03 -0.25772D+03 -0.37131D-14  0.25772D+03
  -0.27956D+03  0.70624D+02  0.21954D+03 -0.25772D+03  0.55629D-14.
```

Apart from rounding error in the diagonal elements of the imaginary part, the reconstructed matrix is identical to the example matrix.

In the SVD factorisation (2.199) of the complex matrix $\mathbf{C}$, the singular values are required to be in descending order down the diagonal of the matrix $\mathbf{W}$. To ensure that this is the case, we sort the singular values and their associated eigenvectors. If $\boldsymbol{u}_1, \boldsymbol{u}_2, \ldots, \boldsymbol{u}_M$ are the column vectors of the matrix $\mathbf{U}$ and if $\boldsymbol{v}_1, \boldsymbol{v}_2, \ldots, \boldsymbol{v}_M$ are the column vectors of $\mathbf{V}$, the product $\mathbf{U} \cdot \mathbf{W} \cdot \mathbf{V}^H$ may be expanded as

$$
s_1 \cdot \begin{pmatrix} u_{11} \\ u_{21} \\ \vdots \\ u_{M1} \end{pmatrix} \otimes \begin{pmatrix} v_{11}^* & v_{21}^* & \cdots & v_{M1}^* \end{pmatrix} + s_2 \cdot \begin{pmatrix} u_{12} \\ u_{22} \\ \vdots \\ u_{M2} \end{pmatrix} \otimes \begin{pmatrix} v_{12}^* & v_{22}^* & \cdots & v_{M2}^* \end{pmatrix}
$$

$$
+ \cdots + s_M \cdot \begin{pmatrix} u_{1M} \\ u_{2M} \\ \vdots \\ u_{MM} \end{pmatrix} \otimes \begin{pmatrix} v_{1M}^* & v_{2M}^* & \cdots & v_{MM}^* \end{pmatrix}. \tag{2.243}
$$

If, in the sorting, the singular values s_i and s_j are exchanged, then the ith and jth column vectors of $\mathbf{U}$, and the ith and jth row vectors of $\mathbf{V}^H$, need to be exchanged.

The subroutine SVD begins with a call to the subroutine BIDIAG to construct the real, positive, upper bidiagonal matrix **B**. It then performs an implicit QR iteration, through a series of Givens rotations, to reduce **B** to diagonal form containing the singular values of **C**. At each step of the iteration, the upper diagonal element of the leading row of the working matrix is tested for negligibility. If it is negligibly small compared to the smallest of its nearest diagonal neighbours, the matrix is deflated and a new leading row of the working matrix is established. Similarly, a new trailing row of the working matrix is established as the first row below the leading row with a negligible upper diagonal element compared to its smallest diagonal neighbour. The QR iteration is then repeated on the smaller split matrix. The final iteration is on a 2×2 matrix. At each stage the unitary matrices **U** and $\mathbf{V}^H$ are updated, and at the end of the iteration their Hermitian transposes are taken to give the unitary matrices appearing in the SVD factored form (2.199).

```
      SUBROUTINE SVD(C,U,S,VH,M,MMAX)
C
C The subroutine SVD calls the subroutine BIDIAG to convert the complex
C matrix C to real, positive, upper bidiagonal form. The diagonal elements
C are stored in the vector S and its upper diagonal is stored in
C the vector UD. An implicit QR iteration with Wilkinson shift is then
C performed on the bidiagonal matrix through a series of Givens rotations
C to chase a rogue element, RE, from the matrix. After each QR iteration,
C the upper diagonal element in the leading row is tested for negligibility.
C The matrix is deflated when this element is found negligible, to
C computational accuracy, compared to its smallest diagonal neighbour,
C establishing a new leading row K. Similarly, a new trailing row, N, is
C established as the first row below the leading row with an upper diagonal
C element that is negligible compared to its smallest diagonal neighbour.
C The final iteration is on a 2 x 2 matrix. The singular values are on
C the diagonal of the final matrix. The singular values and their
C associated eigenvectors are then sorted so that they are in descending
C order down the diagonal of the final matrix. The Hermitian transposes of
C the accumulated matrices U and VH are taken to convert them to
C the matrices appearing in the SVD factorization.
C
      IMPLICIT DOUBLE PRECISION(A-H,O-Z)
      DOUBLE COMPLEX C(MMAX,MMAX),U(MMAX,MMAX),VH(MMAX,MMAX),HOLD
      DIMENSION S(MMAX),UD(MMAX)
C Set maximum number of iterations.
      MAXI=100
      CALL BIDIAG(U,C,VH,M,MMAX)
C Put diagonal of real, positive bidiagonal matrix in S and its upper
C diagonal in UD.
      DO 10 I=1,M
        IP1=I+1
        S(I)=DREAL(C(I,I))
        IF(IP1.GT.M) GO TO 10
        UD(I)=DREAL(C(I,IP1))
 10   CONTINUE
C Begin diagonalization of matrix B.
C Set initial values of iteration number, ITER, leading row, K, and
C trailing row, N, of working matrix.
      ITER=0
      K=1
      N=M
```

```
C Continue reduction of working matrix.
 11   CONTINUE
C Advance iteration counter.
      ITER=ITER+1
      KP1=K+1
      NM1=N-1
C Find Wilkinson shift.
      P=S(N)
      Q=S(NM1)
      R=UD(NM1)
      H1=DSQRT((P+Q)*(P+Q)+R*R)
      H2=DSQRT((P-Q)*(P-Q)+R*R)
      SIG1=(H1-H2)/2.D0
C Find first Givens rotation.
      T11=(S(K)-SIG1)*(S(K)+SIG1)
      T12=S(K)*UD(K)
      CMAG=DSQRT(T11*T11+T12*T12)
      CO=T11/CMAG
      SI=-T12/CMAG
C Update unitary matrix VH.
      DO 12 I=1,M
        HOLD=VH(I,K)
        VH(I,K)=CO*HOLD-SI*VH(I,KP1)
        VH(I,KP1)=SI*HOLD+CO*VH(I,KP1)
 12   CONTINUE
C Post-multiply bidiagonal matrix B by rotation.
      H1=CO*S(K)-SI*UD(K)
      H2=CO*UD(K)+SI*S(K)
      S(K)=H1
      UD(K)=H2
C Find rogue element RE.
      RE=-SI*S(KP1)
      S(KP1)=CO*S(KP1)
C Find remaining Givens rotations.
      DO 13 I=K,NM1
        IP1=I+1
        IP2=I+2
        CMAG=DSQRT(S(I)*S(I)+RE*RE)
        CO=S(I)/CMAG
        SI=-RE/CMAG
C Update unitary matrix U.
        DO 14 J=1,M
          HOLD=U(I,J)
          U(I,J)=CO*HOLD-SI*U(IP1,J)
          U(IP1,J)=SI*HOLD+CO*U(IP1,J)
 14     CONTINUE
C Pre-multiply matrix B by rotation.
        H1=CMAG
        H2=CO*UD(I)-SI*S(IP1)
        H4=CO*S(IP1)+SI*UD(I)
        S(I)=H1
        UD(I)=H2
        S(IP1)=H4
        IF(IP2.GT.N) GO TO 13
        RE=-SI*UD(IP1)
        UD(IP1)=CO*UD(IP1)
C Find next Givens rotation.
        CMAG=DSQRT(UD(I)*UD(I)+RE*RE)
        CO=UD(I)/CMAG
        SI=-RE/CMAG
C Update unitary matrix VH.
```

```
        DO 15 J=1,M
          HOLD=VH(J,IP1)
          VH(J,IP1)=CO*HOLD-SI*VH(J,IP2)
          VH(J,IP2)=SI*HOLD+CO*VH(J,IP2)
 15     CONTINUE
C Post-multiply matrix B by rotation.
        H1=CMAG
        H2=CO*S(IP2)
        H3=CO*S(IP1)-SI*UD(IP1)
        H4=CO*UD(IP1)+SI*S(IP1)
        UD(I)=H1
        S(IP1)=H3
        UD(IP1)=H4
        RE=-SI*S(IP2)
        S(IP2)=H2
 13   CONTINUE
C Find leading row of working matrix.
      J=K
      MM1=M-1
      DO 16 I=K,MM1
        IP1=I+1
C Find smallest neighbouring diagonal element.
        SMIN=DMIN1(DABS(S(I)),DABS(S(IP1)))
C Compare to see if upper diagonal element is non-negligible.
        SIZE=SMIN+DABS(UD(I))
        IF(SIZE.NE.SMIN) GO TO 17
        J=J+1
 16   CONTINUE
 17   CONTINUE
C Set new leading row number.
      K=J
C Test to see if diagonalization is complete.
      IF(K.EQ.M) GO TO 20
C Find trailing row of working matrix.
      J=K+1
      KKP1=K+1
      DO 18 I=KKP1,MM1
        IP1=I+1
C Find smallest neighbouring diagonal element.
        SMIN=DMIN1(DABS(S(I)),DABS(S(IP1)))
C Compare to see if upper diagonal element is negligible.
        SIZE=SMIN+DABS(UD(I))
        IF(SIZE.EQ.SMIN) GO TO 19
        J=J+1
 18   CONTINUE
 19   CONTINUE
C Set new trailing row number.
      N=J
C Test for failure to converge.
      NNM1=N-1
      IF(KKP1.NE.KP1.OR.NNM1.NE.NM1) ITER=0
      IF(ITER.EQ.MAXI) PAUSE 'Maximum iterations reached.'
      GO TO 11
 20   CONTINUE
C Take Hermitian transpose of U and VH to get their final forms.
      DO 21 I=1,M
        DO 21 J=1,I
        HOLD=U(I,J)
        U(I,J)=DCONJG(U(J,I))
        U(J,I)=DCONJG(HOLD)
 21   CONTINUE
```

```
      DO 22 I=1,M
        DO 22 J=1,I
        HOLD=VH(I,J)
        VH(I,J)=DCONJG(VH(J,I))
        VH(J,I)=DCONJG(HOLD)
 22   CONTINUE
C Sort singular values into descending order, exchange columns of U
C and rows of VH.
      N=1
 23   CONTINUE
      IMAX=N
      DO 24 I=N,M
        IF(S(I).GE.S(IMAX)) IMAX=I
 24   CONTINUE
C Exchange singular values.
      HOLD=S(N)
      S(N)=S(IMAX)
      S(IMAX)=HOLD
C Exchange columns of U and rows of VH.
      DO 25 I=1,M
        HOLD=U(I,N)
        U(I,N)=U(I,IMAX)
        U(I,IMAX)=HOLD
        HOLD=VH(N,I)
        VH(N,I)=VH(IMAX,I)
        VH(IMAX,I)=HOLD
 25   CONTINUE
      N=N+1
      IF(N.LE.M) GO TO 23
      RETURN
      END
```

2.4 Fourier series and transforms

In the discrete, equally spaced case, the discrete Fourier transform (DFT) pair, (2.147) and (2.148), provide a one-to-one mapping between the time and frequency domains. If we hold the record length fixed at T but increase the sampling rate indefinitely, $N \to \infty$ and $\Delta t \to 0$, making the sequence g_j a continuous function of time, $g(t)$. Thus,

$$G_k \to \int_{-T/2}^{T/2} g(t)\, e^{-i2\pi f_k t}\, dt, \tag{2.244}$$

and

$$g_j \to g(t) = \Delta f \sum_{k=-\infty}^{\infty} G_k e^{i2\pi f_k t}. \tag{2.245}$$

This is the infinite *Fourier series* representation of the continuous function of time, $g(t)$, defined for $-T/2 < t < T/2$,

$$g(t) = \frac{1}{T} \sum_{k=-\infty}^{\infty} G_k e^{i2\pi \frac{k}{T} t}, \tag{2.246}$$

with coefficients given by the Euler formula

$$G_k = \int_{-T/2}^{T/2} g(t)\, e^{-i2\pi \frac{k}{T} t}\, dt. \tag{2.247}$$

The Fourier series representation is periodic in the record length, as is the representation (2.147) in the DFT pair, seen previously. In the case of the Fourier series representation of $g(t)$, we have,

$$\begin{aligned} g(t+T) &= \frac{1}{T} \sum_{k=-\infty}^{\infty} G_k\, e^{i2\pi \frac{k}{T}(t+T)} \\ &= \frac{1}{T} \sum_{k=-\infty}^{\infty} G_k\, e^{i2\pi \frac{k}{T} t + i2\pi k} \\ &= \frac{1}{T} \sum_{k=-\infty}^{\infty} G_k\, e^{i2\pi \frac{k}{T} t} \\ &= g(t). \end{aligned} \tag{2.248}$$

If we now increase the record length without limit, $T \to \infty$ and $\Delta f = 1/T \to 0$. Thus $k\Delta f$ becomes the continuous frequency variable, f, and G_k becomes the continuous function of frequency, $G(f)$. We then have that

$$g(t) = \int_{-\infty}^{\infty} G(f)\, e^{i2\pi f t}\, df, \tag{2.249}$$

the *Fourier integral* representation of $g(t)$, and

$$G(f) = \int_{-\infty}^{\infty} g(t)\, e^{-i2\pi f t}\, dt, \tag{2.250}$$

the *Fourier transform* of $g(t)$.

The Fourier series converges if $g(t)$ satisfies the *Dirichlet conditions*: $g(t)$ is piecewise continuous and has only a finite number of minima and maxima on the interval $-T/2 \le t \le T/2$. The Fourier transform exists if $g(t)$ is absolutely integrable, or

$$\int_{-\infty}^{\infty} |g(t)|\, dt < \infty, \tag{2.251}$$

for then,

$$\begin{aligned} \int_{-\infty}^{\infty} g(t)\, e^{-i2\pi f t}\, dt &\le \int_{-\infty}^{\infty} \left| g(t)\, e^{-i2\pi f t} \right| dt \\ &= \int_{-\infty}^{\infty} |g(t)| \cdot 1 \cdot dt < \infty. \end{aligned} \tag{2.252}$$

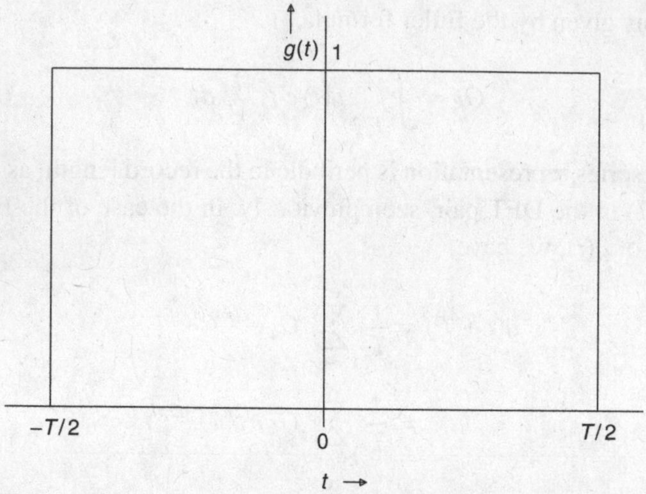

Figure 2.5 The *boxcar* function $g(t)$ to be used in modelling finite length records.

We conclude this section with the Fourier transforms of time functions to be used in subsequent analyses.

First, there is the Fourier transform of the *boxcar* function of time, defined by

$$g(t) = \begin{cases} 0, & |t| > T/2, \\ 1, & |t| \le T/2. \end{cases} \tag{2.253}$$

As illustrated in Figure 2.5, it is of unit height, extending over the interval $-T/2 < t < T/2$.

The Fourier transform of the boxcar is

$$\begin{aligned} G(f) &= \int_{-\infty}^{\infty} g(t)\, e^{-i2\pi f t}\, dt = \int_{-T/2}^{T/2} e^{-i2\pi f t}\, dt \\ &= \left. \frac{e^{-i2\pi f t}}{-i2\pi f} \right|_{-T/2}^{T/2} = \frac{e^{-i2\pi f T/2} - e^{i2\pi f T/2}}{-i2\pi f} = T\,\frac{\sin(\pi f T)}{\pi f T}. \end{aligned} \tag{2.254}$$

Introducing the *sinc* function, defined as

$$\operatorname{sinc}(x) = \frac{\sin(x)}{x}, \tag{2.255}$$

the Fourier transform of the boxcar function of time can be written

$$G(f) = T \operatorname{sinc}(\pi f T). \tag{2.256}$$

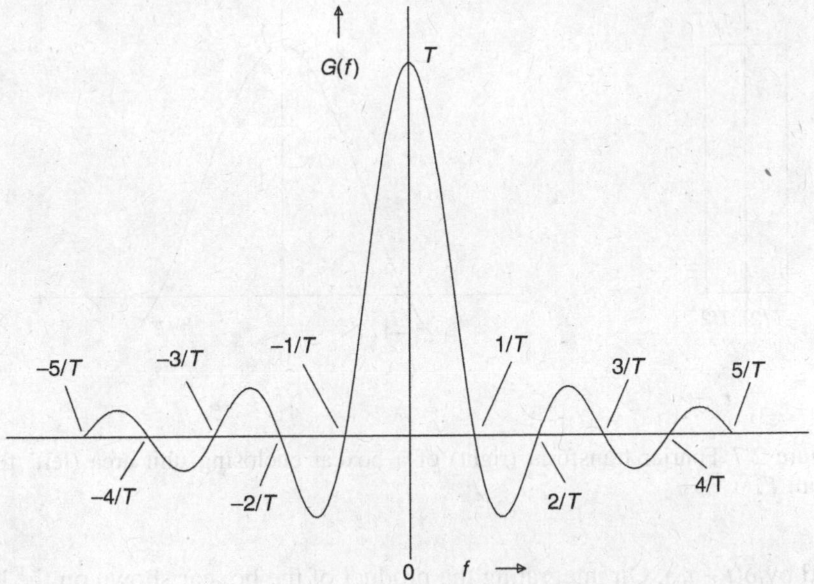

Figure 2.6 The Fourier transform of the boxcar time function, a sinc function of frequency. The sidelobes fall off according to the reciprocal of the distance along the frequency axis from the main lobe.

Figure 2.6 shows a plot of $G(f)$ for the band of frequencies $-5/T < f < 5/T$. The height of the nth sidelobe of the sinc function (2.256) is approximately

$$\frac{(-1)^n}{\pi\,(2n+1)\,/2T} = \frac{(-1)^n 2T}{(2n+1)\,\pi}, \quad n = 1, 2, 3, \ldots. \tag{2.257}$$

The first sidelobe is negative, and 21.2% in magnitude, compared to the main lobe.

Next, we examine the Fourier transform of a different boxcar, shown in Figure 2.7. The area enclosed by this boxcar is unity. It is identical to the previous boxcar, except that it is scaled by $1/T$. It then has the Fourier transform $\mathrm{sinc}(\pi f T)$. This transform pair illustrates a general property of Fourier transforms. A function of short duration in the time domain has a Fourier transform of long duration in the frequency domain. Similarly, a function of long duration in the time domain will have a Fourier transform of short duration in the frequency domain. Now, consider the limit as $T \to 0$. The time domain function vanishes everywhere, except at the origin where it goes to infinity. This infinity is such that the area under the function remains unity. The frequency domain function becomes unity for all frequencies.

In this limit of $T \to 0$, the time domain function becomes a *Dirac delta* function at the time origin. It is denoted by $\delta(t)$. Located at time t_0, a Dirac delta function is

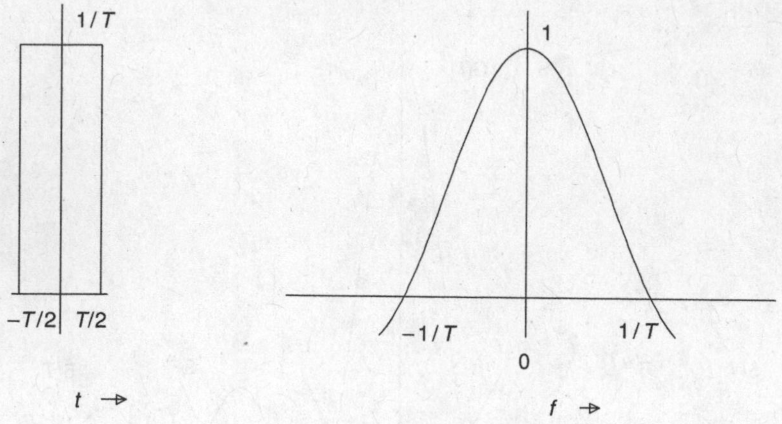

Figure 2.7 Fourier transform (right) of a boxcar enclosing unit area (left) for small T.

denoted by $\delta(t - t_0)$. On integrating the product of the boxcar shown on the left of Figure 2.7, but centred on time $t = t_0$, with a function of time, $f(t)$, we find that

$$\int_{t_0-T/2}^{t_0+T/2} f(t)\, b(t)\, dt = \frac{\bar{f}(t)}{T} \cdot T, \tag{2.258}$$

with $b(t)$ the boxcar, and where $\bar{f}(t)$ lies between the minimum and maximum of $f(t)$ on the closed interval $[t_0 - T/2,\ t_0 + T/2]$, by the mean value theorem for integrals. Then,

$$\lim_{T\to 0} \int_{t_0-T/2}^{t_0+T/2} f(t)\, b(t)\, dt = \int_{t_0-T/2}^{t_0+T/2} f(t)\, \delta(t - t_0)\, dt = f(t_0). \tag{2.259}$$

The Fourier transform of the Dirac delta function, $\delta(t - t_0)$, is

$$\int_{-\infty}^{\infty} \delta(t - t_0)\, e^{-i2\pi f t}\, dt = e^{-i2\pi f t_0}. \tag{2.260}$$

The Fourier integral representation of $\delta(t - t_0)$ is then

$$\delta(t - t_0) = \int_{-\infty}^{\infty} e^{-i2\pi f t_0} \cdot e^{i2\pi f t} df. \tag{2.261}$$

Thus,

$$\delta(t - t_0) = \int_{-\infty}^{\infty} e^{i2\pi f (t - t_0)} df. \tag{2.262}$$

2.4.1 Convolution theorems

Consider a function of time, $k(t)$, whose Fourier transform, $K(f)$, is the product of two Fourier transforms. Then,

$$K(f) = G(f) \cdot H(f), \tag{2.263}$$

where $G(f)$ and $H(f)$ are the Fourier transforms of the functions of time, $g(t)$ and $h(t)$, respectively. The Fourier integral representation of $k(t)$ is

$$k(t) = \int_{-\infty}^{\infty} G(f)\, H(f)\, e^{i2\pi ft}\, df. \tag{2.264}$$

We can write

$$G(f) = \int_{-\infty}^{\infty} g(t_1)\, e^{-i2\pi f t_1}\, dt_1, \quad H(f) = \int_{-\infty}^{\infty} h(t_2)\, e^{-i2\pi f t_2}\, dt_2. \tag{2.265}$$

Thus,

$$\begin{aligned} k(t) &= \int_{-\infty}^{\infty} \cdot \int_{-\infty}^{\infty} g(t_1)\, e^{-i2\pi f t_1}\, dt_1 \cdot \int_{-\infty}^{\infty} h(t_2)\, e^{-i2\pi f t_2}\, dt_2 \cdot e^{i2\pi ft}\, df \\ &= \int_{-\infty}^{\infty} g(t_1) \int_{-\infty}^{\infty} h(t_2) \cdot \int_{-\infty}^{\infty} e^{i2\pi f(t-t_1-t_2)}\, df \cdot dt_2\, dt_1. \end{aligned} \tag{2.266}$$

From (2.262),

$$\int_{-\infty}^{\infty} e^{i2\pi f(t-t_1-t_2)} df = \delta(t - t_1 - t_2). \tag{2.267}$$

Hence,

$$k(t) = \int_{-\infty}^{\infty} g(t_1) \int_{-\infty}^{\infty} h(t_2)\, \delta(t - t_1 - t_2)\, dt_2\, dt_1, \tag{2.268}$$

or

$$k(t) = \int_{-\infty}^{\infty} g(t_1)\, h(t - t_1)\, dt_1. \tag{2.269}$$

$k(t)$ is called the *convolution* of $g(t)$ and $h(t)$. For continuous signals this is the analogue of the convolution of discrete sequences considered in Section 2.1.2. The Fourier transform of the convolution of two functions of time is the product of their Fourier transforms, just as the z-transform of the convolution of two discrete time sequences is the product of their z-transforms, as shown in Section 2.1.2.

In the Fourier integral (2.249) and Fourier transform (2.250) pair, f and t play symmetrical rôles. The Fourier transform of the product of the functions of time, $g(t)\,h(t) = k(t)$, is

$$\begin{aligned}\int_{-\infty}^{\infty} & g(t)\,h(t)\,e^{-i2\pi ft}\,dt \\ &= \int_{-\infty}^{\infty} \cdot \int_{-\infty}^{\infty} G(f_1)\,e^{i2\pi f_1 t}\,df_1 \cdot \int_{-\infty}^{\infty} H(f_2)\,e^{i2\pi f_2 t}\,df_2 \cdot e^{-i2\pi ft}\,dt \\ &= \int_{-\infty}^{\infty} G(f_1) \int_{-\infty}^{\infty} H(f_2) \int_{-\infty}^{\infty} e^{i2\pi(f_1+f_2-f)t}\,dt\,df_2\,df_1. \end{aligned} \tag{2.270}$$

If we replace t by f and t_0 by f_0 in the Fourier integral representation of the Dirac delta function (2.262), we have

$$\delta(f - f_0) = \int_{-\infty}^{\infty} e^{i2\pi t(f-f_0)}\,dt, \tag{2.271}$$

or

$$\delta(f_0 - f) = \int_{-\infty}^{\infty} e^{i2\pi(f_0-f)t}\,dt, \tag{2.272}$$

and

$$\delta(f_1 + f_2 - f) = \int_{-\infty}^{\infty} e^{i2\pi(f_1+f_2-f)t}\,dt. \tag{2.273}$$

Expression (2.270) for the Fourier transform of the product of the two functions of time, $k(t) = g(t)\,h(t)$, becomes

$$\begin{aligned} K(f) &= \int_{-\infty}^{\infty} g(t)\,h(t)\,e^{-i2\pi ft}\,dt \\ &= \int_{-\infty}^{\infty} G(f_1) \int_{-\infty}^{\infty} H(f_2)\,\delta(f_1 + f_2 - f)\,df_2\,df_1 \\ &= \int_{-\infty}^{\infty} G(f_1)\,H(f - f_1)\,df_1. \end{aligned} \tag{2.274}$$

Thus, the Fourier transform of the product of two functions of time is the convolution, in the frequency domain, of their Fourier transforms.

2.4.2 The effect of finite record length

Any real, physical record of a function of time, $g(t)$, is bound to be of finite length T. Yet, to calculate its Fourier transform (2.250), we require a record extending indefinitely in both directions in time. Instead, we have only $g(t)$ for

$-T/2 \le t \le T/2$. The function of time we have access to is not $g(t)$ but $h(t) = b(t)\,g(t)$, where $b(t)$ is the boxcar function (2.253)

$$b(t) = \begin{cases} 0, & |t| > T/2, \\ 1, & |t| \le T/2. \end{cases} \tag{2.275}$$

Thus, in attempting to calculate the Fourier transform $G(f)$ from the finite record, we find instead the Fourier transform of the product, $h(t) = b(t)\,g(t)$, of two functions of time. By our result (2.274), the Fourier transform, found from the finite record, is the convolution of the true transform with the sinc function of frequency, representing the Fourier transform of the boxcar (2.256). It is

$$\begin{aligned} H(f) &= \int_{-\infty}^{\infty} T\,\mathrm{sinc}(\pi f' T)\,G(f - f')\,df' \\ &= \int_{-\infty}^{\infty} T\,\mathrm{sinc}[\pi(f - \lambda)T]\,G(\lambda)\,d\lambda, \end{aligned} \tag{2.276}$$

on changing the integration variable from f' to $\lambda = f - f'$. The Fourier transform found from the finite record for frequency f is the average of the true transform, taken with the sinc averager centred on f. The whole of $G(\lambda)$ contributes to each value of $H(f)$. The only case in which the true transform is returned is that in which $G(\lambda)$ is everywhere unity. In that case, the limit of the sinc integral gives

$$\int_0^{\infty} \frac{\sin x}{x}\,dx = \frac{\pi}{2}, \tag{2.277}$$

or

$$\pi T \int_0^{\infty} \mathrm{sinc}(\pi f T)\,df = \frac{\pi}{2}, \tag{2.278}$$

and

$$\int_{-\infty}^{\infty} T\,\mathrm{sinc}(\pi f T)\,df = 1. \tag{2.279}$$

Otherwise, the finite length of the actual record causes an undesirable mixture of frequency content.

The question then arises as to how to mitigate the frequency mixing effects of finite record length. We can obtain the finite record by multiplying $g(t)$ by a *window* other than the boxcar. Suppose we use the triangular function, $\Delta(t)$, called

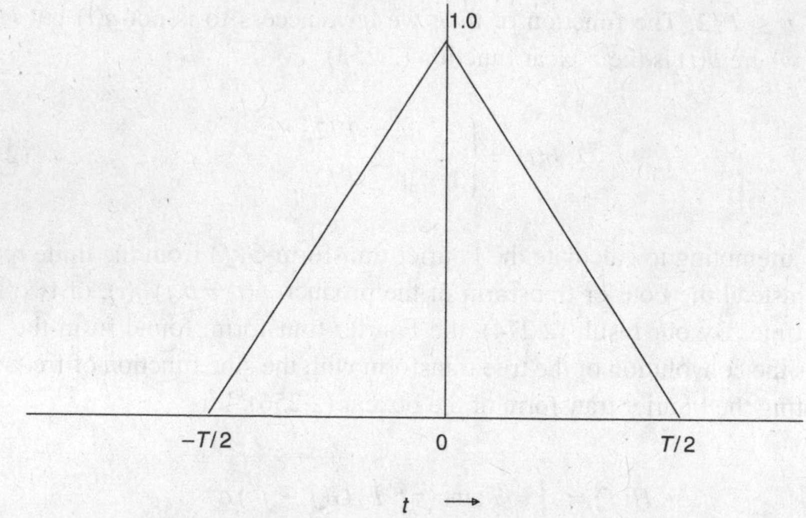

Figure 2.8 The triangular Bartlett window $\Delta(t)$.

the *Bartlett window*, shown in Figure 2.8. The Fourier transform of the Bartlett window is

$$\begin{aligned}
&\int_{-\infty}^{\infty} \Delta(t)\, e^{-i2\pi f t} dt \\
&= \int_{-T/2}^{0} \left(1 + \frac{2t}{T}\right) e^{-i2\pi f t} dt + \int_{0}^{T/2} \left(1 - \frac{2t}{T}\right) e^{-i2\pi f t} dt \\
&= T \operatorname{sinc}(\pi f T) - \frac{4}{T} \int_{0}^{T/2} t \cos(2\pi f t)\, dt \\
&= T \operatorname{sinc}(\pi f T) - \frac{4}{T} \left[T^2 \frac{\sin(\pi f T)}{4\pi f T} + \frac{1}{(2\pi f)^2} \left\{ \cos(\pi f T) - 1 \right\} \right] \\
&= \frac{T}{(\pi f T)^2} \left[1 - \cos(\pi f T) \right] = \frac{T}{(\pi f T)^2} \left[1 - 1 + 2 \sin^2 \left(\frac{\pi f T}{2} \right) \right] \\
&= \frac{T}{2} \left[\operatorname{sinc} \left(\frac{\pi f T}{2} \right) \right]^2 . \qquad (2.280)
\end{aligned}$$

The Bartlett frequency window, the Fourier transform of $\Delta(t)$, is shown in Figure 2.9. The height of the nth sidelobe is approximately

$$\frac{T}{2} \cdot \frac{1}{(T \cdot \pi\, (2n+1)\,/2T)^2} = \frac{2T}{(2n+1)^2\, \pi^2}, \qquad n = 1, 2, 3, \ldots . \qquad (2.281)$$

Although the main lobe is twice as wide as that of the frequency window of the boxcar, shown in Figure 2.6, the height of the first sidelobe of the Bartlett frequency

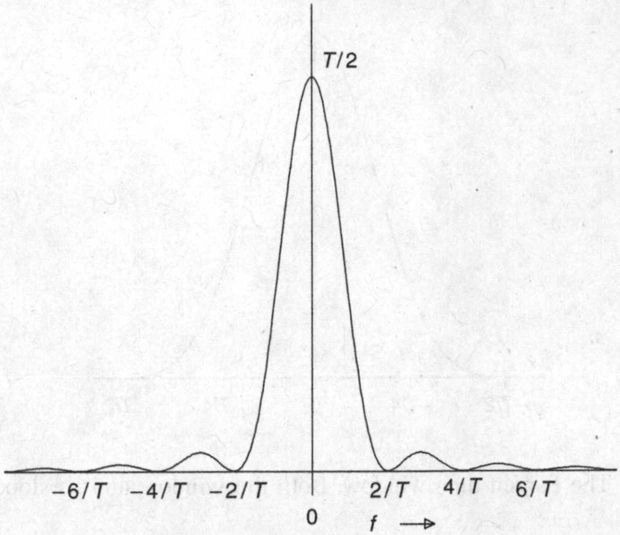

Figure 2.9 The Fourier transform of the Bartlett window as a function of frequency.

window is only 4.5% of the main lobe, instead of 21.2% as is the case for the frequency window of the boxcar. In addition, the sidelobe heights fall off inversely as the square of the distance along the frequency axis from the midpoint of the window, instead of just inversely as in the case of the boxcar frequency window.

The success of the Bartlett window, in mitigating the frequency mixing that arises from calculating Fourier transforms from finite records, raises the question of the possibility of further improvement. We note that the Fourier transform of the Bartlett window (2.280) can be written as the product of two Fourier transforms, both of the form

$$\sqrt{\frac{T}{2}}\mathrm{sinc}\frac{\pi f T}{2}. \tag{2.282}$$

This is the Fourier transform of a boxcar extending over $-T/4 < t < T/4$ of height $\sqrt{2/T}$. Thus, by (2.269), the Bartlett window is the convolution of two of these boxcars. Further improvement is indeed possible, by continuing the process, and taking the convolution of two Bartlett windows to construct the *Parzen window*. The Parzen window is the convolution of two Bartlett windows, extending over $-T/4 < t < T/4$ with height $\sqrt{6/T}$. Its Fourier transform is

$$P(f) = \frac{3T}{8}\mathrm{sinc}^4\left(\frac{\pi f T}{4}\right). \tag{2.283}$$

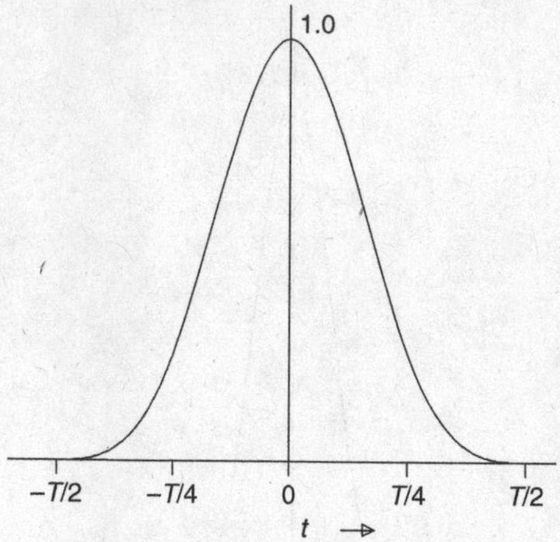

Figure 2.10 The Parzen time window. Both the window and its slope vanish at $t = \pm T/2$.

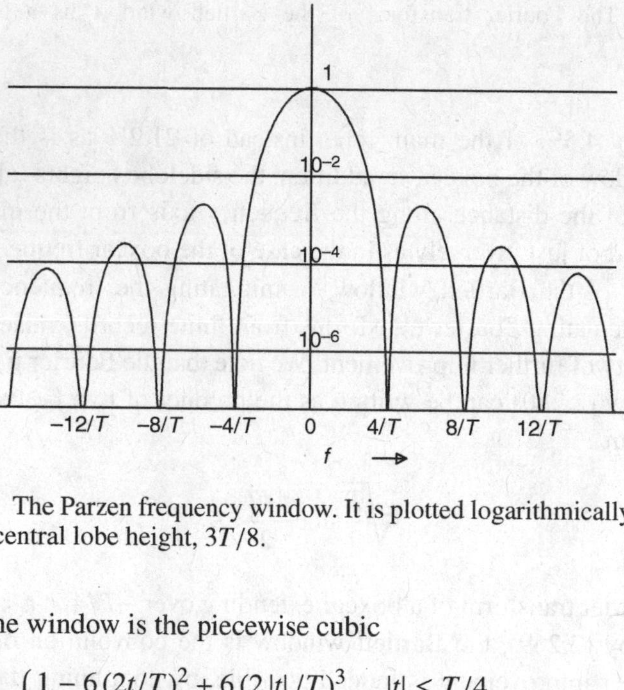

Figure 2.11 The Parzen frequency window. It is plotted logarithmically, as a fraction of the central lobe height, $3T/8$.

The Parzen time window is the piecewise cubic

$$p(t) = \begin{cases} 1 - 6\,(2t/T)^2 + 6\,(2\,|t|\,/T)^3\,, & |t| \leq T/4, \\ 2\,(1 - 2\,|t|\,/T)^3\,, & T/4 < |t| < T/2, \\ 0, & |t| \geq T/2. \end{cases} \tag{2.284}$$

The Parzen time window is shown in Figure 2.10 and the Parzen frequency window, its Fourier transform, is illustrated in Figure 2.11. To show the sidelobes clearly, the Parzen frequency window is plotted logarithmically, as a fraction of the peak height of the main lobe.

The height of the nth sidelobe of the Parzen frequency window is approximately

$$\frac{3T}{8}\left(\frac{2}{(2n+1)\pi}\right)^4, \quad n = 1, 2, 3, \ldots . \tag{2.285}$$

The first sidelobe is only 0.2% of the height of the main lobe, and the sidelobes fall off inversely as the fourth power of the distance along the frequency axis from the centre point of the window. The width of the main lobe is twice that of the Bartlett frequency window, as the width of the main lobe of the Bartlett window is twice that of the boxcar frequency window. The suppression of frequency mixing has been accomplished at the cost of a slight decrease in frequency resolution. The half-power points of the Parzen frequency window are $1.82/T$ apart on the frequency axis, giving an indication of the frequency resolution limit imposed by the use of this window to suppress frequency mixing in finite records.

Window carpentry is a highly developed art, but improvements over the Parzen window are marginal. In the time domain the Parzen window is conveniently a piecewise cubic. We will retain the Parzen window as our commonly used window. The subroutine WINDOW calculates the Parzen window value W, at time T, for a window of length AM, centred at time T0.

```
      SUBROUTINE WINDOW(T,T0,AM,W)
C
C WINDOW finds the value of the PARZEN window, W, at time T,
C for a window of length AM centred at time T0.
C
      IMPLICIT DOUBLE PRECISION(A-H,O-Z)
C Find relative time on (-1,1) time base.
      X=2.D0*(T-T0)/AM
C Find unity minus absolute value of relative time.
      OMX=1.D0-DABS(X)
C If absolute value of relative time is greater than 0.5,
C select alternate cubic.
      IF(DABS(X).GT.0.5D0) GO TO 10
      W=1.D0-6.D0*OMX*X*X
      GO TO 20
10    W=2.D0*OMX*OMX*OMX
20    CONTINUE
      RETURN
      END
```

2.4.3 *The effects of discrete sampling*

Modern data sequences are almost exclusively digital. Usually, the signals are continuous functions of time, reduced to discrete values by a digitiser.

The digitiser operates by integrating the signal over a small but finite length of time. The sampling process can be modelled by taking the product of the modified boxcar, shown on the left of Figure 2.7, with the continuous signal $g(t)$. We take the mean value of $g(t)$ over the interval $[-T/2, T/2]$ about the point to be sampled, and then let $T \to 0$. The modified boxcar, in this limit, becomes the Dirac delta function. Taking $2N + 1$ samples is equivalent to forming the product of $g(t)$ with the *finite Dirac comb*, to obtain the sequence of samples, $h(t)$. Thus,

$$h(t) = g(t) \cdot \sum_{j=-N}^{N} \delta(t - j\Delta t), \tag{2.286}$$

where Δt is the uniform sample interval.

To see the effect of discrete sampling in the frequency domain, we take the Fourier transform of the finite Dirac comb. It is

$$\int_{-\infty}^{\infty} \sum_{j=-N}^{N} \delta(t - j\Delta t)\, e^{-i2\pi ft} dt = \sum_{j=-N}^{N} e^{-i2\pi f j\Delta t}. \tag{2.287}$$

The right-hand side is the sum of a finite geometric progression, with first term

$$a = e^{i2\pi f N\Delta t}, \tag{2.288}$$

and common ratio

$$r = e^{-i2\pi f\Delta t}. \tag{2.289}$$

The sum is then

$$\begin{aligned}
\sum_{j=-N}^{N} e^{-i2\pi f j\Delta t} &= \frac{a\left(1 - r^{2N+1}\right)}{1 - r} \\
&= e^{i2\pi f N\Delta t}\frac{1 - e^{-i2\pi f(2N+1)\Delta t}}{1 - e^{-i2\pi f\Delta t}} \\
&= \frac{e^{i\pi f(2N+1)\Delta t} - e^{-i\pi f(2N+1)\Delta t}}{e^{i\pi f\Delta t} - e^{-i\pi f\Delta t}} \\
&= \frac{\sin[\pi f\,(2N+1)\,\Delta t]}{\sin(\pi f\Delta t)} \\
&= \frac{\sin(\pi f T)}{\sin(\pi f\Delta t)},
\end{aligned} \tag{2.290}$$

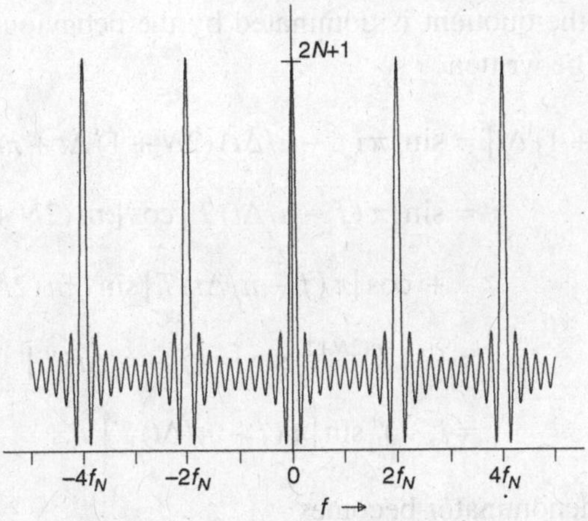

Figure 2.12 The Fourier transform of the finite Dirac comb with 21 sample points and unit sample interval Δt. The transform is periodic in frequency with period equal to twice the Nyquist frequency, $2f_N = 1/\Delta t$.

where $T = (2N + 1)\,\Delta t$ is the record length. A graph of the Fourier transform of the finite Dirac comb sampler is shown in Figure 2.12 for $2N + 1 = 21$ sample points with unit sample interval. It is periodic in frequency, with period equal to twice the *Nyquist frequency*, $2f_N = 1/\Delta t$.

The Fourier transform of the finite Dirac comb sampler (2.290), illustrated in Figure 2.12, requires some interpretation. The numerator has zeros at $fT = 0, \pm1, \pm2, \ldots$. The denominator has zeros at $f\Delta t = 0, \pm1, \pm2, \ldots$, or at $f = 0, \pm1/\Delta t, \pm2/\Delta t, \ldots$. Consider the value of the transform at $f = n/\Delta t$, a zero of the denominator. Since $\sin[\pi(n/\Delta t)T] = \sin[(2N + 1)\,n\pi] = 0$, the numerator has a zero there too. By l'Hôpital's rule,

$$\begin{aligned}
&\lim_{f\to\frac{n}{\Delta t}} \frac{\sin\left[\pi f\,(2N+1)\,\Delta t\right]}{\sin\left(\pi f \Delta t\right)} \\
&= \lim_{f\to\frac{n}{\Delta t}} (2N+1)\,\frac{\cos\left[\pi f\,(2N+1)\,\Delta t\right]}{\cos\left(\pi f \Delta t\right)} \\
&= (2N+1)\,\frac{(-1)^{n(2N+1)}}{(-1)^{n}} \\
&= (-1)^{2Nn}\,(2N+1) = 2N+1.
\end{aligned} \tag{2.291}$$

Both numerator and denominator have zeros and are changing sign at the points $f = n/\Delta t$. Their quotient is therefore not changing sign at these points. The denominator is a much weaker function of frequency than the numerator and, thus, near the points $f = n/\Delta t$ the quotient is dominated by the behaviour of the numerator. The numerator may be written

$$\begin{aligned}\sin[\pi f(2N+1)\Delta t] &= \sin[\pi(f - n/\Delta t)(2N+1)\Delta t + \pi n(2N+1)] \\ &= \sin[\pi(f - n/\Delta t)T]\cos[\pi n(2N+1)] \\ &\quad + \cos[\pi(f - n/\Delta t)T]\sin[\pi n(2N+1)] \\ &= (-1)^{n(2N+1)}\sin[\pi(f - n/\Delta t)T] \\ &= (-1)^n \sin[\pi(f - n/\Delta t)T]. \end{aligned} \tag{2.292}$$

Near $f = n/\Delta t$, the denominator becomes

$$\begin{aligned}\sin[\pi(f - n/\Delta t)\Delta t + n\pi] &= \sin[\pi(f - n/\Delta t)\Delta t]\cos(n\pi) \\ &\quad + \cos[\pi(f - n/\Delta t)\Delta t]\sin(n\pi) \\ &= (-1)^n \sin[\pi(f - n/\Delta t)\Delta t] \\ &\approx (-1)^n \pi(f - n/\Delta t)\Delta t. \end{aligned} \tag{2.293}$$

Thus, near $f = n/\Delta t$, the Fourier transform of the finite Dirac comb sampler is closely

$$\frac{(-1)^n \sin[\pi(f - n/\Delta t)T]}{(-1)^n \pi(f - n/\Delta t)\Delta t} = (2N+1)\,\mathrm{sinc}[\pi(f - n/\Delta t)T]. \tag{2.294}$$

Hence, as illustrated in Figure 2.12, the transform resembles a series of sinc functions of amplitude $2N + 1$, centred on the frequencies $f = n/\Delta t = 2nf_N$.

To isolate the effects of discrete sampling from finite record effects, we take the limit as $N \to \infty$, $T \to \infty$. From equation (2.279), $T\mathrm{sinc}(\pi fT)$ contains unit area. Hence, $(2N+1)\,\mathrm{sinc}[\pi(f - n/\Delta t)T]$ contains area $(2N+1)/T = 1/\Delta t$. As N, $T \to \infty$, with Δt constant, each of the sinc functions in the Fourier transform of the finite Dirac comb sampler become distinct and very narrow. In the limit, each one approaches

$$\frac{1}{\Delta t}\delta(f - n/\Delta T). \tag{2.295}$$

The Fourier transform of the infinite Dirac comb,

$$\sum_{j=-\infty}^{\infty} \delta(t - j\Delta t), \tag{2.296}$$

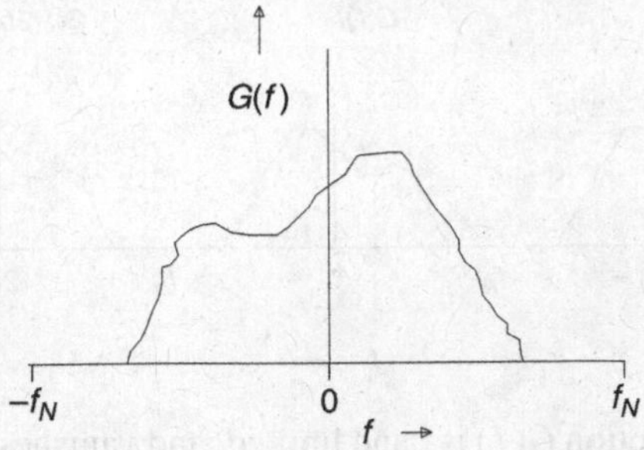

Figure 2.13 Illustration of a *band limited* function of frequency that vanishes outside the Nyquist band, $-f_N < f < f_N$.

is then

$$\frac{1}{\Delta t}\sum_{n=-\infty}^{\infty}\delta(f - n/\Delta t). \tag{2.297}$$

Using the infinite Dirac comb scaled by Δt,

$$\Delta t \sum_{j=-\infty}^{\infty}\delta(t - j\Delta t), \tag{2.298}$$

as the sampler, the Fourier transform of the sampled record is the transform of the product of time functions,

$$h(t) = g(t)\cdot \Delta t \sum_{j=-\infty}^{\infty}\delta(t - j\Delta t). \tag{2.299}$$

The transform is then

$$H(f) = \int_{-\infty}^{\infty} G(f - f')\sum_{n=-\infty}^{\infty}\delta\left(f' - \frac{n}{\Delta t}\right)df'. \tag{2.300}$$

It is the convolution of the true transform, $G(f)$, with the infinite Dirac comb in the frequency domain.

To see the effects of discrete sampling, we first suppose $G(f)$ vanishes outside the band of frequencies $-f_N < f < f_N$, with $f_N = 1/2\Delta t$ the Nyquist frequency, as illustrated in Figure 2.13. Such a function of frequency is said to be *band limited.* If the sampling interval Δt is made sufficiently small, any function of

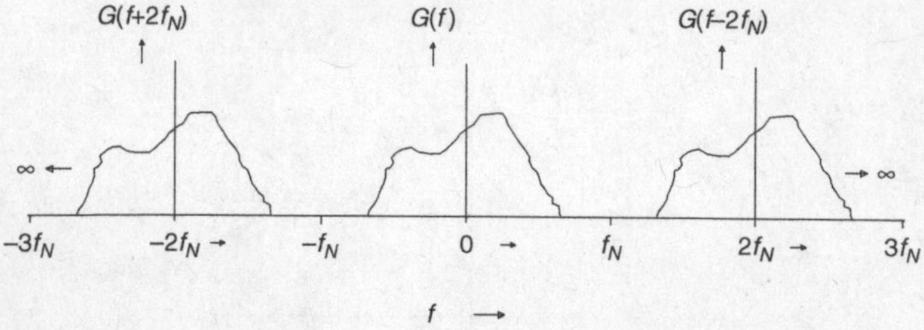

Figure 2.14 If the function $G(f)$ is band limited, and vanishes outside the Nyquist band $-f_N < f < f_N$, then discrete sampling only produces a periodic repetition of the true transform up and down the frequency axis, without any other distortion.

frequency encountered in practice can be made band limited. For a band limited function of frequency, the Fourier transform calculated from the sampled record is simply

$$H(f) = \sum_{n=-\infty}^{\infty} G(f - 2nf_N). \tag{2.301}$$

This represents a periodic repetition of the true transform, $G(f)$, along the frequency axis, as shown in Figure 2.14.

If $G(f)$ is not band limited, and extends beyond the range of frequencies $-f_N < f < f_N$, then

$$H(f) = \cdots + G(f + 2f_N) + G(f) + G(f - 2f_N) + \cdots . \tag{2.302}$$

The transform in the interval $f_N \le f \le 3f_N$ is *folded back* onto the interval $-f_N \le f \le f_N$, and the transform in the interval $-3f_N \le f \le -f_N$ is folded back onto the interval $-f_N \le f \le f_N$, and so on. The folding of the frequency axis is illustrated in Figure 2.15. The contents of portions of the frequency axis, multiples of $2f_N$ away, are *aliased* back onto the principal Nyquist band $-f_N \le f \le f_N$. Thus, if significant portions of $G(f)$ lie outside the principal Nyquist band, $H(f)$ will be seriously distorted by *aliasing*. To prevent distortion by aliasing, it is necessary to choose Δt sufficiently small, so that $f_N = 1/2\Delta t$ is larger than the highest significant frequency, in both the content of the desired signal and the content of the ever present noise. If $T_N = 1/f_N$ is the period of the highest frequency component present, then Δt must be chosen to be no greater than $T_N/2$, or we must sample at least *twice per cycle of the highest frequency* present, in either signal or noise, to prevent distortion by aliasing.

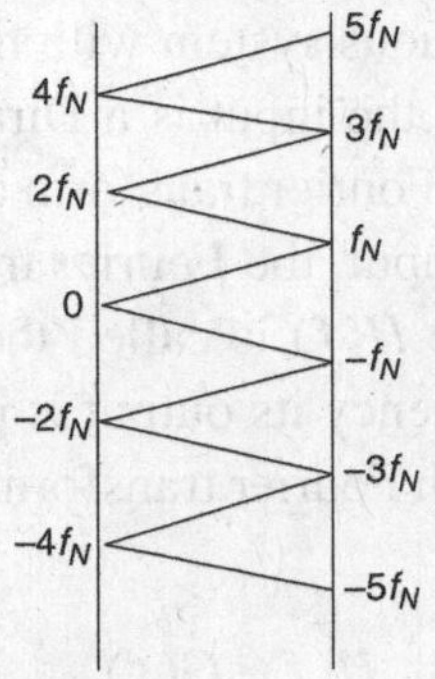

Figure 2.15 Folding of the content on the frequency axis causing *aliasing* in the case of time sequences not limited in their frequency content to the principal Nyquist band, $-f_N \leq f \leq f_N$.

2.5 Power spectral density estimation

The power spectral density, which displays how power in a sequence is distributed over the frequency axis, is a very useful tool in the search for new signals in background noise. Usually, a new signal, to be observed for the first time, will be obscured by background noise, otherwise it would have been found previously. Thus, the estimation of power spectral density is a very important topic for consideration.

2.5.1 Autocorrelation and spectral density

In our earlier discussion of discrete time sequences (Section 2.1.3), we defined the autocorrelation at lag j, time index k, of an equispaced time sequence f, as

$$\phi_{ff}(j,k) = E\left\{f_k f_{k-j}^*\right\}. \tag{2.303}$$

The equivalent definition for a continuous sequence at lag τ, time t, is

$$\phi_{ff}(\tau,t) = E\left\{f(t)\, f^*(t-\tau)\right\}. \tag{2.304}$$

For stationary sequences, the autocorrelation becomes a function of lag alone,

$$\phi_{ff}(\tau) = E\left\{f(t)\, f^*(t-\tau)\right\}. \tag{2.305}$$

Again, for a stationary process, the expectation over an ensemble of realisations may be replaced, by the ergodic hypothesis, with the average over time of an infinite record. Hence,

$$\phi_{ff}(\tau) = \lim_{T\to\infty} \frac{1}{T} \int_{-T/2}^{T/2} f(t)\, f^*(t-\tau)\, dt. \tag{2.306}$$

Now consider a linear, continuous system with impulse response $h(t)$. By definition of the impulse response, if the input is a Dirac delta function, $\delta(t)$, then the output of the system is $h(t)$. The Fourier transform of the Dirac delta function, $\delta(t)$, is unity (2.260). Thus, for this input, the Fourier transform of the output from the linear system is $H(f) \cdot 1$, where $H(f)$ is called the *system function*. Because the system is linear, at a given frequency its output is proportional to the input at that frequency. If the input is $g(t)$ with Fourier transform $G(f)$, the Fourier transform of the output is

$$H(f) \cdot G(f), \tag{2.307}$$

the product of the two Fourier transforms. Therefore, by (2.269), the output $o(t)$ is the convolution

$$o(t) = \int_{-\infty}^{\infty} h(\lambda)\, g(t-\lambda)\, d\lambda. \tag{2.308}$$

The autocorrelation of the output at lag τ is then

$$\begin{aligned}\phi_{oo}(\tau) &= \lim_{T\to\infty} \frac{1}{T} \int_{-T/2}^{T/2} \int_{-\infty}^{\infty} h(\lambda)\, g(t-\lambda)\, d\lambda \\ &\quad \cdot \int_{-\infty}^{\infty} h^*(\lambda')\, g^*(t-\tau-\lambda')\, d\lambda'\, dt \\ &= \int_{-\infty}^{\infty} \int_{-\infty}^{\infty} h(\lambda)\, h^*(\lambda')\, \phi_{gg}(\tau+\lambda'-\lambda)\, d\lambda\, d\lambda'. \end{aligned} \tag{2.309}$$

If the Fourier transform of the autocorrelation, $\phi_{gg}(\tau)$, considered as a lag domain function, is $S_{gg}(f)$, its Fourier integral representation is

$$\phi_{gg}(\tau) = \int_{-\infty}^{\infty} S_{gg}(f)\, e^{i2\pi f\tau}\, df. \tag{2.310}$$

Thus,

$$\phi_{oo}(\tau) = \int_{-\infty}^{\infty} \int_{-\infty}^{\infty} \int_{-\infty}^{\infty} h(\lambda)\, h^*(\lambda')\, S_{gg}(f)\, e^{i2\pi f(\tau+\lambda'-\lambda)} df\, d\lambda\, d\lambda'. \tag{2.311}$$

The Fourier transform of the impulse response is

$$H(f) = \int_{-\infty}^{\infty} h(t)\, e^{-i2\pi ft}\, dt, \tag{2.312}$$

and its conjugate is

$$H^*(f) = \int_{-\infty}^{\infty} h^*(t)\, e^{i2\pi ft}\, dt. \tag{2.313}$$

The autocorrelation of the output then reduces to

$$\begin{aligned}\phi_{oo}(\tau) &= \int_{-\infty}^{\infty} H(f)\, H^*(f)\, S_{gg}(f)\, e^{i2\pi f\tau}\, df \\ &= \int_{-\infty}^{\infty} |H(f)|^2\, S_{gg}(f)\, e^{i2\pi f\tau}\, df.\end{aligned} \tag{2.314}$$

The autocorrelation of the output at zero lag is

$$\phi_{oo}(0) = \lim_{T\to\infty} \frac{1}{T} \int_{-T/2}^{T/2} o(t)\, o^*(t)\, dt = \lim_{T\to\infty} \frac{1}{T} \int_{-T/2}^{T/2} |o(t)|^2\, dt. \tag{2.315}$$

It represents the power of the output. From expression (2.314), it is also

$$\phi_{oo}(0) = \int_{-\infty}^{\infty} |H(f)|^2\, S_{gg}(f)\, df. \tag{2.316}$$

Function $|H(f)|^2$ is the *power transfer function* of the linear system; $S_{gg}(f)$ represents the *power spectrum* or *power spectral density* of the signal $g(t)$. It gives the distribution of power over the frequency axis, and has the units of power per unit frequency.

The power spectrum and autocorrelation are Fourier transform pairs,

$$S_{gg}(f) = \int_{-\infty}^{\infty} \phi_{gg}(\tau)\, e^{-i2\pi f\tau}\, d\tau, \tag{2.317}$$

$$\phi_{gg}(\tau) = \int_{-\infty}^{\infty} S_{gg}(f)\, e^{i2\pi f\tau}\, df. \tag{2.318}$$

This is sometimes referred to as the *Wiener–Khintchine theorem*.

By definition, the autocorrelation of $g(t)$ is

$$\phi_{gg}(\tau) = \lim_{T\to\infty} \int_{-T/2}^{T/2} g(t)\, g^*(t-\tau)\, dt. \tag{2.319}$$

If we define

$$h(t) = \begin{cases} g(t), & |t| \le T/2, \\ 0, & |t| > T/2 \end{cases} \tag{2.320}$$

then we can write

$$\phi_{gg}(\tau) = \lim_{T\to\infty} \frac{1}{T} \int_{-\infty}^{\infty} h(t)\, h^*(t-\tau)\, dt. \tag{2.321}$$

The Fourier integral representation of $h(t)$ is

$$h(t) = \int_{-\infty}^{\infty} H(f)\, e^{i2\pi ft}\, df, \tag{2.322}$$

giving

$$h^*(t-\tau) = \int_{-\infty}^{\infty} H^*(f)\, e^{-i2\pi f(t-\tau)}\, df. \tag{2.323}$$

Thus, the autocorrelation of $g(t)$ becomes

$$\begin{aligned}\phi_{gg}(\tau) &= \lim_{T\to\infty} \frac{1}{T} \int_{-\infty}^{\infty} h(t) \int_{-\infty}^{\infty} H^*(f)\, e^{-i2\pi f(t-\tau)}\, df\, dt \\ &= \lim_{T\to\infty} \frac{1}{T} \int_{-\infty}^{\infty} H^*(f)\, e^{i2\pi f\tau} \cdot \int_{-\infty}^{\infty} h(t)\, e^{-i2\pi ft}\, dt \cdot df,\end{aligned} \tag{2.324}$$

on reversing the order of integration. Now,

$$\int_{-\infty}^{\infty} h(t)\, e^{-i2\pi ft}\, dt = H(f), \tag{2.325}$$

the Fourier transform of $h(t)$. Hence,

$$\begin{aligned}\phi_{gg}(\tau) &= \lim_{T\to\infty} \frac{1}{T} \int_{-\infty}^{\infty} H^*(f)\, H(f)\, e^{i2\pi f\tau}\, df \\ &= \lim_{T\to\infty} \frac{1}{T} \int_{-\infty}^{\infty} |H(f)|^2\, e^{i2\pi f\tau}\, df = \int_{-\infty}^{\infty} S_{gg}(f)\, e^{i2\pi f\tau}\, df,\end{aligned} \tag{2.326}$$

indicating that the power spectral density is

$$\begin{aligned}S_{gg}(f) &= \lim_{T\to\infty} \frac{1}{T} |H(f)|^2 \\ &= \lim_{T\to\infty} \frac{1}{T} \left| \int_{-\infty}^{\infty} h(t)\, e^{-i2\pi ft}\, dt \right|^2 \\ &= \lim_{T\to\infty} \frac{1}{T} \left| \int_{-T/2}^{T/2} g(t)\, e^{-i2\pi ft}\, dt \right|^2 .\end{aligned} \tag{2.327}$$

In any practical case, we are confined to records of finite length and have only the *sample power spectral density estimator*,

$$\begin{aligned}\tilde{S}_{gg}(f) &= \frac{1}{T} \left| \int_{-T/2}^{T/2} g(t)\, e^{-i2\pi ft}\, dt \right|^2 \\ &= \frac{1}{T} \left| \int_{-\infty}^{\infty} h(t)\, e^{-i2\pi ft}\, dt \right|^2 \qquad (2.328) \\ &= \frac{1}{T} |H(f)|^2, \qquad (2.329)\end{aligned}$$

and the *sample autocorrelation estimator*,

$$\tilde{\phi}_{gg}(\tau) = \frac{1}{T} \int_{-\infty}^{\infty} |H(f)|^2\, e^{i2\pi f\tau}\, df. \tag{2.330}$$

The definition (2.320) of $h(t)$ is equivalently the product $h(t) = b(t)\,g(t)$, where $b(t)$ is the familiar boxcar function (2.253). The sample power spectral density estimator can then be written as

$$\begin{aligned}\tilde{S}_{gg}(f) &= \frac{1}{T}\left|\int_{-\infty}^{\infty} b(t)\,g(t)\,e^{-i2\pi f t}\,dt\right|^2 \\ &= \frac{1}{T}\left|\int_{-\infty}^{\infty} T\,\mathrm{sinc}(\pi f_1 T)\,G(f - f_1)\,df_1\right|^2 \qquad (2.331)\end{aligned}$$

by the convolution theorem, (2.274), for the Fourier transform of the product of two functions of time. Thus, finite record length plays the same rôle here as in the calculation of Fourier transforms, considered in Section 2.4.2. The true Fourier transform, $G(f)$, is convolved with the sinc function, with attendant serious frequency mixing. Again, the same procedure is used to suppress frequency mixing; namely, multiplication of the time sequence by a good window, such as the Parzen window, before the transform is calculated.

For discrete, equispaced data sequences, the finite sampled record is given as in (2.286), but with the finite Dirac comb scaled by Δt, and the sample spectral density estimator (2.328) takes the form

$$\tilde{S}_{gg}(f) = \frac{1}{T}\left|\Delta t \sum_{j=-N}^{N} g_j e^{-i2\pi f j \Delta t}\right|^2 . \qquad (2.332)$$

This is just the squared magnitude of the discrete Fourier transform of the sampled record, divided by the record length. In the discrete case, we have only the samples

$$\phi_{gg}(j\Delta\tau) = E\left\{g(k\Delta t)\,g^*(k\Delta t - j\Delta\tau)\right\} \qquad (2.333)$$

of the continuous autocorrelation for stationary sequences. Modelling the sampling process, as before, by multiplication with the infinite Dirac comb in lag,

$$\Delta t \sum_{j=-\infty}^{\infty} \delta(\tau - j\Delta t), \qquad (2.334)$$

the sampled autocorrelation is

$$\Delta t\,\phi_{gg}(\tau) \sum_{j=-\infty}^{\infty} \delta(\tau - j\Delta t). \qquad (2.335)$$

Its Fourier transform, the power spectral density for equispaced, discrete data, is

$$S_{gg}(f) = \Delta t \sum_{j=-\infty}^{\infty} \phi_{gg}(j)\,e^{-i2\pi f j \Delta t}. \qquad (2.336)$$

We recognise this expression as the infinite Fourier series expansion of the periodic function of frequency, $S_{gg}(f)$. The power spectral density of discrete, equispaced data is thus periodic, with period $2f_N = 1/\Delta t$, where f_N is the Nyquist frequency. The corresponding sequence of autocorrelations can be recovered from the Fourier series (2.336) as the Euler coefficients

$$\phi_{gg}(j) = \int_{-f_N}^{f_N} S_{gg}(f)\, e^{i2\pi f j\Delta t}\, df. \tag{2.337}$$

2.5.2 Multiple discrete segment estimate

The variance of a power spectral density estimate is reduced by averaging over multiple individual estimates, based on discrete segments of the record.

For a windowed data segment of length M, we have for analysis,

$$h(t) = w(t)\, g(t), \tag{2.338}$$

where

$$w(t) = 0, \quad |t| > M/2. \tag{2.339}$$

We window the data with a window extending over $-M/2 < t < M/2$. The sample autocorrelation estimator for this data segment is

$$\tilde{\phi}_{hh}(\tau) = \frac{1}{M}\int_{-M/2}^{M/2} h(t)\, h^*(t-\tau)\, dt. \tag{2.340}$$

To determine what this estimator gives, on average, we find its expected value across an infinite ensemble of realisations of the process generating $g(t)$. The expected value is

$$\begin{aligned} E\left\{\tilde{\phi}_{hh}(\tau)\right\} &= E\left\{\frac{1}{M}\int_{-M/2}^{M/2} h(t)\, h^*(t-\tau)\, dt\right\} \\ &= E\left\{\frac{1}{M}\int_{-M/2}^{M/2} w(t)\, g(t)\, w^*(t-\tau)\, g^*(t-\tau)\, dt\right\}. \end{aligned} \tag{2.341}$$

The window, $w(t)$, is deterministic, real and even, and both integration and taking the expected value are linear operations, giving

$$\begin{aligned} E\left\{\tilde{\phi}_{hh}(\tau)\right\} &= \frac{1}{M}\int_{-M/2}^{M/2} w(t)\, w^*(t-\tau)\, E\left\{g(t)\, g^*(t-\tau)\right\} dt \\ &= \frac{1}{M}\int_{-M/2}^{M/2} w(t)\, w(\tau-t)\, \phi_{gg}(\tau)\, dt \\ &= \frac{1}{M}\phi_{gg}(\tau)\int_{-\infty}^{\infty} w(t)\, w(\tau-t)\, dt. \end{aligned} \tag{2.342}$$

The latter integral is the convolution of the window with itself. By (2.269), its Fourier transform is then the square of the Fourier transform, $W(f)$, of the window. The Fourier transform of the expected value of the estimator (2.342), the spectral density estimator, which is the product of two functions of lag, is then by (2.274) the convolution of their Fourier transforms,

$$\frac{1}{M}\int_{-\infty}^{\infty} W^2(f_1)\, S_{gg}(f - f_1)\, df_1, \tag{2.343}$$

instead of the true spectral density $S_{gg}(f)$. The windowing has two effects. First, the true spectral density is smoothed by convolution with $W^2(f)$. Second, its amplitude is distorted.

To remove the amplitude distortion, we need a reference spectrum with known density. White noise of unit power provides such a reference spectrum, for then

$$S_{gg}(f) = 1, \tag{2.344}$$

and our estimator for spectral density gives

$$\frac{1}{M}\int_{-\infty}^{\infty} W^2(f_1)\cdot 1\cdot df_1 = \frac{1}{M}\int_{-\infty}^{\infty} W^2(f_1)\, df_1. \tag{2.345}$$

From the convolution theorem (2.269), we have that

$$\int_{-\infty}^{\infty} W(f)\cdot W(f)\, e^{i2\pi ft}\, df = \int_{-\infty}^{\infty} w(t_1)\, w(t - t_1)\, dt_1. \tag{2.346}$$

Hence, setting $t = 0$, we obtain

$$\int_{-\infty}^{\infty} W^2(f)\, df = \int_{-\infty}^{\infty} w^2(t)\, dt = I. \tag{2.347}$$

This is the continuous form of *Parseval's relation*, equating the integral of the square in the frequency domain to the integral of the square in the time domain. Thus, to remove the amplitude distortion caused by windowing, we need to divide the spectral density estimator by

$$\frac{I}{M} = \frac{1}{M}\int_{-M/2}^{M/2} w^2(t)\, dt. \tag{2.348}$$

The sample spectral density estimator (2.329) becomes

$$\tilde{S}_{gg}(f) = \frac{|H(f)|^2}{I}. \tag{2.349}$$

For discrete, equispaced, windowed data, the spectral density estimator (2.332) becomes

$$\tilde{S}_{gg}(f_k) = \frac{\left| \Delta t \sum_{j=-N}^{N} w_j g_j e^{-i2\pi f_k j \Delta t} \right|^2}{I}. \tag{2.350}$$

This is the squared magnitude of the discrete Fourier transform of the windowed data, divided by I.

When the window being used is the Parzen window (2.284), $w(t) = p(t)$ and

$$I = \int_{-M/2}^{M/2} p^2(t)\, dt = \frac{151}{560} M, \tag{2.351}$$

on performing the elementary integrations of the square of the piecewise cubic. As seen previously (2.343), the true spectrum is convolved, or averaged, with the square of the Parzen frequency window, whose half-power points are separated by a distance $1.82/M$ along the frequency axis. The single segment estimate is an average over a band of frequencies of the same order. Hence, the selection of the segment length, M, controls the frequency resolution of the spectral density estimate.

In order to assess the significance of a spectral density estimate, it is necessary to model the stochastic process generating the time sequence. Usually, we are dealing with time sequences in which the signal we wish to discover is highly corrupted by noise. Thus, a purely random Gaussian process is normally assumed to be generating the sequence, new results being discovered by significant departures from such a process. The coefficients of the discrete Fourier transform, used in the spectral density estimate, are found by linear combinations of independent realisations of the Gaussian process assumed to be generating the time sequence. If the process has zero mean, so will the coefficients, since they are linear, homogeneous combinations of the realisations of the process. If a given Fourier coefficient is $A(f_k)$, then

$$\frac{A^2(f_k)}{\text{var}\,[A(f_k)]} \tag{2.352}$$

is χ^2 distributed with one degree of freedom, with var$[A(f_k)]$ being the variance of the coefficient $A(f_k)$. Since there are two real Fourier coefficients involved in each raw spectral density estimate, each such estimate is χ_2^2 distributed, the subscript 2 indicating two degrees of freedom.

The usefulness of the single segment spectral density estimator (2.350) lies in the fact that the variance of the spectral density estimate can be reduced by breaking the full record, of total length T, into κ multiple discrete segments of length M,

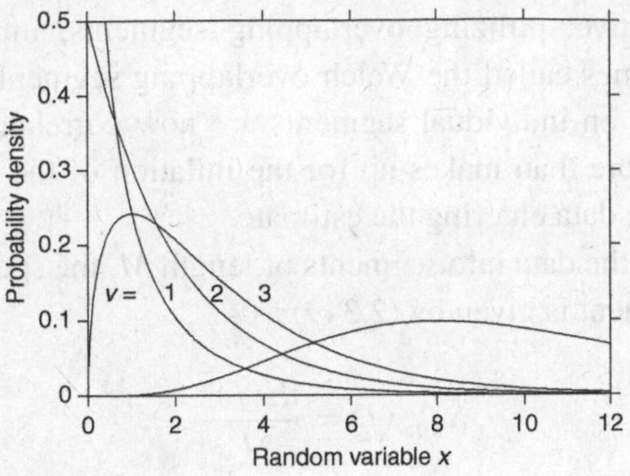

Figure 2.16 The $\chi_\nu^2(x)$ probability density function for degrees of freedom $\nu = 1, 2, 3, 10$. For large ν it approaches that of a normal distribution.

such that $T = \kappa M$. The final spectral density estimate is formed by taking the mean of the κ independent estimates. It is

$$\bar{\tilde{S}}_{gg}(f_k) = \frac{1}{\kappa}\sum_{l=1}^{\kappa} \tilde{S}_{gg,l}(f_k). \tag{2.353}$$

This averaged spectral density estimate has 2κ degrees of freedom, and the variance of an individual estimate is diminished by the same factor. Hence,

$$\frac{\bar{\tilde{S}}_{gg}(f_k)}{S_{gg}(f_k)/2\kappa} \tag{2.354}$$

is $\chi_{2\kappa}^2$ distributed. The probability density for the $\chi_\nu^2(x)$ distribution is shown in Figure 2.16 for degrees of freedom $\nu = 1, 2, 3, 10$ as a function of the random variable x. As the number of degrees of freedom increases without limit, this distribution approaches a normal distribution, as is true of all distributions by the *central limit theorem*.

A confidence interval for the averaged spectral density $\bar{\tilde{S}}_{gg}(f_k)$ can be established from the $\chi_{2\kappa}^2(x)$ probability distribution. For a confidence interval of $(1-\alpha)\times 100\%$, the range is (Jenkins and Watts, 1968, pp. 254–255)

$$\frac{2\kappa\bar{\tilde{S}}_{gg}(f_k)}{x(1-\alpha/2)} \le S_{gg}(f_k) \le \frac{2\kappa\bar{\tilde{S}}_{gg}(f_k)}{x(\alpha/2)}, \tag{2.355}$$

where $x(\zeta)$ is the value of the random variable below which lies a fraction ζ of the area under the cumulative $\chi_{2\kappa}^2$ distribution.

2.5.3 Overlapping segment analysis

Although the average over multiple discrete segments allows the reduction of the variance of the spectral density estimate, it is very wasteful of the data. Here we consider an alternative, utilizing overlapping segments, introduced by Welch (1967). It is sometimes called the Welch overlapping segment analysis (WOSA). While the estimates on individual segments are now correlated, the extra effective record length more than makes up for the inflation of the variance due to the common parts of the data entering the estimate.

If we again break the data into segments of length M, the spectral density estimator for the nth segment is given by (2.349) as

$$\tilde{S}_{gg,n}(f) = \frac{|H(f)|^2}{I}, \tag{2.356}$$

with $H(f)$ the Fourier transform of

$$h(t) = w(t)\, g(t), \tag{2.357}$$

the windowed form of the original time sequence $g(t)$. The expected value of this sample spectral density estimate is

$$\begin{aligned} E\left\{\tilde{S}_{gg,n}(f)\right\} &= \frac{1}{I}\, E\left\{H(f)\, H^*(f)\right\} \\ &= \frac{1}{I}\int_{-\infty}^{\infty}\int_{-\infty}^{\infty} e^{-i2\pi f(t-u)}\, w(t)\, w^*(u)\, E\left\{g(t)\, g^*(u)\right\} dt\, du \\ &= \frac{1}{I}\int_{-\infty}^{\infty}\int_{-\infty}^{\infty} e^{-i2\pi f\tau}\, w(t)\, w^*(t-\tau)\, E\left\{g(t)\, g^*(t-\tau)\right\} dt\, d\tau, \end{aligned} \tag{2.358}$$

on making the variable change $\tau = t - u$. Once again, recognising that the window $w(t)$ is real and even, the expected value of the sample spectral density estimate takes the form

$$E\left\{\tilde{S}_{gg,n}(f)\right\} = \frac{1}{I}\int_{-\infty}^{\infty} e^{-i2\pi f\tau}\, \phi_{gg}(\tau) \cdot \int_{-\infty}^{\infty} w(t)\, w(\tau - t)\, dt \cdot d\tau. \tag{2.359}$$

The latter integral is the Fourier transform of the product of two functions of lag τ, the autocorrelation $\phi_{gg}(\tau)$ and the convolution of the window with itself. Using the convolution results (2.269) and (2.274), the expected value of the sample spectral density estimate is

$$E\left\{\tilde{S}_{gg,n}(f)\right\} = \frac{1}{I}\int_{-\infty}^{\infty} S_{gg}(f - f_1)\, W^2(f_1)\, df_1, \tag{2.360}$$

where $S_{gg}(f)$, the Fourier transform of the autocorrelation $\phi_{gg}(\tau)$, is the true power spectral density of $g(t)$. As before, we see that the estimator, on average, produces a spectral density estimate that is smoothed or convolved with the square of the frequency window, $W^2(f)$. For the Parzen window, the half-power points of $W^2(f)$

are approximately $1.82/M$ apart on the frequency axis. If the true spectral density varies only slightly over this range of frequencies,

$$E\left\{\tilde{S}_{gg,n}(f)\right\} \approx S_{gg}(f), \tag{2.361}$$

and the estimator, on average, gives an *unbiased* estimate of the true spectral density.

In the overlapping segment analysis, the final spectral density estimate, $\tilde{\tilde{S}}_{gg}(f)$, is taken as the average of the estimates (2.360) over κ overlapping segments of length M of a complete record of length T. Thus,

$$\tilde{\tilde{S}}_{gg}(f) = \frac{1}{\kappa}\sum_{j=1}^{\kappa}\tilde{S}_{gg,j}(f) = \frac{1}{\kappa}\sum_{j=1}^{\kappa}\left[\tilde{S}_{gg,j}(f) - S_{gg}(f)\right] + S_{gg}(f). \tag{2.362}$$

The variance of the final spectral density estimate is then

$$\begin{aligned}\operatorname{var}\left\{\tilde{\tilde{S}}_{gg}(f)\right\} &= E\left\{\left[\tilde{\tilde{S}}_{gg}(f) - S_{gg}(f)\right]^2\right\} \\ &= \frac{1}{\kappa}\sum_{j=1}^{\kappa}\sum_{k=1}^{\kappa}E\left\{\left[\tilde{S}_{gg,j}(f) - S_{gg}(f)\right]\left[\tilde{S}_{gg,k}(f) - S_{gg}(f)\right]\right\}.\end{aligned} \tag{2.363}$$

The process generating $g(t)$ is assumed to be stationary, so that the terms in the double sum depend only on the unsigned difference of the indices j and k. Taking m as the non-negative value of this difference, the double sum can be rearranged to give

$$\begin{aligned}\operatorname{var}\left\{\tilde{\tilde{S}}_{gg}(f)\right\} &= \frac{1}{\kappa}\operatorname{var}\left\{\tilde{S}_{gg,n}(f)\right\} \\ &\quad + \frac{2}{\kappa^2}\sum_{m=1}^{\kappa-1}(\kappa - m)\operatorname{cov}\left\{\tilde{S}_{gg,n}(f), \tilde{S}_{gg,n+m}(f)\right\},\end{aligned} \tag{2.364}$$

where the covariance of $\tilde{S}_{gg,n}(f)$ and $\tilde{S}_{gg,n+m}(f)$ is

$$\begin{aligned}\operatorname{cov}\left\{\tilde{S}_{gg,n}(f), \tilde{S}_{gg,n+m}(f)\right\} &= E\left\{\left[\tilde{S}_{gg,n}(f) - S_{gg}(f)\right]\left[\tilde{S}_{gg,n+m}(f) - S_{gg}(f)\right]\right\} \\ &= E\left\{\tilde{S}_{gg,n}(f)\tilde{S}_{gg,n+m}(f)\right\} - \left[E\left\{\tilde{S}_{gg,n}(f)\right\}\right]^2,\end{aligned} \tag{2.365}$$

recognising that both $E\{\tilde{S}_{gg,n}(f)\}$ and $E\{\tilde{S}_{gg,n+m}(f)\}$ are $\approx S_{gg}(f)$.

To evaluate the covariance, we require the expected value of the product of a spectral estimate based on a window $w_n(t)$ of the data centred at b_n on the time axis, with a spectral estimate based on a window $w_{n+m}(t)$ of the data centred at b_{n+m} on the time axis,

$$E\left\{\tilde{S}_{gg,n}(t)\,\tilde{S}_{gg,n+m}(f)\right\} = \frac{1}{I^2}\,E\left\{H_n(f)H_n^*(f)\,H_{n+m}(f)\,H_{n+m}^*(f)\right\}, \qquad (2.366)$$

where $H_n(f)$ and $H_{n+m}(f)$ are Fourier transforms of the respective windowed data segments. Then,

$$\begin{aligned} E\left\{\tilde{S}_{gg,n}(f)\,\tilde{S}_{gg,n+m}(f)\right\} = \frac{1}{I^2}\int_{-\infty}^{\infty}\int_{-\infty}^{\infty}\int_{-\infty}^{\infty}\int_{-\infty}^{\infty} & e^{-i2\pi f(s-t+u-v)} \\ & \cdot w_n(s)\,w_n^*(t)\,w_{n+m}(u)\,w_{n+m}^*(v) \\ & \cdot E\left\{g(s)\,g^*(t)\,g(u)\,g^*(v)\right\} ds\,dt\,du\,dv. \end{aligned} \qquad (2.367)$$

The evaluation of this expected value requires the expected value of a fourth moment of the process generating $g(t)$. The realisations of the process are complex numbers. If they are taken to have real and imaginary parts that jointly are normally distributed with zero mean, at a number of points N along the time axis, the probability density has been given by Wooding (1956) as

$$p\,(g_1, g_2, \ldots, g_N) = \frac{1}{\pi^N\,|\mathbf{C}|}\exp\left(-g_j^* c_{jk}^{-1} g_k\right), \qquad (2.368)$$

where summation is implied over the repeated subscripts j, k, while $\mathbf{C}$ is the Hermitian variance–covariance matrix, $|\mathbf{C}|$ is its determinant and c_{jk}^{-1} are the elements of its inverse. Wooding (1956) also gives the characteristic function for this distribution as

$$\Phi = \exp\left(-\frac{1}{4}\omega_j^* c_{jk}\omega_k\right) \qquad (2.369)$$

$$= \int \exp\left[i\left(\omega_j^* g_j + \omega_j g_j^*\right)/2\right] p\,(g_1, g_2, \ldots, g_N)\,d\mathcal{V}, \qquad (2.370)$$

with the integral over all volume elements $d\mathcal{V}$ of the complex realisation space. Again, summation is implied over repeated subscripts, and c_{jk} are the elements of the variance–covariance matrix $\mathbf{C}$. Labelling the realisations at locations s, t, u, v on the time axis g_1, g_2, g_3, g_4, respectively, power series expansion of the exponential in the integral (2.370) for the characteristic function, and integration term by term, yield

$$\begin{aligned} \Phi &= 1 + \cdots + \frac{1}{4!}\frac{1}{16}\int\left(\cdots + \omega_1^* g_1 + \omega_2 g_2^* + \omega_3^* g_3 + \omega_4 g_4^* + \cdots\right)^4 p\,d\mathcal{V} + \cdots \\ &= 1 + \cdots + \frac{1}{16}\int \omega_1^* g_1 \omega_2 g_2^* \omega_3^* g_3 \omega_4 g_4^*\,p\,d\mathcal{V} + \cdots \\ &= 1 + \cdots + \frac{1}{16}\,\omega_1^*\omega_2\omega_3^*\omega_4\,E\left\{g_1 g_2^* g_3 g_4^*\right\} + \cdots, \end{aligned} \qquad (2.371)$$

showing explicitly only the terms of interest. The fourth moment required in expression (2.367) is then supplied by

$$E\left\{g_1 g_2^* g_3 g_4^*\right\} = 16 \frac{\partial^4 \Phi}{\partial \omega_1^* \, \partial \omega_2 \, \partial \omega_3^* \, \partial \omega_4}, \tag{2.372}$$

evaluated at $\omega_1 = \omega_2 = \omega_3 = \omega_4 = 0$. Differentiation of the first expression (2.369) for the characteristic function, and setting $\omega_1 = \omega_2 = \omega_3 = \omega_4 = 0$, gives the required fourth moment as

$$\begin{aligned} E\left\{g_1 g_2^* g_3 g_4^*\right\} &= c_{12} c_{34} + c_{14} c_{32} \\ &= \phi_{gg}(\tau_1)\, \phi_{gg}(\tau_2) + \phi_{gg}(\tau_3)\, \phi_{gg}(\tau_4), \end{aligned} \tag{2.373}$$

with $\tau_1 = s - t$, $\tau_2 = u - v$, $\tau_3 = s - v$, $\tau_4 = u - t$. Substitution of this expression for the fourth moment in (2.367), and changing variables of integration, produces the formula

$$\begin{aligned} &E\left\{\tilde{S}_{gg,n}(f)\, \tilde{S}_{gg,n+m}(f)\right\} \\ &= \frac{1}{T^2} \int_{-\infty}^{\infty} \exp(-i2\pi f \tau_1)\, \phi_{gg}(\tau_1) \int_{-\infty}^{\infty} w_n(s)\, w_n^*(s - \tau_1)\, ds\, d\tau_1 \\ &\quad \cdot \int_{-\infty}^{\infty} \exp(-i2\pi f \tau_2)\, \phi_{gg}(\tau_2) \int_{-\infty}^{\infty} w_{n+m}(u)\, w_{n+m}^*(u - \tau_2)\, du\, d\tau_2 \\ &\quad + \frac{1}{T^2} \int_{-\infty}^{\infty} \exp(-i2\pi f \tau_3)\, \phi_{gg}(\tau_3) \int_{-\infty}^{\infty} w_n(s)\, w_{n+m}^*(s - \tau_3)\, ds\, d\tau_3 \\ &\quad \cdot \int_{-\infty}^{\infty} \exp(-i2\pi f \tau_4)\, \phi_{gg}(\tau_4) \int_{\infty}^{\infty} w_{n+m}(u)\, w_n^*(u - \tau_4)\, du\, d\tau_4, \end{aligned} \tag{2.374}$$

for the expected value of the product of spectral estimates, based on identical windows $w_n(t)$ and $w_{n+m}(t)$ shifted to centre on b_n and b_{n+m} on the time axis, respectively. Each of the inner integrals on the right-hand side of expression (2.374) represents a convolution of a shifted window with the time reverse (complex conjugate of the function obtained by switching the sign of the argument) of another shifted window. If the Fourier transform of the window $w(t)$ is $W(f)$, the Fourier transform of a window shifted to b on the time axis is

$$\int_{-\infty}^{\infty} w(t-b)\, e^{-i2\pi f t}\, dt = \int_{-\infty}^{\infty} w(t')\, e^{-i2\pi f (t'+b)}\, dt' = e^{-i2\pi f b}\, W(f), \tag{2.375}$$

a general property of Fourier transforms known as the *shifting theorem*. Further, the complex conjugate of $W(f)$ is

$$W^*(f) = \int_{-\infty}^{\infty} w^*(t)\, e^{i2\pi f t}\, dt = \int_{-\infty}^{\infty} w^*(-t')\, e^{-i2\pi f t'}\, dt', \tag{2.376}$$

on changing the integration variable to $t' = -t$. Thus, the Fourier transform of the time reverse of the window $w(t)$ is the complex conjugate of $W(f)$. Given that the window $w(t)$ is real and even, $W(f)$ is real. The convolution satisfies

$$\int_{-\infty}^{\infty} w_n(s)\, w_{n+m}^*(s-\tau)\, ds = \int_{-\infty}^{\infty} w_n(s)\, w_{n+m}(\tau - s)\, ds$$
$$= \int_{-\infty}^{\infty} w_{n+m}(u)\, w_n(\tau-u)\, du = \int_{-\infty}^{\infty} w_{n+m}(u)\, w_n^*(u-\tau)\, du, \qquad (2.377)$$

on changing the integration variable to $u = \tau - s$, and utilising the real and even properties of the windows. Using the foregoing properties and the convolutional relations (2.269) and (2.274), expression (2.374) becomes

$$E\left\{\tilde{S}_{gg,n}(f)\,\tilde{S}_{gg,n+m}(f)\right\}$$
$$= \frac{1}{I^2}\left[\int_{-\infty}^{\infty} S_{gg}(f-f_1)\, W^2(f_1)\, df_1\right]^2$$
$$+ \frac{1}{I^2}\left[\int_{-\infty}^{\infty} S_{gg}(f-f_1)\, \exp\left[i2\pi f_1\,(b_{n+m}-b_n)\right] W^2(f_1)\, df_1\right]^2. \qquad (2.378)$$

From (2.360), we recognise that

$$\frac{1}{I^2}\left[\int_{-\infty}^{\infty} S_{gg}(f-f_1)\, W^2(f_1)\, df_1\right]^2 = \left[E\left\{\tilde{S}_{gg,n}(f)\right\}\right]^2, \qquad (2.379)$$

the square of the expected value of the sample spectral density estimate. Substituting in (2.365), we find for the covariance,

$$\operatorname{cov}\left\{\tilde{S}_{gg,n}(f),\, \tilde{S}_{gg,n+m}(f)\right\}$$
$$= \frac{1}{I^2}\left[\int_{-\infty}^{\infty} S_{gg}(f-f_1) \exp\left[i2\pi f_1\,(b_{n+m}-b_n)\right]\, W^2(f_1)\, df_1\right]^2$$
$$\approx \frac{S_{gg}^2(f)}{I^2}\left[\int_{-\infty}^{\infty} \exp\left[i2\pi f\,(b_{n+m}-b_n)\right]\, W^2(f)\, df\right]^2, \qquad (2.380)$$

for spectral densities assumed to vary only slightly over the passband of $W^2(f)$. The window $w_{n+m}(t)$, centred on b_{n+m}, is displaced along the time axis by $m\,\Delta s$ from the window $w_n(t)$, centred on b_n, where Δs is the separation of successive windows. Now consider the convolution of the window $w_n(t)$ with the time reverse of the window $w_{n+m}(t)$,

$$J_m(\tau) = \int_{-\infty}^{\infty} w_n(t)\, w_{n+m}^*(t-\tau)\, dt. \qquad (2.381)$$

Again, using the shifting theorem and that the Fourier transform of a convolution is the product of the transforms (2.269), the Fourier integral representation of $J_m(\tau)$ is expressed as

$$J_m(\tau) = \int_{-\infty}^{\infty} \exp(i2\pi f\tau) \exp[i2\pi f (b_{n+m} - b_n)]\, W^2(f)\, df. \tag{2.382}$$

For $\tau = 0$, we then have that

$$\begin{aligned} J_m(0) = J_m &= \int_{-\infty}^{\infty} \exp[i2\pi f (b_{n+m} - b_n)]\, W^2(f)\, df \\ &= \int_{-M/2}^{M/2} w(t)\, w(t - m\Delta s)\, dt. \end{aligned} \tag{2.383}$$

Substituting in expression (2.380), we obtain for the covariance,

$$\operatorname{cov}\left\{\tilde{S}_{gg,n}(f), \tilde{S}_{gg,n+m}(f)\right\} = \frac{J_m^2}{I^2} S_{gg}^2(f) = S_{gg}^2(f)\rho(m), \tag{2.384}$$

where $\rho(m) = J_m^2/I^2$. Since $J_0 = I$,

$$\operatorname{cov}\left\{\tilde{S}_{gg,n}(f), \tilde{S}_{gg,n}(f)\right\} = \operatorname{var}\left\{\tilde{S}_{gg,n}(f)\right\} = S_{gg}^2(f). \tag{2.385}$$

Expression (2.364) for the variance of the final spectral density estimate, taken as an average of the sample spectral density estimates over κ overlapping segments of length M, then becomes

$$\operatorname{var}\left\{\bar{\tilde{S}}_{gg}(f)\right\} = \frac{S_{gg}^2(f)}{\kappa}\left[1 + 2\sum_{m=1}^{\kappa-1} \frac{\kappa - m}{\kappa}\rho(m)\right]. \tag{2.386}$$

This formula was quoted and used by Welch (1967) in an analysis using Bartlett windows with 50% overlap.

Instead, we have adopted the Parzen window (2.284), and we use 75% overlap. The Parzen time window is a piecewise cubic, and the integrals, J_m, can be evaluated analytically. We find that

$$J_0 = I = \frac{151}{560}M, \quad J_1 = \frac{397}{40}J_2, \quad J_2 = \frac{3}{224}M, \quad J_3 = \frac{1}{120}J_2, \tag{2.387}$$

giving,

$$\begin{aligned} \rho(1) &= \left(\frac{397}{40}\cdot\frac{3}{224}\cdot\frac{560}{151}\right)^2 = 0.2430131, \\ \rho(2) &= \left(\frac{3}{224}\cdot\frac{560}{151}\right)^2 = 0.0024670, \\ \rho(3) &= \left(\frac{1}{120}\cdot\frac{3}{224}\cdot\frac{560}{151}\right)^2 = 0.0000002, \\ &\rho(1) + \rho(2) + \rho(3) = 0.2454803. \end{aligned} \tag{2.388}$$

Substitution in (2.386) gives the variance of the final spectral density estimate as

$$\begin{aligned}\operatorname{var}\left\{\bar{\bar{S}}_{gg}(f)\right\} &= \frac{S_{gg}^2(f)}{\kappa}\left\{1 + 2\left[\rho(1) + \rho(2) + \rho(3)\right]\right.\\ &\qquad \left. - \frac{2}{\kappa}\left[\rho(1) + 2\rho(2) + 3\rho(3)\right]\right\}\\ &= \frac{S_{gg}^2(f)}{\kappa}\left(1.490961 - \frac{0.495895}{\kappa}\right).\end{aligned} \tag{2.389}$$

For a record of total length T, with 75% window overlap, the number of segments is

$$\kappa = \left(4\frac{T}{M} - 3\right). \tag{2.390}$$

The 75% overlap of the segments inflates the variance, for long records, by just under 50%. This is more than compensated for by the fourfold increase in the number of segments. The equivalent number of degrees of freedom, taking account of the variance inflation due to segment overlap, is

$$\nu = 2\kappa \Big/ \left(1.490961 - \frac{0.495895}{\kappa}\right). \tag{2.391}$$

As in the case of the average over multiple discrete segments (2.355), a confidence interval of $(1 - \alpha) \times 100\%$ for $\bar{\bar{S}}_{gg}(f_k)$ is given by

$$\frac{\nu \bar{\bar{S}}_{gg}(f_k)}{x(1 - \alpha/2)} \le S_{gg}(f_k) \le \frac{\nu \bar{\bar{S}}_{gg}(f_k)}{x(\alpha/2)}, \tag{2.392}$$

where $x(\zeta)$ is the value of the random variable below which lies a fraction ζ of the area under the cumulative χ_ν^2 distribution. On plots of the logarithm of the spectral density, the confidence interval is the fixed length $\log[x(1 - \alpha/2)] - \log[x(\alpha/2)]$, independent of frequency. The vertical confidence interval can then be parallel transported on plots of the logarithm of the spectral density. The lengths of the confidence intervals, on logarithmic plots of spectral density, are shown in Figure 2.17, as functions of the number of degrees of freedom, for confidence levels of 90%, 95% and 99%.

2.5.4 The product spectrum

Often the power spectral density is found from observations at a number of observatories. These, typically, will contain uncorrelated systematic and random errors

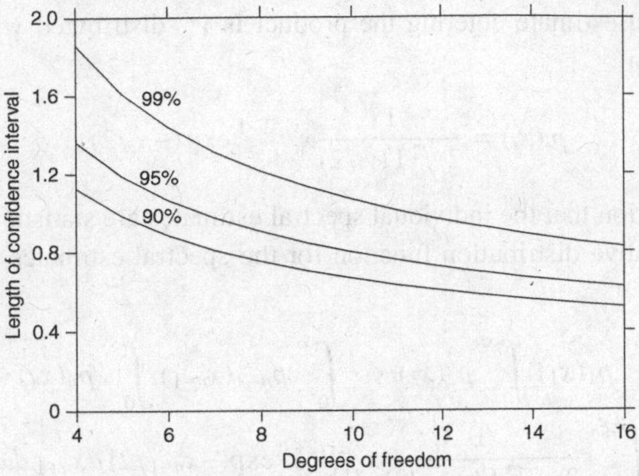

Figure 2.17 Length of the confidence interval, as a function of the number of degrees of freedom, on logarithmic plots of spectral density, for confidence levels of 90%, 95% and 99%.

peculiar to a particular observatory. To bring out features common to all observatories, the average product, or geometric mean, of the individual power spectra has been developed (Smylie *et al.*, 1993). Since it is usual to plot the logarithm of spectral density, the logarithm of the product spectrum is simply the arithmetic mean of the logarithms of the individual spectra. The product spectrum may be regarded as a generalisation of the cross spectrum used to find features common to two records.

In order to establish confidence intervals for the product spectrum, we construct the cumulative distribution function, $F(z)$, for the random variable z that is the product of $n + 1$ random variables $x_1, x_2, \ldots, x_n, x_{n+1}$. Thus,

$$z = x_1 x_2 \cdots x_n x_{n+1}. \tag{2.393}$$

Function $F(z)$ can be found by integrating the joint probability density function $p(x_1, x_2, \ldots, x_n, x_{n+1})$ over all of the realisation space below z. The integration then includes all possible values of the first $n-1$ random variables, but only those values of the last two falling below the rectangular hyperbola

$$x_{n+1} = \frac{z\,\xi_{n-1}}{x_n} \tag{2.394}$$

in the (x_n, x_{n+1}) plane, with

$$\xi_{n-1} = \frac{1}{x_1 x_2 \cdots x_{n-1}}. \tag{2.395}$$

The ith spectral estimate entering the product is $\chi^2_{\nu_i}$ distributed with probability density function

$$p_i(x_i) = \frac{1}{2^{\nu_i/2}\,\Gamma(\nu_i/2)} x_i^{\nu_i/2-1} \exp(-x_i/2). \tag{2.396}$$

On the assumption that the individual spectral estimates are statistically independent, the cumulative distribution function for the spectral estimates in the product spectrum is

$$F(z) = \int_0^\infty p_1(x_1) \int_0^\infty p_2(x_2) \cdots \int_0^\infty p_{n-1}(x_{n-1}) \int_0^\infty p_n(x_n) \cdot \int_0^{z\xi_n} \frac{1}{2^{\nu_{n+1}/2}\,\Gamma(\nu_{n+1}/2)} x_{n+1}^{\nu_{n+1}/2-1} \exp(-x_{n+1}/2)\, dx_{n+1}\, dx_n \cdots dx_1. \tag{2.397}$$

With the change of the variable of integration to $t = x_{n+1}/2$, the innermost integral becomes

$$\frac{1}{\Gamma(\nu_{n+1}/2)} \int_0^{z\xi_n/2} t^{\nu_{n+1}/2-1}\, e^{-t}\, dt = P(\nu_{n+1}/2, z\xi_n/2), \tag{2.398}$$

where

$$P(a, y) = \frac{1}{\Gamma(a)} \int_0^y t^{a-1}\, e^{-t}\, dt \tag{2.399}$$

is the incomplete gamma function and $\Gamma(a)$ is the gamma function. The gamma function has the integral representation

$$\Gamma(a) = \int_0^\infty t^{a-1}\, e^{-t}\, dt. \tag{2.400}$$

Calculation of the incomplete gamma function then requires calculation of the gamma function and the integral

$$\gamma(a, y) = \int_0^y t^{a-1}\, e^{-t}\, dt. \tag{2.401}$$

As a function of y the incomplete gamma function ranges from $P(a, 0) = 0$ to $P(a, \infty) = 1$. Its complement is then

$$\begin{aligned} 1 - P(a, y) &= \frac{1}{\Gamma(a)} \int_0^\infty t^{a-1}\, e^{-t}\, dt - \frac{1}{\Gamma(a)} \int_0^y t^{a-1}\, e^{-t}\, dt \\ &= \frac{1}{\Gamma(a)} \int_y^\infty t^{a-1}\, e^{-t}\, dt = \frac{\Gamma(a, y)}{\Gamma(a)}, \end{aligned} \tag{2.402}$$

where $\Gamma(a, y)$ is defined by the integral

$$\Gamma(a, y) = \int_y^\infty t^{a-1} e^{-t} \, dt. \tag{2.403}$$

Thus, the incomplete gamma function can be expressed as

$$P(a, y) = \frac{\gamma(a, y)}{\Gamma(a)}, \tag{2.404}$$

or,

$$P(a, y) = 1 - \frac{\Gamma(a, y)}{\Gamma(a)}. \tag{2.405}$$

The evaluation of either $\gamma(a, y)$ or $\Gamma(a, y)$ and the gamma function $\Gamma(a)$ allows calculation of the incomplete gamma function. A comprehensive account of numerical methods for the calculation of these and related functions is given by Press *et al.* (1992, pp. 206–213).

For sufficiently small y, the power series expansion of

$$\gamma(a, y) = e^{-y} y^a \sum_{n=0}^{\infty} \frac{\Gamma(a)}{\Gamma(a+1+n)} y^n = e^{-y} y^a \sum_{n=0}^{\infty} \frac{1}{a(a+1)\cdots(a+n)} y^n \tag{2.406}$$

can be used (Erdélyi *et al.*, 1953, p. 135), while for larger y the continued fraction development

$$\Gamma(a, y) = \cfrac{e^{-y} y^a}{y + \cfrac{1-a}{1 + \cfrac{1}{y + \cfrac{2-a}{1 + \cfrac{2}{y + \cfrac{3-a}{1+\cdots}}}}}} \tag{2.407}$$

can be employed (Erdélyi *et al.*, 1953, p. 136). For calculation of the gamma function itself we use the seven-term approximation of Lanczos (1964),

$$\Gamma(a+1) = \left(a + 5 + \frac{1}{2}\right)^{a+\frac{1}{2}} e^{-(a+5+\frac{1}{2})} \sqrt{2\pi}$$
$$\times \left[c_0 + \frac{c_1}{a+1} + \frac{c_2}{a+2} + \cdots + \frac{c_6}{a+6} + \epsilon\right], \tag{2.408}$$

with coefficients,

$$c_0 = 1.000000000178, \quad c_1 = 76.180091729406, \quad c_2 = -86.505320327112,$$
$$c_3 = 24.014098222230, \quad c_4 = -1.231739516140, \quad c_5 = 0.001208580030,$$
$$c_6 = -0.000005363820. \tag{2.409}$$

The relative error term has an absolute value $|\epsilon| < 2 \times 10^{-10}$.

The function subprogramme ALOGAM(X) finds the natural logarithm of the gamma function of x.

```
      DOUBLE PRECISION FUNCTION ALOGAM(X)
C
C A function subprogramme to calculate the natural logarithm of
C the gamma function using the seven-term Lanczos formula.
C
      IMPLICIT DOUBLE PRECISION(A-H,O-Z)
      DIMENSION COEF(7)
C Set values of coefficients.
      COEF(1)=1.000000000178D0
      COEF(2)=76.180091729406D0
      COEF(3)=-86.505320327112D0
      COEF(4)=24.014098222230D0
      COEF(5)=-1.231739516140D0
      COEF(6)=0.001208580030D0
      COEF(7)=-0.000005363820D0
C Set value of square root of 2pi.
      RT2PI=2.506628274631D0
C Change argument to that of factorial.
      Z=X-1.D0
C Set initial term of seven term series.
      SERIES=COEF(1)
C Sum remaining six terms.
      DO 10 I=1,6
        IP1=I+1
        AI=DFLOAT(I)
        DENOM=Z+AI
        SERIES=SERIES+COEF(IP1)/DENOM
 10   CONTINUE
C Multiply sum of series by square root of 2pi.
      SERIES=SERIES*RT2PI
C Set value of gamma for seven term series.
      GAM=5.D0
C Set value of Z+1/2.
      ZPH=Z+0.5D0
C Set value of Z+1/2+GAM
      ZPHG=ZPH+GAM
C Find natural logarithm of coefficient of series sum.
      HOLD=ZPH*DLOG(ZPHG)-ZPHG
C Find natural logarithm of Gamma(x).
      ALOGAM=HOLD+DLOG(SERIES)
      RETURN
      END
```

As Press *et al.* (1992, p. 211) suggest, for y less than $1+a$, the incomplete gamma function is found from (2.404) using the series expansion (2.406). The subroutine SEGAMI returns the incomplete gamma function $P(a, x)$ as the argument SEGI, for positive A and non-negative X, using up to 100 terms in the power series expansion. The expansion is continued until the last term is negligible compared to the series sum to double precision accuracy.

```
      SUBROUTINE SEGAMI(SEGI,A,X)
C
C Subroutine to calculate the incomplete gamma function by power series.
C
      IMPLICIT DOUBLE PRECISION(A-H,O-Z)
C Check if X is negative or A is not positive.
      IF(X.LT.0.D0.OR.A.LE.0.D0) GO TO 10
C Find natural logarithm of gamma(x).
      ALG=ALOGAM(A)
C Set maximum number of terms.
      ITM=100
C Test for zero argument.
      IF(X.EQ.0.D0) GO TO 11
C Set value of first term of series sum.
      SUM=1.D0/A
C Set value of first term.
      TERM=SUM
C Set initial value of inverse of coefficient of X^N.
      COEF=A
C Sum series.
      DO 12 N=1,ITM
C Update inverse of coefficient of X^N.
        COEF=COEF+1.D0
C Update term.
        TERM=TERM*X/COEF
C Store current value of sum.
        SUMOLD=SUM
C Update sum.
        SUM=SUM+TERM
C Check relative size of current term.
        IF(SUM.EQ.SUMOLD) GO TO 13
12    CONTINUE
      PAUSE 'A too small or X too large.'
13    SEGI=SUM*DEXP(-X+A*DLOG(X)-ALG)
      RETURN
10    PAUSE 'X is negative or A is not positive.'
      RETURN
11    SEGI=0.D0
      RETURN
      END
```

For y greater than $1 + a$, the incomplete gamma function is found from (2.405) using the continued fraction expansion (2.407). Evaluation of a continued fraction of the form

$$f = b_0 + \cfrac{a_1}{b_1 + \cfrac{a_2}{b_2 + \cfrac{a_3}{b_3 + \cfrac{a_4}{b_4 + \cfrac{a_5}{b_5 + \cdots}}}}} \qquad (2.410)$$

is a classical problem solved by Wallis in 1655! If f_n denotes the evaluation using terms through to a_n and b_n, then f_n is given by the rational approximation

$$f_n = \frac{A_n}{B_n}, \tag{2.411}$$

with A_n and B_n given by

$$\begin{aligned} A_{-1} &\equiv 1, & B_{-1} &\equiv 0, \\ A_0 &\equiv b_0, & B_0 &\equiv 1, \\ A_j &= b_j A_{j-1} + a_j A_{j-2}, & B_j &= b_j B_{j-1} + a_j B_{j-2}, \quad j = 1, 2, \ldots, n. \end{aligned} \tag{2.412}$$

Applied to the continued fraction (2.407), with the common factor $e^{-y}y^a$ removed, we have $b_0 = 0$ and

$$\begin{array}{llll} a_1 = 1, & a_2 = 1 - a, & b_1 = y, & b_2 = 1, \\ a_3 = 1, & a_4 = 2 - a, & b_3 = y, & b_4 = 1, \\ a_5 = 2, & a_6 = 3 - a, & b_5 = y, & b_6 = 1, \\ a_7 = 3, & a_8 = 4 - a, & b_7 = y, & b_8 = 1, \\ \ldots . \end{array} \tag{2.413}$$

Thus, the even and odd coefficients of the continued fraction show distinct recurrence patterns allowing their simple continuations. The subroutine CFGAMI implements calculation of the incomplete gamma function as the quantity CFGI for arguments A and X, by the continued fraction expansion (2.407) using recurrence relations (2.413) and (2.412). First the odd terms in recurrence (2.412) are computed and the odd terms in recurrence (2.413) are updated, then the even terms in recurrence (2.412) are computed and the even terms in recurrence (2.413) are updated. This sequence of computations is continued until the next term is negligible compared to the total to double precision accuracy.

```
      SUBROUTINE CFGAMI(CFGI,A,X)
C
C Subroutine to calculate the incomplete gamma function
C by continued fraction expansion.
C
      IMPLICIT DOUBLE PRECISION(A-H,O-Z)
C Find natural logarithm of gamma(x).
      ALG=ALOGAM(A)
C Set maximum number of terms.
      ITER=100
C Set initial value of continued fraction.
      CFOLD=0.D0
C Set initial values of even and odd sequences a and b.
      AE=0.D0
      AO=1.D0
      BE=1.D0
      BO=0.D0
```

```
C Set initial values of even and odd sequences a and b.
      AAE=1.D0-A
      AAO=1.D0
      BBE=1.D0
      BBO=X
C Evaluate continued fraction, first by odd sequences, then by even.
      DO 10 N=1,ITER
C Calculate odd a and b.
        AO=BBO*AE+AAO*AO
        BO=BBO*BE+AAO*BO
        CF=AO/BO
C Test for convergence.
        IF(CF.EQ.CFOLD) GO TO 11
C Store old CF.
        CFOLD=CF
C Update odd sequence a.
        IF(N.GT.1)AAO=AAO+1.D0
C Calculate even a and b.
        AE=BBE*AO+AAE*AE
        BE=BBE*BO+AAE*BE
        CF=AE/BE
C Update even sequence a.
        AAE=AAE+1.D0
C Test for convergence.
        IF(CF.EQ.CFOLD) GO TO 11
C Store old CF.
        CFOLD=CF
 10   CONTINUE
      PAUSE 'A too large or X too small.'
 11   CFGI=1.D0-DEXP(-X+A*DLOG(X)-ALG)*CF
      RETURN
      END
```

The subroutine GAMI calls the subroutine SEGAMI or the subroutine CFGAMI to compute the incomplete gamma function for all arguments for which A is positive and X is non-negative.

```
      SUBROUTINE GAMI(GI,A,X)
C
C Subroutine for the calculation of the incomplete gamma function.
C
      IMPLICIT DOUBLE PRECISION(A-H,O-Z)
C Check if X is negative or A is not positive.
      IF(X.LT.0.D0.OR.A.LE.0.D0) GO TO 10
C If X is less than A+1, use series calculation,
C otherwise use continued fraction expansion.
      IF(X.LT.A+1.D0) CALL SEGAMI(GI,A,X)
      IF(X.GE.A+1.D0) CALL CFGAMI(GI,A,X)
      GO TO 11
 10   PAUSE 'X is negative or A is not positive.'
 11   CONTINUE
      RETURN
      END
```

Evaluation of the cumulative distribution function $F(z)$ given by expression (2.397) requires the equivalent degrees of freedom of each individual spectrum entering the product. For 75% overlap, the equivalent number of degrees of

Table 2.1 *Degrees of freedom as a function of the number of overlapping segments for 75% overlap.*

Segments	Degrees
1	2.0099183
2	3.2179875
3	4.5260402
4	5.8522880
5	7.1850360
6	8.5208433
7	9.8583334
8	11.1968478
9	12.5360320
10	13.8756782

freedom is given by (2.391). These are given, as a function of the number of overlapping segments, in Table 2.1.

As a specific example of the computation of the cumulative distribution function (2.397) we consider the case where four separate spectral estimates enter the product. With the change of integration variables to $y_1 = x_1/2$, $y_2 = x_2/2$, $y_3 = x_3/2$, the cumulative distribution function (2.397) becomes

$$F(z) = \frac{1}{\Gamma(\nu_1/2)\Gamma(\nu_2/2)\Gamma(\nu_3/2)} \int_0^\infty y_1^{\nu_1/2-1} e^{-y_1} \int_0^\infty y_2^{\nu_2/2-1} e^{-y_2} \int_0^\infty y_3^{\nu_3/2-1} e^{-y_3} P(\nu_4/2, z\xi_3/2)\, dy_3\, dy_2\, dy_1, \tag{2.414}$$

where

$$z\xi_3/2 = \frac{z}{2 \cdot x_1 \cdot x_2 \cdot x_3} = \frac{z}{16\, y_1 y_2 y_3}. \tag{2.415}$$

A further change of integration variables to $t_i = e^{-y_i}$ brings the three integrals to the range (0, 1) and the cumulative distribution function takes the form

$$F(z) = \frac{1}{\Gamma(\nu_1/2)\,\Gamma(\nu_2/2)\,\Gamma(\nu_3/2)} \int_0^1 (-\log t_1)^{\nu_1/2-1} \int_0^1 (-\log t_2)^{\nu_2/2-1} \cdot \int_0^1 (-\log t_3)^{\nu_3/2-1}\, P\left(\frac{\nu_4}{2}, \frac{-z}{16 \log t_1 \log t_2 \log t_3}\right) dt_3\, dt_2\, dt_1. \tag{2.416}$$

The triple integrations are evaluated numerically by breaking each (0, 1) interval into subintervals, to improve sampling at large values of y, with boundaries

$$\left(1 - \cos\frac{\pi j}{20}\right)^4, \quad j = 0, 1, \ldots, 10 \tag{2.417}$$

and using five-point Gaussian integration on each subinterval.

Gaussian integration (Abramowitz and Stegun, 1964, p. 887) replaces the integral by a weighted sum, as in

$$\int_a^b f(t)\,dt = \frac{b-a}{2}\sum_{i=1}^{n} w_i\, f(t_i), \tag{2.418}$$

with

$$t_i = \left(\frac{b-a}{2}\right)x_i + \left(\frac{b+a}{2}\right). \tag{2.419}$$

For five-point Gaussian integration the weights are

$$w_1 = 0.236926885056189, \quad w_2 = 0.478628670499366,$$
$$w_3 = 0.568888888888889, \quad w_4 = w_2,$$
$$w_5 = w_1, \tag{2.420}$$

and the abscissas are

$$x_1 = -0.906179845938664, \quad x_2 = -0.538469310105683,$$
$$x_3 = 0, \quad x_4 = -x_2,$$
$$x_5 = -x_1. \tag{2.421}$$

The five-point Gaussian integration is ninth-order accurate (Abramowitz and Stegun, 1964, p. 916).

The subroutine CUMDF calculates the cumulative distribution function for the product spectrum of four individual spectra. It calls the subroutine GAMI, which in turn calls either the subroutine SEGAMI, for the series computation of the incomplete gamma function, or the subroutine CFGAMI, for the continued fraction computation, as required. It also uses the function subprogramme ALOGAM(X) for the calculation of the natural logarithm of the gamma function of x.

```
      SUBROUTINE CUMDF(CDF,Z,ANU,N)
C A subroutine for the direct numerical calculation of the cumulative
C distribution function for N, chi-squared distributed random variables
C multiplied together.
      IMPLICIT DOUBLE PRECISION(A-H,O-Z)
      DIMENSION ANU(N),A(N),X(5),W(5),TN(11),XIJ(10,5),BMA(10)
C Take natural logarithm of Z.
      ZLOG=DLOG(Z)
C Construct array of half-degrees of freedom.
      DO 10 I=1,N
        A(I)=ANU(I)/2.D0
10    CONTINUE
C Enter abscissas for 5-point Gaussian integration.
      X(1)=-0.906179845938664D0
      X(2)=-0.538469310105683D0
      X(3)=0.D0
      X(4)=-X(2)
      X(5)=-X(1)
```

```
C Enter weights for 5-point Gaussian integration.
      W(1)=0.236926885056189D0
      W(2)=0.478628670499366D0
      W(3)=0.568888888888889D0
      W(4)=W(2)
      W(5)=W(1)
C Find natural logarithms of weights.
      DO 11 I=1,5
        W(I)=DLOG(W(I))
 11   CONTINUE
C Set node points according to 1-cos(pi*j/20), j=0,1,...,10.
      TN(1)=0.D0
      TN(2)=0.0123117D0
      TN(3)=0.0489435D0
      TN(4)=0.1089935D0
      TN(5)=0.1909830D0
      TN(6)=0.2928932D0
      TN(7)=0.4122148D0
      TN(8)=0.5460095D0
      TN(9)=0.6909830D0
      TN(10)=0.8435655D0
      TN(11)=1.D0
C Raise 1-cos(pi*j/20) to the power 4.
      DO 12 I=1,11
        TN(I)=TN(I)**4
 12   CONTINUE
C Calculate abscissas for ith subinterval and jth integration point.
      DO 13 I=1,10
        IP1=I+1
C Find a.
        AA=TN(I)
C Find b.
        BB=TN(IP1)
C Find (b-a)/2
        BMA(I)=0.5D0*(BB-AA)
C Find (b+a)/2
        BPA=0.5D0*(BB+AA)
        DO 13 J=1,5
C Find abscissa t(i,j).
        TIJ=BMA(I)*X(J)+BPA
C Take logarithm of negative logarithm of abscissa.
        XIJ(I,J)=DLOG(-DLOG(TIJ))
 13   CONTINUE
      DO 14 I=1,10
C Find logarithm of (b-a)/2.
        BMA(I)=DLOG(BMA(I))
 14   CONTINUE
C Begin calculation of cumulative distribution function specialized to N=4.
      CDF=0.D0
C Find contribution of inner integral.
      DO 15 I=1,10
        DO 15 IJ=1,5
C Find contribution of middle integral.
          DO 15 J=1,10
            DO 15 JJ=1,5
C Find contribution of outer integral.
              DO 15 K=1,10
                DO 15 KJ=1,5
C Find argument x (XARG) for incomplete gamma function P(a,x).
                  XARG=ZLOG-XIJ(I,IJ)-XIJ(J,JJ)-XIJ(K,KJ)
                  XARG=DEXP(XARG)/16.D0
```

```
      CALL GAMI(GP,A(N),XARG)
C Find logarithm of current term.
      TERMLN=W(IJ)+W(JJ)+W(KJ)
     1 +(A(1)-1.D0)*XIJ(I,IJ)+(A(2)-1.D0)*XIJ(J,JJ)
     2 +(A(3)-1.D0)*XIJ(K,KJ)+DLOG(GP)
     3 +BMA(I)+BMA(J)+BMA(K)
C Find current term.
      TERM=DEXP(TERMLN)
C Add current term to sum of previous terms.
      CDF=CDF+TERM
15    CONTINUE
C Revert to Z from the logarithm of Z.
      Z=DEXP(ZLOG)
C Divide by triple product of gamma functions in front of integrals.
      CDF=CDF*DEXP(-ALOGAM(A(1))-ALOGAM(A(2))-ALOGAM(A(3)))
      RETURN
      END
```

We will later require the cumulative distribution function for the product of four spectra found from four, five, seven and eight segments with 75% overlap. From Table 2.1 these are found to have 5.8522880, 7.1850360, 9.8583334 and 11.1968478 degrees of freedom, respectively. The resulting cumulative distribution function is shown plotted in Figure 2.18.

Iteration on the cumulative distribution function $F(z)$ allows the calculation of confidence intervals for the product spectrum. For $z = 292.22$, $F(z) = 0.025$, and for $z = 19{,}500$, $F(z) = 0.975$, giving a 95% confidence interval of 1.824 on logarithmic plots of the product spectrum.

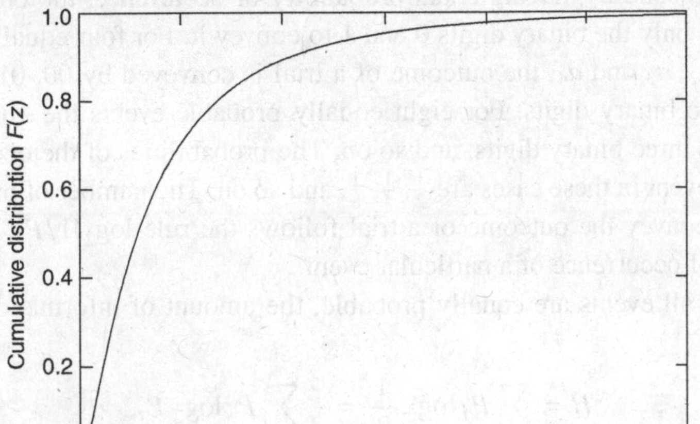

Figure 2.18 The cumulative distribution function $F(z)$ for the product of four spectral density estimates based on four, five, seven and eight segments with 75% overlap, respectively. Iteration on this function gives a 95% confidence interval of 1.824 on logarithmic plots of the product spectrum.

2.6 Maximum entropy spectral analysis

In traditional power spectral density estimation, the signal is assumed to vanish outside the finite record for which it is available for sampling. This limits the frequency resolution to the order of the reciprocal of the record length. The discrete Fourier transform representation of the finite record (2.147), as shown previously, is periodic in the record length, giving a periodic extension of the time sequence outside the finite record. This inserts periodicities in the spectrum, which can only be suppressed by windowing, at the cost of a further reduction in frequency resolution.

A way out of this dilemma was pointed out by John Parker Burg, who suggested that, using entropy as a measure of information, the entropy within the measured finite record be maximised, with no entropy added from any assumption about the information outside the known record (Burg, 1967, 1968). In this case, the signal is neither assumed to vanish outside the measured record nor to be a periodic extension of the signal within the measured record. This has led to the *Burg maximum entropy algorithm* for data adaptive spectral analysis of short records, promising improved resolution compared to conventional methods of spectral analysis (Lacoss, 1971; Smylie *et al.*, 1973).

2.6.1 Information and entropy

A measure of information is the length of the message required to convey it. For two events, a_1 and a_2, having equal probability of occurrence, the outcome of a trial requires only the binary digits 0 and 1 to convey it. For four equally probable events, a_1, a_2, a_3 and a_4, the outcome of a trial is conveyed by 00, 01, 10 or 11, requiring two binary digits. For eight equally probable events the outcome of a trial requires three binary digits, and so on. The probabilities of the occurrence of a particular event in these cases are $\frac{1}{2}$, $\frac{1}{4}$, $\frac{1}{8}$, and so on. The number of binary digits required to convey the outcome of a trial follows the rule $\log_2(1/P)$, with P the probability of occurrence of a particular event.

When not all events are equally probable, the amount of information is measured by

$$H = \sum_j P_j \log_2 \frac{1}{P_j} = -\sum_j P_j \log_2 P_j, \qquad (2.422)$$

a quantity introduced by Shannon (1948), which he called the entropy. The base of the logarithm used in the definition of entropy depends on the encoding scheme. For natural logarithms, a change from base r rescales the entropy as

$$H = -\sum_j P_j \log_r P_j = -\frac{1}{\log r} \sum_j P_j \log P_j. \qquad (2.423)$$

For random variables that can take on a continuum of values, the sum in the definition of entropy is replaced by an integral. For example, for realisations of the time sequence $z_0, z_1, \ldots, z_N$, we would use

$$H = -\int f(z_0, z_1, \ldots, z_N) \log\left\{a^{2N+2} f(z_0, z_1, \ldots, z_N)\right\} dV \tag{2.424}$$

as a measure of the entropy, where a is a constant with the same units as a dimension of the sample space, $f(z_0, z_1, \ldots, z_N)$ is the joint probability density function, and dV is an element of volume in sample space. If the $N+1$ realisations of the time sequence are complex numbers with real and imaginary parts that jointly are normally distributed with zero mean, the joint probability density function is that (2.368) given by Wooding (1956),

$$f(z_0, z_1, \ldots, z_N) = \frac{1}{\pi^{N+1}|\mathbf{C}|} \exp\left(-z_j^* c_{jk}^{-1} z_k\right), \tag{2.425}$$

where $\mathbf{C}$ is the Hermitian variance–covariance matrix, $|\mathbf{C}|$ is its determinant and c_{jk}^{-1} are the elements of its inverse. For a stationary process with zero mean, the elements of the variance–covariance matrix are given by

$$c_{ij} = E\left\{z_i z_j^*\right\} = \phi(i-j), \tag{2.426}$$

with $\phi(i-j)$ being the autocorrelation at lag $i-j$. Then, $\mathbf{C} = \mathbf{T}_N$, where

$$\mathbf{T}_N = \begin{pmatrix} \phi(0) & \cdots & \phi(-N) \\ \vdots & \ddots & \vdots \\ \phi(N) & \cdots & \phi(0) \end{pmatrix} \tag{2.427}$$

is the Toeplitz coefficient matrix of the prediction error equations (2.80). Substitution of the joint probability density function (2.425) for the Gaussian process $\boldsymbol{z}_N = (z_0, z_1, \ldots, z_N)$ into expression (2.424) for the entropy gives

$$\begin{aligned} H &= -\int f(\boldsymbol{z}_N) \log\left[\frac{a^{2N+2}}{\pi^{N+1}|\mathbf{C}|} \exp\left(-z_j^* c_{jk}^{-1} z_k\right)\right] dV \\ &= -\log\left(\frac{a^{2N+2}}{\pi^{N+1}|\mathbf{C}|}\right) \int f(\boldsymbol{z}_N)\, dV - \int f(\boldsymbol{z}_N)\left(-z_j^* c_{jk}^{-1} z_k\right) dV \\ &= \log\left(\frac{\pi^{N+1}|\mathbf{C}|}{a^{2N+2}}\right) + c_{jk}^{-1} \int f(\boldsymbol{z}_N)\, z_j^* z_k\, dV, \end{aligned} \tag{2.428}$$

since $\int f(\boldsymbol{z}_N)\, dV = 1$ by definition. Further,

$$\int f(\boldsymbol{z}_N)\, z_j^* z_k\, dV = E\left\{z_k z_j^*\right\} = c_{kj}. \tag{2.429}$$

Since

$$c_{jk}^{-1} c_{kj} = \delta_j^j = N + 1, \tag{2.430}$$

the entropy becomes

$$H = \log |\mathbf{C}| + (N + 1)(\log \pi - 2 \log a + 1). \tag{2.431}$$

Choosing the constant $a = \sqrt{\pi e}$, the final expression for the entropy is

$$H = \log |\mathbf{C}| = \log (\det \mathbf{T}_N). \tag{2.432}$$

Thus far, we have considered the entropy of processes with finite length $N + 1$. For processes of indefinite duration we use the *entropy density* or *entropy rate* which Middleton (1960, p. 306) defined as

$$h = \lim_{N \to \infty} \frac{H}{N + 1} = \lim_{N \to \infty} \log \left[(\det \mathbf{T}_N)^{1/(N+1)} \right]. \tag{2.433}$$

The Toeplitz matrix $\mathbf{T}_N$ has non-negative definite eigenvalues. For eigenvalue λ and corresponding eigenvector $\boldsymbol{x}$, by definition,

$$\mathbf{T}_N \boldsymbol{x} = \lambda \boldsymbol{x}. \tag{2.434}$$

Left multiplication gives

$$\boldsymbol{x}^H \mathbf{T}_N \boldsymbol{x} = \lambda \boldsymbol{x}^H \boldsymbol{x}, \tag{2.435}$$

where the superscript H indicates the adjoint. Then,

$$\begin{aligned} \boldsymbol{x}^H \mathbf{T}_N \boldsymbol{x} &= \sum_{j=0}^{N} \sum_{k=0}^{N} \phi(j-k)\, x_j^* x_k \\ &= \sum_{j=0}^{N} \sum_{k=0}^{N} E\left\{ z_j z_k^* \right\} x_j^* x_k \\ &= E\left\{ \sum_{j=0}^{N} z_j x_j^* \cdot \sum_{k=0}^{N} z_k^* x_k \right\} \\ &= E\left\{ \left| \sum_{j=0}^{N} z_j x_j^* \right|^2 \right\} \geq 0. \end{aligned} \tag{2.436}$$

Thus, as stated,

$$\lambda = \frac{\boldsymbol{x}^H \mathbf{T}_N \boldsymbol{x}}{\boldsymbol{x}^H \boldsymbol{x}} \geq 0. \tag{2.437}$$

If we form the matrix $\mathbf{X}$ with the eigenvectors of $\mathbf{T}_N$ as columns, then

$$\mathbf{T}_N \mathbf{X} = \mathbf{D}\mathbf{X} \tag{2.438}$$

with

$$\mathbf{D} = \begin{pmatrix} \lambda_0 & & & 0 \\ & \lambda_1 & & \\ & & \ddots & \\ 0 & & & \lambda_N \end{pmatrix} \tag{2.439}$$

the diagonal matrix of eigenvalues. On taking determinants of both sides of (2.438), we have

$$\det \mathbf{T}_N \cdot \det \mathbf{X} = \det \mathbf{D} \cdot \det \mathbf{X} \tag{2.440}$$

or

$$\det \mathbf{T}_N = \det \mathbf{D} = \lambda_0 \lambda_1 \cdots \lambda_N \geq 0. \tag{2.441}$$

Hence, $\mathbf{T}_N$ is non-negative definite.

The determinant of $\mathbf{T}_N$ can be related to the spectral density $S(f)$ through a remarkable theorem of G. Szegö (1920), (Widom, 1965, p. 202), which states that for F a continuous function,

$$\lim_{N\to\infty} \frac{F(\lambda_0) + F(\lambda_1) + \cdots + F(\lambda_N)}{N+1} = \frac{1}{2f_N} \int_{-f_N}^{f_N} F\left[2f_N S(f)\right] df, \tag{2.442}$$

with f_N denoting the Nyquist frequency. Taking F to be the logarithmic function, with substitution from (2.441), we have

$$\begin{aligned} \lim_{N\to\infty} \frac{\log\lambda_0 + \log\lambda_1 + \cdots + \log\lambda_N}{N+1} &= \lim_{N\to\infty} \log\left[(\lambda_0\lambda_1 \cdots \lambda_N)^{1/(N+1)}\right] \\ &= \lim_{N\to\infty} \log\left[(\det \mathbf{T}_N)^{1/(N+1)}\right] \\ &= \frac{1}{2f_N} \int_{-f_N}^{f_N} \log\left[2f_N S(f)\right] df. \end{aligned} \tag{2.443}$$

Expanding the right-hand side produces

$$\begin{aligned} \frac{1}{2f_N}\left[\int_{-f_N}^{f_N} \log(2f_N)\, df + \int_{-f_N}^{f_N} \log\left[S(f)\right] df\right] \\ = \log(2f_N) + \frac{1}{2f_N}\int_{-f_N}^{f_N} \log\left[S(f)\right] df. \end{aligned} \tag{2.444}$$

Taking exponentials produces

$$\lim_{N\to\infty} (\det \mathbf{T}_N)^{1/(N+1)} = 2f_N \exp\left\{\frac{1}{2f_N}\int_{-f_N}^{f_N} \log\left[S(f)\right] df\right\}. \tag{2.445}$$

Expression (2.433) for the entropy rate then becomes

$$h = \lim_{N\to\infty} \log\left[(\det \mathbf{T}_N)^{1/(N+1)}\right]$$
$$= \log(2f_N) + \frac{1}{2f_N}\int_{-f_N}^{f_N} \log\left[S(f)\right] df. \tag{2.446}$$

Substituting the infinite Fourier series expansion of the spectral density (2.336), the entropy rate can be expressed as

$$h = \log(2f_N) + \frac{1}{2f_N}\left(\int_{-f_N}^{f_N} \log(\Delta t)\, df + \int_{-f_N}^{f_N} \log\left\{\sum_{k=-\infty}^{\infty} \phi(k)\, e^{-i2\pi f k \Delta t}\right\} df\right)$$
$$= \log(2f_N) + \frac{1}{2f_N}\left(-\log(2f_N)\cdot 2f_N + \int_{-f_N}^{f_N} \log\left\{\sum_{k=-\infty}^{\infty} \phi(k)\, e^{-i2\pi f k \Delta t}\right\} df\right)$$
$$= \frac{1}{2f_N}\int_{-f_N}^{f_N} \log\left\{\sum_{k=-\infty}^{\infty} \phi(k)\, e^{-i2\pi f k \Delta t}\right\} df, \tag{2.447}$$

where $\phi(k)$ is the autocorrelation at lag k and $\Delta t = 1/2f_N$.

2.6.2 The maximum entropy spectrum

The starting point for the Burg maximum entropy method of spectral analysis is expression (2.447) for the entropy rate. For a finite record of $N + 1$ sample points, only autocorrelations for lags in magnitude less than N are available. Burg suggests that the most reasonable choice for the unknown autocorrelations is the one that adds no information or entropy. Hence, the entropy rate h is made stationary with respect to the unknown autocorrelations. Thus, we set

$$\frac{\partial h}{\partial \phi(k)} = 0, \quad |k| \geq N + 1. \tag{2.448}$$

The conditions that the unknown autocorrelations add no entropy, on taking derivatives, are found to be

$$\int_{-f_N}^{f_N} \frac{e^{-i2\pi f k \Delta t}}{S(f)}\, df = 0, \quad |k| \geq N + 1. \tag{2.449}$$

Thus, the Euler coefficients of the Fourier series expansion of the real, periodic function $1/S(f)$ truncate, and its Fourier series expansion becomes

$$\frac{1}{S(f)} = \frac{1}{2f_N}\sum_{k=-N}^{N} c_k\, e^{i2\pi f k \Delta t}. \tag{2.450}$$

The realness of $1/S(f)$ requires that $c_k = c^*_{-k}$. Then

$$\frac{1}{S(f)} = \frac{1}{2f_N} \sum_{k=-N}^{N} c_{-k}\, e^{-i2\pi f k\Delta t}. \tag{2.451}$$

If $C(z)$ is the z-transform of $c_N, c_{N-1}, \ldots, c_{-N+1}, c_{-N}$, then $C(z)$ can be factored, as in (2.58), to give

$$C(z) = G(z)\, G^*(1/z^*) \tag{2.452}$$

with $G(z)$ a polynomial of degree N. As described in Section 2.1.5, $G(z)$ can be put in minimum delay form so that its roots lie outside the unit circle in the complex z-plane, while those of $G^*(1/z^*)$ lie inside the unit circle. If $\Phi(z)$ is the z-transform of the infinite autocorrelation series in expression (2.336) for the spectral density, as outlined in Section 2.3.2,

$$2f_N S(f) = \Phi(z) \quad \text{for} \quad z = e^{-i2\pi f\Delta t}. \tag{2.453}$$

Then, (2.451) can be replaced with

$$\frac{1}{\Phi(z)} = \frac{1}{4f_N^2} C(z) = \frac{1}{4f_N^2}\, G(z)\, G^*(1/z^*), \quad z = e^{-i2\pi f\Delta t}. \tag{2.454}$$

As well as satisfying the extremum conditions (2.449), the spectral density must also be consistent with the known autocorrelations

$$\phi(-N), \phi(-N+1), \ldots, \phi(0), \ldots, \phi(N-1), \phi(N). \tag{2.455}$$

If these have the z-transform $\Psi(z)$, then like terms in the z-transform $\Phi(z)$ must be identical where the two overlap. $\Phi(z)$ itself is indefinitely long. To apply this restriction we use the prediction error equations (2.80).

The prediction error equations can be expressed in the form

$$\sum_{k=0}^{N} \gamma_k\, \phi(j-k) = P_{N+1}\delta_j^0, \quad j = 0, 1, \ldots, N. \tag{2.456}$$

The sum can be converted to a convolution, giving

$$\sum_{-N}^{N} \gamma_k\, \phi(j-k) = P_{N+1}\delta_j^0 + h_j, \quad j = -N, \ldots, 0, \ldots, N, \tag{2.457}$$

since γ_k vanishes for k negative and the sequence h_j is taken to vanish for $0 \le j \le N$. Then, on taking z-transforms of both sides of this equation, we get

$$\Gamma(z)\, \Psi(z) = P_{N+1} + H(z), \tag{2.458}$$

with

$$\Gamma(z) = 1 + \gamma_1 z + \cdots + \gamma_N z^N \tag{2.459}$$

as the z-transform of the prediction error filter and

$$H(z) = h_{-N} z^{-N} + \cdots + h_1 z^{-1} \tag{2.460}$$

as the z-transform of the sequence h_j. As explained in Section 2.1.5, $\Psi(z)$ can be factored as

$$\Psi(z) = F(z)\, F^*(1/z^*) = F(z)\, R(z)/z^N, \tag{2.461}$$

where $F(z)$ is the z-transform of a wavelet,

$$F(z) = f_0 + f_1 z + \cdots + f_N z^N, \tag{2.462}$$

and

$$R(z) = f_N^* + f_{N-1}^* z + \cdots + f_0^* z^N \tag{2.463}$$

is the z-transform of the reverse wavelet. Multiplying the z-transform of the prediction error equations (2.458) by $z^N/R(z)$ and then performing polynomial division, we find that

$$\begin{aligned} \Gamma(z)\, F(z) &= \frac{z^N P_{N+1}}{R(z)} + \frac{z^N H(z)}{R(z)} \\ &= \frac{P_{N+1}}{f_0^*} - \frac{f_1^*}{f_0^{*2}} P_{N+1}\, z^{-1} + \cdots + \frac{h_1 z^{-1}}{f_0^*} + \cdots . \end{aligned} \tag{2.464}$$

The left-hand side of this expression is a polynomial of degree $2N$, while the right-hand side, except for the first constant term, consists of terms in negative powers of z. Thus,

$$\Gamma(z)\, F(z) = \frac{P_{N+1}}{f_0^*} = f_0. \tag{2.465}$$

The reciprocal of $\Psi(z)$ can then be written

$$\begin{aligned} \frac{1}{\Psi(z)} &= \frac{1}{F(z)\, F^*(1/z^*)} = \frac{|f_0|^2}{P_{N+1}^2} \Gamma(z)\, \Gamma^*(1/z^*) \\ &= \frac{\Gamma(z)\, \Gamma^*(1/z^*)}{P_{N+1}}. \end{aligned} \tag{2.466}$$

The condition (2.454) for the entropy rate to be an extremum (a maximum) with respect to the unknown autocorrelations provides the identification

$$G(z) = \frac{2 f_N \Gamma(z)}{(P_{N+1})^{1/2}} \tag{2.467}$$

in order for the z-transforms $\Psi(z)$ and $\Phi(z)$ to have matching terms where they overlap. Returning to the frequency domain, the truncated form of $1/S(f)$ becomes

$$\frac{1}{S(f)} = \frac{2f_N\,\Gamma(z)\,\Gamma^*(z)}{P_{N+1}}, \quad z = e^{-i2\pi f\Delta t}. \tag{2.468}$$

Hence, the *maximum entropy spectral density* is found to be

$$S(f) = \frac{P_{N+1}}{2f_N\,\Gamma\left[\exp(-i2\pi f\Delta t)\right]\,\Gamma^*\left[\exp(-i2\pi f\Delta t)\right]}. \tag{2.469}$$

In terms of the prediction error filter coefficients it becomes

$$S(f) = \frac{P_{N+1}}{2f_N\left|1+\sum_{j=1}^{N}\gamma_j\,e^{-i2\pi f j\Delta t}\right|^2}. \tag{2.470}$$

Calculation of the maximum entropy spectral density then involves determination of the prediction error filter and the error power of the prediction. Burg (1968) has demonstrated a recursive method of solving the prediction error system of equations (2.80).

2.6.3 The Burg algorithm

We begin with the finite, equispaced data sequence $(f_1, f_2, \ldots, f_{M-1}, f_M)$ as the available record. First, consider the prediction error system of equations for $N = 0$. It is simply

$$\phi_{ff}(0) = P_1. \tag{2.471}$$

For the estimate of the autocorrelation at zero lag we only have

$$\phi_{ff}(0) = \frac{1}{M}\sum_{j=1}^{M} f_j f_j^* = P_1. \tag{2.472}$$

Then, we write the system for $N = 1$ to be compatible with that for $N = 0$,

$$\begin{pmatrix} P_1 & \phi_{ff}(-1) \\ \phi_{ff}(1) & P_1 \end{pmatrix}\begin{pmatrix} 1 \\ \gamma_{1,1} \end{pmatrix} = \begin{pmatrix} P_2 \\ 0 \end{pmatrix}. \tag{2.473}$$

A second subscript 1 is added to γ_1 to indicate that it is the prediction error coefficient for $N = 1$.

There is now the possibility to consider backward prediction as well as forward prediction. Backward prediction is equivalent to forward prediction of the sequence f_{M-j+1}, $j = 1, 2, \ldots, M$. Assuming that the given sequence is stationary, the autocorrelation $\phi_{ff}(k)$ is replaced by $\phi_{ff}(-k)$ for the new sequence. From the Hermitian property of the autocorrelation for stationary sequences, (2.20), $\phi_{ff}(k)$

is replaced by its conjugate, $\phi^*_{ff}(k)$. Then, the coefficient matrix of the prediction error equations for backward prediction becomes the conjugate of that for forward prediction. Taking conjugates of the equations for backward prediction, we find that

$$\begin{pmatrix} \phi_{ff}(0) & \cdots & \phi_{ff}(-N) \\ \vdots & \ddots & \vdots \\ \phi_{ff}(N) & \cdots & \phi_{ff}(0) \end{pmatrix} \begin{pmatrix} 1 \\ \gamma_1^* \\ \vdots \\ \gamma_N^* \end{pmatrix} = \begin{pmatrix} P_{N+1} \\ 0 \\ \vdots \\ 0 \end{pmatrix}. \tag{2.474}$$

By comparison with the equations (2.80) for forward prediction, the prediction error filter for backward prediction is just the complex conjugate of that for forward prediction.

Returning to the equations (2.473) for $N = 1$, forward prediction gives the error sequence

$$\epsilon^F_{j+1,1} = f_{j+1} + \gamma_{1,1} f_j, \quad j = 1, \ldots, M-1, \tag{2.475}$$

while backward prediction gives the error sequence

$$\epsilon^B_{j,1} = f_j + \gamma^*_{1,1} f_{j+1} \quad j = 1, \ldots, M-1. \tag{2.476}$$

For the assumed stationarity of the given sequence, the best estimate of P_2 is the average of the forward and backward prediction values,

$$P_2 = \frac{1}{2(M-1)} \sum_{j=1}^{M-1} \left\{ \left| f_{j+1} + \gamma_{1,1} f_j \right|^2 + \left| f_j + \gamma^*_{1,1} f_{j+1} \right|^2 \right\}. \tag{2.477}$$

The Burg algorithm is data adaptive in the sense that $\gamma_{1,1}$ can be chosen to minimise the value of P_2 calculated from the data. For a minimum, the partial derivatives of P_2 with respect to the real and imaginary parts of $\gamma_{1,1}$ vanish. Then,

$$\begin{aligned} \frac{\partial P_2}{\partial \mathrm{Re}\gamma_{1,1}} = \frac{1}{2(M-1)} \sum_{j=1}^{M-1} \Big\{ & f_j \left(f^*_{j+1} + \gamma^*_{1,1} f^*_j \right) + f^*_j \left(f_{j+1} + \gamma_{1,1} f_j \right) \\ & + f_{j+1} \left(f^*_j + \gamma_{1,1} f^*_{j+1} \right) + f^*_{j+1} \left(f_j + \gamma^*_{1,1} f_{j+1} \right) \Big\} = 0, \end{aligned} \tag{2.478}$$

and

$$\begin{aligned} \frac{\partial P_2}{\partial \mathrm{Im}\gamma_{1,1}} = \frac{i}{2(M-1)} \sum_{j=1}^{M-1} \Big\{ & f_j \left(f^*_{j+1} + \gamma^*_{1,1} f^*_j \right) - f^*_j \left(f_{j+1} + \gamma_{1,1} f_j \right) \\ & - f_{j+1} \left(f^*_j + \gamma_{1,1} f^*_{j+1} \right) + f^*_{j+1} \left(f_j + \gamma^*_{1,1} f_{j+1} \right) \Big\} = 0. \end{aligned} \tag{2.479}$$

Adding $i \times$ (2.479) to (2.478) gives

$$\sum_{j=1}^{M-1} \left\{ f_j^* \left(f_{j+1} + \gamma_{1,1} f_j \right) + f_{j+1} \left(f_j^* + \gamma_{1,1} f_{j+1}^* \right) \right\} = 0. \tag{2.480}$$

Solving for the value of $\gamma_{1,1}$ that minimises P_2, we get

$$\begin{aligned} \gamma_{1,1} &= -2 \sum_{j=1}^{M-1} \left\{ f_{j+1} f_j^* \right\} \Big/ \sum_{j=1}^{M-1} \left\{ f_{j+1} f_{j+1}^* + f_j f_j^* \right\} \\ &= -2 \sum_{j=1}^{M-1} \left\{ \epsilon_{j+1,0}^{F} \epsilon_{j,0}^{B*} \right\} \Big/ \sum_{j=1}^{M-1} \left\{ \left| \epsilon_{j+1,0}^{F} \right|^2 + \left| \epsilon_{j,0}^{B*} \right|^2 \right\}, \end{aligned} \tag{2.481}$$

if we interpret the forward prediction error sequence as

$$\epsilon_{j+1,0}^{F} = f_{j+1}, \tag{2.482}$$

and the backward prediction error sequence as

$$\epsilon_{j,0}^{B} = f_j \tag{2.483}$$

for $N = 0$. In fact, the prediction error filter for $N = 0$ is just the unit impulse and the two prediction error sequences are both just the original data sequence itself. For this value of $\gamma_{1,1}$, equations (2.473) give the recurrence relations

$$\phi_{ff}(1) = -\gamma_{1,1} \phi_{ff}(0) \quad \text{and} \quad P_2 = \left(1 - \gamma_{1,1} \gamma_{1,1}^* \right) P_1. \tag{2.484}$$

The next step is reminiscent of the Levinson algorithm (Section 2.2.3). We begin with the system of equations

$$\begin{pmatrix} \phi_{ff}(0) & \phi_{ff}(-1) & \phi_{ff}(-2) \\ \phi_{ff}(1) & \phi_{ff}(0) & \phi_{ff}(-1) \\ \phi_{ff}(2) & \phi_{ff}(1) & \phi_{ff}(0) \end{pmatrix} \begin{pmatrix} 1 \\ \gamma_{1,1} \\ 0 \end{pmatrix} = \begin{pmatrix} P_2 \\ 0 \\ \Delta_2 \end{pmatrix}, \tag{2.485}$$

where Δ_2 is defined by the third equation, and the first two equations are those (2.473) for $N = 1$. Reversing the order of equations and unknowns yields the system

$$\begin{pmatrix} \phi_{ff}(0) & \phi_{ff}(1) & \phi_{ff}(2) \\ \phi_{ff}(-1) & \phi_{ff}(0) & \phi_{ff}(1) \\ \phi_{ff}(-2) & \phi_{ff}(-1) & \phi_{ff}(0) \end{pmatrix} \begin{pmatrix} 0 \\ \gamma_{1,1} \\ 1 \end{pmatrix} = \begin{pmatrix} \Delta_2 \\ 0 \\ P_2 \end{pmatrix}. \tag{2.486}$$

Taking complex conjugates of both sides and using the Hermitian property of the autocorrelations gives

$$\begin{pmatrix} \phi_{ff}(0) & \phi_{ff}(-1) & \phi_{ff}(-2) \\ \phi_{ff}(1) & \phi_{ff}(0) & \phi_{ff}(-1) \\ \phi_{ff}(2) & \phi_{ff}(1) & \phi_{ff}(0) \end{pmatrix} \begin{pmatrix} 0 \\ \gamma_{1,1}^* \\ 1 \end{pmatrix} = \begin{pmatrix} \Delta_2^* \\ 0 \\ P_2 \end{pmatrix}. \tag{2.487}$$

Now, we multiply the system (2.487) by the parameter $\gamma_{2,2}$ and add it to the system (2.485) to get

$$\begin{pmatrix} \phi_{ff}(0) & \phi_{ff}(-1) & \phi_{ff}(-2) \\ \phi_{ff}(1) & \phi_{ff}(0) & \phi_{ff}(-1) \\ \phi_{ff}(2) & \phi_{ff}(1) & \phi_{ff}(0) \end{pmatrix} \left(\begin{pmatrix} 1 \\ \gamma_{1,1} \\ 0 \end{pmatrix} + \gamma_{2,2} \begin{pmatrix} 0 \\ \gamma_{1,1}^* \\ 1 \end{pmatrix} \right)$$
$$= \left(\begin{pmatrix} P_2 \\ 0 \\ \Delta_2 \end{pmatrix} + \gamma_{2,2} \begin{pmatrix} \Delta_2^* \\ 0 \\ P_2 \end{pmatrix} \right). \tag{2.488}$$

If this system is to match the prediction error equations (2.80) for $N = 2$,

$$P_2 + \gamma_{2,2}\Delta_2^* = P_3, \tag{2.489}$$

$$\gamma_{1,1} + \gamma_{2,2}\gamma_{1,1}^* = \gamma_{1,2}, \tag{2.490}$$

$$\Delta_2 + \gamma_{2,2}P_2 = 0. \tag{2.491}$$

$\gamma_{2,2}$ is then the only free parameter left to be determined. From equation (2.491), $\Delta_2 = -\gamma_{2,2}P_2$. Then, from equation (2.489),

$$P_3 = \left(1 - \gamma_{2,2}\gamma_{2,2}^*\right) P_2. \tag{2.492}$$

The third equation of the system (2.488) yields

$$\phi_{ff}(2) = -\gamma_{1,2}\phi_{ff}(1) - \gamma_{2,2}\phi_{ff}(0). \tag{2.493}$$

Once again we take P_3 to be given by the average of the forward and backward prediction errors,

$$\begin{aligned} \epsilon_{j+2,2}^F &= f_{j+2} + \gamma_{1,2} f_{j+1} + \gamma_{2,2} f_j, \quad j = 1, \ldots, M-2 \\ &= f_{j+2} + \left(\gamma_{1,1} + \gamma_{2,2}\gamma_{1,1}^*\right) f_{j+1} + \gamma_{2,2} f_j \\ &= \epsilon_{j+2,1}^F + \gamma_{2,2}\epsilon_{j,1}^B, \end{aligned} \tag{2.494}$$

$$\begin{aligned} \epsilon_{j,2}^B &= f_j + \gamma_{1,2}^* f_{j+1} + \gamma_{2,2}^* f_{j+2}, \quad j = 1, \ldots, M-2 \\ &= f_j + \left(\gamma_{1,1}^* + \gamma_{2,2}^*\gamma_{1,1}\right) f_{j+1} + \gamma_{2,2}^* f_{j+2} \\ &= \epsilon_{j,1}^B + \gamma_{2,2}^*\epsilon_{j+2,1}^F. \end{aligned} \tag{2.495}$$

Thus,

$$P_3 = \frac{1}{2(M-2)} \sum_{j=1}^{M-2} \left(\left|\epsilon_{j+2,2}^F\right|^2 + \left|\epsilon_{j,2}^B\right|^2 \right)$$

$$= \frac{1}{2(M-2)} \sum_{j=1}^{M-2} \left(\left|\epsilon_{j+2,1}^F + \gamma_{2,2}\epsilon_{j,1}^B\right|^2 + \left|\epsilon_{j,1}^B + \gamma_{2,2}^*\epsilon_{j+2,1}^F\right|^2 \right). \qquad (2.496)$$

As before, the free parameter is chosen to minimise the prediction error power. For a minimum of P_3 the partial derivatives with respect to the real and imaginary parts of $\gamma_{2,2}$ vanish. Then,

$$\frac{\partial P_3}{\partial \mathrm{Re}\gamma_{2,2}} = \frac{1}{2(M-2)} \sum_{j=1}^{M-2} \Big\{ \epsilon_{j,1}^B \left(\epsilon_{j+2,1}^{F*} + \gamma_{2,2}^*\epsilon_{j,1}^{B*}\right) + \epsilon_{j,1}^{B*} \left(\epsilon_{j+2,1}^F + \gamma_{2,2}\epsilon_{j,1}^B\right)$$

$$+ \epsilon_{j+2,1}^F \left(\epsilon_{j,1}^{B*} + \gamma_{2,2}\epsilon_{j+2,1}^{F*}\right) + \epsilon_{j+2,1}^{F*} \left(\epsilon_{j,1}^B + \gamma_{2,2}^*\epsilon_{j+2,1}^F\right) \Big\} = 0, \qquad (2.497)$$

and

$$\frac{\partial P_3}{\partial \mathrm{Im}\gamma_{2,2}} = \frac{i}{2(M-2)} \sum_{j=1}^{M-2} \Big\{ \epsilon_{j,1}^B \left(\epsilon_{j+2,1}^{F*} + \gamma_{2,2}^*\epsilon_{j,1}^{B*}\right) - \epsilon_{j,1}^{B*} \left(\epsilon_{j+2,1}^F + \gamma_{2,2}\epsilon_{j,1}^B\right)$$

$$- \epsilon_{j+2,1}^F \left(\epsilon_{j,1}^{B*} + \gamma_{2,2}\epsilon_{j+2,1}^{F*}\right) + \epsilon_{j+2,1}^{F*} \left(\epsilon_{j,1}^B + \gamma_{2,2}^*\epsilon_{j+2,1}^F\right) \Big\} = 0. \qquad (2.498)$$

Adding $i \times$ (2.498) to (2.497) gives

$$\sum_{j=1}^{m-2} \Big\{ \epsilon_{j,1}^{B*} \left(\epsilon_{j+2,1}^F + \gamma_{2,2}\epsilon_{j,1}^B\right) + \epsilon_{j+2,1}^F \left(\epsilon_{j,1}^{B*} + \gamma_{2,2}\epsilon_{j+2,1}^{F*}\right) \Big\} = 0. \qquad (2.499)$$

Solving for the value of $\gamma_{2,2}$ that minimises P_3, we get

$$\gamma_{2,2} = -2 \sum_{j=1}^{M-2} \left(\epsilon_{j+2,1}^F \epsilon_{j,1}^{B*}\right) \Big/ \sum_{j=1}^{M-2} \left(\left|\epsilon_{j+2,1}^F\right|^2 + \left|\epsilon_{j,1}^B\right|^2 \right). \qquad (2.500)$$

Successive coefficients of the prediction error filter can be found in this manner so that after $N+1$ recursions it is found that

$$\gamma_{N+1,N+1} = -2 \sum_{j=1}^{M-N-1} \left(\epsilon_{j+N+1,N}^F \epsilon_{j,N}^{B*}\right) \Big/ \sum_{j=1}^{M-N-1} \left(\left|\epsilon_{j+N+1,N}^F\right|^2 + \left|\epsilon_{j,N}^B\right|^2 \right). \qquad (2.501)$$

The forward and backward prediction error sequences are

$$\begin{aligned}\epsilon^F_{j+N+1,N+1} &= f_{j+N+1} + \gamma_{1,N+1} f_{j+N} + \cdots + \gamma_{N+1,N+1} f_j \\ &= \epsilon^F_{j+N+1,N} + \gamma_{N+1,N+1}\epsilon^B_{j,N}, \qquad (2.502)\\ \epsilon^B_{j,N+1} &= f_j + \gamma^*_{1,N+1} f_{j+1} + \cdots + \gamma^*_{N+1,N+1} f_{j+N+1} \\ &= \epsilon^B_{j,N} + \gamma^*_{N+1,N+1}\epsilon^F_{j+N+1,N}, \qquad (2.503)\end{aligned}$$

for $j = 1, \ldots, M - N - 1$. The prediction error filter coefficients are updated according to

$$\begin{aligned}\gamma_{1,N+1} &= \gamma_{1,N} + \gamma_{N+1,N+1}\gamma^*_{N,N}, \\ \gamma_{2,N+1} &= \gamma_{2,N} + \gamma_{N+1,N+1}\gamma^*_{N-1,N}, \\ &\vdots \\ \gamma_{N,N+1} &= \gamma_{N,N} + \gamma_{N+1,N+1}\gamma^*_{1,N}. \qquad (2.504)\end{aligned}$$

The new autocorrelation is found from

$$\phi_{ff}(N+1) = -\sum_{j=1}^{N+1} \gamma_{j,N+1}\,\phi_{ff}(N-j+1), \qquad (2.505)$$

and the new prediction error power is found from the recursion

$$P_{N+2} = P_{N+1}\left(1 - \gamma_{N+1,N+1}\gamma^*_{N+1,N+1}\right). \qquad (2.506)$$

The subroutine BPEC performs the recursion to find the prediction error coefficients G of length NP1 from those of length N for the given data sequence F of length M. Current values of the prediction error coefficients, and the forward and backward prediction error sequences, are included as input and their updated values are returned as output.

```
      SUBROUTINE BPEC(M,NP1,F,G,PEF,PER)
C
C Subroutine BPEC finds the prediction error filter coefficients G
C of length NP1 from those of length N for the time sequence F of length M.
C It also returns the forward and backward prediction error sequences.
C
      IMPLICIT DOUBLE COMPLEX (A-H,O-Z)
      DIMENSION F(M),G(M),H(M),O(M),PEF(M),PER(M)
C Find previous recursion number.
      N=NP1-1
      IF(N.NE.0) GO TO 10
C If previous recursion number is zero, set forward and backward
C prediction error sequences to given data sequence.
      DO 11 J=1,M
        PEF(J)=F(J)
        PER(J)=F(J)
11    CONTINUE
```

```
C Find next member of prediction error filter G(NP1).
 10    SN=(0.D0,0.D0)
       SD=(0.D0,0.D0)
C Find upper limit of sum.
       JJ=M-N-1
       DO 12 J=1,JJ
         SN=SN-2.D0*PEF(J+N+1)*DCONJG(PER(J))
         SD=SD+PEF(J+N+1)*DCONJG(PEF(J+N+1))+PER(J)*DCONJG(PER(J))
 12    CONTINUE
       G(NP1)=SN/SD
C If previous recursion number is zero, do not update prediction error
C filter coefficients.
       IF(N.EQ.0) GO TO 13
C Update prediction error filter coefficients and store them in vector H(J).
       DO 14 J=1,N
         K=N-J+1
         H(J)=G(J)+G(NP1)*DCONJG(G(K))
 14    CONTINUE
C Put updated prediction error coefficients in G(J).
       DO 15 J=1,N
         G(J)=H(J)
 15    CONTINUE
C Update forward prediction error sequence and store in H(K).
 13    DO 16 J=1,JJ
       K=J+NP1
       H(K)=PEF(K)+G(NP1)*PER(J)
 16    CONTINUE
C Update backward prediction error sequence and store in O(J).
       DO 17 J=1,JJ
         O(J)=PER(J)+DCONJG(G(NP1))*PEF(J+NP1)
 17    CONTINUE
C Put updated forward and backward prediction error sequences
C in PEF(K) and PER(J),respectively.
       DO 18 J=1,JJ
         K=J+NP1
         PEF(K)=H(K)
         PER(J)=O(J)
 18    CONTINUE
       RETURN
       END
```

3

Earth deformations

The Earth is a rotating, self-gravitating body with significant radial variations in its physical properties. Lateral variations in its physical properties are generally less pronounced but important in certain circumstances.

For the purpose of studying deformations except at very long periods, it is usually sufficient to treat the Earth as perfectly elastic, although even at relatively short periods the small but finite dissipation can be geophysically important.

3.1 Equilibrium equations

As well as allowing for spatial variation in elastic properties and for self-gravitation, the elastic equations, when applied to the Earth, must be modified to take into account pre-existing stress.

We adopt the notation of Cartesian tensors and use the range and summation conventions in our analysis.

In infinitesimal elasticity theory, the stresses are measured at points in the deformed medium, while the displacements and strains are calculated with reference to points in the undeformed medium. For a displacement field u_k, the stresses, τ_{ij}, are related to the dilatation

$$\frac{\partial u_k}{\partial x_k}, \tag{3.1}$$

and the shear strain

$$\frac{1}{2}\left(\frac{\partial u_j}{\partial x_i} + \frac{\partial u_i}{\partial x_j}\right), \tag{3.2}$$

by the generalised Hooke's law

$$\tau_{ij} = \lambda \frac{\partial u_k}{\partial x_k} \delta^i_j + \mu \left(\frac{\partial u_j}{\partial x_i} + \frac{\partial u_i}{\partial x_j}\right), \tag{3.3}$$

where λ and μ are the Lamé coefficients of elasticity.

Strictly speaking, one must use the stress σ_{ij} at points in the undeformed medium, since the co-ordinate system generally refers to such points. If the pre-existing stress field is T_{ij}, the particle carried into the reference point by the deformation was under an initial stress field

$$T_{ij} - u_k \frac{\partial T_{ij}}{\partial x_k}. \tag{3.4}$$

The same particle experienced an additional applied stress τ_{ij}, sufficient to carry it into the point under consideration. In infinitesimal elasticity theory, the second term in (3.4) is usually of second order in small quantities and can be neglected. The same cannot be said in the presence of initial stress, since T_{ij} may be due to physical effects entirely different from those giving rise to the displacement field u_k. In a pre-stressed body, such as the Earth, the total stress at the reference point, in the deformed medium, is then

$$\sigma_{ij} = T_{ij} - u_k \frac{\partial T_{ij}}{\partial x_k} + \tau_{ij}. \tag{3.5}$$

It is usual to take the initial stress in the Earth to be that of hydrostatic equilibrium, due entirely to self-gravitation and the centrifugal force arising from rotation. Thus,

$$T_{ij} = -p_0 \delta^i_j, \quad \frac{\partial p_0}{\partial x_i} = \rho_0 g_{0i}, \tag{3.6}$$

where p_0 is the equilibrium pressure, g_{0i} is the equilibrium force of gravity (including the centrifugal force) per unit mass, and ρ_0 is the equilibrium density.

In the deformed state, the divergence of the total stress, plus the total body force per unit volume, vanishes for force equilibrium, giving the equilibrium equation

$$\frac{\partial \sigma_{ki}}{\partial x_k} = -\rho g_i - F_i, \tag{3.7}$$

with ρ representing the density, g_i gravity and F_i the body force, over and above gravity, per unit volume. The deformation causes small perturbations in density and gravity, ρ_1 and g_{1i}, respectively, and we can write that

$$\rho = \rho_0 + \rho_1, \tag{3.8}$$

and

$$g_i = g_{0i} + g_{1i}. \tag{3.9}$$

The change in density arises from the dilatation, and from transport through the non-uniform equilibrium density field into the reference point. Thus,

$$\rho_1 = -\rho_0 \frac{\partial u_k}{\partial x_k} - u_k \frac{\partial \rho_0}{\partial x_k} = -\frac{\partial}{\partial x_k} (\rho_0 u_k), \tag{3.10}$$

correct to first order in the displacements. To the same degree of approximation, using the Poisson equation (5.4) for the gravitational potential,

$$\frac{\partial^2 V_1}{\partial x_k \partial x_k} = -4\pi G \rho_1 = 4\pi G \frac{\partial}{\partial x_k}(\rho_0 u_k), \tag{3.11}$$

where V_1 is defined as the decrease in gravitational potential arising from the density perturbation ρ_1. Then, the perturbation in gravity is given by

$$g_{1i} = \frac{\partial V_1}{\partial x_i} \tag{3.12}$$

and, from (3.7),

$$\frac{\partial \sigma_{ki}}{\partial x_k} = -\rho_0 g_{0i} - \rho_0 g_{1i} - \rho_1 g_{0i} - F_i. \tag{3.13}$$

Substitution from expressions (3.5) and (3.6) yields a modified equilibrium equation,

$$\frac{\partial \tau_{ki}}{\partial x_k} = -\frac{\partial}{\partial x_i}(\rho_0 u_k g_{0k}) - \rho_0 g_{1i} - \rho_1 g_{0i} - F_i. \tag{3.14}$$

On substitution of the generalised Hooke's law (3.3), we have

$$\begin{aligned}
\frac{\partial}{\partial x_k}&\left[\lambda \frac{\partial u_l}{\partial x_l}\delta_i^k + \mu\left(\frac{\partial u_i}{\partial x_k} + \frac{\partial u_k}{\partial x_i}\right)\right] \\
&= \frac{\partial \lambda}{\partial x_i}\frac{\partial u_l}{\partial x_l} + \lambda \frac{\partial^2 u_l}{\partial x_i \partial x_l} + \frac{\partial \mu}{\partial x_k}\left(\frac{\partial u_i}{\partial x_k} + \frac{\partial u_k}{\partial x_i}\right) + \mu \frac{\partial^2 u_i}{\partial x_k^2} + \mu \frac{\partial^2 u_k}{\partial x_k \partial x_i} \\
&= -\frac{\partial}{\partial x_i}(\rho_0 u_k g_{0k}) - \rho_0 g_{1i} - \rho_1 g_{0i} - F_i.
\end{aligned} \tag{3.15}$$

With some difficulty, the equilibrium equation (3.15) can be converted to symbolic vector notation. Making use of the properties of the alternating tensor ξ_{ijk}, the ith component of $\nabla \times (\nabla \times \boldsymbol{u})$ is

$$[\nabla \times (\nabla \times \boldsymbol{u})]_i = \xi_{ijk} \frac{\partial}{\partial x_j} \xi_{klm} \frac{\partial u_m}{\partial x_l} = \frac{\partial^2 u_k}{\partial x_i \partial x_k} - \frac{\partial^2 u_i}{\partial x_k^2}, \tag{3.16}$$

while the ith component of $\nabla \times (\boldsymbol{u} \times \nabla \mu)$ is

$$\begin{aligned}
[\nabla \times (\boldsymbol{u} \times \nabla \mu)]_i &= \xi_{ijk} \frac{\partial}{\partial x_j} \xi_{klm} u_l \frac{\partial \mu}{\partial x_m} \\
&= \frac{\partial \mu}{\partial x_k}\frac{\partial u_i}{\partial x_k} + u_i \frac{\partial^2 \mu}{\partial x_k^2} - \frac{\partial \mu}{\partial x_i}\frac{\partial u_k}{\partial x_k} - u_k \frac{\partial^2 \mu}{\partial x_k \partial x_i}.
\end{aligned} \tag{3.17}$$

The ith component of the combination $-\mu\nabla\times(\nabla\times\boldsymbol{u})+\nabla\times(\boldsymbol{u}\times\nabla\mu)+\nabla(\boldsymbol{u}\cdot\nabla\mu)$ is then

$$\frac{\partial\mu}{\partial x_k}\left(\frac{\partial u_i}{\partial x_k}+\frac{\partial u_k}{\partial x_i}\right)+\mu\frac{\partial^2 u_i}{\partial x_k^2}-\mu\frac{\partial^2 u_k}{\partial x_k\partial x_i}-\frac{\partial\mu}{\partial x_i}\frac{\partial u_k}{\partial x_k}+u_i\frac{\partial^2\mu}{\partial x_k^2}. \tag{3.18}$$

If we now add $\nabla\lambda\nabla\cdot\boldsymbol{u}+\lambda\nabla(\nabla\cdot\boldsymbol{u})+2\mu\nabla(\nabla\cdot\boldsymbol{u})+\nabla\mu\nabla\cdot\boldsymbol{u}-\boldsymbol{u}\nabla^2\mu$ to this combination, we obtain the left side of the equilibrium equation (3.15). In symbolic vector notation it takes the form

$$\begin{aligned}(\lambda+2\mu)\,\nabla(\nabla\cdot\boldsymbol{u})&-\mu\nabla\times(\nabla\times\boldsymbol{u})\\&+(\nabla\lambda+\nabla\mu)\,\nabla\cdot\boldsymbol{u}+\nabla\times(\boldsymbol{u}\times\nabla\mu)+\nabla(\boldsymbol{u}\cdot\nabla\mu)-\boldsymbol{u}\nabla^2\mu\\&=-\nabla(\rho_0\boldsymbol{u}\cdot\boldsymbol{g}_0)-\boldsymbol{g}_0\rho_1-\boldsymbol{g}_1\rho_0-\boldsymbol{F}.\end{aligned} \tag{3.19}$$

The first two terms on the left side and the last term on the right side appear in the classical Navier equation of equilibrium (1.271) for uniform elastic solids, free of self-gravitation. The extra terms on the left side reflect the spatial variation of elastic properties. The extra terms on the right side reflect, respectively, the effects of the extra pressure arising from transport through the hydrostatic pressure field, the equilibrium gravity acting on the perturbed density and the perturbed gravity acting on the equilibrium density. To complete the description of the elasto-gravitational system, (3.19) must be augmented by the gravitational equation (3.11) expressed in symbolic vector notation,

$$\nabla^2 V_1=4\pi G\nabla\cdot(\rho_0\boldsymbol{u}). \tag{3.20}$$

3.2 The reciprocal theorem of Betti

The reciprocal theorem of Betti (1.266) is one of the basic theorems of classical elasticity theory. It can be regarded as the basis for generating formulae of the Green's theorem type, where the solution at the field point is an integral over point sources in the source region. In this section, we generalise the reciprocal theorem of Betti to elasto-gravitational systems whose elastic properties are spatially varying, and which are subject to hydrostatic pre-stress in the equilibrium reference state, as described in the previous section.

Consider two systems of surface tractions and body forces (primed and unprimed), t_i, F_i, and t_i', F_i', producing, respectively, displacement fields u_i and u_i'. Let these force systems act on material contained in a volume $\mathcal{V}$ by a surface S, which is such as to allow the divergence theorem of Gauss to be applied. The work done by the unprimed tractions and forces, acting through the primed displacements, is

$$\int_S t_i u_i' dS+\int_{\mathcal{V}} F_i u_i' d\mathcal{V}=\int_S \tau_{ji}\nu_j u_i' dS+\int_{\mathcal{V}} F_i u_i' d\mathcal{V}, \tag{3.21}$$

where ν_i is the unit outward normal vector to the surface S. Transforming the surface integral to a volume integral by Gauss's theorem gives, for the work done,

$$\int_{\mathcal{V}} \left[\frac{\partial}{\partial x_j} \left(\tau_{ji} u'_i \right) + F_i u'_i \right] d\mathcal{V} = \int_{\mathcal{V}} \left[\tau_{ji} \frac{\partial u'_i}{\partial x_j} + \frac{\partial \tau_{ji}}{\partial x_j} u'_i + F_i u'_i \right] d\mathcal{V}. \tag{3.22}$$

Substitution from the equilibrium equation (3.14) gives the combination of terms

$$t_c = \frac{\partial \tau_{ji}}{\partial x_j} u'_i + F_i u'_i = \left[-\frac{\partial}{\partial x_i} \left(\rho_0 u_j g_{0j} \right) - \rho_0 g_{1i} - \rho_1 g_{0i} - F_i \right] u'_i + F_i u'_i. \tag{3.23}$$

Using expression (3.10) for ρ_1,

$$t_c = -g_{0j} \frac{\partial \rho_0}{\partial x_i} u_j u'_i - \rho_0 \frac{\partial}{\partial x_i} \left(u_j g_{0j} \right) u'_i + g_{0i} \frac{\partial}{\partial x_j} \left(\rho_0 u_j \right) u'_i - \rho_0 g_{1i} u'_i. \tag{3.24}$$

With the exchange of the repeated subscripts, the second term on the right side becomes $-\rho_0 \partial / \partial x_j \left(u_i g_{0i} \right) u'_j$, while the third term on the right can be replaced identically by

$$\frac{\partial}{\partial x_j} \left(\rho_0 g_{0i} u_j u'_i \right) - \rho_0 u_j \frac{\partial}{\partial x_j} \left(u'_i g_{0i} \right). \tag{3.25}$$

The expression for the work done then takes the form

$$\begin{aligned} \int_{\mathcal{V}} \Bigg[\tau_{ji} \frac{\partial u'_i}{\partial x_j} &- g_{0j} \frac{\partial \rho_0}{\partial x_i} u_j u'_i - \rho_0 u'_j \frac{\partial}{\partial x_j} \left(u_i g_{0i} \right) \\ &- \rho_0 u_j \frac{\partial}{\partial x_j} \left(u'_i g_{0i} \right) + \frac{\partial}{\partial x_j} \left(\rho_0 g_{0i} u_j u'_i \right) - \rho_0 g_{1i} u'_i \Bigg] d\mathcal{V}. \end{aligned} \tag{3.26}$$

On substitution from (3.3), the first term in the integrand can be transformed as

$$\begin{aligned} \lambda \frac{\partial u_k}{\partial x_k} \delta^j_i \frac{\partial u'_i}{\partial x_j} + \mu \left(\frac{\partial u_i}{\partial x_j} + \frac{\partial u_j}{\partial x_i} \right) \frac{\partial u'_i}{\partial x_j} &= \lambda \frac{\partial u_i}{\partial x_i} \frac{\partial u'_j}{\partial x_j} + \mu \frac{\partial u_i}{\partial x_j} \frac{\partial u'_i}{\partial x_j} + \mu \frac{\partial u_j}{\partial x_i} \frac{\partial u'_i}{\partial x_j} \\ &= \lambda \frac{\partial u_i}{\partial x_i} \frac{\partial u'_j}{\partial x_j} + \mu \frac{\partial u_i}{\partial x_j} \frac{\partial u'_i}{\partial x_j} + \mu \frac{\partial u_i}{\partial x_j} \frac{\partial u'_j}{\partial x_i}, \end{aligned} \tag{3.27}$$

the latter form being realised on the interchange of repeated subscripts in the last term. The second term in the integrand can be transformed by the recognition that surfaces of equal density (isopycnic surfaces) coincide with equipotentials of the geopotential in equilibrium, as demonstrated by equation (5.107). If s is the co-ordinate in the orthometric direction (opposite to gravity), then

$$-g_{0j}\frac{\partial \rho_0}{\partial x_i}u_j u_i' = g_0\frac{\partial \rho_0}{\partial s}u_s u_s'. \tag{3.28}$$

The last term in the integrand can be replaced by the identity

$$-\rho_0 g_{1i}u_i' = \frac{1}{4\pi G}\left\{\frac{\partial}{\partial x_i}\left[V_1\left(\frac{\partial V_1'}{\partial x_i} - 4\pi G\rho_0 u_i'\right)\right] - \frac{\partial V_1}{\partial x_i}\frac{\partial V_1'}{\partial x_i}\right\}. \tag{3.29}$$

Finally, with the aid of Gauss's theorem, we are able to write that

$$\begin{aligned}
&\int_S t_i u_i' dS + \int_V F_i u_i' dV \\
&= \int_V \left[\lambda\frac{\partial u_i}{\partial x_i}\frac{\partial u_j'}{\partial x_j} + \mu\left(\frac{\partial u_i}{\partial x_j}\frac{\partial u_i'}{\partial x_j} + \frac{\partial u_j}{\partial x_i}\frac{\partial u_i'}{\partial x_j}\right) + g_0\frac{\partial \rho_0}{\partial s}u_s u_s' \right. \\
&\qquad \left. - \rho_0 u_j\frac{\partial}{\partial x_j}\left(u_i' g_{0i}\right) - \rho_0 u_j'\frac{\partial}{\partial x_j}\left(u_i g_{0i}\right) - \frac{1}{4\pi G}\frac{\partial V_1}{\partial x_i}\frac{\partial V_1'}{\partial x_i}\right] dV \\
&\quad - \int_S \rho_0 g_0 u_i u_s' \nu_i\, dS + \frac{1}{4\pi G}\int_S V_1\left(\frac{\partial V_1'}{\partial x_i} - 4\pi G\rho_0 u_i'\right)\nu_i\, dS.
\end{aligned} \tag{3.30}$$

With the use of the second form of the right side of equation (3.27), on exchange of the primed and unprimed systems of forces and displacements, it is seen that of all the terms on the right side of expression (3.30), only the surface integrals change. We then have the required generalisation of the reciprocal theorem of Betti, expressed as

$$\begin{aligned}
&\int_S \left(t_i - \rho g_0 u_s \nu_i\right) u_i'\, dS + \int_V F_i u_i'\, dV \\
&= \int_S \left(t_i' - \rho_0 g_0 u_s' \nu_i\right) u_i\, dS + \int_V F_i' u_i\, dV \\
&\quad + \frac{1}{4\pi G}\int_S \left[V_1\left(\frac{\partial V_1'}{\partial x_i} - 4\pi G\rho_0 u_i'\right) - V_1'\left(\frac{\partial V_1}{\partial x_i} - 4\pi G\rho_0 u_i\right)\right]\nu_i\, dS.
\end{aligned} \tag{3.31}$$

The surface tractions are modified by the contributions from transport through the initial hydrostatic stress field, and new terms arise to account for the effect of self-gravitation.

3.3 Radial equations: spheroidal and torsional

The Earth departs by only about one part in 300 from spherical shape. The internal surfaces of equal density, geopotential and pressure are close to spherical, and in close approximation the orthometric co-ordinate becomes the radius. In this approximation, the elastic properties are taken to depend on radius alone. The displacement field $\boldsymbol{u}$ can be broken into the sum of lamellar, poloidal and torsional vectors (1.174), dependent on lamellar, poloidal and torsional scalars, which, in turn, can be expanded in spherical harmonics, as in (1.184), for Earth models with spherical symmetry. The divergence of the lamellar vector, and curls of the poloidal and torsional vectors, lead to new scalars and radial coefficients, as described by expression (1.194).

Using the methods of Section 1.3, the equilibrium equation (3.19) can be reduced to a system of radial equations for its radial spheroidal, transverse spheroidal and torsional parts. The elasto-gravitational system, in addition to the equilibrium equation, includes the gravitational relation (3.20). The decrease in gravitational potential, V_1, is expanded in spherical harmonics as

$$V_1 = \sum_{n=0}^{\infty} \sum_{m=-n}^{n} \phi_n^m(r,t)\, P_n^m(\cos\theta)\, e^{im\phi}. \tag{3.32}$$

Although the reduction of terms in the equilibrium equation is generally straightforward, the term $\nabla\times(\boldsymbol{u}\times\nabla\mu)$ requires special consideration. By comparison with expressions (1.195) and (1.187), the vector $\boldsymbol{u}\times\nabla\mu$ is easily shown to be made up of a transverse spheroidal part with radial coefficient $-t_n^m d\mu/dr$ and a torsional part with radial coefficient $v_n^m d\mu/dr$. The transverse spheroidal part, in turn, by (1.186), has a poloidal part such that

$$\frac{1}{r}\frac{\partial}{\partial r}(r p_n^m) = -t_n^m \frac{d\mu}{dr}. \tag{3.33}$$

Since the curl of the lamellar part vanishes, the curl of the spheroidal vector field depends only on the curl of its poloidal part. The curl of a torsional vector is poloidal and the negative of the curl of a poloidal vector is torsional with scalar given by (1.189). Thus, the curl of the vector $\boldsymbol{u}\times\nabla\mu$ is made up of a poloidal part with radial coefficient $v_n^m d\mu/dr$, and a torsional part, by relations (1.189) and (1.198), with radial coefficient

$$\frac{1}{r}\frac{\partial}{\partial r}\left(r\frac{d\mu}{dr}t_n^m\right). \tag{3.34}$$

With the definitions (1.194) of $l_n''^m$, $p_n''^m$ and $t_n''^m$, the radial spheroidal, transverse spheroidal and torsional parts of the equilibrium equation can be identified in terms of their radial coefficients. The radial spheroidal part of the equilibrium equation gives the radial equation

$$(\lambda+2\mu)\frac{\partial l_n^{\prime\prime m}}{\partial r}+\mu\frac{n(n+1)}{r}p_n^{\prime\prime m}+\left(\frac{d\lambda}{dr}+\frac{d\mu}{dr}\right)l_n^{\prime\prime m}$$
$$+\frac{d\mu}{dr}\frac{n(n+1)}{r}v_n^m+\frac{d^2\mu}{dr^2}u_n^m+\frac{d\mu}{dr}\frac{\partial u_n^m}{\partial r}-\frac{1}{r}\frac{d^2}{dr^2}(r\mu)\,u_n^m$$
$$=\frac{\partial}{\partial r}(\rho_0 g_0 u_n^m)-\rho_0 g_0 l_n^{\prime\prime m}-\frac{d\rho_0}{dr}g_0 u_n^m-\rho_0\frac{\partial\phi_n^m}{\partial r}-u_{F\,n}^m, \tag{3.35}$$

with $u_{F\,n}^m$ representing the radial coefficient of the radial spheroidal part of the body force per unit volume. The transverse spheroidal part of the equilibrium equation gives the radial equation

$$\frac{(\lambda+2\mu)}{r}l_n^{\prime\prime m}+\frac{\mu}{r}\frac{\partial}{\partial r}(rp_n^{\prime\prime m})+\frac{1}{r}\frac{\partial}{\partial r}\left(r\frac{d\mu}{dr}v_n^m\right)+\frac{1}{r}\frac{d\mu}{dr}u_n^m-\frac{1}{r}\frac{d^2}{dr^2}(r\mu)\,v_n^m$$
$$=\frac{1}{r}\rho_0 g_0 u_n^m-\frac{1}{r}\rho_0\phi_n^m-v_{F\,n}^m, \tag{3.36}$$

with $v_{F\,n}^m$ representing the radial coefficient of the transverse spheroidal part of the body force per unit volume. The torsional part of the equilibrium equation gives the radial equation

$$\mu t_n^{\prime\prime m}+\frac{1}{r}\frac{\partial}{\partial r}\left(r\frac{d\mu}{dr}t_n^m\right)-\frac{1}{r}\frac{d^2}{dr^2}(r\mu)\,t_n^m=-t_{F\,n}^m, \tag{3.37}$$

with $t_{F\,n}^m$ representing the radial coefficient of the torsional part of the body force per unit volume. A fourth radial equation follows from the gravitational equation (3.20),

$$\frac{1}{r}\frac{\partial^2}{\partial r^2}(r\phi_n^m)-\frac{n(n+1)}{r^2}\phi_n^m=4\pi G\left(\rho_0 l_n^{\prime\prime m}+\frac{d\rho_0}{dr}u_n^m\right). \tag{3.38}$$

Together with the defining relations (1.196), these four radial equations form six equations in the six radial coefficients u_n^m, v_n^m, $l_n^{\prime\prime m}$, $p_n^{\prime\prime m}$, $t_n^{\prime\prime m}$ and ϕ_n^m.

We may recast these equations in terms of more physical variables by considering the stresses associated with the deformations. The stress normal to an internal spherical surface of radius r is

$$\tau_{rr}=\lambda\nabla^2 L+2\mu\frac{\partial u_r}{\partial r}, \tag{3.39}$$

while the tangential shear stress has components

$$\tau_{r\theta}=\mu\left(\frac{\partial u_\theta}{\partial r}+\frac{1}{r}\frac{\partial u_r}{\partial\theta}-\frac{u_\theta}{r}\right),$$
$$\tau_{r\phi}=\mu\left(\frac{\partial u_\phi}{\partial r}+\frac{1}{r\sin\theta}\frac{\partial u_r}{\partial\phi}-\frac{u_\phi}{r}\right). \tag{3.40}$$

The radial coefficient of the normal stress is then

$$y_2 = \lambda l_n''^m + 2\mu \frac{\partial u_n^m}{\partial r} = \lambda \left[\frac{\partial u_n^m}{\partial r} + \frac{(2u_n^m - n(n+1)v_n^m)}{r} \right] + 2\mu \frac{\partial u_n^m}{\partial r}. \tag{3.41}$$

The shear stress arising from the spheroidal part of the deformation field is a transverse spheroidal vector field with radial coefficient

$$y_4 = \mu \left[\frac{\partial v_n^m}{\partial r} - \frac{(v_n^m - u_n^m)}{r} \right], \tag{3.42}$$

while the shear stress arising from the torsional part of the deformation field is a torsional vector field with radial coefficient

$$z_2 = \mu \left[\frac{\partial t_n^m}{\partial r} - \frac{1}{r} t_n^m \right]. \tag{3.43}$$

The gravitational equation (3.20) may be expressed as

$$\nabla \cdot (\nabla V_1 - 4\pi G \rho_0 \boldsymbol{u}) = 0. \tag{3.44}$$

The vector $\nabla V_1 - 4\pi G \rho_0 \boldsymbol{u}$ may be regarded as a gravitational flux vector, with outward normal component to an internal, spherical equipotential surface having the radial coefficient

$$y_6 = \frac{\partial \phi_n^m}{\partial r} - 4\pi G \rho_0 u_n^m. \tag{3.45}$$

In the approximation that the geopotential surfaces are spherical, the equation (5.5) governing the equilibrium geopotential may be recast as

$$-\nabla \cdot g_0 = 4\pi G \rho_0 - 2\Omega^2. \tag{3.46}$$

Then,

$$\frac{dg_0}{dr} = 4\pi G \rho_0 - \frac{2}{r}\left(g_0 + r\Omega^2\right). \tag{3.47}$$

With appropriate substitutions from (1.194), (1.197), (1.198), (3.41), (3.42), (3.43) and (3.47), equation (3.35) is reduced to

$$\begin{aligned} &\frac{\partial y_2}{\partial r} - \frac{n(n+1)}{r} y_4 - \frac{2\mu}{r\lambda}\left[y_2 - (3\lambda + 2\mu)\frac{\partial u_n^m}{\partial r} \right] \\ &= \left[4\pi G \rho_0^2 - \frac{2}{r}\rho_0 \left(g_0 + r\Omega^2\right) \right] u_n^m - \frac{\rho_0 g_0}{\lambda}\left[y_2 - (\lambda + 2\mu)\frac{\partial u_n^m}{\partial r} \right] \\ &- \rho_0 \frac{\partial \phi_n^m}{\partial r} - u_{F\,n}^m, \end{aligned} \tag{3.48}$$

while equation (3.36) is reduced to

$$\frac{\partial y_4}{\partial r} + \frac{1}{r} y_2 + \frac{1}{r} y_4 + \frac{2\mu}{r\lambda}\left[y_2 - (\lambda + 2\mu)\frac{\partial u_n^m}{\partial r}\right] + \frac{2\mu}{r}\left(\frac{\partial v_n^m}{\partial r} - \frac{\partial u_n^m}{\partial r}\right)$$
$$= \frac{1}{r}\rho_0 g_0 u_n^m - \frac{1}{r}\rho_0 \phi_n^m - v_{Fn}^m, \tag{3.49}$$

and equation (3.37) is reduced to

$$\frac{\partial z_2}{\partial r} - \frac{\mu}{r^2}\left[n(n+1) + 1 \right] t_n^m + \frac{3\mu}{r}\frac{\partial t_n^m}{\partial r} = -t_{Fn}^m. \tag{3.50}$$

Recast in this form, the radial spheroidal, transverse spheroidal and torsional parts of the equilibrium equation are free of derivatives of the Earth model properties λ, μ, ρ_0 and g_0. From the numerical point of view, the advantages of this form of the equations were recognised by Alterman, Jarosch and Pekeris (Alterman *et al.*, 1959) and led to their *y-notation* in which $y_1 = u_n^m$ denotes the radial coefficient of the radial spheroidal displacement, $y_3 = v_n^m$ denotes the radial coefficient of the transverse spheroidal displacement and $y_5 = \phi_n^m$ denotes the radial coefficient of the decrease in gravitational potential. The six variables y_1, y_2, y_3, y_4, y_5, y_6 describe the spheroidal deformation field. From equation (3.41), we obtain

$$\frac{\partial y_1}{\partial r} = -\frac{2\lambda}{(\lambda + 2\mu)\, r} y_1 + \frac{1}{\lambda + 2\mu} y_2 + \frac{n(n+1)\lambda}{(\lambda + 2\mu)\, r} y_3. \tag{3.51}$$

Using this equation and (3.45), we can rewrite equation (3.48) as

$$\frac{\partial y_2}{\partial r} = \left[-\frac{2\rho_0}{r}\left(2g_0 + r\Omega^2\right) + 4\mu \frac{(3\lambda + 2\mu)}{(\lambda + 2\mu)\, r^2}\right] y_1 - \frac{4\mu}{(\lambda + 2\mu)\, r} y_2$$
$$+ \left[\frac{n(n+1)}{r}\rho_0 g_0 - \frac{2n(n+1)}{r^2}\mu \frac{(3\lambda + 2\mu)}{(\lambda + 2\mu)}\right] y_3 + \frac{n(n+1)}{r} y_4$$
$$- \rho_0 y_6 - u_{Fn}^m. \tag{3.52}$$

Equation (3.42) gives

$$\frac{\partial y_3}{\partial r} = -\frac{1}{r} y_1 + \frac{1}{r} y_3 + \frac{1}{\mu} y_4. \tag{3.53}$$

Using this equation and (3.51), we can rewrite equation (3.49) as

$$\frac{\partial y_4}{\partial r} = \left[\frac{\rho_0 g_0}{r} - 2\mu \frac{(3\lambda + 2\mu)}{(\lambda + 2\mu)\, r^2}\right] y_1 - \frac{\lambda}{(\lambda + 2\mu)\, r} y_2$$
$$+ \frac{2\mu}{(\lambda + 2\mu)\, r^2}\left[\left(2n^2 + 2n - 1\right)\lambda + 2\left(n^2 + n - 1\right)\mu \right] y_3$$
$$- \frac{3}{r} y_4 - \frac{\rho_0}{r} y_5 - v_{Fn}^m. \tag{3.54}$$

Written in y-notation, equation (3.45) becomes

$$\frac{\partial y_5}{\partial r} = 4\pi G\rho_0 y_1 + y_6. \tag{3.55}$$

A sixth equation is obtained on differentiating (3.45) and substituting the result in the gravitational relation (3.38). With further substitution from (1.197), (3.41) and (3.55), it takes the form

$$\frac{\partial y_6}{\partial r} = -4\pi G\rho_0 \frac{n(n+1)}{r} y_3 + \frac{n(n+1)}{r^2} y_5 - \frac{2}{r} y_6. \tag{3.56}$$

The spheroidal deformation field is then governed by the six first-order linear differential equations, (3.51) through (3.56).

The torsional deformation field may be described similarly by writing $z_1 = t_n^m$. From equation (3.43), we then have

$$\frac{\partial z_1}{\partial r} = \frac{1}{r} z_1 + \frac{1}{\mu} z_2. \tag{3.57}$$

Substitution of this relation in the torsional part of the equilibrium equation (3.50) produces

$$\frac{\partial z_2}{\partial r} = \frac{\mu}{r^2}[n(n+1) - 2] z_1 - \frac{3}{r} z_2 - t_{Fn}^m. \tag{3.58}$$

The torsional part of the deformation field is then described by the variables z_1, z_2, governed by the two first-order linear differential equations (3.57) and (3.58).

For degree $n = 0$, we have $m = 0$ and $P_n^m = 1$. Thus, from (1.195), there is no transverse spheroidal vector field, and, from (1.187), there is no torsional part of the equilibrium equation. Equations (3.36) and (3.42), derived from transverse spheroidal parts, no longer appear, and, for $n = 0$, there is no torsional displacement field. The spheroidal displacement field is entirely in the radial direction, and, from (3.40), the shear stress vanishes. The variables y_3 and y_4 no longer appear and the governing spheroidal system degenerates to the four first-order linear differential equations

$$\frac{\partial y_1}{\partial r} = -\frac{2\lambda}{(\lambda + 2\mu) r} y_1 + \frac{1}{\lambda + 2\mu} y_2, \tag{3.59}$$

$$\frac{\partial y_2}{\partial r} = \left[-\frac{2\rho_0}{r}\left(2g_0 + r\Omega^2\right) + 4\mu \frac{(3\lambda + 2\mu)}{(\lambda + 2\mu) r^2} \right] y_1 - \frac{4\mu}{(\lambda + 2\mu) r} y_2 - \rho_0 y_6 - u_{F0}^0, \tag{3.60}$$

$$\frac{\partial y_5}{\partial r} = 4\pi G\rho_0 y_1 + y_6, \tag{3.61}$$

$$\frac{\partial y_6}{\partial r} = -\frac{2}{r} y_6. \tag{3.62}$$

In the fluid outer core, the modulus of rigidity vanishes and, neglecting viscosity, there is no resistance to shear. From equation (3.42), the variable y_4 then vanishes and, for $n \geq 1$, the governing system (3.51) through (3.56) degenerates to

$$\frac{\partial y_1}{\partial r} = -\frac{2}{r} y_1 + \frac{1}{\lambda} y_2 + \frac{n(n+1)}{r} y_3, \tag{3.63}$$

$$\frac{\partial y_2}{\partial r} = -\frac{2\rho_0}{r}\left(2g_0 + r\Omega^2\right) y_1 + \frac{n(n+1)}{r} \rho_0 g_0 y_3 - \rho_0 y_6 - u^m_{Fn}, \tag{3.64}$$

$$0 = \frac{\rho_0 g_0}{r} y_1 - \frac{1}{r} y_2 - \frac{\rho_0}{r} y_5 - v^m_{Fn}, \tag{3.65}$$

$$\frac{\partial y_5}{\partial r} = 4\pi G \rho_0 y_1 + y_6, \tag{3.66}$$

$$\frac{\partial y_6}{\partial r} = -4\pi G \rho_0 \frac{n(n+1)}{r} y_3 + \frac{n(n+1)}{r^2} y_5 - \frac{2}{r} y_6. \tag{3.67}$$

Equation (3.53), derived from expression (3.42) for the shear stress, no longer holds and is not included.

For $n = 0$, the governing system for the fluid outer core degenerates to

$$\frac{\partial y_1}{\partial r} = -\frac{2}{r} y_1 + \frac{1}{\lambda} y_2, \tag{3.68}$$

$$\frac{\partial y_2}{\partial r} = -\frac{2\rho_0}{r}\left(2g_0 + r\Omega^2\right) y_1 - \rho_0 y_6 - u^0_{F0}, \tag{3.69}$$

$$\frac{\partial y_5}{\partial r} = 4\pi G \rho_0 y_1 + y_6, \tag{3.70}$$

$$\frac{\partial y_6}{\partial r} = -\frac{2}{r} y_6. \tag{3.71}$$

3.4 Dynamical equations

In the equation for force equilibrium (3.7), the body force F_i per unit volume, over and above gravity, was left unspecified. The static equilibrium equation can be converted to one expressing dynamic equilibrium if F_i is replaced by $-\rho a_i$, with ρ representing the mass density and a_i the acceleration. This simple device is sometimes called *d'Alembert's principle*.

Equations of motion are generally expressed with respect to an inertial or space-fixed frame of reference. In a rotating frame of reference, such as that of the Earth, the time rate of change of an arbitrary vector $\boldsymbol{q}$ must be augmented by an extra rate of change caused by the rotation of the reference frame. In a purely kinematic relation, the respective time derivatives are given by

$$\left(\frac{d\boldsymbol{q}}{dt}\right)_{\text{fix}} = \left(\frac{d\boldsymbol{q}}{dt}\right)_{\text{rot}} + \boldsymbol{\omega} \times \boldsymbol{q}, \tag{3.72}$$

with $\boldsymbol{\omega}$ the vector angular velocity of the rotation. Applied to the radius vector $\boldsymbol{r}$, the velocity observed in the fixed frame, $\boldsymbol{v}_{\text{fix}}$, is related to the velocity observed in the rotating frame, $\boldsymbol{v}_{\text{rot}}$, by

$$\boldsymbol{v}_{\text{fix}} = \boldsymbol{v}_{\text{rot}} + \boldsymbol{\omega} \times \boldsymbol{r}. \tag{3.73}$$

Taking time derivatives with respect to the fixed frame on both sides, and using the kinematic relation (3.72) to express those on the right as time derivatives in the rotating frame, the acceleration in the fixed frame, $\boldsymbol{a}_{\text{fix}}$, is related to the acceleration in the rotating frame, $\boldsymbol{a}_{\text{rot}}$, by

$$\begin{aligned} \boldsymbol{a}_{\text{fix}} &= \boldsymbol{a}_{\text{rot}} + \boldsymbol{\omega} \times \boldsymbol{v}_{\text{rot}} + \boldsymbol{\alpha}_{\text{rot}} \times \boldsymbol{r} + \boldsymbol{\omega} \times \boldsymbol{v}_{\text{rot}} + \boldsymbol{\omega} \times (\boldsymbol{\omega} \times \boldsymbol{r}) \\ &= \boldsymbol{a}_{\text{rot}} + 2\boldsymbol{\omega} \times \boldsymbol{v}_{\text{rot}} + \boldsymbol{\omega} \times (\boldsymbol{\omega} \times \boldsymbol{r}) + \boldsymbol{\alpha}_{\text{rot}} \times \boldsymbol{r}. \end{aligned} \tag{3.74}$$

Thus, the acceleration measured in the rotating frame is augmented by the Coriolis acceleration, $2\boldsymbol{\omega} \times \boldsymbol{v}_{\text{rot}}$, the centripetal acceleration, $\boldsymbol{\omega} \times (\boldsymbol{\omega} \times \boldsymbol{r})$ and the Poincaré acceleration, $\boldsymbol{\alpha}_{\text{rot}} \times \boldsymbol{r}$, where $\boldsymbol{\alpha}_{\text{rot}}$ is the vector angular acceleration rate, if any, observed in the rotating frame.

By d'Alembert's principle, the negatives of these accelerations appear as body forces per unit mass in the equations of motion. Usually, the reference frame of the Earth is taken to be in uniform rotation at a fixed rate, $\boldsymbol{\Omega}$, and we are left with the Coriolis force per unit mass,

$$-2\boldsymbol{\Omega} \times \boldsymbol{v}, \tag{3.75}$$

and the centrifugal force per unit mass,

$$-\boldsymbol{\Omega} \times (\boldsymbol{\Omega} \times \boldsymbol{r}), \tag{3.76}$$

where $\boldsymbol{v}$ is the velocity in the Earth frame and $\boldsymbol{r}$ is the radius vector.

The centrifugal force has already been incorporated through the total geopotential, described by equation (5.5), whose negative gradient is the equilibrium gravity $\boldsymbol{g}_0$ appearing in the equilibrium equation (3.19). For small harmonic displacements at angular frequency ω, their time dependence may be represented by the phasor $e^{i\omega t}$, and the body force per unit volume, arising from dynamical terms, becomes

$$\boldsymbol{F} = \rho_0 \left[\omega^2 \boldsymbol{u} - 2i\omega \boldsymbol{\Omega} \times \boldsymbol{u} \right], \tag{3.77}$$

with ρ_0 representing the equilibrium density, and $\boldsymbol{u}$ representing the vector displacement field.

In a spherical polar co-ordinate system (r, θ, ϕ), with the x_3-axis aligned with the rotation vector $\boldsymbol{\Omega}$, the vector $\boldsymbol{\Omega} \times \boldsymbol{u}$ has components

$$\left(-\Omega u_\phi \sin\theta,\ -\Omega u_\phi \cos\theta,\ \Omega u_\theta \cos\theta + \Omega u_r \sin\theta\right), \tag{3.78}$$

the components of the vector displacement field being (u_r, u_θ, u_ϕ). In turn, using expressions (1.195) and (1.187), the components of the vector displacement field can be expanded in spheroidal and torsional vector harmonics as

$$\begin{aligned} u_r &= \sum_{n=0}^{\infty} \sum_{m=-n}^{n} u_n^m P_n^m(\cos\theta)\, e^{im\phi}, \\ u_\theta &= \sum_{n=0}^{\infty} \sum_{m=-n}^{n} \left[v_n^m \frac{dP_n^m(\cos\theta)}{d\theta} + \frac{im}{\sin\theta} t_n^m P_n^m(\cos\theta) \right] e^{im\phi}, \\ u_\phi &= \sum_{n=0}^{\infty} \sum_{m=-n}^{n} \left[\frac{im}{\sin\theta} v_n^m P_n^m(\cos\theta) - t_n^m \frac{dP_n^m}{d\theta} \right] e^{im\phi}, \end{aligned} \tag{3.79}$$

where u_n^m, v_n^m and t_n^m are, respectively, the radial coefficients of the radial spheroidal part, the transverse spheroidal part and the torsional part of the vector displacement field. For $n = 0$, we have $m = 0$ and $P_n^m = 1$. Thus, both the transverse spheroidal part and the torsional part vanish for $n = 0$.

Using the orthogonality relations (1.205) and (1.209), and expression (3.78), the radial coefficients of the radial spheroidal part, the transverse spheroidal part and the torsional part of the vector $\boldsymbol{\Omega} \times \boldsymbol{u}$ can be found.

The radial coefficient of the radial spheroidal part of the vector $\boldsymbol{\Omega} \times \boldsymbol{u}$ is given by

$$\begin{aligned} u_{C\,n}^{-m} = (-1)^m \frac{2n+1}{4\pi} \int_0^{2\pi} \int_0^{\pi} \sum_{l=0}^{\infty} \sum_{k=-l}^{k=l} \Bigg[& -\Omega i k v_l^k P_l^k P_n^m \sin\theta \\ & + \Omega t_l^k \frac{dP_l^k}{d\theta} P_n^m \sin^2\theta \Bigg] e^{i(k+m)\phi}\, d\theta\, d\phi. \end{aligned} \tag{3.80}$$

On using the recurrence relation (B.12), we find that

$$\begin{aligned} u_{C\,n}^{-m} = (-1)^m \frac{2n+1}{4\pi} \sum_{l=0}^{\infty} \sum_{k=-l}^{k=l} \Bigg[& -i\Omega k v_l^k \int_0^{2\pi} \int_0^{\pi} P_l^k P_l^m \sin\theta\, e^{i(k+m)\phi}\, d\theta\, d\phi \\ & + \Omega t_l^k \int_0^{2\pi} \int_0^{\pi} \frac{1}{2l+1} \\ & \times \left\{ l(l-k+1) P_{l+1}^k - (l+1)(l+k) P_{l-1}^k \right\} \\ & \times P_n^m \sin\theta\, e^{i(k+m)\phi}\, d\theta\, d\phi \Bigg]. \end{aligned} \tag{3.81}$$

Applying the orthogonality relation (1.180), this expression reduces to

$$u_{C\,n}^{-m} = i\Omega m v_n^{-m} + \Omega\left[\frac{(n-1)(n+m)}{2n-1}t_{n-1}^{-m} - \frac{(n+2)(n-m+1)}{2n+3}t_{n+1}^{-m}\right]. \tag{3.82}$$

After replacing $-m$ with m, the radial coefficient of the radial spheroidal part of the body force per unit volume becomes

$$\begin{aligned} u_{F\,n}^{m} &= \omega^2\rho_0 u_n^m - 2m\omega\Omega\rho_0 v_n^m \\ &\quad - 2i\omega\Omega\rho_0\left[\frac{(n-1)(n-m)}{2n-1}t_{n-1}^{m} - \frac{(n+2)(n+m+1)}{2n+3}t_{n+1}^{m}\right]. \end{aligned} \tag{3.83}$$

The radial coefficient of the transverse spheroidal part of the vector $\boldsymbol{\Omega}\times\boldsymbol{u}$ is given by

$$\begin{aligned} v_{C\,n}^{-m} &= (-1)^m\frac{2n+1}{4\pi n(n+1)}\int_0^{2\pi}\int_0^{\pi}\sum_{l=0}^{\infty}\sum_{k=-l}^{k=l}\left[-\Omega i k v_l^k P_l^k\frac{dP_n^m}{d\theta}\cos\theta\right. \\ &\quad + im\Omega v_l^k\frac{dP_l^k}{d\theta}P_n^m\cos\theta + im\Omega u_l^k P_l^k P_n^m\sin\theta + \Omega t_l^k\frac{dP_l^k}{d\theta}\frac{dP_n^m}{d\theta}\cos\theta\sin\theta \\ &\quad \left. - \Omega m k t_l^k P_l^k P_n^m\cot\theta\right] e^{i(k+m)\phi}\,d\theta\,d\phi. \end{aligned} \tag{3.84}$$

Consider the first integral arising on the right,

$$k\int_0^{2\pi}\int_0^{\pi} P_l^k\frac{dP_n^m}{d\theta}\cos\theta\, e^{i(k+m)\phi}\,d\theta\,d\phi. \tag{3.85}$$

On integrating by parts, it becomes

$$\begin{aligned} &2\pi k\delta_{-k}^m P_l^k P_n^m\cos\theta\Big|_0^{\pi} - k\int_0^{2\pi}\int_0^{\pi}\left(\frac{dP_l^k}{d\theta}\cos\theta - P_l^k\sin\theta\right)P_n^m e^{i(k+m)\phi}\,d\theta\,d\phi \\ &= 2\pi k\delta_{-k}^m\int_0^{\pi}\left(P_l^k\sin\theta - \frac{dP_l^k}{d\theta}\cos\theta\right)P_n^m\,d\theta, \end{aligned} \tag{3.86}$$

where the first term on the left vanishes for $m = -k = 0$, and for $m = -k \neq 0$ it also vanishes, since $P_l^k(\pm 1) = P_n^m(\pm 1) = 0$. This combines with the second integral arising on the right of (3.84),

$$-2\pi m\delta_{-k}^m\int_0^{\pi}\frac{dP_l^k}{d\theta}P_n^m\cos\theta\,d\theta, \tag{3.87}$$

to give

$$-2\pi(k+m)\,\delta^m_{-k}\int_0^\pi \frac{dP_l^k}{d\theta}P_n^m\cos\theta\,d\theta + 2\pi k\delta^m_{-k}\int_0^\pi P_l^kP_n^m\sin\theta\,d\theta$$
$$= 2\pi k\delta^m_{-k}\int_0^\pi P_l^kP_n^m\sin\theta\,d\theta$$
$$= (-1)^m\frac{4\pi k}{2n+1}\delta^m_{-k}\delta^n_l, \tag{3.88}$$

using the orthogonality relation (1.180). Now consider the fourth integral arising on the right of (3.84),

$$\int_0^\pi \frac{dP_l^k}{d\theta}\frac{dP_n^m}{d\theta}\cos\theta\sin\theta\,d\theta. \tag{3.89}$$

On integrating by parts, it becomes

$$\frac{dP_l^k}{d\theta}P_n^m\cos\theta\sin\theta\Big|_0^\pi - \int_0^\pi\left[\frac{d}{d\theta}\left(\sin\theta\frac{dP_l^k}{d\theta}\right)\cos\theta - \frac{dP_l^k}{d\theta}\sin^2\theta\right]P_n^m\,d\theta$$
$$= \int_0^\pi\left[l(l+1)\,P_l^k\cos\theta\sin\theta - k^2P_l^k\cot\theta + \frac{dP_l^k}{d\theta}\sin^2\theta\right]P_n^m\,d\theta, \tag{3.90}$$

using Legendre's equation (B.1). This combines with the fifth integral arising on the right of (3.84), for $m=-k$, to give

$$\int_0^\pi\left[l(l+1)\sin\theta\cos\theta P_l^k + \sin^2\theta\frac{dP_l^k}{d\theta}\right]P_n^m, \tag{3.91}$$

which is easily evaluated using the recurrence relations (B.11) and (B.12), along with the orthogonality relation (1.180). The third integral on the right of (3.84) is evaluated directly using the orthogonality relation (1.180). Finally, collecting terms, we find for $n \geq 1$,

$$v_{C\,n}^{-m} = \frac{im\Omega}{n(n+1)}\left(u_n^{-m}+v_n^{-m}\right)$$
$$+\frac{\Omega}{n(n+1)}\left[\frac{(n-1)(n+1)(n+m)}{2n-1}t_{n-1}^{-m} + \frac{n(n+2)(n-m+1)}{2n+3}t_{n+1}^{-m}\right]. \tag{3.92}$$

After replacing $-m$ with m, the radial coefficient of the transverse spheroidal part of the body force per unit volume becomes

$$v_{F\,n}^m = \omega^2\rho_0v_n^m - \frac{2m\omega\Omega\rho_0}{n(n+1)}\left(u_n^m+v_n^m\right)$$
$$-\frac{2i\omega\Omega\rho_0}{n(n+1)}\left[\frac{(n-1)(n+1)(n-m)}{2n-1}t_{n-1}^m + \frac{n(n+2)(n+m+1)}{2n+3}t_{n+1}^m\right]. \tag{3.93}$$

The radial coefficient of the torsional part of the vector $\boldsymbol{\Omega} \times \boldsymbol{u}$ is given by

$$t_{C\,n}^{-m} = (-1)^m \frac{2n+1}{4\pi n(n+1)} \int_0^{2\pi} \int_0^{\pi} \sum_{l=0}^{\infty} \sum_{k=-l}^{k=l} \Big[\Omega k m v_l^k P_l^k P_n^m \cot\theta + im\Omega t_l^k \frac{dP_l^k}{d\theta} P_n^m \cos\theta - \Omega v_l^k \frac{dP_l^k}{d\theta} \frac{dP_n^m}{d\theta} \cos\theta \sin\theta - ik\Omega t_l^k P_l^k \frac{dP_n^m}{d\theta} \cos\theta - \Omega u_l^k P_l^k \frac{dP_n^m}{d\theta} \sin^2\theta \Big] e^{i(k+m)\phi}\, d\theta\, d\phi. \tag{3.94}$$

Consider the third integral arising on the right,

$$\int_0^{\pi} \frac{dP_l^k}{d\theta} \frac{dP_n^m}{d\theta} \cos\theta \sin\theta\, d\theta. \tag{3.95}$$

This is identical to the integral (3.89), which, on integrating by parts, led to (3.90), using Legendre's equation (B.1). This combines with the first integral on the right of (3.94), for $m = -k$, to give (3.91), which, as before, is easily evaluated using the recurrence relations (B.11) and (B.12), along with the orthogonality relation (1.180). Consider the fourth integral arising on the right of (3.94),

$$k \int_0^{2\pi} \int_0^{\pi} P_l^k \frac{dP_n^m}{d\theta} \cos\theta\, e^{i(k+m)\phi}\, d\theta\, d\phi. \tag{3.96}$$

This is identical to the integral (3.85), which, on integrating by parts, led to (3.86). This result combines with the second integral arising on the right of (3.94),

$$-2\pi m \delta_{-k}^{m} \int_0^{\pi} \frac{dP_l^k}{d\theta} P_n^m \cos\theta\, d\theta, \tag{3.97}$$

giving, as before, expression (3.88). The remaining integral, fifth on the right of (3.94), is easily evaluated, using the recurrence relation (B.12) and the orthogonality relation (1.180). Finally, collecting terms, we find for $n \geq 1$,

$$t_{C\,n}^{-m} = \frac{im\Omega}{n(n+1)} t_n^{-m} - \frac{\Omega}{n(n+1)} \left[\frac{(n-1)(n+1)(n+m)}{2n-1} v_{n-1}^{-m} + \frac{n(n+2)(n-m+1)}{2n+3} v_{n+1}^{-m} \right] + \frac{\Omega}{n(n+1)} \left[\frac{(n+1)(n+m)}{2n-1} u_{n-1}^{-m} - \frac{n(n-m+1)}{2n+3} u_{n+1}^{-m} \right]. \tag{3.98}$$

After replacing $-m$ with m, the radial coefficient of the torsional part of the body force per unit volume becomes

$$
\begin{aligned}
t_{Fn}^{m} &= \omega^2 \rho_0 t_n^m - \frac{2m\omega\Omega\rho_0}{n(n+1)} t_n^m \\
&- \frac{2i\omega\Omega\rho_0}{n(n+1)} \left[\frac{(n+1)(n-m)}{2n-1} u_{n-1}^m - \frac{n(n+m+1)}{2n+3} u_{n+1}^m \right] \\
&+ \frac{2i\omega\Omega\rho_0}{n(n+1)} \left[\frac{(n-1)(n+1)(n-m)}{2n-1} v_{n-1}^m + \frac{n(n+2)(n+m+1)}{2n+3} v_{n+1}^m \right].
\end{aligned}
\tag{3.99}
$$

Substitution of the body force expressions (3.83) for u_{Fn}^m, (3.93) for v_{Fn}^m and (3.99) for t_{Fn}^m, into equations (3.52), (3.54) and (3.58), respectively, gives a coupled eighth-order differential system comprising equations (3.51) through (3.58), for each harmonic degree n. The magnitude of the Coriolis acceleration compared with the magnitude of the local acceleration is $2\omega\Omega/\omega^2 = 2\Omega/\omega$, or the ratio of the period to 12 sidereal hours. For periods shorter than an hour, in the range of conventional free oscillations, the effect of Coriolis coupling can be handled by first- or second-order perturbation theory. For long-period core oscillations the coupling is much stronger and two infinite coupled chains emerge,

$$\text{the spheroidal chain} \qquad \boldsymbol{S}_m^m \Leftrightarrow \boldsymbol{T}_{m+1}^m \Leftrightarrow \boldsymbol{S}_{m+2}^m \Leftrightarrow \boldsymbol{T}_{m+3}^m \Leftrightarrow \cdots, \tag{3.100}$$

and

$$\text{the torsional chain} \qquad \boldsymbol{T}_m^m \Leftrightarrow \boldsymbol{S}_{m+1}^m \Leftrightarrow \boldsymbol{T}_{m+2}^m \Leftrightarrow \boldsymbol{S}_{m+3}^m \Leftrightarrow \cdots. \tag{3.101}$$

At periods of several hours and longer (Johnson and Smylie, 1977), the convergence of the coupled chains is very slow.

In summary, the sixth-order spheroidal governing system, for $n \geq 1$, expressed by equations (3.51) through (3.56), becomes

$$\frac{dy_1}{dr} = -\frac{2\lambda\beta}{r} y_1 + \beta y_2 + \frac{n_1 \lambda\beta}{r} y_3, \tag{3.102}$$

$$
\begin{aligned}
\frac{dy_2}{dr} &= \left[-\frac{2\rho_0}{r} \left(2g_0 + r\Omega^2 \right) + \frac{2\delta}{r^2} \right] y_1 - \frac{4\mu\beta}{r} y_2 + n_1 \left(\frac{\rho_0 g_0}{r} - \frac{\delta}{r^2} \right) y_3 \\
&+ \frac{n_1}{r} y_4 - \rho_0 y_6 - u_{Fn}^m,
\end{aligned}
\tag{3.103}
$$

$$\frac{dy_3}{dr} = -\frac{1}{r} y_1 + \frac{1}{r} y_3 + \frac{1}{\mu} y_4, \tag{3.104}$$

$$\frac{dy_4}{dr} = \left(\frac{\rho_0 g_0}{r} - \frac{\delta}{r^2} \right) y_1 - \frac{\lambda\beta}{r} y_2 + \frac{\epsilon}{r^2} y_3 - \frac{3}{r} y_4 - \frac{\rho_0}{r} y_5 - v_{Fn}^m, \tag{3.105}$$

$$\frac{dy_5}{dr} = 4\pi G\rho_0 y_1 + y_6, \tag{3.106}$$

$$\frac{dy_6}{dr} = -4\pi G\rho_0 \frac{n_1}{r} y_3 + \frac{n_1}{r^2} y_5 - \frac{2}{r} y_6, \tag{3.107}$$

using the shorthand notations

$$n_1 = n(n+1), \qquad \beta = \frac{1}{\lambda + 2\mu}, \tag{3.108}$$

$$\delta = 2\mu(3\lambda + 2\mu)\beta, \qquad \epsilon = 4n_1\mu(\lambda + \mu)\beta - 2\mu. \tag{3.109}$$

For $n = 0$, the spheroidal system degenerates to fourth order, as expressed by equations (3.59) through (3.62), becoming

$$\frac{dy_1}{dr} = -\frac{2\lambda\beta}{r} y_1 + \beta y_2, \tag{3.110}$$

$$\frac{dy_2}{dr} = \left[-\frac{2\rho_0}{r} \left(2g_0 + r\Omega^2 \right) + \frac{2\delta}{r^2} \right] y_1 - \frac{4\mu\beta}{r} y_2 - \rho_0 y_6 - u^0_{F0}, \tag{3.111}$$

$$\frac{dy_5}{dr} = 4\pi G\rho_0 y_1 + y_6, \tag{3.112}$$

$$\frac{dy_6}{dr} = -\frac{2}{r} y_6. \tag{3.113}$$

In the fluid outer core, for $n \geq 1$, the governing spheroidal system, expressed by equations (3.63) through (3.67), becomes

$$\frac{dy_1}{dr} = -\frac{2}{r} y_1 + \frac{1}{\lambda} y_2 + \frac{n_1}{r} y_3, \tag{3.114}$$

$$\frac{dy_2}{dr} = -\frac{2\rho_0}{r} \left(2g_0 + r\Omega^2 \right) y_1 + \frac{n_1}{r} \rho_0 g_0 y_3 - \rho_0 y_6 - u^m_{Fn}, \tag{3.115}$$

$$0 = \frac{\rho_0 g_0}{r} y_1 - \frac{1}{r} y_2 - \frac{\rho_0}{r} y_5 - v^m_{Fn}, \tag{3.116}$$

$$\frac{dy_5}{dr} = 4\pi G\rho_0 y_1 + y_6, \tag{3.117}$$

$$\frac{dy_6}{dr} = -4\pi G\rho_0 \frac{n_1}{r} y_3 + \frac{n_1}{r^2} y_5 - \frac{2}{r} y_6. \tag{3.118}$$

The torsional deformation, for which $n \geq 1$, described by equations (3.57) and (3.58), is governed by the second-order system

$$\frac{dz_1}{dr} = \frac{1}{r} z_1 + \frac{1}{\mu} z_2, \tag{3.119}$$

$$\frac{dz_2}{dr} = [n_1 - 2] \frac{\mu}{r^2} - \frac{3}{r} z_2 - t^m_{Fn}. \tag{3.120}$$

In the fluid outer core for $n = 0$, the fourth-order system expressed by equations (3.68) through (3.71) becomes

$$\frac{dy_1}{dr} = -\frac{2}{r}y_1 + \frac{1}{\lambda}y_2, \tag{3.121}$$

$$\frac{dy_2}{dr} = -\frac{2\rho_0}{r}\left(2g_0 + r\Omega^2\right)y_1 - \rho_0 y_6 - u^0_{F0}, \tag{3.122}$$

$$\frac{dy_5}{dr} = 4\pi G\rho_0 y_1 + y_6, \tag{3.123}$$

$$\frac{dy_6}{dr} = -\frac{2}{r}y_6. \tag{3.124}$$

While, in the fluid outer core, the spheroidal and torsional deformation fields are strongly coupled across degree by the Coriolis acceleration, in the solid inner core and shell, the inertial terms are dominated by elastic restoring forces and it suffices to take only *self-coupling* into account. The dynamical body force expressions (3.83), (3.93) and (3.99), become, for deformations with angular frequency ω,

$$u^m_{Fn} = \omega^2\rho_0 y_1 - 2m\omega\Omega\rho_0 y_3 \qquad n \geq 0, \tag{3.125}$$

$$v^m_{Fn} = \omega^2\rho_0 y_3 - \frac{2m\omega\Omega\rho_0}{n_1}(y_1 + y_3) \quad n \geq 1, \tag{3.126}$$

$$t^m_{Fn} = \omega^2\rho_0 z_1 - \frac{2m\omega\Omega\rho_0}{n_1}z_1 \qquad n \geq 1. \tag{3.127}$$

For static deformations, the inertial terms vanish and, as we shall see, this modifies the static behaviour in the fluid outer core, compared with the dynamical case.

3.5 Solutions near the geocentre

Near the geocentre, gravity is constrained to vary linearly with radius. If $g_0(r) = \gamma r$, then from equation (3.47),

$$\gamma = \frac{4}{3}\pi G\rho_0 - \frac{2}{3}\Omega^2, \tag{3.128}$$

or

$$4\pi G\rho_0 = 3\gamma + 2\Omega^2. \tag{3.129}$$

Both pressure and temperature gradients vanish at the geocentre, even in the presence of heat sources. Density and the elastic constants there are then taken to be independent of radius, since the thermodynamic state is locally constant. Thus, they are given at zero implied radius. Near the geocentre, taking only self-coupling into account, the sixth-order spheroidal governing system, for $n \geq 1$, expressed by equations (3.102) through (3.107), becomes

$$\frac{dy_1}{dr} = -\frac{2\lambda\beta}{r} y_1 + \beta y_2 + \frac{n_1 \lambda\beta}{r} y_3, \tag{3.130}$$

$$\frac{dy_2}{dr} = \left[-\rho_0 \left(4\gamma + 2\Omega^2 \right) + \frac{2\delta}{r^2} \right] y_1 - \frac{4\mu\beta}{r} y_2 + n_1 \left(\rho_0 \gamma - \frac{\delta}{r^2} \right) y_3$$
$$+ \frac{n_1}{r} y_4 - \rho_0 y_6 - \omega^2 \rho_0 y_1 + 2m\omega\Omega\rho_0 y_3, \tag{3.131}$$

$$\frac{dy_3}{dr} = -\frac{1}{r} y_1 + \frac{1}{r} y_3 + \frac{1}{\mu} y_4, \tag{3.132}$$

$$\frac{dy_4}{dr} = \left(\rho_0 \gamma - \frac{\delta}{r^2} \right) y_1 - \frac{\lambda\beta}{r} y_2 + \frac{\epsilon}{r^2} y_3 - \frac{3}{r} y_4 - \frac{\rho_0}{r} y_5$$
$$- \omega^2 \rho_0 y_3 + \frac{2m\omega\Omega\rho_0}{n_1} \left(y_1 + y_3 \right), \tag{3.133}$$

$$\frac{dy_5}{dr} = 4\pi G \rho_0 y_1 + y_6, \tag{3.134}$$

$$\frac{dy_6}{dr} = -4\pi G \rho_0 \frac{n_1}{r} y_3 + \frac{n_1}{r^2} y_5 - \frac{2}{r} y_6. \tag{3.135}$$

Taking only self-coupling into account, the second-order torsional governing system, (3.119) and (3.120), for arbitrary radius, has the form

$$\frac{dz_1}{dr} = \frac{1}{r} z_1 + \frac{1}{\mu} z_2, \tag{3.136}$$

$$\frac{dz_2}{dr} = (n_1 - 2) \frac{\mu}{r^2} z_1 - \frac{3}{r} z_2 - \omega^2 \rho_0 z_1 + \frac{2m\omega\Omega\rho_0}{n_1} z_1. \tag{3.137}$$

Solutions of the spheroidal system, regular at the geocentre, can be expanded in power series as

$$y_i(r) = r^\alpha \sum_{\nu=0}^{\infty} A_{i,\nu} r^\nu, \tag{3.138}$$

with α non-negative. The coefficients $A_{i,\nu}$ vanish for $\nu < 0$. A series of indicial equations, determining the admissible values of α, are found on substituting these expansions in the governing equations and equating like powers of the radius.

Substitution of the expansions (3.138) in equations (3.130) through (3.133), and equating the coefficients of like powers of the radius, yields the system

$$\left[\mathbf{M}_n + \eta \mathbf{I} \right] \boldsymbol{x}_\nu = \boldsymbol{b}_\nu, \tag{3.139}$$

for $\eta = \alpha + \nu - 1$, with $\mathbf{I}$ representing the unit matrix, and

$$\mathbf{M}_n = \begin{pmatrix} 2\lambda\beta + 1 & -\beta & -n_1\lambda\beta & 0 \\ -2\delta & 4\mu\beta & n_1\delta & -n_1 \\ 1 & 0 & 0 & -1/\mu \\ \delta & \lambda\beta & -\epsilon & 3 \end{pmatrix}, \tag{3.140}$$

while the vectors $\boldsymbol{x}_\nu$ and $\boldsymbol{b}_\nu$ are

$$\boldsymbol{x}_\nu = \begin{pmatrix} A_{1,\nu} \\ A_{2,\nu-1} \\ A_{3,\nu} \\ A_{4,\nu-1} \end{pmatrix}, \tag{3.141}$$

and

$$\boldsymbol{b}_\nu = \rho_0 \begin{pmatrix} 0 \\ -\left(4\gamma + \omega^2 + 2\Omega^2\right) A_{1,\nu-2} + (n_1\gamma + 2m\omega\Omega) A_{3,\nu-2} - A_{6,\nu-2} \\ 0 \\ (\gamma + 2m\omega\Omega/n_1) A_{1,\nu-2} - \left(\omega^2 - 2m\omega\Omega/n_1\right) A_{3,\nu-2} - A_{5,\nu-1} \end{pmatrix}, \tag{3.142}$$

for $\nu = 0, 1, 2, \ldots$.

Similarly, substitution of the expansions (3.138) in equations (3.134) and (3.135), and equating the coefficients of like powers of the radius, yields the system

$$[\mathbf{N}_n + \eta\mathbf{I}]\, \boldsymbol{y}_\nu = \boldsymbol{c}_\nu, \tag{3.143}$$

$$\mathbf{N}_n = \begin{pmatrix} 1 & -1 \\ -n_1 & 2 \end{pmatrix}, \tag{3.144}$$

with the vectors $\boldsymbol{y}_\nu$ and $\boldsymbol{c}_\nu$ given by

$$\boldsymbol{y}_\nu = \begin{pmatrix} A_{5,\nu} \\ A_{6,\nu-1} \end{pmatrix} \quad \text{and} \quad \boldsymbol{c}_\nu = 4\pi G\rho_0 \begin{pmatrix} A_{1,\nu-1} \\ -n_1 A_{3,\nu-1} \end{pmatrix}, \tag{3.145}$$

again for $\nu = 0, 1, 2, \ldots$. For cases where the systems (3.139) or (3.143) are homogeneous, we will refer to the values of η permitting non-trivial solutions as eigenvalues, although by conventional usage they are of opposite sign to the true eigenvalues of the coefficient matrices.

For $\nu = 0$, the systems (3.139) and (3.143) degenerate to

$$\begin{pmatrix} \alpha + 2\lambda\beta & -n_1\lambda\beta \\ -2\delta & n_1\delta \\ 1 & \alpha - 1 \\ \delta & -\epsilon \end{pmatrix} \begin{pmatrix} A_{1,0} \\ A_{3,0} \end{pmatrix} = 0 \quad \text{and} \quad \begin{pmatrix} \alpha \\ -n_1 \end{pmatrix} A_{5,0} = 0, \tag{3.146}$$

respectively. For $n \geq 1$, the first two equations of the first system have a non-trivial solution for $n_1 \delta \alpha = 0$, which implies that $\alpha = 0$. For $\alpha = 0$, the last two equations give the condition for a non-trivial solution as

$$\epsilon - \delta = \frac{4(n-1)(n+2)\mu(\lambda+\mu)}{\lambda + 2\mu} = 0. \tag{3.147}$$

Thus, the first system has the non-trivial solution $A_{1,0} = A_{3,0}$ only for $\alpha = 0$, $n = 1$, when $n \geq 1$. This degree-one displacement field is special, since it describes a rigid-body displacement with respect to the centre of mass, that is involved in any exchange of linear momentum. The second system gives $A_{5,0} = 0$, except when $\alpha = 0$, $n = 0$. This degree-zero, radial spheroidal deformation field is special, since the governing system degenerates to fourth order. The special cases of $n = 0$ and $n = 1$ will be considered separately later. Let us now consider the case where $n \geq 1$.

For $\nu \geq 1$, the homogeneous first and third equations of the system (3.139) may be used to eliminate unknowns. The third equation gives

$$A_{1,\nu} = -\eta A_{3,\nu} + \frac{1}{\mu} A_{4,\nu-1}. \tag{3.148}$$

Eliminating $A_{1,\nu}$ in the first equation with this relation yields

$$A_{2,\nu-1} = -q_1(\eta) A_{3,\nu} + q_2(\eta) A_{4,\nu-1}, \tag{3.149}$$

with

$$q_1(\eta) = [n_1 + \eta(\eta+3)]\lambda + 2\eta(\eta+1)\mu, \tag{3.150}$$

$$q_2(\eta) = 2(\eta+1) + (\eta+3)\frac{\lambda}{\mu}. \tag{3.151}$$

These two relations allow the elimination of $A_{1,\nu}$ and $A_{2,\nu-1}$ in the second equation, giving the new left side

$$-p_1(\eta) A_{3,\nu} + p_2(\eta) A_{4,\nu-1}, \tag{3.152}$$

with

$$p_1(\eta) = \eta\big[n_1 + \eta(\eta+3)\big]\lambda + \big[2\eta(\eta^2 + 3\eta) - 2n_1\big]\mu, \tag{3.153}$$

$$p_2(\eta) = 2\eta(\eta+3) - n_1 + \eta(\eta+3)\frac{\lambda}{\mu}, \tag{3.154}$$

and in the fourth equation, giving the new left side

$$-r_1(\eta) A_{3,\nu} + r_2(\eta) A_{4,\nu-1}, \tag{3.155}$$

with

$$r_1(\eta) = [n_1 + \eta(\eta+3)]\lambda + 2(n_1 + \eta - 1)\mu, \tag{3.156}$$

$$r_2(\eta) = \eta + 5 + (\eta+3)\frac{\lambda}{\mu}. \tag{3.157}$$

The system (3.139) is then reduced to

$$\begin{pmatrix} 0 & 1 & q_1(\eta) & -q_2(\eta) \\ 0 & 0 & -p_1(\eta) & p_2(\eta) \\ 1 & 0 & \eta & -1/\mu \\ 0 & 0 & -r_1(\eta) & r_2(\eta) \end{pmatrix} \boldsymbol{x}_\nu = \boldsymbol{b}_\nu. \tag{3.158}$$

The determinant of the coefficient matrix of the reduced system (3.158) is

$$\begin{vmatrix} -p_1(\eta) & p_2(\eta) \\ -r_1(\eta) & r_2(\eta) \end{vmatrix} = p_2(\eta)\, r_1(\eta) - p_1(\eta)\, r_2(\eta). \tag{3.159}$$

The coefficient matrix is singular for

$$p_1(\eta)\, \mu r_2(\eta) = r_1(\eta)\, \mu p_2(\eta), \tag{3.160}$$

or

$$\begin{aligned} &\left[\eta\,\{n_1 + \eta\,(\eta+3)\}\,\lambda + \left\{2\eta\left(\eta^2+3\eta\right) - 2n_1\right\}\mu\right]\left[(\eta+3)\,\lambda + (\eta+5)\,\mu\right] \\ &= \left[\{n_1 + \eta\,(\eta+3)\}\lambda + 2\{n_1 + \eta - 1\}\mu\right]\left[\eta\,(\eta+3)\,\lambda + \{2\eta\,(\eta+3) - n_1\}\mu\right]. \end{aligned} \tag{3.161}$$

The terms in λ^2 cancel in this condition, and those in μ^2 and $\lambda\mu$ each yield the quartic equation

$$\begin{aligned} &\eta^4 + 6\eta^3 + [11 - 2n\,(n+1)]\,\eta^2 + 6\,[1 - n\,(n+1)]\,\eta \\ &\quad + (n-2)\,n\,(n+1)\,(n+3) = 0, \end{aligned} \tag{3.162}$$

with n_1 replaced by $n\,(n+1)$. In turn, the quartic factors into the product of quadratics,

$$\left[\eta^2 + 3\eta - (n-2)\,(n+1)\right]\left[\eta^2 + 3\eta - n\,(n+3)\right] = 0. \tag{3.163}$$

Finally, this expression factors to

$$(\eta - n + 2)\,(\eta - n)\,(\eta + n + 1)\,(\eta + n + 3) = 0. \tag{3.164}$$

The eigenvalues of the matrix, $\mathbf{M}_n$, in the first system (3.139), are then $\eta = -\,(n+3)$ and $\eta = -\,(n+1)$, as well as $\eta = n - 2$ and $\eta = n$.

Much more directly, the eigenvalues of the matrix, $\mathbf{N}_n$, in the second system (3.143), are $\eta = -\,(n+2)$ and $\eta = n - 1$.

Now consider how the power series solutions propagate for higher values of $\nu \geq 1$ with $n \geq 2$. Then, from (3.146), $A_{1,0} = A_{3,0} = A_{5,0} = 0$.

For $\nu = 1$, we have $\eta = \alpha$ and $\boldsymbol{b}_1$ is a null vector, so the system (3.158) is homogeneous. Since $\alpha \geq 0$, only the eigenvalues $\eta = \alpha = n - 2$ and $\eta = \alpha = n$ are admissible. The corresponding eigenvectors, $\boldsymbol{x}_1$, are

$$\begin{pmatrix} A_{1,1} \\ A_{2,0} \\ A_{3,1} \\ A_{4,0} \end{pmatrix} = \begin{pmatrix} 1 \\ 2\,(n-1)\,\mu \\ 1/n \\ 2\mu\,(n-1)\,/n \end{pmatrix} A_{1,1}, \tag{3.165}$$

and

$$\begin{pmatrix} A_{1,1} \\ A_{2,0} \\ A_{3,1} \\ A_{4,0} \end{pmatrix} = \begin{pmatrix} 1/\mu - np_2(n)/p_1(n) \\ q_2(n) - q_1(n)p_2(n)/p_1(n) \\ p_2(n)/p_1(n) \\ 1 \end{pmatrix} A_{4,0}, \tag{3.166}$$

where $A_{1,1}$ and $A_{4,0}$ are arbitrary free constants, determining the two free solutions for $\alpha = n-2$ and $\alpha = n$, respectively. For $\nu = 1$, the system (3.143) is also homogeneous. Since $\alpha \geq 0$, only the eigenvalue $\eta = n-1$ is admissible, and thus for $\eta = n-2$ and $\eta = n$, only the trivial solution $A_{5,1} = A_{6,0} = 0$ exists.

For $\nu = 2$, the system (3.158) is homogeneous and non-singular for both $\alpha = n-2$, $\eta = n-1$ and $\alpha = n$, $\eta = n+1$, and it has only the trivial solution $A_{1,2} = A_{2,1} = A_{3,2} = A_{4,1} = 0$. The system (3.143) becomes

$$\begin{pmatrix} \eta + 1 & -1 \\ -n_1 & \eta + 2 \end{pmatrix} \begin{pmatrix} A_{5,2} \\ A_{6,1} \end{pmatrix} = 4\pi G\rho_0 \begin{pmatrix} A_{1,1} \\ -n_1 A_{3,1} \end{pmatrix}. \tag{3.167}$$

For $\nu = 2$, we have $\eta = \alpha + 1$. Then, for the value $\alpha = n$ of the free solution (3.166), the coefficient matrix of this system is non-singular, giving $A_{5,2}$ and $A_{6,1}$ in terms of $A_{1,1}$ and $A_{3,1}$ for $\eta = n+1$, allowing continuation of the free solution. Solving for $A_{5,2}$, we have

$$A_{5,2} = \frac{2\pi G\rho_0}{2n+3}\Big[(n+3)\,A_{1,1} - n\,(n+1)\,A_{3,1}\Big]. \tag{3.168}$$

Then,

$$A_{6,1} = (n+2)\,A_{5,2} - 4\pi G\rho_0 A_{1,1}. \tag{3.169}$$

For the eigenvalue $\eta = n-1$ of the coefficient matrix, $\alpha = n-2$. From the system (3.165), $A_{3,1} = A_{1,1}/n$ for $\alpha = n-2$, and both equations of the system (3.167) reduce to

$$A_{6,1} = nA_{5,2} - 4\pi G\rho_0 A_{1,1} = nA_{5,2} - \left(3\gamma + 2\Omega^2\right) A_{1,1}, \tag{3.170}$$

giving rise to a third free solution by choice of $A_{6,1}$.

For $\nu = 3$, we have $\eta = \alpha + 2$ and the eigenvalue $\eta = n$ results in a singular coefficient matrix in the system (3.158) for $\alpha = n - 2$. For $\eta = n$, we have

$$p_1(n) = 2n\left[n(n+2)\lambda + \left(n^2 + 2n - 1\right)\mu\right], \tag{3.171}$$

$$r_1(n) = 2\left[n(n+2)\lambda + \left(n^2 + 2n - 1\right)\mu\right], \tag{3.172}$$

$$p_2(n) = n(n+5) + n(n+3)\lambda/\mu, \tag{3.173}$$

$$r_2(n) = n + 5 + (n+3)\lambda/\mu. \tag{3.174}$$

Thus, the left side of the second equation in the system (3.158) is just n times the left side of the fourth equation. Again, $A_{3,1} = A_{1,1}/n$ for $\alpha = n-2$ and, substituting from (3.170) for $A_{6,1}$, the right side of the second equation becomes

$$\rho_0\left[\left(n\gamma - \omega^2 + 2m\omega\Omega/n\right)A_{1,1} - nA_{5,2}\right], \tag{3.175}$$

while the right side of the fourth equation becomes

$$\rho_0\left[\left(n\gamma - \omega^2 + 2m\omega\Omega/n\right)A_{1,1}/n - A_{5,2}\right]. \tag{3.176}$$

Thus, the whole of the second equation in the system (3.158) is just n times the fourth. Dropping the fourth equation, and replacing $A_{3,1}$ by $A_{1,1}/n$, for $\alpha = n - 2$, the system can be rearranged to

$$\begin{pmatrix} 0 & 1 & q_1(n) \\ 0 & 0 & -p_1(n)/\rho_0 \\ 1 & 0 & n \end{pmatrix}\begin{pmatrix} A_{1,3} \\ A_{2,2} \\ A_{3,3} \end{pmatrix} = \begin{pmatrix} q_2(n)A_{4,2} \\ -\left[(3-n)\gamma + \omega^2 + 2\Omega^2 - 2m\omega\Omega/n\right]A_{1,1} - A_{6,1} - p_2(n)A_{4,2}/\rho_0 \\ A_{4,2}/\mu \end{pmatrix}. \tag{3.177}$$

Given the three free constants $A_{1,1}$ and $A_{6,1}$, for $\alpha = n - 2$ and $A_{4,2}$, for $\alpha = n$, the second equation of this system yields,

$$A_{3,3} = \frac{p_2(n)}{p_1(n)}A_{4,2} + \frac{\rho_0}{p_1(n)}\left\{\left[(3-n)\gamma + \omega^2 + 2\Omega^2 - 2m\omega\Omega/n\right]A_{1,1} + A_{6,1}\right\}. \tag{3.178}$$

Then, the first equation of the system gives

$$A_{2,2} = -q_1(n)A_{3,3} + q_2(n)A_{4,2}, \tag{3.179}$$

and the third equation gives

$$A_{1,3} = -nA_{3,3} + \frac{1}{\mu}A_{4,2}. \tag{3.180}$$

In addition to the free constants $A_{1,1}$, $A_{4,0}$ and $A_{6,1}$ already encountered, this system appears to introduce a fourth free constant, $A_{4,2}$. This fourth free constant generates the solution

$$\begin{pmatrix} A_{1,3} \\ A_{2,2} \\ A_{3,3} \\ A_{4,2} \end{pmatrix} = \begin{pmatrix} 1/\mu - np_2(n)/p_1(n) \\ q_2(n) - q_1(n)p_2(n)/p_1(n) \\ p_2(n)/p_1(n) \\ 1 \end{pmatrix} A_{4,2}, \tag{3.181}$$

for $\alpha = n - 2$. Replacing the free constant $A_{4,2}$ with $A_{4,0}$, this is identical to the second of the eigenvectors (3.166) found for $\alpha = n$. Thus, no new solution arises from the eigenvalue $\eta = n$ giving $\alpha = n - 2$ for $\nu = 3$, and we may set $A_{4,2} = 0$ without loss of generality.

Using relations (3.178) through (3.180), a third eigenvector, $\boldsymbol{x}_3$, for $\alpha = n - 2$, associated with the free constant $A_{6,1}$, is found to be

$$\begin{pmatrix} A_{1,3} \\ A_{2,2} \\ A_{3,3} \\ A_{4,2} \end{pmatrix} = \frac{\rho_0}{p_1(n)} \begin{pmatrix} -n \\ -q_1(n) \\ 1 \\ 0 \end{pmatrix} A_{6,1}. \tag{3.182}$$

Another eigenvector, $\boldsymbol{x}_3$, associated with the free constant $A_{1,1}$, also appears, having the form, determined by relations (3.178) through (3.180),

$$\begin{pmatrix} A_{1,3} \\ A_{2,2} \\ A_{3,3} \\ A_{4,2} \end{pmatrix} = \frac{\rho_0}{p_1(n)} \left[(3-n)\gamma + \omega^2 + 2\Omega^2 - 2m\omega\Omega/n \right] \begin{pmatrix} -n \\ -q_1(n) \\ 1 \\ 0 \end{pmatrix} A_{1,1}. \tag{3.183}$$

This is just the next member of the sequence of eigenvectors beginning with (3.165). To complete preparation for the next recurrence, we consider the system (3.143) for $\nu = 4$, $\eta = \alpha + 3$. With $\alpha = n - 2$ and $\eta = n + 1$, the system (3.143) becomes

$$\begin{pmatrix} n+2 & -1 \\ -n_1 & n+3 \end{pmatrix} \begin{pmatrix} A_{5,4} \\ A_{6,3} \end{pmatrix} = 4\pi G\rho_0 \begin{pmatrix} A_{1,3} \\ -n_1 A_{3,3} \end{pmatrix}. \tag{3.184}$$

This can easily be solved to give

$$A_{5,4} = \frac{4\pi G\rho_0}{2(2n+3)} \left[(n+3) A_{1,3} - n(n+1) A_{3,3} \right], \tag{3.185}$$

and

$$A_{6,3} = (n+2) A_{5,4} - 4\pi G\rho_0 A_{1,3}. \tag{3.186}$$

No further eigenvalues are encountered in the recursions for higher terms in the power series expansions.

For numerical integration, the y-variables are replaced by a system of z-variables defined by

$$\begin{aligned} z_i(r) &= y_i(r)/r^{\alpha}, \qquad i = 2, 4, \\ z_i(r) &= y_i(r)/r^{\alpha+1}, \quad i = 1, 3, 6, \\ z_i(r) &= y_i(r)/r^{\alpha+2}, \quad i = 5. \end{aligned} \tag{3.187}$$

In summary, there are three fundamental solutions, regular at the geocentre, generated by the three free constants: $A_{1,1}$, $A_{6,1}$, for $\alpha = n-2$, and $A_{4,0}$, for $\alpha = n$.

The fundamental solution generated by the free constant $A_{1,1}$ is

$$z_1(r) = A_{1,1} + A_{1,3}r^2 + A_{1,5}r^4 + \cdots, \tag{3.188}$$

$$z_2(r) = 2\,(n-1)\,\mu A_{1,1} + A_{2,2}r^2 + A_{2,4}r^4 + \cdots, \tag{3.189}$$

$$z_3(r) = \frac{1}{n}A_{1,1} + A_{3,3}r^2 + A_{3,5}r^4 + \cdots, \tag{3.190}$$

$$z_4(r) = 2\mu\frac{n-1}{n}A_{1,1} + A_{4,2}r^2 + A_{4,4}r^4 + \cdots, \tag{3.191}$$

$$z_5(r) = \frac{4\pi G\rho_0}{n}A_{1,1} + A_{5,4}r^2 + A_{5,6}r^4 + \cdots, \tag{3.192}$$

$$z_6(r) = A_{6,3}r^2 + A_{6,5}r^4 + \cdots. \tag{3.193}$$

The coefficients $A_{1,3}$, $A_{2,2}$, $A_{3,3}$, $A_{4,2}$ are given in terms of $A_{1,1}$ by relation (3.183). In turn, $A_{5,4}$ is given by (3.185) and $A_{6,3}$ is given by (3.186). The coefficients $A_{1,5}$, $A_{2,4}$, $A_{3,5}$, $A_{4,4}$ are given by the system (3.139) in terms of $A_{1,3}$, $A_{3,3}$, $A_{6,3}$, $A_{5,4}$ for $\nu = 5$, $\eta = n+2$. In turn, $A_{5,6}$, $A_{6,5}$ are given by the system (3.143) in terms of $A_{1,5}$, $A_{3,5}$ for $\nu = 6$, $\eta = n+3$.

The fundamental solution generated by the free constant $A_{6,1}$ is

$$z_1(r) = -n\frac{\rho_0}{p_1(n)}A_{6,1}r^2 + A_{1,5}r^4 + \cdots, \tag{3.194}$$

$$z_2(r) = -\frac{q_1(n)}{p_1(n)}\rho_0 A_{6,1}r^2 + A_{2,4}r^4 + \cdots, \tag{3.195}$$

$$z_3(r) = \frac{\rho_0}{p_1(n)}A_{6,1}r^2 + A_{3,5}r^4 + \cdots, \tag{3.196}$$

$$z_4(r) = A_{4,4}r^4 + \cdots, \tag{3.197}$$

$$z_5(r) = \frac{1}{n}A_{6,1} + A_{5,4}r^2 + A_{5,6}r^4 + \cdots, \tag{3.198}$$

$$z_6(r) = A_{6,1} + A_{6,3}r^2 + A_{6,5}r^4 + \cdots. \tag{3.199}$$

Again $A_{5,4}$ is given by (3.185), $A_{6,3}$ is given by (3.186), and $A_{1,5}, A_{2,4}, A_{3,5}, A_{4,4}$ are given by the system (3.139) in terms of $A_{1,3}, A_{3,3}, A_{6,3}, A_{5,4}$ for $\nu = 5$, $\eta = n+2$, while $A_{5,6}, A_{6,5}$ are given by the system (3.143) in terms of $A_{1,5}, A_{3,5}$ for $\nu = 6$, $\eta = n+3$.

The fundamental solution generated by the free constant $A_{4,0}$ is

$$z_1(r) = \left(\frac{1}{\mu} - n\frac{p_2(n)}{p_1(n)}\right) A_{4,0} + A_{1,3}r^2 + A_{1,5}r^4 + \cdots, \tag{3.200}$$

$$z_2(r) = \left(q_2(n) - q_1(n)\frac{p_2(n)}{p_1(n)}\right) A_{4,0} + A_{2,2}r^2 + A_{2,4}r^4 + \cdots, \tag{3.201}$$

$$z_3(r) = \frac{p_2(n)}{p_1(n)} A_{4,0} + A_{3,3}r^2 + A_{3,5}r^4 + \cdots, \tag{3.202}$$

$$z_4(r) = A_{4,0} + A_{4,2}r^2 + A_{4,4}r^4 + \cdots, \tag{3.203}$$

$$z_5(r) = A_{5,2} + A_{5,4}r^2 + A_{5,6}r^4 + \cdots, \tag{3.204}$$

$$z_6(r) = A_{6,1} + A_{6,3}r^2 + A_{6,5}r^4 + \cdots. \tag{3.205}$$

$A_{5,2}$ and $A_{6,1}$ are given in terms of $A_{1,1}$ and $A_{3,1}$ by (3.168) and (3.169). Next $A_{1,3}$, $A_{2,2}, A_{3,3}, A_{4,2}$ are given by the system (3.139) in terms of $A_{1,1}, A_{3,1}, A_{6,1}, A_{5,2}$ for $\nu = 3$, $\eta = n+2$, while $A_{5,4}, A_{6,3}$ are given by the system (3.143) in terms of $A_{1,3}$, $A_{3,3}$ for $\nu = 4$, $\eta = n+3$. Then $A_{1,5}, A_{2,4}, A_{3,5}, A_{4,4}$ are given by the system (3.139) in terms of $A_{1,3}, A_{3,3}, A_{6,3}, A_{5,4}$ for $\nu = 5$, $\eta = n+4$, while $A_{5,6}, A_{6,5}$ are given by the system (3.143) in terms of $A_{1,5}, A_{3,5}$ for $\nu = 6$, $\eta = n+5$.

Now consider the special case of $n = 1$. From (3.146) and (3.147), $A_{1,0} = A_{3,0}$ for $n = 1$, $\alpha = 0$. For $\nu = 1$, we have $\eta = \alpha = 0$ and from (3.146) $A_{5,0} = 0$. Thus, the system (3.139) is homogeneous and non-singular and $A_{1,1} = A_{2,0} = A_{3,1} = A_{4,0} = 0$. The system (3.143), for $\nu = 1$, becomes

$$\begin{pmatrix} \eta+1 & -1 \\ -n_1 & \eta+2 \end{pmatrix} \begin{pmatrix} A_{5,1} \\ A_{6,0} \end{pmatrix} = 4\pi G\rho_0 \begin{pmatrix} A_{1,0} \\ -n_1 A_{3,0} \end{pmatrix}. \tag{3.206}$$

The coefficient matrix is singular for $\eta = n-1 = 0$. For $n = 1$, we have $A_{1,0} = A_{3,0}$, and for $\eta = 0$ both equations reduce to

$$A_{6,0} = A_{5,1} - 4\pi G\rho_0 A_{1,0} = A_{5,1} - \left(3\gamma + 2\Omega^2\right) A_{1,0}. \tag{3.207}$$

For $\nu = 2$, $\alpha = 0$, $\eta = 1 = n$, the system (3.158) is singular. With $\eta = n = 1$,

$$p_1(1) = r_1(1), \tag{3.208}$$

$$p_2(1) = r_2(1). \tag{3.209}$$

Then, with $A_{1,0} = A_{3,0}$, and substituting from (3.207) for $A_{6,0}$, the second and fourth equations are found to be identical. Solving the first three equations in terms of $A_{4,1}$ yields

$$A_{1,2} = -A_{3,2} + \frac{1}{\mu} A_{4,1}, \tag{3.210}$$

$$A_{2,1} = -q_1(1)A_{3,2} + q_2(1)A_{4,1}, \tag{3.211}$$

$$A_{3,2} = \frac{1}{p_1(1)} \left[\rho_0 \left(2\gamma + \omega^2 + 2\Omega^2 - 2m\omega\Omega\right) A_{1,0} + p_2(1)A_{4,1} + \rho_0 A_{6,0}\right]. \tag{3.212}$$

Now $A_{1,0}$, $A_{6,0}$, $A_{4,1}$ may be regarded as the three free constants determining the three regular solutions at the geocentre for $n = 1$. No further eigenvalues are encountered in the recursions for higher terms in the power series expansions.

For the regular solution generated by the free constant $A_{1,0}$, from equations (3.210) through (3.212), we have

$$\begin{pmatrix} A_{1,2} \\ A_{2,1} \\ A_{3,2} \\ A_{4,1} \end{pmatrix} = \frac{\rho_0}{p_1(1)} \left(2\gamma + \omega^2 + 2\Omega^2 - 2m\omega\Omega\right) \begin{pmatrix} -1 \\ -q_1(1) \\ 1 \\ 0 \end{pmatrix} A_{1,0}. \tag{3.213}$$

From (3.207), $A_{5,1} = 4\pi G\rho_0 A_{1,0}$ for $A_{6,0} = 0$. For numerical integration, the y-variables are again replaced by a system of z-variables for $n = 1$. For the regular solutions generated by the free constants $A_{1,0}$ and $A_{6,0}$ the new z-variables are just those defined in (3.187) for $\alpha = n - 2 = -1$. For the regular solution generated by the free constant $A_{4,1}$ they are those defined in (3.187) for $\alpha = n = 1$. The power series expansions for the regular solution generated by the free constant $A_{1,0}$ then have the forms

$$z_1(r) = y_1(r) = A_{1,0} + A_{1,2}r^2 + A_{1,4}r^4 + \cdots, \tag{3.214}$$

$$z_2(r) = ry_2(r) = A_{2,1}r^2 + A_{2,3}r^4 + \cdots, \tag{3.215}$$

$$z_3(r) = y_3(r) = A_{1,0} + A_{3,2}r^2 + A_{3,4}r^4 + \cdots, \tag{3.216}$$

$$z_4(r) = ry_4(r) = A_{4,3}r^4 + \cdots, \tag{3.217}$$

$$z_5(r) = y_5(r)/r = 4\pi G\rho_0 A_{1,0} + A_{5,3}r^2 + A_{5,5}r^4 + \cdots, \tag{3.218}$$

$$z_6(r) = y_6(r) = A_{6,2}r^2 + A_{6,4}r^4 + \cdots. \tag{3.219}$$

For the regular solution generated by the free constant $A_{6,0}$, from equations (3.210) through (3.212), we have

$$\begin{pmatrix} A_{1,2} \\ A_{2,1} \\ A_{3,2} \\ A_{4,1} \end{pmatrix} = \begin{pmatrix} -\rho_0/p_1(1) \\ -\rho_0 q_1(1)/p_1(1) \\ \rho_0/p_1(1) \\ 0 \end{pmatrix} A_{6,0}. \tag{3.220}$$

From (3.207), $A_{5,1} = A_{6,0}$ for $A_{1,0} = 0$. The power series expansions for this regular solution have the forms

$$z_1(r) = y_1(r) = A_{1,2}r^2 + A_{1,4}r^4 + \cdots, \tag{3.221}$$

$$z_2(r) = ry_2(r) = A_{2,1}r^2 + A_{2,3}r^4 + \cdots, \tag{3.222}$$

$$z_3(r) = y_3(r) = A_{3,2}r^2 + A_{3,4}r^4 + \cdots, \tag{3.223}$$

$$z_4(r) = ry_4(r) = A_{4,3}r^4 + \cdots, \tag{3.224}$$

$$z_5(r) = y_5(r)/r = A_{6,0} + A_{5,3}r^2 + A_{5,5}r^4 + \cdots, \tag{3.225}$$

$$z_6(r) = y_6(r) = A_{6,0} + A_{6,2}r^2 + A_{6,4}r^4 + \cdots. \tag{3.226}$$

For the regular solution generated by the free constant $A_{4,1}$, from equations (3.210) through (3.212), we have

$$\begin{pmatrix} A_{1,2} \\ A_{2,1} \\ A_{3,2} \\ A_{4,1} \end{pmatrix} = \begin{pmatrix} 1/\mu - p_2(1)/p_1(1) \\ q_2(1) - q_1(1)p_2(1)/p_1(1) \\ p_2(1)/p_1(1) \\ 1 \end{pmatrix} A_{4,1}. \tag{3.227}$$

From (3.207), $A_{5,1} = 0$ for $A_{1,0} = A_{6,0} = 0$. The power series expansions for this regular solution have the forms

$$z_1(r) = y_1(r)/r^2 = A_{1,2} + A_{1,4}r^2 + A_{1,6}r^4 + \cdots, \tag{3.228}$$

$$z_2(r) = y_2(r)/r = A_{2,1} + A_{2,3}r^2 + A_{2,5}r^4 + \cdots, \tag{3.229}$$

$$z_3(r) = y_3(r)/r^2 = A_{3,2} + A_{3,4}r^2 + A_{3,6}r^4 + \cdots, \tag{3.230}$$

$$z_4(r) = y_4(r)/r = A_{4,1} + A_{4,3}r^2 + A_{4,5}r^4 + \cdots, \tag{3.231}$$

$$z_5(r) = y_5(r)/r^3 = A_{5,3} + A_{5,5}r^2 + A_{5,7}r^4 + \cdots, \tag{3.232}$$

$$z_6(r) = y_6(r)/r^2 = A_{6,2} + A_{6,4}r^2 + A_{6,6}r^4 + \cdots. \tag{3.233}$$

Higher terms in the power series expansions for all three regular solutions are found from the recurrence relations provided by the systems (3.143) and (3.139). For $\nu = 3$, $\eta = 2$, the system (3.143) gives

$$\begin{pmatrix} 3 & -1 \\ -n_1 & 4 \end{pmatrix} \begin{pmatrix} A_{5,3} \\ A_{6,2} \end{pmatrix} = 4\pi G\rho_0 \begin{pmatrix} A_{1,2} \\ -n_1 A_{3,2} \end{pmatrix}. \tag{3.234}$$

This can be easily solved to give

$$A_{5,3} = \frac{4\pi G\rho_0}{10}\left[4A_{1,2} - 2A_{3,2}\right], \tag{3.235}$$

and

$$A_{6,2} = 3A_{5,3} - 4\pi G\rho_0 A_{1,2}. \tag{3.236}$$

$A_{1,4}$, $A_{2,3}$, $A_{3,4}$, $A_{4,3}$ are given by the system (3.139) in terms of $A_{1,2}$, $A_{3,2}$, $A_{6,2}$, $A_{5,3}$ for $\nu = 4$, $\eta = 3$. In turn, $A_{5,5}$, $A_{6,4}$ are given by the system (3.143) in terms of $A_{1,4}$, $A_{3,4}$ for $\nu = 5$, $\eta = 4$. For the regular solution generated by the free constant $A_{4,1}$, the extra terms $A_{1,6}$, $A_{2,5}$, $A_{3,6}$, $A_{4,5}$ are given by the system (3.139) in terms of $A_{1,4}$, $A_{3,4}$, $A_{6,4}$, $A_{5,5}$ for $\nu = 6$, $\eta = 5$. In turn, the extra terms $A_{5,7}$, $A_{6,6}$ are given by the system (3.143) in terms of $A_{1,6}$, $A_{3,6}$ for $\nu = 7$, $\eta = 6$.

For $n = 0$, near the geocentre, the fourth-order system governing the radial spheroidal deformations, (3.110) through (3.113), takes the form

$$\frac{dy_1}{dr} = -\frac{2\lambda\beta}{r}y_1 + \beta y_2, \tag{3.237}$$

$$\frac{dy_2}{dr} = \left[-\rho_0\left(4\gamma + 2\Omega^2\right) + \frac{2\delta}{r^2}\right]y_1 - \frac{4\mu\beta}{r}y_2 - \rho_0 y_6 - \omega^2\rho_0 y_1, \tag{3.238}$$

$$\frac{dy_5}{dr} = 4\pi G\rho_0 y_1 + y_6, \tag{3.239}$$

$$\frac{dy_6}{dr} = -\frac{2}{r}y_6. \tag{3.240}$$

Substitution of the power series expansions (3.138) in these equations, and equating coefficients of like powers of the radius, yields the systems

$$\begin{pmatrix} 2\lambda\beta + 1 + \eta & -\beta \\ -2\delta & 4\mu\beta + \eta \end{pmatrix}\begin{pmatrix} A_{1,\nu} \\ A_{2,\nu-1} \end{pmatrix} = \rho_0\begin{pmatrix} 0 \\ -\left(4\gamma + \omega^2 + 2\Omega^2\right)A_{1,\nu-2} - A_{6,\nu-2} \end{pmatrix}, \tag{3.241}$$

and

$$\begin{pmatrix} 1 + \eta & -1 \\ 0 & 2 + \eta \end{pmatrix}\begin{pmatrix} A_{5,\nu} \\ A_{6,\nu-1} \end{pmatrix} = \begin{pmatrix} 4\pi G\rho_0 A_{1,\nu-1} \\ 0 \end{pmatrix}, \tag{3.242}$$

for $\nu = 0, 1, 2, \ldots$. The first system has a singular coefficient matrix for $\eta = -3$ and for $\eta = 0$, while the coefficient matrix of the second system is singular for $\eta = -2$ and for $\eta = -1$.

For $\nu = 0$, $\eta = \alpha - 1$, the system (3.241) gives $(\alpha + 2\lambda\beta)A_{1,0} = -2\delta A_{1,0} = 0$ or $A_{1,0} = 0$. The system (3.242) gives $\alpha A_{5,0} = 0$ or $A_{5,0} = 0$, except for $\alpha = 0$.

For $\nu = 1$, $\eta = \alpha$, the system (3.241) is homogeneous and has a non-trivial solution only for $\eta = \alpha = 0$. The solution is $(3\lambda + 2\mu)A_{1,1} = A_{2,0}$. The system (3.242) is homogeneous and non-singular for $\alpha = \eta = 0$, giving $A_{5,1} = A_{6,0} = 0$. No further singular values of η are encountered in either system with increasing ν.

For $\nu = 2$, we have $\eta = \alpha + 1 = 1$ for $\alpha = 0$, so the system (3.241) is homogeneous and non-singular, implying that $A_{1,2} = A_{2,1} = 0$. The system (3.242) is also non-singular, giving $A_{5,2} = 2\pi G\rho_0 A_{1,1}$ and $A_{6,1} = 0$.

The only solution regular at the geocentre is then generated by the free constant $A_{1,1}$ with $\alpha = 0$. Although $A_{5,0}$ can be non-zero, this solution does not propagate and may be regarded as the arbitrary constant determining the reference level for gravitational potential, usually set to zero at infinity.

In summary, the power series solutions, near the geocentre, take the forms

$$y_1 = A_{1,1}r + A_{1,3}r^3 + A_{1,5}r^5 + \cdots , \tag{3.243}$$

$$y_2 = (3\lambda + 2\mu)A_{1,1} + A_{2,2}r^2 + A_{2,4}r^4 + \cdots , \tag{3.244}$$

$$y_5 = A_{5,0} + \frac{1}{2}(4\pi G\rho_0)A_{1,1}r^2 + A_{5,4}r^4 + A_{5,6}r^6 + \cdots , \tag{3.245}$$

$$y_6 = A_{6,3}r^3 + A_{6,5}r^5 + A_{6,7}r^7 + \cdots . \tag{3.246}$$

The solutions, regular at the geocentre, are generated from the two free constants, $A_{1,1}$ and $A_{5,0}$. The latter leads to $y_1 = y_2 = y_6 = 0$, with y_5 constant and equal to $A_{5,0}$. The higher terms in the solutions generated from the free constant $A_{1,1}$, for $\alpha = 0$, are found from the system (3.241) with $\nu = 3$,

$$\begin{pmatrix} 2\lambda\beta + 3 & -\beta \\ -2\delta & 4\mu\beta + 2 \end{pmatrix} \begin{pmatrix} A_{1,3} \\ A_{2,2} \end{pmatrix} = \rho_0 \begin{pmatrix} 0 \\ -\left(4\gamma + \omega^2 + 2\Omega^2\right)A_{1,1} \end{pmatrix}, \tag{3.247}$$

and with $\nu = 5$,

$$\begin{pmatrix} 2\lambda\beta + 5 & -\beta \\ -2\delta & 4\mu\beta + 4 \end{pmatrix} \begin{pmatrix} A_{1,5} \\ A_{2,4} \end{pmatrix} = \rho_0 \begin{pmatrix} 0 \\ -\left(4\gamma + \omega^2 + 2\Omega^2\right)A_{1,3} - A_{6,3} \end{pmatrix}, \tag{3.248}$$

and from the system (3.242) with $\nu = 4$,

$$\begin{pmatrix} 4 & -1 \\ 0 & 5 \end{pmatrix} \begin{pmatrix} A_{5,4} \\ A_{6,3} \end{pmatrix} = \begin{pmatrix} 4\pi G\rho_0 A_{1,3} \\ 0 \end{pmatrix}, \tag{3.249}$$

and with $\nu = 6$,

$$\begin{pmatrix} 6 & -1 \\ 0 & 7 \end{pmatrix} \begin{pmatrix} A_{5,6} \\ A_{6,5} \end{pmatrix} = \begin{pmatrix} 4\pi G\rho_0 A_{1,5} \\ 0 \end{pmatrix}. \tag{3.250}$$

Thus $A_{6,3} = A_{6,5} = \cdots = 0$, $y_6 = 0$, and the solutions generated from $A_{1,1}$, again replacing the y-variables by a system of z-variables defined by (3.187) for $\alpha = n = 0$, become

$$z_1(r) = y_1(r)/r = A_{1,1} + A_{1,3}r^2 + A_{1,5}r^4 + \cdots , \tag{3.251}$$

$$z_2(r) = y_2(r) = (3\lambda + 2\mu)A_{1,1} + A_{2,2}r^2 + A_{2,4}r^4 + \cdots , \tag{3.252}$$

$$z_5(r) = y_5(r)/r^2 = 2\pi G\rho_0 A_{1,1} + A_{5,4}r^2 + A_{5,6}r^4 + \cdots , \tag{3.253}$$

$$z_6(r) = y_6(r)/r = 0. \tag{3.254}$$

The coefficients of the expansions are found easily from the generating free constant $A_{1,1}$. Solving (3.247), we find that

$$A_{1,3} = -\frac{\rho_0\left(4\gamma + \omega^2 + 2\Omega^2\right)}{10\,(\lambda + 2\mu)}A_{1,1}, \tag{3.255}$$

$$A_{2,2} = (5\lambda + 6\mu)\,A_{1,3}. \tag{3.256}$$

Then (3.249) gives

$$A_{5,4} = \pi G\rho_0 A_{1,3}. \tag{3.257}$$

Solving (3.248), we find that

$$A_{1,5} = -\frac{\rho_0\left(4\gamma + \omega^2 + 2\Omega^2\right)}{28\,(\lambda + 2\mu)}A_{1,3}, \tag{3.258}$$

$$A_{2,4} = (7\lambda + 10\mu)\,A_{1,5}. \tag{3.259}$$

Then (3.250) gives

$$A_{5,6} = \frac{2}{3}\pi G\rho_0 A_{1,5}. \tag{3.260}$$

3.6 Numerical integration of the radial equations

The power series solutions regular at the geocentre, considered in the previous section, are useful only for small radii. In general, the radial equations for the spheroidal and torsional deformations, summarised in Section 3.4 for small harmonic oscillations, require integration for arbitrary radii. All of the governing systems are ordinary differential equations.

There are many methods for the numerical integration of ordinary differential equations (see, for example, Press *et al.* (1992), Ch. 6). We focus our attention on one of the most commonly used, the Runge–Kutta method, and implement it to integrate the radial equations.

Let us first consider the numerical integration of a single first-order differential equation,

$$\frac{dy}{dr} = f(r, y), \tag{3.261}$$

where $f(r, y)$ is a continuous, differentiable function of radius r and the solution $y(r)$. The basic step in numerical integration is to extend the solution $y(r)$ from radius $r = r_i$ to radius $r = r_{i+1}$. The integration stepsize is then $h = r_{i+1} - r_i$. Taylor series expansion yields

$$y_{i+1} - y_i = \sum_{n=1}^{\infty} h^n y_i^{(n)}/n!, \tag{3.262}$$

where $y_{i+1} = y(r_{i+1})$, $y_i = y(r_i)$ and $y_i^{(n)}$ is the nth derivative of y evaluated at $r = r_i$. From (3.261), upon differentiating $(n-1)$ times with respect to r,

$$y_i^{(n)} = \frac{d^{n-1}}{dr^{n-1}} f(r_i, y_i) \equiv \left. \frac{d^{n-1}}{dr^{n-1}} f(r, y) \right|_{r=r_i}, \tag{3.263}$$

with $|_{r=r_i}$ indicating that the expression to the left is to be evaluated at $r = r_i$.

By the chain rule for partial derivatives,

$$\frac{d}{dr} \equiv \left(\frac{\partial}{\partial r} + \frac{dy}{dr}\frac{\partial}{\partial y} \right) \equiv \left(\frac{\partial}{\partial r} + f(r,y)\frac{\partial}{\partial y} \right). \tag{3.264}$$

Then, from (3.263),

$$y_i^{(n)} = \left(\frac{\partial}{\partial r} + f\frac{\partial}{\partial y} \right)^{n-1} f(r_i, y_i). \tag{3.265}$$

The Taylor expansion (3.262) becomes

$$y_{i+1} - y_i = \sum_{n=0}^{\infty} \frac{h^{n+1}}{(n+1)!} \left(\frac{\partial}{\partial r} + f\frac{\partial}{\partial y} \right)^n f(r_i, y_i). \tag{3.266}$$

With the abbreviations

$$f_i = f(r_i, y_i) \quad \text{and} \quad D = \frac{\partial}{\partial r} + f_i \frac{\partial}{\partial y}, \tag{3.267}$$

we find that

$$D^2 f = \left(\frac{\partial}{\partial r} + f_i\frac{\partial}{\partial y} \right)^2 f = \frac{\partial^2 f}{\partial r^2} + 2 f_i \frac{\partial^2 f}{\partial r \partial y} + f_i^2 \frac{\partial^2 f}{\partial y^2}, \tag{3.268}$$

and

$$\left. \left\{ \left(\frac{\partial}{\partial r} + f\frac{\partial}{\partial y} \right)^2 f \right\} \right|_{r=r_i} = \left. \left\{ \frac{\partial^2 f}{\partial r^2} + 2f\frac{\partial^2 f}{\partial r \partial y} + f^2 \frac{\partial^2 f}{\partial y^2} + \frac{\partial f}{\partial y}\left(\frac{\partial f}{\partial r} + f\frac{\partial f}{\partial y} \right) \right\} \right|_{r=r_i}$$
$$= \left. \left\{ D^2 f + \frac{\partial f}{\partial y} D f \right\} \right|_{r=r_i}. \tag{3.269}$$

Similarly,

$$D^3 f = \left(\frac{\partial}{\partial r} + f_i\frac{\partial}{\partial y} \right)^3 f = \frac{\partial^3 f}{\partial r^3} + 3 f_i \frac{\partial^3 f}{\partial r^2 \partial y} + 3 f_i^2 \frac{\partial^3 f}{\partial r \partial y^2} + f_i^3 \frac{\partial^3 f}{\partial y^3}, \tag{3.270}$$

and

$$\left\{\left(\frac{\partial}{\partial r}+f\frac{\partial}{\partial y}\right)^3 f\right\}\Bigg|_{r=r_i} = \left\{\frac{\partial^3 f}{\partial r^3}+3f\frac{\partial^3 f}{\partial r^2\partial y}+3f^2\frac{\partial^3 f}{\partial r\partial y^2}+f^3\frac{\partial^3 f}{\partial y^3}\right.$$
$$+\frac{\partial f}{\partial y}\left(\frac{\partial^2 f}{\partial r^2}+2f\frac{\partial^2 f}{\partial r\partial y}+f^2\frac{\partial^2 f}{\partial y^2}\right)+\left(\frac{\partial f}{\partial y}\right)^2\left(\frac{\partial f}{\partial r}+f\frac{\partial f}{\partial y}\right)$$
$$\left.+3\left(\frac{\partial f}{\partial r}+f\frac{\partial f}{\partial y}\right)\left(\frac{\partial^2 f}{\partial r\partial y}+f\frac{\partial^2 f}{\partial y^2}\right)\right\}\Bigg|_{r=r_i}$$
$$=\left\{D^3 f+\frac{\partial f}{\partial y}D^2 f+\left(\frac{\partial f}{\partial y}\right)^2 Df+3Df\,D\frac{\partial f}{\partial y}\right\}\Bigg|_{r=r_i}. \quad (3.271)$$

With the abbreviation $f_y = \partial f/\partial y$, substituting (3.269) and (3.271) in (3.266), it is found that

$$y_{i+1}-y_i = \left\{hf+\frac{h^2}{2!}Df+\frac{h^3}{3!}\left(D^2 f+f_y Df\right)\right.$$
$$\left.+\frac{h^4}{4!}\left(D^3 f+f_y D^2 f+f_y^2 Df+3Df\,Df_y\right)\right\}\Bigg|_{r=r_i}+O(h^5). \quad (3.272)$$

To lowest order, extension of the solution from radius r_i to r_{i+1} is

$$y_{i+1}-y_i = hf(r_i, y_i)+O(h^2). \quad (3.273)$$

This is simply a trapezoidal extension of $y(r)$ assuming constant slope, $dy/dr = f(r_i, y_i)$.

A better approximation would allow for the change of slope from that at the beginning of the interval to an estimate of the slope elsewhere on the interval. Taking a weighted average of the two, we would write

$$y_{i+1}-y_i = w_1 hf_i + w_2 hf\left(r_i+\alpha_2 h,\ y_i+\beta_{21}hf_i\right)$$
$$= w_1 k_1 + w_2 k_2, \quad (3.274)$$

where w_1, w_2, α_2, β_{21} are parameters to be determined. Taylor expansion in two variables gives

$$f(r_i+\alpha_2 h,\ y_i+\beta_{21}hf_i) = \left\{f+\alpha_2 h\frac{\partial f}{\partial r}+\beta_{21}hf_i\frac{\partial f}{\partial y}\right\}\Bigg|_{r=r_i}+\cdots \quad (3.275)$$

Then,

$$y_{i+1}-y_i = \left\{h(w_1+w_2)f+w_2 h^2\left(\alpha_2\frac{\partial f}{\partial r}+\beta_{21}f_i\frac{\partial f}{\partial y}\right)\right\}\Bigg|_{r=r_i}+O\left(h^3\right)$$
$$= \left\{h(w_1+w_2)f+w_2 h^2 D_2 f\right\}\Big|_{r=r_i}+O(h^3), \quad (3.276)$$

with

$$D_2 \equiv \alpha_2 \frac{\partial}{\partial r} + \beta_{21} f_i \frac{\partial}{\partial y}. \tag{3.277}$$

Comparison with the Taylor expansion (3.272) shows that $w_1 + w_2 = 1$, $2w_2\alpha_2 = 1$, $2w_2\beta_{21} = 1$. These three equations are insufficient to determine the four parameters uniquely. If we weight the two slopes equally, $w_1 = w_2 = 1/2$ and $\alpha_2 = 1, \beta_{21} = 1$. Substitution in (3.274) then yields

$$\begin{aligned} y_{i+1} - y_i &= \frac{1}{2}\Big(hf(r_i, y_i) + hf(r_i + h,\ y_i + hf_i)\Big) + O(h^3) \\ &= \frac{1}{2}(k_1 + k_2) + O(h^3), \end{aligned} \tag{3.278}$$

for $k_1 = hf(r_i, y_i)$, $k_2 = hf(r_i + h,\ y_i + k_1)$. This second-order accurate approximation has been obtained without calculation of derivatives of $f(r, y)$. Such schemes of integration, avoiding the calculation of derivatives, were first developed by Runge (1895), with further contributions by Kutta (1901). Instead, two evaluations of $f(r, y)$ are required for second-order accuracy. Higher-order accuracy can be achieved by increasing the number of evaluations. Because of the underdetermined nature of the parameters, a wide variety of integration formulae have been derived, generally referred to as *Runge–Kutta methods*. The scheme (3.278) is one of several second-order Runge–Kutta formulae.

The foregoing analysis can be generalised. For example, for fourth-order accuracy we write

$$y_{i+1} - y_i = w_1 k_1 + w_2 k_2 + w_3 k_3 + w_4 k_4, \tag{3.279}$$

with the four required evaluations,

$$\begin{aligned} k_1 &= hf(r_i, y_i), \\ k_2 &= hf(r_i + \alpha_2 h,\ y_i + \beta_{21} k_1), \\ k_3 &= hf(r_i + \alpha_3 h,\ y_i + \beta_{31} k_1 + \beta_{32} k_2), \\ k_4 &= hf(r_1 + \alpha_4 h,\ y_i + \beta_{41} k_1 + \beta_{42} k_2 + \beta_{43} k_3). \end{aligned} \tag{3.280}$$

The requisite Taylor expansions of $f(r, y)$ in two variables are listed by Ralston and Rabinowitz (1978, Sec. 5.8). Once again, comparison with the expansion (3.272) gives an underdetermined system of equations in the unknown parameters, allowing a wide variety of fourth-order accurate schemes.

Instead, we concentrate on the most commonly used fourth-order Runge–Kutta scheme,

$$y_{i+1} - y_i = \frac{1}{6}(k_1 + 2k_2 + 2k_3 + k_4), \tag{3.281}$$

where

$$\begin{aligned} k_1 &= hf(r_i, y_i), \\ k_2 &= hf\left(r_i + \frac{h}{2}, y_i + \frac{k_1}{2}\right), \\ k_3 &= hf\left(r_i + \frac{h}{2}, y_i + \frac{k_2}{2}\right), \\ k_4 &= hf(r_i + h, y_i + k_3). \end{aligned} \tag{3.282}$$

Since $k_1 = hf_i$, Taylor expansion in two variables gives

$$\begin{aligned} k_2 &= hf\left(r_i + \frac{h}{2}, y_i + \frac{k_1}{2}\right) \\ &= \left\{ hf + \frac{h^2}{2}Df + \frac{h^3}{8}D^2 f + \frac{h^4}{48}D^3 f \right\}\bigg|_{r=r_i} + O(h^5). \end{aligned} \tag{3.283}$$

This expression for k_2 can be used to develop the expansion,

$$\begin{aligned} f\left(r_i + \frac{h}{2}, y_i + \frac{k_2}{2}\right) = \Bigg\{ & f(r,y) + \frac{h}{2}\frac{\partial f}{\partial r} + \frac{k_2}{2}\frac{\partial f}{\partial y} + \frac{1}{2!}\left(\frac{h}{2}\frac{\partial}{\partial r} + \frac{k_2}{2}\frac{\partial}{\partial y}\right)^2 f \\ & + \frac{1}{3!}\left(\frac{h}{2}\frac{\partial}{\partial r} + \frac{k_2}{2}\frac{\partial}{\partial y}\right)^3 f + \cdots \Bigg\}\Bigg|_{r=r_i}, \end{aligned} \tag{3.284}$$

required for the calculation of k_3. Thus,

$$\begin{aligned} \frac{1}{2!}\left(\frac{h}{2}\frac{\partial}{\partial r} + \frac{k_2}{2}\frac{\partial}{\partial y}\right)^2 f &= \frac{1}{2}\left(\frac{h}{2}\frac{\partial}{\partial r} + \frac{k_2}{2}\frac{\partial}{\partial y}\right)\left(\frac{h}{2}\frac{\partial f}{\partial r} + \frac{k_2}{2}\frac{\partial f}{\partial y}\right) \\ &= \frac{1}{8}\left(h^2\frac{\partial^2 f}{\partial r^2} + 2hk_2\frac{\partial^2 f}{\partial r \partial y} + k_2^2\frac{\partial^2 f}{\partial y^2}\right) \\ &= \Bigg\{\frac{1}{8}\Bigg[h^2\frac{\partial^2 f}{\partial r^2} + 2h\left(hf + \frac{h^2}{2}Df\right)\frac{\partial^2 f}{\partial r \partial y} \\ &\quad + \left(h^2 f^2 + h^3 f Df\right)\frac{\partial^2 f}{\partial y^2}\Bigg]\Bigg\}\Bigg|_{r=r_i} + O(h^4). \end{aligned} \tag{3.285}$$

Then,

$$\begin{aligned} \frac{1}{2!}\left(\frac{h}{2}\frac{\partial}{\partial r} + \frac{k_2}{2}\frac{\partial}{\partial y}\right)^2 f &= \left\{\frac{h^2}{8}D^2 f + \frac{h^3}{8}Df\left(\frac{\partial^2 f}{\partial r \partial y} + f\frac{\partial^2 f}{\partial y^2}\right)\right\}\bigg|_{r=r_i} + O(h^4) \\ &= \left\{\frac{h^2}{8}D^2 f + \frac{h^3}{8}Df Df_y\right\}\bigg|_{r=r_i} + O(h^4), \end{aligned} \tag{3.286}$$

using the expression (3.268) for $D^2 f$, and recognising the identity

$$\frac{\partial^2 f}{\partial r \partial y} + f_i \frac{\partial^2 f}{\partial y^2} = D f_y. \tag{3.287}$$

For the term following (3.286), we have

$$\begin{aligned}
&\frac{1}{3!}\left(\frac{h}{2}\frac{\partial}{\partial r} + \frac{k_2}{2}\frac{\partial}{\partial y}\right)^3 f \\
&\quad = \frac{1}{24}\left(\frac{h}{2}\frac{\partial}{\partial r} + \frac{k_2}{2}\frac{\partial}{\partial y}\right)\left(h^2 \frac{\partial^2 f}{\partial r^2} + 2hk_2 \frac{\partial^2 f}{\partial r \partial y} + k_2^2 \frac{\partial^2 f}{\partial y^2}\right) \\
&\quad = \left\{\frac{h^3}{48}\left(\frac{\partial}{\partial r} + f\frac{\partial}{\partial y}\right)\left(\frac{\partial^2 f}{\partial r^2} + 2f\frac{\partial^2 f}{\partial r \partial y} + f^2 \frac{\partial^2 f}{\partial y^2}\right) + O(h^4)\right\}\Bigg|_{r=r_i} \\
&\quad = \left\{\frac{h^3}{48}\left(\frac{\partial^3 f}{\partial r^3} + 3f\frac{\partial^3 f}{\partial r^2 \partial y} + 3f^2 \frac{\partial^3 f}{\partial r \partial y^2} + f^3 \frac{\partial^3 f}{\partial y^3}\right) + O(h^4)\right\}\Bigg|_{r=r_i} \\
&\quad = \left\{\frac{h^3}{48} D^3 f + O(h^4)\right\}\Bigg|_{r=r_i}
\end{aligned} \tag{3.288}$$

upon substitution from (3.270). From (3.284), with substitution from (3.283), (3.286) and (3.288), we find that

$$\begin{aligned}
k_3 = hf\left(r_i + \frac{h}{2}, y_i + \frac{k_2}{2}\right) = \Bigg\{ & hf + \frac{h^2}{2} Df + \frac{h^3}{8}\left(D^2 f + 2f_y Df\right) \\
& + \frac{h^4}{48}\left(D^3 f + 3f_y D^2 f + 6Df\, Df_y\right)\Bigg\}\Bigg|_{r=r_i} + O(h^5).
\end{aligned} \tag{3.289}$$

In turn, this expression for k_3 can be used to develop the expansion,

$$\begin{aligned}
f(r_i + h, y_i + k_3) = f(r_i, y_i) + h\frac{\partial f}{\partial r} + k_3 \frac{\partial f}{\partial y} + \frac{1}{2!}\left(h\frac{\partial}{\partial r} + k_3 \frac{\partial}{\partial y}\right)^2 f \\
+ \frac{1}{3!}\left(h\frac{\partial}{\partial r} + k_3 \frac{\partial}{\partial y}\right)^3 f + \cdots
\end{aligned} \tag{3.290}$$

required for the calculation of k_4. Thus,

$$\begin{aligned}
\frac{1}{2!}\left(h\frac{\partial}{\partial r}+k_3\frac{\partial}{\partial y}\right)^2 f &= \frac{1}{2}\left(h\frac{\partial}{\partial r}+k_3\frac{\partial}{\partial y}\right)\left(h\frac{\partial f}{\partial r}+k_3\frac{\partial f}{\partial y}\right)\\
&= \frac{1}{2}\left(h^2\frac{\partial^2 f}{\partial r}+2hk_3\frac{\partial^2 f}{\partial r\partial y}+k_3^2\frac{\partial^2 f}{\partial y^2}\right)\\
&= \left\{\frac{1}{2}\left[h^2\frac{\partial^2 f}{\partial r^2}+2h\left(hf+\frac{h^2}{2}Df\right)\frac{\partial^2 f}{\partial r\partial y}\right.\right.\\
&\quad\left.\left.+\left(h^2f^2+h^3fDf\right)\frac{\partial^2 f}{\partial y^2}\right]\right\}\Bigg|_{r=r_i}+O(h^4)\\
&= \left\{\frac{h^2}{2}D^2 f+\frac{h^3}{2}Df\left(\frac{\partial^2 f}{\partial r\partial y}+f\frac{\partial^2 f}{\partial y^2}\right)\right\}\Bigg|_{r=r_i}+O(h^4)\\
&= \left\{\frac{h^2}{2}D^2 f+\frac{h^3}{2}Df\,Df_y\right\}\Bigg|_{r=r_i}+O(h^4),
\end{aligned} \tag{3.291}$$

again using (3.268) and the identity

$$\frac{\partial^2 f}{\partial r\partial y}+f_i\frac{\partial^2 f}{\partial y^2}=Df_y. \tag{3.292}$$

Again,

$$\begin{aligned}
\frac{1}{3!}\left(h\frac{\partial}{\partial r}+k_3\frac{\partial}{\partial y}\right)^3 f &= \frac{1}{6}\left(h\frac{\partial}{\partial r}+k_3\frac{\partial}{\partial y}\right)\left(h^2\frac{\partial^2 f}{\partial r^2}+2hk_3\frac{\partial^2 f}{\partial r\partial y}+k_3^2\frac{\partial^2 f}{\partial y^2}\right)\\
&= \left\{\frac{h^3}{6}\left(\frac{\partial}{\partial r}+f\frac{\partial}{\partial y}\right)\left(\frac{\partial^2 f}{\partial r^2}+2f\frac{\partial^2 f}{\partial r\partial y}+f^2\frac{\partial^2 f}{\partial y^2}\right)\right\}\Bigg|_{r=r_i}+O(h^4)\\
&= \left\{\frac{h^3}{6}D^3 f\right\}\Bigg|_{r=r_i}+O(h^4),
\end{aligned} \tag{3.293}$$

using (3.270). From (3.290) with substitution from (3.289), (3.291) and (3.293), we find that

$$\begin{aligned}
k_4 &= hf(r_i+h,\,y_i+k_3)\\
&= \left\{hf+h^2Df+\frac{h^3}{2}\left(D^2f+f_yDf\right)\right.\\
&\quad\left.+\frac{h^4}{6}\left(D^3f+\frac{3}{4}f_yD^2f+\frac{3}{2}f_y^2Df+3Df\,Df_y\right)\right\}\Bigg|_{r=r_i}+O(h^5).
\end{aligned} \tag{3.294}$$

With $k_1 = hf_i$, and using expression (3.283) for k_2, (3.289) for k_3, and (3.294) for k_4, we find the sum

$$\frac{1}{6}(k_1 + 2k_2 + 2k_3 + k_4) = \left\{ hf + \frac{h^2}{2!}Df + \frac{h^3}{3!}\left(D^2 f + f_y Df\right) + \frac{h^4}{4!}\left(D^3 f + f_y D^2 f + f_y^2 Df + 3Df\, Df_y\right)\right\}\Bigg|_{r=r_i} + O(h^5)$$
$$= y_{i+1} - y_i, \tag{3.295}$$

in agreement with the Taylor expansion (3.272), and the Runge–Kutta scheme (3.281).

The governing equations for the spheroidal and torsional deformations, (3.102) to (3.107), (3.110) to (3.113), (3.114) to (3.118), (3.119) and (3.120), and (3.121) to (3.124), summarised in Section 3.4, are all systems of linear, first-order ordinary differential equations. The simple, single, first-order equation (3.261) is replaced by the vector differential equation

$$\frac{d\boldsymbol{y}(r)}{dr} = \boldsymbol{f}(r, \boldsymbol{y}) = \mathbf{A}(r) \cdot \boldsymbol{y}(r), \tag{3.296}$$

where $\boldsymbol{y}(r)$ is the solution vector. For the sixth-order spheroidal system represented by equations (3.102) through (3.107), for example, the vector $\boldsymbol{y}(r)$ is $[y_1(r), y_2(r), y_3(r), y_4(r), y_5(r), y_6(r)]^T$, and $\mathbf{A}(r)$ is a 6×6 coefficient matrix expressing the linear nature of the differential system through the product $\mathbf{A}(r) \cdot \boldsymbol{y}(r)$. The Runge–Kutta scheme (3.281), for continuing the solution vector from radius r_i to radius $r_i + h$, becomes the vector relation

$$\boldsymbol{y}_{i+1} - \boldsymbol{y}_i = \frac{1}{6}(\boldsymbol{k}_1 + 2\boldsymbol{k}_2 + 2\boldsymbol{k}_3 + \boldsymbol{k}_4), \tag{3.297}$$

with

$$\begin{aligned}
\boldsymbol{k}_1 &= h\,\mathbf{A}(r_i) \cdot \boldsymbol{y}(r_i), \\
\boldsymbol{k}_2 &= h\,\mathbf{A}\left(r_i + \frac{h}{2}\right) \cdot \left[\boldsymbol{y}\left(r_i + \frac{h}{2}\right) + \frac{\boldsymbol{k}_1}{2}\right], \\
\boldsymbol{k}_3 &= h\,\mathbf{A}\left(r_i + \frac{h}{2}\right) \cdot \left[\boldsymbol{y}\left(r_i + \frac{h}{2}\right) + \frac{\boldsymbol{k}_2}{2}\right], \\
\boldsymbol{k}_4 &= h\,\mathbf{A}(r_i + h) \cdot [\boldsymbol{y}(r_i + h) + \boldsymbol{k}_3].
\end{aligned} \tag{3.298}$$

More generally, we will be dealing with the fundamental solutions of the differential system. For a system of order n, there are n linearly independent solutions, and every solution of the system is expressible as a linear combination of these n fundamental solutions (Cole (1968), Theorem 3-2.2, p. 43). This leads to the concept of a *propagator matrix* (Gilbert and Backus, 1966), $\mathbf{Y}(r,\rho)$, which is the solution of the differential system

$$\frac{d\mathbf{Y}(r,\rho)}{dr} = \mathbf{A}(r) \cdot \mathbf{Y}(r,\rho) \tag{3.299}$$

that propagates the fundamental solutions from radius ρ to radius r with initial condition $\mathbf{Y}(\rho,\rho) = \mathbf{I}$, where $\mathbf{I}$ is the unit matrix. Thus, each column vector of $\mathbf{Y}$ is a linearly independent solution as guaranteed by the initial condition.

In the inner core, the y-variables are replaced by the z-variables defined by (3.187). The sixth-order spheroidal system, represented by equations (3.102) through (3.107), in terms of the z-variables becomes

$$\frac{dz_1}{dr} + \frac{\alpha+1}{r}z_1 = -\frac{2\lambda\beta}{r} \cdot z_1 + \beta \cdot \frac{z_2}{r} + \frac{n_1\lambda\beta}{r} \cdot z_3, \tag{3.300}$$

$$\begin{aligned} \frac{dz_2}{dr} + \frac{\alpha}{r}z_2 = {} & \left[-\frac{2\rho_0}{r}\left(2g_0 + r\Omega^2\right) + \frac{2\delta}{r^2}\right] \cdot rz_1 - \frac{4\mu\beta}{r} \cdot z_2 \\ & + n_1\left(\frac{\rho_0 g_0}{r} - \frac{\delta}{r^2}\right) \cdot rz_3 + \frac{n_1}{r} \cdot z_4 - \rho_0 \cdot rz_6 - \omega^2\rho_0 \cdot rz_1 \\ & + 2m\omega\Omega\rho_0 \cdot rz_3, \end{aligned} \tag{3.301}$$

$$\frac{dz_3}{dr} + \frac{\alpha+1}{r}z_3 = -\frac{1}{r} \cdot z_1 + \frac{1}{r} \cdot z_3 + \frac{1}{\mu} \cdot \frac{z_4}{r}, \tag{3.302}$$

$$\begin{aligned} \frac{dz_4}{dr} + \frac{\alpha}{r}z_4 = {} & \left(\frac{\rho_0 g_0}{r} - \frac{\delta}{r^2}\right) \cdot rz_1 - \frac{\lambda\beta}{r} \cdot z_2 + \frac{\epsilon}{r^2} \cdot rz_3 - \frac{3}{r} \cdot z_4 \\ & - \frac{\rho_0}{r} \cdot r^2 z_5 - \omega^2\rho_0 \cdot rz_3 + \frac{2m\omega\Omega\rho_0}{n_1} \cdot r\,(z_1 + z_3), \end{aligned} \tag{3.303}$$

$$\frac{dz_5}{dr} + \frac{\alpha+2}{r}z_5 = 4\pi G\rho_0 \cdot \frac{z_1}{r} + \frac{z_6}{r}, \tag{3.304}$$

$$\frac{dz_6}{dr} + \frac{\alpha+1}{r}z_6 = -4\pi G\rho_0 \frac{n_1}{r} \cdot z_3 + \frac{n_1}{r^2} \cdot rz_5 - \frac{2}{r} \cdot z_6. \tag{3.305}$$

This system can be converted to the vector differential equation

$$\frac{dz(r)}{dr} = \mathbf{A}'(r) \cdot z(r) - z'_c, \tag{3.306}$$

where $z(r) = [z_1(r), z_2(r), z_3(r), z_4(r), z_5(r), z_6(r)]^T$ is the new solution vector and z'_c is a correction to the derivative vector given by

$$z'_c = \left[\frac{\alpha+1}{r}z_1,\ \frac{\alpha}{r}z_2,\ \frac{\alpha+1}{r}z_3,\ \frac{\alpha}{r}z_4,\ \frac{\alpha+2}{r}z_5,\ \frac{\alpha+1}{r}z_6\right]^T. \tag{3.307}$$

The matrix $\mathbf{A}'$ is derived from $\mathbf{A}$ by the element modifications

$$
\begin{aligned}
&A'_{12} = \frac{1}{r}A_{12}, \quad A'_{21} = rA_{21}, \quad A'_{23} = rA_{23}, \quad A'_{26} = rA_{26}, \\
&A'_{34} = \frac{1}{r}A_{34}, \quad A'_{41} = rA_{41}, \quad A'_{43} = rA_{43}, \quad A'_{45} = r^2A_{45}, \\
&A'_{51} = \frac{1}{r}A_{51}, \quad A'_{56} = \frac{1}{r}A_{56}, \quad A'_{65} = rA_{65}.
\end{aligned}
\tag{3.308}
$$

The numerical propagation of the fundamental solutions by the Runge–Kutta scheme (3.281) requires the repeated calculation of their derivatives as the product of the matrix $\mathbf{A}(r)$ with the propagator matrix $\mathbf{Y}(r,\rho)$, as in (3.299). The subroutine YPRIME performs this operation, after a call elsewhere to the subroutine SPMAT to enable cubic spline interpolation of a specified Earth model through calls to the subroutine INTPL. The subroutines SPMAT and INTPL are described in Section 1.6.

```
      SUBROUTINE YPRIME(R,Y,A,YP,N1,C,RM,RHOM,MUM,LAMBDAM,GZEROM,N,
     1 PI,G,IR,RMIN,ANGS,COR,WES)
C
C This subroutine finds the derivatives, YP(6,6), of the six
C fundamental solutions represented by the propagator matrix, Y(6,6),
C in the mantle and crust (shell), at an arbitrary radius R
C for degree N>0. For N=0, the third and fourth solution vectors
C can be ignored. In the inner core, it finds the derivatives of
C the fundamental solutions in terms of the z-variables, which are
C transformations of the y-variables.
C
      IMPLICIT DOUBLE PRECISION(A-H,O-Z)
      DIMENSION RM(100),RHOM(100),GZEROM(100),C(100,100),
     1 Y(6,6),YP(6,6),A(6,6)
      DOUBLE PRECISION MUM(100),LAMBDAM(100),MU,LAMBDA
C Zero fill derivative matrix YP(I,J) and coefficient matrix A(I,J).
      DO 10 I=1,6
        DO 11 J=1,6
          YP(I,J)=0.D0
          A(I,J)=0.D0
 11     CONTINUE
 10   CONTINUE
C Set maximum number of points for interpolation.
      M1=100
C If R is less than or equal to RMIN, take properties to be those
C at the geocentre.
      IF(R.LE.RMIN) GO TO 12
C Interpolate Earth properties at radius R.
      CALL INTPL(R,RHO,N1,C,RM,RHOM,M1)
      CALL INTPL(R,MU,N1,C,RM,MUM,M1)
      CALL INTPL(R,LAMBDA,N1,C,RM,LAMBDAM,M1)
      CALL INTPL(R,GZERO,N1,C,RM,GZEROM,M1)
      GO TO 13
C Assign geocentric Earth properties.
 12   RHO=RHOM(1)
      MU=MUM(1)
      LAMBDA=LAMBDAM(1)
      GZERO=(4.D0*PI*G*RHO-2.D0*WES)*R/3.D0
```

```
C Construct matrix A.
 13    AN=DFLOAT(N)
C Define factors.
       F1=AN*(AN+1.D0)
       F2=LAMBDA+2.D0*MU
       F3=3.D0*LAMBDA+2.D0*MU
       F4=RHO*GZERO
       F5=4.D0*PI*G*RHO
C Assign values to matrix elements.
       A(1,1)=-2.D0*LAMBDA/(F2*R)
       A(1,2)=1.D0/F2
       A(1,3)=F1*LAMBDA/(F2*R)
       A(1,4)=0.D0
       A(1,5)=0.D0
       A(1,6)=0.D0
       A(2,1)=4.D0*(-F4+MU*F3/(F2*R))/R-2.D0*RHO*WES-ANGS*RHO
       A(2,2)=-4.D0*MU/(F2*R)
       A(2,3)=F1*(F4-2.D0*MU*F3/(F2*R))/R+COR*RHO
       A(2,4)=F1/R
       A(2,5)=0.D0
       A(2,6)=-RHO
       IF(N.EQ.0.AND.IR.EQ.1) GO TO 14
       A(3,1)=-1.D0/R
       A(3,2)=0.D0
       A(3,3)=1.D0/R
       A(3,4)=1.D0/MU
       A(3,5)=0.D0
       A(3,6)=0.D0
       A(4,1)=(F4-2.D0*MU*F3/(F2*R))/R
C If degree is not zero, add Coriolis term.
       IF(N.NE.0) A(4,1)=A(4,1)+COR*RHO/F1
       A(4,2)=-LAMBDA/(F2*R)
       A(4,3)=2.D0*MU*((2.D0*F1-1.D0)*LAMBDA
      1 +(2.D0*F1-2.D0)*MU)/(F2*R*R)-ANGS*RHO
C If degree is not zero, add Coriolis term.
       IF(N.NE.0) A(4,3)=A(4,3)+COR*RHO/F1
       A(4,4)=-3.D0/R
       A(4,5)=-RHO/R
       A(4,6)=0.D0
 14    CONTINUE
       A(5,1)=F5
       A(5,2)=0.D0
       A(5,3)=0.D0
       A(5,4)=0.D0
       A(5,5)=0.D0
       A(5,6)=1.D0
       IF(N.EQ.0.AND.IR.EQ.1) GO TO 15
       A(6,1)=0.D0
       A(6,2)=0.D0
       A(6,3)=-F5*F1/R
       A(6,4)=0.D0
       A(6,5)=F1/(R*R)
 15    CONTINUE
       A(6,6)=-2.D0/R
C Branch to inner core calculation
       IF(IR.EQ.1) GO TO 16
C Multiply A into Y to find derivative matrix for the shell.
       DO 17 I=1,6
         DO 18 J=1,6
           DO 19 K=1,6
             YP(I,J)=YP(I,J)+A(I,K)*Y(K,J)
```

```
 19        CONTINUE
 18      CONTINUE
 17    CONTINUE
       GO TO 20
C Modify matrix A for inner core calculation.
 16    A(1,2)=A(1,2)/R
       A(2,1)=A(2,1)*R
       A(2,3)=A(2,3)*R
       A(2,6)=A(2,6)*R
       A(3,4)=A(3,4)/R
       A(4,1)=A(4,1)*R
       A(4,3)=A(4,3)*R
       A(4,5)=A(4,5)*R*R
       A(5,1)=A(5,1)/R
       A(5,6)=A(5,6)/R
       A(6,5)=A(6,5)*R
C Multiply A into Y to find YP for inner core.
       DO 21 I=1,6
         DO 22 J=1,3
           DO 23 K=1,6
             YP(I,J)=YP(I,J)+A(I,K)*Y(K,J)
 23        CONTINUE
 22      CONTINUE
 21    CONTINUE
C Correct YP to give derivatives of z-variables.
C Set value of alpha plus one for first two solutions.
       ANM1=AN-1.D0
       DO 24 J=1,3
C Set value of alpha plus one for third solution.
         IF(J.EQ.3) ANM1=AN+1.D0
         YP(1,J)=YP(1,J)-ANM1*Y(1,J)/R
         YP(2,J)=YP(2,J)-(ANM1-1.D0)*Y(2,J)/R
         YP(3,J)=YP(3,J)-ANM1*Y(3,J)/R
         YP(4,J)=YP(4,J)-(ANM1-1.D0)*Y(4,J)/R
         YP(5,J)=YP(5,J)-(ANM1+1.D0)*Y(5,J)/R
         YP(6,J)=YP(6,J)-ANM1*Y(6,J)/R
 24    CONTINUE
 20    CONTINUE
       RETURN
       END
```

3.7 Fundamental, regular solutions in the inner core

The fundamental solutions of the spheroidal system, regular at the geocentre, are found by power series expansions for small radii followed by Runge–Kutta integration out to the inner core boundary.

For $n > 1$, in terms of z-variables, the three fundamental solutions are given by the expansions (3.188) through (3.193), (3.194) through (3.199), and (3.200) through (3.205). The first two terms in the expansions of the first fundamental solution, generated by the free constant $A_{1,1}$, and the second fundamental solution, generated by the free constant $A_{6,1}$, are obtained directly, as are the first terms of the third fundamental solution, generated by the free constant $A_{4,0}$. The second terms of the third fundamental solution are given by the systems (3.139) for

$\nu = 3$, $\eta = n + 2$, and (3.143) for $\nu = 4$, $\eta = n + 3$, which combine to yield the 6×6 system

$$\mathbf{A} \cdot \begin{pmatrix} A_{1,3} \\ A_{2,2} \\ A_{3,3} \\ A_{4,2} \\ A_{5,4} \\ A_{6,3} \end{pmatrix} = \rho_0 \begin{pmatrix} 0 \\ -\left(4\gamma + \omega^2 + 2\Omega^2\right) A_{1,1} + (n_1\gamma + 2m\omega\Omega) A_{3,1} - A_{6,1} \\ 0 \\ (\gamma + 2m\omega\Omega/n_1) A_{1,1} - \left(\omega^2 - 2m\omega\Omega/n_1\right) A_{3,1} - A_{5,2} \\ 0 \\ 0 \end{pmatrix}, \quad (3.309)$$

where the coefficient matrix, $\mathbf{A}$, is

$$\begin{pmatrix} 2\lambda\beta + n + 3 & -\beta & -n_1\lambda\beta & 0 & 0 & 0 \\ -2\delta & 4\mu\beta + n + 2 & n_1\delta & -n_1 & 0 & 0 \\ 1 & 0 & n + 2 & -1/\mu & 0 & 0 \\ \delta & \lambda\beta & -\epsilon & n + 5 & 0 & 0 \\ -3\gamma - 2\Omega^2 & 0 & 0 & 0 & n + 4 & -1 \\ 0 & 0 & \left(3\gamma + 2\Omega^2\right) n_1 & 0 & -n_1 & n + 5 \end{pmatrix}, \quad (3.310)$$

with $4\pi G\rho_0$ replaced by the right side of relation (3.129). The third terms in the first and second fundamental solutions are given by the systems (3.139) for $\nu = 5$, $\eta = n + 2$, and (3.143) for $\nu = 6$, $\eta = n + 3$, which combine to yield the 6×6 system

$$\mathbf{A} \cdot \begin{pmatrix} A_{1,5} \\ A_{2,4} \\ A_{3,5} \\ A_{4,4} \\ A_{5,6} \\ A_{6,5} \end{pmatrix} = \rho_0 \begin{pmatrix} 0 \\ -\left(4\gamma + \omega^2 + 2\Omega^2\right) A_{1,3} + (n_1\gamma + 2m\omega\Omega) A_{3,3} - A_{6,3} \\ 0 \\ (\gamma + 2m\omega\Omega/n_1) A_{1,3} - \left(\omega^2 - 2m\omega\Omega/n_1\right) A_{3,3} - A_{5,4} \\ 0 \\ 0 \end{pmatrix}. \quad (3.311)$$

The third terms in the third fundamental solution are given by the systems (3.139) for $\nu = 5$, $\eta = n + 4$, and (3.143) for $\nu = 6$, $\eta = n + 5$, which combine to yield the 6×6 system

$$\mathbf{A}' \cdot \begin{pmatrix} A_{1,5} \\ A_{2,4} \\ A_{3,5} \\ A_{4,4} \\ A_{5,6} \\ A_{6,5} \end{pmatrix} = \rho_0 \begin{pmatrix} 0 \\ -\left(4\gamma + \omega^2 + 2\Omega^2\right) A_{1,3} + (n_1\gamma + 2m\omega\Omega) A_{3,3} - A_{6,3} \\ 0 \\ (\gamma + 2m\omega\Omega/n_1) A_{1,3} - \left(\omega^2 - 2m\omega\Omega/n_1\right) A_{3,3} - A_{5,4} \\ 0 \\ 0 \end{pmatrix}, \quad (3.312)$$

with the modified coefficient matrix $\mathbf{A}'$ given by

$$\mathbf{A}' = \mathbf{A} + 2\mathbf{I}, \tag{3.313}$$

where $\mathbf{I}$ is the 6×6 unit matrix. This completes the first three terms in the power series expansions of the three fundamental solutions for $n > 1$.

For $n = 1$, in terms of z-variables, the three fundamental solutions are given by the expansions (3.214) through (3.219), (3.221) through (3.226), and (3.228) through (3.233). The first two terms in the expansions of the first fundamental solution, generated by the free constant $A_{1,0}$, and the second fundamental solution, generated by the free constant $A_{6,0}$, are obtained directly, as are the first terms of the third fundamental solution, generated by the free constant $A_{4,1}$. The third terms in the first and second fundamental solutions, and the second terms of the third fundamental solution, are given by the systems (3.139) for $\nu = 4$, $\eta = n + 2 = 3$, and (3.143) for $\nu = 5$, $\eta = n + 3 = 4$, which combine to yield the 6×6 system

$$\mathbf{A} \cdot \begin{pmatrix} A_{1,4} \\ A_{2,3} \\ A_{3,4} \\ A_{4,3} \\ A_{5,5} \\ A_{6,4} \end{pmatrix} = \rho_0 \begin{pmatrix} 0 \\ -\left(4\gamma + \omega^2 + \Omega^2\right) A_{1,2} + (2\gamma + 2m\omega\Omega) A_{3,2} - A_{6,2} \\ 0 \\ (\gamma + m\omega\Omega) A_{1,2} - \left(\omega^2 - m\omega\Omega\right) A_{3,2} - A_{5,3} \\ 0 \\ 0 \end{pmatrix}. \tag{3.314}$$

The third terms in the third fundamental solution are given by the system (3.139) for $\nu = 6$, $\eta = n + 4 = 5$, and (3.143) for $\nu = 7$, $\eta = n + 5 = 6$, which combine to yield the 6×6 system

$$\mathbf{A}' \cdot \begin{pmatrix} A_{1,6} \\ A_{2,5} \\ A_{3,6} \\ A_{4,5} \\ A_{5,7} \\ A_{6,6} \end{pmatrix} = \rho_0 \begin{pmatrix} 0 \\ -\left(4\gamma + \omega^2 + 2\Omega^2\right) A_{1,4} + (2\gamma + 2m\omega\Omega) A_{3,4} - A_{6,4} \\ 0 \\ (\gamma + m\omega\Omega) A_{1,4} - \left(\omega^2 - m\omega\Omega\right) A_{3,4} - A_{5,5} \\ 0 \\ 0 \end{pmatrix}. \tag{3.315}$$

This completes the first three terms in the power series expansions of the three fundamental solutions for $n = 1$.

For $n = 0$, the governing spheroidal system degenerates to fourth order and, apart from the arbitrary constant determining the reference level for gravitational potential, there is only one fundamental solution regular at the geocentre, generated by the free constant $A_{1,1}$. In terms of z-variables, it is given by the expansions (3.251) through (3.254). The first three terms of these expansions are obtained directly.

For $n \geq 1$, completion of the power series expansions requires the solution of the 6×6 linear systems (3.309), (3.311), (3.312), (3.314) and (3.315), with coefficient matrix **A**, or its modification **A**$'$. The subroutine MATRIX calculates the coefficient matrix **A**, allowing the solution of these linear systems with calls elsewhere to the subroutine LINSOL, described in Section 1.5.

```
      SUBROUTINE MATRIX(A,RHO,LAMBDA,MU,PI,G,AN,WE)
C
C This subroutine calculates the 6X6 matrix A required to find
C higher terms in the power series expansions of the three
C fundamental solutions at the geocentre.
C
      IMPLICIT DOUBLE PRECISION(A-H,O-Z)
      DIMENSION A(6,6),RHO(100)
      DOUBLE PRECISION LAMBDA(100),MU(100)
      BETA=1.D0/(LAMBDA(1)+2.D0*MU(1))
      DELTA=2.D0*MU(1)*(3.D0*LAMBDA(1)+2.D0*MU(1))*BETA
      GAMMA=(4.D0*PI*G*RHO(1)-2.D0*WE*WE)/3.D0
      AN1=AN*(AN+1.D0)
      EPSLN=4.D0*AN1*MU(1)*(LAMBDA(1)+MU(1))*BETA-2.D0*MU(1)
      A(1,1)=AN+3.D0+2.D0*LAMBDA(1)*BETA
      A(1,2)=-BETA
      A(1,3)=-AN1*LAMBDA(1)*BETA
      A(1,4)=0.D0
      A(1,5)=0.D0
      A(1,6)=0.D0
      A(2,1)=-2.D0*DELTA
      A(2,2)=AN+2.D0+4.D0*MU(1)*BETA
      A(2,3)=AN1*DELTA
      A(2,4)=-AN1
      A(2,5)=0.D0
      A(2,6)=0.D0
      A(3,1)=1.D0
      A(3,2)=0.D0
      A(3,3)=AN+2.D0
      A(3,4)=-1.D0/MU(1)
      A(3,5)=0.D0
      A(3,6)=0.D0
      A(4,1)=DELTA
      A(4,2)=LAMBDA(1)*BETA
      A(4,3)=-EPSLN
      A(4,4)=AN+5.D0
      A(4,5)=0.D0
      A(4,6)=0.D0
      A(5,1)=-3.D0*GAMMA-2.D0*WE*WE
      A(5,2)=0.D0
      A(5,3)=0.D0
      A(5,4)=0.D0
      A(5,5)=AN+4.D0
      A(5,6)=-1.D0
      A(6,1)=0.D0
      A(6,2)=0.D0
      A(6,3)=AN1*(3.D0*GAMMA+2.D0*WE*WE)
      A(6,4)=0.D0
      A(6,5)=-AN1
      A(6,6)=AN+5.D0
      RETURN
      END
```

As well, for $n \geq 1$, the power series expansions require evaluation of the polynomials $p_1(n)$, $p_2(n)$, defined by (3.153) and (3.154), and the polynomials $q_1(n)$, $q_2(n)$ defined by (3.150) and (3.151). These are evaluated by the double precision function subprogrammes P1, P2, Q1 and Q2, listed below.

```
      DOUBLE PRECISION FUNCTION P1(AN,LAMBDA,MU)
C Finds value of function P1 for series solution at geocentre.
      IMPLICIT DOUBLE PRECISION(A-H,O-Z)
      DOUBLE PRECISION LAMBDA,MU
      P1=2.D0*AN*(AN*(AN+2.D0)*LAMBDA+(AN*(AN+2.D0)-1.D0)*MU)
      RETURN
      END

      DOUBLE PRECISION FUNCTION P2(AN,LAMBDA,MU)
C Finds value of function P2 for series solution at geocentre.
      IMPLICIT DOUBLE PRECISION(A-H,O-Z)
      DOUBLE PRECISION LAMBDA,MU
      P2=AN*(AN+5.D0+(AN+3.D0)*(LAMBDA/MU))
      RETURN
      END

      DOUBLE PRECISION FUNCTION Q1(AN,LAMBDA,MU)
C Finds value of function Q1 for series solution at geocentre.
      IMPLICIT DOUBLE PRECISION(A-H,O-Z)
      DOUBLE PRECISION LAMBDA,MU
      Q1=2.D0*AN*((AN+2.D0)*LAMBDA+(AN+1.D0)*MU)
      RETURN
      END

      DOUBLE PRECISION FUNCTION Q2(AN,LAMBDA,MU)
C Finds value of function Q2 for series solution at geocentre.
      IMPLICIT DOUBLE PRECISION(A-H,O-Z)
      DOUBLE PRECISION LAMBDA,MU
      Q2=2.D0*(AN+1.D0)+(AN+3.D0)*(LAMBDA/MU)
      RETURN
      END
```

The power series expansions of the fundamental solutions, regular at the geocentre, take the density and elastic constants to be independent of radius, out to a minimum radius, RMIN. They are then used there to calculate the initial solutions, and their derivatives, required for Runge–Kutta integration. RMIN is found so that the maximum ratio of the second term to first term in the power series expansions, ERRMAX, is bounded above by the specified relative error, EPS, at radius RMIN. Numerical experimentation shows that a conservative value of RMIN = 1.2 km easily meets this criterion for EPS = 10^{-5}. In addition, an estimate of the initial stepsize for the Runge–Kutta integration is required. Since the Runge–Kutta scheme used is fourth order, the error term is proportional to the stepsize to the fifth power. Thus, the initial stepsize is taken as the fifth root of the ratio EPS/ERRMAX. The evaluation of ERRMAX requires determination of the maximum relative ratio of the elements of the arrays of second term expansion coefficients to the corresponding elements in the solution arrays. This is accomplished by the subroutine REL.

The subroutine REL is also used in monitoring stepsize in the Runge–Kutta integration to ensure that relative error is kept below the specified value of EPS. As outlined in Press *et al.* (1992, pp. 708–712) the Runge–Kutta integration is performed for a single step of size $2h$ and then for two steps of size h. Since the truncation error is $O(h^5)$, the latter gives a truncation error of order $2h^5$, the former a truncation error of order $(2h)^5$. The difference of the second solution minus the first is then of order $30h^5$. The second solution can then be improved by adding 1/15 of this difference, and the subroutine REL is used to determine the maximum relative error. If it is larger than the specified value, EPS, the stepsize needs to be reduced; if smaller, the stepsize can be increased. In the latter case, the stepsize is increased in proportion to the fifth root of the ratio EPS/ERRMAX. If the stepsize is decreased, there is a potential that additional error could accumulate in proportion to the number of extra steps. As a precaution, the stepsize is only decreased in proportion to the fourth root of the ratio EPS/ERRMAX. The constant of proportionality for both increases and decreases of stepsize is taken as the conservative value of 0.9. The subroutine REL, listed below, finds the maximum ratio ERRMAX of elements of the array DELTA(6,3) to those of the array YIC(6,3), for solutions of degree N.

```
      SUBROUTINE REL(ERRMAX,DELTA,YIC,N)
C
C This subroutine is used in the variable stepsize Runge-Kutta
C integration of the fundamental solutions in the inner core.
C It finds the maximum relative error in the elements of
C the difference array, DELTA(6,3), compared to the elements of
C the fundamental solution array, YIC(6,3), for degree N.
C
      IMPLICIT DOUBLE PRECISION(A-H,O-Z)
      DIMENSION DELTA(6,3),YIC(6,3)
C Zero relative error.
      ERRMAX=0.D0
      DO 10 I=1,6
C Branch for degree zero.
        IF(N.EQ.0)GO TO 11
        GO TO 12
C If degree is zero, skip y3 and y4.
 11     IF(I.EQ.3.OR.I.EQ.4)GO TO 10
 12     DO 13 J=1,3
C Third solution is only fundamental solution for degree zero.
          IF(N.EQ.0.AND.J.NE.3)GO TO 13
          IF(YIC(I,J).EQ.0.D0)GO TO 13
          RATIO=DABS(DELTA(I,J)/YIC(I,J))
          ERRMAX=DMAX1(ERRMAX,RATIO)
 13     CONTINUE
 10   CONTINUE
      RETURN
      END
```

The subroutine RK4, listed below, propagates the solution vector, over stepsize H, by the fourth-order Runge–Kutta scheme (3.297), with calls to the subroutine YPRIME, described in Section 3.6, for the calculation of derivatives.

```
      SUBROUTINE RK4(R,Y,A,K1,N1,C,RM,RHOM,MUM,LAMBDAM,GZEROM,N,
     1 PI,G,H,IR,RMIN,ANGS,COR,WES)
C
C This subroutine completes the last three steps of a fourth-order
C Runge-Kutta integration, given the derivatives K1(6,6) at
C the starting radius R.
C
      IMPLICIT DOUBLE PRECISION(A-H,O-Z)
      DIMENSION RM(100),RHOM(100),GZEROM(100),C(100,100),
     1 Y(6,6),YY(6,6),A(6,6)
      DOUBLE PRECISION MUM(100),LAMBDAM(100),MU,LAMBDA,
     1 K1(6,6),K2(6,6),K3(6,6),K4(6,6)
C Set half stepsize, one sixth stepsize.
      HH=0.5D0*H
      H6=H/6.D0
C Set number of solutions for shell.
      NS=6
C If region is inner core, reset number of solutions to three.
      IF(IR.EQ.1)NS=3
C Complete remaining steps of Runge--Kutta integration.
      DO 10 I=1,6
        DO 11 J=1,NS
          YY(I,J)=Y(I,J)+HH*K1(I,J)
 11     CONTINUE
 10   CONTINUE
C Increment radius and solve for new derivatives.
      R=R+HH
      CALL YPRIME(R,YY,A,K2,N1,C,RM,RHOM,MUM,LAMBDAM,GZEROM,N,PI,G,
     1 IR,RMIN,ANGS,COR,WES)
      DO 12 I=1,6
        DO 13 J=1,NS
          YY(I,J)=Y(I,J)+HH*K2(I,J)
 13     CONTINUE
 12   CONTINUE
C Find new derivatives.
      CALL YPRIME(R,YY,A,K3,N1,C,RM,RHOM,MUM,LAMBDAM,GZEROM,N,PI,G,
     1 IR,RMIN,ANGS,COR,WES)
      DO 14 I=1,6
        DO 15 J=1,NS
          YY(I,J)=Y(I,J)+H*K3(I,J)
 15     CONTINUE
 14   CONTINUE
C Increment radius and solve for new derivatives.
      R=R+HH
      CALL YPRIME(R,YY,A,K4,N1,C,RM,RHOM,MUM,LAMBDAM,GZEROM,N,PI,G,
     1 IR,RMIN,ANGS,COR,WES)
      DO 16 I=1,6
        DO 17 J=1,NS
          Y(I,J)=Y(I,J)+H6*(K1(I,J)+2.D0*(K2(I,J)+K3(I,J))+K4(I,J))
 17     CONTINUE
 16   CONTINUE
      RETURN
      END
```

Calculation of the fundamental solutions requires specification of the Newtonian constant of gravitation, G, as well as the mean rate of rotation of the Earth, Ω. A recommended value for the former has been given by the Committee on Data for Science and Technology (CODATA) for 2006 as $G = 6.67428 \times 10^{-11}\ \mathrm{m^3\,kg^{-1}\,s^{-2}}$ (Mohr *et al.*, 2008, pp. 688–691). The mean rate of rotation of the Earth is specified by the World Geodetic System 1984 (WGS84) as $\Omega = 7.292115 \times 10^{-5}$ rad s^{-1} (Hofmann-Wellenhof and Moritz, 2006, p. 88).

The programme ICFS calculates the free solutions in the inner core, regular at the geocentre, for a given Earth model. Of course, the solutions are only meaningful for $n \geq |m|$. The filename for the Earth model is typed in on request. This can be one of the four models listed in Appendix C, or one in a file with the appropriate format. The Earth model is interpolated by cubic splines with calls to the subroutines SPMAT and INTPL, described in Section 1.6. Power series expansions of the free solutions, regular at the geocentre, are performed with the assistance of the double precision function subprogrammes, P1, P2, Q1 and Q2, as well as the subroutine MATRIX, and the subroutine LINSOL, described in Section 1.5. The subroutine REL tracks the relative error in both the power series expansions and the Runge–Kutta integration. Derivatives of the propagator matrix, required for the Runge–Kutta integration, are calculated by the subroutine YPRIME, described in Section 3.6, and the Runge–Kutta integration itself is performed by the subroutine RK4. The programme ICFS outputs the first three terms of the power series expansions of the solutions regular at the geocentre, at radius RMIN, in the file fsolns.dat. For $n > 0$, there are three regular solutions; for $n = 0$, only the third solution is non-zero, with terms in y_1, y_2 and y_5. The fundamental free solutions throughout the inner core are given in the output files, fs1.dat, fs2.dat and fs3.dat. We list the programme ICFS below.

```
C     PROGRAMME ICFS.FOR
C
C ICFS.FOR calculates the three fundamental solutions of
C the sixth-order spheroidal system, regular at the geocentre, for n>0.
C For n=0, it calculates the one solution of the fourth-order
C spheroidal system, regular at the geocentre, other than
C the trivial solution y5 constant. Interpolation of Earth model
C properties is made by calls to the subroutines SPMAT and INTPL to
C perform cubic spline interpolation. Power series expansions at
C the geocentre are continued by fourth-order Runge-Kutta integration
C by the subroutine RK4 with calls to the subroutine YPRIME to
C evaluate derivatives at each step.
C
      IMPLICIT DOUBLE PRECISION(A-H,O-Z)
      DIMENSION R(100),RHO(100),GZERO(100),RI(300),RHOI(300),
     1 GZEROI(300),ENAME(10),NM(4),NI(4),B(98,198),C(100,100),
     2 Y(6,6),YSCAL(6,3),YSCAL2(6,3),YA(6),A(6,6),AS(6,6),CAUG(6,13),
     3 RM1(6,3),RM2(6,3),YP(6,6),YIC(6,3),Y1(6,6),Y2(6,6),DELTA(6,3),
     4 FS(6,3)
      DOUBLE PRECISION MU(100),LAMBDA(100),MUI(30),LAMBDAI(300),
     1 K1(6,6)
      CHARACTER*20 EMODEL
C Enter Earth model name.
      WRITE(6,10)
 10   FORMAT(1X,'Type in Earth model file name.')
      READ(5,11)EMODEL
 11   FORMAT(A20)
```

```
C Open Earth model file.
      OPEN(UNIT=1,FILE=EMODEL,STATUS='OLD')
C Open fundamental solutions files.
      OPEN(UNIT=2,FILE='fsolns.dat',STATUS='UNKNOWN')
      OPEN(UNIT=3,FILE='fs1.dat',STATUS='UNKNOWN')
      OPEN(UNIT=4,FILE='fs2.dat',STATUS='UNKNOWN')
      OPEN(UNIT=7,FILE='fs3.dat',STATUS='UNKNOWN')
C Set angular frequency of Earth's rotation (WGS 84).
      WE=7.292115D-5
C Calculate square of angular frequency of Earth's rotation.
      WES=WE*WE
C Set value of pi.
      PI=3.141592653589793D0
C Set value of universal constant of gravitation (CODATA 2006).
      G=6.67428D-11
C Enter period, azimuthal number, inertial and Coriolis switches.
      WRITE(6,12)
 12   FORMAT(1X,'Enter period in hours, azimuthal number and',1X,
     1 'inertial and Coriolis switches'/'(1 for in, 0 for out).')
      READ(5,*)TPER,MAZ,INERT,ICOR
C Convert period to seconds.
      TPER=3600.D0*TPER
C Calculate dimensionless angular frequency.
      SIG=PI/(TPER*WE)
C Calculate square of dimensionless angular frequency.
      SIGS=SIG*SIG
C Compute square of angular frequency.
      ANGS=DFLOAT(INERT)
      ANGS=2.D0*PI*ANGS/TPER
      ANGS=ANGS*ANGS
C Compute Coriolis term.
      COR=2.D0*DFLOAT(MAZ)*DFLOAT(ICOR)
      COR=2.D0*PI*COR*WE/TPER
C Set maximum relative error tolerance for integrations.
      EPS=1.D-5
C Set minimum starting radius for variable Earth properties.
      RMIN=1200.D0
C Set maximum dimensions for interpolation.
      M1=100
      M2=198
      M3=98
C Read in and write out inner core values for Earth model.
C Read in and write out headers.
C Read in Earth model name.
      READ(1,13)(ENAME(I),I=1,10)
 13   FORMAT(10A7)
      WRITE(2,14)(ENAME(I),I=1,10)
      WRITE(6,14)(ENAME(I),I=1,10)
 14   FORMAT(5X,10A7)
C Read additional headers.
      READ(1,15)NN
 15   FORMAT(I10)
C Read in number of model points in each region NM(I)
C and number of integration steps NI(I).
      READ(1,16)(NM(I),NI(I),I=1,4)
 16   FORMAT(8I10)
C Write out Earth properties.
      WRITE(6,17)
 17   FORMAT(/,/,5X,'Radius',6X,'Rho',4X,'Lambda',7X,'Mu',6X,'Gzero')
      WRITE(6,18)
```

```
 18   FORMAT(6X,'(km)',5X,'(gm/cc)',2X,'(kbars)',4X,'(kbars)',2X,
     1 '(cm/sec/sec)',/)
C Read in inner core model.
      N1=NM(1)
      READ(1,19)(R(I),RHO(I),LAMBDA(I),MU(I),GZERO(I),I=1,N1)
 19   FORMAT(1X,F10.1,F10.2,F10.1,F10.1,F10.1)
C Scale inner core model to SI values and store.
      DO 20 I=1,N1
        RI(I)=R(I)*1.D3
        RHOI(I)=RHO(I)*1.D3
        LAMBDAI(I)=LAMBDA(I)*1.D8
        MUI(I)=MU(I)*1.D8
        GZEROI(I)=GZERO(I)*1.D-2
C Write out inner core model to screen.
        WRITE(6,19)R(I),RHO(I),LAMBDA(I),MU(I),GZERO(I)
 20   CONTINUE
C Store radius of the inner core boundary.
      RIOB=RI(N1)
C Set up interpolation for the inner core.
      N1=NM(1)
      N2=2*N1-2
      N3=N1-2
C Put inner core values in active locations.
      DO 21 I=1,N1
        J=I
        R(I)=RI(J)
        RHO(I)=RHOI(J)
        MU(I)=MUI(J)
        LAMBDA(I)=LAMBDAI(J)
        GZERO(I)=GZEROI(J)
 21   CONTINUE
C Calculate gravity gradient at geocentre.
      GAMMA=(4.D0*PI*G*RHO(1)-2.D0*WES)/3.D0
C Set minimum radius for variable Earth properties.
      R(1)=RMIN
C Set gravity at minimum radius.
      GZERO(1)=GAMMA*RMIN
C Construct interpolation matrix.
      CALL SPMAT(N1,N2,N3,C,R,B,M1,M2,M3)
C Enter degree of fundamental solutions to be computed.
      WRITE(6,22)
 22   FORMAT(1X,'Enter degree N of fundamental solutions to',1X,
     1 'be computed'
      READ(5,*)N
      AN=DFLOAT(N)
C Write degree of fundamental solutions.
      WRITE(2,23)N
 23   FORMAT(/,5X,'Degree of computed fundamental solutions =',I3)
C Write period of fundamental solutions.
      WRITE(2,24)TPER
 24   FORMAT(/,5X,'Period (sec.) of computed fundamental solutions',1X,
     1 '=',F9.1)
C Write azimuthal number of fundamental solutions.
      WRITE(2,25)MAZ
 25   FORMAT(/,5X,'Azimuthal number of computed fundamental solutions',
C Test if N=0.
      IF(N.EQ.0) GO TO 26
C Begin power series expansions for fundamental solutions for N>0.
C Set initial values of fundamental solutions.
C Set initial values of first fundamental solution.
```

```
      Y(1,1)=1.D0
      Y(2,1)=2.D0*(AN-1.D0)*MU(1)
      Y(3,1)=1.D0/AN
      Y(4,1)=Y(2,1)/AN
      Y(5,1)=4.D0*PI*G*RHO(1)/AN
      Y(6,1)=0.D0
C Set initial values of second fundamental solution.
      Y(1,2)=0.D0
      Y(2,2)=0.D0
      Y(3,2)=0.D0
      Y(4,2)=0.D0
      Y(5,2)=1.D0/AN
      Y(6,2)=1.D0
C Set initial values of third fundamental solution.
      Y(3,3)=P2(AN,LAMBDA(1),MU(1))/P1(AN,LAMBDA(1),MU(1))
      Y(2,3)=-Q1(AN,LAMBDA(1),MU(1))*Y(3,3)+Q2(AN,LAMBDA(1),MU(1))
      Y(1,3)=-AN*Y(3,3)+1.D0/MU(1)
      Y(4,3)=1.D0
      Y(5,3)=2.D0*PI*G*RHO(1)*((AN+3.D0)*Y(1,3)-AN*(AN+1.D0)*Y(3,3))/
     1 (2.D0*AN+3.D0)
      Y(6,3)=(AN+2.D0)*Y(5,3)-4.D0*PI*G*RHO(1)*Y(1,3)
C Calculate coefficients of second terms in series expansions.
C Calculate coefficients of second terms
C for first fundamental solution.
      YSCAL(3,1)=RHO(1)*((3.D0-AN)*GAMMA+ANGS+2.D0*WES-COR/AN)/
     1 P1(AN,LAMBDA(1),MU(1))
      YSCAL(2,1)=-Q1(AN,LAMBDA(1),MU(1))*YSCAL(3,1)
      YSCAL(1,1)=-AN*YSCAL(3,1)
      YSCAL(4,1)=0.D0
      YSCAL(5,1)=2.D0*PI*G*RHO(1)*((AN+3.D0)*YSCAL(1,1)
     1 -AN*(AN+1.D0)*YSCAL(3,1))/(2.D0*AN+3.D0)
      YSCAL(6,1)=(AN+2.D0)*YSCAL(5,1)-4.D0*PI*G*RHO(1)*YSCAL(1,1)
C Calculate coefficients of second terms
C for second fundamental solution.
      YSCAL(3,2)=RHO(1)/P1(AN,LAMBDA(1),MU(1))
      YSCAL(2,2)=-Q1(AN,LAMBDA(1),MU(1))*YSCAL(3,2)
      YSCAL(1,2)=-AN*YSCAL(3,2)
      YSCAL(4,2)=0.D0
      YSCAL(5,2)=2.D0*PI*G*RHO(1)*((AN+3.D0)*YSCAL(1,2)
     1 -AN*(AN+1.D0)*YSCAL(3,2))/(2.D0*AN+3.D0)
      YSCAL(6,2)=(AN+2.D0)*YSCAL(5,2)-4.D0*PI*G*RHO(1)*YSCAL(1,2)
C Generate matrix for evaluation of second terms in power series
C expansion of third fundamental solution.
      CALL MATRIX(A,RHO,LAMBDA,MU,PI,G,AN,WE)
C Set right-hand side for calculation of second terms
C in third solution.
      YA(1)=0.D0
      YA(2)=RHO(1)*(-(4.D0*GAMMA+ANGS+2.D0*WES)*Y(1,3)
     1 +(AN*(AN+1.D0)*GAMMA+COR)*Y(3,3)-Y(6,3))
      YA(3)=0.D0
      YA(4)=RHO(1)*((GAMMA+COR/(AN*(AN+1.D0)))*Y(1,3)
     1 -(ANGS-COR/(AN*(AN+1.D0)))*Y(3,3)-Y(5,3))
      YA(5)=0.D0
      YA(6)=0.D0
C Store matrix A in AS.
      DO 27 I=1,6
        DO 28 J=1,6
          AS(I,J)=A(I,J)
28      CONTINUE
27    CONTINUE
```

```
C Solve for second terms in third solution.
      CALL LINSOL(A,YA,6,CAUG,DET,6,13)
      YSCAL(1,3)=YA(1)
      YSCAL(2,3)=YA(2)
      YSCAL(3,3)=YA(3)
      YSCAL(4,3)=YA(4)
      YSCAL(5,3)=YA(5)
      YSCAL(6,3)=YA(6)
C Set right-hand side for calculation of third terms
C in first solution.
      YA(1)=0.D0
      YA(2)=RHO(1)*(-(4.D0*GAMMA+ANGS+2.D0*WES)*YSCAL(1,1)
     1 +(AN*(AN+1.D0)*GAMMA+COR)*YSCAL(3,1)-YSCAL(6,1))
      YA(3)=0.D0
      YA(4)=RHO(1)*((GAMMA+COR/(AN*(AN+1.D0)))*YSCAL(1,1)
     1 -(ANGS-COR/(AN*(AN+1.D0)))*YSCAL(3,1)-YSCAL(5,1))
      YA(5)=0.D0
      YA(6)=0.D0
C Copy stored matrix AS into matrix A.
      DO 29 I=1,6
        DO 30 J=1,6
          A(I,J)=AS(I,J)
 30     CONTINUE
 29   CONTINUE
C Solve for third terms in first solution.
      CALL LINSOL(A,YA,6,CAUG,DET,6,13)
      YSCAL2(1,1)=YA(1)
      YSCAL2(2,1)=YA(2)
      YSCAL2(3,1)=YA(3)
      YSCAL2(4,1)=YA(4)
      YSCAL2(5,1)=YA(5)
      YSCAL2(6,1)=YA(6)
C Set right-hand side for calculation of third terms
C in second solution.
      YA(1)=0.D0
      YA(2)=RHO(1)*(-(4.D0*GAMMA+ANGS+2.D0*WES)*YSCAL(1,2)
     1 +(AN*(AN+1.D0)*GAMMA+COR)*YSCAL(3,2)-YSCAL(6,2))
      YA(3)=0.D0
      YA(4)=RHO(1)*((GAMMA+COR/(AN*(AN+1.D0)))*YSCAL(1,2)
     1 -(ANGS-COR/(AN*(AN+1.D0)))*YSCAL(3,2)-YSCAL(5,2))
      YA(5)=0.D0
      YA(6)=0.D0
C Copy stored matrix AS into matrix A.
      DO 31 I=1,6
        DO 32 J=1,6
          A(I,J)=AS(I,J)
 32     CONTINUE
 31   CONTINUE
C Solve for third terms in second solution.
      CALL LINSOL(A,YA,6,CAUG,DET,6,13)
      YSCAL2(1,2)=YA(1)
      YSCAL2(2,2)=YA(2)
      YSCAL2(3,2)=YA(3)
      YSCAL2(4,2)=YA(4)
      YSCAL2(5,2)=YA(5)
      YSCAL2(6,2)=YA(6)
C Set right-hand side for calculation of third terms
C in third solution.
      YA(1)=0.D0
      YA(2)=RHO(1)*(-(4.D0*GAMMA+ANGS+2.D0*WES)*YSCAL(1,3)
```

```
     1 +(AN*(AN+1.D0)*GAMMA+COR)*YSCAL(3,3)-YSCAL(6,3))
      YA(3)=0.D0
      YA(4)=RHO(1)*((GAMMA+COR/(AN*(AN+1.D0)))*YSCAL(1,3)
     1 -(ANGS-COR/(AN*(AN+1.D0)))*YSCAL(3,3)-YSCAL(5,3))
      YA(5)=0.D0
      YA(6)=0.D0
C Copy stored matrix AS into matrix A and add 2.0 to diagonal of A.
      DO 33 I=1,6
        DO 34 J=1,6
          A(I,J)=AS(I,J)
          IF(I.EQ.J)A(I,J)=A(I,J)+2.D0
 34     CONTINUE
 33   CONTINUE
C Solve for third terms in third solution.
      CALL LINSOL(A,YA,6,CAUG,DET,6,13)
      YSCAL2(1,3)=YA(1)
      YSCAL2(2,3)=YA(2)
      YSCAL2(3,3)=YA(3)
      YSCAL2(4,3)=YA(4)
      YSCAL2(5,3)=YA(5)
      YSCAL2(6,3)=YA(6)
      GO TO 35
C Begin power series expansion of fundamental solution
C for degree N=0.
 26   CONTINUE
C Set first two solutions to zero.
      DO 36 I=1,6
        DO 37 J=1,2
          Y(I,J)=0.D0
          YSCAL(I,J)=0.D0
          YSCAL2(I,J)=0.D0
 37     CONTINUE
 36   CONTINUE
C Set values for first three terms of fundamental solution.
      Y(1,3)=1.D0
      Y(2,3)=3.D0*LAMBDA(1)+2.D0*MU(1)
      Y(3,3)=0.D0
      Y(4,3)=0.D0
      Y(5,3)=2.D0*PI*G*RHO(1)
      Y(6,3)=0.D0
      YSCAL(1,3)=-RHO(1)*(4.D0*GAMMA+ANGS+2.D0*WES)/
     1 (10.D0*(LAMBDA(1)+2.D0*MU(1)))
      YSCAL(2,3)=(5.D0*LAMBDA(1)+6.D0*MU(1))*YSCAL(1,3)
      YSCAL(3,3)=0.D0
      YSCAL(4,3)=0.D0
      YSCAL(5,3)=PI*G*RHO(1)*YSCAL(1,3)
      YSCAL(6,3)=0.D0
      YSCAL2(1,3)=-RHO(1)*(4.D0*GAMMA+ANGS+2.D0*WES)*YSCAL(1,3)/
     1 (4.D0*(7.D0*LAMBDA(1)+14.D0*MU(1)))
      YSCAL2(2,3)=(7.D0*LAMBDA(1)+10.D0*MU(1))*YSCAL2(1,3)
      YSCAL2(3,3)=0.D0
      YSCAL2(4,3)=0.D0
      YSCAL2(5,3)=2.D0*PI*G*RHO(1)*YSCAL2(1,3)/3.D0
      YSCAL2(6,3)=0.D0
 35   CONTINUE
C Calculate terms in power series expansions at minimum radius RMIN.
      XV=RMIN
      XVS=XV*XV
      XV3=VX*XVS
      XV4=XVS*XVS
```

```
      DO 38 I=1,6
        DO 39 J=1,3
          RM1(I,J)=XVS*YSCAL(I,J)
          RM2(I,J)=XV4*YSCAL2(I,J)
 39     CONTINUE
 38   CONTINUE
C Write out terms in power series expansions.
      DO 40 J=1,3
        WRITE(2,41)J
 41     FORMAT(/,5X,'Fundamental solution number ',I3)
        WRITE(2,42)
 42     FORMAT(/,6X,'first term',10X,'second term',9X,'third term')
        WRITE(2,43)(Y(I,J),RM1(I,J),RM2(I,J),I=1,6)
 43     FORMAT(5X,D15.8,5X,D15.8,5X,D15.8)
 40   CONTINUE
C Set last three columns of arrays Y(6,6) and YP(6,6) to zero
C for inner core.
      DO 44 I=1,6
        DO 45 J=4,6
          Y(I,J)=0.D0
          YP(I,J)=0.D0
 45     CONTINUE
 44   CONTINUE
C Put inner core solutions in array YIC(6,3)
C and free solutions array FS(6,3).
      DO 46 I=1,6
        DO 47 J=1,3
          YIC(I,J)=Y(I,J)
          FS(I,J)=Y(I,J)
 47     CONTINUE
 46   CONTINUE
C Write out starting values of free solutions.
      RAD=XV*1.D-3
      WRITE(3,48)RAD,(FS(I,1),I=1,6)
      WRITE(4,48)RAD,(FS(I,2),I=1,6)
      WRITE(7,48)RAD,(FS(I,3),I=1,6)
 48   FORMAT(F7.2,6D14.6)
C Begin Runge-Kutta integration for the inner core.
C Find maximum ratio of second coefficient to first term
C in power series expansions.
      CALL REL(ERRMAX,YSCAL,YIC,N)
C Set initial stepsize so that fourth-order method would have
C relative error bound EPS.
      H=(EPS/ERRMAX)**(0.2D0)
C Find solutions and derivatives at minimum radius RMIN
C by power series expansions.
      DO 49 I=1,6
        DO 50 J=1,3
C Find solutions.
          Y(I,J)=Y(I,J)+XVS*YSCAL(I,J)+XV4*YSCAL2(I,J)
C Find derivatives.
          YP(I,J)=2.D0*XV*YSCAL(I,J)+4.D0*XV3*YSCAL2(I,J)
 50     CONTINUE
 49   CONTINUE
C Set inner core index.
      IR=1
 51   CONTINUE
C Calculate derivatives at beginning of current step.
      CALL YPRIME(XV,Y,A,K1,N1,C,R,RHO,MU,LAMBDA,GZERO,N,PI,G,
     1 IR,RMIN,ANGS,COR,WES)
```

```
C Initialize function values at beginning of integration step.
      DO 52 I=1,6
        DO 53 J=1,6
          Y1(I,J)=Y(I,J)
          Y2(I,J)=Y(I,J)
 53     CONTINUE
 52   CONTINUE
C Integrate solution forward one full step.
      CALL RK4(XV,Y1,A,K1,N1,C,R,RHO,MU,LAMBDA,GZERO,N,PI,G,H,IR,
     1 RMIN,ANGS,COR,WES)
C Integrate solution forward two half steps.
C Halve step.
      HH=0.5D0*H
C Reset radius.
      XV=XV-H
      CALL RK4(XV,Y2,A,K1,N1,C,R,RHO,MU,LAMBDA,GZERO,N,PI,G,HH,IR,
     1 RMIN,ANGS,COR,WES)
      CALL YPRIME(XV,Y2,A,K1,N1,C,R,RHO,MU,LAMBDA,GZERO,N,PI,G,IR,
     1 RMIN,ANGS,COR,WES)
      CALL RK4(XV,Y2,A,K1,N1,C,R,RHO,MU,LAMBDA,GZERO,N,PI,G,HH,IR,
     1 RMIN,ANGS,COR,WES)
C Find error matrix and maximum relative error.
      DO 54 I=1,6
        DO 55 J=1,3
          DELTA(I,J)=Y1(I,J)-Y2(I,J)
 55     CONTINUE
 54   CONTINUE
C Copy Y2(6,6) into inner core solution array YIC(6,3).
      DO 56 I=1,6
        DO 57 J=1,3
          YIC(I,J)=Y2(I,J)
 57     CONTINUE
 56   CONTINUE
C Find maximum relative error.
      CALL REL(ERRMAX,DELTA,YIC,N)
C Scale maximum relative error by specified relative error tolerance.
      ERRMAX=ERRMAX/EPS
C Test if step was successful.
      IF(ERRMAX.LT.1.D0) GO TO 58
C Step too large, reduce stepsize and repeat.
      XV=XV-H
      H=0.9D0*H*(ERRMAX**(-0.25D0))
      GO TO 51
 58   CONTINUE
C Step successful, accept solution and improve truncation error.
      DO 59 I=1,6
        DO 60 J=1,3
          Y(I,J)=Y2(I,J)+DELTA(I,J)/15.D0
          FS(I,J)=Y(I,J)
 60     CONTINUE
 59   CONTINUE
      RAD=XV*1.D-3
      WRITE(3,48)RAD,(FS(I,1),I=1,6)
      WRITE(4,48)RAD,(FS(I,2),I=1,6)
      WRITE(7,48)RAD,(FS(I,3),I=1,6)
C Increase stepsize.
      H=0.9D0*H*(ERRMAX**(-0.20D0))
C Test if solution is finished.
      XVT=XV+H
      IF(XVT.LT.R(N1)) GO TO 51
```

```
C Find remaining distance to the ICB.
      H=R(N1)-XV
C Compute fraction of radius.
      RATIO=H/R(N1)
C Continue integration only if remaining distance is more than
C 1.D-9 of radius.
      IF(RATIO.GT.1.D-9) GO TO 51
C Convert solutions back to y-variables.
C Set value of alpha for first two solutions.
      ANM2=AN-2.D0
C Convert three solutions.
      DO 61 J=1,3
C Set value of alpha for third solution.
        IF(J.EQ.3) ANM2=AN
C Find constant equal to the radius of the ICB to the power of alpha.
        CONST=RIOB**ANM2
        Y(2,J)=CONST*Y(2,J)
        Y(4,J)=CONST*Y(4,J)
        CONST=CONST*RIOB
        Y(1,J)=CONST*Y(1,J)
        Y(3,J)=CONST*Y(3,J)
        Y(6,J)=CONST*Y(6,J)
        CONST=CONST*RIOB
        Y(5,J)=CONST*Y(5,J)
61    CONTINUE
      END
```

As an illustration of output from the programme ICFS, we show the free solutions, regular at the geocentre, for degree $n = 1$, azimuthal number $m = 1$, at a period of 4 hours. They are generated by the free constants $A_{1,0}$, $A_{6,0}$ and $A_{4,1}$, and are shown in Figures 3.1, 3.2 and 3.3. In each case, the plots are shown for unit value of the respective free constant.

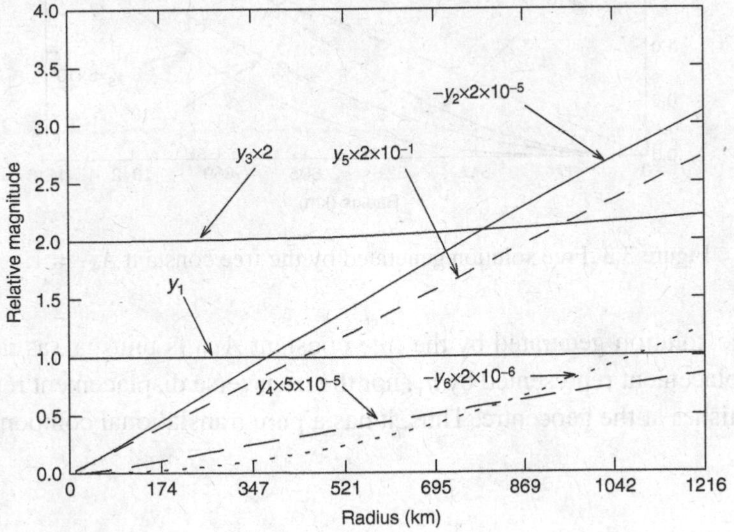

Figure 3.1 Free solution generated by the free constant $A_{1,0} = 1$.

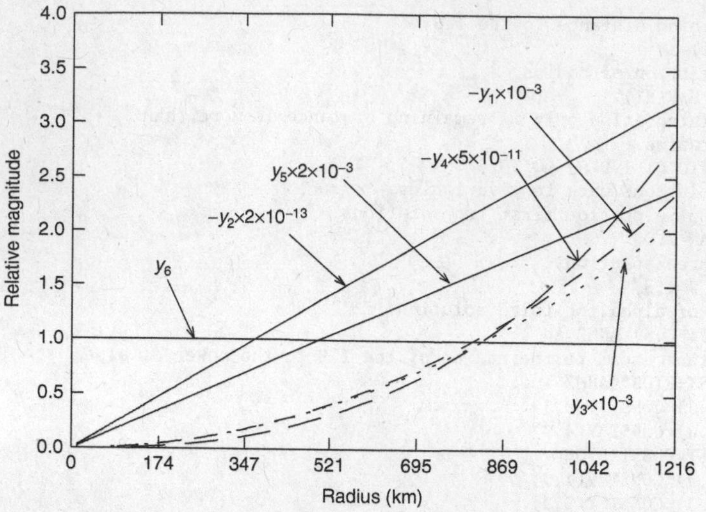

Figure 3.2 Free solution generated by the free constant $A_{6,0} = 1$.

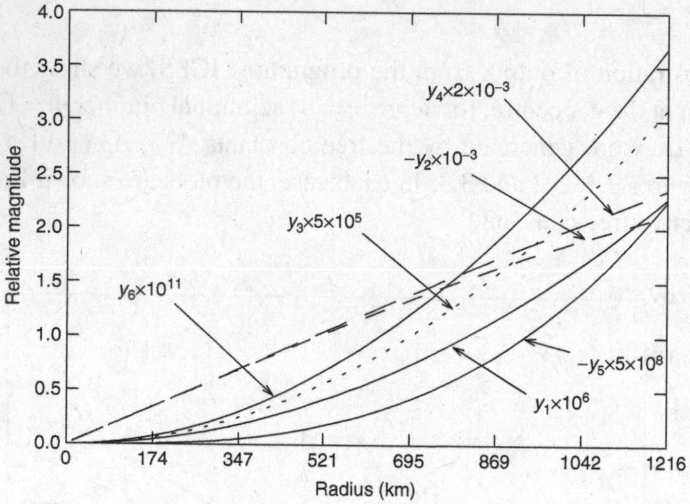

Figure 3.3 Free solution generated by the free constant $A_{4,1} = 1$.

The free solution generated by the free constant $A_{1,0}$ is unusual, as neither the radial displacement represented by y_1, nor the transverse displacement represented by y_3, vanishes at the geocentre. Thus, it has a pure translational component.

4

Earth's rotation: observations and theory

The study of Earth's rotation would not be of much interest if the Earth rotated uniformly about a fixed axis. Variations in the speed of rotation and changes in the orientation of the rotation axis, both within the body of the Earth and in space, make the subject deeply fascinating and rewarding. The subject has a long and interesting history that is well reviewed in the classic treatise *The Rotation of the Earth* (Munk and MacDonald, 1960), which did much to revitalise modern interest. More recently, a second authoritative treatise entitled *The Earth's Variable Rotation*, by Kurt Lambeck, has appeared (Lambeck, 1980), giving a modern overview of the subject.

Observations of the rotation are generally made by observatories attached to the Earth, measuring motions with reference to stars and other celestial objects. Thus, both a terrestrial reference frame and a celestial reference frame need to be defined to make such observations.

4.1 Reference frames

To the lowest order of approximation, observations are made in a rigid, uniformly rotating frame. Of course, the subject is of interest because the actual frame is neither perfectly rigid nor perfectly uniform in its rotation. The observer's frame is usually defined by a prescribed method of adjusting the frame to the mean motion of a set of observatories, in such a way as to approximately minimise the variance of the relative motions over all the observatories. For example, the Bureau International de l'Heure (BIH) defines the 1968 BIH (Guinot and Feissel, 1969) reference system in terms of the latitudes and longitudes assigned to 68 observatories, with each having an assigned weight in latitude and time.

The modern definition of the terrestrial reference frame is given by the International Earth Rotation and Reference Systems Service (IERS) (www.iers.org) and is called the International Terrestrial Reference System (ITRS). This is geocentric, in the sense that it has its origin at the centre of mass of the whole Earth, including the oceans and the atmosphere. Its initial orientation is defined to coincide with that of the BIH at year 1984.0. The time evolution of the orientation is defined by a *no net rotation condition* with respect to horizontal tectonic motions over the whole Earth.

Similarly, the modern definition of the celestial reference frame is given by the IERS and is called the International Celestial Reference System (ICRS). This is based on the co-ordinates of extragalactic radio sources. Its origin is at the barycentre of the solar system. It replaced the previous system, based on the FK5 star catalogue (Fricke *et al.*, 1988), as of 1 January 1998, to provide a reference system with greater accuracy (Feissel and Mignard, 1998).

4.2 Polar motion and wobble

Classically, the Earth's wobble, or the motion of the rotation axis within the Earth, has been measured by observations of the associated latitude variation. Euler had observed in 1765 that a rigid Earth would have a free wobble with period, in sidereal days, near the reciprocal of the dynamical ellipticity, or close to 10 months. The associated latitude variation was confirmed by simultaneous observations at Berlin and Waikiki in 1891. Since these two locations are close to 180° apart in longitude, the variations, as expected, were found to be opposite in phase. Also in 1891, S. C. Chandler, an amateur astronomer, produced his analysis of a series of latitude observations showing that there were two principal components, an annual term and a 14-month variation some 40% longer than Euler's period for a rigid Earth. The lengthened period was quickly shown to be due to the fluidity of the oceans and outer core and the elastic yielding of the solid Earth. This motion is now called the Chandler wobble.

These discoveries led to the establishment of the International Latitude Service (ILS) in 1895, with five stations all operating at latitude 39°08′ N and beginning regular observations in 1899. Details of the early history are given by Munk and MacDonald (1960).

The original ILS stations used visual zenith telescopes (VZTs). The inclusion of additional stations and instruments followed the formation of the International Polar Motion Service (IPMS) as the successor to the ILS in 1962. In the meantime, the Bureau International de l'Heure had established an independent set of observatories to monitor polar motion to be applied to correct for the effect of polar motion on Universal Time (UT). These incorporated instruments such as the photographic zenith tube (PZT), having improved stability and measurement accuracy. These developments are described in more detail by Lambeck (1980).

The co-ordination of polar motion observations using modern space measurement techniques, such as very long baseline interferometry (VLBI), lunar laser

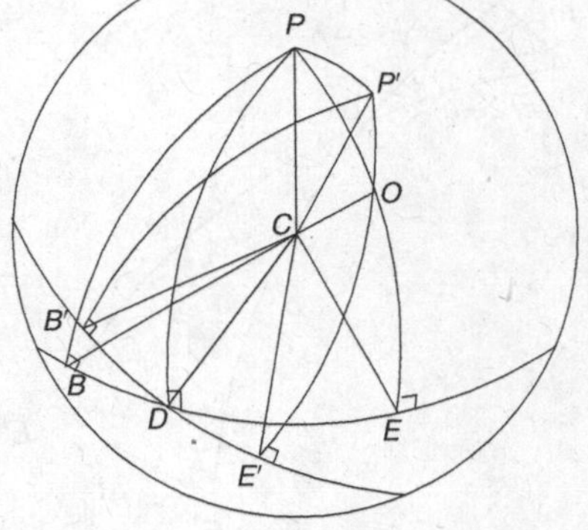

Figure 4.1 The effects of polar motion on the sidereal time and latitude of an observatory.

ranging (LLR) and satellite laser ranging (SLR), began with the formation of the International Earth Rotation Service (IERS) in 1987 as the successor to both the BIH and IPMS. In contrast with the classical astronometric techniques used by the IPMS and BIH, which often yielded pole positions differing by as much as ten centiseconds of arc (one centisecond of arc is close to one foot, or 30 cm, of pole displacement), modern space measurement techniques yield pole positions with error levels decreased by at least three orders of magnitude.

Polar motion changes both the sidereal time and latitude of an observer. These effects of a shift in the pole of rotation are illustrated in Figure 4.1. C is the point at the Earth's centre of mass and O is the position of an observatory on the surface. The meridian circle, through the initial pole position P and the observatory, intersects the equator at the point E, while B is the point on the equator directly below the vernal equinox. Then, the initial latitude of the observatory is given by the angle OCE, and the initial sidereal time is given by the angle BCE.

If the pole of rotation moves from its initial position P to a new position P' on the surface of the Earth, the equator then pivots about an axis through the surface point D on the equator and the centre of mass C, orthogonal to the plane containing PCP'. After the polar shift, the latitude becomes the angle OCE', measured along the new meridian circle through O, intersecting the new equator at E'. The new sidereal time is given by the angle $B'CE'$, where B' is the point on the new equator directly below the vernal equinox. Changes in the spin rate, or speed of axial rotation, also affect sidereal time.

It is usual to adopt a geocentric Cartesian co-ordinate system in which the origin is placed at the centre of mass, the x_3-axis is taken through a fixed reference pole, the x_1-axis is taken toward the Greenwich meridian at 0° longitude, and the x_2-axis

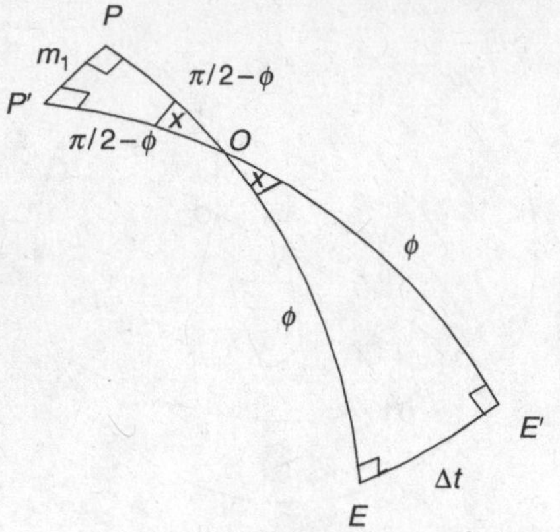

Figure 4.2 The effect of the $m_\perp$ component of polar motion on sidereal time.

is taken toward 90° E longitude. The angular velocity vector of the Earth is then written as

$$\Omega(\boldsymbol{m} + \hat{\mathbf{e}}_3), \tag{4.1}$$

where $\Omega = 7.292115 \times 10^{-5}$ rad/s is the adopted mean rate of rotation of the Earth and $\hat{\mathbf{e}}_3$ is the unit vector in the x_3 direction. The components m_1 and m_2 of the vector $\boldsymbol{m}$ represent the small angular displacement of the rotation axis from the x_3 reference axis, and m_3 gives the relative change in axial rotation rate from the mean rate Ω. Hence, m_1 represents the angular displacement of the rotation axis in radians, toward 0° longitude, and m_2 represents that toward 90° E longitude.

To find the effects of polar motion on latitude and on sidereal time, it is of interest to express the polar motion in terms of its component along the meridian of the observatory, measured positive in the direction of the observatory,

$$m_{\parallel} = m_1 \cos \lambda + m_2 \sin \lambda, \tag{4.2}$$

and its component orthogonal to the meridian of the observatory, measured positive in the direction of decreasing longitude,

$$m_{\perp} = m_1 \sin \lambda - m_2 \cos \lambda, \tag{4.3}$$

with λ representing the east longitude of the observatory.

The component $m_{\parallel}$ gives an increase in the latitude of the observatory. The component $m_{\perp}$ increases the sidereal time angle as illustrated in Figure 4.2. The polar motion is very small compared with Earth's radius and this component of polar motion does not affect the latitude ϕ. The law of sines for the spherical triangle OPP' then gives

$$\frac{\sin m_{\perp}}{\sin x} = \sin(\pi/2 - \phi) = \cos\phi. \tag{4.4}$$

Since the angles $m_{\perp}$ and x are very small, this expression closely approximates to

$$x = \frac{m_{\perp}}{\cos\phi}. \tag{4.5}$$

The law of sines for the spherical triangle OEE' gives

$$\frac{\sin\Delta t}{\sin x} = \sin\phi, \tag{4.6}$$

where Δt is the increase in the sidereal time angle. Again, the angles Δt and x are very small, and thus to close approximation,

$$\Delta t = m_{\perp}\tan\phi = \left(m_1\sin\lambda - m_2\cos\lambda\right)\tan\phi. \tag{4.7}$$

The raw, instantaneous Universal Time, $UT0_i$, is given by the angle measured eastward along the instantaneous equator from the vernal equinox to the Greenwich meridian, while Universal Time corrected for polar motion is $UT1$. Co-ordinated Universal Time, UTC, is the time kept by atomic clocks. Thus, two observables are

$$UT0_i - UTC = UT1 - UTC + \tan\phi\left(m_1\sin\lambda - m_2\cos\lambda\right), \tag{4.8}$$

and

$$\phi_i - \phi = m_1\cos\lambda + m_2\sin\lambda, \tag{4.9}$$

with ϕ_i representing the instantaneous latitude of the observatory.

Traditionally, the co-ordinates of the pole are given as (x, y) with $x = m_1$ and $y = -m_2$. In addition, the longitude is expressed as the west longitude $\psi = 2\pi - \lambda$. Then, the increase in observed latitude is

$$\phi_i - \phi = x\cos(2\pi - \psi) - y\sin(2\pi - \psi) = x\cos\psi + y\sin\psi, \tag{4.10}$$

and

$$UT0_i - UTC = UT1 - UTC + \tan\phi\left(-x\sin\psi + y\cos\psi\right). \tag{4.11}$$

4.2.1 The VLBI pole path

The pole path determined by the very long baseline interferometry (VLBI) technique is listed on the Goddard Space Flight Center website (http://gemini.gsfc.nasa.gov/solutions/2009a). The downloadable file 2009a.eops contains 4443 pole position co-ordinates together with their standard errors. The time tags for each pole position are given in Julian days minus 2,400,000. The programme PP2009A extracts from the file 2009a.eops the time in days from the beginning of the record, the pole co-ordinates, and their standard errors.

```
C           PROGRAMME PP2009A.FOR
C
C PP2009A.FOR extracts pole co-ordinates and their standard errors
C from the file 2009a.eops on the Goddard Space Flight Center website.
C
      IMPLICIT DOUBLE PRECISION(A-H,O-Z)
      OPEN(UNIT=1,FILE='2009a.eops',STATUS='OLD')
      OPEN(UNIT=2,FILE='pp2009a.dat',STATUS='UNKNOWN')
C Read in extracted data.
      DO 10 I=1,4443
        READ(1,11)DAY,XWOB,YWOB,UT1MC,PSI,EPS,EXWOB,EYWOB,
     1  EUT1MC,SEPSI,SEEPS,WMS,C1,C2,C3,C4,N,SC,SD,RX,RY,
     2  DLOD,F1,F2,UX,UY,UL,Q1,Q2,P1,P2
 11     FORMAT(1X,F12.6,2F9.6,F11.7,2F9.3,2F9.6,F10.7,2F8.3,
     1  F8.2,4F7.4,I7,A7,F6.2,2F10.6,F11.7,2A3,2F10.6,F11.7,2A3,2A33)
C Set time origin to zero.
        DAY=DAY-44089.993750D0
C Write out extracted data.
        WRITE(2,12)I,DAY,XWOB,EXWOB,YWOB,EYWOB
 12     FORMAT(1X,I4,F16.6,4F10.6)
 10   CONTINUE
      END
```

Tests reveal that there are 346 null values in the extracted file pp2009a.dat, leaving 4097 pole positions and their standard errors. Of these, 537 pairs were found to have identical time tags, or time tags differing by less than one day. These were averaged along with their time tags, leaving a total of 3560 pole position co-ordinates and their standard errors in the file pp2009a.dat.

Traditionally, a left-handed co-ordinate system is used, with the x_2 co-ordinate toward 90° W. Values are given in seconds of arc subtended at the geocentre. We convert to a right-handed co-ordinate system with the x_1 co-ordinate toward Greenwich and the x_2 co-ordinate toward 90° E. We also convert the pole co-ordinates to centiseconds of arc subtended at the geocentre. Each pole position is represented by the complex quantity $x_1 + ix_2$. The time base is converted to years from the beginning of the record by dividing by 365.25636, the number of mean solar days in one year (Mueller, 1969). It is also useful to have the pole position dates in calendar days. The conversion from Julian days to calendar days is accomplished by the subroutine CALDAT(JULIAN,MM,ID,IYYY) adapted from Press *et al.* (1992, p. 16).

With the Julian day input as JULIAN, the subroutine outputs the corresponding year, month and day as IYYY, MM, ID, given that a specified Julian day starts at noon on a calendar day. All of these conversions are applied by the programme PPCY2009A listed below. The output file ppy2009a.dat contains 3560 pole positions and their standard errors running from 3 August 1979 to 22 October 2009.

```
C            PROGRAMME PPCY2009A.FOR
C
C PPCY2009A.FOR converts pole co-ordinates to centiseconds of arc subtended
C at the geocentre, reverses the sign of the x2 co-ordinate to change
C to a right-handed co-ordinate system, changes the time base to years
C and converts Julian days to calendar days. Input is pp2009a.dat,
C extracted from the file 2009a.eops on the GSFC website with null values
C eliminated and time values differing by less than a day averaged.
C Output is ppy2009a.dat.
      IMPLICIT DOUBLE PRECISION(A-H,O-Z)
      OPEN(UNIT=1,FILE='pp2009a.dat',STATUS='OLD')
      OPEN(UNIT=2,FILE='ppy2009a.dat',STATUS='UNKNOWN')
C Read in VLBI pole co-ordinates with null values eliminated and values
C with identical time tags averaged.
      DO 10 I=1,3560
        READ(1,11)J,DAY,XWOB,EXWOB,YWOB,EYWOB
 11     FORMAT(1X,I4,F16.6,4F10.6)
C Change Julian days from beginning of the record to years.
        TOLD=DAY/365.25636D0
C Change sign of x2 co-ordinate, scale up to centiseconds of arc.
        XWOB=100.D0*XWOB
        EXWOB=100.D0*EXWOB
        YWOB=-100.D0*YWOB
        EYWOB=100.D0*EYWOB
C Change Julian days to calendar days as year, month, day.
C Add initial Julian day to Julian days from beginning of the record.
        DAY=DAY+44089.993750D0+2400000.D0
        IDAY=INT(DAY)
        CALL CALDAT(IDAY,MM,ID,IYYY)
C Write out converted times and co-ordinates.
        WRITE(2,12)J,IYYY,MM,ID,TOLD,XWOB,YWOB,EXWOB,EYWOB
 12     FORMAT(1X,4I5,F10.6,4F11.6)
 10   CONTINUE
      END
```

As an illustration of the converted pole path, we show a plot of the pole path for the year 2004 in Figure 4.3. The standard error bars inflated by a factor of 100 are overplotted to show estimated uncertainties.

4.2.2 Spectral analysis of the VLBI pole path

The co-ordinates of the VLBI pole path are given at unequally spaced times t_j with accompanying standard errors. To calculate the discrete Fourier transform (DFT) of a segment of the VLBI pole path we then apply the methods of Section 2.3.5. If the sequence of complex pole positions is $g_j = x_{1j} + ix_{2j}$, $j = 1, \ldots, L$, it is represented by the sum (2.178) of $2N+1$ complex sinusoids,

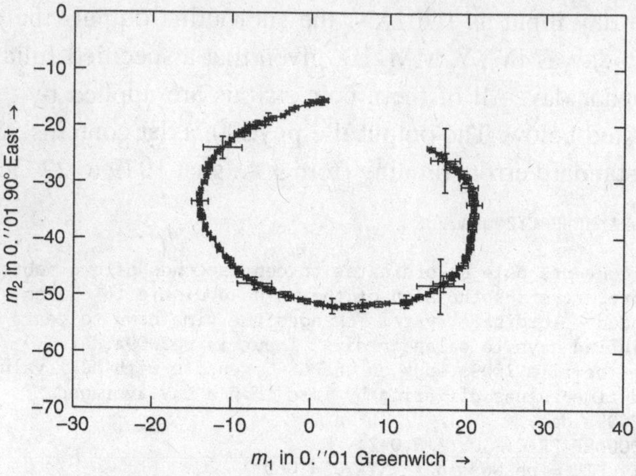

Figure 4.3 The pole path for the year 2004 determined by the VLBI technique. Standard error bars, inflated by a factor of 100 for clarity, are overplotted.

$$g'_j = \frac{1}{M} \sum_{k=-N}^{N} G_k e^{i2\pi(k/M)t_j}, \tag{4.12}$$

where M is the length of the record segment and G_k is the DFT sequence. The DFT is found by least squares adjustment of g_j to g'_j weighting by the inverse of the square of the standard error σ_j. The conditional equations for the DFT sequence (2.187) have a Toeplitz coefficient matrix with elements

$$C_m = \sum_{j=1}^{L} \frac{1}{\sigma_j^2} e^{-i2\pi(m/M)t_j}, \tag{4.13}$$

while the components of the right-hand side vector are

$$d_l = M \sum_{j=1}^{L} \frac{g_j}{\sigma_j^2} e^{-i2\pi(l/M)t_j} \tag{4.14}$$

with the sums over the L sample points.

The conditional equations for the DFT for unequally spaced time sequences can be solved by the methods of Section 2.3.6. The coefficient matrix is first subjected to singular value decomposition (SVD). Singular values are then eliminated, starting with the smallest and working upward, until Parseval's relation is satisfied as closely as possible. The method is quite general, although here we consider its application to VLBI observations of polar motion and nutation.

The programme SVDDFT finds the DFT of an unequally spaced time sequence. The name of the sequence file is entered, as is the name of the output file for the DFT. The first and last point numbers of the data segment are entered, as are the length and midpoint of the segment. The data are reduced to zero mean and Parzen windowed to suppress finite record effects by a call to the subroutine WINDOW. The singular value decomposition is accomplished by the subroutine SVD with calls to the subroutines BIDIAG and HHOLDER. The number of singular values eliminated (NSVE) to satisfy Parseval's theorem as closely as possible is found by iteration. The programme is quite general and could be used to find the DFT of any unequally spaced sequence with suitable adjustment of the input FORMAT statement.

```
          PROGRAMME SVDDFT.FOR
C
C SVDDFT.FOR finds the DFT of an unequally spaced time sequence of VLBI
C polar motion or nutation observations using singular value decomposition
C (SVD) combined with Parseval's theorem to determine the number of
C singular values to eliminate (NSVE). The sequence is reduced to zero mean,
C and it is then Parzen windowed by a call to the subroutine WINDOW
C to suppress finite record effects. The SVD is found by a call to
C the subroutine SVD which in turn calls the subroutines HHOLDER and BIDIAG.
C The programme allows iteration to find the optimum value of NSVE
C to satisfy Parseval's theorem.
C
      IMPLICIT DOUBLE COMPLEX(A-H,O-Z)
      DOUBLE PRECISION SIGS(4000),DAY(4000),S(4000),XWOB,YWOB,EXWOB,
     1 EYWOB,SVL,T,T0,DT,AM,AN,AJ1,AJ2,ARG,ARG1,ARG2,RMS,SUMX,SUMY,
     2 AAI,ERROR,AJ,FREQ,W,PI,SUMS,OMEGA,TD
      DIMENSION G(4000),GP(4000),A(4000,4000),SV(4000),U(4000,4000),
     1 US(4000,4000),VH(4000,4000),B(4000)
      CHARACTER*20 INSEQ,DFTOUT
C Type in name of input sequence file.
      WRITE(6,10)
 10   FORMAT(1X,'Type in name of input sequence file.')
      READ(5,11)INSEQ
 11   FORMAT(A20)
C Type in name of output DFT file.
      WRITE(6,12)
 12   FORMAT(1X,'Type in name of output DFT file.')
      READ(5,11)DFTOUT
      OPEN(UNIT=1,FILE=INSEQ,STATUS='OLD')
      OPEN(UNIT=2,FILE='err.dat',STATUS='UNKNOWN')
      OPEN(UNIT=3,FILE=DFTOUT,STATUS='UNKNOWN')
      OPEN(UNIT=4,FILE='sv.dat',STATUS='UNKNOWN')
      OPEN(UNIT=7,FILE='rec.dat',STATUS='UNKNOWN')
C Set values of pi and 2pi.
      PI=3.141592653589793D0
      OMEGA=2.D0*PI
C Enter segment length and time at segment midpoint, from beginning of record.
      WRITE(6,13)
 13   FORMAT(1X,'Enter segment length and time at segment midpoint.')
      READ(5,*)T,T0
C Enter first and last point numbers of segment data set.
      WRITE(6,14)
 14   FORMAT(1X,'Enter first and last point numbers of segment.')
      READ(5,*)IB,IE
```

```
C Calculate length of data set N.
      N=IE-IB+1
C Find length of DFT, M, an odd integer less than or equal to N.
      M=N
      NB2=N/2
      IF(N.EQ.2*NB2)M=N-1
C Find number of positive harmonics.
      MP=(M-1)/2
C Find equivalent time sample interval.
      AM=DFLOAT(M)
      DT=T/AM
C Read in unused initial record segment.
      IBM1=IB-1
      DO 15 I=1,IBM1
        READ(1,16)K,IYYY,MM,ID,TD,XWOB,YWOB,EXWOB,EYWOB
 16     FORMAT(1X,4I5,F10.6,4F11.6)
 15   CONTINUE
C Set sums of real and imaginary parts of co-ordinates to zero.
      SUMX=0.D0
      SUMY=0.D0
C Read in segment data set.
      DO 17 I=1,N
        READ(1,16)K,IYYY,MM,ID,DAY(I),XWOB,YWOB,EXWOB,EYWOB
C Accumulate sums of real and imaginary parts of co-ordinates.
        SUMX=SUMX+XWOB
        SUMY=SUMY+YWOB
C Convert co-ordinates to complex form.
        G(I)=DCMPLX(XWOB,YWOB)
C Find variance.
        SIGS(I)=EXWOB*EXWOB+EYWOB*EYWOB
 17   CONTINUE
C Find means of real and imaginary parts of co-ordinates.
      AN=DFLOAT(N)
      SUMX=SUMX/AN
      SUMY=SUMY/AN
C Reduce co-ordinates to zero mean and calculate their sum of squares.
      RMS=0.D0
      DO 18 I=1,N
        G(I)=G(I)-DCMPLX(SUMX,SUMY)
        RMS=RMS+G(I)*DCONJG(G(I))
C Window the data segment with the Parzen window.
        CALL WINDOW(DAY(I),T0,T,W)
        G(I)=G(I)*W
 18   CONTINUE
C Find left side of Parseval's relation.
      SUMS=RMS*DT
C Find root mean square of signal.
      RMS=RMS/AN
      RMS=DSQRT(RMS)
C Construct first column, A(J,1), of Toeplitz conditional equations matrix,
C and right-hand side vector B(J).
      DO 19 J=1,M
        AJ1=DFLOAT(J-1)
        AJ2=DFLOAT(J-MP-1)
        A(J,1)=(0.D0,0.D0)
        B(J)=(0.D0,0.D0)
        DO 20 I=1,N
C Form arguments of complex exponentials.
          ARG=-2.D0*PI*DAY(I)/T
          ARG1=ARG*AJ1
          ARG2=ARG*AJ2
```

```
C Perform sums.
          A(J,1)=A(J,1)+DCMPLX(DCOS(ARG1),DSIN(ARG1))/SIGS(I)
          B(J)=B(J)+G(I)*DCMPLX(DCOS(ARG2),DSIN(ARG2))/SIGS(I)
 20     CONTINUE
        B(J)=T*B(J)
 19   CONTINUE
C Complete construction of conditional equations matrix A.
      DO 21 J=2,M
        DO 23 I=1,M
          K=I-J+1
          L=2-K
          IF(K.LE.0)GO TO 22
          A(I,J)=A(K,1)
          GO TO 23
 22       A(I,J)=DCONJG(A(L,1))
 23     CONTINUE
 21   CONTINUE
C Do singular value decomposition of conditional equations matrix A=U*S*VH.
      CALL SVD(A,U,S,VH,M,4000)
C Multiply the Hermitian transpose of U into right-hand side vector B.
      DO 24 I=1,M
C Accumulate product vector in SV.
        SV(I)=(0.D0,0.D0)
        DO 25 J=1,M
          SV(I)=SV(I)+DCONJG(U(J,I))*B(J)
 25     CONTINUE
 24   CONTINUE
C Replace right-hand side vector B by product vector SV.
      DO 26 I=1,M
        B(I)=SV(I)
 26   CONTINUE
C Write out singular values.
      WRITE(4,27)
 27   FORMAT(1X,'Singular values are:')
      WRITE(4,28)(S(I),I=1,M)
 28   FORMAT(1X,D23.15)
C Begin solution of reduced conditional equations.
C Set number of singular values to be eliminated.
      WRITE(6,29)
 29   FORMAT(1X,'Enter number of singular values to be eliminated.')
      READ(5,*)NSVE
C Find number of remaining singular values.
 30   NMAX=M-NSVE
C Multiply V into the inverse of truncated S.
      DO 31 I=1,M
        DO 32 J=1,M
          IF(J.GT.NMAX)GO TO 33
          US(I,J)=DCONJG(VH(J,I))/S(J)
          GO TO 32
 33       US(I,J)=(0.D0,0.D0)
 32     CONTINUE
 31   CONTINUE
C Multiply result into modified right-hand side to find solution vector SV
C and solution vector length SVL.
      SVL=0.D0
      DO 34 I=1,M
        SV(I)=(0.D0,0.D0)
        DO 35 J=1,M
          SV(I)=SV(I)+US(I,J)*B(J)
 35     CONTINUE
        SVL=SVL+SV(I)*DCONJG(SV(I))
```

```
 34    CONTINUE
C Divide by segment length to get right-hand side of Parseval's relation,
C then divide by left side of Parseval's relation to convert SVL to
C Parseval's ratio.
       SVL=SVL/(T*SUMS)
C Construct approximation, GP, to original data set G, from DFT SV.
       DO 36 J=1,N
         GP(J)=(0.D0,0.D0)
         DO 37 I=1,M
           AAI=DFLOAT(I-MP-1)
           ARG=OMEGA*AAI*DAY(J)/T
           GP(J)=GP(J)+SV(I)*DCMPLX(DCOS(ARG),DSIN(ARG))
 37      CONTINUE
C Normalize reconstructed data set.
         GP(J)=GP(J)/T
 36    CONTINUE
C Write out original and reconstructed data sets and calculate mean square
C error in reconstruction ERROR.
       ERROR=0.D0
       DO 38 J=1,N
         ERRORC=G(J)-GP(J)
         ERROR=ERROR+ERRORC*DCONJG(ERRORC)
         WRITE(7,39)DAY(J),G(J),GP(J)
 39      FORMAT(1X,F12.4,4D15.7)
 38    CONTINUE
C Find mean square error.
       ERROR=ERROR/AN
C Find relative reconstruction error.
       ERROR=DSQRT(ERROR)/RMS
C Write out number of singular values eliminated (NSVE),
C relative reconstruction error (RRE), and Parseval's ratio (PR).
       WRITE(2,40)NSVE,ERROR,SVL
       WRITE(6,40)NSVE,ERROR,SVL
 40    FORMAT(1X,I10,2D15.8)
C Decide whether to try new value of NSVE or to end programme.
       WRITE(6,41)
 41    FORMAT(1X,'Enter 1 for new NSVE or 0 to end.')
       READ(5,*)ISWITCH
       IF(ISWITCH.NE.0)WRITE(6,42)
       IF(ISWITCH.EQ.0)GO TO 43
 42    FORMAT(1X,'Enter new value of NSVE.')
       READ(5,*)NSVE
       GO TO 30
 43    CONTINUE
C Write out recovered DFT.
       DO 44 J=1,M
         AJ=DFLOAT(J-MP-1)
         FREQ=AJ/T
         WRITE(3,45)FREQ,SV(J)
 45      FORMAT(1X,D15.5,2D26.15)
 44    CONTINUE
       END
```

The series of 3560 pole positions and their standard errors, running from 3 August 1979 to 22 October 2009, in the file ppy2009a.dat, spans a record length of 30.218 years. The initial part of the record is very sparsely sampled and we take the record to be analysed to begin at 4.0 years into the record at point number 48. The end of the record is set at 30.25 years, the last sample being at point 3560. The record then has 3513 time points and we use the programme SVDDFT to find the

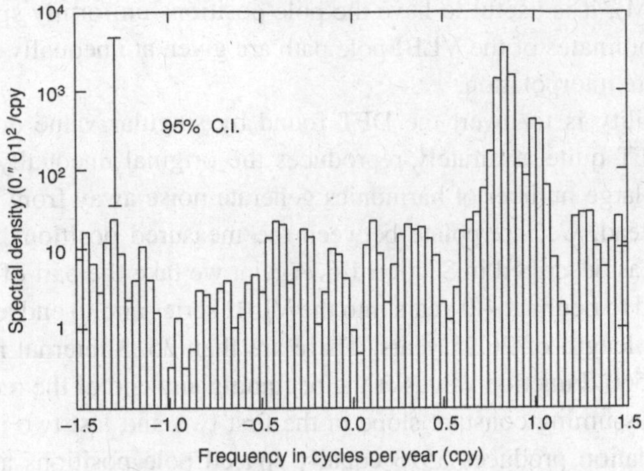

Figure 4.4 Spectral density of the VLBI polar motion based on the DFT found by singular value decomposition (SVD).

discrete Fourier transform series of length 3513. The record segment length is then 26.25 years and the midpoint of the segment is at 17.125 years into the record. The number of singular values eliminated (NSVE) to bring the Parseval ratio, expressed by (2.202), as close as possible to unity, is found to be 320. The relative error in reconstructing the sequence g_j by the representation (4.12), or the relative reconstruction error (RRE), is found to be 0.00081545722 for a Parseval ratio (PR) of 0.69568607.

To maximise frequency resolution we use the single segment of length 26.25 years. The spectral density estimate is then given by the normalised squared magnitude of the discrete Fourier transform (2.350). The spectral density estimate for the single segment is then χ_2^2 distributed with two degrees of freedom (2.355). For two degrees of freedom, the probability density function (2.425) becomes simply $e^{-x/2}/2$, and the cumulative distribution function is $1 - e^{-x/2}$. A fraction α of the realisations will be below the random variable x for $\alpha = 1 - e^{-x/2}$ or $x(\alpha) = -2\log(1-\alpha)$. On logarithmic plots of the spectral density, the confidence interval is the fixed length $\log_{10}[x(1-\alpha/2)] - \log_{10}[x(\alpha/2)]$ (2.355). For a 95% confidence interval the fixed length is $0.8679 + 1.2955 = 2.1634$. The spectral density estimate for the single segment of length 26.25 years is shown plotted in Figure 4.4.

4.2.3 Interpolation of the VLBI pole path

In many applications, such as the use of the fast Fourier transform (FFT) to find the discrete Fourier transform (DFT), or the application of maximum entropy spectral analysis (MEM), it is useful to have the pole positions uniformly spaced in time. Since the co-ordinates of the VLBI pole path are given at unequally spaced times, this will require interpolation.

One possibility is to invert the DFT found by singular value decomposition. While this DFT quite accurately reproduces the original unequally spaced pole positions, the large number of harmonics generate noise away from the measured positions. Instead, we interpolate between the measured positions using natural cubic splines, as described in Section 1.6. Again, we take the part of the record to be interpolated to begin at 4.0 years into the VLBI series and to end at 30.25 years, for a segment length of 26.25 years. There are then 3513 internal nodes at time points 48 to 3560. Function values at the beginning and end of the record segment are found by assuming constant slope in the first two and last two intervals. The spline interpolation produces 3513 equally spaced pole positions at intervals of 26.25/3513 years, or approximately 2.73 days. The programme PPINTP performs natural spline interpolation with calls to the subroutines SPMAT and INTPL. The corresponding calendar dates are found by calls to the subroutine CALDAT. The output file ppintp.dat contains the 3513 equally spaced pole positions as well as their times and calendar dates.

```
C          PROGRAMME PPINTP.FOR
C
C PPINTP.FOR does a natural spline interpolation of the VLBI pole path
C onto 3513 equally spaced time points. The record segment is set
C to begin at 4.0 years into the VLBI series and to end at 30.25 years,
C for a segment length of 26.25 years. There are then 3513 internal nodes
C at time points 48 to 3560. Values at the beginning and end of
C the record segment are found by assuming constant slope in the first two
C and last two intervals. The corresponding calendar dates are found by
C calls to the subroutine CALDAT.
C
      IMPLICIT DOUBLE PRECISION(A-H,O-Z)
      DIMENSION DAY(4000),XWOB(4000),YWOB(4000),C(4000,4000),
     1 B(4000,8000),GTX(4000),GTY(4000)
      OPEN(UNIT=1,FILE='ppy2009a.dat',STATUS='OLD')
      OPEN(UNIT=2,FILE='ppintp.dat',STATUS='UNKNOWN')
C Set number of points, N, and number of interpolates, M.
      N=3513
      M=3513
      AM=DFLOAT(M)
C Set segment length.
      T=26.25D0
C Find time increment.
      DT=T/AM
C Read in unused initial record segment, points 1 to 47.
      DO 10 I=1,47
        READ(1,11)K,IYYY,MM,ID,TD,X,Y,EX,EY
 11     FORMAT(1X,4I5,F10.6,4F11.6)
 10   CONTINUE
C Read in segment to be interpolated, points 48 to 3560.
      DO 12 I=1,N
        IP1=I+1
        READ(1,11)K,IYYY,MM,ID,DAY(IP1),XWOB(IP1),YWOB(IP1),EXWOB,EYWOB
```

```
 12   CONTINUE
C Add nodes at beginning and end by assuming first derivatives are constant
C in first two and last two subintervals.
      DAY(1)=4.D0
C Find slope between second and third nodes.
      SLOPE=(XWOB(3)-XWOB(2))/(DAY(3)-DAY(2))
      XWOB(1)=SLOPE*(DAY(1)-DAY(2))+XWOB(2)
      SLOPE=(YWOB(3)-YWOB(2))/(DAY(3)-DAY(2))
      YWOB(1)=SLOPE*(DAY(1)-DAY(2))+YWOB(2)
      DAY(3515)=30.25D0
C Find slope betwen third last and second last nodes.
      SLOPE=(XWOB(3514)-XWOB(3513))/(DAY(3514)-DAY(3513))
      XWOB(3515)=SLOPE*(DAY(3515)-DAY(3514))+XWOB(3514)
      SLOPE=(YWOB(3514)-YWOB(3513))/(DAY(3514)-DAY(3513))
      YWOB(3515)=SLOPE*(DAY(3515)-DAY(3514))+YWOB(3514)
C Set number of nodes.
      N1=N+1
      N2=2*N1-2
      N3=N1-2
      M1=4000
      M2=8000
      M3=4000
C Construct interpolation matrix C.
      CALL SPMAT(N1,N2,N3,C,DAY,B,M1,M2,M3)
C Find equispaced time sequence.
C Find time from beginning of record.
      TIME=4.D0-DT/2.D0
      DO 13 I=1,M
        TIME=TIME+DT
C Convert time to days from beginning of the record.
        DAYS=365.25636D0*TIME
C Convert to Julian days.
        DAYS=DAYS+44089.993750D0+2400000.D0
C Find calendar date.
        IDAYS=INT(DAYS)
        CALL CALDAT(IDAYS,MM,ID,IYYY)
C Interpolate co-ordinates.
        CALL INTPL(TIME,XX,N1,C,DAY,XWOB,M1)
        CALL INTPL(TIME,YY,N1,C,DAY,YWOB,M1)
        GTX(I)=XX
        GTY(I)=YY
C Write out equispaced time domain sequence.
        WRITE(2,14)I,IYYY,MM,ID,TIME,GTX(I),GTY(I)
 14     FORMAT(1X,4I5,F12.7,2F12.6)
 13   CONTINUE
      END
```

As an example of the interpolated VLBI pole path, we show that for the year 2004 in Figure 4.5. Pole positions, determined by the VLBI technique, are indicated by crosses, and the interpolated path is shown as a continuous curve.

Of interest is the spectral density estimate based on the interpolated VLBI pole path. The file ppintp.dat contains 3513 interpolated pole positions at equally spaced time intervals of approximately 2.73 days. As illustrated in Figure 2.2, the first sample is at half the sample interval from the beginning of the record segment and the last sample is at half the sample interval from the end of the record segment. Once again, the record segment is of length 26.25 years and the spectral

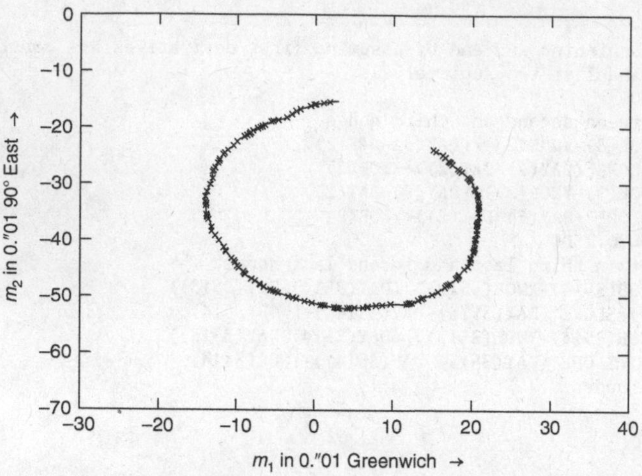

Figure 4.5 Interpolated pole path for 2004. Pole positions measured by the VLBI technique are shown as crosses. Continuous curve is the interpolated path.

density estimate for the single segment is given by the normalised squared magnitude of the discrete Fourier transform (2.350). The programme INTPSPEC reads in the 3513 interpolated pole path positions and reduces them to zero mean. They are then Parzen windowed to suppress finite record effects by a call to the subroutine WINDOW. The generalised DFT (2.150) is found by a call to the subroutine FFTN, which in turn calls the subroutine FFT2N. We use the DFT expressed by (2.148) which is obtained from the generalised DFT through expressions (2.160). The single segment spectral density estimate is given as a function of frequency in the output file intpspec.dat. As before, the confidence interval is of fixed length on logarithmic plots of the spectral density. For a 95% confidence interval the fixed length is 2.1634. A histogram plot of the spectral density is shown in Figure 4.6.

```
C          PROGRAMME INTPSPEC.FOR
C
C INTPSPEC.FOR finds the spectral density estimate for the VLBI pole path,
C interpolated onto 3513 equally spaced time points by natural spline
C interpolation, for a record segment of 26.25 years in length.
C The fast Fourier transform is found by calls to the subroutines FFTN
C and FFT2N and then converted to the DFT, after the sequence is reduced
C to zero mean, and Parzen windowed to prevent finite record effects by
C a call to the subroutine WINDOW. The spectral density is given by
C the normalized squared magnitude of the DFT. The output file intpspec.dat
C gives the spectral density estimate as a function of frequency.
C
      IMPLICIT DOUBLE COMPLEX(A-H,O-Z)
      DOUBLE PRECISION FREQ(4000),SD(4000),DAY(4000),OMEGA,ARG,AI,
     1 PI,T,T0,AN,DT,ANP,W
      DIMENSION GT(4000),WK1(20000),WK2(20000),GF(4000)
      OPEN(UNIT=1,FILE='ppintp.dat',STATUS='OLD')
      OPEN(UNIT=2,FILE='intpspec.dat',STATUS='UNKNOWN')
C Set values of pi and 2pi.
```

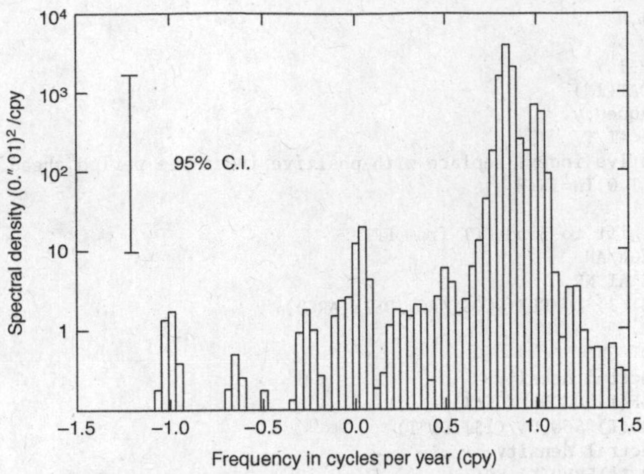

Figure 4.6 Spectral density of the VLBI polar motion, based on the pole path interpolated onto equally spaced time points using natural splines.

```
      PI=3.141592653589793D0
      OMEGA=2.D0*PI
C Set segment length.
      T=26.25D0
C Set midpoint.
      T0=17.125D0
C Set number of interpolates, N.
      N=3513
      AN=DFLOAT(N)
C Find time domain sample interval.
      DT=T/AN
C Find number of positive harmonics.
      NP=(N-1)/2
      ANP=DFLOAT(NP)
C Set mean of pole path to zero.
      GTM=(0.D0,0.D0)
C Read in interpolated pole path.
      DO 10 I=1,N
        READ(1,11)J,IYYY,MM,ID,DAY(I),GT(I)
 11     FORMAT(1X,4I5,F12.7,2F12.6)
C Accumulate sum of co-ordinates.
        GTM=GTM+GT(I)
 10   CONTINUE
C Find mean of pole positions.
      GTM=GTM/AN
C Reduce co-ordinates to zero mean and window sequence.
      DO 12 I=1,N
        GT(I)=GT(I)-GTM
        CALL WINDOW(DAY(I),T0,T,W)
        GT(I)=GT(I)*W
 12   CONTINUE
C Prepare to calculate spectral density.
      IS=-1
C Calculate FFT.
      CALL FFTN(N,GT,IS,L,WK1,WK2)
C Convert to DFT.
```

```
      DO 13 I=1,N
C Find index.
      IN=I-NP-1
      AI=DFLOAT(IN)
C Calculate frequency.
      FREQ(I)=AI/T
C Test for negative index, replace with positive index one period ahead.
      IF(IN.LT.0)IN=IN+N
      J=IN+1
C Apply phase shift to find DFT from FFT.
      ARG=OMEGA/AN
      ARG=ARG*AI*NP
      GF(I)=GT(J)*DCMPLX(DCOS(ARG),DSIN(ARG))
C Normalize DFT.
      GF(I)=DT*GF(I)
C Calculate spectral density.
      SD(I)=GF(I)*DCONJG(GF(I))
      SD(I)=SD(I)*560.D0/(151.D0*T)
C Write out spectral density.
      WRITE(2,14)FREQ(I),SD(I)
 14   FORMAT(1X,D15.5,D26.15)
 13   CONTINUE
      END
```

Both the spectral density estimate of the VLBI pole path found by singular value decomposition (SVD), shown in Figure 4.4, and that found for the interpolated pole path, shown in Figure 4.6, have only the Chandler resonance and the annual line rising above the 95% confidence level. The true spectrum is convolved, or averaged, with the square of the Parzen frequency window, as given in expression (2.343). The half-power points of the Parzen window, in the present case, are separated by a distance of 1.82/26.25 cycles per year, or approximately 0.07 cpy. Thus, the Chandler and annual features are barely separated, suggesting the application of a super-resolution technique, such as the maximum entropy spectral analysis considered in Section 2.6.

4.2.4 Maximum entropy spectral analysis of the VLBI pole path

The maximum entropy spectral density, given by expression (2.470), is proportional to the inverse of the squared magnitude of the discrete Fourier transform (DFT) of the prediction error filter. In turn, the prediction error filter is found using the Burg algorithm described in Section 2.6.3, and implemented by the subroutine BPEC.

The programme MEMSPEC finds the spectral density by the maximum entropy method (MEM), with calls to the subroutine BPEC, and gives it as a function of frequency in cycles per year (cpy) in the output file memspec.dat. The input file is the pole path interpolated by natural splines, ppintp.dat, as generated by the programme PPINTP. The length of the prediction error filter is entered as input on request.

```
C         PROGRAMME MEMSPEC.FOR
C
C MEMSPEC.FOR finds the maximum entropy method (MEM) spectral density for
C the VLBI pole path, interpolated on to 3513 equally spaced time points
C by natural spline interpolation, for a record segment of 26.25 years
C in length. The length of the prediction error filter is entered as input
C on request.
C
      IMPLICIT DOUBLE COMPLEX(A-H,O-Z)
      DIMENSION GT(4000),G(4000),PEF(4000),PER(4000),DFT(4000)
      DOUBLE PRECISION DAY(4000),AM,PEP,PI,OMEGA,T,DT,AIM1,FREQ(4000),
     1 AJ,ARG,SME(4000),DF
      OPEN(UNIT=1,FILE='ppintp.dat',STATUS='OLD')
      OPEN(UNIT=2,FILE='memspec.dat',STATUS='UNKNOWN')
C Set number of interpolated pole positions.
      M=3513
      AM=DFLOAT(M)
C Set values of pi and 2pi.
      PI=3.141592653589793D0
      OMEGA=2.D0*PI
C Set record length.
      T=26.25D0
C Find sample interval.
      DT=T/AM
C Set mean of pole path to zero.
      GTM=(0.D0,0.D0)
C Set prediction error power to zero.
      PEP=0.D0
C Read in interpolated pole path.
      DO 10 I=1,M
        READ(1,11)J,IYYY,IM,ID,DAY(I),GT(I)
 11     FORMAT(1X,4I5,F12.7,2F12.6)
C Accumulate sum of co-ordinates.
        GTM=GTM+GT(I)
 10   CONTINUE
C Find mean of pole positions.
      GTM=GTM/AM
C Reduce co-ordinates to zero mean.
      DO 12 I=1,M
        GT(I)=GT(I)-GTM
C Find sum of squares of pole positions.
        PEP=PEP+GT(I)*DCONJG(GT(I))
 12   CONTINUE
C Find initial value of prediction error power.
      PEP=PEP/AM
C Enter length of prediction error filter.
      WRITE(6,13)
 13   FORMAT(1X,'Enter length of prediction error filter.')
      READ(5,*)K
C Find prediction error filter.
      DO 14 NP1=1,K
        CALL BPEC(M,NP1,GT,G,PEF,PER)
C Update prediction error power.
        PEP=PEP*(1.D0-G(NP1)*DCONJG(G(NP1)))
 14   CONTINUE
C Find DFT of prediction error filter.
      DF=0.01D0
      DO 15 I=1,301
        IM1=I-1
        AIM1=DFLOAT(IM1)
        FREQ(I)=-1.5D0+AIM1*DF
```

```
C Initialize DFT
.        DFT(I)=(1.D0,0.D0)
         DO 16 J=1,K
           AJ=DFLOAT(J)
           ARG=-OMEGA*FREQ(I)
           ARG=ARG*AJ*DT
           DFT(I)=DFT(I)+G(J)*DCMPLX(DCOS(ARG),DSIN(ARG))
 16      CONTINUE
 15    CONTINUE
C Find and write out MEM spectrum.
       DO 17 I=1,301
         SME(I)=DT*PEP/(DFT(I)*DCONJG(DFT(I)))
         WRITE(2,18)FREQ(I),SME(I)
 18      FORMAT(1X,D15.5,D26.15)
 17    CONTINUE
       END
```

The spectral density estimated by the maximum entropy method (MEM), for a prediction error filter of length 500, is shown in Figure 4.7. Both the Chandler resonance and the annual motion are well resolved and there is the suggestion of a weaker retrograde annual motion. Compared with the spectra produced by singular value decomposition, shown in Figure 4.5, and that found for the pole path interpolated by natural splines found by conventional discrete Fourier transform methods, shown in Figure 4.6, the maximum entropy spectrum does indeed show much improved resolution.

A more detailed plot of the Chandler and annual features of the maximum entropy spectral density is shown in Figure 4.8. The annual motion peaks at 0.992 cpy with corresponding period about 3 days longer than a year. Interestingly, the much weaker retrograde annual motion shown in Figure 4.7 peaks at a period a few days shorter than a year.

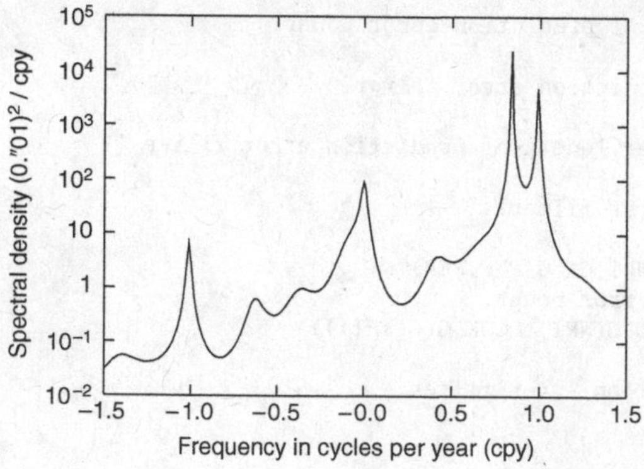

Figure 4.7 The maximum entropy spectral density found by a prediction error filter of length 500.

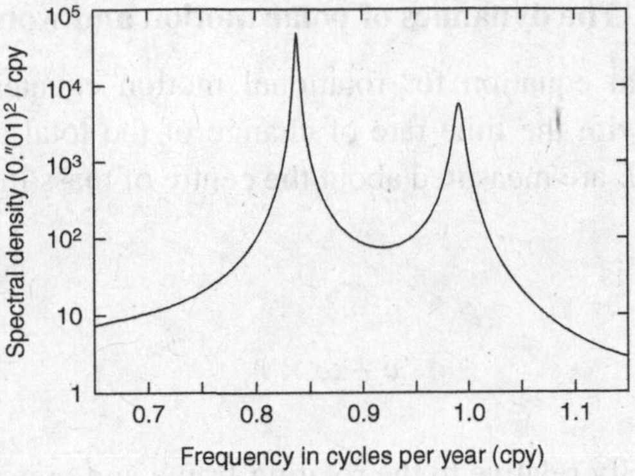

Figure 4.8 Detailed plot of the Chandler resonance and annual features of the maximum entropy spectral density.

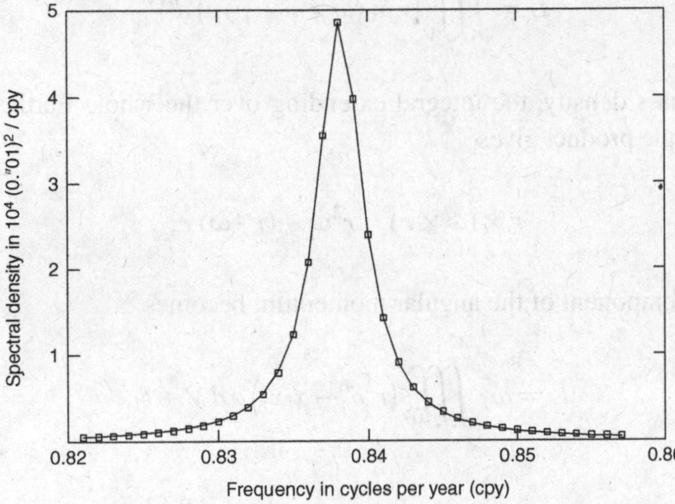

Figure 4.9 Fit of the Chandler resonance to the curve in expression (4.15). Squares mark the locations of the maximum entropy spectral estimates.

In order to recover the parameters of the Chandler wobble resonance, we fit a resonance curve with the form

$$\frac{a^2}{1 + 4Q^2[(f - f_0)/f_0]^2} \tag{4.15}$$

to values of the spectral density above $500\,(0.''01)^2/\text{cpy}$. The resulting fit is shown in Figure 4.9. The recovered peak amplitude is $a^2 = 48,775\,(0.''01)^2/\text{cpy}$. The recovered central frequency is $f_0 = 0.83813$ cpy, equivalent to a period of 435.8 days. The recovered quality factor Q is found to be 228.

4.3 The dynamics of polar motion and wobble

The basic dynamical equation for rotational motion equates the total external applied torque, $\boldsymbol{\Gamma}$, with the time rate of change of the total angular momentum, $\boldsymbol{L}$. Note that $\boldsymbol{\Gamma}$ and $\boldsymbol{L}$ are measured about the centre of mass in an inertial frame of reference.

The total velocity is

$$\boldsymbol{v} + \boldsymbol{\omega} \times \boldsymbol{r}, \tag{4.16}$$

where $\boldsymbol{v}$ is the velocity relative to the rotating frame and $\boldsymbol{\omega}$ is its angular velocity. Then, the total angular momentum is

$$\boldsymbol{L} = \iiint_{\mathcal{V}} [\boldsymbol{r} \times (\boldsymbol{\omega} \times \boldsymbol{r} + \boldsymbol{v})\rho]\, d\mathcal{V}, \tag{4.17}$$

with ρ the mass density, the integral extending over the whole Earth. Expanding the vector triple product gives

$$\boldsymbol{r} \times (\boldsymbol{\omega} \times \boldsymbol{r}) = r^2\boldsymbol{\omega} - (\boldsymbol{r} \cdot \boldsymbol{\omega})\,\boldsymbol{r}, \tag{4.18}$$

and the ith component of the angular momentum becomes

$$L_i = \omega_j \iiint_{\mathcal{V}} \left(r^2\delta^i_j - x_i x_j\right)\rho\, d\mathcal{V} + \ell_i, \tag{4.19}$$

where

$$I_{ij} = \iiint_{\mathcal{V}} \left(r^2\delta^i_j - x_i x_j\right)\rho\, d\mathcal{V} \tag{4.20}$$

are the components of the *inertia tensor* of the Earth and

$$\boldsymbol{\ell} = \iiint_{\mathcal{V}} (\boldsymbol{r} \times \boldsymbol{v})\rho\, d\mathcal{V} \tag{4.21}$$

is the relative angular momentum about the centre of mass relative to the rotating frame. Finally, the total angular momentum can be expressed as

$$\boldsymbol{L} = \boldsymbol{I} \cdot \boldsymbol{\omega} + \boldsymbol{\ell}. \tag{4.22}$$

The time derivative of a vector in the inertial frame is related to that in the rotating frame, by expression (3.72). Thus, the equation for rotational motion becomes

$$\boldsymbol{\Gamma} = \dot{\boldsymbol{L}} + \boldsymbol{\omega} \times \boldsymbol{L}, \tag{4.23}$$

with $\dot{\boldsymbol{L}}$ denoting the time derivative of the total angular momentum in the rotating frame. On substituting expression (4.22) for the total angular momentum into (4.23), we obtain the equation for rotational motion attributed to Liouville in 1858,

$$\boldsymbol{\Gamma} = \frac{d}{dt}(\boldsymbol{I} \cdot \boldsymbol{\omega} + \boldsymbol{\ell}) + \boldsymbol{\omega} \times (\boldsymbol{I} \cdot \boldsymbol{\omega} + \boldsymbol{\ell}) \,. \tag{4.24}$$

Adopting the geocentric Cartesian co-ordinate system of Section 4.2, the angular velocity takes the form (4.1)

$$\boldsymbol{\omega} = \Omega\,(\boldsymbol{m} + \hat{\boldsymbol{e}}_3) \,. \tag{4.25}$$

For most problems, large-scale polar wandering being an obvious exception, the rotation axis does not travel far from the reference axis and the axis of figure does not depart appreciably from the reference axis. Under these conditions,

$$\boldsymbol{I} \cdot \boldsymbol{\omega} = \begin{pmatrix} A + r_{11} + c_{11} & r_{12} + c_{12} & r_{13} + c_{13} \\ r_{12} + c_{12} & A + r_{22} + c_{22} & r_{23} + c_{23} \\ r_{13} + c_{13} & r_{23} + c_{23} & C + r_{33} + c_{33} \end{pmatrix} \times \begin{pmatrix} \Omega m_1 \\ \Omega m_2 \\ \Omega\,(1 + m_3) \end{pmatrix}, \tag{4.26}$$

where A is the equatorial moment of inertia in the absence of disturbance and C is the axial moment of inertia under the same condition. The r_{ij} are contributions to the inertia tensor caused by deformation under the changed centrifugal force, while the c_{ij} represent all other contributions to changes in the inertia tensor. The quantities ℓ_i, m_i, r_{ij} and c_{ij} are then all regarded as small and their squares and products are neglected. The components of the total angular momentum, to this level of approximation, become

$$L_1 = A\Omega m_1 + r_{13}\Omega + c_{13}\Omega + \ell_1, \tag{4.27}$$

$$L_2 = A\Omega m_2 + r_{23}\Omega + c_{23}\Omega + \ell_2, \tag{4.28}$$

$$L_3 = C\Omega + r_{33}\Omega + c_{33}\Omega + C\Omega m_3 + \ell_3. \tag{4.29}$$

To the same level of approximation, the three components of equation (4.23) governing rotational motion about the centre of mass are

$$\Gamma_1 = A\Omega\dot{m}_1 + \dot{r}_{13}\Omega + \dot{c}_{13}\Omega + \dot{\ell}_1 + C\Omega^2 m_2 - A\Omega^2 m_2 - r_{23}\Omega^2 - c_{23}\Omega^2 - \ell_2\Omega, \tag{4.30}$$

$$\Gamma_2 = A\Omega\dot{m}_2 + \dot{r}_{23}\Omega + \dot{c}_{23}\Omega + \dot{\ell}_2 + A\Omega^2 m_1 - C\Omega^2 m_1 + r_{13}\Omega^2 + c_{13}\Omega^2 + \ell_1\Omega, \tag{4.31}$$

$$\Gamma_3 = C\Omega\dot{m}_3 + \dot{r}_{33}\Omega + \dot{c}_{33}\Omega + \dot{\ell}_3. \tag{4.32}$$

Multiplying equation (4.31) by the unit imaginary number i and adding it to equation (4.30) produces

$$A\Omega\dot{\tilde{m}} - i(C-A)\,\Omega^2\tilde{m} + \dot{\tilde{r}}\Omega + i\tilde{r}\Omega^2 = \tilde{\Gamma} - \dot{\tilde{\ell}} - i\tilde{l}\Omega - \dot{\tilde{c}}\Omega - i\tilde{c}\Omega^2, \tag{4.33}$$

with complex phasors defined by $\tilde{m} = m_1 + im_2$, $\tilde{r} = r_{13} + i\tilde{r}_{23}$, $\tilde{\Gamma} = \Gamma_1 + i\Gamma_2$, $\tilde{\ell} = \ell_1 + i\ell_2$, and $\tilde{c} = c_{13} + ic_{23}$.

4.3.1 Response of the Earth to changes in the centrifugal force

When the deformation can be regarded as linear in the forcing function, a very simple description can be given. For example, if U_n is a forcing potential that is a solid harmonic of degree n, then the disturbance in the gravity potential at the surface of the Earth is given by

$$V_n = k_n U_n, \tag{4.34}$$

where k_n is a *Love number*, in general complex, but real for a purely elastic deformation.

This description applies rather well to the crust and mantle, which are elastic solids with quite high Q factors (300 and greater). It does not apply to the oceans and outer core where basic fluid dynamical behaviour cannot be ignored, including possible non-linear response.

The centrifugal force per unit mass,

$$-\boldsymbol{\omega} \times (\boldsymbol{\omega} \times \boldsymbol{r}) = \omega^2\boldsymbol{r} - (\boldsymbol{\omega}\cdot\boldsymbol{r})\,\boldsymbol{\omega}, \tag{4.35}$$

is derivable by taking the gradient of the potential

$$U = \frac{1}{2}\left[\omega^2 r^2 - (\boldsymbol{\omega}\cdot\boldsymbol{r})^2\right]. \tag{4.36}$$

In spherical polar co-ordinates, at the surface of the Earth ($r = d$), this potential becomes

$$
\begin{aligned}
U &= \frac{1}{2}\left(\omega_1^2 + \omega_2^2 + \omega_3^2\right) d^2 \\
&\quad - \frac{1}{2}\left(\omega_1 \sin\theta\cos\phi + \omega_2 \sin\theta\sin\phi + \omega_3\cos\theta\right)^2 d^2 .
\end{aligned}
\tag{4.37}
$$

In terms of the Legendre functions,

$$P_2(\cos\theta) = \frac{1}{2}\left(3\cos^2\theta - 1\right), \tag{4.38}$$

$$P_2^1(\cos\theta) = -3\cos\theta\sin\theta, \tag{4.39}$$

$$P_2^2(\cos\theta) = 3\sin^2\theta, \tag{4.40}$$

the potential is expressible as

$$
\begin{aligned}
U &= \frac{1}{3}\omega^2 d^2 + \frac{1}{6}\left(\omega_1^2 + \omega_2^2 - 2\omega_3^2\right) d^2 P_2(\cos\theta) \\
&\quad + \frac{1}{3}\omega_1\omega_3 d^2 P_2^1(\cos\theta)\cos\phi + \frac{1}{3}\omega_2\omega_3 d^2 P_2^1(\cos\theta)\sin\phi \\
&\quad - \frac{1}{12}\left(\omega_1^2 - \omega_2^2\right) d^2 P_2^2(\cos\theta)\cos 2\phi - \frac{1}{6}\omega_1\omega_2 d^2 P_2^2(\cos\theta)\sin 2\phi ,
\end{aligned}
\tag{4.41}
$$

using the identities

$$
\begin{aligned}
\sin\phi\cos\phi &= \frac{1}{2}\sin 2\phi, \\
\sin^2\phi = \frac{1-\cos 2\phi}{2}, &\quad \cos^2\phi = \frac{1+\cos 2\phi}{2}.
\end{aligned}
\tag{4.42}
$$

A similar expansion can be made for the external gravitational potential,

$$V(\boldsymbol{r}) = -G \iiint_{\mathcal{V}} \frac{\rho(\boldsymbol{r}')}{|\boldsymbol{r} - \boldsymbol{r}'|}\, d\mathcal{V}. \tag{4.43}$$

Expanding $1/|\boldsymbol{r} - \boldsymbol{r}'|$ in a three-dimensional Taylor series around $\boldsymbol{r}' = 0$ gives

$$\frac{1}{|\boldsymbol{r} - \boldsymbol{r}'|} = \frac{1}{r} + \frac{x_j x_j'}{r^3} + \frac{1}{2r^5}\left(3x_i x_j - r^2\delta_j^i\right) x_i' x_j' + \cdots , \tag{4.44}$$

valid for $r > r'$. Thus, we can write that

$$\begin{aligned} V(\boldsymbol{r}) = &-\frac{GM}{r} - \frac{Gx_j}{r^3}\iiint_{\mathcal{V}} x'_j\,\rho(\boldsymbol{r}')\,d\mathcal{V} \\ &-\frac{G}{2r^5}\left(3x_ix_j - r^2\delta^i_j\right)\iiint_{\mathcal{V}} x'_ix'_j\,\rho(\boldsymbol{r}')\,d\mathcal{V} + \cdots . \end{aligned} \tag{4.45}$$

In a reference co-ordinate system with origin at the centre of mass, by definition

$$\iiint_{\mathcal{V}} x'_j\,\rho(\boldsymbol{r}')\,d\mathcal{V} = 0. \tag{4.46}$$

From expression (4.20) for the components of the inertia tensor, we find that

$$\iiint_{\mathcal{V}} x'_ix'_j\,\rho(\boldsymbol{r}')\,d\mathcal{V} = -I_{ij} + \iiint_{\mathcal{V}} r'^2\delta^i_j\,\rho(\boldsymbol{r}')\,d\mathcal{V}. \tag{4.47}$$

Again, from expression (4.20), the *trace* of the inertia tensor is

$$\mathrm{Tr}(I) = I_{11} + I_{22} + I_{33} = 2\iiint_{\mathcal{V}} r'^2\rho(\boldsymbol{r}')\,d\mathcal{V}. \tag{4.48}$$

Thus,

$$\iiint_{\mathcal{V}} x'_ix'_j\,\rho(\boldsymbol{r}')\,d\mathcal{V} = -I_{ij} + \frac{1}{2}\mathrm{Tr}(I)\,\delta^i_j. \tag{4.49}$$

In the expansion (4.45), this expression is multiplied by $3x_ix_j - r^2\delta^i_j$, and the second term on the right side of (4.49) yields

$$\frac{1}{2}\mathrm{Tr}(I)\left(3x_1^2 - r^2 + 3x_2^2 - r^2 + 3x_3^2 - r^2\right) = 0. \tag{4.50}$$

Hence, the expansion (4.45) takes the form

$$\begin{aligned} V(\boldsymbol{r}) = &-\frac{GM}{r} \\ &-\frac{G}{2r^5}\Big[I_{11}\left(x_2^2 + x_3^2 - 2x_1^2\right) + I_{22}\left(x_3^2 + x_1^2 - 2x_2^2\right) + I_{33}\left(x_1^2 + x_2^2 - 2x_3^2\right) \\ &\qquad -6I_{12}x_1x_2 - 6I_{23}x_2x_3 - 6I_{31}x_3x_1\Big] + \cdots . \end{aligned} \tag{4.51}$$

Relations connecting the gravitational potential to the components of the inertia tensor are referred to as MacCullagh's formula (MacCullagh, 1845). Converting to spherical polar co-ordinates, and again using the identities (4.42), at the surface of the Earth ($r = d$), this expression becomes

$$V = -\frac{GM}{d} - \frac{3G}{2d^3}\left[\left(\frac{1}{3}I_{11} + \frac{1}{3}I_{22} - \frac{2}{3}I_{33}\right)P_2(\cos\theta) + \frac{2}{3}I_{13}P_2^1(\cos\theta)\cos\phi + \frac{2}{3}I_{23}P_2^1(\cos\theta)\sin\phi - \frac{1}{6}(I_{11} - I_{22})P_2^2(\cos\theta)\cos 2\phi - \frac{1}{3}I_{12}P_2^2(\cos\theta)\sin 2\phi\right] + \cdots. \tag{4.52}$$

Since the solid spherical harmonics in expressions (4.41) and (4.52) are linearly independent, equating V_2 to k_2U_2 following equation (4.34) gives five equations in the six components of the inertia tensor. A sixth equation is

$$\mathrm{Tr}(I) = I_{11} + I_{22} + I_{33}. \tag{4.53}$$

Solution of these equations for the components of the inertia tensor yields

$$I_{ij} = \frac{k_2 d^5}{3G}\omega_i\omega_j + \frac{1}{3}\left(\mathrm{Tr}(I) - \frac{k_2 d^5\omega^2}{3G}\right)\delta^i_j. \tag{4.54}$$

The term $\omega^2 d^2/3$ in the deforming potential (4.41) will cause a purely radial expansion, contributing only equal amounts to I_{11}, I_{22}, I_{33}, thus altering only $\mathrm{Tr}(I)$ (Rochester and Smylie, 1974). The contributions to the inertia tensor, r_{ij}, caused by deformation under the changed centrifugal force, neglecting squares and products of the small quantities m_1, m_2, m_3, are then

$$\left.\begin{aligned} r_{11} &= \frac{1}{3}\mathrm{Tr}(R) - \frac{2k_2 d^5}{9G}\Omega^2 m_3 \\ r_{22} &= \frac{1}{3}\mathrm{Tr}(R) - \frac{2k_2 d^5}{9G}\Omega^2 m_3 \\ r_{33} &= \frac{1}{3}\mathrm{Tr}(R) + \frac{4k_2 d^5}{9G}\Omega^2 m_3, \end{aligned}\right\} \tag{4.55}$$

where $\mathrm{Tr}(R) = r_{11} + r_{22} + r_{33}$ is the contribution to the trace of the inertia tensor, and

$$r_{12} = 0, \quad r_{23} = \frac{k_2 d^5}{3G}\Omega^2 m_2, \quad r_{13} = \frac{k_2 d^5}{3G}\Omega^2 m_1. \tag{4.56}$$

Thus, the complex phasor,

$$\tilde{r} = r_{13} + ir_{23} = \frac{k_2 d^5}{3G}\Omega^2\tilde{m}, \tag{4.57}$$

and the equation governing polar motion and wobble, (4.33), becomes

$$\begin{aligned}\dot{\tilde{m}}\left(1+\frac{k_2 d^5}{3GA}\Omega^2\right) - i\left\{\frac{C-A}{A}\Omega - \frac{k_2 d^5}{3GA}\Omega^3\right\}\tilde{m} \\ = \frac{\tilde{\Gamma}}{A\Omega} - \frac{\dot{\tilde{\ell}}}{A\Omega} - i\frac{\tilde{\ell}}{A} - \frac{\dot{\tilde{c}}}{A} - i\frac{\tilde{c}\Omega}{A}.\end{aligned} \tag{4.58}$$

From equation (4.32), changes in the spin rate and length of day are governed by

$$\dot{m}_3\left(1+\frac{4k_2 d^5}{9GC}\Omega^2\right) + \frac{1}{3C}\mathrm{Tr}(\dot{R}) = \frac{\Gamma_3}{C\Omega} - \frac{\dot{c}_{33}}{C} - \frac{\dot{\ell}_3}{C\Omega}. \tag{4.59}$$

4.3.2 Free and forced polar motion and dissipation

From equation (4.58), free polar motion is governed by

$$\dot{\tilde{m}} - i\sigma_0\tilde{m} = 0, \tag{4.60}$$

with

$$\sigma_0 = \left(\frac{C-A}{A}\Omega - \frac{k_2 d^5}{3GA}\Omega^3\right)\Bigg/\left(1+\frac{k_2 d^5}{3GA}\Omega^2\right). \tag{4.61}$$

Free polar motion then takes the form

$$\tilde{m} = ce^{i\sigma_0 t}, \tag{4.62}$$

where c is an arbitrary constant. The rotational deformation of the Earth is not perfectly elastic, hence σ_0 is the complex number

$$\sigma_0 = \omega_0 + i/\tau. \tag{4.63}$$

ω_0 is the resonant angular frequency and τ is the damping time of the motion. The free motion is then described by

$$\tilde{m} = ce^{-t/\tau + i\omega_0 t}. \tag{4.64}$$

Another description of the dissipation is through the dimensionless quality factor Q. By definition it is given by

$$Q = -2\pi\frac{E}{\Delta E}, \tag{4.65}$$

where E is the energy of the motion and ΔE is the energy dissipated per cycle. The energy is proportional to the square of the amplitude of the motion and decays according to

$$e^{-2t/\tau}. \tag{4.66}$$

Hence,

$$\frac{dE}{dt} = -\frac{2}{\tau}E, \tag{4.67}$$

and

$$\Delta E = \frac{dE}{dt}T_0 = -\frac{2}{\tau}E/f_0, \tag{4.68}$$

where $T_0 = 2\pi/\omega_0 = 1/f_0$ is the period of the motion and f_0 is the resonant frequency. Thus, the quality factor is

$$Q = \pi f_0 \tau, \tag{4.69}$$

and

$$\sigma_0 = \omega_0 + \frac{i\pi}{Q}f_0 = 2\pi f_0 + \frac{i\pi}{Q}f_0. \tag{4.70}$$

Now consider forced motion. The right-hand side of the polar motion equation (4.58) no longer vanishes. If the excitation is of unit amplitude and angular frequency ω, the right-hand side of equation (4.58) becomes $e^{i\omega t}$. This will excite a motion $re^{i\omega t}$ with

$$r = \frac{1}{i\omega - i\pi f_0\,(2 + i/Q)}. \tag{4.71}$$

The response at frequency $f = \omega/2\pi$ is then

$$r = \frac{Q}{\pi f_0[1 + i2Q\,(f - f_0)\,/f_0]}. \tag{4.72}$$

The power of the response, such as might be observed in a spectral density estimate, is then found to be

$$|r|^2 = \frac{a^2}{1 + 4Q^2[(f - f_0)\,/f_0]^2}, \tag{4.73}$$

with $a^2 = Q^2/\pi^2 f_0^2$. This is just the form (4.15) fitted to the maximum entropy spectrum of the VLBI polar motion path.

4.3.3 Relating theory to observations

A number of fundamental physical constants enter the foregoing theoretical discussion. Among these are the precessional constant, $H = (C - A)\,/C = 3.27379 \times 10^{-3}$, and the equatorial moment of inertia, $A = 8.0100 \times 10^{37}$ kg m^2 (Stacey, 1992, p. 409). As well, the Newtonian constant of gravitation $G = 6.67428 \times 10^{-11}$ m^3 kg^{-1} s^{-2} (Mohr *et al.*, 2008, pp. 688–691), the mean rotation rate of the Earth $\Omega = 7.292115 \times 10^{-5}$ rad s^{-1} (Hofmann-Wellenhof and Moritz, 2006, p. 88), and the mean radius of the Earth $d = 6371.012$ km (Lambeck, 1980, p. 27) enter the discussion.

Neglecting dissipation and rotational deformation,

$$\sigma_0 = \frac{C - A}{A}\Omega. \tag{4.74}$$

Since

$$\frac{C - A}{A} = \frac{C - A}{C + A - C} = \frac{1}{1/H - 1} = \frac{1}{304.456}, \tag{4.75}$$

σ_0 is 1/304.456 of the angular frequency of Earth's rotation. The period of free wobble of a rigid Earth is then 304.456 sidereal days or 305.290 solar days. This is called the *Euler wobble* period. In the real, deformable Earth, the resonant frequency and Q fitted by expression (4.15) to the maximum entropy spectrum of the VLBI pole path are, respectively,

$$f_0 = 0.83813 \text{ cpy} \quad \text{and} \quad Q = 228. \tag{4.76}$$

The resonant frequency is not affected by dissipation, which is reflected only in the value of Q. The observed resonant frequency corresponds to a period of 435.80 solar days or 434.61 sidereal days. Then, from (4.61), the value of the Love number, k_2, required to account for the increase in the observed period of the Chandler wobble by rotational deformation, is 0.2820. This value of k_2 is close to the values computed from seismically determined Earth models (Farrell, 1972), but the agreement is more apparent than real, because the liquid core actually shortens the wobble period by roughly 32 days, while the oceans lengthen the period by roughly 28 days (Lambeck, 1980, p. 202).

The value of Q, determined from the maximum entropy spectrum, is consistent with values for the mantle and crust found from seismic wave attenuation (Lambeck, 1988, p. 63), which are in the range 100 to 600. Conventional spectral analysis has usually given much lower values. This is likely due to the limitations on frequency resolution due to finite record effects, discussed in Section 2.4.2. Even for a 'good' window, such as the Parzen window, frequency resolution is limited by half-power points separated by $1.82/T$ for a record length T. The half-power points of the resonance determined by (4.15) are separated in frequency by $f_0/2Q$. Simply to match this value to that for the Parzen window requires a record length

$$T = \frac{1.82 \times 2Q}{f_0} = 990 \text{ years}. \tag{4.77}$$

The maximum entropy spectral analysis method avoids this limitation of record length, allowing the true value of Q to be determined.

4.3.4 Geometry of free polar motion, wobble and sway

Neglecting the small dissipation reflected in the high value of Q, free polar motion is governed by equation (4.60) with σ_0 real. The solution to this equation has the form

$$\tilde{m} = e^{i(\sigma_0 t+\delta)} \tan\beta, \tag{4.78}$$

with δ an arbitrary phase constant and β the angle the rotation axis makes with the reference axis $\hat{e}_3$. The instantaneous pole of rotation describes a circle of radius $d\tan\beta$ about the reference pole with a period of 435.8 days in the prograde sense. The axis of rotation cuts out a body cone (the polhode) in the Earth every complete revolution. Changes in the trace of the inertia tensor are second order in small quantities. Correct to first order in small quantities, the components of the total angular momentum are

$$L_1 = A\Omega m_1 + r_{13}\Omega = \left(A + \frac{k_2 d^5}{3G}\Omega^2\right)\Omega m_1, \tag{4.79}$$

$$L_2 = A\Omega m_2 + r_{23}\Omega = \left(A + \frac{k_2 d^5}{3G}\Omega^2\right)\Omega m_2, \tag{4.80}$$

$$L_3 = C\Omega = \left(A + \frac{k_2 d^5}{3G}\Omega^2\right)\Omega + \left(C - A - \frac{k_2 d^5}{3G}\Omega^2\right)\Omega. \tag{4.81}$$

Thus, using (4.61), the total angular momentum vector is

$$\boldsymbol{L} = A\left(1 + \frac{k_2 d^5}{3GA}\Omega^2\right)(\boldsymbol{\omega} + \sigma_0\hat{e}_3). \tag{4.82}$$

Since the torques have been assumed to vanish, $\boldsymbol{L}$ is an invariable axis in space. Further, because $\boldsymbol{L}$ is a linear combination of vectors in the directions of $\boldsymbol{\omega}$ and $\hat{e}_3$, the vectors $\boldsymbol{\omega}$, $\hat{e}_3$ and $\boldsymbol{L}$ are coplanar.

Let γ be the angle between the rotation axis and the invariable axis of angular momentum. Then, the angle between the reference axis and the invariable axis is $\beta - \gamma$, as shown in Figure 4.10. Since

$$\boldsymbol{\omega} = \Omega m_1\hat{e}_1 + \Omega m_2\hat{e}_2 + \Omega\hat{e}_3, \tag{4.83}$$

we have that

$$|\boldsymbol{\omega}| = \Omega\sec\beta. \tag{4.84}$$

Writing

$$A' = A\left(1 + \frac{k_2 d^5}{3GA}\Omega^2\right), \tag{4.85}$$

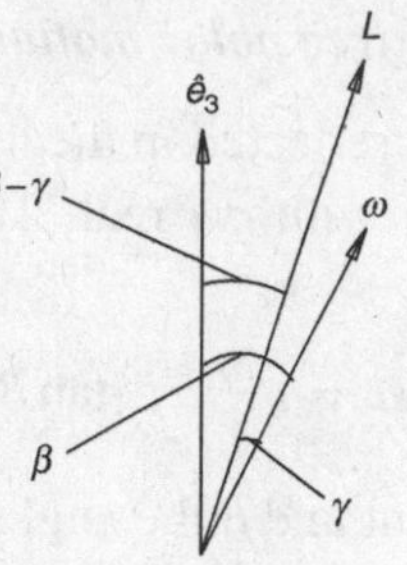

Figure 4.10 The three coplanar vectors: the unit vector $\hat{e}_3$ in the direction of the reference axis, the angular momentum vector $\boldsymbol{L}$ and the angular velocity vector $\boldsymbol{\omega}$ of the rotation.

from equation (4.82), we have that

$$\frac{|\boldsymbol{L}|}{A'}\cos(\beta-\gamma) = \hat{e}_3\cdot\frac{\boldsymbol{L}}{A'} = \Omega+\sigma_0, \tag{4.86}$$

neglecting squares of the small quantities m_1, m_2. We also have that

$$\begin{aligned}\frac{|\boldsymbol{L}|}{A'} &= \left[(\boldsymbol{\omega}+\sigma_0\hat{e}_3)\cdot(\boldsymbol{\omega}+\sigma_0\hat{e}_3)\right]^{1/2}\\ &= \left[\boldsymbol{\omega}\cdot\boldsymbol{\omega}+2\sigma_0\boldsymbol{\omega}\cdot\hat{e}_3+\sigma_0^2\right]^{1/2}.\end{aligned} \tag{4.87}$$

From equation (4.84),

$$|\boldsymbol{\omega}|^2 = \Omega^2\sec^2\beta = \boldsymbol{\omega}\cdot\boldsymbol{\omega} = \Omega^2\left(1+\tan^2\beta\right), \tag{4.88}$$

using the trigonometric identity $\sec^2\beta = 1+\tan^2\beta$. Then,

$$\begin{aligned}\frac{|\boldsymbol{L}|}{A'} &= \left[\Omega^2+\Omega^2\tan^2\beta+2\sigma_0\Omega+\sigma_0^2\right]^{1/2}\\ &= \left[(\Omega+\sigma_0)^2+\Omega^2\tan^2\beta\right]^{1/2} = (\Omega+\sigma_0)\sec(\beta-\gamma),\end{aligned} \tag{4.89}$$

using equation (4.86). We can thus write

$$\begin{aligned}\tan^2(\beta-\gamma) &= \sec^2(\beta-\gamma)-1\\ &= \frac{(\Omega+\sigma_0)^2+\Omega^2\tan^2\beta}{(\Omega+\sigma_0)^2}-1\\ &= \frac{\Omega^2}{(\Omega+\sigma_0)^2}\tan^2\beta.\end{aligned} \tag{4.90}$$

Now,

$$\tan(\beta - \gamma) = \frac{\sin(\beta - \gamma)}{\cos(\beta - \gamma)} = \frac{\sin\beta\cos\gamma - \cos\beta\sin\gamma}{\cos\beta\cos\gamma + \sin\beta\sin\gamma} = \frac{\tan\beta - \tan\gamma}{1 + \tan\beta\tan\gamma}, \tag{4.91}$$

and hence,

$$\frac{\tan\beta - \tan\gamma}{1 + \tan\beta\tan\gamma} = \frac{\Omega}{(\Omega + \sigma_0)}\tan\beta, \tag{4.92}$$

on choosing the positive root. This gives

$$\tan\beta - \tan\gamma\left(1 + \frac{\Omega}{\Omega + \sigma_0}\tan^2\beta\right) = \frac{\Omega}{\Omega + \sigma_0}\tan\beta \tag{4.93}$$

or

$$\frac{\tan\gamma}{\tan\beta} = \frac{1 - \frac{\Omega}{\Omega + \sigma_0}}{1 + \frac{\Omega}{\Omega + \sigma_0}\tan^2\beta} = \frac{\sigma_0}{\sigma_0 + \Omega\sec^2\beta}. \tag{4.94}$$

The quantities β, σ_0, Ω are all constants of the motion, therefore γ is also a constant of the motion. The angular velocity vector $\boldsymbol{\omega}$ then describes a small cone (the herpolhode) about the invariable axis $\boldsymbol{L}$ with semi-apex angle γ.

Of interest is the rate of progression of $\boldsymbol{\omega}$ about $\boldsymbol{L}$. Let ξ measure the angular position of $\boldsymbol{\omega}$ in its path about $\boldsymbol{L}$. In time δt the increase in ξ is $\delta\xi$. The component of $\boldsymbol{\omega}$ normal to $\boldsymbol{L}$ is $|\boldsymbol{\omega}|\sin\gamma$. This acts as a radius vector moving through the angle $\delta\xi$ in time δt to produce the change $\delta\boldsymbol{\omega}$ in time δt. Then,

$$\delta\boldsymbol{\omega} = \delta\xi\,|\boldsymbol{\omega}|\sin\gamma = \delta\xi \cdot \frac{\boldsymbol{L} \times \boldsymbol{\omega}}{|\boldsymbol{L}|} \tag{4.95}$$

and the time derivatives of $\boldsymbol{\omega}$ and ξ are related by

$$\dot{\boldsymbol{\omega}} = \dot{\xi}\frac{\boldsymbol{L} \times \boldsymbol{\omega}}{|\boldsymbol{L}|}. \tag{4.96}$$

We also have that

$$\dot{\boldsymbol{\omega}} = \Omega\dot{m}_1\hat{\mathbf{e}}_1 + \Omega\dot{m}_2\hat{\mathbf{e}}_2. \tag{4.97}$$

From the real and imaginary parts of the free motion equation (4.60), $\dot{m}_1 = -\sigma_0 m_2$ and $\dot{m}_2 = \sigma_0 m_1$, giving

$$\begin{aligned}\dot{\boldsymbol{\omega}} &= -\Omega\sigma_0 m_2 \hat{\mathbf{e}}_1 + \Omega\sigma_0 m_1 \hat{\mathbf{e}}_2 \\ &= \Omega\sigma_0 \hat{\mathbf{e}}_3 \times (m_1\hat{\mathbf{e}}_1 + m_2\hat{\mathbf{e}}_2) \\ &= \sigma_0 \hat{\mathbf{e}}_3 \times (\Omega m_1 \hat{\mathbf{e}}_1 + \Omega m_2 \hat{\mathbf{e}}_2 + \Omega\hat{\mathbf{e}}_3) \\ &= \sigma_0 \hat{\mathbf{e}}_3 \times \boldsymbol{\omega}. \end{aligned} \tag{4.98}$$

Note that $\dot{\boldsymbol{\omega}}$ is the same, whether measured in the space frame or the rotating frame, since $\boldsymbol{\omega} \times \boldsymbol{\omega} = 0$. On equating expressions (4.96) and (4.98) for $\dot{\boldsymbol{\omega}}$, we find that

$$\dot{\xi}\frac{\boldsymbol{L} \times \boldsymbol{\omega}}{|\boldsymbol{L}|} = \sigma_0 \hat{\mathbf{e}}_3 \times \boldsymbol{\omega}. \tag{4.99}$$

Since

$$\boldsymbol{L} \times \boldsymbol{\omega} = |\boldsymbol{L}|\,|\boldsymbol{\omega}| \sin\gamma \quad \text{and} \quad \hat{\mathbf{e}}_3 \times \boldsymbol{\omega} = |\boldsymbol{\omega}| \sin\beta, \tag{4.100}$$

we obtain that

$$\dot{\xi} = \sigma_0 \frac{\sin\beta}{\sin\gamma} = \sigma_0 \frac{\cos\beta \tan\beta}{\cos\gamma \tan\gamma} = \frac{\cos\beta}{\cos\gamma}\left(\sigma_0 + \Omega \sec^2\beta\right), \tag{4.101}$$

using expression (4.94).The angular rate σ_0 at which $\boldsymbol{\omega}$ moves about $\hat{\mathbf{e}}_3$, and the angular rate $\dot{\xi}$ at which $\boldsymbol{\omega}$ moves about $\boldsymbol{L}$, are in inverse proportion to the sines of the semi-apex angles of the cones. Thus, the body cone (polhode) rolls once per day without slipping on the space cone (herpolhode), as illustrated in Figure 4.11. This geometrical representation originated with Poinsot in 1834 and is called the *Poinsot construction*. The resulting polar motion is the free Chandler wobble.

The nearly diurnal motion of the rotation axis in space is called *sway*. For the free polar motion or Chandler wobble, the angle β is only parts in 10^6 and the angle γ is even smaller. The tangents are then very close to the angles themselves and their ratio is closely

$$\frac{\gamma}{\beta} = \frac{\sigma_0}{\Omega + \sigma_0}. \tag{4.102}$$

Similarly, the sines of the angles are very close to the angles themselves and

$$\dot{\xi} = \sigma_0 \frac{\beta}{\gamma} = \Omega + \sigma_0. \tag{4.103}$$

The motion of $\boldsymbol{\omega}$ about $\boldsymbol{L}$ is faster than the diurnal rotation by the factor $1 + \sigma_0/\Omega$, which at 1.00229 is just in excess of unity.

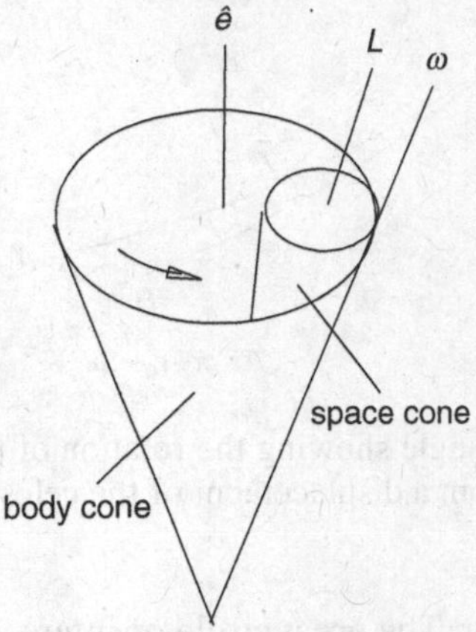

Figure 4.11 In the Poinsot construction, the body cone rolls once per day around the space cone. The ensuing polar motion is the free Chandler wobble, while the nearly diurnal motion of the rotation axis in space is called *sway*.

4.4 Nutation and motion of the celestial pole

Motion of Earth's rotation axis in space is described with reference to the *celestial sphere*, a sphere of indefinite radius defined by the International Celestial Reference System (ICRS) fixed to the co-ordinates of extragalactic radio sources, as illustrated in Figure 4.12.

The celestial reference pole at point P is the point where the reference rotation axis pierces the celestial sphere. The reference equatorial plane is orthogonal to the reference rotation axis and intersects the celestial sphere in the great circle

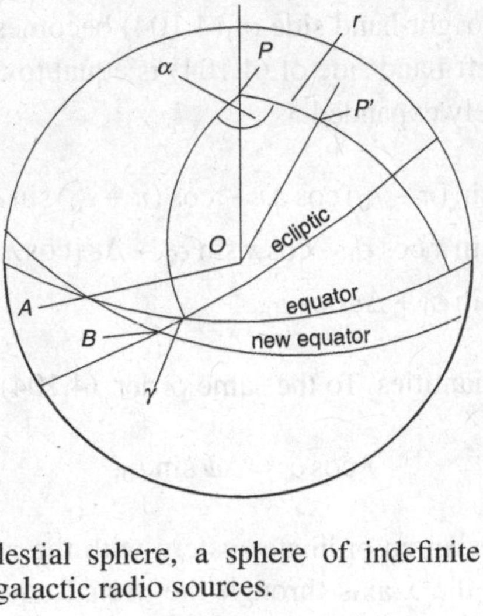

Figure 4.12 The celestial sphere, a sphere of indefinite radius fixed to the co-ordinates of extragalactic radio sources.

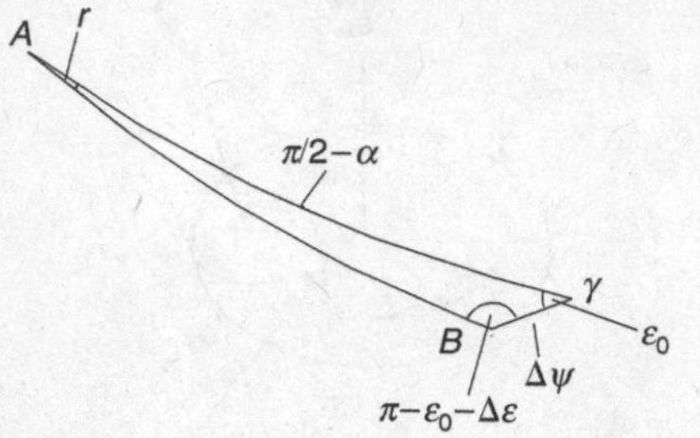

Figure 4.13 Spherical triangle showing the relation of the increases in longitude and obliquity resulting from a displacement of the celestial pole.

marking the celestial equator. The great circle of intersection of the plane of Earth's orbit with the celestial sphere is the celestial ecliptic. The celestial equator and the celestial ecliptic intersect at γ, the vernal equinox or *the first point of Aries*.

If the rotation axis moves through the angle r at ecliptic longitude α measured along the ecliptic from the vernal equinox γ, an increase $\Delta\psi$ in the ecliptic longitude results, given by the side γB of the spherical triangle $A\gamma B$ illustrated in Figure 4.13. At the same time, the obliquity of the ecliptic increases from the reference value ϵ_0 to $\epsilon_0 + \Delta\epsilon$, giving the angle $AB\gamma$ of the spherical triangle as $\pi - \epsilon_0 - \Delta\epsilon$. The equatorial plane rotates about the axis OA orthogonal to the plane POP', giving the side $A\gamma$ of the spherical triangle as $\pi/2 - \alpha$. By the law of sines for spherical triangles,

$$\frac{\sin(\pi/2 - \alpha)}{\sin(\pi - \epsilon_0 - \Delta\epsilon)} = \frac{\sin\Delta\psi}{\sin r}. \tag{4.104}$$

The angles r, $\Delta\psi$ and $\Delta\epsilon$ are all small quantities whose squares and products can be neglected. Thus, the right-hand side of (4.104) becomes the simple ratio $\Delta\psi/r$. The numerator of the left-hand side of (4.104) is equal to $\cos\alpha$, while the denominator can be successively expanded as

$$\begin{aligned}\sin(\pi - \epsilon_0 - \Delta\epsilon) &= \sin(\pi - \epsilon_0)\cos\Delta\epsilon - \cos(\pi - \epsilon_0)\sin\Delta\epsilon \\ &= \sin\pi\cos\epsilon_0 - \cos\pi\sin\epsilon_0 - \Delta\epsilon\,[\cos\pi\cos\epsilon_0 + \sin\pi\sin\epsilon_0] \\ &= \sin\epsilon_0 + \Delta\epsilon\cos\epsilon_0,\end{aligned} \tag{4.105}$$

to first order in small quantities. To the same order, (4.104) then yields

$$r\cos\alpha = \Delta\psi\sin\epsilon_0. \tag{4.106}$$

In a right-handed Cartesian co-ordinate system with the z-axis through the celestial reference pole and the x-axis through the vernal equinox with origin at the barycentre of the solar system, displacement of the rotation axis involved in nutation is described by co-ordinates

$$x = r\cos\alpha = \Delta\psi\sin\epsilon_0, \quad y = r\sin\alpha = \Delta\epsilon, \tag{4.107}$$

using (4.106).

4.4.1 VLBI nutation observations

As well as the pole path, the very long baseline interferometry (VLBI) technique is used to measure motions of the celestial pole or nutations. The observations are expressed as residuals of longitude and obliquity compared to a specific theoretical model. Two principal nutation models have been developed. The first is the 1980 IAU theory developed by John Wahr (Wahr, 1981). A later nutation model, the MHB2000 theory, developed by Mathews, Herring and Buffet (Mathews *et al.*, 2002), fits a model of known nutations to observations and is therefore not independent of the data or the model. Because this procedure might suppress new, unmodelled signals, we choose to examine residuals with respect to the 1980 IAU theory.

The residuals in longitude, $\Delta\psi$, and obliquity, $\Delta\epsilon$, compared to the 1980 IAU nutation series, and their standard errors, in milliarcseconds, are available on the Goddard Space Flight Center website (http://gemini.gsfc.nasa.gov/solutions/2003b/2003b.eops). The downloadable file 2003b.eops contains 3370 points running from Julian day 2,444,089.993750 (3 August 1979) to Julian day 2,452,705.270139 (6 March 2003). The time tags are given in Julian days minus 2,400,000. The programme NUT2003B extracts the residuals of longitude and obliquity with their standard errors and converts them to the Cartesian co-ordinates given by (4.107) using $\sin\epsilon_0 = 0.39777716$ for the year 2000.0. The resulting nutation residuals x, y and their standard errors, in terms of the time in days from the beginning of the record, are returned in the file nut2003b.dat.

```
C         PROGRAMME NUT2003B.FOR
C
C NUT2003B.FOR extracts residuals of ecliptic longitude and
C obliquity of the ecliptic with their standard errors,
C compared to the 1980 IAU model of Wahr, from the file
C 2003b.eops on the Goddard Space Flight Center website.
C They are then converted to Cartesian co-ordinates of
C the celestial pole before output.
C
      IMPLICIT DOUBLE PRECISION(A-H,O-Z)
      OPEN(UNIT=1,FILE='2003b.eops',STATUS='OLD')
      OPEN(UNIT=2,FILE='nut2003b.dat',STATUS='UNKNOWN')
C Read in extracted residuals of longitude and obliquity.
      DO 10,I=1,3369
        READ(1,11)DAY,XWOB,YWOB,UT1MC,PSI,EPS,EXWOB,EYWOB,
     1  EUT1MC,SEPSI,SEEPS,WMS,C1,C2,C3,C4,N,SC,SD,RX,RY,
```

```
     2    DLOD,F1,F2,UX,UY,UL,Q1,Q2
 11     FORMAT(1X,F12.6,2F9.6,F11.7,2F9.3,2F9.6,F10.7,2F8.3,
     1    F8.2,4F7.4,I7,A7,F6.2,2F10.6,F11.7,2A3,2F10.6,F11.7,2A3)
C Set time origin to zero.
        DAY=DAY-44089.99375OD0
C Find Cartesian co-ordinates of the celestial pole and their
C standard errors.
        XNUT=0.39777716D0*PSI
        SEXNUT=0.39777716D0*SEPSI
        YNUT=EPS
        SEYNUT=SEEPS
C Write out extracted co-ordinates of the celestial pole.
        WRITE(2,12)I,DAY,XNUT,SEXNUT,YNUT,SEYNUT
 12     FORMAT(1X,I4,F15.6,4F10.3)
 10   CONTINUE
      END
```

Tests reveal that four of the points have null values and 23 pairs of points have identical time tags. The former were removed and the values of the latter were averaged with weights in inverse proportion to the squares of their standard errors. Series of 3343 points were left for analysis in the file nut.dat.

Before performing spectral analysis, the series were reduced to zero mean. The linear trends were removed, and 18.6-year (6798.58 day), 9.3-year (6798.58/2 day), annual (365.25971 day), and semi-annual (365.2597/2 day) periodic terms were removed by least squares fits to

$$x = i_x t + j_x + a_x \cos(\omega_1 t) + b_x \sin(\omega_1 t) + c_x \cos(\omega_2 t) + d_x \sin(\omega_2 t) + e_x \cos(\omega_3 t) + f_x \sin(\omega_3 t) + g_x \cos(\omega_4 t) + h_x \sin(\omega_4 t)\,, \tag{4.108}$$

$$y = i_y t + j_y + a_y \cos(\omega_1 t) + b_y \sin(\omega_1 t) + c_y \cos(\omega_2 t) + d_y \sin(\omega_2 t) + e_y \cos(\omega_3 t) + f_y \sin(\omega_3 t) + g_y \cos(\omega_4 t) + h_y \sin(\omega_4 t)\,, \tag{4.109}$$

where $\omega_1 = 2\pi/6798.58$, $\omega_2 = 2\omega_1$, $\omega_3 = 2\pi/365.25971$ and $\omega_4 = 2\omega_3$. Coefficients of the extracted terms are listed in Table 4.1.

The programme NUTRES restores the original time tags from the beginning of the record at 44089.99375 days (Julian days minus 2,400,000) and subtracts

Table 4.1 *Coefficients of terms extracted from the GSFC VLBI nutation series.*

a_x	b_x	c_x	d_x	e_x	f_x
2.4496	1.9307	−0.37214	−0.24779	−1.0785	1.7798
g_x	h_x	i_x	j_x		
0.82405	0.19983	−0.00324337	150.02		
a_y	b_y	c_y	d_y	e_y	f_y
1.9026	−2.3741	−0.18244	0.13710	1.6863	1.2533
g_y	h_y	i_y	j_y		
−0.17113	0.71118	−0.000665044	29.144		

the 18.6-year, 9.3-year, annual and semi-annual periodic terms as well as the linear and constant terms, as determined by least squares fits with coefficients given in Table 4.1. The nutation residuals cover a time span of 8615.276389 days. We take the record length to be 8617 days, the first point to be 0.8618055 days into the record, and the last point to be 0.8618055 days before the end of the record. After subtraction of the foregoing terms, the time base is reset to have its origin 0.8618055 days before the first point. The resulting nutation residuals are returned in the file nutres.dat.

```
C           PROGRAMME NUTRES.FOR
C
C Programme NUTRES removes 18.6 yr term, 9.3 yr term, annual,
C and semi-annual terms, and the linear and constant terms from
C the nutation co-ordinates, taking the time origin to be the original
C 44,089.99375 days.
C
      IMPLICIT DOUBLE PRECISION(A-H,O-Z)
      OPEN(UNIT=1,FILE='nut.dat',STATUS='OLD')
      OPEN(UNIT=2,FILE='nutres.dat',STATUS='UNKNOWN')
C Set constants.
      PI=3.14159265358979D0
      W=2.D0*PI/6798.58D0
      W9=2.D0*W
      WA=2.D0*PI/365.25971D0
      WSA=2.D0*WA
      AX=2.4496D0
      AY=1.9026D0
      BX=1.9307D0
      BY=-2.3741D0
      CX=-0.37214D0
      CY=-0.18244D0
      DX=-0.24779D0
      DY=0.13710D0
      EX=-1.0785D0
      EY=1.6863D0
      FX=1.7798D0
      FY=1.2533D0
      GX=0.82405D0
      GY=-0.17113D0
      HX=0.19983D0
      HY=0.71118D0
      AIX=-0.00324337D0
      AIY=-0.000665044D0
      AJX=150.02D0
      AJY=29.144D0
      DO 10,I=1,3343
        READ(1,11)J,DAY,XNUT,SEX,YNUT,SEY
 11     FORMAT(1X,I4,F15.6,4F10.3)
C Reset time origin.
        DAY=DAY+44089.99375D0
C Subtract 18.6 yr term, 9.3 yr term, annual, and semi-annual terms,
C as well as linear and constant terms.
        XNUT=XNUT-AX*DCOS(W*DAY)-BX*DSIN(W*DAY)
     1  -CX*DCOS(W9*DAY)-DX*DSIN(W9*DAY)
     2  -EX*DCOS(WA*DAY)-FX*DSIN(WA*DAY)
     3  -GX*DCOS(WSA*DAY)-HX*DSIN(WSA*DAY)
     4  -AIX*DAY-AJX
```

```
      YNUT=YNUT-AY*DCOS(W*DAY)-BY*DSIN(W*DAY)
     1   -CY*DCOS(W9*DAY)-DY*DSIN(W9*DAY)
     2   -EY*DCOS(WA*DAY)-FY*DSIN(WA*DAY)
     3   -GY*DCOS(WSA*DAY)-HY*DSIN(WSA*DAY)
     4   -AIY*DAY-AJY
C Reset time origin for total record length of 8617 days.
      DAY=DAY-44089.99375D0+0.8618055D0
 10   WRITE(2,11)J,DAY,XNUT,SEX,YNUT,SEY
      END
```

4.4.2 Spectral analysis of the nutation residuals

We apply the overlapping segment analysis described in Section 2.5.3 and break the record into four overlapping segments of length 4924 days, each with 75% overlap. The first segment consists of points 1 through 1866, the second segment consists of points 105 through 2438, the third segment consists of points 443 through 2889 and the fourth and last segment consists of points from 1046 through to the last point 3343.

The discrete Fourier transform (DFT) for each segment is found for the corresponding series of unequally spaced nutation residuals by singular value decomposition (SVD), as outlined for the polar motion in Section 4.2.2. The results for each segment in terms of the number of singular values eliminated (NSVE) giving the value of Parseval's ratio R closest to unity, the relative reconstruction error (RRE), the number of sample points, n, and the odd number of terms in the DFT representation, m, are given in Table 4.2.

As described in Section 2.5.3, the four segments with 75% overlap have 5.85 equivalent degrees of freedom. On plots of the logarithm of spectral density, the 95% confidence interval is of fixed length 1.189. From expression (2.350), the spectral density estimate for each segment is given by the squared magnitude of the discrete Fourier transform, normalised for a Parzen window by the factor $560/(151M)$ for segment length M (in this case $M = 4924$).

Table 4.2 *Number of singular values eliminated (NSVE), Parseval's ratio (R), relative reconstruction error (RRE), number of sample points, n, and odd number of terms in the DFT representation, m, for the four segments analysed by the SVD technique.*

Segment no.	NSVE	R	RRE	n	m
1	543	0.9813	0.2670	1866	1865
2	465	1.0363	0.3881	2334	2333
3	509	0.9932	0.2751	2447	2447
4	462	0.9105	0.1471	2298	2297

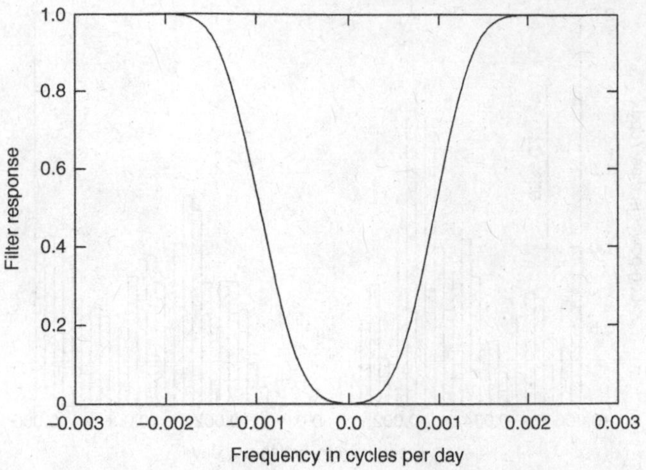

Figure 4.14 High-pass filter response for removal of spurious long-period nutations.

In order to suppress potentially spurious long-period terms in the nutation residuals, a high-pass filter was constructed by subtracting the convolution in the frequency domain of a boxcar function $B(f)$ and a Parzen frequency window, $P(f)$, given by expression (2.283), from an all-pass filter, giving the form

$$1 - B(f) \otimes P(f) . \tag{4.110}$$

Filter parameters are chosen so that the roll off for long periods begins at ±490.6 days and is 90% complete at 1867 days. The response of the high-pass filter is shown in Figure 4.14.

The discrete Fourier transform of each segment is subjected to this filter. The overall spectral density estimate is given by the average over the four segments. The logarithm of the power spectral density of the nutation residuals is plotted in Figure 4.15 complete with the 95% confidence interval.

Two features stand out in the power spectrum of the nutation residuals. The first is a retrograde resonance near -2.4×10^{-3} cycles per day. The second is a similar, but somewhat smaller, prograde resonance near 2.6×10^{-3} cycles per day. The retrograde resonance was first predicted by Poincaré (1910) for a completely fluid, incompressible core bounded by a rigid shell. In a variational calculation of wobble–nutation modes in realistic Earth models, Jiang (1993) found not only the classical retrograde free core nutation (RFCN) but also a prograde free core nutation (PFCN). He was able to find the RFCN in the spectrum of very long baseline interferometry (VLBI) nutation observations well above the 95% confidence level, but the PFCN could not be definitively identified due to the short length of the VLBI record available at the time. Later, both modes were found in the VLBI

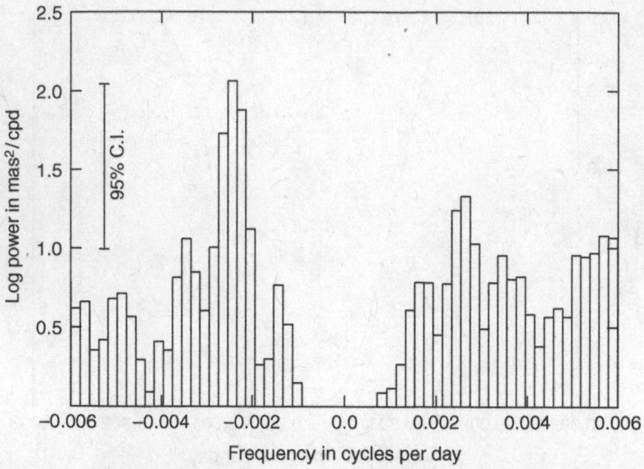

Figure 4.15 Average spectral density of the nutation residuals.

records from the Goddard Space Flight Center and the United States Naval Observatory, both in excess of 23 years in length (Palmer and Smylie, 2005).

Resonances of the form

$$\frac{a^2}{1 + 4Q^2\left[(f - f_0)/f_0\right]^2}, \tag{4.111}$$

used previously in the study of polar motion (4.15), derived from the equation of motion for polar wobble (4.73), are fitted to the PFCN and the RFCN resonances.

For the PFCN, a peak power of $a^2 = 23.36$ milliarcseconds squared (mas^2) was found with $Q = 5.8912$ and central frequency $f_0 = 2.57614 \times 10^{-3}$ cycles per day, equivalent to a period of 388.178 days. The fitted resonance curve is shown plotted over the local power spectrum in Figure 4.16. In order to obtain an estimate of the amplitude of the PFCN, we integrate the fitted resonance curve (4.111) between its half-power points to get the square of the amplitude as

$$A^2 = a^2 \int_{f_0(1-1/2Q)}^{f_0(1+1/2Q)} \frac{df}{1 + 4Q^2\left[(f - f_0)/f_0\right]^2}. \tag{4.112}$$

On changing the integration variable to

$$f' = 2Q\frac{f - f_0}{f_0}, \tag{4.113}$$

we find that

$$A^2 = \frac{a^2 f_0}{2Q}\int_{-1}^{1} \frac{df'}{1 + f'^2} = \pi\frac{a^2 f_0}{4Q}. \tag{4.114}$$

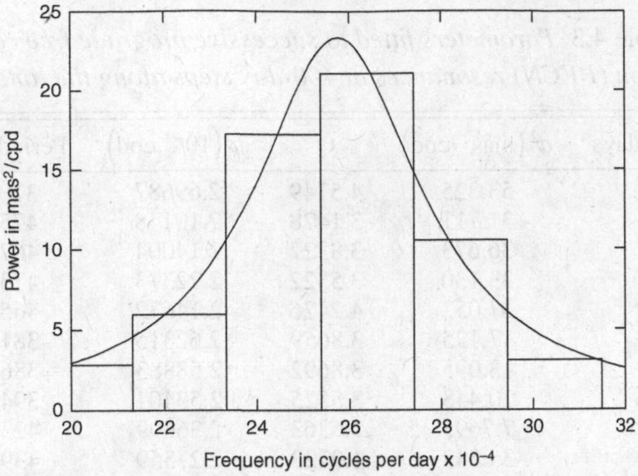

Figure 4.16 Prograde free core nutation (PFCN) spectral resonance.

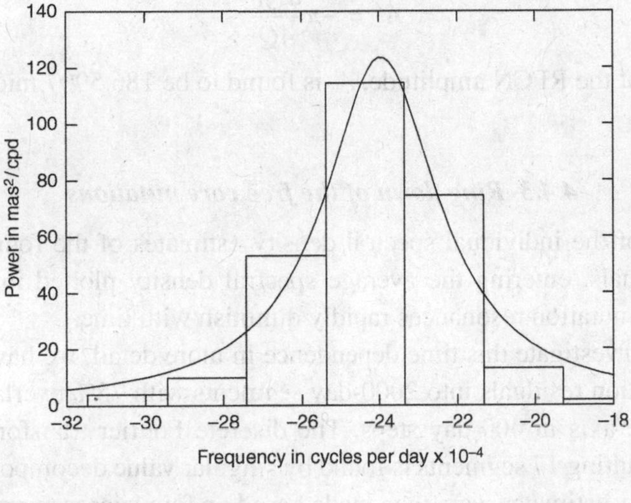

Figure 4.17 Retrograde free core nutation (RFCN) spectral resonance.

The estimate of the PFCN amplitude, A, is found to be 89.5723 microarcseconds (μas).

For the RFCN, a peak power of $a^2 = 123.51$ milliarcseconds squared (mas^2) was found with $Q = 6.6747$ and central frequency $f_0 = -2.39563 \times 10^{-3}$ cycles per day, equivalent to a period of 417.427 days. The fitted resonance curve is shown plotted over the local power spectrum in Figure 4.17.

As for the PFCN, the fitted resonance curve (4.111) is integrated between its half-power points to get an estimate of the square of the amplitude as

Table 4.3 *Parameters fitted to successive prograde free core nutation (PFCN) resonances at 400-day steps along the time axis.*

Time in days	$a^2\left(\text{mas}^2/\text{cpd}\right)$	Q	$f_0\left(10^{-3}\text{ cpd}\right)$	Period (days)
1600	53.325	4.5749	2.69687	370.800
2000	31.711	3.1478	2.10148	475.855
2400	36.673	3.8327	2.14004	467.202
2800	33.130	3.5322	2.22373	473.097
3200	37.052	4.2476	2.14632	465.914
3600	17.125	3.8659	2.62315	381.221
4000	13.095	3.8692	2.58883	386.275
4400	10.443	3.5525	2.53401	394.631
4800	7.7695	2.6363	2.36099	423.551
5200	3.6543	1.9723	2.27559	439.446
5600	3.7166	2.1370	1.79344	557.588
6000	3.9700	2.0140	1.80939	552.672

$$A^2 = -\pi \frac{a^2 f_0}{4Q}. \tag{4.115}$$

The estimate of the RFCN amplitude, A, is found to be 186.5907 microarcseconds (μas).

4.4.3 Ring down of the free core nutations

Examination of the individual spectral density estimates of the four segments of nutation residuals, entering the average spectral density plotted in Figure 4.15, shows that the nutation resonances rapidly diminish with time.

In order to investigate this time dependence in more detail, we have divided the record of nutation residuals into 2000-day segments with 75% overlap, advancing along the time axis in 400-day steps. The discrete Fourier transform (DFT) for each of the resulting 17 segments is found by singular value decomposition (SVD). Spectral density estimates were then made based on four successive segments with 75% overlap. This leads to 14 spectral density estimates centred at 1600 days into the record and advancing along the time axis in 400-day steps. Resonance curves of the form (4.111) were then fitted to both the prograde free core nutation (PFCN) and the retrograde free core nutation (RFCN) features to recover values of a^2, Q and f_0. The results for the PFCN are given in Table 4.3, and those for the RFCN are given in Table 4.4.

The rapid decrease with time of both the PFCN and RFCN modes shown in Tables 4.3 and 4.4 suggests that they may be in free decay. In free decay, the amplitudes of the free core nutations, with squares given by expressions (4.114) and (4.115), follow the exponential decay schemes

Table 4.4 *Parameters fitted to successive retrograde free core nutation (RFCN) resonances at 400-day steps along the time axis.*

Time in days	$a^2\left(\text{mas}^2/\text{cpd}\right)$	Q	$f_0\left(10^{-3}\ \text{cpd}\right)$	Period (days)
1600	168.91	3.4934	−2.00415	−498.965
2000	217.10	8.3631	−2.26549	−441.406
2400	204.58	8.3887	−2.27840	−438.904
2800	180.64	8.2660	−2.26412	−441.673
3200	132.81	8.3479	−2.27136	−440.265
3600	53.808	6.4096	−2.38518	−419.256
4000	34.536	5.7286	−2.38140	−419.921
4400	24.125	5.1931	−2.42864	−411.753
4800	19.160	5.2115	−2.41370	−414.302
5200	15.594	6.1112	−2.31616	−431.749
5600	14.098	5.5330	−2.30438	−433.956
6000	8.5950	3.6275	−2.27682	−439.209
6400	7.8865	3.5702	−2.25907	−442.660
6800	6.8884	2.9255	−2.14030	−467.224

$$A = A_0 e^{\pm \pi t/QT}, \tag{4.116}$$

where A_0 is the amplitude at time $t = 0$ and T is the period (positive for the PFCN, negative for the RFCN). The positive sign in the exponent refers to the RFCN, the negative sign in the exponent refers to the PFCN. In free decay, the logarithms of the amplitudes of the free core nutations then follow the linear relation

$$\log A = ct + d, \tag{4.117}$$

with the slope given by

$$c = \pm \frac{\pi \log e}{QT} = -\frac{\log e}{\tau}, \tag{4.118}$$

where τ is the e-folding time, and the intercept is $d = \log A_0$. The half-life is then

$$t_{1/2} = \tau \ln 2. \tag{4.119}$$

A plot of the logarithm of the amplitude of the PFCN in microarcseconds (μas) against time is shown in Figure 4.18. The straight line, fitted by least squares, returns the values $c = -1.14877 \pm 0.0741224 \times 10^{-4}$ and $d = 2.3881 \pm 0.0299684$.

A similar plot of the logarithm of the amplitude of the RFCN in microarcseconds (μas) against time is shown in Figure 4.19. The straight line, fitted by least squares, returns the values $c = -1.35076 \pm 0.0896257 \times 10^{-4}$ and $d = 2.6179 \pm 0.0403216$.

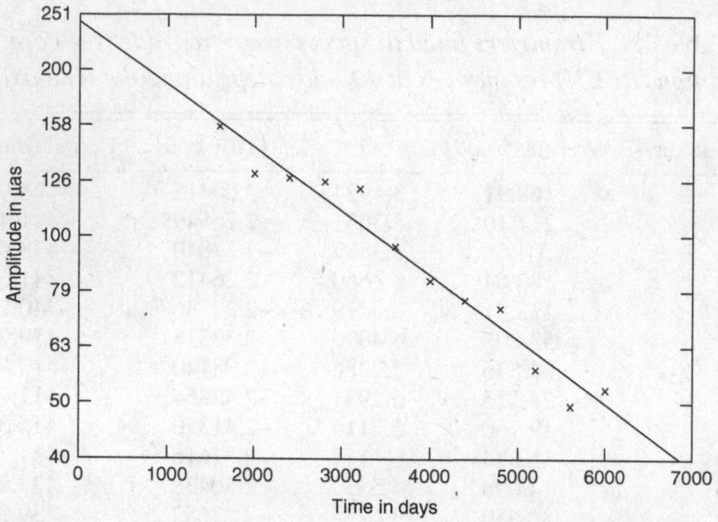

Figure 4.18 Ring down of the prograde free core nutation (PFCN).

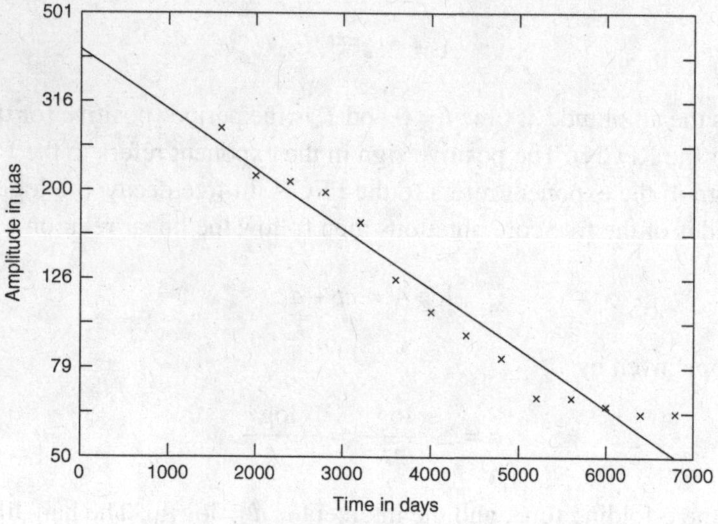

Figure 4.19 Ring down of the retrograde free core nutation (RFCN).

4.4.4 Geometry of the free core nutations and wobbles

Even in realistic Earth models, the free core nutations are closely rigid-body rotations of the fluid outer core with respect to the nearly stationary mantle and crust (shell) (Jiang, 1993; Smylie and Jiang, 1993). In the Earth frame, both the prograde free core nutation (PFCN) and the retrograde free core nutation (RFCN) involve

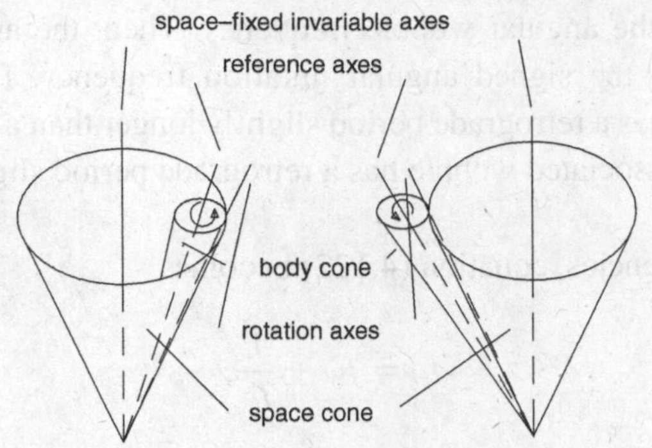

Figure 4.20 Poinsot construction for the PFCN (right) and the RFCN (left).

nearly diurnal retrograde wobbles with angular frequencies close to $-\Omega$, where Ω is the angular frequency of Earth's rotation. The relation between the nutations and wobbles is illustrated by the Poinsot constructions in Figure 4.20. For the PFCN, a small body cone rolls without slipping, once per sidereal day, on the outside of a large space-fixed cone, as shown on the right of Figure 4.20. The RFCN can be described as a small body cone rolling, without slipping, once per sidereal day on the inside of a large space-fixed cone, as shown on the left of Figure 4.20. With slipping, the body cone in the RFCN would make one revolution backwards for each traverse of the space cone. For nutation amplitude A_N and wobble amplitude A_W, without slipping the body cone rotates forward A_N/A_W times for each traverse of the space cone. Then,

$$\frac{A_N}{A_W} - 1 = -\frac{\Omega}{\sigma_N}, \tag{4.120}$$

with σ_N denoting the signed angular nutation frequency (negative for the RFCN). For the PFCN, shown on the right of Figure 4.20, with slipping, the body cone makes one revolution forward for each traverse of the space cone. Without slipping, the body cone rotates forward A_N/A_W times for each traverse of the space cone. Then,

$$\frac{A_N}{A_W} + 1 = \frac{\Omega}{\sigma_N}, \tag{4.121}$$

with σ_N denoting the positive angular nutation frequency. For both modes, the rotation axis is along the line of contact of the two cones. Its motion in space is the sum of the motion of the reference axis and its motion with respect to the reference axis. Thus,

$$\sigma_N = \sigma_W + \Omega, \tag{4.122}$$

with σ_W denoting the angular wobble frequency. Then, the angular wobble frequency is $-\Omega$ plus the signed angular nutation frequency. For the PFCN, the associated wobble has a retrograde period slightly longer than a sidereal day, while for the RFCN the associated wobble has a retrograde period slightly shorter than a sidereal day.

In terms of frequencies, equation (4.122) becomes

$$f_N = f_W + \frac{1}{T_s}, \tag{4.123}$$

with f_N, f_W denoting the nutation and wobble frequencies and T_s representing the length of the sidereal day. The length of the sidereal day is $T_s = 2\pi/\Omega$. For the adopted mean rate of rotation of the Earth, $\Omega = 7.292115 \times 10^{-5}$ rad/s and $T_s = 0.9972696$ days. Combining relations (4.120) and (4.122) gives the ratio of the nutation amplitude to the wobble amplitude for the RFCN as

$$\frac{A_N}{A_W} = \frac{\sigma_W}{\sigma_W + \Omega}, \tag{4.124}$$

while combining relations (4.121) and (4.122) gives the ratio of the nutation amplitude to the wobble amplitude for the PFCN as

$$\frac{A_N}{A_W} = -\frac{\sigma_W}{\sigma_W + \Omega}. \tag{4.125}$$

In the space frame of reference, in which the nutations are observed, the resonances are of the form (4.111). To indicate that they refer to nutations, we write them as

$$\frac{a_N^2}{1 + Q_N^2\left[(f_N - f_{N_0})/f_{N_0}\right]^2}. \tag{4.126}$$

Of more direct geophysical interest are the resonances of the associated nearly diurnal wobbles, which we write as

$$\frac{a_W^2}{1 + Q_W^2\left[(f_W - f_{W_0})/f_{W_0}\right]^2}. \tag{4.127}$$

Expressing the right sides of equations (4.124) and (4.125) in terms of frequencies via relation (4.123), we find that

$$A_W = \pm\frac{f_{N_0}}{f_{N_0} - 1/T_s} A_N, \tag{4.128}$$

with the positive sign referring to the RFCN and the negative sign referring to the PFCN. Comparing the denominators of the resonances expressed by the forms (4.126) and (4.127), and again using the frequency relation (4.123), the wobble Q_W is found to be related to the nutation Q_N by

$$Q_W = \pm \frac{f_{N_0} - 1/T_s}{f_{N_0}} Q_N. \tag{4.129}$$

Again the positive sign refers to the RFCN and the negative sign refers to the PFCN.

The wobble and nutation frequencies, from expression (4.123), are related by

$$f_W = f_N - \frac{1}{T_s}. \tag{4.130}$$

The wobble period is then related to the nutation period by

$$\frac{1}{T_W} = \frac{1}{T_N} - \frac{1}{T_s}, \tag{4.131}$$

or

$$T_W = \frac{T_s T_N}{T_s - T_N}. \tag{4.132}$$

It is of interest to find the amplitude, period and Q of the nearly diurnal wobbles associated with each of the two nutation resonances, fitted to the average spectral density, in Figures 4.16 and 4.17. For the nearly diurnal wobble associated with the PFCN, the amplitude $A_W = 0.230713$ microarcseconds (µas), the period is $T_W = -0.999838$ days and $Q_W = 2287.20$. The nearly diurnal wobble associated with the RFCN has amplitude $A_W = 0.444719$ microarcseconds (µas), period $T_W = -0.994893$ days and $Q_W = 2800.50$.

The nutation Q_N can be recovered from the straight lines fitted to the ring downs of the PFCN and RFCN, shown in Figures 4.18 and 4.19, using the values of the nutation periods recovered from the resonances fitted to the average spectral density of the nutation residuals, shown in Figures 4.16 and 4.17. The slope of the fitted straight lines is

$$c = \pm \frac{\pi \log e}{Q_N T_N} = -\frac{\log e}{\tau}, \tag{4.133}$$

Table 4.5 *Free decay parameters of the PFCN and RFCN ring downs.*

$c\ (10^{-4})$	d	A_{N_0} (µas)	Q_N	Q_W	$t_{1/2}$ (days)
PFCN					
−1.14877	2.3881	244.40	30.596	11879	2620.45
RFCN					
−1.35076	2.6179	414.86	24.198	10152	2228.60

allowing solution for Q_N and τ, and the half-life $t_{1/2} = \tau \ln 2$. The intercepts of the fitted straight lines yield

$$d = \log A_{N_0}. \tag{4.134}$$

The wobble Q_W, reflecting the viscosity at the top of the fluid outer core, can then be recovered via relation (4.129). The free decay parameters recovered from the ring downs of the PFCN and the RFCN are listed in Table 4.5.

5
Earth's figure and gravitation

The study of Earth's figure and gravitation goes back to the very roots of the physical sciences. We begin with a description of its very interesting historical roots.

5.1 Historical development

Although ancient Greek and Egyptian philosophers and astronomers believed the Earth to be spherical and had made rough measurements of its size, the modern theory of the Earth's figure originates with the work of Newton. Using his newly developed laws of dynamics and gravitation, and making the remarkable assumptions that the Earth's figure is nearly an oblate spheroid (the surface generated by an ellipse revolved about its minor axis) and that the Earth behaves as a fluid, he was able to show that the ellipticity of figure is directly given by the ratio of centrifugal force to gravitational force at the equator. As a result of the neglect of the concentration of mass towards the Earth's centre, Newton's estimate of the ellipticity was more than 30% too large. However, the oblate spheroid continues as the figure of reference for geodesy, and the assumption of hydrostatic equilibrium remains as the basis of much of the modern theory. Perhaps Newton's most important contribution, though, was his unequivocal demonstration that any theory of the Earth's figure must take account of both its gravitation and its rotation.

Newton published his calculation in the first edition of the *Principia* in 1687, but the meridian arc length per angular unit of geographical latitude, as measured on European baselines, appeared to increase with increasing latitude, implying that the Earth was prolate rather than oblate. It was not until expeditions had been dispatched by the French Academy to Peru and Lapland that the matter was settled in Newton's favour in the 1740s.

In 1743, a few years after his return as a member of the Lapland expedition, Clairaut published a theory that related the latitudinal increase of gravity (gravitational plus centrifugal force) to the ellipticity and the ratio of centrifugal to gravitational force at the equator. He also gave, in the form of a differential equation, the law of variation of the ellipticity of level surfaces in the interior of the Earth, provided that the law of density variation with depth is prescribed. His theory remains largely unaltered to this day as the first-order theory of the Earth's figure.

It soon became apparent from astronomic measurements that real level surfaces are far from regular and depart markedly in curvature from the ellipsoid that describes the figure as a whole. In 1828, the suggestion was made by Gauss that, in a mathematical sense, the surface of the Earth is nothing but that surface 'which everywhere intersects the direction of gravity at right angles, and of which the surface of the oceans is a part'. This was the forerunner to the definition of the *geoid* in modern physical geodesy as 'that equipotential surface of the Earth's gravity field, which, on the average, coincides with mean sea level in the open undisturbed ocean' (Fischer, 1975), although the name geoid was not applied to this surface until 1872, by J. B. Listing.

The usefulness of the geoid as a reference surface was enhanced by a theorem of Stokes in 1849, which gave the local geoidal height above the reference ellipsoid solely in terms of the magnitude of the anomalous gravity, measured on the geoid, compared to that expected on the ellipsoid. Although practical difficulties are encountered in applying the theorem, it showed that gravity measurements alone could, in principle, be used to determine the geoid.

While geodesy was developing into a flourishing applied science, the problem of the equilibrium figure of rotating, gravitating fluid bodies was attracting the attention of a series of distinguished mathematicians. First, in 1742, Maclaurin demonstrated that the oblate spheroid was exactly the equilibrium figure of a uniform fluid mass rotating as a solid body and that it remained a possible figure of equilibrium no matter how great the rotation speed. An interesting feature of Maclaurin's solution, subsequently noticed by Simpson and d'Alembert, is that for low rotation speeds two oblata are possible: one of small ellipticity, as might be expected; but, surprisingly, another of great ellipticity, which tends to disk shape as the angular velocity tends to zero. Another surprising result was obtained in 1834 by Jacobi, who showed that as the rotation speed increases the Maclaurin spheroids are not the only possible figures of equilibrium and that triaxial ellipsoids are permissible as well. Indeed, in the presence of dissipation, above a critical angular velocity, the Maclaurin spheroids branch over to Jacobi ellipsoids. Finally, Poincaré demonstrated in 1885 that, at still greater angular velocity, pear-shaped figures emerge, but in 1916 these were shown by Jeans to be unstable. The subject has had a modern rebirth (Chandrasekhar, 1969), but most of its results apply to bodies with equal density surfaces of perfectly ellipsoidal shape, with relative rotation speeds far in excess of that of the Earth and, therefore, not of great geophysical consequence.

For more than a century Clairaut's differential equation, governing the variation of the ellipticity of the level surfaces in the Earth, remained unused, owing to the apparent need to first specify the density variation with depth. In 1885, Radau found a transformation of the second-order Clairaut equation that reduced it to first order. G. H. Darwin (1899), in the course of extending the theory of the Earth's figure to second-degree terms in small quantities, used the Radau transformation to show that, to good approximation, the internal density distribution enters the theory only through the ratio of the radius of gyration to the equatorial radius. Therefore, little about the density distribution was to be learned from geodetic measurements, but the approximation allowed the ellipticity of figure to be determined from the precessional constant.

The advent of the artificial Earth satellite has put the study of the figure and gravitational field on a new level of sophistication. Since the orbit of a satellite is determined by the gravitational field of the Earth, apart from perturbing effects such as atmospheric drag and radiation pressure, inversion of tracking data can be performed to produce a very detailed description of the gravitational field. Soon after the first satellites were launched, a much more accurate direct measure of the dynamical ellipticity became available (O'Keefe *et al.*, 1958). This led to a value for the ellipticity of figure, independent of the hydrostatic assumption, with far greater accuracy than had been available from conventional geodesy. It also altered the moment of inertia constraint on Earth models computed from seismological information, requiring a consequent revision of the density profile (Bullen, 1975). Further, the ellipticity of figure, determined through solution of the Clairaut equation on the hydrostatic assumption, was shown to be significantly smaller than that based on the directly observed dynamical ellipticity. Satellites, in addition, have permitted the techniques of geometrical geodesy to be used over great distances, tying together reference systems between continents; and, since their orbits are nearly about the geocentre, they give direct access to true geocentric co-ordinates.

5.2 External gravity and figure

Description of the Earth's figure in mathematical terms is usually accomplished, as Gauss suggested, by specifying the shape of a level reference surface. It is possible in principle to measure topography directly, and it may even become practical in the future to describe topography in detail in geocentric co-ordinates, through the application of evolving space measurement techniques; however, the level reference surface is not only more convenient in theoretical development but also arises naturally in surveying, gravimetry and astronomic observations.

The reference level surface is an equipotential U_0 of the total geopotential, or gravity potential,

$$U = V + W, \tag{5.1}$$

where V is the gravitational potential and W is the centrifugal potential. With Ω as the Earth's mean rate of rotation, r the radius from the geocentre and θ the geocentric co-latitude, the centrifugal force per unit mass (3.76) becomes $\Omega^2 r \sin\theta$. In turn, this can be expressed as the negative gradient of the centrifugal potential

$$\begin{aligned} W &= -\frac{1}{2}\Omega^2\left(x_1^2 + x_2^2\right) \\ &= -\frac{1}{2}\Omega^2 r^2 \sin^2\theta \\ &= -\frac{1}{3}\Omega^2 r^2 + \frac{1}{3}\Omega^2 r^2 P_2(\cos\theta), \end{aligned} \tag{5.2}$$

$P_2 = \left(3\cos^2\theta - 1\right)/2$ being the second-degree Legendre polynomial. The gravitational potential is

$$V(\boldsymbol{r}) = -G \iiint \frac{\rho(\boldsymbol{r}')}{|\boldsymbol{r} - \boldsymbol{r}'|} d\mathcal{V}', \tag{5.3}$$

where the potential at the field point at $\boldsymbol{r}$ is found by integrating over all source points at $\boldsymbol{r}'$. Further, G is the universal constant of gravitation and ρ is the mass density. Just as expression (1.168) is the solution of the Poisson equation (1.163), the gravitational potential obeys the Poisson equation

$$\nabla^2 V = 4\pi G\rho. \tag{5.4}$$

Thus, the equation governing the geopotential is easily shown to be

$$\nabla^2 U = 4\pi G\rho - 2\Omega^2. \tag{5.5}$$

Identification of the particular equipotential surface, U_0, being used for reference can be made through a number of geometrical parameters, but the *mean equivolumetric radius*, d, appears most convenient to use. The mean equivolumetric radius, or more simply the *mean radius*, is the radius of a sphere containing the same volume as the reference surface. In considering the global figure problem, it is illuminating to make (5.5) non-dimensional by expressing the variables in terms of d and the gravitational attraction GM/d^2 on the surface of a spherically symmetric mass distribution of radius d enclosing the same mass M and volume $\mathcal{V}$ as the Earth's surface. The geopotential U is then expressed in units of the gravitational potential GM/d, and ρ is expressed in units of the mean density $3M/4\pi d^3$, while the co-ordinates are given in units of d. The dimensionless version of (5.5) is then

$$\nabla^2 U = 3\rho - 2\frac{\Omega^2 d^3}{GM}. \tag{5.6}$$

The problem is now expressed solely in terms of the dimensionless parameter,

$$m = \frac{\Omega^2 d^3}{GM} \approx 3.45 \times 10^{-3} = \frac{1}{290}, \tag{5.7}$$

the ratio of centrifugal force to gravitational force on the surface of a hypothetical, spherically symmetric mass distribution with the same mean radius and mean density as the Earth. We shall see later that it is very close to the parameter introduced by Newton in the first physical discussion of the Earth's figure. It would now be called a Froude number in hydrodynamics. The utility of expressing the problem in dimensionless form lies in the fact that, for the Earth, m is only about 1/290, so expansion of the dimensionless variables in powers of m permits uniform approximation by rapidly converging series.

We expect the reference level surface to be a surface of revolution and to be symmetrical under reflection in the equatorial plane. At least, there is no known theoretical reason, at the relatively low rotation rate of the Earth, to expect a departure of the overall figure from these symmetries. Thus, the dimensionless radius of the reference level surface, R, is an even function of the vertical distance to the equatorial plane, proportional to $\cos\theta$, and has the expansion

$$R(\theta) = R_0(\theta) + mR_1(\theta) + m^2R_2(\theta) + \cdots, \tag{5.8}$$

where the functions of θ are expressible in series of Legendre polynomials of even degree.

The function $R_0(\theta)$ describes the figure free of the influence of rotation. A proof that a sphere is the unique figure of equilibrium for a self-gravitating fluid at rest was given by Liapounov in 1918 (Liapounov, 1930). Thus, in dimensionless form,

$$R_0 = 1 \tag{5.9}$$

describes our zeroth-order approximation to the Earth's figure. With λ as the east longitude, the volume enclosed by the true reference level surface is

$$\begin{aligned} \mathcal{V} &= \int_0^{2\pi}\int_0^{\pi}\int_0^{R(\theta)} r^2 dr \sin\theta \, d\theta \, d\lambda \\ &= \frac{2\pi}{3}\int_0^{\pi} R^3(\theta)\sin\theta \, d\theta. \end{aligned} \tag{5.10}$$

Now,

$$R(\theta) = 1 + mR_1(\theta) + m^2R_2(\theta), \tag{5.11}$$

so with binomial expansion of R^3, we have

$$\begin{aligned}\mathcal{V} &= \frac{2\pi}{3}\int_0^\pi \left\{1 + 3mR_1 + 3m^2\left(R_2 + R_1^2\right) + \cdots\right\}\sin\theta\, d\theta \\ &= \frac{4\pi}{3} + 2\pi m\int_0^\pi R_1 \sin\theta\, d\theta \\ &\quad + 2\pi m^2 \int_0^\pi \left(R_2 + R_1^2\right)\sin\theta\, d\theta + \cdots,\end{aligned} \tag{5.12}$$

showing that the volume is given correctly by the term of zeroth order in m. The corresponding equipotential expressed in dimensionless variables is -1, and dimensionless gravity, everywhere uniform on the spherical surface (5.9), is $+1$.

A theory of the figure, correct to terms of first order in m, must include Legendre polynomials of at least second degree in $\cos\theta$, since in dimensionless variables expression (5.2) for the centrifugal potential becomes

$$W = -\frac{1}{3}mr^2 + \frac{1}{3}mr^2 P_2(\cos\theta). \tag{5.13}$$

Thus, $R_1(\theta)$ in (5.8) has terms up to and including $P_2(\cos\theta)$. However, since it must be even in $\cos\theta$, its $P_1(\cos\theta)$ term is missing, and its $P_0(\cos\theta)$ term must be zero, in order to correctly give $\mathcal{V}$ in the expansion (5.12). If a is the equatorial radius of the true reference surface, and c its polar radius,

$$a = 1 + mR_1(\pi/2) + m^2R_2(\pi/2) + \cdots, \tag{5.14}$$

$$c = 1 + mR_1(0) + m^2R_2(0) + \cdots. \tag{5.15}$$

By binomial expansion,

$$\frac{1}{a} = 1 - mR_1(\pi/2) + m^2\left\{R_1^2(\pi/2) - R_2(\pi/2)\right\} + \cdots. \tag{5.16}$$

The remaining coefficient of $P_2(\cos\theta)$ in $R_1(\theta)$ is then determined by the flattening

$$\begin{aligned} f &= \frac{a-c}{a} \\ &= m\{R_1(\pi/2) - R_1(0)\} \\ &\quad + m^2\{R_2(\pi/2) - R_2(0) - R_1^2(\pi/2) + R_1(\pi/2)R_1(0)\} + \cdots. \end{aligned} \tag{5.17}$$

For the flattening to be given correctly to first order in m, we must have

$$R_1(\theta) = -\frac{2}{3}\frac{f}{m}P_2(\cos\theta), \tag{5.18}$$

showing that f is of order m.

To complete the first-order theory, the equipotential is now augmented by the second-degree term of order m in the gravitational potential and by the centrifugal potential expressed in (5.13). Outside the Earth, the gravitational potential obeys Laplace's equation. Hence, for a reference level surface that is a surface of revolution, symmetrical under reflection in the equatorial plane, it is represented by a series with terms $P_n(\cos\theta)/r^{n+1}$, where the Legendre polynomials are of even degree n. With the expression of R in powers of m (5.8), and again using the binomial expansion, in dimensionless variables we have

$$\begin{aligned} U_0 &= -\frac{1}{R} + \frac{\alpha m}{R^3} P_2(\cos\theta) - \frac{1}{3} m R^2 + \frac{1}{3} m R^2 P_2(\cos\theta) \\ &= -\{1 - mR_1 - m^2\left(R_2 - R_1^2\right) + \cdots\} \\ &\quad + \alpha m\{1 - 3mR_1 + \cdots\} P_2(\cos\theta) \\ &\quad - \frac{1}{3} m\{1 + 2mR_1 + \cdots\} + \frac{1}{3} m\{1 + 2mR_1 + \cdots\} P_2(\cos\theta), \end{aligned} \tag{5.19}$$

with α a coefficient to be determined. With substitution for R_1 from (5.18), correct to first order in m we have that

$$\begin{aligned} U_0 = &- \left\{1 + m \cdot \frac{2}{3} \frac{f}{m} P_2(\cos\theta)\right\} + \alpha m P_2(\cos\theta) \\ &- \frac{1}{3} m + \frac{1}{3} m P_2(\cos\theta). \end{aligned} \tag{5.20}$$

Recognising that U_0 is a constant, this equation, on equating coefficients of Legendre polynomials to terms correct to first order in m, yields

$$U_0 = -1 - \frac{1}{3} m, \tag{5.21}$$

with α found to be

$$\alpha = -\frac{1}{3}\left\{1 - 2\frac{f}{m}\right\}. \tag{5.22}$$

The reference surface (5.19) is an equipotential of the gravity potential outside the Earth, given by

$$U = -\frac{1}{r} + \frac{\alpha m}{r^3} P_2(\cos\theta) + \frac{1}{3} m r^2 (P_2(\cos\theta) - 1), \tag{5.23}$$

correct to first order in m. Normally, ground-based gravimeters measure normal gravity, or the gravity intensity,

$$g = |\nabla U| = \sqrt{\left(\frac{\partial U}{\partial r}\right)^2 + \frac{1}{r^2}\left(\frac{\partial U}{\partial \theta}\right)^2}, \tag{5.24}$$

with

$$\frac{\partial U}{\partial r} = \frac{1}{r^2} - 3\frac{\alpha m}{r^4} P_2 + \frac{2}{3} m r (P_2 - 1), \tag{5.25}$$

and

$$\frac{1}{r}\frac{\partial U}{\partial \theta} = \left(\frac{\alpha m}{r^4} + \frac{1}{3}mr\right)\frac{dP_2}{d\theta}. \tag{5.26}$$

On the level reference surface $U = U_0$ and $r = R$. By binomial expansion,

$$\begin{aligned}\frac{\partial U}{\partial r} &= 1 - m\left\{2R_1 + 3\alpha P_2 - \frac{2}{3}(P_2 - 1)\right\} \\ &\quad - m^2\left\{2R_1 - 3R_1^2 - 12\alpha R_1 P_2 - \frac{2}{3}R_1(P_2 - 1)\right\} + \cdots,\end{aligned} \tag{5.27}$$

and

$$\frac{1}{r}\frac{\partial U}{\partial \theta} = \left\{m\left(\alpha + \frac{1}{3}\right) + m^2\left(\frac{1}{3}R_1 - 4\alpha R_1\right)\right\}\frac{dP_2}{d\theta} + \cdots. \tag{5.28}$$

By binomial expansion, the intensity of gravity (5.24) becomes

$$\begin{aligned}|\nabla U| &= \frac{\partial U}{\partial r}\left(1 + \frac{1}{2r^2}\left(\frac{\partial U}{\partial \theta}\right)^2 \Big/ \left(\frac{\partial U}{\partial r}\right)^2 + \cdots\right) \\ &= \frac{\partial U}{\partial r} + \frac{1}{2r^2}\left(\frac{\partial U}{\partial \theta}\right)^2 + \cdots \\ &= \frac{\partial U}{\partial r} + \frac{1}{2}m^2\left(\alpha + \frac{1}{3}\right)^2\left(\frac{dP_2}{d\theta}\right)^2 + \cdots.\end{aligned} \tag{5.29}$$

The intensity of gravity, $|\nabla U|$, on the equipotential described by (5.19) is then

$$\begin{aligned}g &= 1 - m\left\{2R_1 + 3\alpha P_2 - \frac{2}{3}(P_2 - 1)\right\} \\ &\quad - m^2\left\{2R_2 - 3R_1^2 - 12\alpha R_1 P_2 - \frac{2}{3}R_1(P_2 - 1)\right. \\ &\quad \left. + \left(\frac{1}{2}\alpha^2 + \frac{1}{3}\alpha + \frac{1}{18}\right)\left(\frac{dP_2}{d\theta}\right)^2\right\} + \cdots.\end{aligned} \tag{5.30}$$

Substituting for R_1 and α from (5.18) and (5.22), respectively, and retaining terms to first order in m,

$$g = 1 - \frac{m}{3}\left\{2 - \left(5 - 2\frac{f}{m}\right)P_2\right\} + \cdots. \tag{5.31}$$

Evaluated at the equator, (5.31) gives

$$g_e = 1 - \frac{m}{3}\left\{\frac{9}{2} - \frac{f}{m}\right\} + \cdots, \tag{5.32}$$

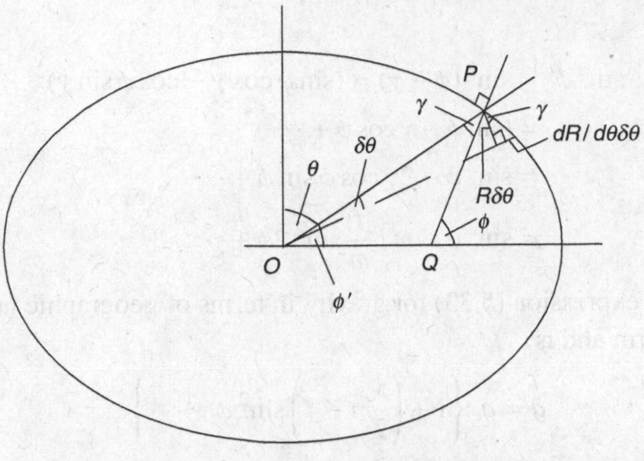

Figure 5.1 An observer at P on the reference level surface with radius $R(\theta)$ measures geographic latitude ϕ with respect to the zenith line QP normal to the level surface. Geocentric latitude ϕ' is measured with respect to the radius line from the geocentre at O.

allowing g to be expressed in terms of g_e, its value at the equator, to first order in m as

$$g = g_e \left\{ 1 + \left(\frac{5}{2} m - f \right) \sin^2 \phi' + \cdots \right\}, \tag{5.33}$$

ϕ' being the geocentric latitude.

Geographic latitude ϕ is often more accessible observationally than ϕ'. Their relationship is illustrated in Figure 5.1.

This shows directly that

$$\phi = \phi' + \gamma, \tag{5.34}$$

where γ is determined to be

$$\gamma = \arctan \left\{ \frac{1}{R} \frac{dR}{d\theta} \right\} \tag{5.35}$$

by considering the effect of a small increment, $\delta\theta$, in the geocentric co-latitude shown by broken lines in Figure 5.1. To first order in m, we have

$$\begin{aligned} \gamma &= m \frac{dR_1}{d\theta} + \cdots = 2m \cdot \frac{f}{m} \cos\theta \sin\theta + \cdots \\ &= 2m \cdot \frac{f}{m} \cos\left(\frac{\pi}{2} - \phi' \right) \sin\left(\frac{\pi}{2} - \phi' \right) + \cdots = 2m \cdot \frac{f}{m} \sin\phi' \cos\phi' + \cdots \\ &= 2m \cdot \frac{f}{m} \sin\phi \cos\phi + \cdots \end{aligned} \tag{5.36}$$

and

$$\begin{aligned}\sin^2\phi' &= \sin^2(\phi-\gamma) = (\sin\phi\cos\gamma - \cos\phi\sin\gamma)^2\\ &= (\sin\phi - \gamma\cos\phi + \cdots)^2\\ &= \sin^2\phi - 2\gamma\cos\phi\sin\phi + \cdots\\ &= \sin^2\phi - m\cdot\frac{f}{m}\sin^2 2\phi + \cdots.\end{aligned} \tag{5.37}$$

The first-order expression (5.33) for gravity in terms of geographic latitude is then unaltered in form and is

$$g = g_e\left\{1 + \left(\frac{5}{2}m - f\right)\sin^2\phi + \cdots\right\}. \tag{5.38}$$

The first-order description of the figure contains Legendre polynomials up to and including second degree. We can then expect $R_2(\theta)$ in the figure expression (5.8) to include Legendre polynomials of fourth and lesser degrees. Odd degree polynomials are missing because $R_2(\theta)$ is an even function of $\cos\theta$. For $z = \cos\theta$,

$$(P_2)^2 = \frac{9}{4}z^4 - \frac{3}{2}z^2 + \frac{1}{4}. \tag{5.39}$$

Since

$$P_4 = \frac{1}{8}\left(35z^4 - 30z^2 + 3\right), \tag{5.40}$$

we have the identity

$$(P_2)^2 = \frac{18}{35}P_4 + \frac{2}{7}P_2 + \frac{1}{5}. \tag{5.41}$$

Description of the volume $\mathcal{V}$ in (5.12), correct to terms of second order in m, requires the constant term in $R_2(\theta)$ to be $(-4/45)(f^2/m^2)$. The coefficients of the remaining polynomials, of second and fourth degree, are related by the requirement that the flattening given by (5.17) has a vanishing second-order term in m. Only one coefficient, δ, remains to be determined, and we have that

$$R_2(\theta) = -\frac{4}{45}\frac{f^2}{m^2}\left\{1 + \frac{5}{2}\left(1 - \frac{1}{6}\delta\right)P_2 + \delta P_4\right\}. \tag{5.42}$$

To complete the second-order theory, we add the fourth-degree term in the gravitational potential,

$$\frac{\beta m^2}{R^5}P_4(\cos\theta) = \beta m^2 P_4(\cos\theta) + \cdots \tag{5.43}$$

to expression (5.19) for the equipotential, and the corresponding term,

$$-5\beta m^2 P_4(\cos\theta) + \cdots, \tag{5.44}$$

to formula (5.30) for the gravity intensity. Equating to second order in m the coefficients of the zeroth-, second- and fourth-degree Legendre polynomials, respectively, in the new expression for the equipotential, after substitution from (5.18), (5.22), (5.41) and (5.42), yields

$$U_0 = -1 - \frac{1}{3}m - \frac{2}{9}m^2 \cdot \frac{f}{m}\left(1 - \frac{2}{5}\frac{f}{m}\right), \tag{5.45}$$

$$\alpha = -\frac{1}{3}\left(1 - 2\frac{f}{m}\right) - \frac{8}{63}m \cdot \frac{f}{m}\left\{1 + \frac{1}{4}\left(1 + \frac{7}{6}\delta\right)\frac{f}{m}\right\}, \tag{5.46}$$

$$\beta = \frac{4}{7}\frac{f}{m}\left\{1 - \frac{4}{5}\left(1 - \frac{7}{36}\delta\right)\frac{f}{m}\right\}. \tag{5.47}$$

For $z = \cos\theta$,

$$\left(\frac{dP_2}{d\theta}\right)^2 = 9\cos^2\theta\sin^2\theta = 9z^2\left(1 - z^2\right) = -9z^4 + 9z^2. \tag{5.48}$$

This leads to the identity

$$\left(\frac{dP_2}{d\theta}\right)^2 = -\frac{72}{35}P_4 + \frac{6}{7}P_2 + \frac{6}{5}, \tag{5.49}$$

which, combined with substitutions from (5.18), (5.22), (5.41), (5.42), (5.46) and (5.47), allows formula (5.30) for the gravity intensity, augmented by (5.44), to be reduced to

$$\begin{aligned} g = 1 &- \frac{m}{3}\left\{2 - \left(5 - 2\frac{f}{m}\right)P_2\right\} - \frac{m^2}{9}\left\{-4\frac{f}{m} + \frac{16}{5}\frac{f^2}{m^2}\right. \\ &+ \left(-\frac{92}{7}\frac{f}{m} + \frac{26}{7}\frac{f^2}{m^2} - \frac{1}{3}\frac{f^2}{m^2}\delta\right)P_2 \\ &\left. + \left(\frac{108}{7}\frac{f}{m} + \frac{72}{35}\frac{f^2}{m^2} + \frac{12}{5}\frac{f^2}{m^2}\delta\right)P_4\right\} + \cdots, \end{aligned} \tag{5.50}$$

correct to second order.

Geocentric latitude, ϕ', is the complement of the co-latitude, θ, thus $\cos\theta = \sin\phi'$ and

$$P_2(\cos\theta) = \frac{3}{2}\sin^2\phi' - \frac{1}{2}. \tag{5.51}$$

Using the half-angle formula,

$$\sin^2\phi' = \frac{1 - \cos 2\phi'}{2}, \tag{5.52}$$

we have

$$\sin^4\phi' = \sin^2\phi' - \frac{1}{4}\sin^2 2\phi'. \tag{5.53}$$

Then P_4 can be expressed as

$$P_4(\cos\theta) = -\frac{35}{32}\sin^4 2\phi' + \frac{5}{8}\sin^2\phi' + \frac{3}{8}. \tag{5.54}$$

The gravity intensity, as a function of geocentric latitude, is then found to be

$$\begin{aligned} g = 1 &- \frac{m}{3}\left\{\frac{9}{2} - \frac{f}{m} - \left(\frac{15}{2} - 3\frac{f}{m}\right)\sin^2\phi'\right\} \\ &- \frac{m^2}{9}\left\{\frac{117}{14}\frac{f}{m} + \frac{74}{35}\frac{f^2}{m^2} + \frac{16}{15}\frac{f^2}{m^2}\delta\right. \\ &+ \left(-\frac{141}{14}\frac{f}{m} + \frac{48}{7}\frac{f^2}{m^2} + \frac{f^2}{m^2}\delta\right)\sin^2\phi' \\ &+ \left.\left(-\frac{135}{8}\frac{f}{m} - \frac{9}{4}\frac{f^2}{m^2} - \frac{21}{8}\frac{f^2}{m^2}\delta\right)\sin^2 2\phi'\right\} + \cdots, \end{aligned} \tag{5.55}$$

correct to second order. To the same order, gravity at the equator is

$$g_e = 1 - \frac{m}{3}\left\{\frac{9}{2} - \frac{f}{m}\right\} - \frac{m^2}{9}\left\{\frac{117}{14}\frac{f}{m} + \frac{74}{35}\frac{f^2}{m^2} + \frac{16}{15}\frac{f^2}{m^2}\delta\right\} + \cdots. \tag{5.56}$$

Expressed in terms of g_e, we then have that

$$\begin{aligned} g = g_e &\left\{1 + \left(\frac{5}{2}m - f + \frac{15}{4}m^2 - \frac{17}{14}mf - \frac{3}{7}f^2 - \frac{1}{9}f^2\delta\right)\sin^2\phi'\right. \\ &+ \left.\left(\frac{15}{8}mf + \frac{1}{4}f^2 + \frac{7}{24}f^2\delta\right)\sin^2 2\phi' + \cdots\right\}. \end{aligned} \tag{5.57}$$

Using relation (5.37), gravity can be given in terms of geographic latitude as

$$\begin{aligned} g = g_e &\left\{1 + \left(\frac{5}{2}m - f + \frac{15}{4}m^2 - \frac{17}{14}mf - \frac{3}{7}f^2 - \frac{1}{9}f^2\delta\right)\sin^2\phi\right. \\ &- \left.\left(\frac{5}{8}mf - \frac{5}{4}f^2 - \frac{7}{24}f^2\delta\right)\sin^2 2\phi + \cdots\right\}. \end{aligned} \tag{5.58}$$

Because of its mathematical simplicity and its closeness to the true reference surface, the oblate spheroid, or ellipsoid of revolution, is generally used in geodesy as the form of the Earth's figure, even though it cannot be an exact equilibrium form in a body with non-uniform density, such as the Earth (Volterra, 1903). For geophysical purposes, it is therefore necessary to allow for the departure of the reference surface from ellipsoidal shape. The volume of the ellipsoid is

$$\mathcal{V} = \frac{4}{3}\pi a^2 c = \frac{4}{3}\pi a^3(1 - f). \tag{5.59}$$

By definition, the equivolumetric radius d is the radius of a sphere with the same volume, so,

$$a^3(1-f) = d^3. \tag{5.60}$$

In dimensionless variables, the equivolumetric radius d is unity, hence,

$$a = \frac{1}{(1-f)^{1/3}}, \tag{5.61}$$

immediately giving the expansion

$$a = 1 + \frac{1}{3}f + \frac{2}{9}f^2 + \cdots \tag{5.62}$$

for the equatorial radius. The ellipsoidal surface obeys the equation

$$\frac{R^2 \sin^2\theta}{a^2} + \frac{R^2\cos^2\theta}{c^2} = 1. \tag{5.63}$$

Solving for R^2, we find that

$$\begin{aligned} R^2(\theta) &= \frac{a^2}{1-\cos^2\theta + \cos^2\theta\,(1-f)^{-2}} \\ &= \frac{a^2}{1-\cos^2\theta + \cos^2(1+2f+3f^2+\cdots)} \\ &= \frac{a^2}{1+2f\cos^2\theta + 3f^2\cos^2\theta}, \end{aligned} \tag{5.64}$$

giving

$$R(\theta) = a\left[1 - f\cos^2\theta - \frac{3}{2}f^2\left(\cos^2\theta - \cos^4\theta\right) + \cdots\right] \tag{5.65}$$

$$= a\left[1 - f\cos^2\theta - \frac{3}{2}f^2\cos^2\theta\sin^2\theta + \cdots\right] \tag{5.66}$$

$$= a\left[1 - f\cos^2\theta - \frac{3}{8}f^2\sin^2 2\theta + \cdots\right]. \tag{5.67}$$

Using the expansion (5.62) for a then yields

$$\begin{aligned} R(\theta) &= 1 + \frac{1}{3}f\left(1 - 3\cos^2\theta\right) + \frac{1}{9}f^2\left(2 - \frac{33}{2}\cos^2\theta + \frac{27}{2}\cos^4\theta\right) + \cdots \\ &= 1 - \frac{2}{3}fP_2 + f^2\left(-\frac{4}{45} - \frac{23}{63}P_2 + \frac{12}{35}P_4\right) + \cdots. \end{aligned} \tag{5.68}$$

Comparison with the general expression (5.8), given (5.9), (5.18) and (5.42), shows that for the value $\delta = -27/7$ the true reference surface becomes an ellipsoid.

In order to allow for an unknown departure from ellipsoidal shape, we rewrite (5.67) as

$$R(\theta) = a\left[1 - f\cos^2\theta - \left(\frac{3}{8}f^2 + \kappa\right)\sin^2 2\theta + \cdots\right]. \tag{5.69}$$

Then, (5.65) takes the form

$$R(\theta) = a\left[1 - f\cos^2\theta - \left(\frac{3}{2}f^2 + 4\kappa\right)\left(\cos^2\theta - \cos^4\theta\right) + \cdots\right]. \tag{5.70}$$

The expansion (5.62) for a will then have an extra second-order term ϵ, giving

$$a = 1 + \frac{1}{3}f + \frac{2}{9} + \epsilon + \cdots. \tag{5.71}$$

With this expansion of a, we then find that

$$\begin{aligned} R(\theta) &= 1 + \frac{1}{3}f\left(1 - 3\cos^2\theta\right) + \frac{1}{9}f^2\left[2 + 9\frac{\epsilon}{f^2} - \left(\frac{33}{2} + 36\frac{\kappa}{f^2}\right)\cos^2\theta \right. \\ &\quad \left. + \left(\frac{27}{2} + 36\frac{\kappa}{f^2}\right)\cos^4\theta\right] + \cdots \\ &= 1 - \frac{2}{3}fP_2 + f^2\left[\frac{\epsilon}{f^2} - \frac{4}{45}\left(1 + 6\frac{\kappa}{f^2}\right) - \frac{1}{63}\left(23 + 24\frac{\kappa}{f^2}\right)P_2 \right. \\ &\quad \left. + \frac{1}{35}\left(12 + 32\frac{\kappa}{f^2}\right)P_4\right] + \cdots. \end{aligned} \tag{5.72}$$

Comparison with (5.42) shows that

$$\delta = -\frac{27}{7} - \frac{72}{7}\frac{\kappa}{f^2}. \tag{5.73}$$

Expression (5.42) then has the alternative form,

$$R_2(\theta) = -\frac{4}{45}\frac{f^2}{m^2}\left\{1 + \frac{5}{7}\left(\frac{23}{4} + 6\frac{\kappa}{f^2}\right)P_2 - \frac{9}{7}\left(3 + 8\frac{\kappa}{f^2}\right)P_4\right\}. \tag{5.74}$$

The expression for the true reference surface, (5.8), then becomes

$$\begin{aligned} R(\theta) &= 1 - \frac{2}{3}fP_2 \\ &\quad - \frac{4}{45}f^2\left[1 + \frac{5}{7}\left(\frac{23}{4} + 6\frac{\kappa}{f^2}\right)P_2 - \frac{9}{7}\left(3 + 8\frac{\kappa}{f^2}\right)P_4\right] + \cdots, \end{aligned} \tag{5.75}$$

showing that the expansion (5.71) for the dimensionless equatorial radius takes the form

$$a = 1 + \frac{1}{3}f + \frac{2}{9}f^2 + \frac{8}{15}\kappa + \cdots, \tag{5.76}$$

giving the extra second-order term ϵ as

$$\epsilon = \frac{8}{15}\kappa. \tag{5.77}$$

With substitution for δ from (5.73), the coefficients of the second- and fourth-degree terms in the gravitational potential, (5.46) and (5.47), become

$$\alpha = -\frac{1}{3}\left(1 - 2\frac{f}{m}\right) - \frac{8}{63}m \cdot \frac{f}{m}\left\{1 - \left(\frac{7}{8} + 3\frac{\kappa}{f^2}\right)\frac{f}{m}\right\}, \tag{5.78}$$

$$\beta = \frac{4}{7}\frac{f}{m}\left\{1 - \frac{1}{5}\left(7 + 8\frac{\kappa}{f^2}\right)\frac{f}{m}\right\}. \tag{5.79}$$

The geocentric latitude, ϕ', is the complement of the co-latitude, θ; hence, $\cos\theta = \sin\phi'$ and $\sin\theta = \cos\phi'$. Equation (5.66) for the figure is transformed to

$$\begin{aligned} R &= a\left[1 - f\cos^2\theta - \left(\frac{3}{2}f^2 + 4\kappa\right)\cos^2\theta\sin^2\theta + \cdots\right] \\ &= a\left[1 - f\sin^2\phi' - \left(\frac{3}{2}f^2 + 4\kappa\right)\sin^2\phi'\cos^2\phi' + \cdots\right] \\ &= a\left[1 - f\sin^2\phi' - \left(\frac{3}{8}f^2 + \kappa\right)\sin^2 2\phi' + \cdots\right], \end{aligned} \tag{5.80}$$

showing that, with the figure expressed in terms of geocentric latitude, κ again represents a perturbation to the second-order term of an ellipsoid.

Returning to dimensional variables, it is usual in geodesy to write the gravitational potential on the reference surface as

$$V = -\frac{GM}{R}\left\{1 - J_2\left(\frac{a}{R}\right)^2 P_2(\cos\theta) - J_4\left(\frac{a}{R}\right)^4 P_4(\cos\theta) + \cdots\right\}. \tag{5.81}$$

We then have

$$\alpha m = \left(\frac{a}{d}\right)^2 J_2, \quad \beta m^2 = \left(\frac{a}{d}\right)^4 J_4. \tag{5.82}$$

For the true reference surface, (5.76) gives

$$\frac{a}{d} = 1 + \frac{1}{3}f + \frac{2}{9}f^2 + \frac{8}{15}\kappa \cdots. \tag{5.83}$$

Using this relation to expand the ratio a/d in equations (5.82), correct to second order in small quantities, (5.78) and (5.79) respectively produce

$$J_2 = -\frac{1}{3}m + \frac{2}{3}f + \frac{2}{21}mf - \frac{1}{3}f^2 + \frac{8}{21}\kappa + \cdots, \tag{5.84}$$

$$J_4 = \frac{4}{7}mf - \frac{4}{5}f^2 - \frac{32}{35}\kappa + \cdots. \tag{5.85}$$

The second-order terms in (5.83) are not required to obtain (5.84), and substitution of d for a is sufficient to obtain (5.85). Geodetic formulae have traditionally been expressed in terms of the ratio of centrifugal force to total gravity at the equator,

$$m_1 = \frac{a\Omega^2}{g_e}, \tag{5.86}$$

the parameter used by Newton, rather than m. Their connection can easily be derived from the dimensional form of expression (5.56) for equatorial gravity and the expansion (5.83). These give

$$\begin{aligned} m_1 &= \frac{\Omega^2 d\,[1 + (1/3)f + \cdots]}{[GM/d^2]\,[1 - (3/2)m + (1/3)f + \cdots]} \\ &= m\left(1 + \frac{3}{2}m + \cdots\right), \end{aligned} \tag{5.87}$$

first-order approximation being adequate for subsequent applications. Again to first order,

$$m = m_1\left(1 - \frac{3}{2}m + \cdots\right) = m_1\left(1 - \frac{3}{2}m_1 + \cdots\right). \tag{5.88}$$

With m_1 as a replacement for m, relations equivalent to (5.84) and (5.85) are

$$J_2 = -\frac{1}{3}m_1 + \frac{2}{3}f + \frac{1}{2}m_1^2 + \frac{2}{21}m_1 f - \frac{1}{3}f^2 + \frac{8}{21}\kappa + \cdots, \tag{5.89}$$

$$J_4 = \frac{4}{7}m_1 f - \frac{4}{5}f^2 - \frac{32}{35}\kappa + \cdots. \tag{5.90}$$

Equatorial gravity is neither calculable nor easily observed, and it is desirable to introduce

$$m_1' = \frac{\Omega^2 a^3}{GM} = \frac{\Omega^2 d^3}{GM}(1 + f + \cdots) = m(1 + f + \cdots), \tag{5.91}$$

or

$$m = m_1'(1 - f + \cdots). \tag{5.92}$$

The first-order form (5.92) is sufficient to convert (5.84) and (5.85) to

$$J_2 = -\frac{1}{3}m_1' + \frac{2}{3}f + \frac{3}{7}m_1' f - \frac{1}{3}f^2 + \frac{8}{21}\kappa + \cdots, \tag{5.93}$$

$$J_4 = \frac{4}{7}m_1' f - \frac{4}{5}f^2 - \frac{32}{35}\kappa + \cdots. \tag{5.94}$$

J_2, J_4, a and GM are well determined by tracking measurements on satellites and spacecraft, while Ω is very well known from centuries of astronomical observations. Thus, equations (5.93) and (5.94) can be solved for the remaining unknowns

f and κ. Solving (5.94) for κ, and substituting the solution in equation (5.93), yields the quadratic equation in the flattening,

$$f^2 - \left(1 + m_1'\right) f + \frac{1}{2}m_1' + \frac{3}{2}J_2 + \frac{5}{8}J_4 = 0. \tag{5.95}$$

Solving the quadratic, we find that

$$2f = 1 + m_1' - \left(1 - 6J_2 + m_1'^2 - \frac{5}{2}J_4\right)^{\frac{1}{2}}, \tag{5.96}$$

selecting the negative root as the only one admissible, since f is first order in small quantities. To second order, we have

$$f = \frac{m_1'}{2}\left(1 - \frac{1}{2}m_1'\right) + \frac{3}{2}J_2\left(1 + \frac{3}{2}J_2\right) + \frac{5}{8}J_4. \tag{5.97}$$

κ is then computed directly from (5.94). Neglecting the departure from ellipsoidal shape, as is conventional in geodesy, permits determination of the flattening, f_e, from J_2 alone, through the quadratic

$$f_e^2 - \left(2 + \frac{9}{7}m_1'\right) f_e + m_1' + 3J_2 = 0. \tag{5.98}$$

Again, selecting the negative root as the only one admissible, we obtain the second-order formula

$$f_e = \frac{m_1'}{2}\left(1 - \frac{11}{28}m_1'\right) + \frac{3}{2}J_2\left(1 + \frac{3}{4}J_2 - \frac{1}{7}m_1'\right). \tag{5.99}$$

The history of the development of geodetic reference systems is described by Hofmann-Wellenhof and Moritz (2006, pp. 83–90). The Geodetic Reference System 1967 (G.R.S. 1967) of the International Association of Geodesy (IAG, 1971) is based on adopted values of J_2, a, GM and Ω, as is its successor, the Geodetic Reference System 1980 (G.R.S. 1980). Both of these systems of adopted values are based on an ellipsoidal reference surface. The GRS 1980 was the basis of the World Geodetic System 1984 (WGS 84), a Conventional Terrestrial Reference System (CTRS). A further terrestrial reference frame called the International Terrestrial Reference System (ITRS) (see Section 4.1) was developed by the International Earth Rotation and Reference Systems Service (IERS). Table 5.1 shows a comparison between the adopted values and results compiled by Rapp (1974). Also shown are values of f, f_e, κ and other parameters computed from the three basic data sets.

Once f and κ have been computed, the mean equivolumetric radius can be calculated using the series (5.83). Then m can be computed from its definition (5.7) or

Table 5.1 *Flattening, ellipsoidal departure and other parameters computed from three sets of data.*

	Data sets		
	(Rapp, 1974)	G. R. S. 1980	G. R. S. 1967
$10^6 J_2$	$1,082.635 \pm 0.011$	$1,082.63$	$1,082.7$
$10^6 J_4$	-1.6410 ± 0.0174	...	...
a (m)	$6,378,128 \pm 6$	$6,378,137$	$6,378,160$
$10^{-10} GM$ ($\mathrm{m^3\,s^{-2}}$)	$39,860.046 \pm 0.025$	$39,860.05$	$39,860.265$
$10^5 \Omega$ ($\mathrm{rad\,s^{-1}}$)	7.2921151467	7.292115	7.2921151467
	Computed values		
$10^3 m_1'$	3.461377	3.461391	3.461410
f	1/298.2175	f_e	f_e
$10^8 \kappa$	−78.96	0	0
f_e	1/298.2579	1/298.2579	1/298.2477
$10^6 J_4$	−1.641040	−2.361394	−2.361747
d (m)	6,370,994	6,371,000.8	6,371,024
$10^3 m$	3.449775	3.449786	3.449805

from m_1' and f via expression (5.92). Equatorial gravity is accessible through the dimensional form of (5.56), which on substituting for δ from (5.73) reduces to

$$g_e = \frac{GM}{d^2}\left\{1 - \frac{3}{2}m\left(1 + \frac{13}{21}f\right) + \frac{1}{3}f\left(1 + \frac{2}{3}f\right) + \frac{128}{105}\kappa + \cdots\right\}. \tag{5.100}$$

A gravity formula is supplied by substituting for δ in equation (5.58). This yields

$$g = g_e\left\{1 + \left[\frac{5}{2}m\left(1 + \frac{3}{2}m\right) - f\left(1 + \frac{17}{14}m\right) + \frac{8}{7}\kappa\right]\sin^2\phi \right.$$
$$\left. + \left[\frac{1}{8}f\,(f - 5m) - 3\kappa\right]\sin^2 2\phi + \cdots\right\}. \tag{5.101}$$

Through the series (5.83) and expression (5.92), we can convert (5.100) and (5.101) to depend on a and m_1' instead of d and m, giving

$$g_e = \frac{GM}{a^2}\left\{1 - \frac{3}{2}m_1'\left(1 + \frac{2}{7}f\right) + f\,(1 + f) + \frac{16}{7}\kappa + \cdots\right\}, \tag{5.102}$$

$$g = g_e\left\{1 + \left[\frac{5}{2}m_1'\left(1 + \frac{3}{2}m_1'\right) - f\left(1 + \frac{26}{7}m_1'\right) + \frac{8}{7}\kappa\right]\sin^2\phi \right.$$
$$\left. + \left[\frac{1}{8}f\left(f - 5m_1'\right) - 3\kappa\right]\sin^2 2\phi + \cdots\right\}. \tag{5.103}$$

Table 5.2 *Comparison of gravity formulae.*

Data sets	g_e (m s^{-2})	Coefficient of $\sin^2\phi$	Coefficient of $\sin^2 2\phi$
Rapp (1974)	9.780340	$5,301.1 \times 10^{-6}$	-5.8×10^{-6}
G.R.S.1980	9.780326	$5,302.5 \times 10^{-6}$	-5.8×10^{-6}
G.R.S.1967	9.780310	$5,302.4 \times 10^{-6}$	-5.8×10^{-6}
1924/30 formula	9.780490	$5,288.4 \times 10^{-6}$	-5.9×10^{-6}

These allow construction of gravity formulae from the three basic data sets given in Table 5.1. Table 5.2 shows the three resulting formulae in contrast with the International Gravity Formula adopted in 1924 and 1930.

The level reference surface described in this section is not, of course, the geoid. In ground-based geodesy, the geoidal surface is referred to mean sea level. Satellite geodesy uses as geoid the equipotential surface with the same gravity potential value as the reference surface described here. Displacement of the satellite geoid from the reference surface is then easily calculated using the satellite determined gravitational coefficients, permitting a map to be made of geoidal undulations. For geophysical purposes, though, it is more useful to display the geoidal undulations with respect to the equilibrium equipotential surface, considered in the next section.

5.3 Equilibrium theory of the internal figure

There are many reasons why the internal figure of the Earth should depart from fluid equilibrium form. The most obvious is that the solid parts of the Earth are capable of sustaining non-hydrostatic stress for long periods of time. It is unlikely that they are capable of sustaining such stresses indefinitely, but dynamical processes could easily be acting to continually renew the disequilibrium. Similarly, flow in the fluid parts of the Earth prevents these more mobile regions from achieving equilibrium. At a more subtle level, it can be shown theoretically that, even in the absence of flow, a rotating fluid body can have no static equilibrium unless it has a particular distribution of heat sources unlikely to be met exactly in the Earth. Nonetheless, it is important to determine just how far the Earth does depart from equilibrium form, for any theory of its static or dynamic behaviour must be in accord with the observed departure.

If the adjustment to fluid equilibrium is complete, the pressure p is related to the density and the geopotential by

$$\nabla p = -\rho \nabla U. \tag{5.104}$$

In general, the density is related to the pressure and temperature T by an equation of state,

$$\rho = \rho\,(p, T). \tag{5.105}$$

Ignoring radiative transfer, the energy equation for static equilibrium is

$$\nabla \cdot (k\nabla T) = -H, \tag{5.106}$$

where k is the thermal conductivity and H is the rate of heat production per unit volume. Taking the curl of (5.104) yields

$$\nabla\rho \times \nabla U = 0, \tag{5.107}$$

showing that, in equilibrium, the surfaces of equal density coincide with equipotentials, which by (5.104) are also isobaric surfaces. Equation (5.105) then implies that the material on such surfaces has the same temperature, and is in the same thermodynamic state, provided that the composition and phase are assumed uniform, and implies that we may take ρ, p, T, k and H all to be functions of U alone. Thus,

$$\begin{aligned} k\nabla T &= k\frac{dT}{dU}\nabla U \\ &= f(U)\nabla U, \end{aligned} \tag{5.108}$$

where $f(U)$ is a function of U. Substitution into (5.106) yields

$$\begin{aligned} \nabla \cdot (f(U)\nabla U) &= f(U)\nabla^2 U + f'(U)\nabla U \cdot \nabla U \\ &= -H, \end{aligned} \tag{5.109}$$

the prime indicating differentiation. With replacement of $\nabla^2 U$ from the governing equation (5.5) for U, and with recognition of ∇U as the negative of the gravity vector, (5.109) becomes

$$f(U)\left(4\pi G\rho - 2\Omega^2\right) + g^2 f'(U) = -H. \tag{5.110}$$

In equilibrium, H depends on U alone, and therefore the left side of (5.110) must depend only on U. From the analysis in Section 5.2, it is clear that, in general, g^2 varies over any equipotential in a rotating Earth. Its coefficient $f'(U)$ must therefore vanish, so $f(U)$ is constant, giving

$$H = C\rho\left(1 - \frac{2}{3}m\bar{\rho}/\rho\right), \tag{5.111}$$

the constant C being

$$C = -3\frac{f(U)}{\bar{\rho}}\frac{GM}{d^3}, \tag{5.112}$$

and $\bar{\rho}$ being the mean density of the Earth,

$$\bar{\rho} = \frac{3M}{4\pi d^3}. \tag{5.113}$$

It is reasonable to expect the volume rate of heat generation to depend on the mass density. Since m is only about 1/290 for the Earth, we can expect (5.111) to be satisfied approximately, but not exactly. A similar demonstration that no exact static equilibrium is possible in a rotating system, for the case where radiative transfer is important, was given by Von Zeipel (1924), and the theorem is sometimes associated with his name.

When the density is supposed to depend only on the pressure, the equation of state is said to be *barotropic* and the difficulties raised by the theorem are avoided, since the heat energy equation (5.106) is then excluded from explicit consideration. For example, if only adiabatic or isothermal processes are contemplated, the equation of state is barotropic and static equilibrium is possible. Much of the classical or modern literature on the equilibrium theory of rotating, gravitating fluid bodies is based on the barotropic assumption.

While the centrifugal potential retains its representation (5.2) in Legendre polynomials in the interior of the Earth, it becomes more difficult to give such a representation for the gravitational potential there. In integral form the latter is

$$V(\boldsymbol{r}) = -G \int \frac{\rho(\boldsymbol{r}')}{|\boldsymbol{r} - \boldsymbol{r}'|}\, d\mathcal{V}, \tag{5.114}$$

the integration with variable $\boldsymbol{r}'$ extending over the whole of the Earth's interior.

The *generating function* for Legendre polynomials derives from the binomial expansion

$$\left(1 + h^2 - 2hz\right)^{-1/2} = \sum_{n=0}^{\infty} h^n P_n(z) \quad \text{for} \quad |h| < \left| z \pm \left(z^2 - 1\right)^{1/2} \right| \tag{5.115}$$

(Copson (1955)[pp. 277–278]), where the Legendre polynomials are generated as the coefficients of the expansion. For $-1 \le z \le 1$, the radius of convergence of the expansion is unity. By the law of cosines, the denominator of the integrand in (5.114) can be expressed as

$$\frac{1}{|\boldsymbol{r} - \boldsymbol{r}'|} = \frac{1}{D} = \frac{1}{(r^2 + r'^2 - 2rr' \cos\Theta)^{1/2}}. \tag{5.116}$$

Applying the generating function, we have the uniformly convergent expansions

$$\frac{1}{D} = \frac{1}{r\left(1 + \left(\frac{r'}{r}\right)^2 - 2\frac{r'}{r}\cos\Theta\right)^{1/2}} = \frac{1}{r}\sum_{n=0}^{\infty}\left(\frac{r'}{r}\right)^n P_n(\cos\Theta), \text{ for } r' < r, \tag{5.117}$$

$$\frac{1}{D} = \frac{1}{r'\left(1 + \left(\frac{r}{r'}\right)^2 - 2\frac{r}{r'}\cos\Theta\right)^{1/2}} = \frac{1}{r'}\sum_{n=0}^{\infty}\left(\frac{r}{r'}\right)^n P_n(\cos\Theta), \text{ for } r < r', \tag{5.118}$$

where Θ is the angle between $\boldsymbol{r}$ and $\boldsymbol{r}'$.

For $\boldsymbol{r}$, the radius vector of an arbitrary point on an internal equipotential of equatorial radius a and polar radius c, there will be a region,

$$c < r' < a, \tag{5.119}$$

where neither of the expansions (5.118) is convergent for all points on the equipotential. To circumvent this difficulty, we adopt the procedure of Wavre (1932).

Let S be the equipotential surface bounding the volume $\mathcal{V}$ of the Earth, with S_i an internal surface of equipotential U_i bounding the volume $\mathcal{V}_i$. Green's second identity, or Green's theorem (A.24), applied to the region $\mathcal{V}_i$ gives

$$\int_{\mathcal{V}_i} \left(\Phi\nabla^2\Psi - \Psi\nabla^2\Phi\right) d\mathcal{V} = \int_{S_i} (\Phi\nabla\Psi - \Psi\nabla\Phi)\cdot\hat{\boldsymbol{\nu}}\,dS, \tag{5.120}$$

$\hat{\boldsymbol{\nu}}$ being the outward unit normal vector. With

$$\Phi = \frac{1}{D}, \quad \Psi = U, \tag{5.121}$$

and noting that (see arguments following (1.172))

$$\nabla^2\left(\frac{1}{D}\right) = -4\pi\delta\left(\boldsymbol{r} - \boldsymbol{r}'\right), \tag{5.122}$$

we have that

$$\int_{\mathcal{V}_i} \frac{\nabla^2 U}{D}\,d\mathcal{V} = \int_{S_i} \frac{\nabla U}{D}\cdot\hat{\boldsymbol{\nu}}\,dS - U_i\int_{S_i} \nabla\left(\frac{1}{D}\right)\cdot\hat{\boldsymbol{\nu}}\,dS, \tag{5.123}$$

for $\boldsymbol{r}$ outside S_i, and

$$\int_{\mathcal{V}_i} \frac{\nabla^2 U}{D}\,d\mathcal{V} + 4\pi U(\boldsymbol{r}) = \int_{S_i} \frac{\nabla U}{D}\cdot\hat{\boldsymbol{\nu}}\,dS - U_i\int_{S_i} \nabla\left(\frac{1}{D}\right)\cdot\hat{\boldsymbol{\nu}}\,dS, \tag{5.124}$$

for $\boldsymbol{r}$ inside S_i. Now by Gauss's theorem,

$$\int_{S_i} \nabla\left(\frac{1}{D}\right)\cdot\hat{\boldsymbol{\nu}}\,dS = \int_{\mathcal{V}_i} \nabla^2\left(\frac{1}{D}\right) d\mathcal{V} = \begin{cases} 0, & \boldsymbol{r} \text{ outside } S_i, \\ -4\pi, & \boldsymbol{r} \text{ inside } S_i. \end{cases} \tag{5.125}$$

Substituting $4\pi G\rho - 2\Omega^2$ for $\nabla^2 U$ from the expression for the geopotential (5.5), and substituting $-\boldsymbol{g}$ for ∇U, equations (5.123) and (5.124) then produce

$$4\pi G\int_{\mathcal{V}_i} \frac{\rho}{D}\,d\mathcal{V} - 2\Omega^2\int_{\mathcal{V}_i} \frac{1}{D}\,d\mathcal{V} - \int_{S_i} \frac{g}{D}\,dS = \begin{cases} 0, & \boldsymbol{r} \text{ outside } S_i, \\ 4\pi(U_i - U(\boldsymbol{r})), & \boldsymbol{r} \text{ inside } S_i. \end{cases} \tag{5.126}$$

Returning to expression (5.114) for the gravitational potential we have

$$V(\mathbf{r}) = -G \int_{\mathcal{V}} \frac{\rho}{D}\, d\mathcal{V}, \tag{5.127}$$

where the integral is carried out throughout the entire volume $\mathcal{V}$ of the Earth. We can then write the integral

$$4\pi G \int_{\mathcal{V}_i} \frac{\rho}{D}\, d\mathcal{V} \tag{5.128}$$

over the volume $\mathcal{V}_i$ enclosed by the internal equipotential S_i as

$$4\pi G \int_{\mathcal{V}_i} \frac{\rho}{D}\, d\mathcal{V} = 4\pi G \int_{\mathcal{V}} \frac{\rho}{D}\, d\mathcal{V} - 4\pi G \int_{\mathcal{V}-\mathcal{V}_i} \frac{\rho}{D}\, d\mathcal{V}, \tag{5.129}$$

where the integral

$$\int_{\mathcal{V}-\mathcal{V}_i} \frac{\rho}{D}\, d\mathcal{V} \tag{5.130}$$

is over the volume bounded on the inside by S_i, and on the outside by the reference surface equipotential S. We can then write

$$4\pi G \int_{\mathcal{V}_i} \frac{\rho}{D}\, d\mathcal{V} = -4\pi V(\mathbf{r}) - 4\pi G \int_{\mathcal{V}-\mathcal{V}_i} \frac{\rho}{D}\, d\mathcal{V}. \tag{5.131}$$

The integral in the transformed expression (5.131) is then over a region exterior to S_i.

Of the integrals in (5.3), only the second now involves integration over the region interior to S_i. We can write it as

$$\int_{\mathcal{V}_i} \frac{1}{D}\, d\mathcal{V} = \int_{\mathcal{V}_i-\mathcal{V}_0} \frac{1}{D}\, d\mathcal{V} + \int_{\mathcal{V}_o} \frac{1}{D}\, d\mathcal{V}, \tag{5.132}$$

where $\mathcal{V}_o$ is the region bounded by the sphere S_o with radius r_o, equal to the equivolumetric radius of the surface S_i. The last integral in (5.132) can be evaluated exactly using the geometry shown in Figure 5.2.

From the law of cosines,

$$r_0^2 = r^2 + E^2 - 2rE\cos(\pi - t) \tag{5.133}$$

or

$$E^2 + 2rE\cos t + r^2 - r_0^2 = 0. \tag{5.134}$$

Since E must be positive, the only admissible root of this quadratic equation is

$$E = -r\cos t + \sqrt{r^2\cos^2 t + r_0^2 - r^2} = -r\cos t + \sqrt{r_0^2 - r^2\sin^2 t}. \tag{5.135}$$

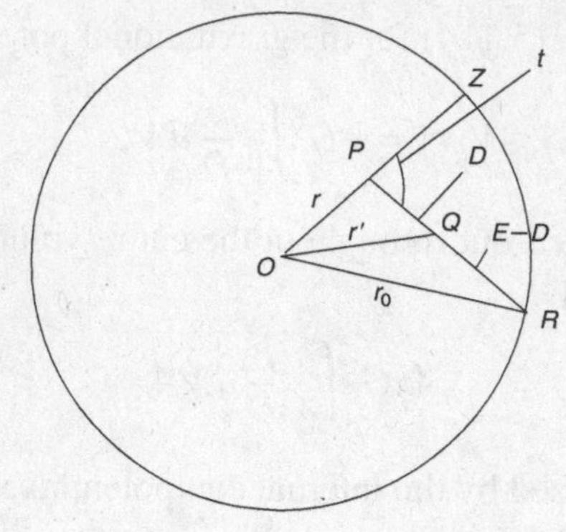

Figure 5.2 The integration point Q is a distance D from the field point P. PQ makes an angle t with the fixed line OP. The length of PR is E.

Then, using OZ as the axis of a system of spherical polar co-ordinates (D, t, p) with P as origin,

$$\begin{aligned}\int_{\mathcal{V}_o} \frac{1}{D}\, d\mathcal{V} &= \int_0^{2\pi}\int_0^{\pi}\int_0^{E} \frac{D^2 \sin t}{D}\, dD\, dt\, dp \\ &= \pi \int_0^{\pi} \left(-r\cos t + \sqrt{r_0^2 - r^2 \sin^2 t}\right)^2 \sin t\, dt \\ &= \pi \left[r^2 \int_0^{\pi} \cos^2 t \sin t\, dt - 2r \int_0^{\pi} \cos t \sqrt{r_0^2 - r^2 \sin^2 t} \sin t\, dt \right. \\ &\quad \left. + \int_0^{\pi} \left(r_0^2 - r^2 \sin^2 t\right) \sin t\, dt \right] \\ &= 2\pi \left(r_0^2 - \frac{1}{3} r^2 \right). \end{aligned} \tag{5.136}$$

Collecting the results of the foregoing integral transformations, we can write (5.3) in the form

$$4\pi G \int_{\mathcal{V}-\mathcal{V}_i} \frac{\rho}{D}\, d\mathcal{V} + 2\Omega^2 \int_{\mathcal{V}_i - \mathcal{V}_o} \frac{1}{D}\, d\mathcal{V} + \int_{S_i} \frac{g}{D}\, dS + 4\pi\Omega^2 \left(r_0^2 - \frac{1}{3} r^2 P_2(\cos\theta) \right) = \begin{cases} -4\pi U(\boldsymbol{r}), & \boldsymbol{r} \text{ outside } S_i, \\ -4\pi U_i, & \boldsymbol{r} \text{ inside } S_i, \end{cases} \tag{5.137}$$

after substitution for $W(\boldsymbol{r})$ from equation (5.2). The advantage of this expression is that none of the integrations include the region $r' < c$ inside the sphere with radius equal to the polar radius of the equipotential S_i. Taking $r < c$, the latter of the expansions (5.118) provides a uniformly convergent series for $1/D$, which can be integrated term by term in each of the integrals of expression (5.137). Equating like powers of r in the result yields

$$4\pi G \int_{\mathcal{V}-\mathcal{V}_i} \frac{\rho P_n(\cos\Theta)}{(r')^{n+1}}\, d\mathcal{V} + 2\Omega^2 \int_{\mathcal{V}_i-\mathcal{V}_o} \frac{P_n(\cos\Theta)}{(r')^{n+1}}\, d\mathcal{V} + \int_{S_i} \frac{g P_n(\cos\Theta)}{(r')^{n+1}}\, dS$$

$$= \begin{cases} 4\pi\left(GM - 2/3\Omega^2 r_0^3\right), & n = -1, \\ -4\pi\left(U_i + \Omega^2 r_0^2\right), & n = 0, \\ (4/3)\pi\Omega^2 P_2(\cos\theta), & n = 2, \\ 0, & n = 1, 3, 4, 5, \cdots . \end{cases} \tag{5.138}$$

The relation for $n = -1$ is found by defining $P_{-1}(\cos\Theta) = 1$ and integrating the main gravity equation (5.5) throughout $\mathcal{V}_i$ to obtain

$$\int_{\mathcal{V}_i} \nabla^2 U\, d\mathcal{V} = \int_{S_i} \nabla U \cdot \hat{\nu}\, dS = \int_{S_i} g\, dS$$
$$= 4\pi G \int_{\mathcal{V}} \rho\, d\mathcal{V} - 4\pi G \int_{\mathcal{V}-\mathcal{V}_i} \rho\, d\mathcal{V} - 2\Omega^2 \cdot \frac{4}{3}\pi r_0^3, \tag{5.139}$$

since by definition the volume $\mathcal{V}_i$ is the same as that of a sphere of radius r_0, denoted by $\mathcal{V}_0$. This then yields

$$4\pi G \int_{\mathcal{V}-\mathcal{V}_i} \rho\, d\mathcal{V} + \int_{S_i} g\, dS = 4\pi(GM - 2/3\Omega^2 r_0^3). \tag{5.140}$$

Hence, if we agree to make P_{-1} unity, (5.140) can be incorporated into the scheme (5.138) with $n = -1$.

By the same sequence of arguments used in Section 5.2 to deduce equation (5.8) for the external equipotential surface, the equation for the internal surfaces is found to be

$$R(r_0, \theta) = r_0\{1 + mR_1(r_0, \theta) + m^2 R_2(r_0, \theta) + \cdots\}, \tag{5.141}$$

where R_1 and R_2 have identical functional forms to those given before, except that the quantities f and κ are now allowed to depend on r_0. Thus, expression (5.75) for $R(\theta)$ now takes the form

$$R(r_0, \theta) = r_0 \left[1 - \frac{2}{3} f(r_0) P_2(\cos\theta) - \frac{4}{45} f^2(r_0) - \frac{23}{63} f^2(r_0) P_2(\cos\theta) \right.$$
$$- \frac{8}{21}\kappa(r_0) P_2(\cos\theta) + \frac{12}{35} f^2(r_0) P_4(\cos\theta)$$
$$\left. + \frac{32}{35}\kappa(r_0) P_4(\cos\theta) + \cdots \right]. \tag{5.142}$$

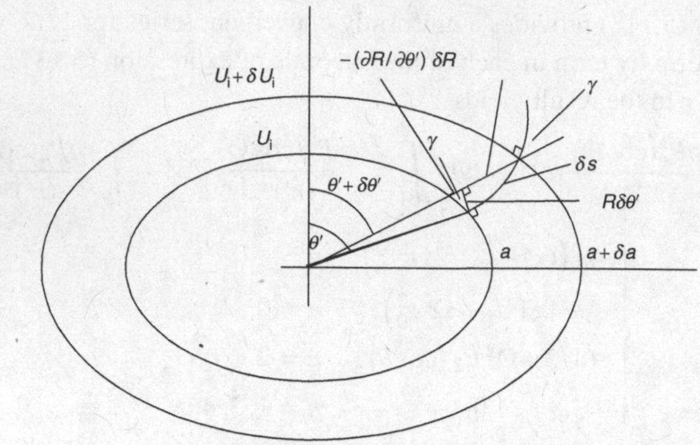

Figure 5.3 Two neighbouring equipotentials with orthometric separation δs.

Consider two neighbouring internal equipotential surfaces, characterised by mean radii r_0 and $r_0 + \delta r_0$, as shown in Figure 5.3.

At an arbitrary point, their orthometric (measured perpendicular to the surfaces) separation is δs. Take (r', θ', λ') as spherical polar co-ordinates of the integration point. We have

$$\delta R = \frac{\partial R}{\partial r_0}\delta r_0 + \frac{\partial R}{\partial \theta'}\delta\theta', \tag{5.143}$$

and from Figure 5.3, it can be deduced that

$$\delta R = \delta s \cos\gamma, \quad \delta\theta' = -\frac{1}{R}\delta s \sin\gamma, \tag{5.144}$$

with

$$\cos\gamma = \frac{1}{\sqrt{1 + \frac{1}{R^2}\left(\frac{\partial R}{\partial\theta'}\right)^2}}, \quad \sin\gamma = \frac{1}{R}\frac{\partial R}{\partial\theta'}\frac{1}{\sqrt{1 + \frac{1}{R^2}\left(\frac{\partial R}{\partial\theta'}\right)^2}}, \tag{5.145}$$

giving

$$\frac{\partial R}{\partial r_0}\delta r_0 = \sqrt{1 + \frac{1}{R^2}\left(\frac{\partial R}{\partial\theta'}\right)^2}\,\delta s \quad \text{or} \quad \frac{\delta s}{\delta r_0} = \frac{\frac{\partial R}{\partial r_0}}{\sqrt{1 + \frac{1}{R^2}\left(\frac{\partial R}{\partial\theta'}\right)^2}}. \tag{5.146}$$

Again from Figure 5.3, the element of surface area can be seen to be

$$dS = \sqrt{1 + \frac{1}{R^2}\left(\frac{\partial R}{\partial\theta'}\right)^2}\, R^2 \sin\theta'\, d\theta'\, d\lambda' \tag{5.147}$$

so that

$$\rho d\mathcal{V} \approx \rho(r_0)\frac{\delta s}{\delta r_0}\, dr_0\, dS$$
$$= \rho(r_0)\frac{\partial R}{\partial r_0} R^2 \sin\theta'\, d\theta'\, d\lambda'\, dr_0, \tag{5.148}$$

after substitution for $\delta s/\delta r_0$ from (5.146). $\rho(r_0)$ is to be interpreted as the density on the equipotential with mean radius r_0. Similarly, since the intensity of gravity is inversely proportional to the orthometric separation of the equipotentials,

$$gdS \approx \frac{\delta U_i}{\delta s} dS = \frac{\delta U_i}{\delta r_0}\frac{\delta r_0}{\delta s} dS$$
$$= \frac{dU_i}{dr_0}\frac{\left(1 + (1/R^2)\,(\partial R/\partial\theta')^2\right)}{\partial R/\partial r_0} R^2 \sin\theta'\, d\theta'\, d\lambda'. \tag{5.149}$$

Together with the addition theorem (B.9),

$$P_n(\cos\Theta) = \sum_{k=-n}^{n} (-1)^k P_n^k(\cos\theta) e^{ik\lambda} P_n^{-k}(\cos\theta') e^{-ik\lambda'}, \tag{5.150}$$

(5.148) and (5.149) can be used to reduce (5.138), including (5.140), to the single scheme

$$4\pi G P_n(\cos\theta)\int_{r_0}^{d} \rho(r_0)\int_0^{\pi} \frac{1}{R^{n-1}}\frac{\partial R}{\partial r_0} P_n(\cos\theta')\sin\theta'\, d\theta'\, dr_0$$
$$+ 2\Omega^2 P_n(\cos\theta)\int_0^{\pi} P_n(\cos\theta')\sin\theta' \int_{r_0}^{R}\frac{dr'}{(r')^{n-1}}\, d\theta'$$
$$+ \frac{dU_i}{dr_0} P_n(\cos\theta)\int_0^{\pi}\frac{1}{R^{n-1}}\frac{1 + (1/R^2)\,(\partial R/\partial\theta')^2}{\partial R/\partial r_0} P_n(\cos\theta')\sin\theta'\, d\theta'$$
$$= \begin{cases} 2\left(GM - \frac{2}{3}\Omega^2 r_0^3\right), & n = -1, \\ -2\left(U_i + \Omega^2 r_0^2\right), & n = 0, \\ (2/3)\Omega^2 P_2(\cos\theta), & n = 2, \\ 0, & n = 1, 3, 4, 5, \cdots. \end{cases} \tag{5.151}$$

In arriving at the relations (5.151), the following should be noted: all the terms in the summation in the addition theorem for $k \neq 0$ vanish on integration over azimuth; and the second integral on the left of (5.138) over the volume $\mathcal{V}_i - \mathcal{V}_0$ involves integration over radius from r_0 to R, with R a function of θ', so the integration over radius is to be completed before the integration over θ'.

Equation (5.141) for the internal equipotential surfaces provides the expansions

$$
\begin{aligned}
\frac{1}{R^{n-1}}\frac{\partial R}{\partial r_0} = \frac{1}{r_0^{n-1}}\Bigg[1 &- m\left\{(n-2)\,R_1 - r_0\frac{\partial R_1}{\partial r_0}\right\} \\
&+ m^2\left\{r_0\frac{\partial R_2}{\partial r_0} - \frac{(n-1)}{2}r_0\frac{\partial}{\partial r_0}\left(R_1^2\right) - (n-2)\,R_2\right. \\
&\left. + \frac{(n-1)\,(n-2)}{2}R_1^2\right\} + \cdots\Bigg],
\end{aligned}
\tag{5.152}
$$

$$
\begin{aligned}
\int_{r_0}^{R}\frac{dr'}{(r')^{n-1}} &= \begin{cases} -\left(R^{-n+2} - r^{-n+2}\right)/\,(n-2), & n \neq 2, \\ \ln R - \ln r_0, & n = 2, \end{cases} \\
&= \frac{1}{r_0^{n-2}}\left[mR_1 + m^2\left\{R_2 - \frac{n-1}{2}R_1^2\right\} + \cdots\right],
\end{aligned}
\tag{5.153}
$$

$$
\begin{aligned}
&\frac{1}{R^{n-1}}\,\frac{1 + (1/R^2)\,(\partial R/\partial\theta')^2}{\partial R/\partial r_0} \\
&\quad = \frac{1}{r_0^{n-1}}\Bigg[1 - m\left\{nR_1 + r_0\frac{\partial R_1}{\partial r_0}\right\} \\
&\qquad + m^2\left\{r_0^2\left(\frac{\partial R_1}{\partial r_0}\right)^2 - r_0\frac{\partial R_2}{\partial r_0} + \frac{(n+1)}{2}r_0\frac{\partial}{\partial r_0}\left(R_1^2\right)\right. \\
&\qquad \left. - nR_2 + \frac{n\,(n+1)}{2}R_1^2 + \left(\frac{\partial R_1}{\partial\theta'}\right)^2\right\} + \cdots\Bigg],
\end{aligned}
\tag{5.154}
$$

for the integrands in (5.151), correct to second order. Only the first-order term in the expansion (5.153) is needed. To second order, the scheme of equations (5.151) contributes relations only for $n = -1, 0, 2$ and 4. These may be found by substituting the functional forms (5.18) and (5.74) for R_1 and R_2, with f and κ dependent on r_0, into the expansions (5.152) through (5.154) and carrying out the integrations in (5.151) using the identities (5.41) and (5.49), as well as the orthogonality among Legendre polynomials.

In stating the results, it is convenient to introduce the mean density within the volume $\mathcal{V}_i$, bounded by the equipotential U_i. The mass enclosed by U_i is

$$\int_{V_i} \rho dV = 2\pi \int_0^{r_0} \rho(r_0) \int_0^{\pi} \frac{\partial R}{\partial r_0} R^2 \sin\theta' d\theta' dr_0$$
$$= 4\pi \int_0^{r_0} r_0^2 \rho(r_0) \left[1 + m^2 \left\{ -\frac{8r_0}{45m^2} f f' + \frac{8r_0}{45m^2} f f' \right. \right.$$
$$\left. \left. - \frac{4}{15m^2} f^2 + \frac{4}{15m^2} f^2 \right\} + \cdots \right] dr_0$$
$$= 4\pi \int_0^{r_0} r_0^2 \rho(r_0)\, dr_0, \tag{5.155}$$

where the expansion (5.152) has been used with $n = -1$, and the prime indicates differentiation. Division of (5.155) by $(4/3)\pi r_0^3$, the volume of $\mathcal{V}_i$, yields the mean density of material enclosed by U_i as

$$\bar{\rho}(r_0) = \frac{3}{r_0^3} \int_0^{r_0} r_0^2 \rho(r_0)\, dr_0. \tag{5.156}$$

For $n = -1$, we then have that

$$g_0(r_0) = \frac{dU_i}{dr_0}$$
$$= \frac{GM_i}{r_0^2} \left[1 - \frac{2}{3} m(r_0) - \frac{4}{45} \left\{ r_0^2 f'^2 + 2r_0 f f' + 5f^2 \right\} + \cdots \right], \tag{5.157}$$

where $g_0(r_0)$ is introduced as a mean gravity intensity and $m(r_0) = \Omega^2 r_0^3 / GM_i$. After substitution from (5.157), the relation for $n = 0$ becomes

$$U_i = V(0) - \frac{GM_i}{r_0} \left[1 + \frac{1}{3} m(r_0) + \frac{4}{45} \left\{ r_0 f f' + f^2 \right\} + \cdots \right], \tag{5.158}$$

$V(0)$ being the gravitational potential at the geocentre, due to mass external to S_i, which by (5.127) is

$$V(0) = -G \int_{V-V_i} \frac{\rho}{r'} dV = -2\pi G \int_{r_0}^{d} \rho(r_0) \int_0^{\pi} \frac{\partial R}{\partial r_0} R \sin\theta' d\theta' dr_0$$
$$= -4\pi G \int_{r_0}^{d} r_0 \rho(r_0) \left[1 - \frac{4}{45} \left\{ r_0 f f' + f^2 \right\} + \cdots \right] dr_0. \tag{5.159}$$

With substitution for $g_0(r_0)$ from (5.157), the relations for $n = 2$ and $n = 4$ become, respectively,

$$\int_{r_0}^{d} \rho(r_0)\left[f' + \frac{9}{7}ff' + \frac{4}{7}\kappa'\right]dr_0 + \bar{\rho}(r_0)\,m(r_0)\left[\frac{10}{9}f + \frac{2}{9}r_0 f'\right]$$
$$-\frac{2}{3}\bar{\rho}(r_0)\left[f + \frac{1}{2}r_0 f' + \frac{2}{21}\left\{r_0^2 f'^2 + \frac{35}{4}r_0 ff' + \frac{47}{4}f^2 + 3r_0\kappa' + 6\kappa\right\}\right]$$
$$= -\frac{5}{6}\bar{\rho}(r_0)\,m(r_0) \tag{5.160}$$

and

$$\int_{r_0}^{d} \frac{1}{r_0^3}\rho(r_0)\,[r_0\kappa' - 2\kappa]\,dr_0$$
$$+\frac{1}{12r_0^2}\bar{\rho}\,(r_0)\left[r_0^2 f'^2 + 2r_0 ff' - 4r_0\kappa' - 16\kappa\right] = 0. \tag{5.161}$$

Differentiation of the mean density expression (5.156) yields

$$\frac{d\bar{\rho}}{dr_0} = -\frac{3\bar{\rho}}{r_0}(1 - \rho/\bar{\rho}). \tag{5.162}$$

Differentiation of expression (5.160) with respect to r_0, and substitution from relation (5.162), gives

$$-\rho(r_0)\left[f' + \frac{9}{7}ff' + \frac{4}{7}\kappa'\right] + \bar{\rho}(r_0)\,m(r_0)\left[\frac{12}{9}f' + \frac{2}{9}r_0 f''\right]$$
$$-\frac{2}{3}\bar{\rho}(r_0)\left[\frac{1}{2}(3f' + r_0 f'') + \frac{2}{21}\left\{2r_0\left(f'^2 + r_0 f' f''\right)\right.\right.$$
$$\left.\left.+\frac{35}{4}\left(r_0 f'^2 + r_0 ff'' + ff'\right) + \frac{47}{2}ff' + 3\,(\kappa' + r_0\kappa'') + 6\kappa'\right\}\right] \tag{5.163}$$
$$+\frac{2}{r_0}\bar{\rho}\,(1 - \rho/\bar{\rho})\left[f + \frac{1}{2}r_0 f' + \frac{2}{21}\left\{r_0^2 f'^2 + \frac{35}{4}r_0 ff' + \frac{47}{4}f^2 + 3r_0\kappa' + 6\kappa\right\}\right]$$
$$= 0.$$

Note that the product $\bar{\rho}(r_0)\,m(r_0)$ is independent of r_0, since $m(r_0) = \Omega^2 r_0^3/GM_i$ and $M_i = 4\pi r_0^3\bar{\rho}\,(r_0)\,/3$, hence $\bar{\rho}(r_0)\,m(r_0) = 3\Omega^2/4\pi G$.

Similarly, differentiating (5.161) with respect to r_0, and again substituting from (5.162), produces

$$
\begin{aligned}
& - \frac{1}{r_0^3}\rho(r_0)\left[r_0\kappa' - 2\kappa\right] - \frac{1}{6r_0^3}\bar{\rho}(r_0)\left[r_0^2 f'^2 + 2r_0 f f' - 4r_0\kappa' - 16\kappa\right] \\
& - \frac{1}{4r_0^3}\bar{\rho}\,(1-\rho/\bar{\rho})\left[r_0^2 f'^2 + 2r_0 f f' - 4r_0\kappa' - 16\kappa\right] \\
& + \frac{1}{12r_0^2}\bar{\rho}(r_0)\Big[2r_0\left(f'^2 + r_0 f' f''\right) + 2\left(f f' + r_0 f'^2 + r_0 f f''\right) \\
& - 4\left(\kappa' + r_0\kappa''\right) - 16\kappa'\Big] = 0.
\end{aligned}
\tag{5.164}
$$

Correct to first order, equation (5.163) reduces to

$$
f'' + \frac{6}{r_0}\frac{\rho}{\bar{\rho}}f' - \frac{6}{r_0^2}(1-\rho/\bar{\rho})\,f = 0,
\tag{5.165}
$$

the first-order *Clairaut equation*. Now, equation (5.164) is entirely of second order. Thus, we can use the first-order expression (5.165) to replace f'' and commit, at most, a third-order error in small quantities. With this substitution, equation (5.164) produces

$$
\begin{aligned}
& \kappa'' + \frac{6}{r_0}\frac{\rho}{\bar{\rho}}\kappa' - \frac{6}{r_0^2}\left(\frac{10}{3} - \rho/\bar{\rho}\right)\kappa \\
& = -\frac{9}{4}\left(\frac{1}{9} + \rho/\bar{\rho}\right)f'^2 + \frac{9}{2}\frac{1}{r_0}\left(\frac{2}{9} - \rho/\bar{\rho}\right)f f' + \frac{3}{r_0^2}(1-\rho/\bar{\rho})\,f^2.
\end{aligned}
\tag{5.166}
$$

Equation (5.166) governs the departure of the internal equipotentials from ellipsoidal shape, as measured by the second-order quantity κ.

Equation (5.166) can be used to eliminate the derivatives of κ from relation (5.163), and, once again, the first-order expression (5.165) can be used to replace f'' in the second-order terms of relation (5.163). After collection of terms, we obtain

$$
\begin{aligned}
& f'' + \frac{6}{r_0}\frac{\rho}{\bar{\rho}}f' - \frac{6}{r_0^2}(1-\rho/\bar{\rho})\,f \\
& = \frac{4}{r_0^2}(1-\rho/\bar{\rho})\,m f + \frac{4}{r_0}(1-\rho/\bar{\rho})\,m f' - 3\left(\frac{4}{9} - \rho/\bar{\rho}\right)f'^2 \\
& - \frac{6}{r_0}\left(\frac{2}{3} - \rho/\bar{\rho}\right)f f' - \frac{5}{r_0^2}(1-\rho/\bar{\rho})\,f^2 - \frac{8}{r_0^2}\kappa.
\end{aligned}
\tag{5.167}
$$

Expression (5.167) is the *Clairaut differential equation* carried to second-order accuracy.

Following Darwin (1899) and de Sitter (1924) it is usual to replace the flattening, f, by the new variable

$$e = f - \frac{5}{42}f^2 + \frac{4}{7}\kappa. \tag{5.168}$$

Then,

$$f = e + \frac{5}{42}f^2 - \frac{4}{7}\kappa, \tag{5.169}$$

$$f' = e' + \frac{5}{21}ff' - \frac{4}{7}\kappa', \tag{5.170}$$

$$f'' = e'' + \frac{5}{21}f'^2 + \frac{5}{21}ff'' - \frac{4}{7}\kappa''. \tag{5.171}$$

As before, f'' in the second-order term on the right side of (5.171) may be replaced by its first-order approximation given by equation (5.165), and the derivatives of κ in (5.171) and (5.170) can be replaced through expression (5.166). Further, since e differs from f only in second-order quantities, we can use it in place of f in any second-order term and commit at most a third-order error. Substituting expressions (5.169) through (5.171) into the second-order Clairaut equation (5.167) and collecting terms, we find, correct to second order, that it becomes

$$\begin{aligned} &e'' + \frac{6}{r_0}\frac{\rho}{\bar{\rho}}e' - \frac{6}{r_0^2}(1 - \rho/\bar{\rho})\,e \\ &= -\frac{4}{7}(1 - \rho/\bar{\rho})\left\{3e'^2 + \frac{6}{r_0}ee' + \frac{7}{r_0^2}e^2 - \frac{7m}{r_0^2}(e + r_0 e')\right\}. \end{aligned} \tag{5.172}$$

The effect of this change of variable is to absorb κ into the new variable e and to make $(1 - \rho/\bar{\rho})$ a common factor of the second-order terms. A further transformation, due to Radau, is to make a variable change of the type used to linearise the Riccati equation,

$$\eta = \frac{r_0 e'}{e}. \tag{5.173}$$

With this change of independent variable, we have

$$e' = \frac{e}{r_0}\eta \quad \text{and} \quad e'' = \frac{e}{r_0}\left(\eta' + \frac{\eta^2}{r_0} - \frac{\eta}{r_0}\right). \tag{5.174}$$

Further, rearranging (5.162) we obtain

$$\rho/\bar{\rho} = 1 + \frac{r_0}{3\bar{\rho}}\frac{d\bar{\rho}}{dr_0} \quad \text{and} \quad 1 - \rho/\bar{\rho} = -\frac{r_0}{3\bar{\rho}}\frac{d\bar{\rho}}{dr_0}. \tag{5.175}$$

Substituting relations (5.174) and (5.175) into the transformed Clairaut equation (5.172), we obtain

$$\begin{aligned} & r_0\eta' + 5\eta + \eta^2 + 2\frac{r_0}{\bar{\rho}}\frac{d\bar{\rho}}{dr_0}(1+\eta) \\ & - \frac{4}{21}r_0\frac{e}{\bar{\rho}}\frac{d\bar{\rho}}{dr_0}\left[3\eta^2 + 6\eta + 7 - 7\frac{m}{e}(1+\eta)\right] = 0. \end{aligned} \tag{5.176}$$

Making use of the easily established identity

$$\frac{2\sqrt{1+\eta}}{\bar{\rho}r_0^4}\frac{d}{dr_0}\left[\bar{\rho}r_0^5\sqrt{1+\eta}\right] - 2\left(5 + \frac{r_0}{\bar{\rho}}\frac{d\bar{\rho}}{dr_0}\right)(1+\eta) = r_0\eta' \tag{5.177}$$

to replace $r_0\eta'$, and using the second of relations (5.175), this equation becomes

$$\begin{aligned} \frac{d}{dr_0}\left[\bar{\rho}r_0^5\sqrt{1+\eta}\right] = \frac{5\bar{\rho}r_0^4}{\sqrt{1+\eta}}\Big[& 1 + \frac{1}{2}\eta - \frac{1}{10}\eta^2 \\ & + \frac{2}{35}(1-\rho/\bar{\rho})\left\{7m(1+\eta) - 3e(1+\eta)^2 - 4e\right\}\Big]. \end{aligned} \tag{5.178}$$

With one integration, (5.172) is converted into the integral equation for η,

$$\sqrt{1+\eta} = \frac{5}{\bar{\rho}r_0^5}\int_0^{r_0}\bar{\rho}r_0^4F(\eta)dr_0, \tag{5.179}$$

with

$$\begin{aligned} F(\eta) = \frac{1}{\sqrt{1+\eta}}\Big[& 1 + \frac{1}{2}\eta - \frac{1}{10}\eta^2 \\ & + \frac{2}{35}(1-\rho/\bar{\rho})\left\{7m(1+\eta) - 3e(1+\eta)^2 - 4e\right\}\Big]. \end{aligned} \tag{5.180}$$

$F(\eta)$ is a very weak function of η and is nearly unity for any reasonable value of η. By its definition (5.173), η is itself of order unity. $F(\eta)$ derives its nearly stationary property from the fact that the terms in curly brackets in (5.180) are of order m and the rest of the function for $\eta < 1$ has the expansion $1 + (1/40)\eta^2 + \cdots$. Prior to the advent of computers, the Radau transformation reduced enormously the labour involved in solving the Clairaut equation, as a good first approximation to the solution could be obtained by taking $F(\eta) = 1$.

Equations (5.172) and (5.166) are second-order ordinary differential equations, and if solutions regular at the geocentre are sought, only one surface condition is required for each equation. These are provided by (5.160) and (5.161), evaluated at the surface $r_0 = d$. From (5.160), we obtain

$$de' + 2e - \frac{5}{2}m + \frac{4}{21}\left\{d^2e'^2 + 10dee' + 13e^2\right\} - \frac{2}{3}m\left\{de' + 5e\right\} = 0, \tag{5.181}$$

after replacing f and f' by expressions (5.169) and (5.170) in the first-order terms, respectively, and by e and e' in the second-order terms. To first order, equation (5.181) reduces to

$$de' + 2e = \frac{5}{2}m, \tag{5.182}$$

which, on squaring, yields the second-order relation

$$d^2e'^2 + 4dee' + 4e^2 = \frac{25}{4}m^2, \tag{5.183}$$

and, on multiplication through by e, yields another second-order relation

$$dee' + 2e^2 = \frac{5}{2}me. \tag{5.184}$$

These permit the reduction of (5.181) to the condition

$$de' + 2e - \frac{5}{2}m - \frac{10}{21}m^2 + \frac{6}{7}me - \frac{4}{7}e^2 = 0. \tag{5.185}$$

Similarly, replacing f and f' by e and e' in second-order terms, (5.161) gives

$$d^2e'^2 + 2dee' - 4\kappa' - 16\kappa = 0, \tag{5.186}$$

which reduces to

$$d\kappa' + 4\kappa + \frac{5}{4}me - \frac{25}{16}m^2 = 0, \tag{5.187}$$

after substitution from (5.183) and (5.184).

There are then two ways of solving for e. We can take the homogeneous first-order form of (5.172) and integrate it to find the solution that is regular at the geocentre and satisfies the first-order condition (5.182) at the surface. This gives a first-order estimate of e, sufficient for calculation of the second-order non-homogeneous terms in (5.172). The full equation can be integrated to find the solution that is regular at the geocentre and satisfies the full second-order condition (5.185). Alternatively, we can use the integral equation for η that resulted from the Radau transformation. First, η is calculated assuming $F(\eta) = 1$. Then, the first-order equation (5.173) is integrated to give

$$e(r_0) = e(0)\exp\left\{\int_0^{r_0} \frac{\eta}{r_0}\, dr_0\right\}. \tag{5.188}$$

It is shown later that $\eta(r_0)$ is linear in r_0 for small r_0, so the integral in the exponent is proper. Substituting $e\eta$ for de', the surface condition (5.185) becomes the quadratic equation

$$e^2 - \frac{1}{4}(14 + 7\eta + 6m)\, e + \frac{35}{8}m + \frac{5}{6}m^2 = 0. \tag{5.189}$$

for the surface value of e, given the surface value of η resulting from the integration in (5.179). The surface value of e is given by the root

$$e(d) = -\frac{B + \sqrt{B^2 - 4C}}{2}, \tag{5.190}$$

with

$$B = -\frac{1}{4}(14 + 7\eta + 6m), \quad C = \frac{35}{8}m + \frac{5}{6}m^2. \tag{5.191}$$

The smaller root is chosen, since the larger root is presumably associated with a figure that is analogous to the highly oblate Maclaurin ellipsoid found as a second solution for homogeneous bodies. A full integration of the exponent in (5.188) provides a value for $e(0)$ in terms of the surface value found from condition (5.185). At all interior points, e is then determined from (5.188). An improved table of values of $F(\eta)$ is then constructed from formula (5.180). The process of calculating η and e is then repeated until the desired accuracy is obtained. One repetition gives second-order accuracy.

It is possible to perform a Radau transformation on equation (5.166) for κ, but a rapidly converging iterative scheme does not result because κ is itself a second-order quantity and the homogeneous equation does not provide a good first approximation. Instead, f and its derivative are replaced by e and its derivative in the second-order terms, to obtain

$$\kappa'' + \frac{6}{r_0}\frac{\rho}{\bar{\rho}}\kappa' + \frac{1}{r_0^2}(6\rho/\bar{\rho} - 20)\,\kappa$$
$$= \frac{3e^2}{4r_0^2}(1 - \rho/\bar{\rho})\left(4 + 6\eta + 3\eta^2\right) - \frac{e^2}{2r_0^2}(7 + 5\eta)\,\eta, \tag{5.192}$$

subject to (5.187). Thus, with values of η and e available, the solution for κ, regular at the geocentre, is straightforward once starting values have been obtained.

Near the geocentre the density may be expanded as

$$\rho(r_0) = \rho(0) + r_0\rho'(0) + \cdots. \tag{5.193}$$

Thus, for small r_0 we have

$$\bar{\rho}(r_0) = \rho(0) + \frac{3}{4}r_0\rho'(0) + \cdots, \tag{5.194}$$

$$\frac{\rho(r_0)}{\bar{\rho}(r_0)} = 1 + \frac{1}{4}\frac{\rho'(0)}{\rho(0)}r_0 + \cdots, \tag{5.195}$$

and

$$1 - \frac{\rho(r_0)}{\bar{\rho}(r_0)} = -\frac{1}{4}\frac{\rho'(0)}{\rho(0)}r_0 + \cdots. \tag{5.196}$$

Using (5.179) with $F(\eta) = 1$, for small r_0 we find that

$$\eta(r_0) = -\frac{1}{4}\frac{\rho'(0)}{\rho(0)}r_0 + \cdots . \tag{5.197}$$

Performing the integration in expression (5.188) and expanding the exponential for small exponent, we obtain

$$e(r_0) = e(0)\left(1 - \frac{1}{4}\frac{\rho'(0)}{\rho(0)}r_0 + \cdots\right). \tag{5.198}$$

Recalculating $F(\eta)$ from (5.180), we now find that

$$F(\eta) = 1 - \frac{1}{10}\frac{\rho'(0)}{\rho(0)}mr_0 + \cdots . \tag{5.199}$$

Thus, $\eta(r_0)$ is linear in r_0 for small r_0 but has a coefficient which differs fractionally from that given in (5.197), by a term of order m. The integrand in the exponent of (5.188) is then obviously proper.

From relation (5.195), the homogeneous form of the κ equation (5.192) for small r_0 becomes

$$\kappa'' + \frac{6}{r_0}\kappa' - \frac{14}{r_0^2}\kappa = 0. \tag{5.200}$$

This equation has the two complementary functions r_0^2 and $1/r_0^7$ with Wronskian $-9/r_0^6$. For small r_0, the non-homogeneous right side of the κ equation (5.192) is

$$I(r_0) = \frac{1}{8}\frac{\rho'(0)}{\rho(0)}\frac{e^2(0)}{r_0} + \cdots . \tag{5.201}$$

By the usual formula for linear differential equations (Kreyszig, 1967, pp. 147–149), the particular integral is

$$\frac{r_0^2}{9}\int \frac{1}{r_0}I(r_0)\,dr_0 - \frac{1}{9r_0^7}\int r_0^8 I(r_0)\,dr_0 = -\frac{1}{64}\frac{\rho'(0)}{\rho(0)}e^2(0)r_0 + \cdots . \tag{5.202}$$

Thus, we may write the full solution for κ, regular at the geocentre, as the sum of the regular complementary function and the particular integral,

$$\kappa(r_0) = Dr_0^2 + \cdots - \frac{1}{64}\frac{\rho'(0)}{\rho(0)}e^2(0)r_0 + \cdots , \tag{5.203}$$

valid for small r_0, with D an arbitrary constant to be determined.

The numerical solution of the κ equation (5.192) follows the Runge–Kutta methods outlined in Section 3.6. The κ equation is first converted to a pair of first-order, coupled, ordinary differential equations by the dependent variable changes, $y_1 = \kappa$ and $y_2 = d\kappa/dr_0$. We then have the pair of equations

$$\frac{dy_1}{dr_0} = y_2,$$

$$\frac{dy_2}{dr_0} = -\frac{1}{r_0^2}\left(6\rho/\bar{\rho} - 20\right) y_1 - \frac{6}{r_0}\frac{\rho}{\bar{\rho}} y_2$$

$$+ \frac{3e^2}{4r_0^2}\left(1 - \rho/\bar{\rho}\right)\left(4 + 6\eta + 3\eta^2\right) - \frac{e^2}{2r_0^2}\left(7 + 5\eta\right)\eta. \tag{5.204}$$

Two separate pairs of numerical solutions are required, one for the regular complementary function of the homogeneous equation, (y_{c_1}, y_{c_2}), and a second for the particular integral of the non-homogeneous equation, $\left(y_{p_1}, y_{p_2}\right)$. From expression (5.203), at the geocentre, the complementary function obeys

$$y_{c_1} = 0, \qquad y_{c_2} = 0, \tag{5.205}$$

$$\frac{dy_{c_1}}{dr_0} = 0, \qquad \frac{dy_{c_2}}{dr_0} = 2D, \tag{5.206}$$

while the particular integral obeys

$$y_{p_1} = 0, \qquad y_{p_2} = -\frac{1}{64}\frac{\rho'(0)}{\rho(0)} e^2(0), \tag{5.207}$$

$$\frac{dy_{p_1}}{dr_0} = -\frac{1}{64}\frac{\rho'(0)}{\rho(0)} e^2(0), \qquad \frac{dy_{p_2}}{dr_0} = 0. \tag{5.208}$$

Since the complementary function is a solution of the linear homogeneous equation, it is proportional to the constant D. If (y_{c_1}, y_{c_2}) is the complementary function found for $D = 1$, then the solution for $D \neq 1$ is (Dy_{c_1}, Dy_{c_2}). The full solution of the kappa equation (5.192) is then $\left(Dy_{c_1} + y_{p_1}, Dy_{c_2} + y_{p_2}\right)$. The constant D is determined by the surface condition (5.187), which becomes

$$d\left(Dy_{c_2} + y_{p_2}\right) + 4\left(Dy_{c_1} + y_{p_1}\right) + \frac{5}{4}me - \frac{25}{16}m^2 = 0. \tag{5.209}$$

Thus, D is given by

$$D = -\frac{dy_{p_2} + 4y_{p_1} + 5/4\, me - 25/16\, m^2}{dy_{c_2} + 4y_{c_1}}. \tag{5.210}$$

The programme FIGURE.FOR computes the mean density $\bar{\rho}$, the reciprocal of the flattening $1/f$, and the departure of the internal equipotentials from ellipsoidal shape, κ. Input to the programme is the Earth model file, the name of which is typed in on request. This can be one of the four models listed in Appendix C, or one in a file with the appropriate format. The Clairaut equation is solved using the Radau transformation, which reduces its solution to two successive quadratures.

The quadratures required for the calculation of the mean density, and the solution of the Clairaut equation, are performed by the subroutine QUAD. The Earth model is interpolated by cubic splines with calls to the subroutines SPMAT and INTPL, described in Section 1.6. The solution of the second-order equation for κ (5.192) requires both the complementary function and the particular integral. These are found by the fourth-order Runge–Kutta scheme, described in Section 3.6, by the subroutine RK. The subroutine DYDR provides the derivatives required by the Runge–Kutta integration scheme. The starting values of the particular integral and its derivative at the geocentre require the derivative of the density there. This is found by cubic spline interpolation by a call to the subroutine DERIV, described in Section 1.6. The number of steps between Earth model points, for both the quadratures and the Runge–Kutta integrations, is specified by the integer variable NIS, input at the request of the programme. As well as output to the screen, the file figure.dat is generated by the programme.

```
C                   PROGRAMME FIGURE.FOR
C
C FIGURE.FOR calculates the mean density, the 'flattening', f, and
C departure from ellipticity, kappa, of the internal level surfaces for
C an input Earth model, under the hydrostatic assumption. Related
C second- and fourth-degree coefficients of the gravitational potential,
C J2 and J4, are also calculated.
C
      IMPLICIT DOUBLE PRECISION(A-H,O-Z)
      DIMENSION Q(100),R(100),RHO(100),GZERO(100),RI(300),RHOI(300),
     1 GZEROI(300),ENAME(10),NM(4),NI(4),NK(4),RHOB(100),
     2 B(98,198), C(100,100),D(100,100,4),RHOBI(300),ETA(100),ETAI(300),
     3 E(100), EI(300),W(6),X(6),YYC(2,100),YYCI(2,300),YYP(2,100),
     4 YYPI(2,300),FI(300),YYI(2,300),RHS(100)
      DOUBLE PRECISION MU(100),LAMBDA(100),MUI(300),LAMBDAI(300),
     1 INT(100,4),INTI(5,5),INITYC(2,5),INITYP(2,5),KAPPA,J2,J4
      CHARACTER*20 EMODEL
C Enter Earth model file name.
      WRITE(6,10)
 10   FORMAT(1X,'Type in Earth model file name.')
      READ(5,11)EMODEL
 11   FORMAT(A20)
C Open Earth model file.
      OPEN(UNIT=1,FILE=EMODEL,STATUS='OLD')
C Open figure file.
      OPEN(UNIT=2,FILE='figure.dat',STATUS='UNKNOWN')
C Open inner core data file.
      OPEN(UNIT=3,FILE='innerc.dat',STATUS='UNKNOWN')
C Set value of pi.
      PI=3.141592653589793D0
C Set value of universal constant of gravitation (CODATA 2006).
      G=6.67428D-11
C Set angular frequency of Earth's rotation (WGS84).
      WE=7.292115D-5
C Set maximum dimensions for interpolation.
      M1=100
      M2=198
      M3=98
C Read in and write out Earth model.
```

```
C Read in and write out headers.
C Earth model name.
      READ(1,12)(ENAME(I),I=1,10)
 12   FORMAT(10A7)
      WRITE(2,13)(ENAME(I),I=1,10)
      WRITE(3,12)(ENAME(I),I=1,10)
      WRITE(6,13)(ENAME(I),I=1,10)
 13   FORMAT(11X,10A7)
      READ(1,14)NN
 14   FORMAT(I10)
C NN is not used.
C Number of model points (NI is not used).
      READ(1,15)(NM(I),NI(I),I=1,4)
 15   FORMAT(8I10)
      WRITE(2,16)
      WRITE(6,16)
 16   FORMAT(//5X,'Number of model points for inner core, outer core, ',
     1 'mantle and crust:'/)
      WRITE(2,15)(NM(I),I=1,4)
      WRITE(6,15)(NM(I),I=1,4)
C Enter number of steps in each region for integrations.
      WRITE(6,17)
 17   FORMAT(//1X,'Select number of steps per subinterval for',
     1 ' integrations.')
      READ(5,*)NIS
      WRITE(2,18)NIS
 18   FORMAT(/5X,'Number of steps per subinterval for integrations =',
     1 I3)
C Initialize model point count.
      K=0
C Read in Earth model.
      DO 19 M=1,4
        N1=NM(M)
        READ(1,20)(R(I),RHO(I),LAMBDA(I),MU(I),GZERO(I),I=1,N1)
 20     FORMAT(1X,F10.1,F10.2,F10.1,F10.1,F10.1)
        DO 21 I=1,N1
C Scale Earth model to SI values and store.
          J=K+I
          RI(J)=R(I)*1.D3
          RHOI(J)=RHO(I)*1.D3
          LAMBDAI(J)=LAMBDA(I)*1.D8
          MUI(J)=MU(I)*1.D8
          GZEROI(J)=GZERO(I)*1.D-2
 21     CONTINUE
C Record number of model points up to end of previous region.
        NK(M)=K
        K=K+N1
 19   CONTINUE
C Begin quadratures for solution of the Clairaut equation by
C the Radau transformation.
C Set weights and abscissae for six-point Gaussian integration.
      W(1)=0.171324492379170D0
      W(2)=0.360761573048139D0
      W(3)=0.467913934572691D0
      W(4)=W(3)
      W(5)=W(2)
      W(6)=W(1)
      X(1)=-0.9324695142031521D0
      X(2)=-0.6612093864662649D0
      X(3)=-0.2386191860831970D0
      X(4)=-X(3)
```

```
      X(5)=-X(2)
      X(6)=-X(1)
C Initialize starting values of the five quadratures at the geocentre.
      INTI(1,1)=0.D0
      INTI(2,1)=0.D0
      INTI(3,1)=0.D0
      INTI(4,1)=0.D0
      INTI(5,1)=0.D0
      DO 22 M=1,4
        MP1=M+1
C Set up interpolation for region M.
        N1=NM(M)
        N2=2*N1-2
        N3=N1-2
C Put radius and density for region M in active locations.
        DO 23 I=1,N1
          J=NK(M)+I
          R(I)=RI(J)
          RHO(I)=RHOI(J)
 23     CONTINUE
C Construct interpolation matrix.
        CALL SPMAT(N1,N2,N3,C,R,B,M1,M2,M3)
C Store interpolation matrices.
        DO 24 I=1,N1
          DO 25 J=1,N1
            D(I,J,M)=C(I,J)
 25       CONTINUE
 24     CONTINUE
C Begin quadrature for mean density.
C Set exponent in integrand.
        EX=2.D0
C Initialize starting value of integral INT.
        INT(1,M)=INTI(1,M)
C Form integrand function Q.
        DO 26 I=1,N1
          Q(I)=RHO(I)
 26     CONTINUE
C Perform quadrature.
        CALL QUAD(INT,EX,Q,NIS,W,X,N1,C,R,M,M1)
C Store starting value for next region.
        INTI(1,MP1)=INT(N1,M)
C Form mean density and store in RHOBI.
        DO 27 I=1,N1
          IF(I.EQ.1.AND.M.EQ.1)GO TO 28
          RHOB(I)=3.D0*INT(I,M)/(R(I)**3)
          GO TO 29
 28       RHOB(I)=RHO(I)
C Store mean density.
 29       J=NK(M)+I
          RHOBI(J)=RHOB(I)
 27     CONTINUE
C Perform first estimate of Radau variable ETA.
C Set exponent in integrand.
        EX=4.D0
C Initialize starting value of integral INT.
        INT(1,M)=INTI(2,M)
C Form integrand function Q.
        DO 30 I=1,N1
          Q(I)=RHOB(I)
 30     CONTINUE
C Perform quadrature.
```

```
      CALL QUAD(INT,EX,Q,NIS,W,X,N1,C,R,M,M1)
C Store starting value for next region.
      INTI(2,MP1)=INT(N1,M)
C Form first estimate of ETA and store in ETAI.
      DO 31 I=1,N1
        IF(I.EQ.1.AND.M.EQ.1)GO TO 32
        ETA(I)=5.D0*INT(I,M)/(RHOB(I)*(R(I)**5))
        ETA(I)=ETA(I)*ETA(I)-1.D0
        GO TO 33
 32     ETA(I)=0.D0
C Store ETA.
 33     J=NK(M)+I
        ETAI(J)=ETA(I)
 31   CONTINUE
C Perform first estimate of figure parameter e.
C Set exponent in integral.
      EX=-1.D0
C Initialize starting value of integral INT.
      INT(1,M)=INTI(3,M)
C Form integrand function Q.
      DO 34 I=1,N1
        Q(I)=ETA(I)
 34   CONTINUE
C Perform quadrature.
      CALL QUAD(INT,EX,Q,NIS,W,X,N1,C,R,M,M1)
C Store starting value for next region.
      INTI(3,MP1)=INT(N1,M)
C Form first estimate of e(r)/e(0) and store in EI.
      DO 35 I=1,N1
        IF(I.EQ.1.AND.M.EQ.1)GO TO 36
        E(I)=DEXP(INT(I,M))
        GO TO 37
 36     E(I)=1.D0
C Store e(r)/e(0) in EI.
 37     J=NK(M)+I
        EI(J)=E(I)
 35   CONTINUE
 22   CONTINUE
C Find total number of model points.
      NT=NK(4)+NM(4)
C Find surface value of Froude number or dimensionless parameter m(d).
      FRD=3.D0*WE*WE/(4.D0*PI*G*RHOBI(NT))
C Find surface value e(d).
C Generate coefficients of quadratic.
      BB=-(14.D0+7.D0*ETAI(NT)+6.D0*FRD)/4.D0
      CC=FRD*(35.D0/8.D0+5.D0*FRD/6.D0)
      ED=-(BB+DSQRT(BB*BB-4.D0*CC))/2.D0
C Find e(0).
      E0=ED/EI(NT)
C Find e(r).
      DO 38 I=1,NT
        EI(I)=E0*EI(I)
 38   CONTINUE
C Begin second pass on integration of the Clairaut equation
C by the Radau transformation.
      DO 39 M=1,4
        MP1=M+1
C Fix number of points for region M.
        N1=NM(M)
C Restore interpolation matrix for region M.
        DO 40 I=1,N1
```

```
        DO 41 J=1,N1
          C(I,J)=D(I,J,M)
 41       CONTINUE
 40     CONTINUE
C Put radius, density, mean density, ETA and e in
C active locations.
        DO 42 I=1,N1
          J=NK(M)+I
          R(I)=RI(J)
          RHO(I)=RHOI(J)
          RHOB(I)=RHOBI(J)
          ETA(I)=ETAI(J)
          E(I)=EI(J)
 42     CONTINUE
C Perform final estimate of Radau variable ETA.
C Set exponent in integrand.
        EX=4.D0
C Initialize starting value of integral INT.
        INT(1,M)=INTI(4,M)
C Form integrand function Q.
        DO 43 I=1,N1
C Calculate Froude number or dimensionless parameter m(r).
          FR=3.D0*WE*WE/(4.D0*PI*G*RHOB(I))
          Q(I)=7.D0*FR*(1.D0+ETA(I))-3.D0*E(I)*(1.D0+ETA(I))**2
     1 -4.D0*E(I)
          Q(I)=Q(I)*2.D0*(1.D0-RHO(I)/RHOB(I))/35.D0
          Q(I)=Q(I)+1.D0+ETA(I)/2.D0-ETA(I)*ETA(I)/10.D0
          Q(I)=Q(I)*RHOB(I)/DSQRT(1.D0+ETA(I))
 43     CONTINUE
C Perform quadrature.
        CALL QUAD(INT,EX,Q,NIS,W,X,N1,C,R,M,M1)
C Store starting value for next region.
        INTI(4,MP1)=INT(N1,M)
C Form final estimate of ETA and store in ETA1.
        DO 44 I=1,N1
          IF(I.EQ.1.AND.M.EQ.1)GO TO 45
          ETA(I)=5.D0*INT(I,M)/(RHOB(I)*(R(I)**5))
          ETA(I)=ETA(I)*ETA(I)-1.D0
          GO TO 46
 45       ETA(I)=0.D0
C Store ETA.
 46       J=NK(M)+I
          ETAI(J)=ETA(I)
 44     CONTINUE
C Form final estimate of figure parameter e.
C Set exponent in integral.
        EX=-1.D0
C Initialize starting value of integral INT.
        INT(1,M)=INTI(5,M)
C Form integrand function Q.
        DO 47 I=1,N1
          Q(I)=ETA(I)
 47     CONTINUE
C Perform quadrature.
        CALL QUAD(INT,EX,Q,NIS,W,X,N1,C,R,M,M1)
C Store starting value for next region.
        INTI(5,MP1)=INT(N1,M)
C Form final estimate of e(r)/e(0) and store in EI.
        DO 48 I=1,N1
          IF(I.EQ.1.AND.M.EQ.1)GO TO 49
          E(I)=DEXP(INT(I,M))
```

```
      GO TO 50
49    E(I)=1.D0
C Store e(r)/e(0) IN EI.
50    J=NK(M)+I
      EI(J)=E(I)
48    CONTINUE
39    CONTINUE
C Find surface value e(d).
C Generate new coefficient of quadratic.
      BB=-(14.D0+7.D0*ETAI(NT)+6.D0*FRD)/4.D0
      ED=-(BB+DSQRT(BB*BB-4.D0*CC))/2.D0
C Find e(0).
      E0=ED/EI(NT)
C Find final e(r).
      DO 51 I=1,NT
        EI(I)=E0*EI(I)
51    CONTINUE
C Begin calculation of kappa by Runge-Kutta integration.
      DO 52 M=1,4
        MP1=M+1
C Fix number of model points for region M.
        N1=NM(M)
C Restore interpolation matrix for region M.
        DO 53 I=1,N1
          DO 54 J=1,N1
            C(I,J)=D(I,J,M)
54        CONTINUE
53      CONTINUE
C Put radius, density, mean density, ETA and e in
C active locations.
        DO 55 I=1,N1
          J=NK(M)+I
          R(I)=RI(J)
          RHO(I)=RHOI(J)
          RHOB(I)=RHOBI(J)
          ETA(I)=ETAI(J)
          E(I)=EI(J)
55      CONTINUE
C Begin calculation of the complementary function YYC.
C Set IC switch to unity for complementary function.
        IC=1
C Set initial value of YYC vector at beginning of region.
        IF(M.EQ.1)GO TO 56
        YYC(1,1)=INITYC(1,M)
        YYC(2,1)=INITYC(2,M)
        GO TO 57
56      YYC(1,1)=0.D0
        YYC(2,1)=0.D0
57      CONTINUE
C Do Runge-Kutta extension of the complementary function through region.
        CALL RK(YYC,IC,RHS,RHO,RHOB,E,NIS,N1,C,R,M1)
C Store starting value for next region.
        INITYC(1,MP1)=YYC(1,N1)
        INITYC(2,MP1)=YYC(2,N1)
C Store YYC in YYCI.
        DO 58 I=1,N1
          J=NK(M)+I
          YYCI(1,J)=YYC(1,I)
          YYCI(2,J)=YYC(2,I)
58      CONTINUE
C Begin calculation of the particular integral YYP.
```

```
C Set switch IC to zero for particular integral.
      IC=0
C Set initial value of YYP vector at beginning of region.
      IF(M.EQ.1)GO TO 59
      YYP(1,1)=INITYP(1,M)
      YYP(2,1)=INITYP(2,M)
      GO TO 60
 59   YYP(1,1)=0.D0
      CALL DERIV(R(1),RHOP,N1,C,R,RHO,M1)
      YYP(2,1)=-RHOP*E(1)*E(1)/(64.D0*RHO(1))
 60   CONTINUE
C Begin Runge-Kutta extension of the particular integral through region.
C Construct right-hand side.
      DO 61 I=1,N1
        IF(I.EQ.1.AND.M.EQ.1)GO TO 62
        RHS(I)=3.D0*E(I)*E(I)*(1.D0-RHO(I)/RHOB(I))*
     1 (4.D0+6.D0*ETA(I)+3.D0*ETA(I)*ETA(I))/(4.D0*R(I)*R(I))
     2 -E(I)*E(I)*(7.D0+5.D0*ETA(I))*ETA(I)/(2.D0*R(I)*R(I))
        GO TO 61
 62     RHS(1)=0.D0
 61   CONTINUE
C Do Runge-Kutta extension.
      CALL RK(YYP,IC,RHS,RHO,RHOB,E,NIS,N1,C,R,M1)
C Store starting value for next region.
      INITYP(1,MP1)=YYP(1,N1)
      INITYP(2,MP1)=YYP(2,N1)
C Store YYP in YYPI.
      DO 63 I=1,N1
        J=NK(M)+I
        YYPI(1,J)=YYP(1,I)
        YYPI(2,J)=YYP(2,I)
 63   CONTINUE
 52   CONTINUE
C Apply boundary condition to YYCI and YYPI.
C determine amplitude DD of YYCI at the surface.
      DD=-RI(NT)*YYPI(2,NT)-4.D0*YYPI(1,NT)
     1 -5.D0*FRD*(EI(NT)-5.D0*FRD/4.D0)/4.D0
      DD=DD/(RI(NT)*YYCI(2,NT)+4.D0*YYCI(1,NT))
C Find YYI and flattening FI.
      DO 64 I=1,NT
        YYI(1,I)=DD*YYCI(1,I)+YYPI(1,I)
        YYI(2,I)=DD*YYCI(2,I)+YYPI(2,I)
        FI(I)=EI(I)+5.D0*EI(I)*EI(I)/42.D0-4.D0*YYI(1,I)/7.D0
 64   CONTINUE
C Write out inner core data file, innerc.dat.
C Write out number of model points.
      WRITE(3,65)NM(1)
 65   FORMAT(1X,I9)
C Write out density at the bottom of the outer core.
      NM1P1=NM(1)+1
      WRITE(3,66)RHOI(NM1P1)
 66   FORMAT(1X,D15.8)
      WRITE(3,67)(RI(I),RHOI(I),RHOBI(I),FI(I),I=1,NM(1))
 67   FORMAT(1X,4D15.8)
C Write out radius, density, mean density, 1/f and kappa.
C Write out headings.
      WRITE(2,68)
      WRITE(6,68)
 68   FORMAT(//10X,'Radius',8X,'Density',6X,'Mean Density',6X,'1/f',
     1 11X,'kappa' )
      WRITE(2,69)
```

```
      WRITE(6,69)
 69   FORMAT(11X,'(km)',9X,'(gm/cc)',8X,'(gm/cc)',4X/)
C Scale and write out values.
      DO 70 I=1,NT
        RAD=RI(I)/1000.D0
        DEN=RHOI(I)/1000.D0
        DENB=RHOBI(I)/1000.D0
        RECIPF=1/FI(I)
        KAPPA=YYI(1,I)
        WRITE(2,71)RAD,DEN,DENB,RECIPF,KAPPA
        WRITE(6,71)RAD,DEN,DENB,RECIPF,KAPPA
 71     FORMAT(1X,F15.1,3F15.3,0PD15.3)
 70   CONTINUE
C Calculate values of second- and fourth-degree gravitational coefficients,
C J2, J4, and write out values.
      J2=-FRD/3.D0+2.D0*FI(NT)/3.D0+2.D0*FRD*FI(NT)/21.D0
     1 -FI(NT)*FI(NT)/3.D0+8.D0*YYI(1,NT)/21.D0
      J4=4.D0*FRD*FI(NT)/7.D0-4.D0*FI(NT)*FI(NT)/5.D0
     1 -32.D0*YYI(1,NT)/35.D0
      WRITE(6,72)J2
      WRITE(2,72)J2
 72   FORMAT(//5X,'J2=',4PD14.7/)
      WRITE(6,73)J4
      WRITE(2,73)J4
 73   FORMAT(//5X,'J4=',1PD11.4/)
      END
C
      SUBROUTINE QUAD(INT,EX,Q,NIS,W,X,N1,C,R,M,M1)
C
C This subroutine performs the quadrature integral of a function, Q, of
C mean radius, from the centre of the Earth through the inner core for
C M=1, or the outer core for M=2, or the mantle for M=3, or the crust
C for M=4. The result is returned in INT. The integration uses six-point
C Gaussian integration which is eleventh-order accurate (exact for
C polynomials up to and including degree eleven). Interpolation of Q to
C the Gaussian integration points is performed by the spline
C interpolator subroutine INTPL. The number of integration steps between
C Earth model points is specified by NIS.
C
      IMPLICIT DOUBLE PRECISION(A-H,O-Z)
      DIMENSION Q(M1),R(M1),C(M1,M1),W(6),X(6),RI(6),QI(6)
      DOUBLE PRECISION INT(M1,4)
      N1M1=N1-1
      DO 10 I=1,N1M1
        IP1=I+1
C Find half-increment in radius.
        H=(R(IP1)-R(I))/(2.D0*DFLOAT(NIS))
C Set sum to zero.
        SUM=0.D0
C Integrate over subinterval.
        DO 11 J=1,NIS
          JM1=J-1
C Find mean radius.
          RM=R(I)+2.D0*DFLOAT(JM1)*H+H
C Find radii for Gaussian integration.
          DO 12 K=1,6
            RI(K)=H*X(K)+RM
C Interpolate integrand at Gaussian abscissae.
            CALL INTPL(RI(K),QI(K),N1,C,R,Q,M1)
C Accumulate sum.
            SUM=SUM+W(K)*QI(K)*((RI(K))**EX)
```

```
 12        CONTINUE
 11      CONTINUE
C Find contribution of current increment.
        INT(IP1,M)=INT(I,M)+H*SUM
 10   CONTINUE
      RETURN
      END
C
      SUBROUTINE DYDR(YP,RAD,Y,IC,RHS,RHO,RHOB,E,N1,C,R,M1)
C
C DYDR calculates the derivative YP of the vector Y at radius RAD.
C For IC=1, the derivative of the complementary function for the kappa
C equation is found. For IC=0, the derivative of the particular integral
C solution of the kappa equation is found. The special cases of
C the derivatives at the geocentre are treated separately.
C
      IMPLICIT DOUBLE PRECISION(A-H,O-Z)
      DIMENSION YP(2),Y(2),R(M1),C(M1,M1),RHS(M1),RHO(M1),
     1 RHOB(M1),E(M1)
      IF(RAD.EQ.0.D0)GO TO 10
      YP(1)=Y(2)
C Test complementary function.
      IF(IC.EQ.1)GO TO 11
C Interpolate RHS to get right-hand side RS at radius RAD.
      CALL INTPL(RAD,RS,N1,C,R,RHS,M1)
C Interpolate RHO and RHOB at radius RAD.
 11   CALL INTPL(RAD,DEN,N1,C,R,RHO,M1)
      CALL INTPL(RAD,DENB,N1,C,R,RHOB,M1)
      YP(2)=-(6.D0*DEN/DENB-20.D0)*Y(1)/(RAD*RAD)
     1 -6.D0*DEN*Y(2)/(RAD*DENB)
      IF(IC.EQ.0)YP(2)=YP(2)+RS
      GO TO 12
C Test for particular integral.
 10   IF(IC.EQ.0)GO TO 13
      YP(1)=0.D0
      YP(2)=2.D0
      GO TO 12
 13   CALL DERIV(RAD,RHOP,N1,C,R,RHO,M1)
      YP(1)=-RHOP*E(1)*E(1)/(64.D0*RHO(1))
      YP(2)=0.D0
 12   CONTINUE
      RETURN
      END
C
      SUBROUTINE RK(YY,IC,RHS,RHO,RHOB,E,NIS,N1,C,R,M1)
C
C This subroutine carries out a fourth-order Runge-Kutta integration of
C the second-order equation for kappa, giving the departure of
C the internal figures of equilibrium from the ellipsoid. The number of
C integration steps between Earth model points is specified by NIS.
C
      IMPLICIT DOUBLE PRECISION(A-H,O-Z)
      DIMENSION RHS(M1),YY(2,M1),R(M1),C(M1,M1),Y(2),YP(2),SUM(2),
     1 RHO(M1),RHOB(M1),E(M1)
      DOUBLE PRECISION K1(2),K2(2),K3(2),K4(2)
      N1M1=N1-1
      DO 10 I=1,N1M1
        IP1=I+1
C Determine step size.
        H=(R(IP1)-R(I))/(DFLOAT(NIS))
C Determine initial radius of step.
```

```
      RAD=R(I)
C Set cumulative sums to zero.
      SUM(1)=0.D0
      SUM(2)=0.D0
      DO 11 J=1,NIS
C Determine current Y vector.
        Y(1)=YY(1,I)+SUM(1)
        Y(2)=YY(2,I)+SUM(2)
        CALL DYDR(YP,RAD,Y,IC,RHS,RHO,RHOB,E,N1,C,R,M1)
C Find vector K1.
        K1(1)=H*YP(1)
        K1(2)=H*YP(2)
C Increment RAD and Y.
        RAD=RAD+H/2.D0
        Y(1)=Y(1)+K1(1)/2.D0
        Y(2)=Y(2)+K1(2)/2.D0
        CALL DYDR(YP,RAD,Y,IC,RHS,RHO,RHOB,E,N1,C,R,M1)
C Find vector K2.
        K2(1)=H*YP(1)
        K2(2)=H*YP(2)
C Increment Y.
        Y(1)=Y(1)+K2(1)/2.D0
        Y(2)=Y(2)+K2(2)/2.D0
        CALL DYDR(YP,RAD,Y,IC,RHS,RHO,RHOB,E,N1,C,R,M1)
C Find vector K3.
        K3(1)=H*YP(1)
        K3(2)=H*YP(2)
C Increment RAD and Y.
        RAD=RAD+H/2.D0
        Y(1)=Y(1)+K3(1)
        Y(2)=Y(2)+K3(2)
        CALL DYDR(YP,RAD,Y,IC,RHS,RHO,RHOB,E,N1,C,R,M1)
C Find vector K4.
        K4(1)=H*YP(1)
        K4(2)=H*YP(2)
C Accumulate sums.
        SUM(1)=SUM(1)+(K1(1)+2.D0*K2(1)+2.D0*K3(1)+K4(1))/6.D0
        SUM(2)=SUM(2)+(K1(2)+2.D0*K2(2)+2.D0*K3(2)+K4(2))/6.D0
 11   CONTINUE
C Increment YY.
      YY(1,IP1)=YY(1,I)+SUM(1)
      YY(2,IP1)=YY(2,I)+SUM(2)
 10 CONTINUE
    RETURN
    END
```

To illustrate the output from the programme FIGURE.FOR, we show the results for Earth model Cal8. The radius, density, mean density, reciprocal of the flattening, $1/f$, and the departure of the internal equipotentials from ellipsoidal shape, κ, are listed for each point of the Earth model. To ensure convergence, the number of integration steps per Earth model subinterval is chosen as 100. The results are shown in Table 5.3.

The second- and fourth-degree coefficients of the external gravitational potential, J_2 and J_4, are also computed. For Earth model Cal8 they are,

$$J_2 = 1071.3219 \times 10^{-6}, \tag{5.211}$$

$$J_4 = -2.9605 \times 10^{-6}. \tag{5.212}$$

Table 5.3 *Figure parameters for Earth model Cal8.*

Radius (km)	Density (g/cm³)	Mean density (g/cm³)	$1/f$	κ $\times 10^6$
0.0	13.580	13.580	422.172	0.0
171.0	13.590	13.588	422.255	0.0
771.0	13.550	13.576	422.229	0.002
971.0	13.490	13.548	422.030	0.004
1171.0	13.380	13.501	421.691	0.006
1216.0	13.340	13.486	421.588	0.007
1216.0	12.170	13.486	421.588	0.007
1371.0	12.140	13.085	420.003	0.016
1571.0	12.030	12.751	416.627	0.028
1821.0	11.840	12.458	412.649	0.038
2171.0	11.520	12.137	408.030	0.045
2571.0	11.110	11.807	403.599	0.054
2971.0	10.620	11.474	399.490	0.065
3171.0	10.330	11.297	397.426	0.072
3371.0	10.010	11.108	395.296	0.080
3486.0	9.820	10.994	394.027	0.086
3486.0	5.920	10.994	394.027	0.086
3521.0	5.740	10.841	393.588	0.088
3591.0	5.520	10.541	392.434	0.094
3671.0	5.430	10.217	390.708	0.105
3971.0	5.230	9.189	381.459	0.170
4371.0	5.060	8.175	365.855	0.279
4771.0	4.920	7.439	349.977	0.384
4971.0	4.830	7.142	342.502	0.431
5171.0	4.740	6.879	335.443	0.473
5371.0	4.610	6.643	328.818	0.511
5571.0	4.430	6.421	322.595	0.546
5701.0	4.220	6.284	318.742	0.568
5731.0	4.090	6.250	317.871	0.573
5771.0	3.980	6.205	316.716	0.579
5871.0	3.840	6.089	313.857	0.596
5971.0	3.610	5.973	311.028	0.614
6071.0	3.470	5.854	308.216	0.633
6151.0	3.420	5.761	305.975	0.648
6221.0	3.400	5.682	304.023	0.663
6291.0	3.360	5.606	302.081	0.678
6331.0	3.350	5.564	300.977	0.686
6351.0	3.340	5.543	300.426	0.691
6351.0	3.260	5.543	300.426	0.691
6356.0	3.260	5.538	300.289	0.692
6361.0	2.160	5.531	300.152	0.693
6366.0	2.160	5.523	300.014	0.694
6371.0	2.160	5.516	299.876	0.695

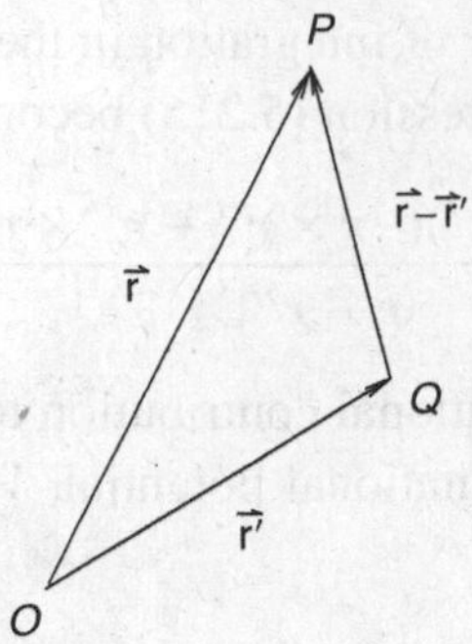

Figure 5.4 Field point P and source point Q.

5.4 Gravity coupling

When the symmetry axis of the inner core is inclined with respect to that of the rest of the Earth, a large gravity restoring torque results.

The evaluation of such coupling begins with elementary Newtonian gravitational relations. With reference to Figure 5.4, an element of mass $\rho(\boldsymbol{r}')d\mathcal{V}'$ at the source point Q at radius $\boldsymbol{r}'$ from the origin O, produces a gravitational force per unit mass at the field point P at radius $\boldsymbol{r}$ from the origin O, given by

$$-G\frac{\boldsymbol{r}-\boldsymbol{r}'}{|\boldsymbol{r}-\boldsymbol{r}'|^3}d\mathcal{V}'. \tag{5.213}$$

If $\mathcal{S}_i$ is the surface of the inner core, bounding the volume $\mathcal{V}_i$, the rest of the Earth is contained in the volume $\mathcal{V}_0$, bounded by the surface $\mathcal{S}_i$ on the inside and the Earth's surface $\mathcal{S}_0$ on the outside. The total gravitational torque acting on the inner core is then

$$\begin{aligned} &-G\int_{\mathcal{V}_i}\boldsymbol{r}\times\rho(\boldsymbol{r})\int_{\mathcal{V}_i}\frac{\boldsymbol{r}-\boldsymbol{r}''}{|\boldsymbol{r}-\boldsymbol{r}''|^3}\rho(\boldsymbol{r}'')d\mathcal{V}''d\mathcal{V} \\ &-G\int_{\mathcal{V}_i}\boldsymbol{r}\times\rho(\boldsymbol{r})\int_{\mathcal{V}_0}\frac{\boldsymbol{r}-\boldsymbol{r}'}{|\boldsymbol{r}-\boldsymbol{r}'|^3}\rho(\boldsymbol{r}')d\mathcal{V}'d\mathcal{V}. \end{aligned} \tag{5.214}$$

The first integral (5.214) represents the torque generated on the inner core by self-gravitation. With the exchange of dummy integration variables, it can be written

$$\begin{aligned} \frac{1}{2}\Bigg\{&-G\int_{\mathcal{V}_i}\boldsymbol{r}\times\rho(\boldsymbol{r})\int_{\mathcal{V}_i}\frac{\boldsymbol{r}-\boldsymbol{r}''}{|\boldsymbol{r}-\boldsymbol{r}''|^3}\rho(\boldsymbol{r}'')d\mathcal{V}''d\mathcal{V} \\ &-G\int_{\mathcal{V}_i}\boldsymbol{r}''\times\rho(\boldsymbol{r}'')\int_{\mathcal{V}_i}\frac{\boldsymbol{r}''-\boldsymbol{r}}{|\boldsymbol{r}''-\boldsymbol{r}|^3}\rho(\boldsymbol{r})d\mathcal{V}\,d\mathcal{V}''\Bigg\}. \end{aligned} \tag{5.215}$$

On the interchange of the order of integration in the second integral, the combination of the two integrals in expression (5.215) becomes

$$\frac{1}{2}\left\{-G\int_{\mathcal{V}_i}\int_{\mathcal{V}_i}\frac{\rho(\boldsymbol{r})\rho(\boldsymbol{r}'')(-\boldsymbol{r}\times\boldsymbol{r}''-\boldsymbol{r}''\times\boldsymbol{r})}{|\boldsymbol{r}-\boldsymbol{r}''|^3}d\mathcal{V}''d\mathcal{V}\right\}=0. \tag{5.216}$$

Thus, in calculating the gravitational contribution to the torque on the inner core, we need only include the gravitational potential, V_0, arising from the mass contained in the volume $\mathcal{V}_0$, or

$$V_0=-G\int_{\mathcal{V}_0}\frac{\rho(\boldsymbol{r}')}{|\boldsymbol{r}-\boldsymbol{r}'|}d\mathcal{V}'. \tag{5.217}$$

In the rotating Earth frame, the calculation of the complete gravity torque requires the inclusion of the centrifugal potential W, as expressed by (5.2), in the total geopotential. Then, the geopotential to be used is

$$U_t=V_0+W, \tag{5.218}$$

with the gravitational potential V_0 being that contributed by the mass outside the inner core. The total gravity torque acting on the inner core is then

$$\begin{aligned}\boldsymbol{\Gamma}&=-\int_{\mathcal{V}_i}(\boldsymbol{r}\times\nabla U_t(\boldsymbol{r}))\rho(\boldsymbol{r})d\mathcal{V}\\&=\int_{\mathcal{V}_i}\nabla\times(\boldsymbol{r}\rho(\boldsymbol{r})U_t(\boldsymbol{r}))d\mathcal{V}+\int_{\mathcal{V}_i}U_t(\boldsymbol{r})(\boldsymbol{r}\times\nabla\rho(\boldsymbol{r}))d\mathcal{V}\\&=-\int_{S_i}U_t(\boldsymbol{r})\rho(\boldsymbol{r})\boldsymbol{r}\times\hat{\boldsymbol{\nu}}dS-\int_{\mathcal{V}_i}U_t(\boldsymbol{r})(\nabla\times\boldsymbol{r}\rho(\boldsymbol{r}))d\mathcal{V},\end{aligned} \tag{5.219}$$

where integral theorem (A.19) has been used to convert one of the volume integrals into a surface integral with $\hat{\boldsymbol{\nu}}$ the outward unit normal vector. For evaluation of the gravity torque, it remains to calculate the gravitational potential V_0 arising from the mass in the volume $\mathcal{V}_0$ lying outside the inner core.

As shown in Figure 5.5, we divide the volume $\mathcal{V}_0$ outside the inner core into three regions. Region ① is bounded on the inside by the inner core and on the outside by a sphere with radius equal to the equatorial radius a_i of the inner core. The polar radius of the inner core is c_i. Region ② lies between region ① and the equipotential with polar radius a_i and thus equatorial radius $a_i(1+f_i)$, where f_i is the flattening of the inner core. Region ③ is exterior to region ①. The surface flattening of the inner core is then

$$f_i=\frac{a_i-c_i}{a_i}. \tag{5.220}$$

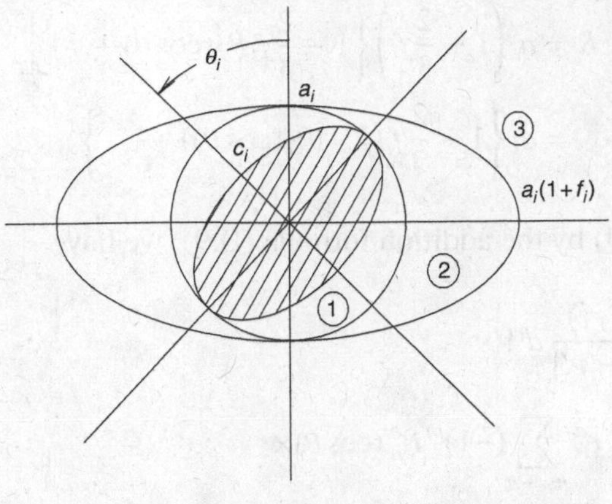

Figure 5.5 The inner core is tilted at angle θ_i with respect to the rest of the Earth. We divide the volume $\mathcal{V}_0$ outside the inner core into three distinct regions. Region ① is bounded on the inside by the inner core and on the outside by a sphere with radius equal to the equatorial radius a_i of the inner core. The polar radius of the inner core is c_i. Region ② lies between region ① and the equipotential with polar radius a_i and thus equatorial radius $a_i(1 + f_i)$, where f_i is the flattening of the inner core. Region ③ is exterior to region ①.

Because $f_i \lesssim 1/400$, our calculations will only be carried out to first order in the flattening. For the same reason, regions ① and ② can be treated as surface densities, using the density just outside the inner core denoted by ρ_i.

We begin with the evaluations of the contributions of regions ① and ② to the gravitational potential V_0. For region ①, entirely within the sphere of radius a_i, we can use the expansion (5.117) for

$$\frac{1}{|\boldsymbol{r} - \boldsymbol{r}'|} = \frac{1}{D}. \tag{5.221}$$

For region ②, entirely outside the sphere of radius a_i, the expansion (5.118) can be used. Since the mass distribution in region ① is symmetrical with respect to the rotated inner core, no contribution to the restoring torque results, and we focus on the contribution of region ② to the gravitational potential V_0. This contribution is

$$-G \int_{②} \frac{\rho(\boldsymbol{r}')}{|\boldsymbol{r} - \boldsymbol{r}'|}\, d\mathcal{V}' \approx -G\rho_i \int_{②} \frac{1}{r'} \sum_{n=0}^{\infty} \left(\frac{r}{r'}\right)^n P_n(\cos\Theta)\, d\mathcal{V}'. \tag{5.222}$$

The equipotential forming the outer boundary of region ②, with polar radius a_i, has mean radius $a_i = (1 + 2f_i/3)$. The equipotential surface then is given by

$$
\begin{aligned}
R &= a_i\left(1+\frac{2}{3}f_i\right)\left\{1-\frac{2}{3}f_iP_2(\cos\theta)+\cdots\right\} \\
&= a_i\left\{1+\frac{2}{3}f_i(1-P_2(\cos\theta))+\cdots\right\}.
\end{aligned}
\tag{5.223}
$$

Replacing $P_n(\cos\Theta)$ by the addition formula (B.9), we have

$$
\begin{aligned}
&-G\int_{②}\frac{\rho(\boldsymbol{r}')}{|\boldsymbol{r}-\boldsymbol{r}'|}d\mathcal{V} \\
&\approx -G\rho_i\sum_{n=0}^{\infty}r^n\sum_{m=-n}^{n}(-1)^m P_n^m(\cos\theta)\times \\
&\int_0^{2\pi}e^{im(\phi-\phi')}\int_0^{\pi}\int_{a_i}^{R}\frac{1}{(r')^{n-1}}dr'\,P_n^{-m}(\cos\theta')\sin\theta'\,d\theta'\,d\phi'.
\end{aligned}
\tag{5.224}
$$

The inner integral becomes

$$
\begin{aligned}
\int_{a_i}^{R}\frac{1}{(r')^{n-1}}dr' &= \left.\frac{(r')^{2-n}}{2-n}\right|_{a_i}^{a_i\left\{1+\frac{2}{3}f_i\left(1-P_2(\cos\theta')\right)\right\}} \\
&= \frac{a_i^{2-n}[1+\frac{2}{3}f_i(1-P_2(\cos\theta'))]^{2-n}-a_i^{2-n}}{2-n} \\
&= \frac{a_i^{2-n}[1+\frac{2}{3}(2-n)f_i(1-P_2(\cos\theta'))]-a_i^{2-n}}{2-n} \\
&= \frac{2}{3}a_i^{2-n}f_i(1-P_2(\cos\theta')).
\end{aligned}
\tag{5.225}
$$

Using the orthogonality relation (B.7), we find the contribution of region ② to the gravitational potential V_0 to be

$$
-G\int_{②}\frac{\rho(\boldsymbol{r}')}{|\boldsymbol{r}-\boldsymbol{r}'|}d\mathcal{V}' \approx -\frac{8\pi}{3}G\rho_i a_i^2 f_i + \frac{8\pi}{15}G\rho_i f_i r^2 P_2(\cos\theta),
\tag{5.226}
$$

correct to first order in the flattening.

For region ③, entirely outside the sphere of radius a_i, once again the expansion (5.118) for $1/D$ can be used. The contribution of this region to the gravitational potential V_0 is then

$$
-G\int_{③}\frac{\rho(\boldsymbol{r}')}{|\boldsymbol{r}-\boldsymbol{r}'|}d\mathcal{V}' = -G\sum_{n=0}^{\infty}r^n\int_{a_i(1+\frac{2}{3}f_i)}^{d}\int_0^{2\pi}\int_0^{\pi}\rho(r_0)\frac{P_n(\cos\Theta)}{R^{n+1}}d\mathcal{V}'.
\tag{5.227}
$$

Using the addition theorem (B.9) for $P_n(\cos\Theta)$ as well as expression (5.148) for the product of the density and the element of volume, the contribution of region ③ to the gravitational potential becomes

$$-G\sum_{n=0}^{\infty}\sum_{m=-n}^{n}(-1)^m r^n P_n^m(\cos\theta)e^{im\phi}\times$$
$$\int_{a_i(1+\frac{2}{3}f_i)}^{d}\int_0^{2\pi}\int_0^{\pi}P_n^{-m}(\cos\theta')e^{-im\phi}\rho(r_0)\frac{1}{R^{n-1}}\frac{\partial R}{\partial r_0}\sin\theta'\,d\theta'\,d\phi'\,dr_0. \tag{5.228}$$

The inner boundary of region ③ is the equipotential

$$R = r_0\left\{1-\frac{2}{3}f(r_0)P_2(\cos\theta)+\cdots\right\}. \tag{5.229}$$

Thus,

$$\frac{1}{R^{n-1}} = \frac{1}{r_0^{n-1}}\left\{1+\frac{2}{3}(n-1)f(r_0)P_2(\cos\theta)+\cdots\right\}, \tag{5.230}$$

and

$$\frac{\partial R}{\partial r_0} = 1-\frac{2}{3}(f(r_0)+r_0f'(r_0))P_2(\cos\theta)+\cdots. \tag{5.231}$$

Then,

$$\frac{1}{R^{n-1}}\frac{\partial R}{\partial r_0} = \frac{1}{r_0^{n-1}}\left\{1+\frac{2}{3}(n-2)f(r_0)P_2(\cos\theta)\right.$$
$$\left.-\frac{2}{3}r_0f'(r_0)P_2(\cos\theta)+\cdots\right\}. \tag{5.232}$$

Substituting this expression in (5.228), and using the orthogonality of spherical harmonics (B.7), the contribution of region ③ to the gravitational potential is reduced to

$$-4\pi G\int_{a_i(1+\frac{2}{3}f_i)}^{d}r_0\rho(r_0)dr_0+\frac{8}{15}\pi Gr^2P_2(\cos\theta)\int_{a_i(1+\frac{2}{3}f_i)}^{d}\rho(r_0)f'(r_0)dr_0+\cdots. \tag{5.233}$$

To interpret this result, we use the uniform method of Wavre, as described in Section 5.3. The mean equivolumetric radius of the equipotential forming the inner boundary of region ③ is $r_0 = a_i(1 + 2f_i/3)$. Replacing $m(r_0)$ and $\bar{\rho}(r_0)$ by their alternative expressions

$$m(r_0) = \frac{\Omega^2 r_0^3}{GM_i} \quad \text{and} \quad \bar{\rho}(r_0) = \frac{3M_i}{4\pi r_0^3}, \tag{5.234}$$

for $n = 0$, expressions (5.158) and (5.159) combine to give

$$U_i = -4\pi G \int_{r_0}^{d} r_0 \rho(r_0) dr_0 - \frac{GM_i}{r_0} - \frac{1}{3}\Omega^2 r_0^2, \tag{5.235}$$

while for $n = 2$, expression (5.160) gives

$$r_0 \int_{r_0}^{d} \rho(r_0) f'(r_0) dr_0 = \frac{1}{4\pi G}\left\{\frac{GM_i}{r_0^2}(2f_i + r_0 f_i') - \frac{5}{2}\Omega^2 r_0\right\}, \tag{5.236}$$

both correct to first order in the flattening. Replacing the integrals in expression (5.233) by those given by the uniform method of Wavre in (5.235) and (5.236), we find the contribution of region ③ to the gravitational potential V_0 to be

$$\begin{aligned} &-G\int \frac{\rho(\boldsymbol{r}')}{|\boldsymbol{r} - \boldsymbol{r}'|} d\mathcal{V}' \\ &\approx U_i + \frac{GM_i}{r_0} + \frac{1}{3}\Omega^2 r_0^2 + \frac{2}{15} r^2 P_2(\cos\theta)\left\{\frac{GM_i}{r_0^3}(2f_i + r_0 f_i') - \frac{5}{2}\Omega^2\right\}, \end{aligned} \tag{5.237}$$

with r_0 the mean equivolumetric radius of the inner boundary of the region.

It remains only to add the centrifugal potential,

$$W = -\frac{1}{3}\Omega^2 r^2 + \frac{1}{3}\Omega^2 r^2 P_2(\cos\theta), \tag{5.238}$$

to the gravitational potential contributions (5.226) and (5.237) to obtain the total geopotential U_t entering the torque integrals (5.219). For evaluation of the torque integrals, we require expressions for $\boldsymbol{r} \times \hat{\boldsymbol{\nu}}$ and $\nabla \times \boldsymbol{r}\rho(\boldsymbol{r})$, correct to first order.

We choose to perform the evaluations of torque integrals in the inner core frame of reference, with the x_3-axis aligned with the axis of symmetry of the inner core. For emphasis, we use double primes for variables in this frame. The surface integral requires the cross product

$$\boldsymbol{r}'' \times \hat{\boldsymbol{\nu}}'' = -\hat{\boldsymbol{\phi}}'' r'' \sin\gamma, \tag{5.239}$$

where $\boldsymbol{r}''$ is the radius vector, $\hat{\boldsymbol{\nu}}''$ is the outward normal vector and $\hat{\boldsymbol{\phi}}''$ is the unit vector in the direction of increasing longitude. γ is the angle between $\boldsymbol{r}''$ and $\hat{\boldsymbol{\nu}}''$. As illustrated in Figure 5.6, the small angle γ is given by the ratio

$$\frac{dR}{d\theta}\delta\theta / R\delta\theta \tag{5.240}$$

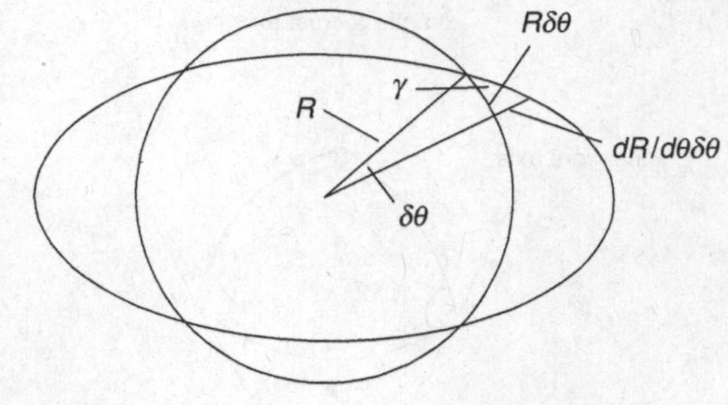

Figure 5.6 A small increment $\delta\theta$ in the angle θ results in an increase in the equipotential surface radius, R, equal to $(dR/d\theta)\delta\theta$. A constant radius vector scribes a circular arc of length $R\delta\theta$. The ratio of these is equal to the small angle γ between $\mathbf{r}''$ and the outward normal to the equipotential surface $\hat{\boldsymbol{\nu}}''$.

with R the equipotential surface given by (5.229). Thus,

$$\gamma = \frac{1}{R}\frac{dR}{d\theta} = 2f_i \cos\theta \sin\theta. \tag{5.241}$$

We adopt both the system of unit vectors $(\hat{\mathbf{r}}'', \hat{\boldsymbol{\theta}}'', \hat{\boldsymbol{\phi}}'')$ for spherical polar co-ordinates in the inner core frame and the system of Cartesian co-ordinate unit vectors $(\hat{\boldsymbol{\imath}}'', \hat{\boldsymbol{\jmath}}'', \hat{\mathbf{k}}'')$ in the same frame. Because γ is only a first-order quantity,

$$\mathbf{r}'' \times \hat{\boldsymbol{\nu}}'' \approx -2\hat{\boldsymbol{\phi}}'' r'' f_i \cos\theta \sin\theta. \tag{5.242}$$

Since

$$\hat{\boldsymbol{\phi}}'' = -\hat{\boldsymbol{\imath}}'' \sin\phi + \hat{\boldsymbol{\jmath}}'' \cos\phi, \tag{5.243}$$

we have

$$\begin{aligned} \mathbf{r}'' \times \hat{\boldsymbol{\nu}}'' &= 2\hat{\boldsymbol{\imath}}'' r'' f_i \cos\theta'' \sin\theta'' \sin\phi'' - 2\hat{\boldsymbol{\jmath}}'' r'' f_i \cos\theta'' \sin\theta'' \cos\phi'' \\ &= \hat{\boldsymbol{\imath}}'' r'' f_i \left[\frac{i}{3} P_2^1(\cos\theta'') e^{i\phi''} + 2i P_2^{-1}(\cos\theta'') e^{-i\phi''} \right] \\ &\quad - \hat{\boldsymbol{\jmath}}'' r'' f_i \left[-\frac{1}{3} P_2^1(\cos\theta'') e^{i\phi''} + 2P_2^{-1}(\cos\theta'') e^{-i\phi''} \right], \end{aligned} \tag{5.244}$$

using the associated Legendre functions,

$$P_2^1(\cos\theta'') = -3\cos\theta'' \sin\theta'' \quad \text{and} \quad P_2^{-1} = \frac{1}{2}\cos\theta'' \sin\theta''. \tag{5.245}$$

Similarly, the volume integral requires

$$\nabla'' \times (\mathbf{r}'' \rho(\mathbf{r}'')) = -\mathbf{r}'' \times \nabla'' \rho(\mathbf{r}'') = \hat{\boldsymbol{\phi}}'' r'' \frac{d\rho_0}{dr_0} \sin\gamma. \tag{5.246}$$

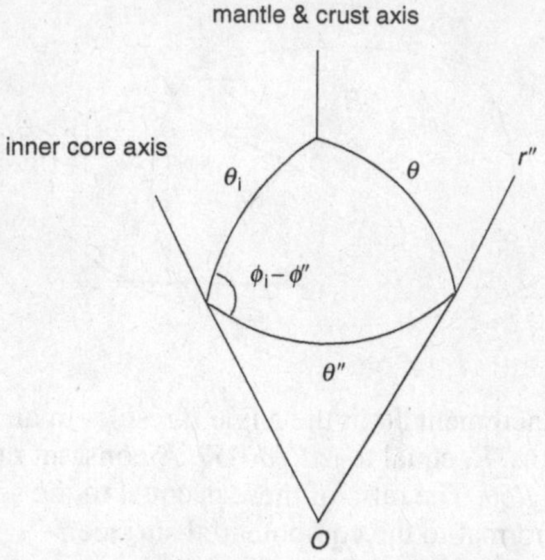

Figure 5.7 Spherical triangle formed by the three axes; namely, the mantle and crust axis, the inner core axis and the axis of the radius vector $\boldsymbol{r}''$ in the inner core reference frame. By the law of cosines for spherical triangles $\cos\theta = \cos\theta_i \cos\theta'' + \sin\theta_i \sin\theta'' \cos(\phi_i - \phi'')$.

Again, because γ is only a first-order quantity,

$$\nabla'' \times (\boldsymbol{r}''\rho(\boldsymbol{r}'')) = 2\hat{\phi}'' r'' \frac{d\rho_0}{dr_0} f \cos\theta'' \sin\theta''. \tag{5.247}$$

Following the same arguments as before,

$$\begin{aligned} &\nabla'' \times (\boldsymbol{r}''\rho(\boldsymbol{r}'')) \\ &= -\hat{\imath}'' r'' \frac{d\rho_0}{dr_0} f \left[\frac{i}{3} P_2^1(\cos\theta'') e^{i\phi''} + 2iP_2^{-1}(\cos\theta'') e^{-i\phi''}\right] \\ &\quad + \hat{\jmath}'' r'' \frac{d\rho_o}{dr_0} f \left[-\frac{1}{3} P_2^1(\cos\theta'') e^{i\phi''} + 2P_2^{-1}(\cos\theta'') e^{-i\phi''}\right]. \end{aligned} \tag{5.248}$$

In evaluating the torque integrals (5.219) in the inner core frame of reference, we require the representation of $P_2(\cos\theta)$ in that frame. As shown in Figure 5.7, the axis of the mantle and crust is taken to pass through co-latitude θ_i and longitude ϕ_i in that frame. The radius vector $\boldsymbol{r}''$ passes through co-latitude θ in the mantle and crust frame and through co-latitude θ'' and longitude ϕ'' in the inner core frame of reference. As shown in Figure 5.7 a spherical triangle is then formed with sides θ_i and θ'' enclosing the angle $\phi_i - \phi''$, with θ forming the third side. By the addition formula (B.9), the required expression for $P_2(\cos\theta)$ is given as

$$P_2(\cos\theta) = \sum_{m=-2}^{2} (-1)^m P_2^m(\cos\theta_i) P_2^{-m}(\cos\theta'') e^{im(\phi_i - \phi'')}. \tag{5.249}$$

In view of the expansions (5.244) and (5.248), only the terms in the geopotential, U_t, proportional to $P_2(\cos\theta)$ will contribute to the torque integrals (5.219). Adding the contribution of the centrifugal potential (5.238) to the contributions (5.226) and (5.237) to the gravitational potential, the total contribution to the geopotential, U_t, proportional to $P_2(\cos\theta)$, becomes

$$\frac{2}{15}\left\{\frac{GM_i}{r_0^3}(2f_i + r_0 f_i') + 3\frac{GM_i'}{r_0^3}f_i + \frac{5}{2}\Omega^2\right\} r^2 P_2(\cos\theta), \tag{5.250}$$

with ρ_i, the density just outside the inner core, replaced by

$$\rho_i = \frac{3M_i'}{4\pi r_0^3}. \tag{5.251}$$

The surface integral in the torque expression (5.219) can then be evaluated using (5.244), the expansion (5.249) of $P_2(\cos\theta)$, and the orthogonality relation (B.7) for spherical harmonics. It reduces to

$$\frac{16}{75}\pi r_0^4 \rho_s f_i\left\{\frac{GM_i}{r_0^2}(2f_i + r_0 f_i') + 3\frac{GM_i'}{r_0^2}f_i + \frac{5}{2}\Omega^2 r_0\right\}\cos\theta_i \sin\theta_i \hat{\phi}_i, \tag{5.252}$$

where ρ_s is the density at the top of the inner core and $\hat{\phi}_i$ is the unit vector in the direction of increasing longitude of the mantle and crust axis. Similarly, the volume integral in the torque expression (5.219) can be evaluated using (5.248), the expansion (5.249) of $P_2(\cos\theta)$, and the orthogonality relation (B.7) for spherical harmonics. In the same way, it reduces to

$$\begin{aligned} &-\frac{16}{75 r_0}\pi\left\{\frac{GM_i}{r_0^2}(2f_i + r_0 f_i') + 3\frac{GM_i'}{r_0^2}f_i + \frac{5}{2}\Omega^2 r_0\right\}\times \\ &\int_0^{r_0} r_0^5 \frac{d\rho_0}{dr_0} f(r_0) dr_0 \ \cos\theta_i \sin\theta_i \hat{\phi}_i. \end{aligned} \tag{5.253}$$

Integrating by parts, we find that

$$\int_0^{r_0} r_0^5 \frac{d\rho_0}{dr_0} f(r_0) dr_0 = \rho_s r_0^5 f_i - \int_0^{r_0} \rho_0 \frac{d}{dr_0}\left\{r_0^5 f(r_0)\right\} dr_0. \tag{5.254}$$

Using this result, the total torque, the sum of (5.253) and (5.252), is found to be

$$\mathbf{\Gamma} = \frac{16}{75r_0}\pi\left\{\frac{GM_i}{r_0^2}(2f_i + r_0 f_i') + 3\frac{GM_i'}{r_0^2}f_i + \frac{5}{2}\Omega^2 r_0\right\}\times$$
$$\int_0^{r_0} \rho_0 \frac{d}{dr_0}\left\{r_0^5 f(r_0)\right\} dr_0 \,\cos\theta_i \sin\theta_i \hat{\phi}_i. \tag{5.255}$$

The integral in expression (5.255) for the torque can be related to the difference between the axial and equatorial moments of inertia of the inner core. In the inner core frame of reference, the axial moment of inertia is

$$C = \int_{\mathcal{V}_i} \rho(\mathbf{r}'')\left(x_1''^2 + x_2''^2\right) d\mathcal{V}. \tag{5.256}$$

The product of the density and the volume element, in terms of the equipotential radius R, is given by (5.148), and the axial moment of inertia then becomes

$$C = \int_0^{r_0}\int_0^{2\pi}\int_0^{\pi} R^2 \sin^2\theta'' \rho(r_0)\frac{\partial R}{\partial r_0} R^2 \sin\theta'' \, d\theta'' \, d\phi'' \, dr_0$$
$$= \frac{2\pi}{5}\int_0^{r_0}\int_0^{\pi} \frac{\partial}{\partial r_0}(R^5)\rho(r_0)\sin^3\theta'' \, d\theta'' \, dr_0. \tag{5.257}$$

To first order, the internal equipotentials (5.142) are

$$R(r_0, \theta'') = r_0\left[1 - \frac{2}{3}f(r_0)P_2(\cos\theta'') + \cdots\right]. \tag{5.258}$$

Then,

$$R^5 = r_0^5\left[1 - \frac{10}{3}f(r_0)P_2(\cos\theta'') + \cdots\right], \tag{5.259}$$

and carrying out the integration in (5.257) over θ'', we obtain

$$C = \frac{8\pi}{15}\int_0^{r_0} \rho(r_0)\frac{d}{dr_0}\left\{r_0^5\left[1 + \frac{2}{3}f(r_0) + \cdots\right]\right\} dr_0. \tag{5.260}$$

In the inner core frame of reference, the equatorial moment of inertia is

$$A = \int_{\mathcal{V}_i} \rho(\mathbf{r}'')\left(x_1''^2 + x_3''^2\right) d\mathcal{V} = \int_{\mathcal{V}_i} \rho(\mathbf{r}'')\left(x_2''^2 + x_3''^2\right) d\mathcal{V}, \tag{5.261}$$

taking the inner core to be axisymmetric. Again, replacing the product of the density and the volume element with (5.148), the equatorial moment of inertia becomes

$$A = \int_0^{r_0}\int_0^{2\pi}\int_0^{\pi} \left(R^2 \sin^2\theta'' \cos\phi'' + R^2 \cos^2\theta''\right)\rho_0(r_0)\frac{\partial R}{\partial r_0}R^2 \sin\theta'' d\theta'' d\phi'' dr_0$$

$$= \frac{2\pi}{5}\int_0^{r_0}\int_0^{\pi} \frac{\partial}{\partial r_0}(R^5)\rho(r_0)\left(\frac{1}{2}\sin^2\theta'' + \cos^2\theta''\right)\sin\theta'' d\theta'' dr_0. \tag{5.262}$$

Substituting for R^5 from (5.259), and carrying out the integration over θ'', we find

$$A = \frac{8\pi}{15}\int_0^{r_0} \rho(r_0)\frac{d}{dr_0}\left\{r_0^5\left[1 - \frac{1}{3}f(r_0) + \cdots\right]\right\} dr_0. \tag{5.263}$$

The difference between the axial and equatorial moments of inertia of the inner core is then

$$C - A = \frac{8\pi}{15}\int_0^{r_0} \rho(r_0)\frac{d}{dr_0}\left\{r_0^5 f(r_0)\right\} dr_0, \tag{5.264}$$

correct to first order in the flattening. The gravity restoring torque on the inner core, tilted at angle θ_i with respect to the mantle and crust, is

$$\boldsymbol{\Gamma} = \Gamma \cos\theta_i \sin\theta_i\, \hat{\boldsymbol{\phi}}_i, \tag{5.265}$$

with the coefficient of the gravity torque, Γ, given by

$$\Gamma = \frac{2}{5}(C - A)\left\{\frac{GM_i}{r_0^3}(2f_i + r_0 f_i') + 3\frac{GM_i'}{r_0^3}f_i + \frac{5}{2}\Omega^2\right\}. \tag{5.266}$$

The programme TORQUE.FOR uses as input the file innerc.dat, generated by the programme FIGURE.FOR described in Section 5.3, containing profiles across the inner core for the density, the mean density and the flattening as functions of mean radius. The input file innerc.dat also provides the number of model points across the inner core for the given Earth model, as well as the density at the bottom of the outer core required for the torque calculation. The inner core values are then interpolated onto a new set of equally spaced model points using the subroutine SPMAT and the subroutine INTPL from Section 1.6. The number of model points in the new profile is specified as input at the request of the programme TORQUE.FOR, which calculates the difference between the axial and equatorial moments of inertia across the inner core using the subroutine QUAD from Section 5.3. The radial derivative of the flattening, required in the integration for the difference in the moments of inertia, is found by the subroutine DERIV from Section 1.6. The remaining quantities, the masses M_i and M_i', required for the calculation of the coefficient of the gravity torque (5.266) are also found. The results are contained in the output file torque.dat.

```
C                     PROGRAMME TORQUE.FOR
C
C TORQUE.FOR calculates the quantities needed to find the gravity
C restoring torque on a tilted inner core. These include
C the difference between the axial and equatorial moments of inertia
C and the radial derivative of the flattening.
C
      IMPLICIT DOUBLE PRECISION(A-H,O-Z)
      DIMENSION R(100),RHO(100),RHOB(100),F(100),RI(100),RHOI(100),
     1 RHOBI(100),FI(100),FPI(100),B(98,198),C(100,100),Q(100),CMA(100),
     2 ENAME(10),W(6),X(6)
      DOUBLE PRECISION MASS(100),INT(100,4),MASSIP,MASSC
      CHARACTER*7 ENAME
C Open inner core data file.
      OPEN(UNIT=1,FILE='innerc.dat',STATUS='OLD')
C Open torque file.
      OPEN(UNIT=2,FILE='torque.dat',STATUS='UNKNOWN')
C Set value of pi.
      PI=3.141592653589793D0
C Set value of universal constant of gravitation (CODATA 2006).
      G=6.67428D-11
C Set angular frequency of Earth's rotation (WGS84).
      OMEGA=7.292115D-5
C Set maximum dimensions for interpolation.
      M1=100
      M2=198
      M3=98
C Read in and write out Earth model name.
      READ(1,10)(ENAME(I),I=1,10)
 10   FORMAT(10A7)
      WRITE(2,11)(ENAME(I),I=1,10)
 11   FORMAT(11X,10A7)
C Read in number of inner core model points.
      READ(1,12)NM
 12   FORMAT(1X,I9)
C Read in density at the bottom of the outer core.
      READ(1,13)RHOBOC
 13   FORMAT(1X,D15.8)
C Read in inner core model data.
      READ(1,14)(R(I),RHO(I),RHOB(I),F(I),I=1,NM)
 14   FORMAT(1X,4D15.8)
C Set up interpolation for inner core.
      N1=NM
      N2=2*N1-2
      N3=N1-2
C Enter number points in model of the inner core.
      WRITE(6,15)
 15   FORMAT(//'Select number of model points for inner core.')
      READ(5,*)NS
      CALL SPMAT(N1,N2,N3,C,R,B,M1,M2,M3)
C Interpolate onto equally spaced inner core model points.
      NSM1=NS-1
C Find space between model points.
      H=R(NM)/DFLOAT(NSM1)
```

```
C Set geocentre values.
      RI(1)=R(1)
      RHOI(1)=RHO(1)
      RHOBI(1)=RHOB(1)
      FI(1)=F(1)
C Construct new equally spaced model.
      DO 16 I=1,NSM1
        IP1=I+1
        RI(IP1)=H*DFLOAT(I)
        CALL INTPL(RI(IP1),RHOI(IP1),NM,C,R,RHO,M1)
        CALL INTPL(RI(IP1),RHOBI(IP1),NM,C,R,RHOB,M1)
        CALL INTPL(RI(IP1),FI(IP1),NM,C,R,F,M1)
        CALL DERIV(RI(IP1),FPI(IP1),NM,C,R,F,M1)
 16   CONTINUE
C Begin integration for the calculation of the difference between
C the axial and equatorial moments of inertia of the inner core.
C Set weights and abscissae for six-point Gaussian integration.
      W(1)=0.171324492379170D0
      W(2)=0.360761573048139D0
      W(3)=0.467913934572691D0
      W(4)=W(3)
      W(5)=W(2)
      W(6)=W(1)
      X(1)=-0.9324695142031521D0
      X(2)=-0.6612093864662649D0
      X(3)=-0.2386191860831970D0
      X(4)=-X(3)
      X(5)=-X(2)
      X(6)=-X(1)
C Set exponent in integrand.
      EX=4.D0
C Initialize starting value of integral INT.
      INT(1,1)=0.D0
C Form integrand function Q.
      DO 17 I=1,NS
        Q(I)=RHOI(I)*(5.D0*FI(I)+RI(I)*FPI(I))
 17   CONTINUE
C Perform quadrature.
      CALL QUAD(INT,EX,Q,1,W,X,NS,C,RI,1,M1)
C Calculate difference between axial and equatorial moments of inertia.
      DO 18 I=1,NS
        CMA(I)=8.D0*PI*INT(I,1)/15.D0
 18   CONTINUE
C Calculate enclosed mass as a function of radius.
      DO 19 I=1,NS
        MASS(I)=4.D0*PI*(RI(I)**3)*RHOBI(I)/3.D0
 19   CONTINUE
C Calculate mass of the inner core as though its density was that
C at the bottom of the outer core.
      MASSIP=4.D0*PI*(RI(NS)**3)*RHOBOC/3.D0
C Find coefficient of gravity restoring torque.
      GAMMA=2.D0*CMA(NS)*(G*MASS(NS)*(2.D0*FI(NS)+RI(NS)*
     1 FPI(NS))/(RI(NS)**3)+3.D0*G*MASSIP*FI(NS)/(RI(NS)**3)
     2 +5.D0*(OMEGA**2)/2.D0)/5.D0
C Write out radius, density, 1/f, f', C-A, and mass for inner core.
C Write out headings.
      WRITE(2,20)
      WRITE(6,20)
 20   FORMAT(//10X,'Radius',8X,'Density',5X,'1/f',5X,'df/dr ',5X,
     1 'C-A',7X,'Mass')
      WRITE(2,21)
```

```
      WRITE(6,21)
 21   FORMAT(11X,'(km)',9X,'(gm/cc)',11X,'(10^-11 ',4X,
     1 '(10^30 ',3X,'(10^21 kg)')
      WRITE(2,22)
      WRITE(6,22)
 22   FORMAT(47X,'/m)',3X,'kg.m^2)')
C Scale and write out values.
      DO 23 I=1,NS
        RAD=RI(I)/1000.D0
        DEN=RHOI(I)/1000.D0
        RECIPF=1.D0/FI(I)
        FP=FPI(I)*1.D11
        CMAS=CMA(I)*1.D-30
        MASSC=MASS(I)*1.D-21
        WRITE(2,24)RAD,DEN,RECIPF,FP,CMAS,MASSC
        WRITE(6,24)RAD,DEN,RECIPF,FP,CMAS,MASSC
 24     FORMAT(1X,F15.1,F15.3,F10.2,F8.3,2F10.2)
 23   CONTINUE
C Write out gravity torque coefficient GAMMA.
      WRITE(2,25)GAMMA
      WRITE(6,25)GAMMA
 25   FORMAT(//5X,'GAMMA=',1PD8.2,1X,'N.m')
C Write out equivalent mass of the inner core as though density is that
C at bottom of the outer core.
      WRITE(2,26)MASSIP
      WRITE(6,26)MASSIP
 26   FORMAT(//5X,'MASSIP=',1PD8.2,1X,'kg')
C Write out density at the bottom of the outer core.
      RHOBOC=RHOBOC/1000.D0
      WRITE(2,27)RHOBOC
      WRITE(6,27)RHOBOC
 27   FORMAT(//5X,'density at bottom of outer core=',2PD10.4,1X,'gm/cc')
      END
```

Table 5.4 *Profiles of parameters for the inner core as computed for Earth model Cal8.*

Radius (km)	Density (g/cm^3)	$1/f$	f' ($\times 10^{-11}$ m^{-1})	$C-A$ ($\times 10^{30}$ kg m^2)	M_i ($\times 10^{21}$ kg)
0.0	13.580	422.17	0.0	0.0	0.0
110.5	13.587	422.23	−0.266	0.0	0.08
221.1	13.592	422.27	−0.210	0.03	0.62
331.6	13.594	422.31	−0.132	0.22	2.08
442.2	13.592	422.33	−0.032	0.91	4.92
552.7	13.584	422.32	0.089	2.78	9.61
663.3	13.570	422.29	0.233	6.92	16.60
773.8	13.549	422.23	0.398	14.94	26.35
884.4	13.520	422.13	0.580	29.10	39.29
994.9	13.481	422.00	0.776	52.36	55.87
1105.5	13.426	421.82	1.029	88.50	76.50
1216.0	13.340	421.59	1.361	142.15	101.57

The profiles of the density, reciprocal of the flattening, radial derivative of the flattening, difference in the axial and equatorial moments of inertia, and enclosed mass are shown in Table 5.4 for Earth model Cal8, as computed by the programme TORQUE.

The mass, M'_I, the inner core would have if its density had the uniform value ρ_i, equal to that at the bottom of the outer core, is found to be

$$M'_i = \frac{4}{3}\pi r_0^3 \rho_i = 9.17 \times 10^{22} \text{ kg}, \tag{5.267}$$

corresponding to density $\rho_i = 12.17$ g/cm^3.

The coefficient of the gravity restoring torque is found to have a very large value of $\Gamma = 2.40 \times 10^{24}$ N m. The large value of the restoring torque led to an investigation of the possible precessional modes of the solid inner core (Smylie *et al.*, 1984). Of particular interest was the possible relation of the observed motion of the dipole axis of the geomagnetic field to precession of the inner core (Szeto and Smylie, 1984a,b).

6

Rotating fluids and the outer core

The fluid outer core is bounded by the shell (mantle and crust) on the outside, and by the solid inner core on the inside. It is thus a contained, rotating fluid.

One of the earliest studies of the dynamics of contained, rotating fluids was that of Poincaré (1885). In these early models the container was assumed rigid, the fluid incompressible, uniform and not self-gravitating, with no inner body. This subject developed rapidly, both theoretically (Greenspan, 1969) and through laboratory experiments (Aldridge and Toomre, 1969). The governing equation is the inertial wave equation. The first suggestion that it might have an analogue in the dynamics of the Earth's liquid core appears to be due to Pekeris and Accad (1972). Of course, in the real Earth the container is deformable, the fluid is compressible, stratified and self-gravitating, and there is an inner body.

We begin with the derivation and study of the inertial wave equation, and then proceed to a scale analysis of long-period motions in the real Earth, finishing with the subseismic condition and the decompression factor.

6.1 The inertial wave equation

For a mode with time dependence $\exp i\omega t$, for small oscillations, the vector displacement field $\boldsymbol{u}$ obeys the equations

$$-\omega^2 \boldsymbol{u} + 2i\omega \boldsymbol{\Omega} \times \boldsymbol{u} = -\nabla \chi, \tag{6.1}$$

$$\nabla \cdot \boldsymbol{u} = 0, \tag{6.2}$$

where ω is the angular frequency of oscillation, and $\boldsymbol{\Omega}$ is the vector rotation rate of the reference frame, taken to be Earth's mean rotation rate around a fixed spatial direction. Here, χ is a reduced pressure potential given by

$$\chi = \frac{p_1}{\rho_0} + W. \tag{6.3}$$

p_1 is the perturbation in pressure, W is the centrifugal potential (5.2), and ρ_0 is the uniform mass density. Equation (6.1) equates the acceleration in the rotating frame (3.77) to the body force per unit mass, while equation (6.2) expresses the incompressibility of the fluid, or more precisely, the assumption that the flow in the uniform fluid is sufficiently slow that its compressibility can be ignored.

The left side of equation (6.1) is linear in the displacement field. It is the vector

$$\boldsymbol{\ell} = -\omega^2 \boldsymbol{u} + 2i\omega\boldsymbol{\Omega} \times \boldsymbol{u} = -\nabla\chi, \tag{6.4}$$

which gives

$$\boldsymbol{\Omega} \times \boldsymbol{u} = \frac{1}{2i\omega}\left(\boldsymbol{\ell} + \omega^2 \boldsymbol{u}\right). \tag{6.5}$$

Then,

$$\boldsymbol{\Omega} \cdot \boldsymbol{\ell} = -\omega^2 \boldsymbol{\Omega} \cdot \boldsymbol{u}, \tag{6.6}$$

and

$$\begin{aligned}\boldsymbol{\Omega} \times \boldsymbol{\ell} &= -\omega^2 (\boldsymbol{\Omega} \times \boldsymbol{u}) + 2i\omega\boldsymbol{\Omega} \times (\boldsymbol{\Omega} \times \boldsymbol{u}) \\ &= -\omega^2 (\boldsymbol{\Omega} \times \boldsymbol{u}) + 2i\omega[(\boldsymbol{\Omega} \cdot \boldsymbol{u})\,\boldsymbol{\Omega} - \Omega^2 \boldsymbol{u}] \\ &= \frac{1}{2i\omega}\left\{-\omega^2(\boldsymbol{\ell} + \omega^2 \boldsymbol{u}) + 4[(\boldsymbol{\Omega} \cdot \boldsymbol{\ell})\,\boldsymbol{\Omega} + \omega^2\Omega^2 \boldsymbol{u}]\right\},\end{aligned} \tag{6.7}$$

on substitution for $\boldsymbol{\Omega} \times \boldsymbol{u}$ from equation (6.5), and for $\boldsymbol{\Omega} \cdot \boldsymbol{u}$ from equation (6.6). Solving (6.7) for the displacement field gives

$$\boldsymbol{u} = -\frac{1}{\omega^2 (\omega^2 - 4\Omega^2)}\left[\omega^2 \boldsymbol{\ell} - 4\boldsymbol{\Omega}\,(\boldsymbol{\Omega} \cdot \boldsymbol{\ell}) + 2i\omega\boldsymbol{\Omega} \times \boldsymbol{\ell}\right]. \tag{6.8}$$

Potential χ then becomes a generalised displacement potential, giving the displacement field in terms of its gradient, through

$$\boldsymbol{u} = \frac{1}{\omega^2 (\omega^2 - 4\Omega^2)}\left[\omega^2 \nabla\chi - 4\boldsymbol{\Omega}\,(\boldsymbol{\Omega} \cdot \nabla\chi) + 2i\omega\boldsymbol{\Omega} \times \nabla\chi\right]. \tag{6.9}$$

In order for this displacement field to satisfy the condition of incompressible flow (6.2), its divergence must vanish. Using the vector identities,

$$\nabla \cdot [\boldsymbol{\Omega}\,(\boldsymbol{\Omega} \cdot \nabla\chi)] = (\boldsymbol{\Omega} \cdot \nabla\chi)\,\nabla \cdot \boldsymbol{\Omega} + \boldsymbol{\Omega} \cdot \nabla\,(\boldsymbol{\Omega} \cdot \nabla\chi) = (\boldsymbol{\Omega} \cdot \nabla)^2\,\chi, \tag{6.10}$$

and

$$\nabla \cdot (\boldsymbol{\Omega} \times \nabla\chi) = \nabla\chi \cdot (\nabla \times \boldsymbol{\Omega}) - \boldsymbol{\Omega} \cdot \nabla \times \nabla\chi = 0, \tag{6.11}$$

on taking the divergence of expression (6.9), and setting it to zero, we find that

$$\omega^2 \nabla^2 \chi - 4\,(\boldsymbol{\Omega} \cdot \nabla)^2\,\chi = 0, \tag{6.12}$$

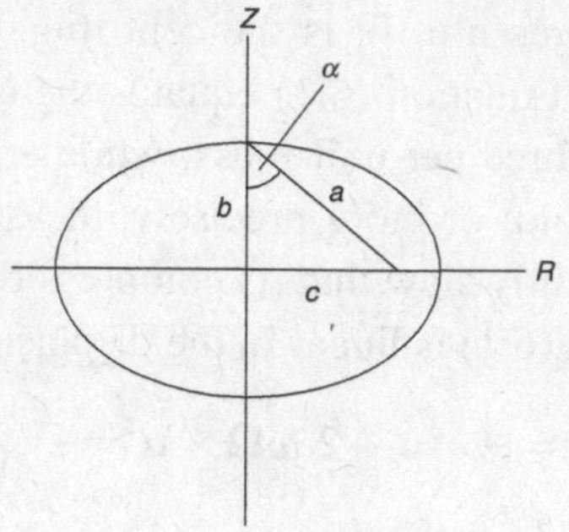

Figure 6.1 Cross-section of the container surface, where a is the equatorial radius, b is the polar radius and c is the focal distance.

the *inertial wave equation* of Poincaré. In a Cartesian co-ordinate system (x, y, z) with the z-axis aligned with the rotation axis, the Poincaré equation becomes

$$\frac{\partial^2 \chi}{\partial x^2} + \frac{\partial^2 \chi}{\partial y^2} + \left(1 - \frac{1}{\sigma^2}\right)\frac{\partial^2 \chi}{\partial z^2} = 0, \tag{6.13}$$

with the angular frequency expressed by the dimensionless Coriolis frequency,

$$\sigma = \frac{\omega}{2\Omega}. \tag{6.14}$$

For modes with periods less than 12 sidereal hours, the square of the Coriolis frequency satisfies $\sigma^2 > 1$. In this case, the Poincaré inertial wave equation can be transformed to Laplace's equation (Bryan, 1889) in new auxiliary co-ordinates (x, y, z') in which the z-axis is stretched by the factor $1/\tau$, where $0 < \tau < 1$, giving $z = \tau z'$, with

$$\tau^2 = 1 - \frac{1}{\sigma^2}. \tag{6.15}$$

If the container is an ellipsoid of revolution, or an oblate spheroid, then in physical cylindrical co-ordinates (R, ϕ, z), with R the cylindrical radius and ϕ the east longitude, as illustrated in Figure 6.1, its surface is described by the equation

$$\frac{R^2}{a^2} + \frac{z^2}{b^2} = \frac{R^2}{c^2 \csc^2 \alpha} + \frac{z^2}{c^2 \cot^2 \alpha} = 1. \tag{6.16}$$

The eccentricity, e, of the cross-section of the container surface is defined as $e = c/a$. Since

$$a^2 = b^2 + c^2, \tag{6.17}$$

we have

$$b^2 = a^2\left(1 - e^2\right). \tag{6.18}$$

Then, in auxiliary co-ordinates, the container surface obeys the equation

$$\frac{R^2}{a^2} + \frac{\tau^2 z'^2}{a^2\,(1-e^2)} = 1. \tag{6.19}$$

We then adopt the prolate spheroidal co-ordinates, (ν, μ), in the auxiliary system. The co-ordinate surfaces of the 'radial-like' co-ordinate ν are the prolate spheroids

$$\frac{R^2}{k^2\,(\nu^2-1)} + \frac{z'^2}{k^2\nu^2} = 1, \quad 1 < \nu < \infty, \tag{6.20}$$

while the co-ordinate surfaces of the 'angular-like' co-ordinate μ are the hyperboloids of two sheets,

$$\frac{R^2}{k^2\,(1-\mu^2)} - \frac{z'^2}{k^2\mu^2} = -1, \quad -1 < \mu < 1. \tag{6.21}$$

If the container surface is given by $\nu = \nu_0$, by comparison of (6.20), for $\nu = \nu_0$, with (6.19), we find that

$$a^2 = k^2\left(\nu_0^2 - 1\right), \quad a^2\frac{(1-e^2)}{\tau^2} = k^2\nu_0^2. \tag{6.22}$$

On substitution for τ from (6.15), the parameters ν_0 and k of the container surface are found to be

$$\nu_0 = \sqrt{\frac{(1-e^2)\,\sigma^2}{1-e^2\sigma^2}}, \quad k = a\sqrt{\frac{1-e^2\sigma^2}{\sigma^2-1}}. \tag{6.23}$$

Solving (6.20) and (6.21) for R^2 and z'^2 in terms of the prolate spheroidal co-ordinates (ν, μ), we find

$$R^2 = k^2\left(\nu^2-1\right)\left(1-\mu^2\right), \quad z'^2 = k^2\nu^2\mu^2. \tag{6.24}$$

We then have the co-ordinate relations

$$x = k\sqrt{(\nu^2-1)\,(1-\mu^2)}\cos\phi = R\cos\phi, \tag{6.25}$$

$$y = k\sqrt{(\nu^2-1)\,(1-\mu^2)}\sin\phi = R\sin\phi, \tag{6.26}$$

$$z = k\tau\nu\mu = \tau z'. \tag{6.27}$$

The co-ordinate surfaces (6.20) and (6.21) may be shown to be orthogonal. Differentiating (6.20) with respect to R, for fixed ν, gives

$$\frac{dz'}{dR} = -\frac{\nu^2}{\nu^2-1}\frac{R}{z'}. \tag{6.28}$$

From (6.24),

$$\frac{R}{z'} = \frac{\sqrt{(\nu^2 - 1)(1 - \mu^2)}}{\nu\mu}, \tag{6.29}$$

giving

$$\frac{dz'}{dR} = -\frac{\nu}{\mu}\sqrt{\frac{1 - \mu^2}{\nu^2 - 1}}. \tag{6.30}$$

Similarly, differentiating (6.21) with respect to R, for fixed μ, gives

$$\frac{dz'}{dR} = \frac{\mu^2}{1 - \mu^2}\frac{R}{z'} = \frac{\mu}{\nu}\sqrt{\frac{\nu^2 - 1}{1 - \mu^2}}. \tag{6.31}$$

Thus, at the points of intersection, the slopes of the co-ordinate surfaces are negative reciprocals and hence they are orthogonal.

Elimination of μ^2 from the pair of equations (6.24) yields

$$\nu^4 - \frac{1}{k^2}\left(R^2 + z'^2 + k^2\right)\nu^2 + \frac{z'^2}{k^2} = 0, \tag{6.32}$$

while elimination of ν^2 yields

$$\mu^4 - \frac{1}{k^2}\left(R^2 + z'^2 + k^2\right)\mu^2 + \frac{z'^2}{k^2} = 0. \tag{6.33}$$

Given R^2 and z'^2, these two equations may be regarded as quadratic equations in ν^2 and μ^2, respectively. Taking the azimuthal dependence to be proportional to $\exp im\phi$, where m is the azimuthal number, in the auxiliary cylindrical co-ordinates (R, ϕ, z'), on multiplying through by R^2, the governing Laplace equation becomes

$$R\frac{\partial}{\partial R}\left(R\frac{\partial\chi}{\partial R}\right) + R^2\frac{\partial^2\chi}{\partial z'^2} - m^2\chi = 0. \tag{6.34}$$

To convert to prolate spheroidal co-ordinates we require

$$R\frac{\partial\chi}{\partial R} = R\frac{\partial\chi}{\partial\nu}\frac{\partial\nu}{\partial R} + R\frac{\partial\chi}{\partial\mu}\frac{\partial\mu}{\partial R}, \tag{6.35}$$

and

$$\frac{\partial\chi}{\partial z'} = \frac{\partial\chi}{\partial\nu}\frac{\partial\nu}{\partial z'} + \frac{\partial\chi}{\partial\mu}\frac{\partial\mu}{\partial z'}, \tag{6.36}$$

and then

$$\begin{aligned}\frac{\partial}{\partial R}\left(R\frac{\partial\chi}{\partial R}\right) = {} & \frac{\partial}{\partial\nu}\left(R\frac{\partial\chi}{\partial\nu}\frac{\partial\nu}{\partial R}\right)\frac{\partial\nu}{\partial R} + \frac{\partial}{\partial\nu}\left(R\frac{\partial\chi}{\partial\mu}\frac{\partial\mu}{\partial R}\right)\frac{\partial\nu}{\partial R} \\ & + \frac{\partial}{\partial\mu}\left(R\frac{\partial\chi}{\partial\mu}\frac{\partial\mu}{\partial R}\right)\frac{\partial\mu}{\partial R} + \frac{\partial}{\partial\mu}\left(R\frac{\partial\chi}{\partial\nu}\frac{\partial\nu}{\partial R}\right)\frac{\partial\mu}{\partial R},\end{aligned} \tag{6.37}$$

and

$$\frac{\partial^2\chi}{\partial z'^2} = \frac{\partial}{\partial\nu}\left(\frac{\partial\chi}{\partial\nu}\frac{\partial\nu}{\partial z'}\right)\frac{\partial\nu}{\partial z'} + \frac{\partial}{\partial\nu}\left(\frac{\partial\chi}{\partial\mu}\frac{\partial\mu}{\partial z'}\right)\frac{\partial\nu}{\partial z'} + \frac{\partial}{\partial\mu}\left(\frac{\partial\chi}{\partial\mu}\frac{\partial\mu}{\partial z'}\right)\frac{\partial\mu}{\partial z'} + \frac{\partial}{\partial\mu}\left(\frac{\partial\chi}{\partial\nu}\frac{\partial\nu}{\partial z'}\right)\frac{\partial\mu}{\partial z'}. \tag{6.38}$$

Differentiation of (6.32) yields

$$R\frac{\partial\nu}{\partial R} = \frac{\nu(\nu^2-1)(1-\mu^2)}{\nu^2-\mu^2}, \qquad \frac{\partial\nu}{\partial z'} = \frac{\mu(\nu^2-1)}{k(\nu^2-\mu^2)}, \tag{6.39}$$

while differentiation of (6.33) yields

$$R\frac{\partial\mu}{\partial R} = -\frac{\mu(\nu^2-1)(1-\mu^2)}{\nu^2-\mu^2}, \qquad \frac{\partial\mu}{\partial z'} = \frac{\nu(1-\mu^2)}{k(\nu^2-\mu^2)}. \tag{6.40}$$

With substitution from (6.39) and (6.40), and from (6.24) for R^2, it is found that

$$\begin{aligned}
&R\frac{\partial}{\partial R}\left(R\frac{\partial\chi}{\partial R}\right) + R^2\frac{\partial^2\chi}{\partial z'^2} \\
&= \frac{\nu(\nu^2-1)(1-\mu^2)^2}{\nu^2-\mu^2}\frac{\partial}{\partial\nu}\left(\frac{\nu(\nu^2-1)}{\nu^2-\mu^2}\frac{\partial\chi}{\partial\nu}\right) - \frac{\nu\mu(\nu^2-1)(1-\mu^2)^2}{\nu^2-\mu^2}\frac{\partial}{\partial\nu}\left(\frac{\nu^2-1}{\nu^2-\mu^2}\frac{\partial\chi}{\partial\mu}\right) \\
&+ \frac{\mu(\nu^2-1)^2(1-\mu^2)}{\nu^2-\mu^2}\frac{\partial}{\partial\mu}\left(\frac{\mu(1-\mu^2)}{\nu^2-\mu^2}\frac{\partial\chi}{\partial\mu}\right) - \frac{\nu\mu(\nu^2-1)^2(1-\mu^2)}{\nu^2-\mu^2}\frac{\partial}{\partial\mu}\left(\frac{1-\mu^2}{\nu^2-\mu^2}\frac{\partial\chi}{\partial\nu}\right) \\
&+ \frac{\mu^2(\nu^2-1)^2(1-\mu^2)}{\nu^2-\mu^2}\frac{\partial}{\partial\nu}\left(\frac{\nu^2-1}{\nu^2-\mu^2}\frac{\partial\chi}{\partial\nu}\right) + \frac{\mu(\nu^2-1)^2(1-\mu^2)^2}{\nu^2-\mu^2}\frac{\partial}{\partial\nu}\left(\frac{\nu}{\nu^2-\mu^2}\frac{\partial\chi}{\partial\mu}\right) \\
&+ \frac{\nu^2(\nu^2-1)(1-\mu^2)^2}{\nu^2-\mu^2}\frac{\partial}{\partial\mu}\left(\frac{1-\mu^2}{\nu^2-\mu^2}\frac{\partial\chi}{\partial\mu}\right) + \frac{\nu(\nu^2-1)^2(1-\mu^2)^2}{\nu^2-\mu^2}\frac{\partial}{\partial\mu}\left(\frac{\mu}{\nu^2-\mu^2}\frac{\partial\chi}{\partial\nu}\right) \\
&= \frac{(\nu^2-1)(1-\mu^2)}{\nu^2-\mu^2}\left[\frac{\partial}{\partial\nu}\left((\nu^2-1)\frac{\partial\chi}{\partial\nu}\right) + \frac{\partial}{\partial\mu}\left((1-\mu^2)\frac{\partial\chi}{\partial\mu}\right)\right].
\end{aligned} \tag{6.41}$$

In the auxiliary prolate spheroidal co-ordinates, the governing Laplace form of the Poincaré equation then becomes

$$\frac{(\nu^2-1)(1-\mu^2)}{\nu^2-\mu^2}\left[\frac{\partial}{\partial\nu}\left((\nu^2-1)\frac{\partial\chi}{\partial\nu}\right) + \frac{\partial}{\partial\mu}\left((1-\mu^2)\frac{\partial\chi}{\partial\mu}\right)\right] - m^2\chi = 0. \tag{6.42}$$

If we take $\chi = f(\nu)\,g(\mu)\exp i(m\phi + \omega t)$, the variables separate, giving

$$\frac{1}{f}\frac{d}{d\nu}\left[(\nu^2-1)\frac{df}{d\nu}\right] - \frac{m^2}{\nu^2-1} = -\frac{1}{g}\frac{d}{d\mu}\left[(1-\mu^2)\frac{dg}{d\mu}\right] + \frac{m^2}{1-\mu^2} = n(n+1), \tag{6.43}$$

where $n(n+1)$ is the separation constant. Thus,

$$\left(1-\nu^2\right)\frac{d^2 f}{d\nu^2}-2\nu\frac{df}{d\nu}+\left\{n(n+1)-\frac{m^2}{1-\nu^2}\right\}f=0, \tag{6.44}$$

and

$$\left(1-\mu^2\right)\frac{d^2 g}{d\mu^2}-2\mu\frac{dg}{d\mu}+\left\{n(n+1)-\frac{m^2}{1-\mu^2}\right\}g=0. \tag{6.45}$$

Both f and g therefore obey the associated Legendre equation and χ has the form

$$P_n^m(\nu)\,P_n^m(\mu)\exp i(m\phi+\omega t). \tag{6.46}$$

For modes with periods less than 12 sidereal hours, $\nu^2 > 1$ and the Legendre function of the second kind, $Q_n^m(\nu)$, is an acceptable solution since the singularities at $\nu = \pm 1$ are excluded, so that χ can also have the form

$$Q_n^m(\nu)\,P_n^m(\mu)\exp i(m\phi+\omega t). \tag{6.47}$$

For modes with periods greater than 12 sidereal hours, the square of the Coriolis frequency satisfies $\sigma^2 < 1$. In order to avoid imaginary co-ordinates, we define

$$\tau^2=\frac{1}{\sigma^2}-1, \tag{6.48}$$

and again scale the z-axis by a factor $1/\tau$. Then, for $\sigma^2 < 1$, we have $0 < \tau < \infty$, with $z = \tau z'$. In the new auxiliary co-ordinates (x, y, z'), the Poincaré inertial wave equation becomes

$$\frac{\partial^2\chi}{\partial x^2}+\frac{\partial^2\chi}{\partial y^2}-\frac{\partial^2\chi}{\partial z'^2}=0, \tag{6.49}$$

rather than Laplace's equation. In auxiliary cylindrical co-ordinates, the equation of the container surface is once again

$$\frac{R^2}{a^2}+\frac{\tau^2 z'^2}{a^2(1-e^2)}=1. \tag{6.50}$$

In auxiliary spheroidal co-ordinates (ξ, η), we adopt the co-ordinate surfaces

$$\frac{R^2}{k^2(1-\xi^2)}+\frac{z'^2}{k^2\xi^2}=1, \quad -1<\xi<1, \tag{6.51}$$

and

$$\frac{R^2}{k^2(1-\eta^2)}+\frac{z'^2}{k^2\eta^2}=1, \quad -1<\eta<1. \tag{6.52}$$

If the container surface is given by $\xi = \xi_0$, by comparison of (6.51) for $\xi = \xi_0$, with (6.50), we find

$$\xi_0^2 = \frac{1 - e^2}{1 - e^2\sigma^2}\sigma^2, \quad k^2 = \frac{1 - e^2\sigma^2}{1 - \sigma^2}a^2. \tag{6.53}$$

Since $e^2 < 1$, we have $\xi_0^2 < 1$, and $k^2 > 0$. Solving (6.51) and (6.52) for R^2 and z'^2 in terms of the spheroidal co-ordinates (ξ, η), we find

$$R^2 = k^2\left(1 - \xi^2\right)\left(1 - \eta^2\right), \quad z'^2 = k^2\xi^2\eta^2. \tag{6.54}$$

As before, the co-ordinate relations are

$$x = k\sqrt{(1 - \xi^2)(1 - \eta^2)}\cos\phi = R\cos\phi, \tag{6.55}$$

$$y = k\sqrt{(1 - \xi^2)(1 - \eta^2)}\sin\phi = R\sin\phi, \tag{6.56}$$

$$z = k\tau\xi\eta = \tau z'. \tag{6.57}$$

The slope of the co-ordinate surface (6.51) may be found by differentiation with respect to R for fixed ξ, giving

$$\frac{dz'}{dR} = -\frac{\xi^2}{1 - \xi^2}\frac{R}{z'}. \tag{6.58}$$

From (6.54),

$$\frac{R}{z'} = \frac{\sqrt{(1 - \xi^2)(1 - \eta^2)}}{\xi\eta}, \tag{6.59}$$

then,

$$\frac{dz'}{dR} = -\frac{\xi}{\eta}\sqrt{\frac{1 - \eta^2}{1 - \xi^2}}. \tag{6.60}$$

Similarly, differentiating (6.52) with respect to R, for fixed η, gives

$$\frac{dz'}{dR} = -\frac{\eta^2}{1 - \eta^2}\frac{R}{z'} = -\frac{\eta}{\xi}\sqrt{\frac{1 - \xi^2}{1 - \eta^2}}. \tag{6.61}$$

Thus, at the points of intersection, the slopes of the co-ordinate surfaces are reciprocals, rather than negative reciprocals, and hence they are not orthogonal.

Elimination of η from the pair of equations (6.54) yields

$$\xi^4 + \frac{1}{k^2}\left(R^2 - z'^2 - k^2\right)\xi^2 + \frac{z'^2}{k^2} = 0, \tag{6.62}$$

while elimination of ξ yields

$$\eta^4 + \frac{1}{k^2}\left(R^2 - z'^2 - k^2\right)\eta^2 + \frac{z'^2}{k^2} = 0. \tag{6.63}$$

In auxiliary cylindrical co-ordinates (R, ϕ, z'), the governing Poincaré inertial wave equation (6.49), taking the azimuthal dependence again as $\exp im\phi$ and multiplying through by R^2, becomes

$$R\frac{\partial}{\partial R}\left(R\frac{\partial \chi}{\partial R}\right) - R^2\frac{\partial^2 \chi}{\partial z'^2} - m^2\chi = 0. \tag{6.64}$$

Differentiation of (6.62) yields

$$R\frac{\partial \xi}{\partial R} = -\xi\frac{(1-\xi^2)(1-\eta^2)}{\xi^2-\eta^2}, \qquad \frac{\partial \xi}{\partial z'} = -\frac{\eta(1-\xi^2)}{k(\xi^2-\eta^2)}, \tag{6.65}$$

while differentiation of (6.63) yields

$$R\frac{\partial \eta}{\partial R} = \eta\frac{(1-\xi^2)(1-\eta^2)}{\xi^2-\eta^2}, \qquad \frac{\partial \eta}{\partial z'} = \frac{\xi(1-\eta^2)}{k(\xi^2-\eta^2)}. \tag{6.66}$$

Following the same procedure as before in the derivation of equation (6.42), we find equation (6.64), in the auxiliary spheroidal co-ordinates (ξ, η), becomes

$$\frac{(1-\xi^2)(1-\eta^2)}{\xi^2-\eta^2}\left[\frac{\partial}{\partial \xi}\left((1-\xi^2)\frac{\partial \chi}{\partial \xi}\right) - \frac{\partial}{\partial \eta}\left((1-\eta^2)\frac{\partial \chi}{\partial \eta}\right)\right] - m^2\chi = 0. \tag{6.67}$$

Again, the variables separate and χ is found to be the product of associated Legendre functions in the co-ordinates ξ and η.

For a rigid container, the normal component of displacement must vanish. In physical cylindrical co-ordinates,

$$\nabla\chi = \hat{\boldsymbol{R}}\frac{\partial \chi}{\partial R} + \hat{\boldsymbol{\phi}}\frac{im}{R}\chi + \hat{\boldsymbol{k}}\frac{\partial \chi}{\partial z}, \tag{6.68}$$

where $(\hat{\boldsymbol{R}}, \hat{\boldsymbol{\phi}}, \hat{\boldsymbol{k}})$ are unit vectors in the directions of the co-ordinates (R, ϕ, z), respectively. Thus, from expression (6.9), the displacement in cylindrical co-ordinates is proportional to the vector

$$\begin{aligned} &\omega^2\nabla\chi - 4\boldsymbol{\Omega}(\boldsymbol{\Omega}\cdot\nabla\chi) + 2i\omega\boldsymbol{\Omega}\times\nabla\chi \\ &= \hat{\boldsymbol{R}}\omega\left(\omega\frac{\partial \chi}{\partial R} + \frac{2m\Omega}{R}\chi\right) + \hat{\boldsymbol{\phi}}i\omega\left(\frac{m\omega}{R}\chi + 2\Omega\frac{\partial \chi}{\partial R}\right) + \hat{\boldsymbol{k}}\left(\omega^2 - 4\Omega^2\right)\frac{\partial \chi}{\partial z}. \end{aligned} \tag{6.69}$$

For the container illustrated in Figure 6.1, the vector

$$\boldsymbol{t} = \hat{\boldsymbol{R}} + \hat{\boldsymbol{k}}\frac{dz}{dR} \tag{6.70}$$

is tangent to the surface. Hence, the vector

$$\boldsymbol{n} = \hat{\boldsymbol{R}} - \hat{\boldsymbol{k}}\frac{dR}{dz} \tag{6.71}$$

is normal to the surface. Differentiating (6.16), we find

$$\frac{dR}{dz} = -\frac{a^2}{b^2}\frac{z}{R}, \tag{6.72}$$

giving, for the normal vector,

$$\boldsymbol{n} = \hat{\boldsymbol{R}} + \hat{\boldsymbol{k}}\frac{a^2}{b^2}\frac{z}{R} = \hat{\boldsymbol{R}} + \hat{\boldsymbol{k}}\frac{1}{1-e^2}\frac{z}{R}. \tag{6.73}$$

Then, from expression (6.69), the condition for the normal component of displacement to vanish at the rigid boundary is

$$\sigma^2 R\frac{\partial \chi}{\partial R} + m\sigma\chi + \frac{\sigma^2 - 1}{1 - e^2}z\frac{\partial \chi}{\partial z} = 0. \tag{6.74}$$

From the derivatives (6.65) and (6.66), we find

$$\begin{aligned} R\frac{\partial \chi}{\partial R} &= R\frac{\partial \chi}{\partial \xi}\frac{\partial \xi}{\partial R} + R\frac{\partial \chi}{\partial \eta}\frac{\partial \eta}{\partial R} \\ &= -\xi\frac{(1-\xi^2)(1-\eta^2)}{\xi^2 - \eta^2}\frac{\partial \chi}{\partial \xi} + \eta\frac{(1-\xi^2)(1-\eta^2)}{\xi^2-\eta^2}\frac{\partial \chi}{\partial \eta}, \end{aligned} \tag{6.75}$$

and

$$\begin{aligned} z\frac{\partial \chi}{\partial z} &= z'\frac{\partial \chi}{\partial z'} = z'\left(\frac{\partial \chi}{\partial \xi}\frac{\partial \xi}{\partial z'} + \frac{\partial \chi}{\partial \eta}\frac{\partial \eta}{\partial z'}\right) \\ &= -\xi\eta^2\frac{1-\xi^2}{\xi^2-\eta^2}\frac{\partial \chi}{\partial \xi} + \xi^2\eta\frac{1-\eta^2}{\xi^2-\eta^2}\frac{\partial \chi}{\partial \eta}. \end{aligned} \tag{6.76}$$

Substitution of these expressions into the boundary condition (6.74) produces

$$\begin{aligned} &\frac{\xi(1-\xi^2)}{\xi^2-\eta^2}\left(\sigma^2 - \frac{\eta^2(1-e^2\sigma^2)}{1-e^2}\right)\frac{\partial \chi}{\partial \xi} \\ &\quad - \frac{\eta(1-\eta^2)}{\xi^2-\eta^2}\left(\sigma^2 - \frac{\xi^2(1-e^2\sigma^2)}{1-e^2}\right)\frac{\partial \chi}{\partial \eta} - m\sigma\chi = 0 \end{aligned} \tag{6.77}$$

as the condition for the vanishing of the normal displacement at the rigid container boundary. From (6.53),

$$\frac{1-e^2\sigma^2}{1-e^2} = \frac{\sigma^2}{\xi_0^2}. \tag{6.78}$$

The variables then separate to give the boundary condition as

$$\begin{aligned} &\frac{\xi(1-\xi^2)}{\sigma^2(1-\xi^2/\xi_0^2)}\frac{\partial \chi}{\partial \xi} - \frac{\eta(1-\eta^2)}{\sigma^2(1-\eta^2/\xi_0^2)}\frac{\partial \chi}{\partial \eta} \\ &\quad - \frac{m\xi_0^2}{\sigma}\left(\frac{1}{\sigma^2(1-\xi^2/\xi_0^2)} - \frac{1}{\sigma^2(1-\eta^2/\xi_0^2)}\right)\chi = 0 \end{aligned} \tag{6.79}$$

on the container surface, given by $\xi = \xi_0$. Multiplying by $\sigma^2(1 - \xi^2/\xi_0^2)$ before letting $\xi \to \xi_0$, we find that

$$\xi\left(1 - \xi^2\right)\frac{\partial\chi}{\partial\xi} = \frac{m\xi_0^2}{\sigma}\chi \tag{6.80}$$

on the surface $\xi = \xi_0$.

For modes of degree n, azimuthal number m, the secular equation for the eigenfrequencies, σ, is then

$$\left(1 - \xi_0^2\right)\frac{dP_n^m(\xi_0)}{d\xi_0} = m\frac{\xi_0}{\sigma}P_n^m(\xi_0), \tag{6.81}$$

with

$$\xi_0 = \sigma\sqrt{\frac{1 - e^2}{1 - e^2\sigma^2}}. \tag{6.82}$$

Replacing the derivative of the Legendre function using the recurrence relation (B.12), the secular equation becomes

$$(n+1)(n+m)P_{n-1}^m(\xi_0) - n(n-m+1)P_{n+1}^m(\xi_0) = (2n+1)m\frac{\xi_0}{\sigma}P_n^m(\xi_0). \tag{6.83}$$

Applied to the Earth's core, we might ignore the inner body and take the container surface to be the core–mantle boundary. The flattening of the boundary is defined as

$$f = \frac{a-b}{a}, \tag{6.84}$$

which can be related to the eccentricity, e, of the spheroidal container. We find that

$$e^2 = 1 - (1-f)^2\,. \tag{6.85}$$

The value of ξ_0, appearing in the secular equation, is given by (6.82). The programme POINCARE calculates the residual of the secular equation through the function subprogramme ERR(Z,M,N) for a range of Coriolis frequencies, and iterates to find a specific root, given two starting periods. The function subprogramme ERR(Z,M,N) calls the function subprogramme PMN(Z,M,N) of Section B.2 for the evaluation of Legendre functions of the first kind.

```
C          PROGRAMME POINCARE.FOR
C
C POINCARE.FOR is an interactive programme to calculate the
C eigenperiods of the Poincare inertial wave equation for an
C ellipsoidal container. The programme allows a range of values of
C the Coriolis frequency to be searched at specified intervals,
C or an iteration to be performed, starting with two initial
C trial periods.
```

```
C
      IMPLICIT DOUBLE PRECISION(A-H,O-Z)
C Specify value of pi.
      PI=3.141592653589793D0
C Set angular frequency of Earth's rotation.
      WE=7.292115D-5
C Calculate number of hours in a semi-sidereal day.
      HSSD=PI/(WE*3600.D0)
C Set flattening to hydrostatic value at the core-mantle boundary
C for Earth model Cal8.
      FLAT=1.D0/394.03D0
C Calculate square of the eccentricity from flattening.
      ESQD=1.D0-(1.D0-FLAT)*(1.D0-FLAT)
C Enter azimuthal number M and degree N.
 10   CONTINUE
      WRITE(6,11)
 11   FORMAT(1X,'Enter M and N')
      READ(5,*)M,N
C Choose between searching a range of Coriolis frequencies or
C iterating on two trial periods.
 12   CONTINUE
      WRITE(6,13)
 13   FORMAT(1X,'Enter -1 for search a range, 0 for iterate')
      READ(5,*)ISW
      IF(ISW.GE.0) GO TO 14
C Enter bottom and top of range of Coriolis frequencies
C and number of increments.
 21   CONTINUE
      WRITE(6,15)
 15   FORMAT(1X,'Enter bottom and top of Coriolis frequency range',1X,
     1 'and number of increments')
      READ(5,*)BOT,TOP,NT
C Find increment in period.
      DSIG=(TOP-BOT)/DFLOAT(NT)
      NTP1=NT+1
C Find residual ERR1 of secular equation at each of trial periods.
      DO 16 I=1,NTP1
C Find Coriolis frequency.
        SIG1=(BOT+DFLOAT(I-1)*DSIG)
        Z=SIG1*DSQRT((1.D0-ESQD)/(1.D0-ESQD*SIG1*SIG1))
C Find error in satisfying secular equation.
        ERR1=ERR(Z,M,N,SIG1)
        TS=12.D0/SIG1
        T=HSSD/SIG1
        WRITE(6,17)SIG1,TS,ERR1,T
 16   CONTINUE
 17   FORMAT(1X,F7.4,2X,'Sidereal period=',F8.4,2X,'Error=',E10.3,2X,
     1 'Period',F7.3,1X,'hours')
C Decide whether to do new search or to iterate.
      GO TO 12
C Enter two trial periods for iteration.
 14   CONTINUE
      WRITE(6,18)
 18   FORMAT(1X,'Enter trial periods')
      READ(5,*)T1,T2
      SIG1=12.D0/T1
      Z=SIG1*DSQRT((1.D0-ESQD)/(1.D0-ESQD*SIG1*SIG1))
      ERR1=ERR(Z,M,N,SIG1)
      SIG2=12.D0/T2
      Z=SIG2*DSQRT((1.D0-ESQD)/(1.D0-ESQD*SIG2*SIG2))
      ERR2=ERR(Z,M,N,SIG2)
```

```
C Find next Coriolis frequency.
 19     CONTINUE
        SIG=(SIG1*ERR2-SIG2*ERR1)/(ERR2-ERR1)
C Calculate new sidereal period.
        TS=12.D0/SIG
C Find new period.
        T=HSSD/SIG
        Z=SIG*DSQRT((1.D0-ESQD)/(1.D0-ESQD*SIG*SIG))
C Find error for new iterate.
        ERRI=ERR(Z,M,N,SIG)
        WRITE(6,17)SIG,TS,ERRI,T
C Decide whether to do new search, do another iterate, or end.
        WRITE(6,20)
 20     FORMAT(1X,'Enter -1 for new search, 0 for another iterate,',1X,
     1 'or 1 for new harmonic')
        READ(5,*)IBR
        IF(IBR.LT.0) GO TO 21
        IF(IBR.GT.0) GO TO 23
C Prepare for another iteration.
        SIG1=SIG2
        ERR1=ERR2
        SIG2=SIG
        ERR2=ERRI
        GO TO 19
C Decide to search new harmonic or go to end.
 23     CONTINUE
        WRITE(6,22)
 22     FORMAT(1X,'Enter -1 for new M and N, 0 for end')
        READ(5,*)ISW
        IF(ISW.LT.0) GO TO 10
        END
C
        DOUBLE PRECISION FUNCTION ERR(Z,M,N,SIG)
C Calculates root function for Poincare inertial wave equation
C in a spheroid.
        IMPLICIT DOUBLE PRECISION(A-H,O-Z)
        AM=DFLOAT(M)
        AN=DFLOAT(N)
        ERR=(AN+1.D0)*(AN+AM)*PMN(Z,M,N-1)-AN*(AN-AM+1.D0)*PMN(Z,M,N+1)
     1 -(2.D0*AN+1.D0)*AM*Z*PMN(Z,M,N)/SIG
        RETURN
        END
```

Eigenperiods, computed by the foregoing codes for a core–mantle boundary with hydrostatic flattening $f = 1/394.03$ (corresponding to that for Earth model Cal8), for some of the shorter period modes, are given in Table 6.1.

6.2 Dynamics of the fluid outer core

The solutions of the Poincaré inertial wave equation show that a contained, rotating fluid is capable of a great variety of free oscillations with periods greater than 12 sidereal hours.

Table 6.1 *Some eigenperiods of the Poincaré inertial wave equation found by the programme POINCARE.*

m	n	σ	Period in sidereal hours	Period in hours
1	2	0.5013	23.9392	23.874
0	4	0.6556	18.3037	18.254
1	4	0.8546	14.0411	14.003
1	4	0.3068	39.1118	39.005
2	4	0.6171	19.4461	19.393
3	4	0.2510	47.8175	47.687
1	6	0.9312	12.8871	12.852
2	6	0.8224	14.5908	14.551
1	8	0.7946	15.1023	15.061
2	8	0.8974	13.3727	13.336
3	8	0.8126	14.7682	14.728

While the outer core is a contained, rotating fluid, the container is deformable, the fluid is compressible, stratified and self-gravitating, and there is an inner body. To consider these realistic properties in its dynamics, we examine the equations for conservation of mass, momentum and energy, in addition to those for gravitation.

The outer core is also the seat of the Earth's main magnetic field and electromagnetic effects must be considered. Fortunately, for motions with wavelengths large compared to the electromagnetic skin depth, core oscillations can be treated as a purely mechanical phenomenon (Crossley and Smylie, 1975; Acheson, 1975). In this approximation, the induced field is roughly uB/L, where u is the typical particle displacement, B is the magnetic field strength and L is a common length scale for both the oscillation and the field. The Lorentz force density is then approximately $uB^2/\mu_0 L^2$, with μ_0 the permeability of free space. For motions with periods of a fraction of a day and longer, the Coriolis acceleration is usually dominant and the Lorentz force density measured against it is $B^2/\mu_0\rho\Omega L^2\omega$, where ρ is the mass density, ω is the angular frequency of the oscillation and Ω is Earth's angular velocity of rotation. For a field strength of 10^{-2} T, length scale 10^3 km, mass density 10^4 kg m^{-3} and an oscillation period of one day, the ratio amounts to only about 1.5×10^{-6}.

The outer core is a compressible, stratified fluid, and its motions are subject to the law of mass conservation. Suppose a surface S, moving with the fluid, encloses a volume $\mathcal{V}$, containing the total mass

$$\int_{\mathcal{V}} \rho(x_i, t)\, d\mathcal{V}, \tag{6.86}$$

where the mass density $\rho(x_i, t)$ is taken to be a function of both the co-ordinates, x_i, and time, t. The surface $\mathcal{S}$, moving with the fluid, sweeps through the density field, increasing the enclosed mass at the rate

$$\int_{\mathcal{S}} \rho v_k n_k \, d\mathcal{S}, \tag{6.87}$$

for surface velocity components (v_1, v_2, v_3) and unit outward normal vector $\hat{\boldsymbol{n}}$, with components (n_1, n_2, n_3). Since the density is also a function of time, the total rate of change of the enclosed mass is then

$$\int_{\mathcal{V}} \frac{\partial \rho}{\partial t} d\mathcal{V} + \int_{\mathcal{S}} \rho v_k n_k \, d\mathcal{S}. \tag{6.88}$$

Applying the theorem of Gauss to the surface integral, the total rate of change of the enclosed mass is

$$\int_{\mathcal{V}} \left[\frac{\partial \rho}{\partial t} + \frac{\partial}{\partial x_k} (\rho v_k) \right] d\mathcal{V}. \tag{6.89}$$

Since the surface moves with the fluid, no mass enters or escapes the enclosed volume, and conservation of mass implies that the integral (6.89) vanishes. The volume is arbitrary and hence the integrand of (6.89) must vanish everywhere, giving

$$\frac{\partial \rho}{\partial t} + \frac{\partial}{\partial x_k} (\rho v_k) = 0 \tag{6.90}$$

as the law of mass conservation. In symbolic notation, conservation of mass yields the *equation of continuity*,

$$\frac{\partial \rho}{\partial t} + \nabla \cdot (\rho \boldsymbol{v}) = 0. \tag{6.91}$$

Expanding the divergence, the equation of continuity becomes

$$\frac{\partial \rho}{\partial t} + \boldsymbol{v} \cdot \nabla \rho + \rho \nabla \cdot \boldsymbol{v} = 0. \tag{6.92}$$

The first two terms on the left side measure the time rate of change of density, moving with the fluid, or the *substantial* time derivative

$$\frac{D}{Dt} \equiv \frac{\partial}{\partial t} + \boldsymbol{v} \cdot \nabla. \tag{6.93}$$

The equation of continuity is then expressible as

$$\frac{D\rho}{Dt} + \rho \nabla \cdot \boldsymbol{v} = 0, \tag{6.94}$$

or

$$\frac{D}{Dt}(\log \rho) + \nabla \cdot v = 0. \tag{6.95}$$

Thus, departure from the solenoidal flow condition (6.2) of the ideal fluid arises from changes in density from local flow pressure fluctuations and from transport through the stratified core fluid. In the Earth's outer core, we will find the former negligible at long periods but the latter very important for motions with substantial radial component.

Now consider the equation for linear momentum. Again, we take a surface S moving with the fluid and enclosing a volume $\mathcal{V}$. The linear momentum of the enclosed fluid is

$$\int_{\mathcal{V}} \rho v_i \, d\mathcal{V}. \tag{6.96}$$

Let f_i be the body force per unit mass and τ_{ji} be the stress field. The total force acting on the fluid within S is then

$$\int_S \tau_{ji} n_j \, dS + \int_{\mathcal{V}} \rho f_i \, d\mathcal{V}. \tag{6.97}$$

Applying the theorem of Gauss to the surface integral, the total force acting is expressible as

$$\int_{\mathcal{V}} \left(\frac{\partial \tau_{ji}}{\partial x_j} + \rho f_i \right) d\mathcal{V}. \tag{6.98}$$

The rate of change of the linear momentum enclosed by the surface S is expressible as the sum of the rate of change within S and the rate at which it is being swept up as S moves through the density and velocity fields, and is thus

$$\int_{\mathcal{V}} \frac{\partial}{\partial t}(\rho v_i) \, d\mathcal{V} + \int_S (\rho v_i) \, v_j n_j \, dS. \tag{6.99}$$

Again, applying Gauss's theorem to the surface integral, the total time rate of change of linear momentum becomes

$$\int_{\mathcal{V}} \left[\frac{\partial}{\partial t}(\rho v_i) + \frac{\partial}{\partial x_j}\left(\rho v_i v_j\right) \right] d\mathcal{V}. \tag{6.100}$$

Equating this to the total force (6.98), for arbitrary $\mathcal{V}$, yields the equation of motion as

$$\frac{\partial}{\partial t}(\rho v_i) + \frac{\partial}{\partial x_j}\left(\rho v_i v_j\right) = \rho f_i + \frac{\partial \tau_{ji}}{\partial x_j}. \tag{6.101}$$

The left side may be expanded as

$$\rho \frac{\partial v_i}{\partial t} + \rho v_j \frac{\partial v_i}{\partial x_j} + v_i \left[\frac{\partial \rho}{\partial t} + \frac{\partial}{\partial x_j}\left(\rho v_j\right) \right]. \tag{6.102}$$

From the equation (6.90) for the conservation of mass, the quantity in square brackets vanishes, and after dividing by ρ the *equation of motion* becomes

$$\frac{\partial v_i}{\partial t} + v_j \frac{\partial v_i}{\partial x_j} = f_i + \frac{1}{\rho}\frac{\partial \tau_{ji}}{\partial x_j}. \tag{6.103}$$

The stress field thus far has been left unspecified. In a fluid, the stress field arises from pressure and the resistance of the fluid to deformation. While a fluid, by definition, can support no permanent shear stress, it does in general resist a finite deformation rate.

Consideration of the surface forces and moments on an arbitrary volume element, as in Section 1.4.1, as the volume element becomes vanishingly small, shows the stress τ_{ji} to be symmetric. The diagonal components of the stress, $\tau_{11}, \tau_{22}, \tau_{33}$, as before are the *normal stresses*, while $\tau_{12}, \tau_{13}, \tau_{21}, \tau_{23}, \tau_{31}, \tau_{32}$ are *shear stresses*. In a static fluid, the shear stresses vanish by definition of a fluid and the normal stresses are all equal to $-p$, where p is the scalar pressure field.

The analysis of the rate of deformation of a fluid closely follows that for a solid in Section 1.4.3, with velocities replacing displacements. The velocity of a fluid particle at $P'(x_k + dx_k)$ relative to a neighbouring fluid particle at $P(x_k)$ is

$$dv_i = \frac{\partial v_i}{\partial x_j} dx_j. \tag{6.104}$$

Once again, the vector gradient can be split into its symmetric and antisymmetric parts as

$$\frac{\partial v_i}{\partial x_j} = e_{ji} + \omega_{ji}, \tag{6.105}$$

where

$$e_{ij} = \frac{1}{2}\left(\frac{\partial v_j}{\partial x_i} + \frac{\partial v_i}{\partial x_j}\right), \quad \omega_{ij} = \frac{1}{2}\left(\frac{\partial v_j}{\partial x_i} - \frac{\partial v_i}{\partial x_j}\right) \tag{6.106}$$

are both second-order tensors. e_{ij} is the *strain rate tensor*. As in the case of the deformation of a solid, e_{ij} represents the true rate of deformation of the fluid, while ω_{ij} can be shown to represent a rigid-body rotation of P' around P at the angular velocity

$$\xi = \frac{1}{2}(\nabla \times v); \tag{6.107}$$

ξ is called the *vorticity* of the flow.

We now require a law relating the stresses produced in resistance to the deformation rate, as measured by the strain rate tensor. The most widely used assumption is that these quantities are linearly related, the law of Newtonian viscosity. For an isotropic viscous fluid, analogous to Hooke's law for an isotropic elastic solid, we write the viscous stress as

$$\lambda' e_{kk}\delta^i_j + 2\mu' e_{ij}, \tag{6.108}$$

where λ' and μ' are two constants of proportionality and

$$e_{kk} = \frac{\partial v_1}{\partial x_1} + \frac{\partial v_2}{\partial x_2} + \frac{\partial v_3}{\partial x_3} = \nabla \cdot \boldsymbol{v} \tag{6.109}$$

is the flow divergence rate. The mean normal viscous stress is

$$\frac{1}{3}(3\lambda'\nabla \cdot \boldsymbol{v} + 2\mu'\nabla \cdot \boldsymbol{v}) = \left(\lambda' + \frac{2}{3}\mu'\right)\nabla \cdot \boldsymbol{v}. \tag{6.110}$$

The viscous stress over and above the mean normal stress is

$$\lambda'\nabla \cdot \boldsymbol{v}\delta^i_j + \mu'\left(\frac{\partial v_j}{\partial x_i} + \frac{\partial v_i}{\partial x_j}\right) - \left(\lambda' + \frac{2}{3}\mu'\right)\nabla \cdot \boldsymbol{v}\delta^i_j = \mu'\left(\frac{\partial v_j}{\partial x_i} + \frac{\partial v_i}{\partial x_j} - \frac{2}{3}\nabla \cdot \boldsymbol{v}\delta^i_j\right), \tag{6.111}$$

which is called the stress deviator.

Similarly, the mean normal strain rate is

$$\frac{1}{3}\nabla \cdot \boldsymbol{v}, \tag{6.112}$$

while the rate of strain over and above the mean normal rate of strain is

$$\frac{1}{2}\left(\frac{\partial v_j}{\partial x_i} + \frac{\partial v_i}{\partial x_j}\right) - \frac{1}{3}\nabla \cdot \boldsymbol{v}\delta^i_j = \frac{1}{2}\left(\frac{\partial v_j}{\partial x_i} + \frac{\partial v_i}{\partial x_j} - \frac{2}{3}\nabla \cdot \boldsymbol{v}\delta^i_j\right), \tag{6.113}$$

which is called the strain rate deviator. The total viscous stress can be written as the sum of the stress deviator plus the mean normal stress, or

$$\mu'\left(\frac{\partial v_j}{\partial x_i} + \frac{\partial v_i}{\partial x_j} - \frac{2}{3}\nabla \cdot \boldsymbol{v}\delta^i_j\right) + \left(\lambda' + \frac{2}{3}\mu'\right)\nabla \cdot \boldsymbol{v}\delta^i_j. \tag{6.114}$$

Hence, the viscous stress deviator is proportional to the rate of strain deviator, and the mean normal stress is proportional to the mean normal rate of strain. The constants of proportionality define the first and second coefficients of viscosity, η and ζ. The total viscous stress is then

$$\eta\left(\frac{\partial v_j}{\partial x_i} + \frac{\partial v_i}{\partial x_j} - \frac{2}{3}\nabla \cdot \boldsymbol{v}\delta^i_j\right) + \zeta\nabla \cdot \boldsymbol{v}\delta^i_j. \tag{6.115}$$

The viscous stress arising through the *bulk viscosity*, ζ, depends on the compression or dilatation rate. For its determination, one requires either a very compressible fluid or very high flow pressures. Few satisfactory measurements of ζ have been made for any fluid. For the Earth's core, the damping of compressional waves indicates that ζ is quite small. The *dynamic viscosity*, η, has traditionally been estimated on the basis of laboratory studies to be in the order of 10^{-2} Pa s (Poirier, 1988). Such extrapolation to core pressures and temperatures has been challenged by Brazhkin (1998) and Brazhkin and Lyapin (2000), who find 10^{11} Pa s at the bottom of the outer core and 10^2 Pa s at the top. These values are closely confirmed by the reduction in the rotational splitting of the equatorial translational modes of the inner core (Smylie, 1999), giving 1.22×10^{11} Pa s at the bottom, and by the free decay of the free core nutations (Palmer and Smylie, 2005; Smylie and Palmer, 2007), leading to 890–3900 Pa s at the top of the outer core. Direct observations have recently been reconciled with laboratory measurements by Smylie *et al.* (2009), taking into account the pressure dependence of the activation volume, which had been ignored in previous extrapolations of laboratory measurements.

In addition to the contribution of viscous terms to the stress field, we must add pressure forces, which contribute the diagonal terms $-p\delta^i_{\ j}$ to the stress tensor. Substituting these and expression (6.115) for the viscous stress in the equation of motion (6.103) yields

$$\begin{aligned} \frac{\partial v_i}{\partial t} + v_j \frac{\partial v_i}{\partial x_j} &= f_i + \frac{1}{\rho}\frac{\partial}{\partial x_j}\left[-p\delta^i_{\ j} + \eta\left(\frac{\partial v_j}{\partial x_i} + \frac{\partial v_i}{\partial x_j} - \frac{2}{3}\nabla\cdot v\delta^i_{\ j}\right) + \zeta\nabla\cdot v\delta^i_{\ j}\right] \\ &= f_i - \frac{1}{\rho}\frac{\partial p}{\partial x_i} + \frac{\eta}{\rho}\frac{\partial^2 v_i}{\partial x_j \partial x_j} + \frac{\zeta + \eta/3}{\rho}\frac{\partial}{\partial x_i}(\nabla\cdot v), \end{aligned} \tag{6.116}$$

ignoring spatial gradients of the coefficients of viscosity. Also left unspecified is the body force per unit mass, f_i. In the uniformly rotating reference frame, it will include the negatives of the Coriolis acceleration (3.75) and the centrifugal acceleration (3.76). As well, in the self-gravitating fluid outer core, it will include the gravitational acceleration g. In symbolic notation, the equation of motion becomes

$$\begin{aligned} &\frac{\partial v}{\partial t} + (v\cdot\nabla)v + 2\Omega\times v \\ &\quad = -\frac{1}{\rho}\nabla p + g - \Omega\times(\Omega\times r) + \frac{\eta}{\rho}\nabla^2 v + \frac{\zeta+\eta/3}{\rho}\nabla(\nabla\cdot v). \end{aligned} \tag{6.117}$$

We next examine the energy equation. Again, we enclose the volume $\mathcal{V}$ by the surface $\mathcal{S}$ moving with the fluid. The change in the internal energy per unit mass is $Tds - pdv$, where T is the temperature, s is the entropy per unit mass or the specific entropy, p is the pressure and $\mathrm{v} = 1/\rho$ is the volume per unit mass or the specific volume. The rate of increase of the total internal energy in the volume $\mathcal{V}$ is then

$$\begin{aligned}
&\int_{\mathcal{V}} \rho \left[T \frac{Ds}{Dt} - p \frac{D}{Dt}\left(\frac{1}{\rho}\right) \right] d\mathcal{V} \\
&\quad = \int_{\mathcal{V}} \rho T \left(\frac{\partial s}{\partial t} + \boldsymbol{v} \cdot \nabla s \right) d\mathcal{V} + \int_{\mathcal{V}} \frac{p}{\rho} \left(\frac{\partial \rho}{\partial t} + \boldsymbol{v} \cdot \nabla \rho \right) d\mathcal{V} \\
&\quad = \int_{\mathcal{V}} \left[\rho T \left(\frac{\partial s}{\partial t} + \boldsymbol{v} \cdot \nabla s \right) - p \nabla \cdot \boldsymbol{v} \right] d\mathcal{V}, \qquad (6.118)
\end{aligned}$$

on using the equation of continuity (6.92). To get the total rate of increase of internal energy of the fluid contained in $\mathcal{S}$, we must add the rate of gain of kinetic energy associated with the fluid motion,

$$\int_{\mathcal{V}} \frac{\partial}{\partial t} \left(\frac{1}{2} \rho v_k v_k \right) d\mathcal{V} + \int_{\mathcal{S}} \left(\frac{1}{2} \rho v_k v_k \right) v_j n_j \, d\mathcal{S}. \qquad (6.119)$$

Using the theorem of Gauss to transform the surface integral, this becomes

$$\int_{\mathcal{V}} \left[\frac{\partial}{\partial t} \left(\frac{1}{2} \rho v_k v_k \right) + \frac{\partial}{\partial x_j} \left(v_j \cdot \frac{1}{2} \rho v_k v_k \right) \right] d\mathcal{V}. \qquad (6.120)$$

At the same time, the fluid contained in $\mathcal{S}$ loses energy by doing work against the body force and on its surroundings at the rate

$$-\int_{\mathcal{V}} \rho f_j v_j \, d\mathcal{V} - \int_{\mathcal{S}} \tau_{ji} n_j v_i \, d\mathcal{S} = -\int_{\mathcal{V}} \left[\rho f_j v_j + \frac{\partial}{\partial x_j} (\tau_{ji} v_i) \right] d\mathcal{V}, \qquad (6.121)$$

on transforming the surface integral by Gauss's theorem. In addition, the fluid contained by the surface, $\mathcal{S}$, gains heat energy by thermal conduction relative to the fluid motion at the rate

$$\int_{\mathcal{S}} k \frac{\partial T}{\partial x_j} n_j \, d\mathcal{S} = \int_{\mathcal{V}} \frac{\partial}{\partial x_j} \left(k \frac{\partial T}{\partial x_j} \right) d\mathcal{V}, \qquad (6.122)$$

with k representing the thermal conductivity, and where once again the surface integral has been transformed to a volume integral by Gauss's theorem. Ignoring radiative transfer, internal heat sources add energy at the rate H per unit volume, giving the total rate of energy entering the system as

$$\int_{\mathcal{V}} \left[\frac{\partial}{\partial x_j} \left(k \frac{\partial T}{\partial x_j} \right) + H \right] d\mathcal{V}. \qquad (6.123)$$

The law of conservation of energy equates the gain of energy within the surface S, given by the sum of (6.118), (6.120) and (6.121), to the energy entering the system (6.123). Since the volume is arbitrary, integrands are equated, giving

$$\rho T\left(\frac{\partial s}{\partial t}+\boldsymbol{v}\cdot\nabla s\right)-p\nabla\cdot\boldsymbol{v}+\frac{\partial}{\partial t}\left(\frac{1}{2}\rho v_k v_k\right)+\frac{\partial}{\partial x_j}\left(v_j\cdot\frac{1}{2}\rho v_k v_k\right)$$
$$-\rho f_j v_j-\frac{\partial}{\partial x_j}\left(\tau_{ji}v_i\right)=\frac{\partial}{\partial x_j}\left(k\frac{\partial T}{\partial x_j}\right)+H. \qquad (6.124)$$

Expanding derivatives, this equation can be rearranged to

$$\rho T\left(\frac{\partial s}{\partial t}+\boldsymbol{v}\cdot\nabla s\right)-p\nabla\cdot\boldsymbol{v}+\frac{1}{2}v_k v_k\left(\frac{\partial\rho}{\partial t}+\frac{\partial}{\partial x_j}\left(\rho v_j\right)\right)+\rho v_k\left(\frac{\partial v_k}{\partial t}+v_j\frac{\partial v_k}{\partial x_j}\right)$$
$$-\rho f_j v_j-v_i\frac{\partial\tau_{ji}}{\partial x_j}-\tau_{ji}\frac{\partial v_i}{\partial x_j}=\frac{\partial}{\partial x_j}\left(k\frac{\partial T}{\partial x_j}\right)+H. \qquad (6.125)$$

It can be further simplified using the mass conservation equation (6.90) and the equation of motion (6.103) to give

$$\rho T\left(\frac{\partial s}{\partial t}+\boldsymbol{v}\cdot\nabla s\right)=\frac{\partial}{\partial x_j}\left(k\frac{\partial T}{\partial x_j}\right)+H+p\nabla\cdot\boldsymbol{v}+\tau_{ji}\frac{\partial v_i}{\partial x_j}. \qquad (6.126)$$

The left side of the equation of energy conservation (6.126) can be further transformed using a Tds equation from thermodynamics,

$$Tds=c_p\,dT-\frac{\alpha'}{\rho}Tdp, \qquad (6.127)$$

where c_p is the specific heat at constant pressure and

$$\alpha'=-\frac{1}{\rho}\left(\frac{\partial\rho}{\partial T}\right)_p \qquad (6.128)$$

is the volume coefficient of thermal expansion. Then, we have

$$\rho T\left(\frac{\partial s}{\partial t}+\boldsymbol{v}\cdot\nabla s\right)=\rho c_p\left(\frac{\partial T}{\partial t}+\boldsymbol{v}\cdot\nabla T\right)-\alpha' T\left(\frac{\partial p}{\partial t}+\boldsymbol{v}\cdot\nabla p\right). \qquad (6.129)$$

For an isotropic viscous fluid, the viscous stress resisting deformation is given by (6.115), and we have

$$\tau_{ji}\frac{\partial v_i}{\partial x_j}=\frac{\partial v_i}{\partial x_j}\left[-p\delta^i_j+\eta\left(\frac{\partial v_j}{\partial x_i}+\frac{\partial v_i}{\partial x_j}-\frac{2}{3}\nabla\cdot\boldsymbol{v}\delta^i_j\right)+\zeta\nabla\cdot\boldsymbol{v}\delta^i_j\right]$$
$$=-p\nabla\cdot\boldsymbol{v}+\eta\frac{\partial v_i}{\partial x_j}\left(\frac{\partial v_j}{\partial x_i}+\frac{\partial v_i}{\partial x_j}-\frac{2}{3}\nabla\cdot\boldsymbol{v}\delta^i_j\right)+\zeta\left(\nabla\cdot\boldsymbol{v}\right)^2. \qquad (6.130)$$

A further transformation can be made for the viscous terms. Expanding, we find that

$$\left(\frac{\partial v_j}{\partial x_i}+\frac{\partial v_i}{\partial x_j}-\frac{2}{3}\nabla\cdot v\delta^i_j\right)^2$$
$$=\frac{\partial v_j}{\partial x_i}\left(\frac{\partial v_j}{\partial x_i}+\frac{\partial v_i}{\partial x_j}-\frac{2}{3}\nabla\cdot v\delta^i_j\right)+\frac{\partial v_i}{\partial x_j}\left(\frac{\partial v_j}{\partial x_i}+\frac{\partial v_i}{\partial x_j}-\frac{2}{3}\nabla\cdot v\delta^i_j\right)$$
$$-\frac{2}{3}\nabla\cdot v\delta^i_j\frac{\partial v_j}{\partial x_i}-\frac{2}{3}\nabla\cdot v\delta^i_j\frac{\partial v_i}{\partial x_j}+\frac{4}{9}(\nabla\cdot v)^2\,\delta^i_j\delta^i_j. \tag{6.131}$$

Since

$$\delta^i_j\frac{\partial v_j}{\partial x_i}=\delta^i_j\frac{\partial v_i}{\partial x_j}=\nabla\cdot v,\quad\text{and}\quad \delta^i_j\delta^i_j=3, \tag{6.132}$$

on the interchange of dummy summation indices, we have the transformation identity

$$\frac{\partial v_i}{\partial x_j}\left(\frac{\partial v_j}{\partial x_i}+\frac{\partial v_i}{\partial x_j}-\frac{2}{3}\nabla\cdot v\delta^i_j\right)=\frac{1}{2}\left(\frac{\partial v_j}{\partial x_i}+\frac{\partial v_i}{\partial x_j}-\frac{2}{3}\nabla\cdot v\delta^i_j\right)^2. \tag{6.133}$$

On substitution from (6.129), (6.130) and (6.133), the equation of energy conservation (6.126) takes the final form

$$\frac{\partial T}{\partial t}+v\cdot\nabla T-\frac{\alpha' T}{\rho c_p}\left(\frac{\partial p}{\partial t}+v\cdot\nabla p\right)$$
$$=\frac{1}{\rho c_p}\left[\nabla\cdot(k\nabla T)+H+\frac{1}{2}\eta\left(\frac{\partial v_j}{\partial x_i}+\frac{\partial v_i}{\partial x_j}-\frac{2}{3}\nabla\cdot v\delta^i_j\right)^2+\zeta(\nabla\cdot v)^2\right]. \tag{6.134}$$

In addition to the conservation equations for mass (6.92), momentum (6.117) and energy (6.134), the dynamics of the fluid outer core are governed by gravitation, expressed by the gravitational acceleration

$$g=-\nabla V, \tag{6.135}$$

where $V(r)$ is the gravitational potential (5.114), related to the mass density through the Poisson equation (5.4),

$$\nabla^2 V=4\pi G\rho. \tag{6.136}$$

6.3 Scaling of the core equations

At the periods of several hours and longer that are considered here, the Coriolis acceleration is taken to be the dominant dynamical term. In a rotating fluid exact static equilibrium is impossible, as demonstrated by the Von Zeipel style of argument as expressed by (5.111). With substitution from (5.7) and (5.113) we see that a near equilibrium is possible if the quantity $\Omega^2/2\pi G\rho$ is small. In Earth's core this dimensionless number is only about 10^{-3}.

The outer core is a fluid stratified in both density and temperature. Taking the density stratification $\Delta\rho$ to be the result of a mean gravity $\bar{g}_0$ acting on a mean density $\bar{\rho}_0$, a flow characterised by velocity scale U would be expected to modify near-equilibrium quantities by the ratio of the Coriolis acceleration to the relative action of gravity on the density stratification, or

$$\frac{\Omega U}{\Delta\rho\,\bar{g}_0/\bar{\rho}_0}. \tag{6.137}$$

Using Ω^{-1} as the time scale and L as the length scale, the velocity scale is $U = L\Omega$, and the ratio (6.137) may be expressed as

$$\frac{\epsilon\Omega^2 L}{\Delta\rho\,\bar{g}_0/\bar{\rho}_0}, \tag{6.138}$$

with $\epsilon = U/\Omega L$ denoting the Rossby number (the ratio of the momentum advection to the Coriolis acceleration). If $\bar{T}_0$ is the mean temperature of the outer core and ΔT characterises its stratification, the stratified density and temperature fields may be written

$$\rho = \bar{\rho}_0 + \Delta\rho\left(\rho_e(r) + \frac{\epsilon\Omega^2 L}{\Delta\rho\,\bar{g}_0/\bar{\rho}_0}\rho_1\right), \tag{6.139}$$

$$T = \bar{T}_0 + \Delta T\left(T_e(r) + \frac{\epsilon\Omega^2 L}{\Delta\rho\,\bar{g}_0/\bar{\rho}_0}T_1\right), \tag{6.140}$$

where $\rho_e(r)$ and $T_e(r)$ are dimensionless functions of radius giving the near-equilibrium variation of density and temperature. ρ_1 and T_1 are the corresponding flow-induced dimensionless quantities. The pressure and gravitational potential fields may be similarly cast in dimensionless forms as

$$p = \bar{\rho}_0\bar{g}_0 L\left(p_e(r) + \frac{\epsilon\Omega^2 L}{\bar{g}_0}p_1\right), \tag{6.141}$$

$$V = \bar{g}_0 L\left(V_e(r) + \frac{\epsilon\Omega^2 L}{\bar{g}_0}V_1\right), \tag{6.142}$$

where $p_e(r)$ and $V_e(r)$ are dimensionless functions of radius giving their near-equilibrium variations and p_1 and V_1 are their dimensionless flow-induced perturbations. For scaled values of velocity $\boldsymbol{v}'$, time t' and gradient ∇', we have that

$$\boldsymbol{v} = U\boldsymbol{v}', \quad \frac{\partial}{\partial t} = \Omega\frac{\partial}{\partial t'}, \quad \text{and} \quad \nabla = \frac{1}{L}\nabla'. \tag{6.143}$$

Table 6.2 *Dimensionless numbers governing the dynamics of the fluid outer core.*

Self-gravitation number $\Gamma = 4\pi G\bar{\rho}_0 L/\bar{g}_0 = 1.2$
Adiabatic to actual temperature gradient ratio $A_R = (\alpha'\bar{g}_0\bar{T}_0/c_p)(\Delta T/L)^{-1} \approx 1$
Density ratio $D_R = \Delta\rho/\bar{\rho}_0 = 0.21$
Temperature ratio $\tau_R = \Delta T/\bar{T}_0 = 0.27$
Compressibility number $C = \bar{\rho}_0\bar{g}_0 L/\bar{\lambda} = \bar{g}_0 L/\alpha^2 = 8.8 \times 10^{-2}$
Internal Froude to Rossby number ratio $f_R = \Omega^2 L(\bar{g}_0\Delta\rho/\bar{\rho}_0)^{-1} = 3.1 \times 10^{-3}$
Froude number to Rossby number ratio $F_R = \Omega^2 L/\bar{g}_0 = 6.6 \times 10^{-4}$
Prandtl number $\sigma_P = \nu/\kappa = 2.4 \times 10^3$ to 2.4×10^{12}
Rossby number $\epsilon = U/\Omega L = 4.1 \times 10^{-6}$
Ekman number $E = \nu/\Omega L^2 = 1.4 \times 10^{-10}$ to 1.4×10^{-1}
Diffusivity coefficient $\delta = \nu\Omega/c_p\Delta T = 8.7 \times 10^{-13}$ to 8.7×10^{-4}
Bulk to dynamic viscosity ratio $\zeta/\eta \approx 1$

Dropping primes on these scaled quantities, mass conservation, expressed by the equation of continuity (6.92), becomes

$$F_R\frac{\partial\rho_1}{\partial t} + D_R\boldsymbol{v}\cdot\nabla\rho_e + \epsilon F_R\boldsymbol{v}\cdot\nabla\rho_1 + \left(1 + D_R\rho_e + \epsilon F_R\rho_1\right)\nabla\cdot\boldsymbol{v} = 0. \qquad (6.144)$$

The dimensionless numbers, F_R, D_R and ϵ, governing the relative importance of the terms in this equation, are defined in Table 6.2.

At zero velocity, the momentum equation (6.117) shows that the scaled near-equilibrium quantities are related by

$$\nabla p_e = (1 + D_R\rho_e)\,\boldsymbol{g}_e - (1 + D_R\rho_e)F_R\left(\hat{\boldsymbol{k}}\times(\hat{\boldsymbol{k}}\times\boldsymbol{r})\right), \qquad (6.145)$$

with $\hat{\boldsymbol{k}}$ as the unit vector along the axis of rotation. Using the near-equilibrium relation (6.145), the full equation of motion (6.117) scales to

$$\left(1 + D_R\rho_e + \epsilon F_R\rho_1\right)\left(\frac{\partial\boldsymbol{v}}{\partial t} + \epsilon\boldsymbol{v}\cdot\nabla\boldsymbol{v} + 2\hat{\boldsymbol{k}}\times\boldsymbol{v} - \boldsymbol{g}_1\right)$$
$$= -\nabla p_1 + \rho_1\boldsymbol{g}_e - F_R\rho_1\left(\hat{\boldsymbol{k}}\times(\hat{\boldsymbol{k}}\times\boldsymbol{r})\right) + E\left[\nabla^2\boldsymbol{v} + \left(\frac{\zeta}{\eta} + \frac{1}{3}\right)\nabla(\nabla\cdot\boldsymbol{v})\right], \qquad (6.146)$$

where, in view of the scaling (6.142) of the gravitational potential, the gravitational acceleration has been written

$$\boldsymbol{g} = \bar{g}_0\left(\boldsymbol{g}_e(\boldsymbol{r}) + \frac{\epsilon\Omega^2 L}{\bar{g}_0}\boldsymbol{g}_1\right), \qquad (6.147)$$

and, again, the dimensionless numbers, D_R, ϵ, F_R and E, governing the relative importance of the various terms, are defined in Table 6.2.

Table 6.3 *Numerical values assigned to outer core parameters in the scaling of the dynamical equations.*

Rotation rate $\Omega = 7.29 \times 10^{-5}$ rad s^{-1}
Characteristic mean density $\bar{\rho}_0 = 1.1 \times 10^4$ kg m^{-3}
Characteristic mean temperature $\bar{T}_0 = 4400$ K
Characteristic mean gravity $\bar{g}_0 = 8$ m s^{-2}
Characteristic length scale $L = 10^6$ m
Density stratification $\Delta\rho = 2.3 \times 10^3$ kg m^{-3}
Temperature stratification $\Delta T = 1200$ K
Flow speed $U = 3 \times 10^{-4}$ m s^{-1}
Mean adiabatic bulk modulus $\bar{\lambda} = 10^{12}$ N m^{-2}
Square of P wave velocity $\alpha^2 = \bar{\lambda}/\bar{\rho}_0 = 91$ km^2 s^{-2}
Coefficient of thermal expansion $\alpha' = 1 \times 10^{-5}$ K^{-1}
Specific heat at constant pressure $c_p = 7 \times 10^2$ J kg^{-1} K^{-1}
Universal constant of gravitation $G = 6.67 \times 10^{-11}$ N m^2 kg^{-2}
Kinematic viscosity $\nu = \eta/\bar{\rho}_0 = 10^{-2}$ to 10^7 m^2 s^{-1}
Thermal diffusivity $\kappa = k/(\bar{\rho}_0 c_p) = 5.1 \times 10^{-6}$ m^2 s^{-1}

In the absence of flow, the energy conservation equation (6.134) shows that the near-equilibrium temperature $T_e(r)$ obeys (see also (5.106))

$$\nabla \cdot (k \nabla T_e) = -H. \tag{6.148}$$

The full equation of energy conservation (6.134), on ignoring the spatial variation of the thermal conductivity, takes the scaled form

$$\begin{aligned}&\left(1 + D_R\rho_e + \epsilon F_R \rho_1\right)\left(f_R \frac{\partial T_1}{\partial t} + \boldsymbol{v} \cdot \nabla T_e + \epsilon f_R \boldsymbol{v} \cdot \nabla T_1\right)\\ &\quad - A_R\left(1 + \tau_R T_e + \tau_R \epsilon f_R T_1\right)\left(F_R \frac{\partial p_1}{\partial t} + \boldsymbol{v} \cdot \nabla p_e + \epsilon F_R \boldsymbol{v} \cdot \nabla p_1\right)\\ &= \frac{E f_R}{\sigma_P} \nabla^2 T_1 + \epsilon\delta\left[\frac{1}{2}\left(\frac{\partial v_j}{\partial x_i} + \frac{\partial v_i}{\partial x_j} - \frac{2}{3}\nabla \cdot \boldsymbol{v}\delta^i_j\right)^2 + \frac{\zeta}{\eta}(\nabla \cdot \boldsymbol{v})^2\right],\end{aligned} \tag{6.149}$$

where, once again, the dimensionless numbers, D_R, ϵ, F_R, f_R, A_R, τ_R, E, σ_P and δ, indicating the relative importance of terms, are defined in Table 6.2.

The values of the dimensionless numbers indicated in Table 6.2 have been computed using the numerical values assigned to outer core parameters listed in Table 6.3.

In the unperturbed, near-equilibrium case, the Poisson equation (6.136) for the gravitational potential scales to

$$\nabla^2 V_e = \Gamma(1 + D_R \rho_e), \tag{6.150}$$

and the scaled, flow-induced gravitational potential obeys

$$\nabla^2 V_1 = \Gamma \rho_1. \tag{6.151}$$

The dimensionless self-gravitation number Γ is defined in Table 6.2.

From the numerical values in Table 6.2, to parts in 10^9, the equation of mass conservation (6.144) takes the dimensional form

$$\frac{\partial \rho_1}{\partial t} + \boldsymbol{v} \cdot \nabla \rho_0 + \rho_0 \nabla \cdot \boldsymbol{v} = 0, \tag{6.152}$$

with subscripted zero indicating near-equilibrium quantities, and subscripted ones indicating perturbation quantities arising from the flow. Similarly, outside boundary layers, to parts in 10^6, the equation of motion (6.146) takes the dimensional form

$$\rho_0 \left(\frac{\partial \boldsymbol{v}}{\partial t} + 2\boldsymbol{\Omega} \times \boldsymbol{v} \right) = -\nabla p_1 + \rho_0 \boldsymbol{g}_1 - \rho_1 \big(\boldsymbol{g}_0 - \boldsymbol{\Omega} \times (\boldsymbol{\Omega} \times \boldsymbol{r}) \big). \tag{6.153}$$

The centrifugal acceleration $\boldsymbol{\Omega} \times (\boldsymbol{\Omega} \times \boldsymbol{r})$ can be expressed as $-\nabla W$ with

$$W = -\frac{1}{2} \Omega^2 r^2 \sin^2 \theta, \tag{6.154}$$

a function of radius r and co-latitude θ. Similarly, the near-equilibrium gravitational acceleration, $\boldsymbol{g}_0$, can be expressed as $-\nabla V_0$, with V_0 the near-equilibrium gravitational potential. Combining the two potentials, as customary, into the total geopotential U_0 (5.1), the equation of motion becomes

$$\rho_0 \left(\frac{\partial \boldsymbol{v}}{\partial t} + 2\boldsymbol{\Omega} \times \boldsymbol{v} \right) = -\nabla p_1 + \rho_0 \boldsymbol{g}_1 - \rho_1 \nabla U_0. \tag{6.155}$$

It is also customary to combine the near-equilibrium gravitational acceleration with the centrifugal acceleration to form the total near-equilibrium gravity vector, denoted by $\boldsymbol{g}_0$, with

$$\boldsymbol{g}_0 = -\nabla U_0. \tag{6.156}$$

The near-equilibrium equation for pressure gradient, (6.145), takes the dimensional form of the hydrostatic condition

$$\nabla p_0 = \rho_0 \boldsymbol{g}_0. \tag{6.157}$$

In the energy equation (6.149), thermal diffusion effects amount to only parts in 10^{16} and viscous dissipation effects amount to only a few parts in 10^9 at most, so that the flow is highly isentropic and the dimensional form of the energy equation becomes

$$\frac{\partial s}{\partial t} + \boldsymbol{v} \cdot \nabla s = \frac{Ds}{Dt} = 0. \tag{6.158}$$

The equation of state relates the state variables density, pressure and entropy, and since entropy is constant, the equation of state is *barotropic*, or the density is a function of pressure alone. The material derivative of the pressure is then

$$\frac{Dp}{Dt} = \left(\frac{\partial p}{\partial \rho}\right)_s \frac{D\rho}{Dt}. \tag{6.159}$$

The adiabatic bulk modulus or *incompressibility* in the fluid outer core is

$$\lambda = \rho\left(\frac{\partial p}{\partial \rho}\right)_s. \tag{6.160}$$

The velocity of compressional waves there is $\alpha = \sqrt{\lambda/\rho}$, and thus the equation of state becomes

$$\frac{Dp}{Dt} = \frac{\partial p}{\partial t} + \boldsymbol{v}\cdot\nabla p = \alpha^2\frac{D\rho}{Dt} = \alpha^2\left(\frac{\partial \rho}{\partial t} + \boldsymbol{v}\cdot\nabla\rho\right). \tag{6.161}$$

Substituting from the continuity equation (6.94), the equation of state leads to the relation

$$\frac{Dp}{Dt} = \frac{\partial p}{\partial t} + \boldsymbol{v}\cdot\nabla p = -\rho\alpha^2\nabla\cdot\boldsymbol{v}. \tag{6.162}$$

With the same scaling as before, this relation takes the scaled form,

$$\left(1 + D_R\rho_e + \epsilon F_R\rho_1\right)\nabla\cdot\boldsymbol{v} = -C\left(\boldsymbol{v}\cdot\nabla p_e + F_R\frac{\partial p_1}{\partial t} + \epsilon F_R\boldsymbol{v}\cdot\nabla p_1\right), \tag{6.163}$$

while the scaled equation of state itself becomes

$$\frac{\partial \rho_1}{\partial t} = -\frac{D_R}{F_R}\boldsymbol{v}\cdot\nabla\rho_e + \frac{C}{F_R}\boldsymbol{v}\cdot\nabla p_e + C\frac{\partial p_1}{\partial t} - \epsilon\left(\boldsymbol{v}\cdot\nabla\rho_1 - C\boldsymbol{v}\cdot\nabla p_1\right). \tag{6.164}$$

6.4 Compressibility and density stratification

The density of the fluid outer core increases by almost 24% from top to bottom. In addition, it has a compressibility well determined from the study of seismic waves. These properties are fundamental to understanding its dynamics.

Since the compressions and dilatations associated with the transmission of P waves are very closely adiabatic, the square of the P-wave velocity gives the ratio λ/ρ_0 of the bulk modulus or incompressibility to the density. A fluid particle suddenly displaced in radius by δr undergoes an adiabatic expansion, giving a decrease of its density,

$$\delta\rho = \rho_0\frac{\rho_0 g_0}{\lambda}\delta r = \frac{\rho_0 g_0}{\alpha^2}\delta r, \tag{6.165}$$

where $\rho_0 g_0\delta r$ is the decrease in pressure resulting from the radial displacement δr. Thus, the adiabatic lapse rate for density is $\rho_0 g_0/\alpha^2$. The actual lapse rate for density is $-d\rho_0/dr$. If the actual lapse rate exceeds the adiabatic lapse rate, the displaced fluid particle will experience a restoring force; if not, it will be subject to a force in the direction of its displacement. Hence, the stability or otherwise of the density stratification turns on the sign of the stability factor

$$\beta = 1 + \frac{\alpha^2}{\rho_0 g_0}\frac{d\rho_0}{dr}. \tag{6.166}$$

If $\beta < 0$, the density stratification is stable; if $\beta > 0$, the density stratification is unstable. The ratio of the actual lapse rate for density to the adiabatic lapse rate, given by $1 - \beta$, is referred to as the stratification parameter. If the stratification is stable, the displaced fluid particle will experience a restoring force $\delta r \beta \rho_0 g_0^2/\alpha^2$ per unit volume. It will, therefore, oscillate around its equilibrium position at angular frequency ω_v with

$$\omega_v^2 = -\beta\left(\frac{g_0}{\alpha}\right)^2. \tag{6.167}$$

ω_v is called the Brunt–Väisälä frequency. For an unstable density stratification, ω_v becomes purely imaginary and w_v can be used as a stability parameter with $w_v^2 = -\omega_v^2$.

At long periods, flow pressure fluctuations have only a small effect on the density. In the scaled equation of state (6.164), terms on the right have been arranged in order of decreasing magnitude. They represent the rate of density decrease due to transport along the near-equilibrium density gradient (of order D_R/F_R), the effect of adiabatic compression by the near-equilibrium pressure field (of order C/F_R), the effects of dilatation–compression by the flow pressure fluctuations (of order C), and non-linear terms in the flow variables (of order ϵ). In comparison to the first term, taken as unity, the others have magnitudes of roughly 0.41, 2.8×10^{-4}, and 1.3×10^{-8}. Neglect of the smallest term is part of the linearisation for slow flows or small amplitude oscillations. The effects of flow pressure fluctuations on the density are small, and depend inversely on the square of the period. At periods of several hours they amount to only parts in 10^3. Neglecting this term leads to the *subseismic approximation* (Smylie and Rochester, 1981). With these approximations, the equation of state (6.164) takes the dimensional form

$$\frac{\partial \rho_1}{\partial t} = -\boldsymbol{v}\cdot\nabla\rho_0 + \frac{1}{\alpha^2}\boldsymbol{v}\cdot\nabla p_0. \tag{6.168}$$

Neglecting only the non-linear term on the right of (6.163), (of order 10^{-9} compared to the largest term on the right), it takes the dimensional form

$$\rho_0\alpha^2\nabla\cdot\boldsymbol{v} = -\boldsymbol{v}\cdot\nabla p_0 - \frac{\partial p_1}{\partial t}. \tag{6.169}$$

With substitution from (6.156), dividing by ρ_0 and differentiating with respect to time, the equation of motion (6.155) becomes

$$\frac{\partial^2 \boldsymbol{v}}{\partial t^2} + 2\boldsymbol{\Omega} \times \frac{\partial \boldsymbol{v}}{\partial t} = -\nabla\left(\frac{1}{\rho_0}\frac{\partial p_1}{\partial t} - \frac{\partial V_1}{\partial t}\right) - \frac{1}{\rho_0^2}\nabla\rho_0\frac{\partial p_1}{\partial t} + \frac{1}{\rho_0}\frac{\partial \rho_1}{\partial t}\boldsymbol{g}_0, \tag{6.170}$$

where V_1 is the decrease in gravitational potential associated with the flow, giving $\boldsymbol{g}_1 = \nabla V_1$ (3.12). Again, at periods of several hours the effects of flow pressure fluctuations amount to only parts in 10^3, and depend inversely on the square of the period, and to this degree of approximation, $\partial p_1/\partial t$ may be neglected along with the second term on the right side of the equation of motion. The third term on the right side of the equation of motion, on substitution from the equation of state (6.168), becomes

$$\begin{aligned}\frac{1}{\rho_0}\frac{\partial \rho_1}{\partial t}\boldsymbol{g}_0 &= \frac{1}{\rho_0}\boldsymbol{g}_0\left(-\boldsymbol{v}\cdot\nabla\rho_0 + \frac{1}{\alpha^2}\boldsymbol{v}\cdot\nabla p_0\right)\\ &= \frac{1}{\rho_0}\boldsymbol{g}_0\left(\frac{\boldsymbol{v}\cdot\boldsymbol{g}_0}{g_0}\frac{d\rho_0}{dr} + \frac{\rho_0}{\alpha^2}\boldsymbol{v}\cdot\boldsymbol{g}_0\right)\\ &= \frac{1}{\alpha^2}\boldsymbol{v}\cdot\boldsymbol{g}_0\left(1 + \frac{\alpha^2}{\rho_0 g_0}\frac{d\rho_0}{dr}\right)\boldsymbol{g}_0,\end{aligned} \tag{6.171}$$

taking gravity to be in the direction opposite to radius, and substituting from the hydrostatic condition (6.157). Again, neglecting the effect of flow pressure, and substituting from the hydrostatic condition, equation (6.169) yields

$$\boldsymbol{v}\cdot\boldsymbol{g}_0 = -\alpha^2\nabla\cdot\boldsymbol{v}. \tag{6.172}$$

Using this relation and expression (6.166) for the stability factor β, the equation of motion is transformed to

$$\frac{\partial^2 \boldsymbol{v}}{\partial^2 t} + 2\boldsymbol{\Omega} \times \frac{\partial \boldsymbol{v}}{\partial t} = -\nabla\left(\frac{\partial \chi}{\partial t}\right) - \beta\boldsymbol{g}_0\nabla\cdot\boldsymbol{v}, \tag{6.173}$$

with

$$\frac{\partial \chi}{\partial t} = \frac{1}{\rho_0}\frac{\partial p_1}{\partial t} - \frac{\partial V_1}{\partial t}. \tag{6.174}$$

At frequencies below seismic frequencies, flow pressure fluctuations have little effect on density compared to transport through the stratified density field or through the hydrostatic pressure field. At periods of several hours, departures from these subseismic conditions amount to only parts in 10^3, and decrease inversely as the square of the period.

For an oscillation with time dependence exp $i\omega t$, from equation (6.173), the vector displacement field $\boldsymbol{u}$ is found to follow the equation of motion

$$-\omega^2 \boldsymbol{u} + 2i\omega \boldsymbol{\Omega} \times \boldsymbol{u} = -\nabla \chi - \beta g_0 \nabla \cdot \boldsymbol{u}, \tag{6.175}$$

with the generalised potential given as

$$\chi = \frac{p_1}{\rho_0} - V_1. \tag{6.176}$$

From the equation of state (6.168), the density perturbation is given by

$$\rho_1 = -\boldsymbol{u} \cdot \nabla \rho_0 + \frac{1}{\alpha^2} \boldsymbol{u} \cdot \nabla p_0. \tag{6.177}$$

Flow pressure fluctuations are given by equation (6.169) as

$$p_1 = -\boldsymbol{u} \cdot \nabla p_0 - \rho_0 \alpha^2 \nabla \cdot \boldsymbol{u}. \tag{6.178}$$

Again neglecting the effects of flow pressure fluctuations, we have

$$\frac{1}{\alpha^2} \boldsymbol{u} \cdot \nabla p_0 = -\rho_0 \nabla \cdot \boldsymbol{u}, \tag{6.179}$$

which, on substitution in the equation of state, yields

$$\rho_1 = -\boldsymbol{u} \cdot \nabla \rho_0 - \rho_0 \nabla \cdot \boldsymbol{u} = -\nabla \cdot (\rho_0 \boldsymbol{u}). \tag{6.180}$$

With substitution from the hydrostatic condition (6.157), expression (6.179) provides *the subseismic form of the equation of continuity*,

$$\nabla \cdot \boldsymbol{u} = -\frac{1}{\alpha^2} g_0 \cdot \boldsymbol{u}. \tag{6.181}$$

The displacement field obeying (6.181) can be shown to be reduced to an equivalent solenoidal vector field by multiplication by a scalar *decompression factor* f (Friedlander, 1985). Thus,

$$\nabla \cdot (f \boldsymbol{u}) = \boldsymbol{u} \cdot \nabla f + f \nabla \cdot \boldsymbol{u} = 0, \tag{6.182}$$

provided that scalar f obeys

$$\frac{\nabla f}{f} = \frac{g_0}{\alpha^2} = -\frac{\nabla U_0}{\alpha^2}, \tag{6.183}$$

where U_0 is the geopotential. Scalar f then obeys the ordinary differential equation in the radius,

$$\frac{df}{f} = -C_{\mathrm{loc}} \frac{dr}{r}, \tag{6.184}$$

with

$$C_{\mathrm{loc}} = \frac{dU_0}{dr} \frac{r}{\alpha^2} = \frac{\rho_0 g_0 r}{\lambda}, \tag{6.185}$$

a *local compressibility number*, λ again representing the adiabatic bulk modulus or incompressibility. The decompression factor, as a function of radius, can be found by solution of the differential equation (6.184). For an initial value of the decompression factor, f_i, at an initial radius r_i, its value at radius r is

$$f(r) = f_i \exp\left(-\int_{r_i}^{r} \frac{C_{\text{loc}}}{r'}\, dr'\right) = f_i \exp\left(-\int_{r_i}^{r} \frac{\rho_0(r')\, g_0(r')}{\lambda(r')}\, dr'\right). \tag{6.186}$$

If the initial radius is chosen as b, the radius of the core–mantle boundary, and the value of the decompression factor there is taken as unity, then the decompression factor as a function of radius becomes

$$f(r) = \exp\left(\int_{r}^{b} \frac{\rho_0(r')\, g_0(r')}{\lambda(r')}\right) dr'. \tag{6.187}$$

The programme DECOMP calculates the decompression factor for the outer core at each of the Earth model integration points for a specific Earth model. The Earth model file name is entered at the request of the programme, and the output, for use in subsequent calculations, is stored in the file decomp.dat. The integral in the exponent of expression (6.187) for the decompression factor is evaluated using Simpson's rule. For two successive model integration points at radii r_1 and r_3, the integrand $g(r)$ has the values $g(r_1) = g_1$ and $g(r_3) = g_3$. Through calls to the subroutines SPMAT and INTPL, described in Section 1.6, the value of the integrand at the midpoint $(r_3 - r_1)/2 = r_2 = r_1 + h$ is found as g_2 by cubic spline interpolation. By Simpson's rule, the contribution to the integral is

$$\frac{h}{3}\,(g_1 + 4g_2 + g_3)\,. \tag{6.188}$$

As can be shown by successive substitution of the functions 1, r, r^2 and r^3, Simpson's rule is third-order accurate; in other words, exact for the cubic spline used in the approximation of the integrand.

```
C          PROGRAMME DECOMP.FOR
C
C DECOMP.FOR calculates the scalar decompression factor f for
C a particular Earth model. It first reads in the Earth model,
C converts it to SI units, and uses the subroutines SPMAT and INTPL
C to interpolate outer core values, using spline interpolation.
C The required quadrature for the calculation of the decompression
C factor as a function of radius is performed using Simpson's rule.
C
      IMPLICIT DOUBLE PRECISION(A-H,O-Z)
      DIMENSION R(100),RHO(100),GZERO(100),RI(300),RHOI(300),F(100),
     1 GZEROI(300),ENAME(10),NM(4),NI(4),NK(4),B(98,198),C(100,100)
      DOUBLE PRECISION MU(100),LAMBDA(100),MUI(300),LAMBDAI(300),
     1 LAMBDAM
      CHARACTER*20 EMODEL
C Read in Earth model file name.
      WRITE(6,10)
```

```
 10   FORMAT(1X,'Type in Earth model file name.')
      READ(5,11)EMODEL
 11   FORMAT(A20)
C Open Earth model file.
      OPEN(UNIT=1,FILE=EMODEL,STATUS='OLD')
      OPEN(UNIT=2,FILE='decomp.dat',STATUS='UNKNOWN')
C Set maximum dimensions for interpolation.
      M1=100
      M2=198
      M3=98
C Read in and write out Earth model.
C Read in and write out headers.
      READ(1,12)(ENAME(I),I=1,10)
 12   FORMAT(10A8)
      WRITE(6,13)(ENAME(I),I=1,10)
 13   FORMAT(11X,10A8)
      WRITE(2,12)(ENAME(I),I=1,10)
      READ(1,14)NN
 14   FORMAT(I10)
      READ(1,15)(NM(I),NI(I),I=1,4)
 15   FORMAT(8I10)
      WRITE(6,16)
 16   FORMAT(/16X,'Number of model points',1X,
     1 'and number of integration steps.'/)
      WRITE(6,15)(NM(I),NI(I),I=1,4)
      WRITE(6,17)
 17   FORMAT(//5X,'Radius',6X,'Rho',4X,'Lambda',7X,'Mu',6X,'Gzero')
      WRITE(6,18)
 18   FORMAT(6X,'(km)',5X,'(gm/cc)',2X,'(kbars)',4X,'(kbars)',2X,
     1'(cm/sec)'/)
C Read in Earth model, inner core, outer core, mantle and crust.
      K=0
      DO 19 M=1,4
        N1=NM(M)
        READ(1,20)(R(I),RHO(I),LAMBDA(I),MU(I),GZERO(I),I=1,N1)
 20     FORMAT(1X,F10.1,F10.2,F10.1,F10.1,F10.1)
        DO 21 I=1,N1
C Scale Earth model to SI values and store.
          J=K+I
          RI(J)=R(I)*1.D3
          RHOI(J)=RHO(I)*1.D3
          LAMBDAI(J)=LAMBDA(I)*1.D8
          MUI(J)=MU(I)*1.D8
          GZEROI(J)=GZERO(I)*1.D-2
C Write Earth model out to screen.
          WRITE(6,20)R(I),RHO(I),LAMBDA(I),MU(I),GZERO(I)
 21     CONTINUE
        NK(M)=K
        K=K+N1
 19   CONTINUE
C Begin calculation of decompression factor f.
C Set up interpolation for the outer core.
      N1=NM(2)
      N2=2*N1-2
      N3=N1-2
C Put outer core values in active locations.
      DO 22 I=1,N1
        J=NK(2)+I
        R(I)=RI(J)
        RHO(I)=RHOI(J)
        MU(I)=MUI(J)
```

```
        LAMBDA(I)=LAMBDAI(J)
        GZERO(I)=GZEROI(J)
 22   CONTINUE
C Construct interpolation matrix.
      CALL SPMAT(N1,N2,N3,C,R,B,M1,M2,M3)
C Begin integration by Simpson's rule.
      WRITE(6,23)
 23   FORMAT(//9X,'radius',9X,'exponent',7X,'f factor')
      EINT=0.D0
      F(N1)=1.D0
      RKM=R(N1)*1.D-3
      WRITE(2,14)N1
      WRITE(2,24)R(N1),F(N1),RHO(N1)
 24   FORMAT(3D23.15)
      WRITE(6,25)RKM,EINT,F(N1)
 25   FORMAT(/1X,F15.1,2F15.3)
      N1M1=N1-1
      DO 26 I=1,N1M1
C Interpolate rho, g, lambda to midpoint of interval.
        J=N1-I
        JP1=J+1
C Find stepsize for integration.
        H=0.5D0*(R(JP1)-R(J))
C Find midpoint of interval.
        RM=R(J)+H
        CALL INTPL(RM,RHOM,N1,C,R,RHO,M1)
        CALL INTPL(RM,GMEAN,N1,C,R,GZERO,M1)
        CALL INTPL(RM,LAMBDAM,N1,C,R,LAMBDA,M1)
C Apply Simpson's rule.
        F1=RHO(J)*GZERO(J)/LAMBDA(J)
        F2=RHOM*GMEAN/LAMBDAM
        F3=RHO(JP1)*GZERO(JP1)/LAMBDA(JP1)
        EINT=EINT+H*(F1+4.D0*F2+F3)/3.D0
        F(J)=DEXP(EINT)
        RKM=R(J)*1.D-3
C Write out results.
        WRITE(2,24)R(J),F(J),RHO(J)
        WRITE(6,25)RKM,EINT,F(J)
 26   CONTINUE
C Interpolate decompression factor onto integration points.
C Find number of integration steps and interval size.
      M=NI(2)
      AM=DFLOAT(M)
      H=(R(N1)-R(1))/AM
      RKM=R(1)*1.D-3
      EINT=DLOG(F(1))
      FI=F(1)
      WRITE(2,25)RKM,EINT,FI
      DO 27 I=1,M
        AI=DFLOAT(I)
        XV=R(1)+AI*H
        CALL INTPL(XV,FI,N1,C,R,F,M1)
        EINT=DLOG(FI)
        RKM=XV*1.D-3
        WRITE(2,25)RKM,EINT,FI
 27   CONTINUE
      END
```

The programme DECOMP can be used to calculate the decompression factor for the Earth models Cal8, 1066A, PREM and Core11, listed in Appendix C, simply

Table 6.4 *The decompression factor f as a function of radius for Earth models Cal8 and Core11.*

Cal8		Core11	
Radius (km)	f	Radius (km)	f
1216.0	1.232	1218.8	1.229
1443.0	1.218	1445.2	1.217
1670.0	1.204	1671.5	1.202
1897.0	1.187	1897.9	1.186
2124.0	1.168	2124.2	1.167
2351.0	1.147	2350.6	1.146
2578.0	1.123	2577.0	1.122
2805.0	1.097	2803.3	1.096
3032.0	1.068	3029.7	1.067
3259.0	1.035	3256.0	1.035
3486.0	1.000	3482.4	1.000

by entering the file names, cal8.dat, 1066a.dat, prem.dat and corell.dat, or that for another Earth model in a file with appropriate format, successively, at the request of the programme during execution. The decompression factor as a function of radius in the outer core is given in Table 6.4 for Earth models Cal8 and Core11. It is seen not to be strongly dependent on Earth model.

7

The subseismic equation and boundary conditions

In contrast with the inertial waves studied in the previous chapter, long-period motions in the outer core of the real Earth take place in a deformable container with an inner boundary, where the fluid is compressible, stratified and self-gravitating.

We first derive the governing subseismic wave equation (Smylie and Rochester, 1981). Then, to incorporate the elasto-gravitational boundary conditions, we develop a system of internal load Love numbers for the shell (mantle and crust), and for the inner core.

7.1 The subseismic wave equation

Expression (6.181) allows the equation of motion (6.175) to be transformed to

$$-\omega^2 \boldsymbol{u} + 2i\omega \boldsymbol{\Omega} \times \boldsymbol{u} = -\nabla\chi + \frac{\beta}{\alpha^2} g_0 \left(g_0 \cdot \boldsymbol{u}\right) = \boldsymbol{\ell}, \tag{7.1}$$

showing that the displacement field $\boldsymbol{u}$ is linearly related to the gradient of the generalised potential.

The displacement field $\boldsymbol{u}$ is given in terms of the vector $\boldsymbol{\ell}$ by equation (6.8). With $\boldsymbol{\Omega} = \Omega\hat{\boldsymbol{k}}$, and $\hat{\boldsymbol{k}}$ a unit vector, we have that

$$\boldsymbol{u} = \frac{1}{4\Omega^2\sigma^2\left(\sigma^2 - 1\right)}\left[-\sigma^2\boldsymbol{\ell} + \hat{\boldsymbol{k}}\left(\hat{\boldsymbol{k}} \cdot \boldsymbol{\ell}\right) - i\sigma\hat{\boldsymbol{k}} \times \boldsymbol{\ell}\right]. \tag{7.2}$$

Substituting the right side of equation (7.1) for $\boldsymbol{\ell}$, we find that

$$\boldsymbol{u} = \frac{1}{4\Omega^2\sigma^2\left(\sigma^2 - 1\right)}\left[\sigma^2\nabla\chi - \hat{\boldsymbol{k}}\left(\hat{\boldsymbol{k}} \cdot \nabla\chi\right) + i\sigma\hat{\boldsymbol{k}} \times \nabla\chi + \frac{\beta}{\alpha^2}\left(g_0 \cdot \boldsymbol{u}\right)\boldsymbol{C}^*\right], \tag{7.3}$$

with $\boldsymbol{C}^*$ the complex conjugate of the vector

$$\boldsymbol{C} = -\sigma^2 g_0 + \left(\hat{\boldsymbol{k}} \cdot g_0\right)\hat{\boldsymbol{k}} + i\sigma\hat{\boldsymbol{k}} \times g_0. \tag{7.4}$$

The scalar product $g_0 \cdot u$ can be expressed entirely in terms of the generalised potential χ. Taking the scalar product of $\hat{k}$ with the equation of motion (7.1), scaled by $4\Omega^2$, we find

$$-\sigma^2 \hat{k} \cdot u = -\frac{1}{4\Omega^2}\hat{k} \cdot \nabla\chi + \frac{\beta}{4\Omega^2\alpha^2}\left(\hat{k} \cdot g_0\right) g_0 \cdot u. \tag{7.5}$$

Taking the scalar product with g_0 gives

$$-\sigma^2 g_0 \cdot u + i\sigma g_0 \cdot \left(\hat{k} \times u\right) = -\frac{1}{4\Omega^2} g_0 \cdot \nabla\chi + \frac{\beta g_0^2}{4\Omega^2\alpha^2} g_0 \cdot u. \tag{7.6}$$

Using the identity

$$g_0 \cdot \left(\hat{k} \times u\right) = -\left(\hat{k} \times g_0\right) \cdot u, \tag{7.7}$$

for the triple scalar product, (7.6) becomes

$$-\left(\frac{\beta g_0^2}{4\Omega^2\alpha^2} + \sigma^2\right) g_0 \cdot u - i\sigma\left(\hat{k} \times g_0\right) \cdot u = -\frac{1}{4\Omega^2} g_0 \cdot \nabla\chi. \tag{7.8}$$

Finally, taking the scalar product of $(\hat{k} \times g_0)$ with (7.1), scaled by $4\Omega^2$, yields

$$-\sigma^2\left(\hat{k} \times g_0\right) \cdot u + i\sigma\left(\hat{k} \times g_0\right) \cdot \left(\hat{k} \times u\right) = -\frac{1}{4\Omega^2}\left(\hat{k} \times g_0\right) \cdot \nabla\chi, \tag{7.9}$$

which can be transformed with the identity

$$\left(\hat{k} \times g_0\right) \cdot \left(\hat{k} \times u\right) = g_0 \cdot u - \left(\hat{k} \cdot g_0\right) \hat{k} \cdot u \tag{7.10}$$

to

$$-\sigma^2\left(\hat{k} \times g_0\right) \cdot u + i\sigma g_0 \cdot u - i\sigma\left(\hat{k} \cdot g_0\right) \hat{k} \cdot u = -\frac{1}{4\Omega^2}\left(\hat{k} \times g_0\right) \cdot \nabla\chi. \tag{7.11}$$

Equations (7.5), (7.8) and (7.11) form the system of linear equations

$$\begin{pmatrix} -\sigma^2 & -\beta(\hat{k} \cdot g_0)/4\Omega^2\alpha^2 & 0 \\ 0 & -\beta g_0^2/4\Omega^2\alpha^2 - \sigma^2 & -i\sigma \\ -i\sigma(\hat{k} \cdot g_0) & i\sigma & -\sigma^2 \end{pmatrix} \begin{pmatrix} \hat{k} \cdot u \\ g_0 \cdot u \\ (\hat{k} \times g_0) \cdot u \end{pmatrix}$$
$$= -\frac{1}{4\Omega^2} \begin{pmatrix} \hat{k} \cdot \nabla\chi \\ g_0 \cdot \nabla\chi \\ (\hat{k} \times g_0) \cdot \nabla\chi \end{pmatrix}. \tag{7.12}$$

Multiplying the first equation through by $\hat{k} \cdot g_0$, the second by $-\sigma^2$, the third by $i\sigma$ and adding all three gives

$$\frac{\beta}{4\Omega^2\alpha^2}\left[\frac{4\Omega^2\alpha^2}{\beta}\sigma^2\left(\sigma^2 - 1\right) + \sigma^2 g_0^2 - \left(\hat{k} \cdot g_0\right)^2\right] g_0 \cdot u = -\frac{1}{4\Omega^2} C \cdot \nabla\chi, \tag{7.13}$$

or

$$g_0 \cdot u = -\frac{\alpha^2}{\beta}\frac{C \cdot \nabla\chi}{B}, \tag{7.14}$$

with

$$B = \frac{4\Omega^2\alpha^2}{\beta}\sigma^2\left(\sigma^2 - 1\right) + \sigma^2 g_0^2 - \left(\hat{k} \cdot g_0\right)^2. \tag{7.15}$$

Substitution for $g_0 \cdot u$ from expression (7.14) in the displacement field formula (7.3) leads to

$$u = \frac{1}{4\Omega^2\sigma^2\left(\sigma^2 - 1\right)}\left[\sigma^2\nabla\chi - \hat{k}\left(\hat{k} \cdot \nabla\chi\right) + i\sigma\hat{k} \times \nabla\chi - C^*\frac{C \cdot \nabla\chi}{B}\right], \tag{7.16}$$

expressing the vector displacement field u entirely in terms of the gradient of the generalised potential χ. For this reason, χ can be referred to as the *generalised displacement potential*. The displacement field must satisfy the subseismic form (6.181) of the equation of continuity, or

$$\nabla \cdot \left[\sigma^2\nabla\chi - \hat{k}\left(\hat{k} \cdot \nabla\chi\right) + i\sigma\hat{k} \times \nabla\chi - C^*\frac{C \cdot \nabla\chi}{B}\right] = -4\Omega^2\sigma^2\left(\sigma^2 - 1\right)\frac{1}{\alpha^2}g_0 \cdot u. \tag{7.17}$$

Once again using the vector identities (6.10) and (6.11), with $\hat{k}$ replacing Ω, but also using the identity (A.9), and substituting from (7.14), we find

$$\sigma^2\nabla^2\chi - \left(\hat{k} \cdot \nabla\right)^2\chi - C^* \cdot \nabla\left(\frac{C \cdot \nabla\chi}{B}\right) = \frac{C \cdot \nabla\chi}{B}\left[\nabla \cdot C^* + \frac{4\Omega^2}{\beta}\sigma^2\left(\sigma^2 - 1\right)\right]. \tag{7.18}$$

The divergence of C^* is

$$\nabla \cdot C^* = -\sigma^2\nabla \cdot g_0 + \nabla \cdot \left[\left(\hat{k} \cdot g_0\right)\hat{k}\right] - i\sigma\nabla \cdot \left(\hat{k} \times g_0\right). \tag{7.19}$$

This expression can be transformed using the vector identities

$$\nabla \cdot \left[\left(\hat{k} \cdot g_0\right)\hat{k}\right] = \left(\hat{k} \cdot g_0\right)\nabla \cdot \hat{k} + \hat{k} \cdot \nabla\left(\hat{k} \cdot g_0\right) = \hat{k} \cdot \nabla\left(\hat{k} \cdot g_0\right) \tag{7.20}$$

and

$$\nabla \cdot \left(\hat{k} \times g_0\right) = g_0 \cdot \left(\nabla \times \hat{k}\right) - \hat{k} \cdot \left(\nabla \times g_0\right) = 0, \tag{7.21}$$

since $\boldsymbol{g}_0 = -\nabla U_0$ and $\nabla \times \boldsymbol{g}_0 = -\nabla \times \nabla U_0 \equiv 0$. Further, from (5.5),

$$\nabla \cdot \boldsymbol{g}_0 = -\nabla^2 U_0 = -\left(4\pi G \rho_0 - 2\Omega^2\right). \tag{7.22}$$

Finally, substituting (7.20), (7.21) and (7.22) in (7.19) yields

$$\nabla \cdot \boldsymbol{C}^* = \sigma^2\left(4\pi G \rho_0 - 2\Omega^2\right) + \hat{\boldsymbol{k}} \cdot \nabla\left(\hat{\boldsymbol{k}} \cdot \boldsymbol{g}_0\right). \tag{7.23}$$

Substituting this expression in (7.18), we arrive at

$$\sigma^2 \nabla^2 \chi - \left(\hat{\boldsymbol{k}} \cdot \nabla\right)^2 \chi - A \frac{\boldsymbol{C} \cdot \nabla \chi}{B} - \boldsymbol{C}^* \cdot \nabla\left(\frac{\boldsymbol{C} \cdot \nabla \chi}{B}\right) = 0, \tag{7.24}$$

with the scalar A defined by

$$A = \sigma^2\left(4\pi G \rho_0 - 2\Omega^2\right) + \hat{\boldsymbol{k}} \cdot \nabla\left(\hat{\boldsymbol{k}} \cdot \boldsymbol{g}_0\right) + \frac{4\Omega^2}{\beta} \sigma^2\left(\sigma^2 - 1\right). \tag{7.25}$$

7.2 Deformation of the shell and inner core

At long periods of several hours and more, for which the subseismic approximation applies, the elastic forces resisting deformation dominate the acceleration terms. There is one exception, in the case of a pure translation of the inner core, where there is little elastic deformation. Although, in principle, a compensating translation of the shell (mantle and crust) will accompany translation of the inner core, the motion is small due to the much larger mass of the shell compared to the inner core.

Neglecting shear stresses exerted by the fluid outer core, deformations of the shell and inner core are governed by the sixth-order spheroidal system of differential equations developed in Section 3.3, for each spherical harmonic constituent of degree $n \geq 1$. For degree $n = 0$, the governing spheroidal system degenerates to fourth order, the terms representing transverse displacement and shear stress no longer appearing. In each case, the frequency dependence is included through the self-coupling body force expressions (3.125) and (3.126).

For $n \geq 1$, in the shell, there are six linearly independent solutions, and the general solution can be written as their linear combination

$$\boldsymbol{y} = C_1 \boldsymbol{y}_1 + C_2 \boldsymbol{y}_2 + C_3 \boldsymbol{y}_3 + C_4 \boldsymbol{y}_4 + C_5 \boldsymbol{y}_5 + C_6 \boldsymbol{y}_6, \tag{7.26}$$

where each $\boldsymbol{y}_j$ is a six-vector with components representing the values of the variables $(y_1, y_2, y_3, y_4, y_5, y_6)$ describing the spheroidal deformation field, and each C_j is a combination coefficient. If $\boldsymbol{y}_j = (y_{1j}, y_{2j}, y_{3j}, y_{4j}, y_{5j}, y_{6j})$, the six linearly independent fundamental solutions are generated by setting $y_{ij}(b) = \delta^i_{\ j}$, with b the radius of the core–mantle boundary, and integrating forward to the surface at radius d. Thus, at the core–mantle boundary $y_i(b) = C_i$.

At the free surface of the Earth, normal and shear stresses vanish and the perturbation in gravitational potential becomes harmonic, while at the core–mantle boundary the shear stress vanishes and the normal displacement, normal stress, the perturbation in gravitational potential and the gravitational flux vector are all continuous. On the fluid side of the boundary, the normal stress is $\lambda\nabla\cdot\boldsymbol{u}$. From equation (6.178), with substitution from the hydrostatic condition (6.157), and replacement of α^2 by λ/ρ_0, we find that

$$p_1 = -\rho_0\boldsymbol{u}\cdot\boldsymbol{g}_0 - \lambda\nabla\cdot\boldsymbol{u}. \tag{7.27}$$

Hence, on the fluid side of the core–mantle boundary ($r = b^-$), the normal stress is

$$\lambda\nabla\cdot\boldsymbol{u} = -p_1 - \rho_0\boldsymbol{u}\cdot\boldsymbol{g}_0 = -\rho_0\,(V_1 + \chi) - \rho_0\boldsymbol{u}\cdot\boldsymbol{g}_0. \tag{7.28}$$

In terms of the radial coefficients of degree-n vector spherical harmonics, this equation becomes

$$y_2(b) = -\rho_0(b^-)\,(y_5(b) + \chi_n(b^-)) + \rho_0(b^-)\,g_0(b)\,y_1(b), \tag{7.29}$$

where we have used the continuity of y_1, y_2, y_5 and g_0. Vanishing shear stress there requires $C_4 = 0$. By rearrangement of (7.28), the perturbation of gravitational potential on the fluid side of the boundary is found as

$$V_1 = -\frac{\lambda}{\rho_0}\nabla\cdot\boldsymbol{u} - \boldsymbol{u}\cdot\boldsymbol{g}_0 - \chi. \tag{7.30}$$

Thus, continuity of y_5 is also expressed by (7.29). The gravitational flux vector,

$$\nabla V_1 - 4\pi G\rho_0\boldsymbol{u}, \tag{7.31}$$

is also continuous across the boundary. In terms of radial coefficients of degree-n vector spherical harmonics, this leads to

$$y_6(b) = \frac{dV_{1,n}}{dr}(b^-) - 4\pi G\rho_0(b^-)y_1(b). \tag{7.32}$$

Just outside the surface of the Earth ($r = d^+$), the perturbation in gravitational potential becomes harmonic, and continuity of y_6 becomes

$$y_6(d) = \frac{dV_{1,n}}{dr}(d^+) = -\frac{n+1}{d}V_{1,n}(d^+) = -\frac{n+1}{d}y_5(d), \tag{7.33}$$

since y_5 is also continuous there. For $n \geq 1$, application of the boundary conditions for the shell (excepting continuity of normal displacement at the core–mantle boundary) leads to the system of linear equations,

$$\begin{gathered}
C_1 y_{21}(d) + C_2 y_{22}(d) + C_3 y_{23}(d) + C_5 y_{25}(d) + C_6 y_{26}(d) = 0, \\
C_1 y_{41}(d) + C_2 y_{42}(d) + C_3 y_{43}(d) + C_5 y_{45}(d) + C_6 y_{46}(d) = 0, \\
C_1 z_1(d) + C_2 z_2(d) + C_3 z_3(d) + C_5 z_5(d) + C_6 z_6(d) = 0, \\
g_0(b) C_1 - \frac{1}{\rho_0(b^-)} C_2 - C_5 = \chi_n(b^-), \\
4\pi G \rho_0(b^-) C_1 + C_6 = \frac{dV_{1,n}}{dr}(b^-),
\end{gathered} \tag{7.34}$$

with the shorthand,

$$z_i(d) = \frac{d}{n+1} y_{6i}(d) + y_{5i}(d). \tag{7.35}$$

For $n = 0$, the spheroidal system degenerates to fourth order. Since there is no transverse displacement or shear stress, $y_3 = y_4 = 0$ and $\boldsymbol{y}_i = (y_{1i}, y_{2i}, y_{5i}, y_{6i})$. The system (7.34) then takes the reduced form,

$$\begin{gathered}
C_1 y_{21}(d) + C_2 y_{22}(d) + C_5 y_{25}(d) + C_6 y_{26}(d) = 0, \\
C_1 z_1(d) + C_2 z_2(d) + C_5 z_5(d) + C_6 z_6(d) = 0, \\
g_0(b) C_1 - \frac{1}{\rho_0(b^-)} C_2 - C_5 = \chi_0(b^-), \\
4\pi G \rho_0(b^-) C_1 + C_6 = \frac{dV_{1,0}}{dr}(b^-),
\end{gathered} \tag{7.36}$$

where

$$z_i(d) = d y_{6i}(d) + y_{5i}(d). \tag{7.37}$$

For $n \geq 1$, in the inner core, there are only three fundamental, linearly independent solutions regular at the geocentre, as shown in Section 3.5. If they are denoted by $\boldsymbol{y}_1, \boldsymbol{y}_2$ and $\boldsymbol{y}_3$, the most general solution in the inner core, regular at the geocentre, is then formed by the linear combination

$$\boldsymbol{y} = D_1\boldsymbol{y}_1 + D_2\boldsymbol{y}_2 + D_3\boldsymbol{y}_3, \tag{7.38}$$

where, once again, each $\boldsymbol{y}_j$ is a six-vector with components representing the values of the variables $(y_1, y_2, y_3, y_4, y_5, y_6)$ describing the spheroidal deformation field, and each D_j is a linear combination coefficient.

At the inner core boundary of radius a the shear stress vanishes and the normal displacement, normal stress, the perturbation in gravitational potential and the gravitational flux vector are all continuous. Following the same arguments as used at the core–mantle boundary, continuity of both the normal stress and the perturbation in gravitational potential imply that

$$y_2(a) = -\rho_0(a^+)\,(y_5(a) + \chi_n(a^+)) + \rho_0(a^+)\,g_0(a)\,y_1(a), \tag{7.39}$$

while continuity of the gravitational flux vector implies

$$y_6(a) = \frac{dV_{1,n}}{dr}(a^+) - 4\pi G\rho_0(a^+)\,y_1(a), \tag{7.40}$$

with $r = a^+$ indicating the fluid side of the inner core boundary. Application of the conditions at the inner core boundary (again excepting continuity of normal displacement) leads to the system of linear equations,

$$\begin{aligned} D_1y_{41}(a) + D_2y_{42}(a) + D_3y_{43}(a) &= 0,\\ D_1v_1(a) + D_2v_2(a) + D_3v_3(a) &= \chi_n(a^+),\\ D_1w_1(a) + D_2w_2(a) + D_3w_3(a) &= \frac{dV_{1,n}}{dr}(a^+), \end{aligned} \tag{7.41}$$

with

$$v_i(a) = g_0(a)y_{1i}(a) - \frac{1}{\rho_0(a^+)}y_{2i}(a) - y_{5i}(a), \tag{7.42}$$

and

$$w_i(a) = 4\pi G\rho_0(a^+)y_{1i}(a) + y_{6i}(a). \tag{7.43}$$

For $n = 0$, as shown in Section 3.5, there is only one solution in the inner core, regular at the geocentre, which corresponds to a true deformation. This involves only the variables y_1, y_2 and y_5. A second solution, regular at the geocentre, is the trivial constant solution for which only y_5 is a non-zero constant. For this second solution, $v_2(a) = -y_{52}(a)$, a constant, and $w_2(a) = 0$. In both solutions, $y_6 = 0$ everywhere, while y_3 and y_4 do not appear. The system (7.41) then degenerates to

$$D_1 v_1(a) + D_2 v_2(a) = \chi_0(a^+),$$
$$D_1 w_1(a) = \frac{dV_{1,0}}{dr}(a^+). \quad (7.44)$$

Without loss of generality, the constant $v_2(a)$ can be absorbed into the linear combination constant D_2, giving

$$D_1 v_1(a) + D_2 = \chi_0(a^+),$$
$$D_1 w_1(a) = \frac{dV_{1,0}}{dr}(a^+). \quad (7.45)$$

The systems (7.34), (7.36), (7.41) and (7.45) are all seen to depend only on the boundary values of χ_n and the derivatives of $V_{1,n}$. Thus, the entire deformation field in either the shell or the inner core is determined by the values of these two quantities on the boundaries. Following the traditional definition of the Love number h as a dimensionless number giving the radial displacement as a proportion of the ratio of the disturbing potential to gravity, we write the radial coefficients of the degree-n parts of the respective radial displacement components on the two boundaries as

$$y_{1,n}(b) = \frac{1}{g_0(b)}\left[h_n^{1S}\chi_n(b^-) - bh_n^{2S}\frac{dV_{1,n}}{dr}(b^-)\right] \quad (7.46)$$

and

$$y_{1,n}(a) = \frac{1}{g_0(a)}\left[h_n^{1I}\chi_n(a^+) - ah_n^{2I}\frac{dV_{1,n}}{dr}(a^+)\right]. \quad (7.47)$$

For each spherical harmonic of degree n, the Love numbers h_n^{1S}, h_n^{2S} for the shell, and the Love numbers h_n^{1I}, h_n^{2I} for the inner core, are found by numerical integration of the sixth-order spheroidal system of differential equations, (3.102) through (3.107), for $n \geq 1$, or the fourth-order spheroidal system, (3.110) through (3.113), for $n = 0$. Only the conditions of continuity of normal displacement at the two solid–fluid interfaces remain to be applied.

Application of the subseismic form (6.181) of the equation of continuity provides an additional constraint. In terms of the radial coefficients of the degree-n vector spherical harmonics, it becomes

$$y_2(r) = \rho_0(r)\, g_0(r)\, y_1(r), \quad (7.48)$$

for $a \leq r \leq b$. At the core–mantle boundary this yields the constraint

$$g_0(b)C_1 - \frac{1}{\rho_0(b^-)}C_2 = 0, \quad (7.49)$$

while at the inner core boundary it gives the constraint

$$D_1 x_1(a) + D_2 x_2(a) + D_3 x_3(a) = 0, \tag{7.50}$$

for $n \geq 1$ with $x_i(a) = v_i(a) + y_{5i}(a)$. For $n = 0$, the second solution does not affect the radial displacement and the second equation of the system (7.45) gives

$$D_1 = \frac{1}{w_1(a)} \frac{dV_{1,0}}{dr}(a^+) = \frac{1}{4\pi G\rho_0(a^+) y_{11}(a)} \frac{dV_{1,0}}{dr}(a^+). \tag{7.51}$$

The radial coefficient of the radial displacement at the inner core boundary is then

$$y_{1,0}(a) = D_1 y_{11}(a) = \frac{1}{4\pi G\rho_0(a^+)} \frac{dV_{1,0}}{dr}(a^+). \tag{7.52}$$

Comparison with expression (7.47) gives $h_0^{1I} = 0$ and

$$h_0^{2I} = -\frac{g_0(a)}{4\pi G\rho_0(a^+) a}. \tag{7.53}$$

Under the subseismic condition, (7.50) is replaced by

$$D_1 x_1(a) + D_2 x_2(a) = 0, \tag{7.54}$$

where

$$x_1(a) = g_0(a) y_{11}(a) - \frac{y_{21}(a)}{\rho_0(a^+)} \neq 0,$$
$$x_2(a) = 0. \tag{7.55}$$

Thus, $D_1 = 0$ and there is no degree-zero deformation field in the inner core, so $dV_{1,0}(a^+)/dr = 0$ in agreement with (7.52). D_2 may be non-zero, representing the usual freedom of choice for the reference level of a potential.

The neglect of flow pressure fluctuations involved in the subseismic condition leads to $\chi = -V_1$, from (6.176), so the generalised potential is equal to the change in gravitational potential. Further, the boundary values of χ_n and the derivatives of $V_{1,n}$ are no longer independent but are linearly related. If the shell were rigid, we would expect $V_{1,n}$ to be proportional to $1/r^{n+1}$ at the core–mantle boundary, while if the inner core were rigid, we would expect $V_{1,n}$ to be proportional to r^n at the inner core boundary. Thus, we would expect the linear relations

$$(n+1)\chi_n(b^-) - b_n \frac{dV_{1,n}}{dr}(b^-) = 0 \tag{7.56}$$

$$n\chi_n(a^+) + a_n \frac{dV_{1,n}}{dr}(a^+) = 0, \tag{7.57}$$

with the quantities b_n and a_n representing *equivalent rigid-boundary* radii computed from the systems (7.34), (7.36) and (7.41), augmented by the subseismic condition. Again, the case $n = 0$ at the inner core boundary is an exception. Since $dV_{1,0}(a^+)/dr$ vanishes there, a_0 becomes indeterminate.

Under the subseismic approximation, single *effective* internal load Love numbers for the shell and inner core, from relations (7.46)–(7.47) and (7.56)–(7.57), can be defined as

$$h_n^S = \left[h_n^{1S} - (n+1)\frac{b}{b_n} h_n^{2S} \right], \tag{7.58}$$

and

$$h_n^I = \left[h_n^{1I} + \frac{a}{a_n} h_n^{2I} \right]. \tag{7.59}$$

The radial coefficients of the degree-n parts of the radial displacements on the two boundaries can be expressed as

$$y_{1,n}(b) = \frac{h_n^S}{g_0(b)} \chi_n(b^-), \tag{7.60}$$

and

$$y_{1,n}(a) = \frac{h_n^I}{g_0(a)} \chi_n(a^+). \tag{7.61}$$

Since, under the subseismic approximation, there is no degree-zero deformation field in the inner core, we have $h_0^I = 0$.

The programme LOVE.FOR computes the internal load Love numbers for a given Earth model. The Love numbers are only physically meaningful for degree $n \geq |m|$, where m is the azimuthal number. Input to the programme is the Earth model file, the name of which is typed in at the request of the programme. The first line of this file is the name of the Earth model, in format (10A7). The second line contains the number of Love numbers to be computed, in format (I10). The third line gives the numbers of model points and integration steps, in format (8I10), for each region specified by the integer variable IR = 1, 2, 3, 4 for the inner core, outer core, mantle and crust, respectively. The Earth model follows in format (1X,F10.1,F10.2,F10.1,F10.1,F10.1), in traditional units: radius in km, density in g/cm^3, Lamé coefficients in kilobars and gravity in cm/s^2. The input Earth model file is written out to the screen for verification. For each degree n, the Love numbers h_n^{1I}, h_n^{2I} for the inner core, defined by expression (7.47), and the Love numbers h_n^{1S}, h_n^{2S} for the shell, defined by expression (7.46), are written out to the screen as well.

Output from the programme is contained in the file love.dat. The first line of this file gives the Earth model name, the second line the number of Love numbers computed, the third line the radii of the inner core boundary and the core–mantle boundary, and the fourth line the gravity at the inner core boundary and the core–mantle boundary. These are followed by the effective load Love numbers for the inner core, h_n^I, the equivalent rigid-boundary radius a_n, the effective load Love numbers for the shell, h_n^S, and the equivalent rigid-boundary radius b_n, for each degree n. The final line of the output file gives the coefficients a, b, c of the expression

$$\frac{1}{h_1^I} = a\sigma^2 + b\sigma + c, \tag{7.62}$$

for the reciprocal of the degree-one Love number of the inner core as a function of dimensionless angular frequency $\sigma = \omega/2\Omega$. Since the effective Love numbers represent admittances, we expect the inertial drag in the degree-one translational modes to show up as a quadratic term and the resistance to the Coriolis acceleration to register as a linear term of opposite sign, for translations in the equatorial plane represented by azimuthal number $m = \pm 1$ (Smylie and Jiang, 1993, pp. 1354–1355).

The programme allows the interactive input of the period in hours, the azimuthal number m and the switching on or off of the inertial and Coriolis terms. The computation in the inner core begins with the power series expansions of the fundamental solutions regular at the geocentre, followed by variable stepsize Runge–Kutta integration, as in the programme ICFS.FOR described in Section 3.7. The number of steps in the Runge–Kutta integrations in the mantle and crust are those specified in the input Earth model file. As in the case of ICFS.FOR, the programme proceeds with calls to the subroutines SPMAT and INTPL for cubic spline interpolation of the Earth model, as described in Section 1.6. Power series expansions of the free solutions, regular at the geocentre, are performed with the assistance of the double precision function subprogrammes P1, P2, Q1 and Q2, as well as the subroutine MATRIX, described in Section 3.7, and the subroutine LINSOL, described in Section 1.5. The subroutine REL, described in Section 3.7, tracks the relative error in both the power series expansions and the Runge–Kutta integration. Derivatives of the propagator matrix, required for the Runge–Kutta integration, are calculated by the subroutine YPRIME, described in Section 3.6, and the Runge–Kutta integration is performed by the subroutine RK4, described in Section 3.7.

```
C             PROGRAMME LOVE.FOR
C LOVE.FOR computes the 'effective' internal Love numbers for the shell
C (mantle plus crust) and inner core for an input Earth model.
C These Love numbers allow the radial displacement at the boundaries of
C the fluid outer core to be expressed in terms of the generalised
C displacement potential there.
C
      IMPLICIT DOUBLE PRECISION(A-H,O-Z)
      DIMENSION R(100),RHO(100),GZERO(100),RI(300),RHOI(300),
     1 GZEROI(300),ENAME(10),NM(4),NI(4),NK(4),B(98,198),C(100,100),
```

```
    2 Y(6,6),YSCAL(6,3),YSCAL2(6,3),YA(6),A(6,6),AS(6,6),CAUG(6,13),
    3 YIC(6,3),Y1(6,6),Y2(6,6),DELTA(6,3),CM1(5,5),CM2(5,5),
    4 CM3(5,5),BV1(5),BV2(5),BV3(5),CAUGC(3,7),CMC1(3,3),CMC2(3,3),
    5 CMC3(3,3),BVC1(3),BVC2(3),BVC3(3),CAUGS(5,11),CAUGS0(4,9),
    6 CM01(4,4),CM02(4,4),CM03(4,4),BV01(4),BV02(4),BV03(4)
      DOUBLE PRECISION MU(100),LAMBDA(100),MUI(300),LAMBDAI(300),K1(6,6)
      CHARACTER*7 ENAME
      CHARACTER*20 EMODEL
C Enter Earth model file name.
      WRITE(6,10)
 10   FORMAT(1X,'Type in Earth model file name.')
      READ(5,11)EMODEL
 11   FORMAT(A20)
C Open Earth model file.
      OPEN(UNIT=1,FILE=EMODEL,STATUS='OLD')
C Open Love number file.
      OPEN(UNIT=2,FILE='love.dat',STATUS='UNKNOWN')
C Set angular frequency of Earth's rotation (WGS 84).
      WE=7.292115D-5
C Calculate square of angular frequency of Earth's rotation.
      WES=WE*WE
C Set value of PI.
      PI=3.141592653589793D0
C Set value of universal constant of gravitation (CODATA 2006).
      G=6.67428D-11
C Enter period in hours, azimuthal number, inertial and Coriolis switches.
      WRITE(6,12)
 12   FORMAT(1X,'Enter period in hours, azimuthal number and inertial',
     1 ' and Coriolis switches (1 for in, 0 for out).')
      READ(5,*)TPER,MAZ,INERT,ICOR
C Convert period to seconds.
      TPER=3600.D0*TPER
C Calculate dimensionless angular frequency.
      SIG=PI/(TPER*WE)
C Calculate square of dimensionless angular frequency.
      SIGS=SIG*SIG
C Compute square of angular frequency.
      ANGS=DFLOAT(INERT)
      ANGS=2.D0*PI*ANGS/TPER
      ANGS=ANGS*ANGS
C Compute Coriolis term.
      COR=2.D0*DFLOAT(MAZ)*DFLOAT(ICOR)
      COR=2.D0*PI*COR*WE/TPER
C Set maximum relative error tolerance for integrations.
      EPS=1.D-5
C Set minimum starting radius for variable Earth properties.
      RMIN=1200.D1
C Set maximum dimensions for interpolation.
      M1=100
      M2=198
      M3=98
C Read in and write out Earth model.
C Read in and write out headers.
C Earth model name.
      READ(1,13)(ENAME(I),I=1,10)
 13   FORMAT(10A7)
      WRITE(2,13)(ENAME(I),I=1,10)
      WRITE(6,14)(ENAME(I),I=1,10)
 14   FORMAT(11X,10A7)
C Number of Love numbers to be computed.
      READ(1,15)NN
```

```
   15   FORMAT(I10)
        WRITE(2,15)NN
C Number of model points and number of integration steps.
        READ(1,16)(NM(I),NI(I),I=1,4)
   16   FORMAT(8I10)
        WRITE(6,17)
   17   FORMAT(//16X,'Number of model points and number of integration',
      1 ' steps.'/)
        WRITE(6,16)(NM(I),NI(I),I=1,4)
        WRITE(6,18)
   18   FORMAT(//5X,'Radius',6X,'Rho',4X,'Lambda',7X,'Mu',6X,'Gzero')
        WRITE(6,19)
   19   FORMAT(6X,'(km)',5X,'(gm/cc)',2X,'(kbars)',4X,'(kbars)',2X,
      1 '(cm/sec/sec)'/)
C Read in Earth model, inner core, outer core, mantle and crust.
        K=0
        DO 20 M=1,4
        N1=NM(M)
        READ(1,21)(R(I),RHO(I),LAMBDA(I),MU(I),GZERO(I),I=1,N1)
   21   FORMAT(1X,F10.1,F10.2,F10.1,F10.1,F10.1)
        DO 22 I=1,N1
C Scale Earth model to SI values and store.
        J=K+I
        RI(J)=R(I)*1.D3
        RHOI(J)=RHO(I)*1.D3
        LAMBDAI(J)=LAMBDA(I)*1.D8
        MUI(J)=MU(I)*1.D8
        GZEROI(J)=GZERO(I)*1.D-2
C Write Earth model out to screen.
        WRITE(6,21)R(I),RHO(I),LAMBDA(I),MU(I),GZERO(I)
   22   CONTINUE
        NK(M)=K
        K=K+N1
   20   CONTINUE
C Find density above, lambda above, gravity at the inner core boundary.
        RHOAC=RHOI(NK(2)+1)
        GZEROC=GZEROI(NK(2)+1)
C Find density below, lambda below, gravity at the core-mantle boundary.
        RHOBM=RHOI(NK(3))
        GZEROM=GZEROI(NK(3))
C Store radii of inner core and core-mantle boundaries
C and surface radius.
        RIOB=RI(NK(2))
        RCMB=RI(NK(3))
        D=RI(NK(4)+NM(4))
C Write out radii of ICB and CMB.
        WRITE(2,23) RIOB,RCMB
C Write out gravity at the ICB and the CMB.
        WRITE(2,23) GZEROC,GZEROM
   23   FORMAT(2D23.15)
C Begin calculation of Love numbers for the inner core.
C Set up interpolation for the inner core.
        N1=NM(1)
        N2=2*N1-2
        N3=N1-2
C Put inner core values in active locations.
        DO 24 I=1,N1
          J=NK(1)+I
          R(I)=RI(J)
          RHO(I)=RHOI(J)
          MU(I)=MUI(J)
```

```
      LAMBDA(I)=LAMBDAI(J)
      GZERO(I)=GZEROI(J)
 24   CONTINUE
C Calculate gravity gradient at geocentre.
      GAMMA=(4.D0*PI*G*RHO(1)-2.D0*WES)/3.D0
C Set minimum radius for variable Earth properties.
      R(1)=RMIN
C Set gravity at minimum radius.
      GZERO(1)=GAMMA*RMIN
C Construct interpolation matrix.
      CALL SPMAT(N1,N2,N3,C,R,B,M1,M2,M3)
C Begin loop over degree for inner core.
      WRITE(6,25)
 25   FORMAT(//11X,'Degree',14X,'Love number h1',14X,'Love number h2')
      DO 26 NP1=1,NN
        N=NP1-1
        AN=DFLOAT(N)
C Test if N=0.
        IF(N.EQ.0) GO TO 27
C Test if N=1 and set counter.
        IF(N.NE.1) GO TO 28
        ICOUNT=1
C Store ANGS and COR
        ANGSS=ANGS
        CORS=COR
C Set ANGS and COR to zero for calculation of DC value.
        ANGS=0.D0
        COR=0.D0
C Begin power series expansions of fundamental solutions for N>0.
C Set initial values of fundamental solutions.
C Set initial values of first fundamental solution.
 28     CONTINUE
        Y(1,1)=1.D0
        Y(2,1)=2.D0*(AN-1.D0)*MU(1)
        Y(3,1)=1.D0/AN
        Y(4,1)=Y(2,1)/AN
        Y(5,1)=4.D0*PI*G*RHO(1)/AN
        Y(6,1)=0.D0
C Set initial values of second fundamental solution.
        Y(1,2)=0.D0
        Y(2,2)=0.D0
        Y(3,2)=0.D0
        Y(4,2)=0.D0
        Y(5,2)=1.D0/AN
        Y(6,2)=1.D0
C Set initial values of third fundamental solution.
        Y(3,3)=P2(AN,LAMBDA(1),MU(1))/P1(AN,LAMBDA(1),MU(1))
        Y(2,3)=-Q1(AN,LAMBDA(1),MU(1))*Y(3,3)+Q2(AN,LAMBDA(1),MU(1))
        Y(1,3)=-AN*Y(3,3)+1.D0/MU(1)
        Y(4,3)=1.D0
        Y(5,3)=2.D0*PI*G*RHO(1)*((AN+3.D0)*Y(1,3)-AN*(AN+1.D0)*Y(3,3))/
     1  (2.D0*AN+3.D0)
        Y(6,3)=(AN+2.D0)*Y(5,3)-4.D0*PI*G*RHO(1)*Y(1,3)
C Calculate coefficients of second terms in series expansions.
C Calculate coefficients of second terms for first fundamental solution.
        YSCAL(3,1)=RHO(1)*((3.D0-AN)*GAMMA+ANGS+2.D0*WES-COR/AN)/
     1  P1(AN,LAMBDA(1),MU(1))
        YSCAL(2,1)=-Q1(AN,LAMBDA(1),MU(1))*YSCAL(3,1)
        YSCAL(1,1)=-AN*YSCAL(3,1)
        YSCAL(4,1)=0.D0
        YSCAL(5,1)=2.D0*PI*G*RHO(1)*((AN+3.D0)*YSCAL(1,1)
```

```
     1  -AN*(AN+1.D0)*YSCAL(3,1))/(2.D0*AN+3.D0)
        YSCAL(6,1)=(AN+2.D0)*YSCAL(5,1)-4.D0*PI*G*RHO(1)*YSCAL(1,1)
C Calculate coefficients of second terms for second fundamental solution.
        YSCAL(3,2)=RHO(1)/P1(AN,LAMBDA(1),MU(1))
        YSCAL(2,2)=-Q1(AN,LAMBDA(1),MU(1))*YSCAL(3,2)
        YSCAL(1,2)=-AN*YSCAL(3,2)
        YSCAL(4,2)=0.D0
        YSCAL(5,2)=2.D0*PI*G*RHO(1)*((AN+3.D0)*YSCAL(1,2)
     1  -AN*(AN+1.D0)*YSCAL(3,2))/(2.D0*AN+3.D0)
        YSCAL(6,2)=(AN+2.D0)*YSCAL(5,2)-4.D0*PI*G*RHO(1)*YSCAL(1,2)
C Generate matrix for evaluation of second terms in power series
C expansion of third fundamental solution.
        CALL MATRIX(A,RHO,LAMBDA,MU,PI,G,AN,WE)
C Set right-hand side for calculation of second terms in third solution.
        YA(1)=0.D0
        YA(2)=RHO(1)*(-(4.D0*GAMMA+ANGS+2.D0*WES)*Y(1,3)
     1  +(AN*(AN+1.D0)*GAMMA+COR)*Y(3,3)-Y(6,3))
        YA(3)=0.D0
        YA(4)=RHO(1)*((GAMMA+COR/(AN*(AN+1.D0)))*Y(1,3)
     1  -(ANGS-COR/(AN*(AN+1.D0)))*Y(3,3)-Y(5,3))
        YA(5)=0.D0
        YA(6)=0.D0
C Store matrix A in AS.
        DO 29 I=1,6
          DO 30 J=1,6
            AS(I,J)=A(I,J)
 30       CONTINUE
 29     CONTINUE
C Solve for second terms in third solution.
        CALL LINSOL(A,YA,6,CAUG,DET,6,13)
        YSCAL(1,3)=YA(1)
        YSCAL(2,3)=YA(2)
        YSCAL(3,3)=YA(3)
        YSCAL(4,3)=YA(4)
        YSCAL(5,3)=YA(5)
        YSCAL(6,3)=YA(6)
C Set right-hand side for calculation of third terms in first solution.
        YA(1)=0.D0
        YA(2)=RHO(1)*(-(4.D0*GAMMA+ANGS+2.D0*WES)*YSCAL(1,1)
     1  +(AN*(AN+1.D0)*GAMMA+COR)*YSCAL(3,1)-YSCAL(6,1))
        YA(3)=0.D0
        YA(4)=RHO(1)*((GAMMA+COR/(AN*(AN+1.D0)))*YSCAL(1,1)
     1  -(ANGS-COR/(AN*(AN+1.D0)))*YSCAL(3,1)-YSCAL(5,1))
        YA(5)=0.D0
        YA(6)=0.D0
C Copy stored matrix AS into matrix A.
        DO 31 I=1,6
          DO 32 J=1,6
            A(I,J)=AS(I,J)
 32       CONTINUE
 31     CONTINUE
C Solve for third terms in first solution.
        CALL LINSOL(A,YA,6,CAUG,DET,6,13)
        YSCAL2(1,1)=YA(1)
        YSCAL2(2,1)=YA(2)
        YSCAL2(3,1)=YA(3)
        YSCAL2(4,1)=YA(4)
        YSCAL2(5,1)=YA(5)
        YSCAL2(6,1)=YA(6)
C Set right-hand side for calculation of third terms in second solution.
        YA(1)=0.D0
```

```
      YA(2)=RHO(1)*(-(4.D0*GAMMA+ANGS+2.D0*WES)*YSCAL(1,2)
     1 +(AN*(AN+1.D0)*GAMMA+COR)*YSCAL(3,2)-YSCAL(6,2))
      YA(3)=0.D0
      YA(4)=RHO(1)*((GAMMA+COR/(AN*(AN+1.D0)))*YSCAL(1,2)
     1 -(ANGS-COR/(AN*(AN+1.D0)))*YSCAL(3,2)-YSCAL(5,2))
      YA(5)=0.D0
      YA(6)=0.D0
C Copy stored matrix AS into matrix A.
      DO 33 I=1,6
        DO 34 J=1,6
          A(I,J)=AS(I,J)
 34     CONTINUE
 33   CONTINUE
C Solve for third terms in second solution.
      CALL LINSOL(A,YA,6,CAUG,DET,6,13)
      YSCAL2(1,2)=YA(1)
      YSCAL2(2,2)=YA(2)
      YSCAL2(3,2)=YA(3)
      YSCAL2(4,2)=YA(4)
      YSCAL2(5,2)=YA(5)
      YSCAL2(6,2)=YA(6)
C Set right-hand side for calculation of third terms in third solution.
      YA(1)=0.D0
      YA(2)=RHO(1)*(-(4.D0*GAMMA+ANGS+2.D0*WES)*YSCAL(1,3)
     1 +(AN*(AN+1.D0)*GAMMA+COR)*YSCAL(3,3)-YSCAL(6,3))
      YA(3)=0.D0
      YA(4)=RHO(1)*((GAMMA+COR/(AN*(AN+1.D0)))*YSCAL(1,3)
     1 -(ANGS-COR/(AN*(AN+1.D0)))*YSCAL(3,3)-YSCAL(5,3))
      YA(5)=0.D0
      YA(6)=0.D0
C Copy stored matrix AS into matrix A and add 2.0 to diagonal of A.
      DO 35 I=1,6
        DO 36 J=1,6
          A(I,J)=AS(I,J)
          IF(I.EQ.J)A(I,J)=A(I,J)+2.D0
 36     CONTINUE
 35   CONTINUE
C Solve for third terms in third solution.
      CALL LINSOL(A,YA,6,CAUG,DET,6,13)
      YSCAL2(1,3)=YA(1)
      YSCAL2(2,3)=YA(2)
      YSCAL2(3,3)=YA(3)
      YSCAL2(4,3)=YA(4)
      YSCAL2(5,3)=YA(5)
      YSCAL2(6,3)=YA(6)
      GO TO 37
 27   CONTINUE
C Begin power series expansion of fundamental solution for degree N=0.
C Set first two solutions to zero.
      DO 38 I=1,6
        DO 39 J=1,2
          Y(I,J)=0.D0
          YSCAL(I,J)=0.D0
          YSCAL2(I,J)=0.D0
 39     CONTINUE
 38   CONTINUE
C Set values for first three terms of fundamental solution.
      Y(1,3)=1.D0
      Y(2,3)=3.D0*LAMBDA(1)+2.D0*MU(1)
      Y(3,3)=0.D0
      Y(4,3)=0.D0
```

```
      Y(5,3)=2.D0*PI*G*RHO(1)
      Y(6,3)=0.D0
      YSCAL(1,3)=-RHO(1)*(4.D0*GAMMA+ANGS+2.D0*WES)/
     1 (10.D0*(LAMBDA(1)+2.D0*MU(1)))
      YSCAL(2,3)=(5.D0*LAMBDA(1)+6.D0*MU(1))*YSCAL(1,3)
      YSCAL(3,3)=0.D0
      YSCAL(4,3)=0.D0
      YSCAL(5,3)=PI*G*RHO(1)*YSCAL(1,3)
      YSCAL(6,3)=0.D0
      YSCAL2(1,3)=-RHO(1)*(4.D0*GAMMA+ANGS+2.D0*WES)*YSCAL(1,3)/
     1 (4.D0*(7.D0*LAMBDA(1)+14.D0*MU(1)))
      YSCAL2(2,3)=(7.D0*LAMBDA(1)+10.D0*MU(1))*YSCAL2(1,3)
      YSCAL2(3,3)=0.D0
      YSCAL2(4,3)=0.D0
      YSCAL2(5,3)=2.D0*PI*G*RHO(1)*YSCAL2(1,3)/3.D0
      YSCAL2(6,3)=0.D0
 37   CONTINUE
C Set last three columns of array Y(6,6) to zero for inner core.
      DO 40 I=1,6
        DO 41 J=4,6
          Y(I,J)=0.D0
 41     CONTINUE
 40   CONTINUE
C Put inner core solutions in array YIC(6,3).
      DO 42 I=1,6
        DO 43 J=1,3
          YIC(I,J)=Y(I,J)
 43     CONTINUE
 42   CONTINUE
C Begin Runge-Kutta integration for the inner core.
C Find maximum ratio of second to first coefficient in power series
C expansion at the geocentre.
      CALL REL(ERRMAX,YSCAL,YIC,N)
C Set initial stepsize so that fifth-order method would have relative
C error bound EPS.
      H=(EPS/ERRMAX)**(0.2D0)
C Find solutions and derivatives at minimum radius RMIN by power series
C expansions.
      XV=RMIN
      XVS=XV*XV
      XV3=XV*XVS
      XV4=XVS*XVS
C Find solutions.
      DO 44 I=1,6
        DO 45 J=1,3
          Y(I,J)=Y(I,J)+XVS*YSCAL(I,J)+XV4*YSCAL2(I,J)
 45     CONTINUE
 44   CONTINUE
C Set inner core index.
      IR=1
 46   CONTINUE
C Calculate derivatives at beginning of current step.
      CALL YPRIME(XV,Y,A,K1,N1,C,R,RHO,MU,LAMBDA,GZERO,N,PI,G,
     1 IR,RMIN,ANGS,COR,WES)
C Initialize function values at beginning of integration step.
      DO 47 I=1,6
        DO 48 J=1,6
          Y1(I,J)=Y(I,J)
          Y2(I,J)=Y(I,J)
 48     CONTINUE
 47   CONTINUE
```

```
C Integrate solution forward one full step.
      CALL RK4(XV,Y1,A,K1,N1,C,R,RHO,MU,LAMBDA,GZERO,N,PI,G,H,IR,
     1   RMIN,ANGS,COR,WES)
C Integrate solution forward two half-steps.
C Halve step.
      HH=0.5D0*H
C Reset radius.
      XV=XV-H
      CALL RK4(XV,Y2,A,K1,N1,C,R,RHO,MU,LAMBDA,GZERO,N,PI,G,HH,IR,
     1   RMIN,ANGS,COR,WES)
      CALL YPRIME(XV,Y2,A,K1,N1,C,R,RHO,MU,LAMBDA,GZERO,N,PI,G,IR,
     1   RMIN,ANGS,COR,WES)
      CALL RK4(XV,Y2,A,K1,N1,C,R,RHO,MU,LAMBDA,GZERO,N,PI,G,HH,IR,
     1   RMIN,ANGS,COR,WES)
C Find error matrix and maximum relative error.
      DO 49 I=1,6
        DO 50 J=1,3
          DELTA(I,J)=Y1(I,J)-Y2(I,J)
 50     CONTINUE
 49   CONTINUE
C Copy Y2(6,6) into inner core solution array YIC(6,3).
      DO 51 I=1,6
        DO 52 J=1,3
          YIC(I,J)=Y2(I,J)
 52     CONTINUE
 51   CONTINUE
C Find maximum relative error.
      CALL REL(ERRMAX,DELTA,YIC,N)
C Scale maximum relative error by specified relative error tolerance.
      ERRMAX=ERRMAX/EPS
C Test if step was successful.
      IF(ERRMAX.LT.1.D0) GO TO 53
C Step too large, reduce stepsize and repeat.
      XV=XV-H
      H=0.9D0*H*(ERRMAX**(-0.25D0))
      GO TO 46
 53   CONTINUE
C Step successful, accept solution and improve truncation error.
      DO 54 J=1,6
        DO 55 K=1,3
          Y(J,K)=Y2(J,K)+DELTA(J,K)/15.D0
 55     CONTINUE
 54   CONTINUE
C Increase stepsize.
      H=0.9D0*H*(ERRMAX**(-0.20D0))
C Test if solution is finished.
      XVT=XV+H
      IF(XVT.LT.R(N1)) GO TO 46
C Find remaining distance to the ICB.
      H=R(N1)-XV
C Compute fraction of radius.
      RATIO=H/R(N1)
C Continue integration only if remaining distance is more than
C 1.D-9 of radius.
      IF(RATIO.GT.1.D-9) GO TO 46
C Convert solutions back to y-variables.
C Set value of alpha for first two solutions.
      ANM2=AN-2.D0
C Convert three solutions.
      DO 56 J=1,3
C Set value of alpha for third solution.
```

```
          IF(J.EQ.3) ANM2=AN
C Find constant equal to the radius of the ICB to the power alpha.
          CONST=RIOB**ANM2
          Y(2,J)=CONST*Y(2,J)
          Y(4,J)=CONST*Y(4,J)
          CONST=CONST*RIOB
          Y(1,J)=CONST*Y(1,J)
          Y(3,J)=CONST*Y(3,J)
          Y(6,J)=CONST*Y(6,J)
          CONST=CONST*RIOB
          Y(5,J)=CONST*Y(5,J)
 56     CONTINUE
C Test if N=0.
        IF(N.EQ.0) GO TO 57
C Begin construction of coefficient matrix.
        DO 58 I=1,3
          CMC1(1,I)=Y(4,I)
          CMC1(2,I)=GZEROC*Y(1,I)-Y(2,I)/RHOAC-Y(5,I)
          CMC1(3,I)=4.D0*PI*G*RHOAC*Y(1,I)+Y(6,I)
 58     CONTINUE
C Set coefficient matrices equal.
        DO 59 I=1,3
          DO 60 J=1,3
            CMC2(I,J)=CMC1(I,J)
            CMC3(I,J)=CMC2(I,J)
 60       CONTINUE
 59     CONTINUE
C Set constant vectors.
        BVC1(1)=0.D0
        BVC1(2)=1.D0
        BVC1(3)=0.D0
        BVC2(1)=0.D0
        BVC2(2)=0.D0
        BVC2(3)=-1.D0
C Store first constant vector.
        DO 61 I=1,3
          BVC3(I)=BVC1(I)
 61     CONTINUE
C Solve for linear combination coefficients.
        CALL LINSOL(CMC1,BVC1,3,CAUGC,DET,3,7)
        CALL LINSOL(CMC2,BVC2,3,CAUGC,DET,3,7)
C Alter coefficient matrix to apply subseismic condition.
        DO 62 I=1,3
          CMC3(3,I)=GZEROC*Y(1,I)-Y(2,I)/RHOAC
 62     CONTINUE
C Solve for subseismic coefficients.
        CALL LINSOL(CMC3,BVC3,3,CAUGC,DET,3,7)
C Calculate unscaled Love numbers.
        SUM1=0.D0
        SUM2=0.D0
        SUM3=0.D0
        DO 63 I=1,3
          SUM1=SUM1+BVC1(I)*Y(1,I)
          SUM2=SUM2+BVC2(I)*Y(1,I)
          SUM3=SUM3+BVC3(I)*Y(1,I)
 63     CONTINUE
C Calculate radial derivative of V1n at inner core boundary.
        SUM=0.D0
        DO 64 I=1,3
          SUM=SUM+(4.D0*PI*G*RHOAC*Y(1,I)+Y(6,I))*BVC3(I)
 64     CONTINUE
```

```
C Calculate potential condition parameter AND.
      AND=-AN/SUM
      GO TO 65
C Construct coefficient matrix for N=0.
 57   CONTINUE
      CMC1(1,1)=GZEROC*Y(1,3)-Y(2,3)/RHOAC-Y(5,3)
      CMC1(1,2)=1.D0
      CMC1(2,1)=4.D0*PI*G*RHOAC*Y(1,3)
      CMC1(2,2)=0.D0
C Solve for linear combination coefficients and AND.
      BVC1(1)=0.D0
      BVC2(1)=-1.D0/CMC1(2,1)
      AND=0.D0
C Calculate unscaled Love numbers.
      SUM1=BVC1(1)*Y(1,3)
      SUM2=BVC2(1)*Y(1,3)
      SUM3=0.D0
 65   CONTINUE
C Scale Love numbers to dimensionless values.
      BVC1(1)=GZEROC*SUM1
      BVC2(1)=GZEROC*SUM2/RIOB
      BVC3(1)=GZEROC*SUM3
C Test if N=1.
      IF(N.NE.1) GO TO 66
      IF(ICOUNT.EQ.3) GO TO 67
      IF(ICOUNT.EQ.2) GO TO 68
      GAMA=1.D0/BVC3(1)
      COR=CORS
      ICOUNT=2
      GO TO 28
 68   CONTINUE
      BETA=(1.D0/BVC3(1)-GAMA)/SIG
      ANGS=ANGSS
      ICOUNT=3
      GO TO 28
 67   CONTINUE
      ALPHA=(1.D0/BVC3(1)-BETA*SIG-GAMA)/SIGS
C Write out internal Love numbers.
 66   CONTINUE
      WRITE(6,69)N,BVC1(1),BVC2(1)
 69   FORMAT(/I15,11X,F16.8,F28.8)
      WRITE(2,70)BVC3(1),AND
 70   FORMAT(2D23.15)
C End loop over degree for inner core.
 26   CONTINUE
C Begin loop over degree for the shell (mantle + crust).
     WRITE(6,25)
     DO 71 NP1=1,NN
      N=NP1-1
C Set up interpolation for the mantle.
      N1=NM(3)
      N2=2*N1-2
      N3=N1-2
C Put mantle values in active locations.
      DO 72 I=1,N1
        J=NK(3)+I
        R(I)=RI(J)
        RHO(I)=RHOI(J)
        MU(I)=MUI(J)
        LAMBDA(I)=LAMBDAI(J)
        GZERO(I)=GZEROI(J)
```

```
 72     CONTINUE
C Prepare for spline interpolation.
        CALL SPMAT(N1,N2,N3,C,R,B,M1,M2,M3)
C Set initial values for fundamental solutions.
        DO 73 I=1,6
          Y(I,I)=1.D0
          DO 74 J=1,6
            IF(I.NE.J) Y(I,J)=0.D0
 74       CONTINUE
 73     CONTINUE
C Begin Runge-Kutta integration for the mantle.
        M=NI(3)
        AM=DFLOAT(M)
C Find stepsize.
        H=(R(N1)-R(1))/AM
C Set initial radius.
        XV=R(1)
C Set integration region counter.
        IR=3
        DO 75 I=1,M
          CALL YPRIME(XV,Y,A,K1,N1,C,R,RHO,MU,LAMBDA,GZERO,N,PI,G,
     1     IR,RMIN,ANGS,COR,WES)
          CALL RK4(XV,Y,A,K1,N1,C,R,RHO,MU,LAMBDA,GZERO,N,PI,G,H,IR,
     1     RMIN,ANGS,COR,WES)
 75     CONTINUE
C Set up interpolation for the crust.
        N1=NM(4)
        N2=2*N1-2
        N3=N1-2
C Put crustal values in active locations.
        DO 76 I=1,N1
          J=NK(4)+I
          R(I)=RI(J)
          RHO(I)=RHOI(J)
          MU(I)=MUI(J)
          LAMBDA(I)=LAMBDAI(J)
          GZERO(I)=GZEROI(J)
 76     CONTINUE
C Prepare for spline interpolation.
        CALL SPMAT(N1,N2,N3,C,R,B,M1,M2,M3)
C Begin Runge-Kutta integration for the crust.
        M=NI(4)
        AM=DFLOAT(M)
C Find stepsize.
        H=(R(N1)-R(1))/AM
C Set initial radius.
        XV=R(1)
C Set integration region counter.
        IR=4
        DO 77 I=1,M
          CALL YPRIME(XV,Y,A,K1,N1,C,R,RHO,MU,LAMBDA,GZERO,N,PI,G,
     1     IR,RMIN,ANGS,COR,WES)
          CALL RK4(XV,Y,A,K1,N1,C,R,RHO,MU,LAMBDA,GZERO,N,PI,G,H,IR,
     1     RMIN,ANGS,COR,WES)
 77     CONTINUE
C Begin construction of coefficient matrix.
        AN=DFLOAT(N)
C Test if N=0.
        IF(N.EQ.0) GO TO 78
        DO 79 I=1,5
          J=I
```

```
        IF(I.GT.3)J=J+1
        CM1(1,I)=Y(2,J)
        CM1(2,I)=Y(4,J)
        CM1(3,I)=Y(5,J)+D*Y(6,J)/(AN+1.D0)
        CM1(4,I)=0.D0
        CM1(5,I)=0.D0
 79   CONTINUE
      CM1(4,1)=GZEROM
      CM1(4,2)=-1.D0/RHOBM
      CM1(4,4)=-1.D0
      CM1(5,1)=4.D0*PI*G*RHOBM
      CM1(5,2)=0.D0
      CM1(5,5)=1.D0
C Set coefficient matrices equal.
      DO 80 I=1,5
        DO 81 J=1,5
          CM2(I,J)=CM1(I,J)
          CM3(I,J)=CM2(I,J)
 81     CONTINUE
 80   CONTINUE
C Set values of constant vectors.
      BV1(1)=0.D0
      BV1(2)=0.D0
      BV1(3)=0.D0
      BV1(4)=1.D0
      BV1(5)=0.D0
      DO 82 I=1,4
        BV2(I)=0.D0
 82   CONTINUE
      BV2(5)=-1.D0
C Store first constant vector.
      DO 83 I=1,5
        BV3(I)=BV1(I)
 83   CONTINUE
C Solve for linear combination coefficients.
      CALL LINSOL(CM1,BV1,5,CAUGS,DET,5,11)
      CALL LINSOL(CM2,BV2,5,CAUGS,DET,5,11)
C Alter coefficient matrix to apply subseismic condition.
      CM3(5,1)=GZEROM
      CM3(5,2)=-1.D0/RHOBM
      CM3(5,5)=0.D0
C Solve for subseismic coefficients.
      CALL LINSOL(CM3,BV3,5,CAUGS,DET,5,11)
C Calculate potential condition parameter BN.
      BN=(AN+1.D0)/(4.D0*PI*G*RHOBM*BV3(1)+BV3(5))
      GO TO 84
C Construct coefficient matrix for N=0.
C Reduce Y-matrix to 4x4.
 78   CONTINUE
      DO 85 I=1,4
        K=I
        IF(K.GT.2) K=K+2
        DO 86 J=1,4
          L=J
          IF(L.GT.2) L=L+2
          Y(I,J)=Y(K,L)
 86     CONTINUE
 85   CONTINUE
      DO 87 I=1,4
        CM01(1,I)=Y(2,I)
        CM01(2,I)=Y(3,I)+D*Y(4,I)/(AN+1.D0)
```

```
          CM01(3,I)=0.D0
          CM01(4,I)=0.D0
 87     CONTINUE
        CM01(3,1)=GZEROM
        CM01(3,2)=-1.D0/RHOBM
        CM01(3,3)=-1.D0
        CM01(4,1)=4.D0*PI*G*RHOBM
        CM01(4,2)=0.D0
        CM01(4,4)=1.D0
C Set coefficient matrices equal.
        DO 88 I=1,4
          DO 89 J=1,4
            CM02(I,J)=CM01(I,J)
            CM03(I,J)=CM02(I,J)
 89       CONTINUE
 88     CONTINUE
C Set values of constant vectors.
        BV01(1)=0.D0
        BV01(2)=0.D0
        BV01(3)=1.D0
        BV01(4)=0.D0
        DO 90 I=1,3
          BV02(I)=0.D0
 90     CONTINUE
        BV02(4)=-1.D0
C Store first constant vector.
        DO 91 I=1,4
          BV03(I)=BV01(I)
 91     CONTINUE
C Solve for linear combination coefficients.
        CALL LINSOL(CM01,BV01,4,CAUGS0,DET,4,9)
        BV1(1)=BV01(1)
        CALL LINSOL(CM02,BV02,4,CAUGS0,DET,4,9)
        BV2(1)=BV02(1)
C Alter coefficient matrix to apply subseismic condition.
        CM03(4,1)=GZEROM
        CM03(4,2)=-1.D0/RHOBM
        CM03(4,4)=0.D0
C Solve for subseismic coefficients.
        CALL LINSOL(CM03,BV03,4,CAUGS0,DET,4,9)
        BV3(1)=BV03(1)
        BV3(4)=BV03(4)
C Calculate potential condition parameter.
        BN=(AN+1.D0)/(4.D0*PI*G*RHOBM*BV3(1)+BV3(4))
C Scale Love numbers to dimensionless values.
 84     CONTINUE
        BV1(1)=GZEROM*BV1(1)
        BV2(1)=GZEROM*BV2(1)/RCMB
        BV3(1)=GZEROM*BV3(1)
C Write out internal Love numbers.
        WRITE(6,69)N,BV1(1),BV2(1)
        WRITE(2,70)BV3(1),BN
C End loop over degree.
 71   CONTINUE
C Write out dimensionless angular frequency coefficients of degree-one
C reciprocal Love number.
      WRITE(2,92)ALPHA,BETA,GAMA
 92   FORMAT(3D23.15)
      END
```

Table 7.1 *Internal load Love numbers at period* 4.0 *hr, with azimuthal number* $m = 1$, *based on Earth model Cal8.*

n	h_n^{1I}	h_n^{2I}	h_n^{1S}	h_n^{2S}
0	0.000000	−0.369806	1.464194	1.066759
1	−31.748834	−3.174732	−2.129650	−1.130747
2	−0.112801	−0.001361	0.730491	0.158910
3	−0.069769	−0.000151	0.396377	0.058026
4	−0.052181	0.000011	0.253257	0.029156
5	−0.042075	0.000039	0.183732	0.017747
6	−0.035384	0.000042	0.145664	0.012157
7	−0.030586	0.000038	0.122066	0.008958
8	−0.026962	0.000033	0.105849	0.006919
9	−0.024119	0.000028	0.093827	0.005523
10	−0.021827	0.000024	0.084436	0.004518
11	−0.019938	0.000021	0.076832	0.003766
12	−0.018353	0.000019	0.070519	0.003188
13	−0.017004	0.000016	0.065180	0.002734
14	−0.015840	0.000014	0.060600	0.002370
15	−0.014827	0.000013	0.056624	0.002075
16	−0.013936	0.000011	0.053139	0.001831
17	−0.013146	0.000010	0.050060	0.001628
18	−0.012442	0.000009	0.047319	0.001457
19	−0.011809	0.000009	0.044863	0.001311
20	−0.011238	0.000008	0.042651	0.001186
21	−0.010720	0.000007	0.040647	0.001078
22	−0.010247	0.000007	0.038823	0.000985
23	−0.009815	0.000006	0.037157	0.000902

To illustrate the output from the programme LOVE.FOR, we show the first 24 internal load Love numbers for Earth model Cal8, with azimuthal number $m = 1$. The results are given in Table 7.1 for a period of 4 hours. Except for the inner core Love number of degree one, the internal load Love numbers are only weakly dependent on period (Smylie and Jiang, 1993, p. 1355).

Single effective load Love numbers for the inner core and the shell, along with their equivalent rigid boundary radii, under the subseismic approximation, are shown in Table 7.2. Again, they are shown for a period of 4 hours, azimuthal number $m = 1$, and Earth model Cal8.

As mentioned, the internal load Love numbers are only weakly dependent on period, except for the degree-one number reflecting the translation of the inner core, as expressed by the quadratic dependence of the reciprocal of the effective

Table 7.2 *Effective internal load Love numbers and equivalent rigid-boundary radii found under the subseismic approximation at period 4.0 hr, with azimuthal number m = 1, for Earth model Cal8.*

n	h_n^l	a_n (km)	h_n^S	b_n (km)
0	0.0		0.137979	2804.0
1	−19.015768	−303.2	0.111106	3518.3
2	−0.115485	1233.5	0.213218	3212.8
3	−0.070219	1223.0	0.154227	3341.4
4	−0.052138	1219.9	0.104171	3408.7
5	−0.041878	1218.5	0.075792	3438.9
6	−0.035135	1217.8	0.059770	3453.8
7	−0.030323	1217.3	0.049913	3462.2
8	−0.026700	1217.0	0.043242	3467.6
9	−0.023865	1216.8	0.038357	3471.2
10	−0.021583	1216.7	0.034567	3473.8
11	−0.019705	1216.5	0.031507	3475.8
12	−0.018131	1216.5	0.028969	3477.3
13	−0.016792	1216.4	0.026821	3478.5
14	−0.015638	1216.3	0.024976	3479.5
15	−0.014634	1216.3	0.023372	3480.3
16	−0.013752	1216.3	0.021965	3480.9
17	−0.012970	1216.2	0.020719	3481.5
18	−0.012273	1216.2	0.019609	3482.0
19	−0.011647	1216.2	0.018614	3482.4
20	−0.011083	1216.2	0.017716	3482.7
21	−0.010570	1216.2	0.016901	3483.0
22	−0.010103	1216.1	0.016160	3483.3
23	−0.009676	1216.1	0.015482	3483.5

load Love number, h_1^l, given in relation (7.62). The coefficients of this relation are computed to be

$$a = 5.5807 \times 10^{-3}, \quad b = -5.5804 \times 10^{-3}, \quad c = -85.84 \times 10^{-3}. \qquad (7.63)$$

The coefficients a and b are nearly equal in magnitude but opposite in sign, indicating that the motion of the inner core is closely a pure translation for azimuthal number $m = 1$.

8

Variational methods and core modes

In the subseismic description of core dynamics, presented in the previous chapter, the changes in density caused by adiabatic compression or expansion due to transport through the hydrostatic pressure field are included, but those arising from flow pressure fluctuations are ignored in comparison. While this leads to the subseismic wave equation governing long-period core oscillations, the geophysicist's favourite analytical tool, the representation of solutions by expansion in spherical harmonics, is poorly convergent for such modes, due to tight Coriolis coupling between harmonics of like azimuthal number but differing zonal number (Johnson and Smylie, 1977). Instead, we use local polynomial basis functions to represent the generalised displacement potential. We show that solutions of the governing subseismic wave equation and boundary conditions have either purely even or purely odd symmetry across the equatorial plane. The basis functions then take the forms developed in Section 1.6.3. Solutions to the subseismic wave equation are found through the development of a variational principle, which includes the continuity of the normal component of displacement as a *natural boundary condition* of the problem (Smylie *et al.*, 1992). The remaining elasto-gravitational boundary conditions are incorporated through the load Love numbers described in the previous chapter.

8.1 A subseismic variational principle

A variational principle for the subseismic wave equation, (7.24), will involve stationarity of the functional with respect to variations in the generalised displacement potential, χ, subject to boundary conditions. The vector displacement field is given by expression (7.16), entirely in terms of the gradient of the generalised displacement potential. As we have seen, substitution of this expression in the subseismic form of the equation of continuity, (6.182), leads to the subseismic wave equation.

Expression (7.16) for the displacement field may be written as

$$\boldsymbol{u} = \boldsymbol{T} \cdot \nabla\chi. \tag{8.1}$$

$\boldsymbol{T}$ is a second-order tensor with Cartesian components,

$$T_{ij} = \frac{1}{4\Omega^2\sigma^2\left(\sigma^2 - 1\right)}\left[\sigma^2\delta_{ij} - k_i k_j + i\sigma\xi_{iqj}k_q - \frac{C_i^* C_j}{B}\right], \tag{8.2}$$

where $k_1,\ k_2,\ k_3$ are the components of the unit vector $\hat{\boldsymbol{k}}$. The components, u_i, of the displacement field, are then given by the contraction

$$u_i = T_{ij}\frac{\partial\chi}{\partial x_j}. \tag{8.3}$$

The tensor $\boldsymbol{T}$ is seen to be Hermitian. On interchange of the indices i and j, the alternating tensor ξ_{iqj} is replaced by $\xi_{jqi} = -\xi_{iqj}$, and the product $C_i^* C_j$ is replaced by its conjugate $C_j^* C_i$.

Consider the functional

$$\mathcal{F} = 4\Omega^2\sigma^2\left(\sigma^2 - 1\right)\int_{\mathcal{V}} \chi^* \nabla \cdot (f\boldsymbol{u})\, d\mathcal{V}, \tag{8.4}$$

where χ^* and $\boldsymbol{u}$ are solutions of the subseismic equations obeying all boundary conditions of the problem, including continuity of the normal component of displacement at the boundaries of the outer core. The integral is over the volume of the outer core. Using the identity

$$\nabla \cdot (f\chi^*\boldsymbol{u}) = \chi^*\nabla \cdot (f\boldsymbol{u}) + f\boldsymbol{u} \cdot \nabla\chi^* \tag{8.5}$$

and the divergence theorem of Gauss, the functional is transformed to

$$\begin{aligned} \mathcal{F} = &-4\Omega^2\sigma^2\left(\sigma^2 - 1\right)\int_{\mathcal{V}} f\boldsymbol{u} \cdot \nabla\chi^* d\mathcal{V} \\ &+ 4\Omega^2\sigma^2\left(\sigma^2 - 1\right)\int_{S} f\chi^*\boldsymbol{u}_B \cdot \hat{\boldsymbol{n}} dS, \end{aligned} \tag{8.6}$$

where $\boldsymbol{u}_B$ is used to denote the vector displacement fields, with continuous normal components, on the solid sides of the two elastic boundaries of the fluid outer core. Substituting for $\boldsymbol{u}$ from (8.1), the functional becomes

$$\begin{aligned} \mathcal{F} = &-4\Omega^2\sigma^2\left(\sigma^2 - 1\right)\int_{\mathcal{V}} f\nabla\chi^* \cdot \boldsymbol{T} \cdot \nabla\chi\, d\mathcal{V} \\ &+ 4\Omega^2\sigma^2\left(\sigma^2 - 1\right)\int_{S} f\chi^*\boldsymbol{u}_B \cdot \hat{\boldsymbol{n}} dS. \end{aligned} \tag{8.7}$$

We now examine the surface integral in (8.7). Since there is no coupling of modes across azimuthal number (see Section 3.4), expansion of χ in spherical harmonics gives

$$\chi = e^{im\phi} \sum_{n=m}^{\infty} \chi_n(r) P_n^m(\cos\theta). \tag{8.8}$$

Similar expansion of $\boldsymbol{u}_B \cdot \hat{\boldsymbol{n}}$ at the core–mantle boundary gives

$$\begin{aligned} \boldsymbol{u}_B \cdot \hat{\boldsymbol{n}} = \boldsymbol{u}_B \cdot \hat{\boldsymbol{s}} &= e^{im\phi} \sum_{n=m}^{\infty} y_{1,n}(b) P_n^m(\cos\theta) \\ &= e^{im\phi} \frac{1}{g_0(b)} \sum_{n=m}^{\infty} h_n^S \chi_n(b^-) P_n^m(\cos\theta), \end{aligned} \tag{8.9}$$

using the internal load Love number for the shell defined by (7.60), while expansion of $\boldsymbol{u}_B \cdot \hat{\boldsymbol{n}}$ at the inner core boundary gives

$$\begin{aligned} \boldsymbol{u}_B \cdot \hat{\boldsymbol{n}} = -\boldsymbol{u}_B \cdot \hat{\boldsymbol{s}} &= -e^{im\phi} \sum_{n=m}^{\infty} y_{1,n}(a) P_n^m(\cos\theta) \\ &= -e^{im\phi} \frac{1}{g_0(a)} \sum_{n=m}^{\infty} h_n^I \chi_n(a^+) P_n^m(\cos\theta), \end{aligned} \tag{8.10}$$

using the internal load Love number defined by (7.61) for the inner core. Spherical harmonics obey the orthogonality relation (1.180) under integration over a sphere. Thus, the surface integral has the expansion

$$\begin{aligned} \int_S f\chi^* \boldsymbol{u}_B \cdot \hat{\boldsymbol{n}} dS = &\frac{4\pi b^2}{g_0(b)} \sum_{n=m}^{\infty} \frac{1}{2n+1} \frac{(n+m)!}{(n-m)!} \Big[h_n^S \chi_n^*(b^-) \chi_n(b^-) \\ &- f(a) \frac{a^2 g_0(b)}{b^2 g_0(a)} h_n^I \chi_n^*(a^+) \chi_n(a^+) \Big] \\ = &\frac{\pi b}{\Omega^2} \Sigma_S , \end{aligned} \tag{8.11}$$

with Σ_S defined by

$$\begin{aligned} \Sigma_S = &\frac{4b\Omega^2}{g_0(b)} \sum_{n=m}^{\infty} \frac{1}{2n+1} \frac{(n+m)!}{(n-m)!} \Big[h_n^S \chi_n^*(b^-) \chi_n(b^-) \\ &- f(a) \frac{a^2 g_0(b)}{b^2 g_0(a)} h_n^I \chi_n^*(a^+) \chi_n(a^+) \Big], \end{aligned} \tag{8.12}$$

having set the decompression factor to unity at the core–mantle boundary. The surface integral is then seen to be real. Because $\boldsymbol{T}$ is Hermitian,

$$(\nabla\chi^* \cdot \boldsymbol{T} \cdot \nabla\chi)^* = \nabla\chi \cdot \boldsymbol{T}^* \cdot \nabla\chi^* = \nabla\chi^* \cdot \boldsymbol{T} \cdot \nabla\chi, \tag{8.13}$$

so the volume integral in the functional (8.7) is seen to be real as well.

Now consider variations, $\delta\chi$, in the generalised displacement potential χ, subject only to *essential boundary conditions*, excluding continuity of the normal component of displacement at the boundaries of the fluid outer core. The variation in the functional $\mathcal{F}$, caused by a variation $\delta\chi$ in χ, is

$$\begin{aligned}\delta\mathcal{F} &= -4\Omega^2\sigma^2\left(\sigma^2-1\right)\int_{\mathcal{V}} f\left[\nabla\delta\chi^*\cdot\boldsymbol{T}\cdot\nabla\chi+\nabla\chi^*\cdot\boldsymbol{T}\cdot\nabla\delta\chi\right]d\mathcal{V}\\ &+4\Omega^2\sigma^2\left(\sigma^2-1\right)\int_{\mathcal{S}} f\left[\delta\chi^*\boldsymbol{u}_B\cdot\hat{\boldsymbol{n}}+\delta\chi\boldsymbol{u}_B^*\cdot\hat{\boldsymbol{n}}\right]d\mathcal{S},\end{aligned} \tag{8.14}$$

using the symmetry of χ_n and χ_n^* in expression (8.11). The volume integral can be transformed with the vector identity,

$$f\nabla\delta\chi^*\cdot\boldsymbol{T}\cdot\nabla\chi=-\delta\chi^*\nabla\cdot(f\boldsymbol{T}\cdot\nabla\chi)+\nabla\cdot(f\delta\chi^*\boldsymbol{T}\cdot\nabla\chi). \tag{8.15}$$

Since $\boldsymbol{T}$ is Hermitian, $\nabla\chi^*\cdot\boldsymbol{T}=(\boldsymbol{T}\cdot\nabla\chi)^*=\boldsymbol{T}^*\cdot\nabla\chi^*=\boldsymbol{u}^*$, and a second vector identity emerges,

$$f\nabla\chi^*\cdot\boldsymbol{T}\cdot\nabla\delta\chi=-\delta\chi\nabla\cdot(f\boldsymbol{T}^*\cdot\nabla\chi^*)+\nabla\cdot(f\delta\chi\boldsymbol{T}^*\cdot\nabla\chi^*). \tag{8.16}$$

Using both vector identities, we find

$$\begin{aligned}\delta\mathcal{F} &= 4\Omega^2\sigma^2\left(\sigma^2-1\right)\int_{\mathcal{V}}\left[\delta\chi^*\nabla\cdot(f\boldsymbol{T}\cdot\nabla\chi)+\delta\chi\nabla\cdot(f\boldsymbol{T}^*\cdot\nabla\chi^*)\right]d\mathcal{V}\\ &-4\Omega^2\sigma^2\left(\sigma^2-1\right)\int_{\mathcal{V}}\left[\nabla\cdot(f\delta\chi^*\boldsymbol{T}\cdot\nabla\chi)+\nabla\cdot(f\delta\chi\boldsymbol{T}^*\cdot\nabla\chi^*)\right]d\mathcal{V}\\ &+4\Omega^2\sigma^2\left(\sigma^2-1\right)\int_{\mathcal{S}} f\left[\delta\chi^*\boldsymbol{u}_B\cdot\hat{\boldsymbol{n}}+\delta\chi\boldsymbol{u}_B^*\cdot\hat{\boldsymbol{n}}\right]d\mathcal{S}\\ &= 4\Omega^2\sigma^2\left(\sigma^2-1\right)\int_{\mathcal{V}}\left[\delta\chi^*\nabla\cdot(f\boldsymbol{T}\cdot\nabla\chi)+\delta\chi\nabla\cdot(f\boldsymbol{T}^*\cdot\nabla\chi^*)\right]d\mathcal{V}\\ &-4\Omega^2\sigma^2\left(\sigma^2-1\right)\int_{\mathcal{S}} f\left[\delta\chi^*\boldsymbol{T}\cdot\nabla\chi+\delta\chi\boldsymbol{T}^*\cdot\nabla\chi^*\right]\cdot\hat{\boldsymbol{n}}d\mathcal{S}\\ &+4\Omega^2\sigma^2\left(\sigma^2-1\right)\int_{\mathcal{S}} f\left[\delta\chi^*\boldsymbol{u}_B\cdot\hat{\boldsymbol{n}}+\delta\chi\boldsymbol{u}_B^*\cdot\hat{\boldsymbol{n}}\right]d\mathcal{S},\end{aligned} \tag{8.17}$$

on using the divergence theorem of Gauss to transform the second volume integral to a surface integral. The vector displacement field in the interior of the outer core is given by equation (8.1) as $\boldsymbol{u}=\boldsymbol{T}\cdot\nabla\chi$. Thus, the variation in the functional $\mathcal{F}$, caused by the variation $\delta\chi$ in χ, becomes

$$\begin{aligned}\delta\mathcal{F} &= 4\Omega^2\sigma^2\left(\sigma^2-1\right)\int_{\mathcal{V}}\left[\nabla\cdot(f\boldsymbol{u})\,\delta\chi^*+\nabla\cdot(f\boldsymbol{u}^*)\,\delta\chi\right]d\mathcal{V}\\ &-4\Omega^2\sigma^2\left(\sigma^2-1\right)\int_{\mathcal{S}} f\left[(\boldsymbol{u}-\boldsymbol{u}_B)\cdot\hat{\boldsymbol{n}}\delta\chi^*+(\boldsymbol{u}^*-\boldsymbol{u}_B^*)\cdot\hat{\boldsymbol{n}}\delta\chi\right]d\mathcal{S}.\end{aligned} \tag{8.18}$$

The variations in χ are unrestricted, except by the essential boundary conditions of the problem. Stationarity of the functional $\mathcal{F}$ then implies that $\nabla\cdot(f\boldsymbol{u})=0$ and that the *natural boundary condition*, $(\boldsymbol{u}-\boldsymbol{u}_B)\cdot\hat{\boldsymbol{n}}=0$, holds as well.

The functional in the form (8.7) may be expanded, by substitution for $\boldsymbol{T}\cdot\nabla\chi=\boldsymbol{u}$ from (7.16), to give

$$\mathcal{F}=-\sigma^2\int_{\mathcal{V}} f\nabla\chi\cdot\nabla\chi^* d\mathcal{V}+\int_{\mathcal{V}} f\left(\hat{\boldsymbol{k}}\cdot\nabla\chi\right)\left(\hat{\boldsymbol{k}}\cdot\nabla\chi^*\right)d\mathcal{V}+\int_{\mathcal{V}} fB\psi\psi^* d\mathcal{V}$$
$$-i\sigma\int_{\mathcal{V}} f\hat{\boldsymbol{k}}\cdot(\nabla\chi\times\nabla\chi^*)\,d\mathcal{V}+4\pi b\sigma^2\left(\sigma^2-1\right)\Sigma_S, \tag{8.19}$$

where $\psi=\boldsymbol{C}\cdot\nabla\chi/B$ and we have used the vector identity

$$\nabla\chi^*\cdot\left(\hat{\boldsymbol{k}}\times\nabla\chi\right)=\hat{\boldsymbol{k}}\cdot(\nabla\chi\times\nabla\chi^*). \tag{8.20}$$

The stability, or otherwise, of the density profile is represented by the Brunt–Väisälä frequency defined by (6.167). It is convenient to work with the dimensionless form of this quantity, $N=\omega_v/2\Omega$. Expression (7.15) for B then takes the form

$$B=-g_0^2\left[\frac{1}{N^2}\sigma^2\left(\sigma^2-1\right)+\left(\hat{\boldsymbol{k}}\cdot\hat{\boldsymbol{s}}\right)^2-\sigma^2\right], \tag{8.21}$$

with the unit vector $\hat{\boldsymbol{s}}$ in the orthometric direction (opposite to gravity, see Section 5.3). The vector quantity $\boldsymbol{C}$, defined by expression (7.4), becomes

$$\boldsymbol{C}=-g_0\left[\left(\hat{\boldsymbol{k}}\cdot\hat{\boldsymbol{s}}\right)\hat{\boldsymbol{k}}+i\sigma\hat{\boldsymbol{k}}\times\hat{\boldsymbol{s}}-\sigma^2\hat{\boldsymbol{s}}\right]. \tag{8.22}$$

The scalar product $\hat{\boldsymbol{k}}\cdot\hat{\boldsymbol{s}}=\cos\theta$, where θ is the geographic co-latitude. For $z=\cos\theta$, we have $B=g_0^2\left(\zeta_1^2-z^2\right)$, where

$$\zeta_1^2=\sigma^2\left(1-\frac{\sigma^2-1}{N^2}\right). \tag{8.23}$$

The unit vector $\hat{\boldsymbol{k}}=\hat{\boldsymbol{s}}\cos\theta-\hat{\boldsymbol{\theta}}\sin\theta=\hat{\boldsymbol{s}}z-\hat{\boldsymbol{\theta}}(1-z^2)^{1/2}$, and $\hat{\boldsymbol{k}}\times\hat{\boldsymbol{s}}=\hat{\boldsymbol{\phi}}\sin\theta=\hat{\boldsymbol{\phi}}(1-z^2)^{1/2}$, for unit vectors $\hat{\boldsymbol{\theta}}$ and $\hat{\boldsymbol{\phi}}$ in the direction of increasing geographic co-latitude and in the azimuthal direction, respectively. Neglecting corrections of the order of the flattening (see Section 5.2) in expressing the gradient in geographical co-ordinates, we have

$$\nabla\chi=\hat{\boldsymbol{s}}\frac{\partial\chi}{\partial r_0}-\hat{\boldsymbol{\theta}}\frac{\left(1-z^2\right)^{1/2}}{r_0}\frac{\partial\chi}{\partial z}+\hat{\boldsymbol{\phi}}\frac{im}{r_0\,(1-z^2)^{1/2}}\chi, \tag{8.24}$$

with r_0 representing the equivolumetric radius (see Section 5.3). We then find that

$$\psi=\frac{\boldsymbol{C}\cdot\nabla\chi}{B}=\frac{1}{g_0\left(\zeta_1^2-z^2\right)}\left[\left(\sigma^2-z^2\right)\frac{\partial\chi}{\partial r_0}-\frac{z\left(1-z^2\right)}{r_0}\frac{\partial\chi}{\partial z}+\frac{m\sigma}{r_0}\chi\right]. \tag{8.25}$$

The functional, given by (8.19), on integrating over the volume of the outer core, becomes

$$
\begin{aligned}
\mathcal{F} = & -2\pi\sigma^2 \int_{-1}^{1}\int_a^b f\left[\frac{m^2}{1-z^2}\chi\chi^* + r_0^2 \frac{\partial\chi}{\partial r_0}\frac{\partial\chi}{\partial r_0}^* + \left(1-z^2\right)\frac{\partial\chi}{\partial z}\frac{\partial\chi}{\partial z}^*\right] dr_0\, dz \\
& + 2\pi \int_{-1}^{1}\int_a^b f\left[r_0^2 z^2 \frac{\partial\chi}{\partial r_0}\frac{\partial\chi}{\partial r_0}^* + r_0 z\left(1-z^2\right)\left(\frac{\partial\chi}{\partial r_0}\frac{\partial\chi}{\partial z}^* + \frac{\partial\chi}{\partial z}\frac{\partial\chi}{\partial r_0}^*\right)\right. \\
& \left. + \left(1-z^2\right)^2 \frac{\partial\chi}{\partial z}\frac{\partial\chi}{\partial z}^*\right] dr_0\, dz \\
& + 2\pi \int_{-1}^{1}\int_a^b \frac{f}{\zeta_1^2 - z^2}\left[m^2\sigma^2\chi\chi^* + m\sigma r_0\left(\sigma^2 - z^2\right)\left(\chi\frac{\partial\chi}{\partial r_0}^* + \frac{\partial\chi}{\partial r_0}\chi^*\right)\right. \\
& + r_0^2\left(\sigma^2 - z^2\right)^2 \frac{\partial\chi}{\partial r_0}\frac{\partial\chi}{\partial r_0}^* - r_0 z\left(\sigma^2 - z^2\right)\left(1-z^2\right)\left(\frac{\partial\chi}{\partial r_0}\frac{\partial\chi}{\partial z}^* + \frac{\partial\chi}{\partial z}\frac{\partial\chi}{\partial r_0}^*\right) \\
& \left. - m\sigma z\left(1-z^2\right)\left(\chi\frac{\partial\chi}{\partial z}^* + \frac{\partial\chi}{\partial z}\chi^*\right) + z^2\left(1-z^2\right)^2 \frac{\partial\chi}{\partial z}\frac{\partial\chi}{\partial z}^*\right] dr_0\, dz \\
& - 2\pi m\sigma \int_{-1}^{1}\int_a^b f\left[r_0\left(\chi\frac{\partial\chi}{\partial r_0}^* + \frac{\partial\chi}{\partial r_0}\chi^*\right) - z\left(\chi\frac{\partial\chi}{\partial z}^* + \frac{\partial\chi}{\partial z}\chi^*\right)\right] dr_0\, dz \\
& + 4\pi b\sigma^2\left(\sigma^2 - 1\right)\Sigma_S .
\end{aligned}
\tag{8.26}
$$

Again neglecting corrections of the order of the flattening, the normal component of the displacement is given by (7.14) as

$$
\boldsymbol{u}\cdot\hat{\boldsymbol{n}} = \frac{\alpha^2}{\beta g_0}\frac{\boldsymbol{C}\cdot\nabla\chi}{B}. \tag{8.27}
$$

From (6.167), the stability factor is expressed as

$$
\beta = -4\Omega^2 N^2 \frac{\alpha^2}{g_0^2}. \tag{8.28}
$$

Substituting for β and using (8.25), the normal component of displacement to be used at the boundaries becomes

$$
\boldsymbol{u}\cdot\hat{\boldsymbol{n}} = -\frac{1}{4\Omega^2 N^2\left(\zeta_1^2 - z^2\right)}\left[\left(\sigma^2 - z^2\right)\frac{\partial\chi}{\partial r_0} - \frac{z\left(1-z^2\right)}{r_0}\frac{\partial\chi}{\partial z} + \frac{m\sigma}{r_0}\chi\right]. \tag{8.29}
$$

8.2 Representation of the functional

The functional in the form (8.26) and the normal component of displacement (8.29) are expressed in terms of equivolumetric radius r_0 and the cosine of geographic co-latitude $z = \cos\theta$. Some general properties of the solution for χ can be deduced from these expressions. First, modes of a given azimuthal number m are independent of modes with other azimuthal numbers. This absence of coupling over different azimuthal numbers derives from the axisymmetric nature of the Earth model assumed in the formulation. In addition, it is evident from expression (8.26) for the functional that only real trial functions are required and that it is unnecessary to include their azimuthal dependence in their representation. Second, reversing the signs of both m and the angular frequency σ results in the same functional and boundary conditions. Thus, we can restrict m to zero and the positive integers, if σ is allowed to range over both positive and negative values. Axisymmetric modes are standing waves, determined by σ^2 alone. Non-axisymmetric modes drift westward at the angular rate $2\Omega\sigma/m$ for positive σ and drift eastward at the rate $-2\Omega\sigma/m$ for negative σ. Finally, modes that are purely even with respect to the equatorial plane are decoupled from modes that are purely odd, since z appears only as a product with $\partial\chi/\partial z$, or as a square, in both the functional (8.26) and the boundary conditions (8.29). Thus, contributions arising from cross products of even and odd parts of solutions will be odd in the equatorial plane and their integral will vanish. Support functions that are purely even or purely odd in the equatorial plane are therefore used in the representation of trial functions.

The functional (8.26) is conveniently displayed in the matrix representation

$$
\begin{aligned}
\mathcal{F} = &-2\pi\sigma^2 \int_{-1}^{1}\int_a^b f\boldsymbol{U}^{T*}\begin{pmatrix} m^2/(1-z^2) & 0 & 0 \\ 0 & 1 & 0 \\ 0 & 0 & (1-z^2) \end{pmatrix}\boldsymbol{U}\, dr_0\, dz \\
&+ 2\pi \int_{-1}^{1}\int_a^b f\boldsymbol{U}^{T*}\begin{pmatrix} 0 \\ z \\ 1-z^2 \end{pmatrix} \otimes \left(0, z, 1-z^2\right)\boldsymbol{U}\, dr_0\, dz \\
&+ 2\pi \int_{-1}^{1}\int_a^b \frac{f}{\zeta_1^2 - z^2}\boldsymbol{U}^{T*}\begin{pmatrix} m\sigma \\ \sigma^2 - z^2 \\ z^3 - z \end{pmatrix} \otimes \left(m\sigma, \sigma^2 - z^2, -z\left(1-z^2\right)\right)\boldsymbol{U}\, dr_0\, dz \\
&- 2\pi m\sigma \int_{-1}^{1}\int_a^b f\boldsymbol{U}^{T*}\begin{pmatrix} 0 & 1 & -z \\ 1 & 0 & 0 \\ -z & 0 & 0 \end{pmatrix}\boldsymbol{U}\, dr_0\, dz \\
&+ 4\pi b\sigma^2\left(\sigma^2 - 1\right)\Sigma_S,
\end{aligned}
\tag{8.30}
$$

where the symbol $\otimes$ denotes the outer vector product and $\boldsymbol{U}^T = (\chi, r_0\chi_{r_0}, \chi_z)$ with $\chi_{r_0} = \partial\chi/\partial r_0$ and $\chi_z = \partial\chi/\partial z$.

In addition to the properties of the functional just described, the factor $\sin^m\theta$ is a general feature of solutions of the subseismic wave equation (Smylie and

Rochester, 1986). This suggests that we make the change of independent variable to Φ with

$$\chi = \left(1 - z^2\right)^{m/2} \Phi. \tag{8.31}$$

The domain of the problem is reduced to a quarter annulus with z ranging over the interval $(0, 1)$. Scaling the radius by the radius b of the core–mantle boundary, the dimensionless radius $r = r_0/b$ ranges over the interval $(a/b, 1)$. We can then divide expression (8.30) by $-4\pi b$ and work with the reduced functional

$$\begin{aligned}
\mathcal{F}_R = {} & \sigma^2 \int_0^1 \int_{a/b}^1 f\left(1 - z^2\right)^m \boldsymbol{V}^T \begin{pmatrix} m^2\left(1+z^2\right)/\left(1-z^2\right) & 0 & -mz \\ 0 & 1 & 0 \\ -mz & 0 & \left(1-z^2\right) \end{pmatrix} \boldsymbol{V}\, dr\, dz \\
& - \int_0^1 \int_{a/b}^1 f\left(1 - z^2\right)^m \boldsymbol{V}^T \begin{pmatrix} mz \\ -z \\ z^2 - 1 \end{pmatrix} \otimes \left(mz, -z, z^2 - 1\right) \boldsymbol{V}\, dr\, dz \\
& - \int_0^1 \int_{a/b}^1 f \frac{\left(1 - z^2\right)^m}{\zeta_1^2 - z^2} \boldsymbol{V}^T \begin{pmatrix} m\left(\sigma + z^2\right) \\ \sigma^2 - z^2 \\ z^3 - z \end{pmatrix} \otimes \left(m\left(\sigma + z^2\right), \sigma^2 - z^2, z^3 - z\right) \boldsymbol{V}\, dr\, dz \\
& + m\sigma \int_0^1 \int_{a/b}^1 f\left(1 - z^2\right)^m \boldsymbol{V}^T \begin{pmatrix} 2mz^2/\left(1-z^2\right) & 1 & -z \\ 1 & 0 & 0 \\ -z & 0 & 0 \end{pmatrix} \boldsymbol{V}\, dr\, dz \\
& - \sigma^2\left(\sigma^2 - 1\right)\Sigma_S,
\end{aligned} \tag{8.32}$$

where $\boldsymbol{V}^T = (\Phi, r\Phi_r, \Phi_z)$ with $\Phi_r = \partial\Phi/\partial r$ and $\Phi_z = \partial\Phi/\partial z$. The normal component of displacement (8.29) is then expressed as

$$\boldsymbol{u} \cdot \hat{\boldsymbol{n}} = -\frac{\hat{\boldsymbol{n}} \cdot \hat{\boldsymbol{s}} \left(1 - z^2\right)^{m/2}}{4\Omega^2 N^2 r_0 \left(\zeta_1^2 - z^2\right)} \left(m\left(\sigma + z^2\right), \sigma^2 - z^2, z^3 - z\right) \cdot \boldsymbol{V}. \tag{8.33}$$

The factor $\zeta_1^2 - z^2$, appearing in the denominator of the third integrand of the functional and in the denominator of the expression of the normal component of displacement, requires special consideration. From the definition (8.23) of ζ_1^2, we have

$$\zeta_1^2 - z^2 = \sigma^2 - z^2 - \frac{\sigma^2\left(\sigma^2 - 1\right)}{N^2}. \tag{8.34}$$

For modes with periods neither close to one-half sidereal day ($\sigma = 1$) nor extremely long, and for small departures from neutral stratification,

$$\left|\frac{N^2}{\sigma^2 - 1}\right| \ll \sigma^2, \tag{8.35}$$

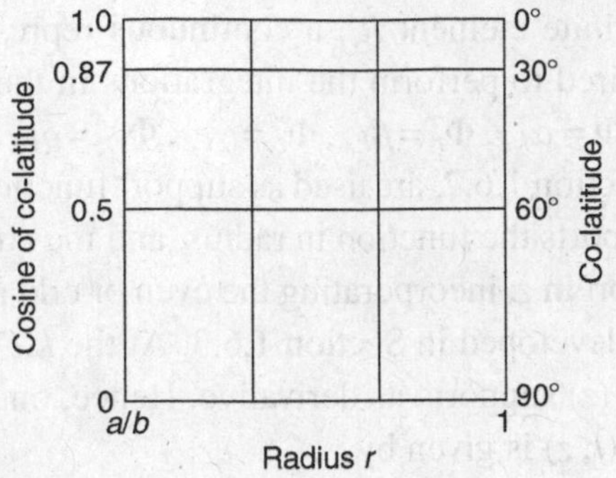

Figure 8.1 The domain in the (r, z) plane broken into a 3×3 finite element grid, equispaced in radius and co-latitude θ, with $z = \cos\theta$.

yielding the expansion

$$\frac{1}{\zeta_1^2 - z^2} = -\frac{N^2}{\sigma^2(\sigma^2-1)}\left[1 + \frac{N^2}{\sigma^2(\sigma^2-1)}\left(\sigma^2 - z^2\right) + \cdots\right]. \tag{8.36}$$

For example, a period of 4 sidereal hours gives $\sigma^2 = 9$, and for a weak departure from neutral stratification we might have $N^2 = 0.01$, giving $N^2/(\sigma^2-1) = 0.00125$. Thus, neglect of the second term in the expansion would lead to an error of slightly over one part in a thousand. To this level of approximation, we can multiply the functional by $\sigma^2(\sigma^2 - 1)$ to make it an eighth-degree polynomial in σ. Similarly, continuity of normal displacement becomes a quadratic in σ.

We can then define a finite element grid for support functions in the (r, z) plane. The domain of the problem is shown in Figure 8.1 as a 3×3 finite element grid, equispaced in radius and co-latitude.

8.3 Finite element support functions

A particular finite element R_{ij} occupies the interval $(i, i+1)$ in radius r and the interval $(j, j+1)$ in the cosine of co-latitude z, as illustrated in Figure 8.2.

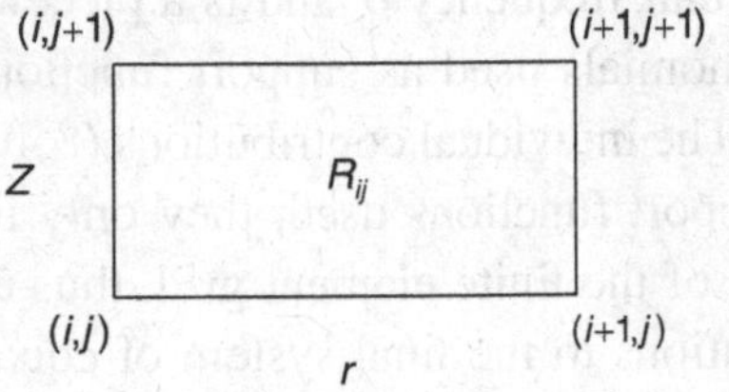

Figure 8.2 The rectangular finite element R_{ij} and its four nodes.

On the rectangular finite element R_{ij} a continuous representation of Φ and its first derivatives is required to perform the integrations in the functional (8.32). At the k,ℓ vertex we write $\Phi = \alpha_{k,\ell}$, $\Phi_r = \beta_{k,\ell}$, $\Phi_z = \gamma_{k,\ell}$, $\Phi_{rz} = \delta_{k,\ell}$. Local Hermite spline cubics, developed in Section 1.6.2, are used as support functions in r. At the k,ℓ vertex, the cubic $\eta_k(r)$ supports the function in radius, and the cubic $\psi_k(r)$ supports the radial derivative. Support in z, incorporating the even or odd properties, is provided by the basis functions developed in Section 1.6.3. At the k,ℓ vertex, $\eta_\ell(z)$ supports the function in z and $\psi_\ell(z)$ supports its derivative. Hence, on the finite element R_{ij}, the representation of $\Phi(r,z)$ is given by

$$\Phi_{i,j}(r,z) = \sum_{k=i}^{k=i+1} \sum_{\ell=j}^{\ell=j+1} \Big[\alpha_{k,\ell}\eta_k(r)\eta_\ell(z) + \beta_{k,\ell}\psi_k(r)\eta_\ell(z) + \gamma_{k,\ell}\eta_k(r)\psi_\ell(z) + \delta_{k,\ell}\psi_k(r)\psi_\ell(z) \Big]. \tag{8.37}$$

A four-vector can be defined at each of the nodes of the finite element grid. For the node at the k,ℓ vertex, the four-vector has the transpose $\boldsymbol{x}_{k,\ell}^T = (\alpha_{k,\ell}, \beta_{k,\ell}, \gamma_{k,\ell}, \delta_{k,\ell})$. On the finite element R_{ij}, the vector $\boldsymbol{V}$ entering the functional (8.32) and the expression (8.33) for the normal component of displacement then has the support

$$\boldsymbol{V} = \sum_{k=i}^{k=i+1} \sum_{\ell=j}^{\ell=j+1} D_{k,\ell}\boldsymbol{x}_{k,\ell}, \tag{8.38}$$

where

$$D_{k,\ell} = \begin{pmatrix} \eta_k(r)\eta_\ell(z) & \psi_k(r)\eta_\ell(z) & \eta_k(r)\psi_\ell(z) & \psi_k(r)\psi_\ell(z) \\ r\eta_k'(r)\eta_\ell(z) & r\psi_k'(r)\eta_\ell(z) & r\eta_k'(r)\psi_\ell(z) & r\psi_k'(r)\psi_\ell(z) \\ \eta_k(r)\eta_\ell'(z) & \psi_k(r)\eta_\ell'(z) & \eta_k(r)\psi_\ell'(z) & \psi_k(r)\psi_\ell'(z) \end{pmatrix}. \tag{8.39}$$

Substitution of the expression (8.38) for $\boldsymbol{V}$ in the functional $\mathcal{F}_R$ (8.32) gives a symmetric quadratic form and setting its variation to zero gives the contribution

$$\int_0^1 \int_{a/b}^1 D_{p,q}^T A(r,z,\sigma) D_{k,\ell}\, dr\, dz \tag{8.40}$$

to the p,q row and k,ℓ column of the final matrix. The 3×3 matrix A is a polynomial in the dimensionless angular frequency σ and is a piecewise polynomial in r and z for the piecewise polynomials used as support functions and to interpolate the decompression factor f. The individual contributions (8.40) are 4×4 matrices. For the local polynomial support functions used, they only interact with the support at the four nearest nodes of the finite element grid, thus each element produces a 16×16 array of contributions to the final system of equations. If the grid has M nodes in r and N nodes in z, the final system of equations is generated by summing the contributions from all $(M-1)\times(N-1)$ elements.

Finally, we consider the support for the sum Σ_s (8.12) appearing in the functional and representing integrals over the boundaries of the outer core. To evaluate this sum, we require the radial coefficients $\chi_n(b^-)$ and $\chi_n(a^+)$ of the spherical harmonic expansions of χ on the outer core boundaries. From expressions (8.38) and (8.39) on the core–mantle boundary, we have

$$\boldsymbol{V} = \sum_{j=1}^{N-1} [D_{M,j}(1,z)\boldsymbol{x}_{M,j} + D_{M,j+1}(1,z)\boldsymbol{x}_{M,j+1}], \tag{8.41}$$

with

$$D_{M,\ell}(1,z) = \begin{pmatrix} \eta_\ell(z) & 0 & \psi_\ell(z) & 0 \\ 0 & \eta_\ell(z) & 0 & \psi_\ell(z) \\ \eta'_\ell(z) & 0 & \psi'_\ell(z) & 0 \end{pmatrix}, \tag{8.42}$$

while on the inner core boundary we have

$$\boldsymbol{V} = \sum_{j=1}^{N-1} [D_{1,j}(a/b,z)\boldsymbol{x}_{1,j} + D_{1,j+1}(a/b,z)\boldsymbol{x}_{1,j+1}], \tag{8.43}$$

with

$$D_{1,\ell}(a/b,z) = \begin{pmatrix} \eta_\ell(z) & 0 & \psi_\ell(z) & 0 \\ 0 & a\eta_\ell(z)/b & 0 & a\psi_\ell(z)/b \\ \eta'_\ell(z) & 0 & \psi'_\ell(z) & 0 \end{pmatrix}. \tag{8.44}$$

The coefficients $\chi_n(r)$ in the spherical harmonic expansion (8.8) of χ can be recovered, using the formula (1.182) and the relation (1.179), as

$$\chi_n(r) = (2n+1)\frac{(n-m)!}{(n+m)!}\int_0^1 \chi P_n^m(z)\,dz. \tag{8.45}$$

In terms of the vector $\boldsymbol{V}$, the coefficients may be expressed as

$$\chi_n(r) = (2n+1)\frac{(n-m)!}{(n+m)!}\int_0^1 (1-z^2)^{m/2}\big(1,0,0\big)\boldsymbol{V} P_n^m(z)\,dz. \tag{8.46}$$

In terms of the four-vector with transpose

$$\boldsymbol{d}_{n,\ell}^T = (2n+1)\frac{(n-m)!}{(n+m)!}\int_0^1 \left(1-z^2\right)^{m/2}\big(\eta_\ell(z),0,\psi_\ell(z),0\big)P_n^m(z)\,dz, \tag{8.47}$$

with substitution from (8.41) and (8.43), the radial coefficients on the boundaries of the outer core are found to be

$$\chi_n(b) = \sum_{j=1}^{N-1}\left(\boldsymbol{d}_{n,j+1}^T\boldsymbol{x}_{M,j} + \boldsymbol{d}_{n,j+1}^T\boldsymbol{x}_{M,j+1}\right) \tag{8.48}$$

and

$$\chi_n(a) = \sum_{j=1}^{N-1} \left(\boldsymbol{d}_{n,j}^T \boldsymbol{x}_{1,j} + \boldsymbol{d}_{n,j+1}^T \boldsymbol{x}_{1,j+1} \right). \tag{8.49}$$

These lead to the squared magnitudes, given by the sums

$$|\chi_n(b)|^2 = \sum_{j=1}^{N-1} \sum_{\ell=1}^{N-1} \sum_{q=0}^{1} \sum_{s=0}^{1} \boldsymbol{x}_{M,j+q}^T \boldsymbol{d}_{n,j+q} \boldsymbol{d}_{n,\ell+s}^T \boldsymbol{x}_{M,\ell+s} \tag{8.50}$$

and

$$|\chi_n(a)|^2 = \sum_{j=1}^{N-1} \sum_{\ell=1}^{N-1} \sum_{q=0}^{1} \sum_{s=0}^{1} \boldsymbol{x}_{1,j+q}^T \boldsymbol{d}_{n,j+q} \boldsymbol{d}_{n,\ell+s}^T \boldsymbol{x}_{1,\ell+s}. \tag{8.51}$$

The full contributions of Σ_S to the partitioned matrix then consist of contributions to the $M,j+q$ row and $M,\ell+s$ column from the core–mantle boundary through (8.50) and to the $1,j+q$ row and $1,\ell+s$ column from the inner core boundary through (8.51).

8.4 Boundary conditions and constraints

In addition to satisfying the known symmetry properties of the problem, the support functions must also satisfy the boundary conditions at the core–mantle boundary and at the inner core boundary. We also include the special case of the Poincaré inertial wave equation (Section 6.1), which has known analytic solutions and can be used as a test of the accuracy of the numerical solutions. For this model of the oscillations of a uniform, not self-gravitating fluid in a rigid container, there is no inner body and conditions need to be imposed at the geocentre.

The remaining boundary condition to be applied to the support functions is that of continuity of the normal component of displacement at the boundaries of the fluid outer core. The normal component of displacement is given by expression (8.33) in terms of the vector $\boldsymbol{V}$. In analogy to the expansion (8.45), for small departures from neutral stratification, the normal component of displacement has the spherical harmonic expansion

$$\boldsymbol{u} \cdot \hat{\boldsymbol{n}} = e^{im\phi} \sum_{n=m}^{\infty} y_{1,n}(r) P_n^m(\cos\theta), \tag{8.52}$$

with

$$y_{1,n}(r) = (2n+1)\frac{(n-m)!}{(n+m)!}\int_0^1 \boldsymbol{u}\cdot\hat{\boldsymbol{n}}P_n^m(z)\,dz$$
$$= \frac{\hat{\boldsymbol{n}}\cdot\hat{\boldsymbol{s}}}{4\Omega^2\sigma^2(\sigma^2-1)\,r_0}(2n+1)\frac{(n-m)!}{(n+m)!}$$
$$\times\int_0^1\left(1-z^2\right)^{m/2}\left(m\left(\sigma+z^2\right),\sigma^2-z^2,z^3-z\right)\boldsymbol{V}P_n^m(z)\,dz. \tag{8.53}$$

With a four-vector defined as

$$e_{n,\ell}(r,\sigma) = (2n+1)\frac{(n-m)!}{(n+m)!}$$
$$\times\int_0^1\left(1-z^2\right)^{m/2}\begin{pmatrix} m(\sigma+z^2)\eta_\ell(z)+z(z^2-1)\eta_\ell'(z)\\ (\sigma^2-z^2)r\eta_\ell(z)\\ m(\sigma+z^2)\psi_\ell(z)+z(z^2-1)\psi_\ell'\\ (\sigma^2-z^2)r\psi_\ell(z)\end{pmatrix}P_n^m(z)\,dz, \tag{8.54}$$

substitution for $\boldsymbol{V}$ in (8.53) from (8.41) gives, for the core–mantle boundary,

$$y_{1,n}(b) = \frac{1}{4\Omega^2\sigma^2(\sigma^2-1)\,b}\sum_{j=1}^{N-1}\left[\boldsymbol{e}_{n,j}^T(1,\sigma)\boldsymbol{x}_{M,j}+\boldsymbol{e}_{n,j+1}^T(1,\sigma)\boldsymbol{x}_{M,j+1}\right], \tag{8.55}$$

while substitution for $\boldsymbol{V}$ in (8.53) from (8.43) gives, for the inner core boundary,

$$y_{1,n}(a) = \frac{1}{4\Omega^2\sigma^2(\sigma^2-1)\,a}\sum_{j=1}^{N-1}\left[\boldsymbol{e}_{n,j}^T(a/b,\sigma)\boldsymbol{x}_{1,j}+\boldsymbol{e}_{n,j+1}^T(a/b,\sigma)\boldsymbol{x}_{1,j+1}\right]. \tag{8.56}$$

Using the internal load Love number for the shell (7.60), $y_{1,n}(b)$ may be related to $\chi_n(b)$, which in turn is expressed by (8.48). Similarly, using the internal load Love number for the inner core (7.61), $y_{1,n}(a)$ may be related to $\chi_n(a)$, which in turn is expressed by (8.49). Continuity of normal displacement at the boundaries of the outer core then yields two linear, homogeneous constraint equations for each harmonic degree n:

$$\sum_{j=1}^{N-1}\left[\boldsymbol{c}_{n,j}^T(1,\sigma)\boldsymbol{x}_{M,j}+\boldsymbol{c}_{n,j+1}^T(1,\sigma)\boldsymbol{x}_{M,j+1}\right]=0, \tag{8.57}$$

for the core–mantle boundary, and

$$\sum_{j=1}^{N-1}\left[\boldsymbol{c}_{n,j}^T(a/b,\sigma)\boldsymbol{x}_{1,j}+\boldsymbol{c}_{n,j+1}^T(a/b,\sigma)\boldsymbol{x}_{1,j+1}\right]=0, \tag{8.58}$$

for the inner core boundary, with

$$\boldsymbol{c}^T_{n,\ell}(1,\sigma) = \boldsymbol{e}^T_{n,\ell}(1,\sigma) - 4\Omega^2\sigma^2\left(\sigma^2-1\right)\frac{bh^S_n}{g_0(b)}\boldsymbol{d}^T_{n,\ell} \tag{8.59}$$

and

$$\boldsymbol{c}^T_{n,\ell}(a/b,\sigma) = \boldsymbol{e}^T_{n,\ell}(a/b,\sigma) - 4\Omega^2\sigma^2\left(\sigma^2-1\right)\frac{ah^I_n}{g_0(a)}\boldsymbol{d}^T_{n,\ell}. \tag{8.60}$$

The integrals (8.47) and (8.54) for $\boldsymbol{d}_{n,\ell}$ and $\boldsymbol{e}_{n,\ell}(r,\sigma)$, respectively, take into account that they are non-vanishing only for integrands that are even across the equatorial plane. For modes even across the equatorial plane, n in the constraint equations (8.57) and (8.58) ranges over $n = m,\ m+2,\ m+4,\ldots$. While for modes odd across the equatorial plane n ranges over $n = m+1,\ m+3,\ m+5,\ldots$. For L spherical harmonics constrained to satisfy the boundary conditions, n ranges over the values

$$n - m - \mathcal{P} = 0,\ 2,\ 4,\ldots,\ 2(L-1), \tag{8.61}$$

where the binary number $\mathcal{P}$ is zero for even modes and unity for odd modes.

With the application of the range and summation conventions, the functional multiplied by $\sigma^2\left(\sigma^2-1\right)$ is the symmetric bilinear form

$$\sigma^2\left(\sigma^2-1\right)\mathcal{F}_R = x_i a_{ij} x_j, \quad a_{ij} = a_{ji}, \tag{8.62}$$

the a_{ij} being the elements of a matrix that is an eighth-degree polynomial in σ. The variables x_i can be divided into those that are constrained, x''_p, and those that are free to be chosen to make the functional stationary, x'_k. The pth constraint equation can then be solved for x''_p to give

$$x''_p = c_{pk}x'_k. \tag{8.63}$$

This allows the symmetric bilinear form (8.62) to be expressed entirely in terms of unconstrained variables as

$$x'_k a_{k\ell} x'_\ell + x'_k a_{kp} c_{p\ell} x'_\ell + x'_k a_{\ell p} c_{pk} x'_\ell + x'_k c_{pk} a_{pq} c_{q\ell} x'_\ell, \tag{8.64}$$

using the symmetry of the elements a_{ij}. Stationarity of this form leads to a modified yet symmetric coefficient matrix with $k\ell$ element given by

$$a_{k\ell} + a_{kp}c_{p\ell} + a_{\ell p}c_{pk} + c_{pk}a_{pq}c_{q\ell}. \tag{8.65}$$

The new matrix is easily generated from the unconstrained matrix by sequences of row and column operations.

In the case of the Poincaré inertial wave equation, Φ and its derivatives Φ_r, Φ_z and Φ_{rz} can be shown to vanish at the geocentre, except for the axisymmetric modes

$(m=0)$, where Φ may be a non-zero constant for all z, and the 'flow-through' modes, which involve translational motion either along the axis of rotation or in the equatorial plane. The 'flow-through' modes are the analogue of the translational modes in realistic Earth models, where the inner core and shell have opposing translational motions. The 'flow-through' mode along the axis of rotation is also axisymmetric but is odd in the equatorial plane. For this mode $\Phi_r = \chi_r$ is linear in z at the geocentre and Φ_{rz} is constant there. The 'flow-through' modes in the equatorial plane have azimuthal number $m = 1$ and Φ_r is a non-zero constant at the geocentre. All of these special constraints can be cast in the linear form (8.63) and implemented by the matrix modification described by (8.65).

8.5 Numerical implementation and results

Construction of the coefficient matrix requires evaluation of the volume integral (8.40) for each finite element, the boundary integral (8.47) required for the application of the elasto-gravitational boundary conditions, and (8.54) for ensuring the continuity of the normal component of displacement. With spline interpolation of the decompression factor, the integrands are polynomials. We use 12-point Gaussian integration in both r and z, giving complete accuracy for integrands expressed by polynomials of degree up to and including 23. The coefficient matrix is a polynomial of degree eight in the dimensionless angular frequency σ, while the matrix constraining continuity of normal displacement at the boundaries of the outer core is a quadratic function of σ. Thus, ten matrices are required to be calculated for each different value of σ, in addition to the matrix independent of σ.

The programme MAT.FOR computes the eleven required matrices and puts them in the output file matres.dat. Input to the programme MAT.FOR is the file decomp.dat containing the decompression factor f (Section 6.4) and love.dat containing the required internal Love numbers for the inner core and shell (Section 7.2).

The final eigenvalue–eigenvector problem then takes the form

$$A(\sigma_i)\boldsymbol{x}_i = 0, \tag{8.66}$$

for eigenfrequency σ_i and eigenvector $\boldsymbol{x}_i$. The matrix $A(\sigma)$ is unusual in that it is a λ-matrix (Lancaster, 1966) or polynomial matrix of degree eight, expressible as

$$A(\sigma) = A_8\sigma^8 + A_7\sigma^7 + \cdots + A_1\sigma + A_0. \tag{8.67}$$

A specially adapted form of inverse iteration is employed to solve the eigenvalue–eigenvector problem (8.66). An initial estimate of the eigenfrequency σ_r is found from the vanishing of the determinant of $A(\sigma_r)$. The derivative of $A(\sigma)$ is easy to calculate because A is a polynomial in σ, so for σ close to σ_r we have

$$A(\sigma_r) = A(\sigma) + (\sigma_r - \sigma)A'(\sigma) + \cdots, \tag{8.68}$$

where $A'(\sigma)$ is the derivative of $A(\sigma)$. Then

$$A'(\sigma) \approx \frac{A(\sigma_r) - A(\sigma)}{\sigma_r - \sigma}. \tag{8.69}$$

Now consider the problem

$$A(\sigma)\boldsymbol{u} = A'(\sigma)\boldsymbol{x}_r \approx \frac{A(\sigma_r) - A(\sigma)}{\sigma_r - \sigma}\boldsymbol{x}_r = -\frac{A(\sigma)\boldsymbol{x}_r}{\sigma_r - \sigma}. \tag{8.70}$$

Thus, the solution to this problem is

$$\boldsymbol{u} = \frac{\boldsymbol{x}_r}{\sigma - \sigma_r}. \tag{8.71}$$

A more complicated problem is

$$A(\sigma)\boldsymbol{v} = A'(\sigma)\boldsymbol{b}, \tag{8.72}$$

where $\boldsymbol{b}$ is the linear combination of all the eigenvectors,

$$\boldsymbol{b} = \sum_i \gamma_i \boldsymbol{x}_i. \tag{8.73}$$

We can verify that

$$\boldsymbol{v} = \sum_i \frac{\gamma_i \boldsymbol{x}_i}{\sigma - \sigma_i} \tag{8.74}$$

is the solution of this more complicated problem. We then have

$$\begin{aligned} A(\sigma)\boldsymbol{v} = A(\sigma)\sum_i \frac{\gamma_i \boldsymbol{x}_i}{\sigma - \sigma_i} &= \sum_i \gamma_i \frac{[A(\sigma_i) + (\sigma - \sigma_i)A'(\sigma) + \cdots]\,\boldsymbol{x}_i}{\sigma - \sigma_i} \\ &= A'(\sigma)\sum_i \gamma_i \boldsymbol{x}_i = A'(\sigma)\boldsymbol{b}, \end{aligned} \tag{8.75}$$

completing the verification. This suggests that if we have an initial value of the eigenfrequency, σ_r, the solution $\boldsymbol{v}$ will enhance the corresponding eigenvector $\boldsymbol{x}_r$ above all others. The iterative scheme then becomes

$$A(\sigma_r)\boldsymbol{x}_{r+1} = A'(\sigma_r)\boldsymbol{x}_r. \tag{8.76}$$

After satisfactory convergence to the eigenvector has been obtained, an improved estimate of the eigenfrequency can be found, satisfying the equation

$$A(\sigma_{r+1})\boldsymbol{x}_{r+1} = 0. \tag{8.77}$$

Then

$$A(\sigma_{r+1}) = A(\sigma_r) + (\sigma_{r+1} - \sigma_r)A'(\sigma_r) + \cdots \tag{8.78}$$

and

$$[A(\sigma_r) + (\sigma_{r+1} - \sigma_r)A'(\sigma_r) + \cdots]\,\boldsymbol{x}_{r+1} = 0. \tag{8.79}$$

Since the current estimate of the eigenvector satisfies (8.76), we have

$$A'(\sigma_r)[\boldsymbol{x}_r + (\sigma_{r+1} - \sigma_r)\boldsymbol{x}_{r+1}] \approx 0. \tag{8.80}$$

The iterative scheme (8.76) allows large changes in the magnitude and direction of the eigenvector, so that at each step the eigenvector is normalised to unit magnitude, giving $\boldsymbol{x}_r \cdot \boldsymbol{x}_r = 1$. Thus, if $\det A' \neq 0$,

$$\boldsymbol{x}_r + (\sigma_{r+1} - \sigma_r)\boldsymbol{x}_{r+1} \approx 0 \tag{8.81}$$

and

$$\sigma_{r+1} = \sigma_r - \frac{1}{\boldsymbol{x}_{r+1} \cdot \boldsymbol{x}_r}, \tag{8.82}$$

giving the improved estimate of the eigenfrequency.

Given the solution for the generalised displacement potential χ, the displacement vector is found from relation (7.16). In spherical polar co-ordinates, the vector displacement $\boldsymbol{u}$ has components $(u_r,\ u_\theta,\ u_\phi)$ given by,

$$ru_r = F^m\left[m\left(\sigma + z^2\right)\Phi + \left(\sigma^2 - z^2\right)r\Phi_r - z\left(1 - z^2\right)\Phi_z\right]e^{im\phi}, \tag{8.83}$$

$$\begin{aligned}\sigma^2 r\sin\theta\, u_\theta = F^m\Big[&\left\{\sigma\left(z^2 - \zeta_2^2\right) + \left(1 - \zeta_2^2\right)\right\} m\sigma z\Phi + z\left(1 - z^2\right)\sigma^2 r\Phi_r \\ &- \left(z^2 - \zeta_2^2\right)\left(1 - z^2\right)\sigma^2\Phi_z\Big]e^{im\phi},\end{aligned} \tag{8.84}$$

$$\begin{aligned}\sigma r\sin\theta\, u_\phi = F^m\Big[&m\left(\sigma + z^2\right)\left(1 - \zeta_2^2\right)\Phi + \sigma^2\left(1 - z^2\right)r\Phi_r \\ &- \left(1 - \zeta_2^2\right)z\left(1 - z^2\right)\Phi_z\Big]e^{i(m\phi + \pi/2)},\end{aligned} \tag{8.85}$$

with the common factor F^m defined by

$$F^m = \frac{\left(1 - z^2\right)^{m/2}}{4\Omega^2 N^2\left(z^2 - \zeta_1^2\right)} \tag{8.86}$$

and the parameter ζ_2^2 defined by

$$\zeta_2^2 = 1 - \sigma^2 + N^2 = \frac{N^2}{\sigma^2}\zeta_1^2. \tag{8.87}$$

Using only the first term of the expansion (8.36) as an approximation, the factor F^m becomes

$$F^m = \frac{\left(1 - z^2\right)^{m/2}}{4\Omega^2\sigma^2\left(\sigma^2 - 1\right)}. \tag{8.88}$$

In the case of the Poincaré inertial wave equation, the parameter ζ_2^2 reduces to

$$\zeta_2^2 = 1 - \sigma^2. \tag{8.89}$$

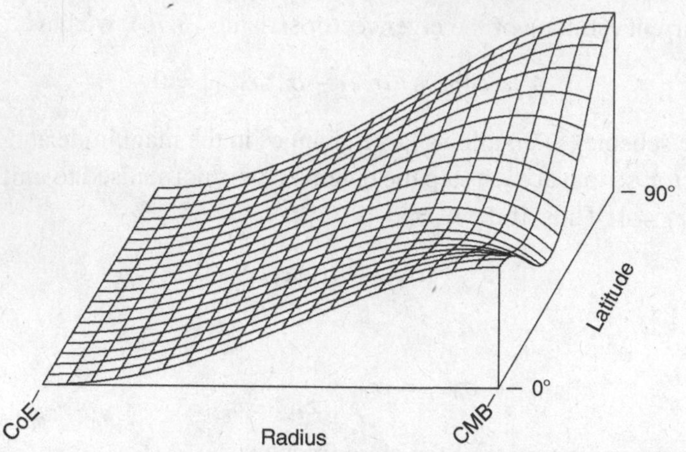

Figure 8.3 The generalised displacement potential for the axisymmetric mode designated (4,1,0) in the notation of Greenspan (1969, p. 64). It is even in the equatorial plane and has a period of 18.254 hours.

The programme EIGENS.FOR finds the eigenvectors and eigenfrequencies, with input matres.dat containing the eleven matrices required for the computation and produced by MAT.FOR. Also input is the file decomp.dat containing the decompression factor. As well as searching for vanishing determinant of the coefficient matrix as a function of frequency and iterating on eigenvectors as outlined, EIGENS.FOR also produces the graphics files THREED and FLOWV as output. THREED is a perspective plot of the generalised displacement potential χ, while FLOWV is a plot of the flow vector field projected on the meridional plane and with the azimuthal component shown in perspective orthogonal to the meridional plane.

As a first example of the computation, we show an axisymmetric solution of the Poincaré inertial wave equation known as the (4,1,0) mode in the notation of Greenspan (1969, p. 64). Solutions of the Poincaré inertial wave equation are included in the code because it has analytical solutions as presented in Section 6.1, and these can be used as a check on the numerical procedures. The programme POINCARE.FOR calculates these analytical solutions and results for a variety of modes are tabulated in Table 6.1. The generalised displacement potential for this mode is shown as a perspective plot in Figure 8.3. The flow vector field for this mode is shown projected on the meridional plane, and the azimuthal component of the flow is shown in perspective orthogonal to the meridional plane in Figure 8.4. Comparison with the analytical solution gives agreement with the eigenfunction to three significant figures, while the eigenfrequency is reproduced to seven significant figures, for the 3×3 finite element grid used in the calculation. Since the eigenfrequency is at a stationary point of the functional, the error in the eigenfrequency is proportional to the square of the error in the eigenfunction.

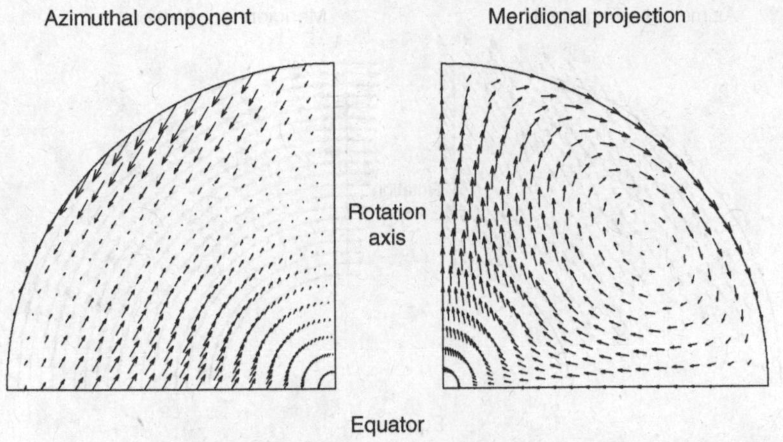

Figure 8.4 Flow vector field for the axisymmetric solution of the Poincaré inertial wave equation known as the (4,1,0) mode in the notation of Greenspan (1969, p. 64). On the right the velocity vector is projected on the meridional plane, and on the left the azimuthal component of velocity is shown in perspective on the plane orthogonal to the meridional plane.

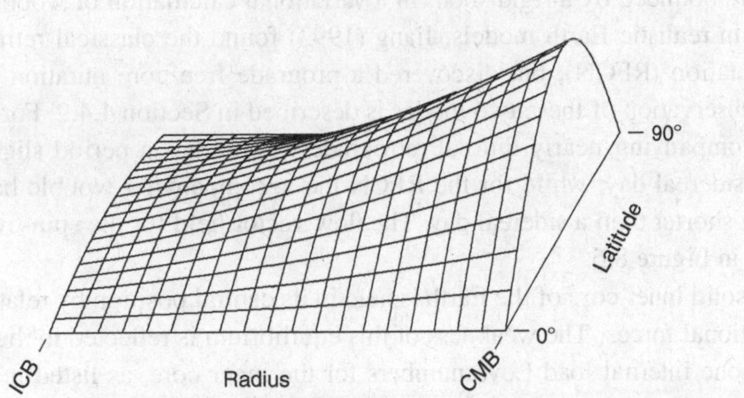

Figure 8.5 Generalised displacement potential for the spin-over mode in a spherical Earth model. It has a period of 23.939 hours.

One of the most interesting core modes is the *spin-over mode*, which reflects the ability of the outer core to rotate about an axis different from that of either the inner core or the shell. It has a nearly diurnal period. The generalised potential for this mode is illustrated in Figure 8.5. In the Earth frame of reference, this mode produces the *nearly diurnal retrograde wobble*. In the space frame of reference it is accompanied by the *free core nutations*. When the flattening of the boundaries of the fluid outer core and the figure–figure gravitational coupling are taken into account, as well as the deformability of the boundaries, both a *retrograde free core nutation* and a *prograde free core nutation* are found. The retrograde free core

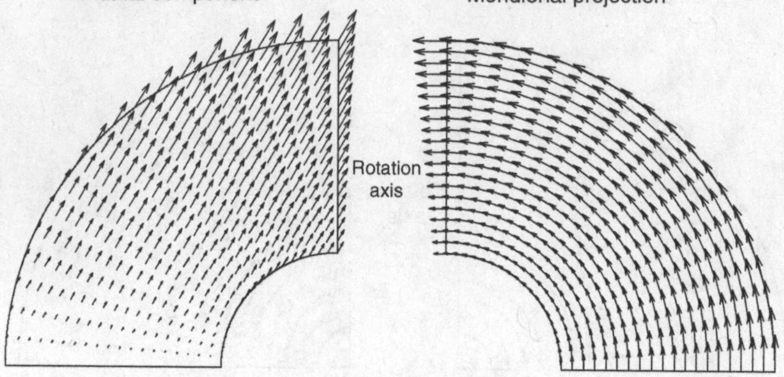

Figure 8.6 Flow vector field for the spin-over mode in a spherical Earth model.

nutation was first predicted by Poincaré (1910) for a completely fluid, incompressible core bounded by a rigid shell. In a variational calculation of wobble–nutation modes in realistic Earth models, Jiang (1993) found the classical retrograde free core nutation (RFCN), but discovered a prograde free core nutation (PFCN) as well. Observation of these two modes is described in Section 4.4.2. For the PFCN, the accompanying nearly diurnal retrograde wobble has a period slightly longer than a sidereal day, while for the RFCN the accompanying wobble has a period slightly shorter than a sidereal day. The flow vector field for the spin-over mode is plotted in Figure 8.6.

The solid inner core of the Earth is held in its central position by relatively weak gravitational forces. The weakness of this equilibrium is reflected in the very large degree-one internal load Love numbers for the inner core, as listed in Tables 7.1 and 7.2. This results in an axial translational mode for the inner core. The generalised displacement potential for this mode is shown in Figure 8.7. It is axially symmetric and odd with respect to the equatorial plane. The motion in this mode is principally in the direction of the rotation axis, so it is not much affected by the Coriolis acceleration. The displacement vector field is plotted in Figure 8.8. The azimuthal component for this mode is quite small, as shown on the left side of the plot. The motion is largely in the meridional plane, as shown on the right side of the plot.

The motion of the inner core can also be parallel to the equatorial plane. In this case, the Coriolis acceleration is a maximum and the mode is split into two motions, one rotating in the equatorial plane in the prograde sense and the other rotating in the equatorial plane in the retrograde sense. The azimuthal number for these two modes is unity and the generalised displacement potential is even in the

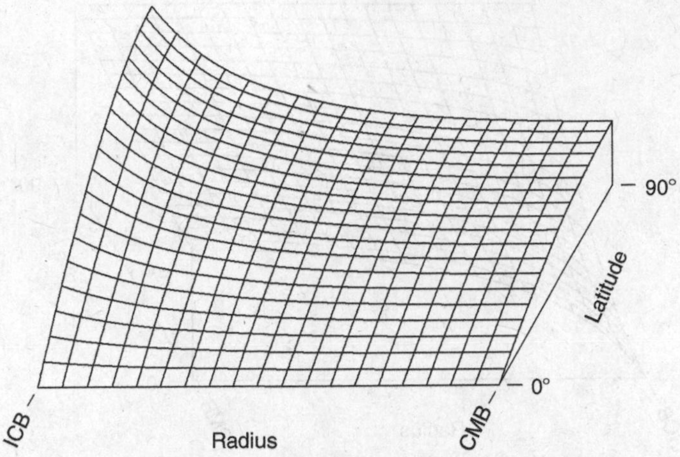

Figure 8.7 Generalised displacement potential for the axial inner core translational mode. For Earth model Cal8 it has a period of 3.7926 hours.

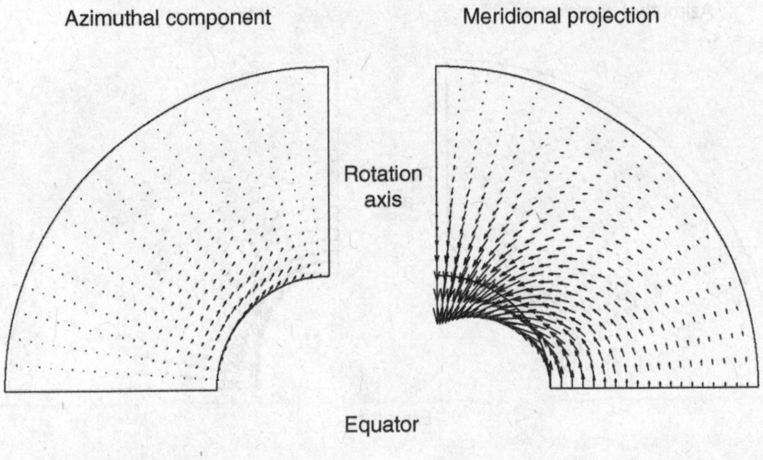

Figure 8.8 Displacement vector field for the axial translational mode of the inner core. On the left the small azimuthal component is shown in projection, while on the right the relatively large axial motion is shown projected on the meridional plane.

equatorial plane. The angular frequency is negative for the prograde mode, which travels westward at the rate $2\Omega\sigma$. The generalised displacement potential for the prograde equatorial mode is shown in Figure 8.9. The displacement vector field for this mode is plotted in Figure 8.10. As well as the dominant motion parallel to the equatorial plane, as shown on the right, there is a large azimuthal component shown on the left in perspective.

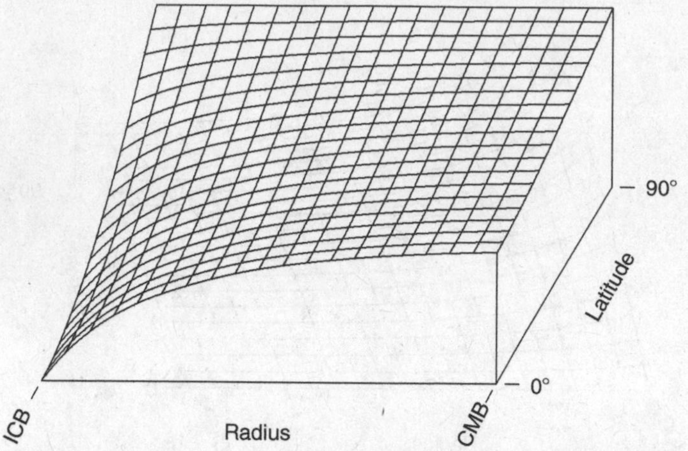

Figure 8.9 The generalised displacement potential for the prograde equatorial translational mode. For Earth model Cal8 it has a period of −4.1118 hours.

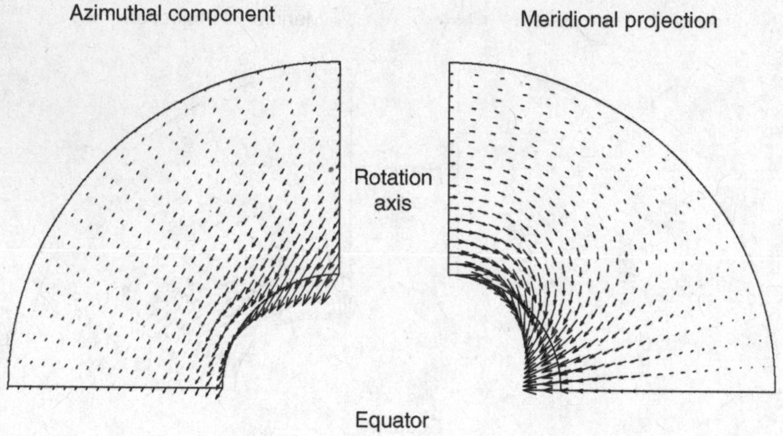

Figure 8.10 Displacement vector field for the prograde equatorial translational mode. The motion parallel to the equatorial plane and projected on the meridional plane is shown on the right, while the azimuthal motion is shown on the left in perspective.

The angular frequency is positive for the retrograde equatorial mode, which travels eastward at the rate $-2\Omega\sigma$. The generalised displacement potential for the retrograde equatorial mode is shown in Figure 8.11. The displacement vector field for this mode is plotted in Figure 8.12. As well as the dominant motion parallel to the equatorial plane as shown on the right, there is a large azimuthal component shown on the left in perspective.

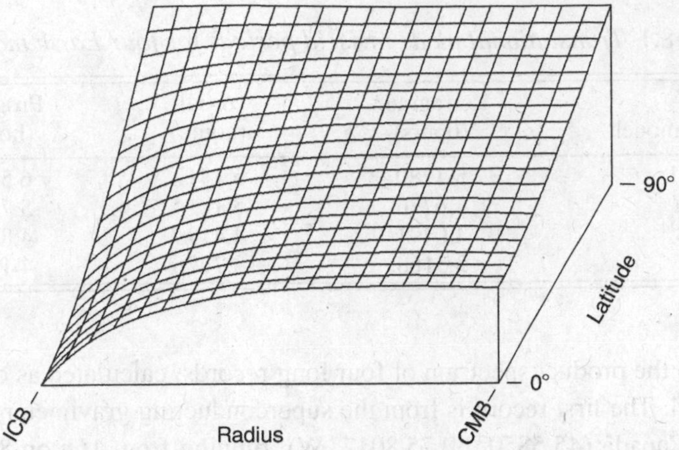

Figure 8.11 The generalised displacement potential for the retrograde equatorial translational mode. For Earth model Cal8 it has a period of 3.5168 hours.

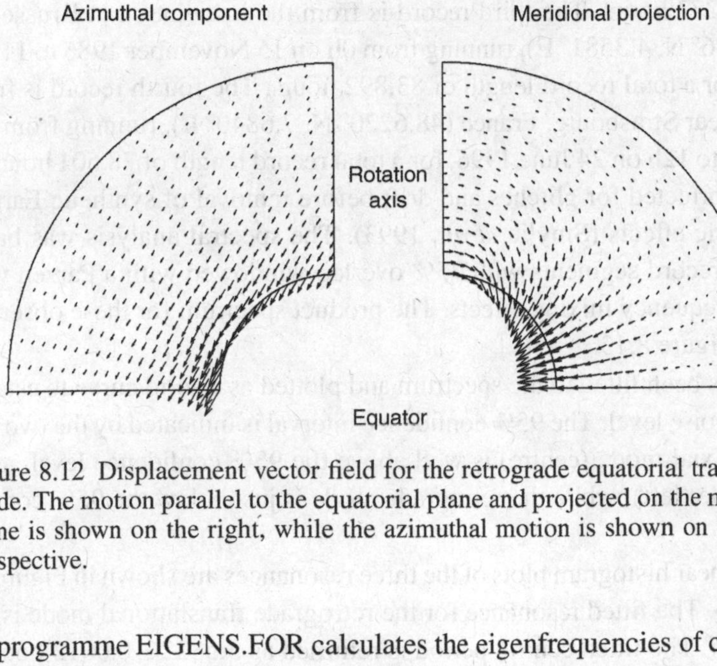

Figure 8.12 Displacement vector field for the retrograde equatorial translational mode. The motion parallel to the equatorial plane and projected on the meridional plane is shown on the right, while the azimuthal motion is shown on the left in perspective.

The programme EIGENS.FOR calculates the eigenfrequencies of core modes in the absence of viscosity. The periods in hours for the translational triplet of modes are listed in Table 8.1 for four Earth models.

8.6 Rotational splitting and viscosity

The translational triplet of inner core modes has been observed in a number of spectra from superconducting gravimeter observations (Courtier *et al.*, 2000).

Table 8.1 *Translational mode inviscid periods for four Earth models.*

Earth model	Retrograde (hours)	Axial (hours)	Prograde (hours)
Core11	5.1280	5.7412	6.5114
PREM	4.6776	5.1814	5.7991
1066A	4.0491	4.4199	4.8603
Cal8	3.5168	3.7926	4.1118

We focus on the product spectrum of four long records, calculated as described in Section 2.5.4. The first record is from the superconducting gravimeter installation at Cantley, Canada (45.5850° N, 75.8017° W), running from 16h on 8 November 1989 to 7h on 14 August 1993, for a total record length of 32,992 hours. The second record is from the installation at Bad Homburg, Germany (50.2285° N, 8.6113° E), running from 0h on 22 March 1986 to 7h on 27 December 1988, for a total record length of 24,272 hours. The third record is from the installation at Brussels, Belgium (50.7986° N, 4.3581° E), running from 0h on 15 November 1986 to 11h on 10 June 1996, for a total record length of 83,892 hours. The fourth record is from the installation near Strasbourg, France (48.6220° N, 7.6840° E), running from 16h on 11 July 1987 to 12h on 24 June 1996, for a total record length of 78,501 hours. Each record was corrected for glitches and drift before removal of synthetic Earth tides and barometric effects (Smylie *et al.*, 1993). The spectral analysis was based on 12,000-hour record segments with 75% overlap windowed with a Parzen window to suppress frequency mixing effects. The product spectrum for these observations is shown in Figure 8.13.

A cubic has been fitted to the spectrum and plotted as a solid curve to establish a background noise level. The 95% confidence interval is indicated by the two dashed curves. The axial mode (centre) is well above the 95% confidence level, as is the prograde mode (left), while the retrograde mode is just below the 95% confidence level.

Detailed linear histogram plots of the three resonances are shown in Figures 8.14, 8.15 and 8.16. The fitted resonance for the retrograde translational mode is shown in Figure 8.14. The fitted central period is returned as 3.5822 ± 0.0012 hours. The fitted resonance for the axial translational mode is shown in Figure 8.15. The fitted central period is returned as 3.7656 ± 0.0015 hours. The fitted resonance for the prograde translational mode is shown in Figure 8.16. It is dominated by the resonance of the S_6 solar heating tide centred on exactly a 4-hour period. To accommodate the solar heating tide, a double resonance has been fitted, giving a central period for the prograde mode of 4.0150 ± 0.0010 hours.

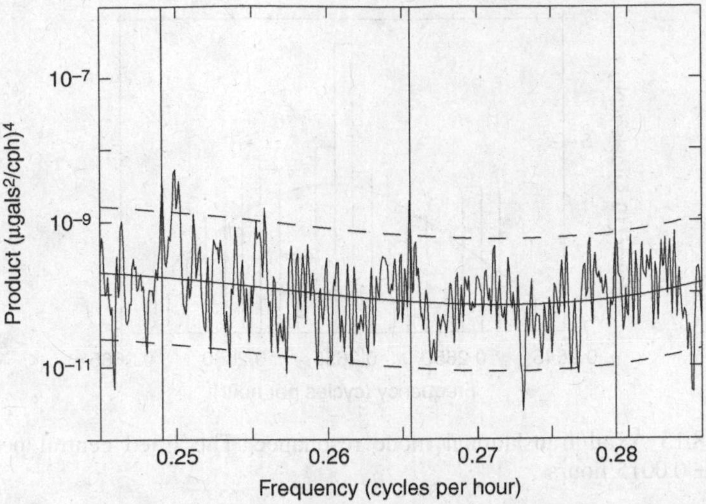

Figure 8.13 Product spectrum of superconducting gravimeter records at Cantley, Bad Homburg, Brussels and Strasbourg. The solid curve is a cubic fitted to give a reference noise level. The two dashed curves indicate the 95% confidence interval. Vertical lines indicate the locations of the central periods of the three translational modes. The period of the prograde mode is recovered as 4.0150 hours (left), the period of the axial mode is recovered as 3.7656 hours (centre) and the period of the retrograde mode is recovered as 3.5822 hours (right).

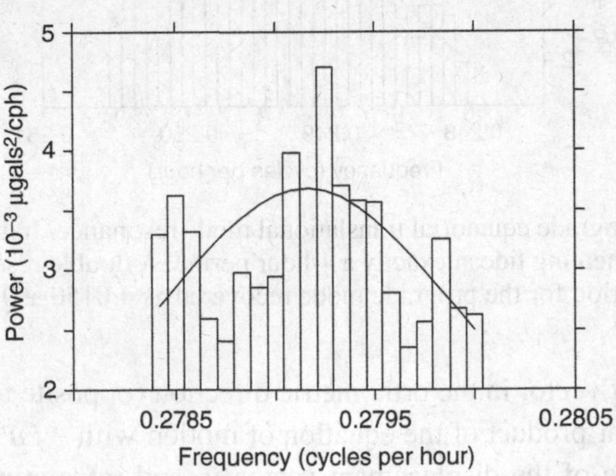

Figure 8.14 Retrograde equatorial translational mode resonance. The recovered central period is 3.5822 ± 0.0012 hours.

The relative locations of the central periods of the three translational mode resonances are governed by a *splitting law*. The equation of motion (7.1), with the Brunt–Väisälä angular frequency (6.167), is transformed to

$$-\omega^2 \boldsymbol{u} + 2i\omega\boldsymbol{\Omega} \times \boldsymbol{u} = -\nabla\chi - \omega_v^2 \hat{\boldsymbol{s}}\,(\hat{\boldsymbol{s}} \cdot \boldsymbol{u})\,, \tag{8.90}$$

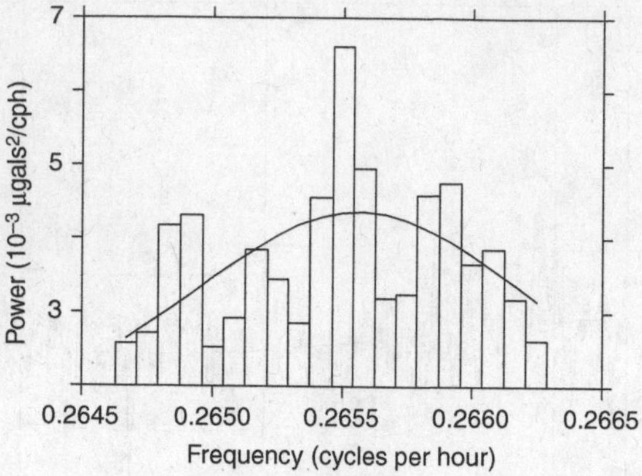

Figure 8.15 Axial translational mode resonance. The fitted central period is 3.7656 ± 0.0015 hours.

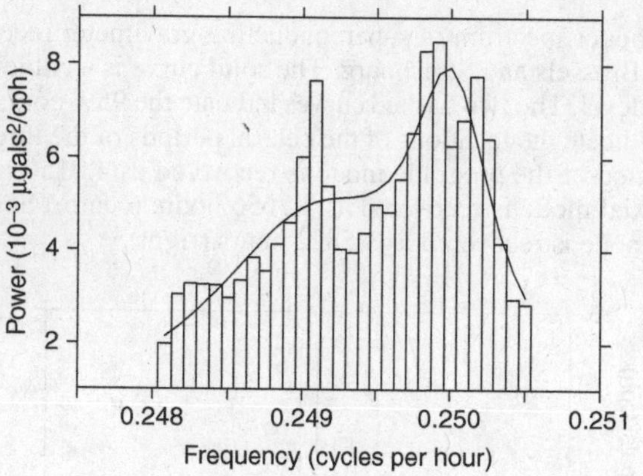

Figure 8.16 Prograde equatorial translational mode resonance. It is just to the left of the S_6 solar heating tide at exactly a 4-hour period. A double resonance is fitted with central period for the prograde mode recovered as 4.0150 ± 0.0010 hours.

where $\hat{\boldsymbol{s}}$ is the unit vector in the orthometric direction (opposite to gravity).

Taking the scalar product of the equation of motion with $-f\boldsymbol{u}^*$, where $\boldsymbol{u}^*$ is the complex conjugate of the displacement vector $\boldsymbol{u}$, and integrating throughout the outer core, we find that

$$\begin{aligned}\omega^2 \int_{\mathcal{V}} f\,|\boldsymbol{u}|^2\, d\mathcal{V} - 2i\omega \int_{\mathcal{V}} f\boldsymbol{u}^* \cdot (\boldsymbol{\Omega} \times \boldsymbol{u})\, d\mathcal{V} \\ = \int_{\mathcal{V}} f\boldsymbol{u}^* \cdot \nabla\chi\, d\mathcal{V} + \int_{\mathcal{V}} f\omega_v^2 \boldsymbol{u}^* \cdot \hat{\boldsymbol{s}}\,(\hat{\boldsymbol{s}} \cdot \boldsymbol{u})\, d\mathcal{V}.\end{aligned} \tag{8.91}$$

Breaking the displacement vector into its real and imaginary parts,

$$\boldsymbol{u} = \boldsymbol{u}_{\mathcal{R}} + i\boldsymbol{u}_{\mathcal{I}}, \tag{8.92}$$

the triple scalar product $\boldsymbol{u}^* \cdot (\boldsymbol{\Omega} \times \boldsymbol{u})$ becomes

$$\begin{aligned}\boldsymbol{u}^* \cdot (\boldsymbol{\Omega} \times \boldsymbol{u}) = \boldsymbol{\Omega} \cdot (\boldsymbol{u} \times \boldsymbol{u}^*) &= \boldsymbol{\Omega} \cdot \left\{(\boldsymbol{u}_{\mathcal{R}} + i\boldsymbol{u}_{\mathcal{I}}) \times (\boldsymbol{u}_{\mathcal{R}} - i\boldsymbol{u}_{\mathcal{I}})\right\} \\ &= -2i\boldsymbol{\Omega} \cdot (\boldsymbol{u}_{\mathcal{R}} \times \boldsymbol{u}_{\mathcal{I}}). \end{aligned} \tag{8.93}$$

The identity

$$f\boldsymbol{u}^* \cdot \nabla\chi = \nabla \cdot (f\boldsymbol{u}^*\chi) - \chi\nabla \cdot (f\boldsymbol{u}^*) \tag{8.94}$$

is transformed to

$$f\boldsymbol{u}^* \cdot \nabla\chi = \nabla \cdot (f\boldsymbol{u}^*\chi) \tag{8.95}$$

using the subseismic condition (6.182). The volume integral

$$\int_{\mathcal{V}} f\boldsymbol{u}^* \cdot \nabla\chi \, d\mathcal{V} \tag{8.96}$$

then becomes, with the use of Gauss's theorem, the surface integral over the boundaries of the outer core,

$$\int_{\mathcal{S}} f\chi\boldsymbol{u}^* \cdot \hat{\boldsymbol{n}} dS. \tag{8.97}$$

With these transformations, equation (8.91) becomes

$$\begin{aligned}\omega^2 \int_{\mathcal{V}} f\,|\boldsymbol{u}|^2 \, d\mathcal{V} - 4\omega \int_{\mathcal{V}} f\boldsymbol{\Omega} \cdot (\boldsymbol{u}_{\mathcal{R}} \times \boldsymbol{u}_{\mathcal{I}}) \, d\mathcal{V} \\ = \int_{\mathcal{S}} f\chi\boldsymbol{u}^* \cdot \hat{\boldsymbol{n}} dS + \int_{\mathcal{V}} f\omega_v^2 \,|\boldsymbol{u} \cdot \hat{\boldsymbol{n}}|^2 \, d\mathcal{V}. \end{aligned} \tag{8.98}$$

The modal intensity is defined by

$$I = \int_{\mathcal{V}} f\,|\boldsymbol{u}|^2 \, d\mathcal{V} > 0, \tag{8.99}$$

while we might define

$$SI = 2 \int_{\mathcal{V}} f\boldsymbol{\Omega} \cdot (\boldsymbol{u}_{\mathcal{R}} \times \boldsymbol{u}_{\mathcal{I}}) \, d\mathcal{V} \tag{8.100}$$

as the product of I with the spin term S. The boundary deformation energy E is represented by the surface integral, expressed by equation (8.11), for which we use the shorthand EI. Finally, the last term on the right side represents the work done in displacement against the non-neutrally stratified density profile and is defined as

$$FI = \int_{\mathcal{V}} f\omega_v^2 \,|\boldsymbol{u} \cdot \hat{\boldsymbol{s}}|^2 \, d\mathcal{V}. \tag{8.101}$$

Dividing through by the intensity gives a simple equation for angular eigenfrequency ω_j,

$$\omega_j^2 - 2\omega_j S = \omega_0^2 = E + F, \tag{8.102}$$

where ω_0 is the angular frequency free of the spin term arising from rotation. In cylindrical co-ordinates, the radial component of displacement is

$$u_R e^{im\phi}, \tag{8.103}$$

while the azimuthal component is

$$u_\phi e^{i(m\phi+\pi/2)}. \tag{8.104}$$

From the identities (8.93),

$$\begin{aligned}\boldsymbol{\Omega}\cdot(\boldsymbol{u}_{\mathcal{R}}\times\boldsymbol{u}_{\mathcal{I}}) &= \frac{1}{-2i}\boldsymbol{u}^*\cdot(\boldsymbol{\Omega}\times\boldsymbol{u})\\ &= \frac{1}{-2i}\left(-u_R e^{-im\phi}\Omega u_\phi e^{i(m\phi+\pi/2)} + u_\phi e^{-i(m\phi+\pi/2)}\Omega u_R e^{im\phi}\right)\\ &= \Omega u_R u_\phi.\end{aligned} \tag{8.105}$$

This allows the spin term to be written simply as $S = g\Omega$, with g a dimensionless geometrical factor expressed as

$$g = 2\int_{\mathcal{V}} f u_r u_\phi \, d\mathcal{V}/I. \tag{8.106}$$

Dividing by ω_j^2, the angular eigenfrequency equation becomes

$$\left(\frac{\omega_0}{\omega_j}\right)^2 + 2g\frac{\Omega}{\omega_j} - 1 = 0. \tag{8.107}$$

In terms of period, with $T_s = 2\pi/\Omega$ the length of the sidereal day and $T_0 = 2\pi/\omega_0$ the non-rotating period, we then have the *quadratic splitting formula*

$$\left(\frac{T_j}{T_0}\right)^2 + 2g\left(\frac{T_j}{T_0}\right)\frac{T_0}{T_s} - 1 = 0 \tag{8.108}$$

for a mode with period T_j. The dimensionless splitting parameter g is found to be

$$g = \frac{T_s}{2T_j}\left(1 - \frac{T_j^2}{T_0^2}\right), \tag{8.109}$$

Table 8.2 *Inviscid splitting parameters for four Earth models.*

Earth model	T_0 (hours)	g_R	g_A	g_P
Core11	5.8197	0.5218	0.0559	−0.4628
PREM	5.2384	0.5184	0.0499	−0.4655
1066A	4.4547	0.5137	0.0421	−0.4688
Cal8	3.8143	0.5101	0.0358	−0.4717

for a given T_0. Solving the quadratic splitting equation (8.108), the three translational mode periods are given by

$$T_R = -g_R \frac{T_0^2}{T_s} + \sqrt{1 + g_R^2 \frac{T_0^2}{T_s^2}}, \tag{8.110}$$

$$T_A = -g_A \frac{T_0^2}{T_s} + \sqrt{1 + g_A^2 \frac{T_0^2}{T_s^2}}, \tag{8.111}$$

$$T_P = -g_P \frac{T_0^2}{T_s} - \sqrt{1 + g_P^2 \frac{T_0^2}{T_s^2}}, \tag{8.112}$$

with T_R representing the period of the retrograde equatorial mode, T_A the axial mode period and T_P the period of the prograde equatorial mode counted negative. For graphing purposes, it is more convenient to count the prograde period as positive, while counting the splitting parameter g_P as negative. Starting with an initial estimate of T_0 as the mean of T_R and T_P, the splitting parameters g_R, g_A and g_P are given by equation (8.109). An improved value of the offset $T_0 - T_A$ is found using equation (8.111). By adding the actual value of T_A to the offset, an improved value of T_0 is obtained. Repetition of this cycle leads to a rapid convergence for the parameters T_0, g_r, g_A and g_P for given T_R, T_A and T_P. Results for the inviscid periods for the four Earth models given in Table 8.1 are shown in Table 8.2. The inviscid splitting curves are shown dashed in Figure 8.17, using the splitting parameters for Earth model Cal8. Inviscid periods for the other Earth models fall close to these curves.

Consideration of pressure and viscous drags on the inner core (Smylie and McMillan, 2000) leads to a quadratic splitting law of the form

$$\left(\frac{T}{T_0}\right)^2 + 2g^{\nu} \frac{T_0}{T_s} \left(\frac{T}{T_0}\right) - 1 = 0, \tag{8.113}$$

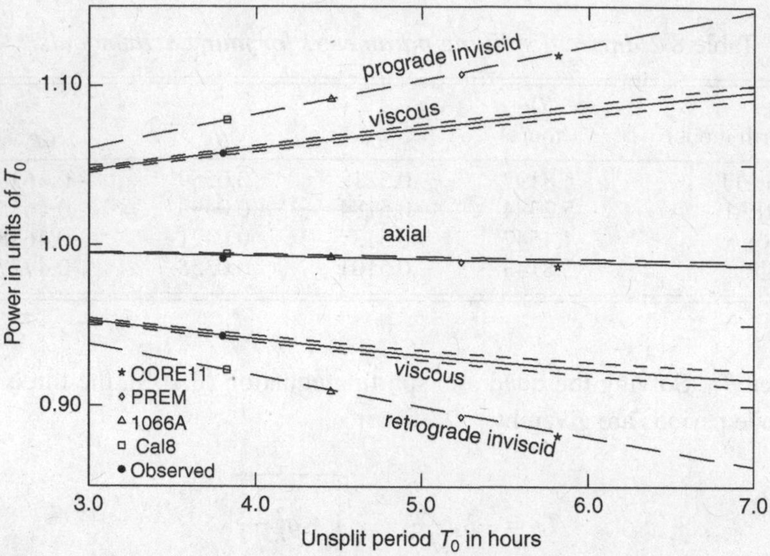

Figure 8.17 Splitting curves for the three translational modes. The inviscid curves for the three modes are shown as dashed, based on the splitting parameters for Earth model Cal8 (open squares). Inviscid periods are overplotted for Earth models Core11 (open stars), PREM (open diamonds) and 1066A (open triangles). Solid viscous splitting curves are for a single viscosity. The observed periods (solid circles) fall on the viscous splitting curves to better than 5 parts in 10^4 for a single viscosity of 1.225×10^{11} Pa s. The dashed curves surrounding the viscous curves are those for a ±20% variation of viscosity.

where T is the period, T_0 is the unsplit period, T_s is the length of the sidereal day and g^ν is a dimensionless viscous splitting parameter. For the axial mode, the viscous splitting parameter is related to the inviscid splitting parameter g^i by

$$g^\nu = g^i \left[1 + \frac{1}{4} \frac{M_I - M_I'}{M_I + \alpha} \sqrt{E_k} f^a(\sigma) \right], \tag{8.114}$$

and for the equatorial modes by

$$g^\nu = g^i \left[1 - \frac{1}{8} \left(\frac{M_I' - \beta}{M_I + \beta} + \frac{M_I' + \alpha}{M_I + \alpha} \right) \sqrt{E_k} f^e(\sigma) \right], \tag{8.115}$$

where E_k is the Ekman number while α and β are coefficients of the pressure drag on the inner core given by

$$\alpha = M_I' \left(\frac{1}{2} + \frac{3}{2} \frac{M_I + (a/b)^3 M_S}{M_O + M_S \left(1 - (a/b)^3\right)} \right), \tag{8.116}$$

and

$$\beta = M_I'\left(\frac{1}{4} - \frac{3}{4}\frac{M_I + (a/b)^3 M_S}{M_O + M_S\left(1 - (a/b)^3\right)}\right). \tag{8.117}$$

M_I is the mass of the inner core, M_O is the mass of the outer core, M_S is the mass of the shell and $M_I' = 4/3\pi a^3 \rho_0(a)$ is the displaced mass, where $\rho_0(a)$ is the density at the bottom of the outer core. $\sigma = \omega/2\Omega$ is the dimensionless Coriolis frequency corresponding to angular frequency ω. Finally, $f^a(\sigma)$ and $f^e(\sigma)$ are dimensionless functions of σ given by

$$\begin{aligned} f^a(\sigma) = \Big\{ & 8\left[(\sigma + 1)^{3/2} + (\sigma - 1)^{3/2}\right] \\ & - \frac{16}{5}\left[(\sigma + 1)^{5/2} - (\sigma - 1)^{5/2}\right]\Big\}, \end{aligned} \tag{8.118}$$

and

$$\begin{aligned} f^e(\sigma) = \Big\{ & \mp 24\,(\pm\sigma \mp 1)^{1/2} - 16\,(\pm\sigma \mp 1)^{3/2} \\ & - \frac{16}{5}\left[(\pm\sigma - 1)^{5/2} - (\pm\sigma + 1)^{5/2}\right]\Big\}, \end{aligned} \tag{8.119}$$

with the upper sign referring to the retrograde mode, for which σ is positive, and the lower sign referring to the prograde mode, for which σ is negative. The viscous splitting curves are shown in Figure 8.17, together with the observed periods. Two independent measures of viscosity are produced, as the reduction in rotational splitting at fixed viscosity is larger for the retrograde mode than for the prograde equatorial mode. The retrograde equatorial mode gives $1.190 \pm 0.035 \times 10^{11}$ Pa s, while the prograde equatorial mode gives $1.304 \pm 0.034 \times 10^{11}$Pa s. A balanced error value of 1.247×10^{11}Pa s yields viscous periods that are only 6.5 s longer than the observed periods. The solid viscous splitting curves in Figure 8.17 are drawn for this value of viscosity. The dashed curves surrounding the solid viscous splitting curves are those for a ±20% variation in viscosity.

A very stringent test of the significance of the identification of the translational triplet in the product spectrum, shown in Figure 8.13, derives from the viscous splitting equation (8.113). For three candidate periods, T_R (retrograde), T_A (axial) and T_P (prograde), the splitting equation (8.113) provides the corresponding values of the dimensionless viscous splitting parameter, g_R^ν, g_A^ν and g_P^ν, for a given value of T_0. Thus, the whole frequency axis can be searched for correctly split resonances. For a resonance centred on frequency f_j, its form at neighbouring frequencies f_i is

$$r_{ij} = \frac{a_j^2}{1 + 4Q\left[\left(f_i - f_j\right)/f_j\right]^2}. \tag{8.120}$$

For record segments of 12,000-hour length, product spectral estimates s_i are spaced at intervals of 1/12,000 cycles/hour along the frequency axis. In the subtidal band, between the 2-hour and 8-hour period, there are 4,501 spectral estimates. For 25 spectral estimates centred on frequency f_j with $Q = 100$, the misfit of (8.120) to spectral estimate s_i is

$$\epsilon_{ij} = A_j r_{ij} - s_i. \tag{8.121}$$

The error energy of the fit is

$$I_j = \sum_{i=j-12}^{j+12} \epsilon_{ij}^2. \tag{8.122}$$

Minimising the error energy of the fit gives

$$A_j = \sum_{i=j-12}^{j+12} r_{ij} s_i \Big/ \sum_{i=j-12}^{j+12} r_{ij}^2, \tag{8.123}$$

with minimum error energy

$$I_{\min} = \sum_{i=j-12}^{j+12} s_i^2 - A_j^2 \sum_{i=j-12}^{j+12} r_{ij}^2. \tag{8.124}$$

As a measure of the strength of a potential resonance of the form (8.120),we use the parameter $S_j^2 = A_j^2/I_{\min}$. When a large well-fitted resonance is found, we expect S_j^2 to be large, and if a small poorly fitted spectral feature is found, we expect S_j^2 to be small. For each of the available 4477 frequencies, f_j, in the subtidal band, we set $T_0 = 1/f_j$ and compute $f_R = 1/T_R$, $f_C = 1/T_A$ and $f_P = -1/T_P$ from equation (8.113). The values of S_R^2, S_A^2 and S_P^2 of the resonance parameter S^2 at the discrete frequencies nearest f_R, f_A and f_P, respectively, are then multiplied together to form the splitting product P_j as an indicator of the presence of correctly split resonances. In Figure 8.18 we show the resulting probability density function (PDF) for the splitting product computed at 4119 points along the frequency axis in the subtidal band. The PDF shown in Figure 8.18 allows evaluation of the significance of translational triplets along the frequency axis. In Figure 8.19 we show the splitting products found between the 2-hour and 10-hour periods. A very large value of P is found at $T_0 = 3.7975$ hours. From the PDF it is found that the probability of a realisation of P_j larger than the largest shown in Figure 8.19 is only 1 in 6.8×10^{38}. This appears to confirm the identification of the translational triplet in the product spectrum.

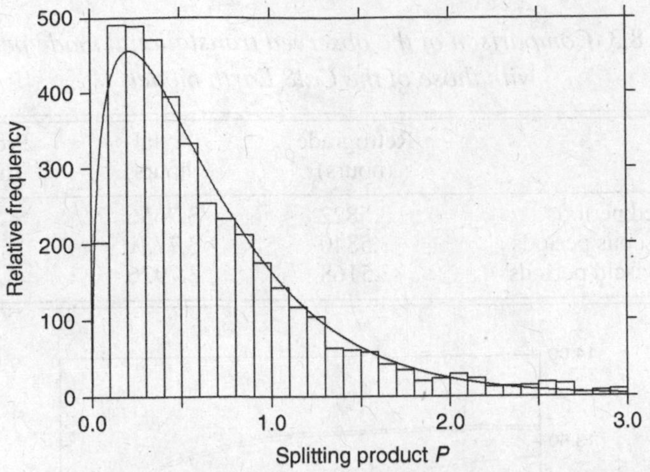

Figure 8.18 Probability density function (PDF) for the splitting product P. Bins in P are 0.1 wide. The fitted PDF is for a χ^2_ν distribution with $\nu = 2.97614$ for the random variable $4.5314P$.

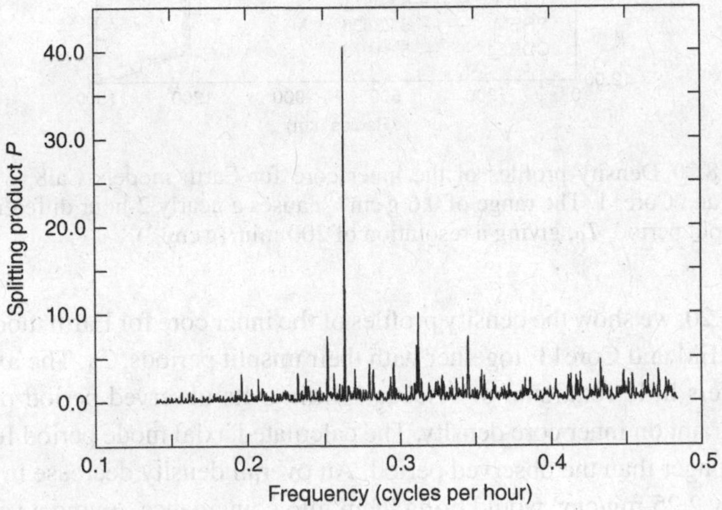

Figure 8.19 Splitting product P as a function of frequency. The large spike at $T_0 = 3.7985$ hours corresponds to the triplet of resonances shown in Figures 8.14, 8.15 and 8.16.

From Figure 8.13, we see that the observed periods are close to those for the Cal8 Earth model. In Table 8.3 a detailed comparison of the Cal8 periods with those observed is shown.

The close match of the observed periods with those of the Cal8 Earth model is due to the sensitivity of the translational mode periods to inner core density.

Table 8.3 *Comparison of the observed translational mode periods with those of the Cal8 Earth model.*

Periods	Retrograde (hours)	Axial (hours)	Prograde (hours)
Observed periods	3.5822	3.7656	4.0150
Cal8 viscous periods	3.5840	3.7731	4.0168
Cal8 inviscid periods	3.5168	3.7926	4.1118

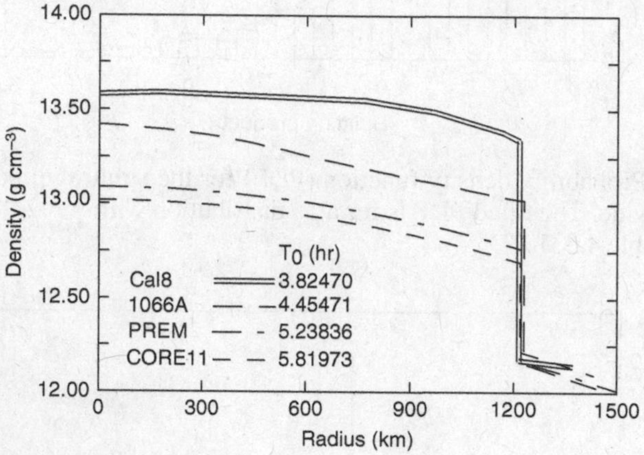

Figure 8.20 Density profiles of the inner core for Earth models Cal8, 1066A, PREM and Core11. The range of 0.6 $\mathrm{g\,cm^{-3}}$ causes a nearly 2-hour difference in the unsplit period, T_0, giving a resolution of 200 $\mathrm{min/(g\,cm^{-3})}$.

In Figure 8.20, we show the density profiles of the inner core for Earth models Cal8, 1066A, PREM and Core11, together with their unsplit periods, T_0. The axial mode period suffers little rotational or viscous splitting. Its observed period provides a strict constraint on inner core density. The calculated axial mode period for Cal8 is only 27 s longer than the observed period. An overall density decrease in the inner core of only 2.25 $\mathrm{mg/cm^3}$ would bring them into coincidence, giving a very strong confirmation of Cal8. In Table 8.4 we show a comparison between the unsplit periods, T_0, for the four Earth models and the observed value.

8.7 A viscosity profile for the outer core

In addition to the viscosity estimates at the bottom of the fluid outer core, from the reduction in the rotational splitting of the two equatorial translational modes, the ring downs of the free core nutations, found from their spectral analysis (Section 4.4.2), lead to viscosity estimates near the top of the fluid outer core. A detailed

Table 8.4 *Comparison between the unsplit periods, T_0, for four Earth models and the observed value.*

Earth model	Unsplit period T_0 (hours)	Deviation ΔT_0 (hours)	Error %
Observed	3.7985		
Càl8	3.82470	0.0262	0.69
1066A	4.45471	0.65621	17.28
PREM	5.23836	1.43986	37.91
Core11	5.81973	2.02123	53.21

Table 8.5 *Profiles of pressure, density, melting temperature and radial pressure gradient.*

Radius (km)	P (10^{11} Pa)	ρ (10^3 kg m^{-3})	T_m (K)	dP/dr (10^4 Pa/m)
1,216	3.300	12.20	4,961	−5.600
1,371	3.223	12.14	4,905	−6.094
1,571	3.094	12.03	4,824	−6.737
1,821	2.916	11.84	4,710	−7.507
2,171	2.636	11.52	4,521	−8.479
2,571	2.278	11.11	4,258	−9.421
2,971	1.886	10.62	3,936	−10.12
3,171	1.681	10.33	3,751	−10.34
3,371	1.473	10.01	3,551	−10.47
3,486	1.350	9.860	3,429	−10.56

analysis of Ekman boundary layer theory (Smylie and Palmer, 2007; Smylie *et al.*, 2009), leads to an expression for the kinematic viscosity, ν_b, at the top of the fluid outer core as

$$\nu_b = \frac{1225 I_c^2 \Omega}{8\pi^2 \rho_0^2(b)\, b^8 \left(9\sqrt{3} + 19\right)^2 Q_W^2}, \tag{8.125}$$

with $I_c = 917.95 \times 10^{34}$ kg m^2 representing the total moment of inertia of the inner and outer cores, $\rho_0(b) = 9.82 \times 10^3$kg m^{-3} the density at the top of the outer core and $b = 3486$ km its radius. From Table 4.5, Q_W, the Q of the associated nearly diurnal wobble of the prograde free core nutation (PFCN) is 11,879, leading to a kinematic viscosity of 0.2685m^2 s^{-1}, while the Q of the associated nearly diurnal wobble of the retrograde free core nutation (RFCN) is 10,152, leading to a kinematic viscosity of 0.3677m^2 s^{-1}. The corresponding values of the dynamic viscosities are 2637 Pa s and 3611 Pa s, respectively, with mean value 3124 Pa s.

The boundary values of viscosity we have found are in very close agreement with an Arrhenius extrapolation of laboratory experiments by Brazhkin (1998) and by Brazhkin and Lyapin (2000), who find 10^{11} Pa s at the bottom of the outer core and 10^2 Pa s at the top. We are prompted, by the very close agreement of the viscosity measures at the boundaries with those provided by the Arrhenius extrapolation, to interpolate between the boundary values in order to obtain a viscosity profile across the entire liquid outer core.

The Arrhenius description of the temperature and pressure dependence of the dynamic viscosity η is (Brazhkin, 1998)

$$\eta \sim \exp\left(\frac{E_{\mathrm{act}_0} + PV_{\mathrm{act}}}{kT}\right), \tag{8.126}$$

with E_{act_0} representing the activation energy at normal pressure, P the pressure, V_{act} the activation volume, k Boltzmann's constant and T the Kelvin temperature. V_{act} is proportional to the atomic volume, which in turn is inversely proportional to the density ρ. While the activation volume for liquid metals at atmospheric pressure is very small, Brazhkin (1998) and Brazhkin and Lyapin (2000) report experimental results on pure iron at the melting temperature, T_m, that show it to be strongly dependent on pressure up to 95 kbar. The strong pressure dependence requires integration of the differential form of the Arrhenius expression. For dominant pressure dependence, from expression (8.126), the differential increment in viscosity is proportional to

$$\frac{D}{\rho T_m} \exp\left(D \frac{P}{\rho T_m}\right) dP, \tag{8.127}$$

with D a pressure-dependent parameter, allowing for the pressure dependence of the activation volume, and dP the differential increment in pressure. The integral of (8.127) over pressure is easily converted to an integral over radius r since $dP/dr = -\rho g$, where g is the gravitational acceleration at radius r. The viscosity at radius r is then

$$\eta(r) = \eta_b + \eta_b \int_b^r \frac{D}{\rho T_m} \exp\left(D \frac{P}{\rho T_m}\right) \frac{dP}{dr} dr, \tag{8.128}$$

with b the radius of the core–mantle boundary and $\eta_b = 3368$ Pa s, the dynamic viscosity at the top of the core. To perform the integration in (8.128), we require profiles of pressure, density, melting temperature and pressure gradient. The pressure profile can be found by integrating the product of gravity and density for an Earth model (we use Cal8). The melting temperatures are found by spline interpolation onto the Cal8 radii from those tabulated by Stacey (1992, p. 459). The required profiles are shown in Table 8.5.

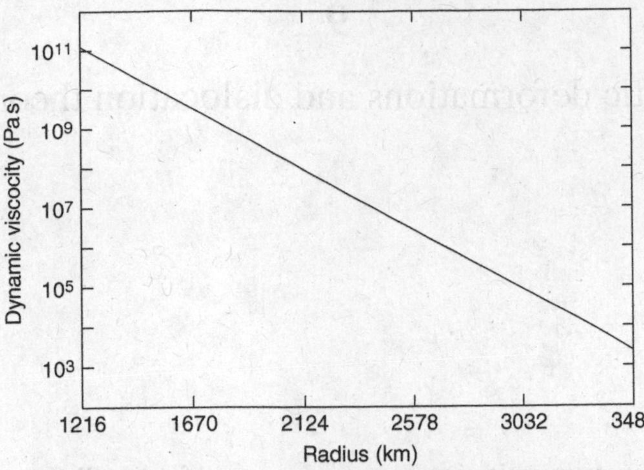

Figure 8.21 Viscosity profile for Earth's outer core.

Since the activation volume increases strongly with pressure (Brazhkin, 1998) as represented by our parameter D in equation (8.127), we allow for a linear variation with depth through

$$D = D_b + \frac{b-r}{b-a} D_a, \tag{8.129}$$

where a is the radius of the inner core while D_b and D_a are constants. The integration in (8.128) is carried out by Simpson's rule over 100 steps with spline interpolation across the whole outer core. It is found that the constant, D_b, controls the curvature of the viscosity profile near the core–mantle boundary, while the curvature otherwise departs only slightly from log-linear. Some numerical experimentation shows that the profile is closely log-linear, even near the core–mantle boundary, for a value $D_b = 4.5 \times 10^{-4}\ \mathrm{m^{-2}\,s^2\,K}$. Moreover, for $D_a = 2.976 \times 10^{-3}\ \mathrm{m^{-2}\,s^2\,K}$, the viscosity at the bottom of the outer core, $\eta_a = 1.247 \times 10^{11}$ Pa s, is closely matched. The resulting viscosity profile is shown in Figure 8.21.

9

Static deformations and dislocation theory

In addition to generating seismic waves, tsunamis and free oscillations of the Earth, earthquakes generate static displacements, strains and tilts both locally and at teleseismic distances. The nearly global extent of these static deformations was brought out clearly by Press (1965) following the great Alaska earthquake of 27 March 1964. The basis for modelling the displacement fields is Volterra's theory of elastic dislocations, published in 1907, dealing with the elasticity theory of surfaces across which displacements are discontinuous. The theory was revived by Steketee (1958a), and applied to geophysical problems by him (Steketee, 1958b) and his student at the time (Rochester, 1956). The surprising extent of the displacement fields demonstrated by Press led to a re-evaluation of their effect on the polar motion (Mansinha and Smylie, 1967). In this chapter, we first present the *elasticity theory of dislocations*, as it is now known, in an infinite uniform elastic half-space, and then in realistic Earth models. We conclude with the calculation of the effects of earthquakes on the polar motion.

9.1 The elasticity theory of dislocations

The starting point for the elasticity theory of dislocations is the reciprocal theorem of Betti (1.266) (Sokolnikoff, 1956, pp. 390–391). This states that for two systems of surface tractions and body forces the work done by the first system, acting through the displacements caused by the second system, is equal to the work done by the second system, acting through the displacements caused by the first system. With t_i and t'_i representing the surface tractions per unit area, F_i and F'_i representing the body forces per unit volume, and u_i and u'_i representing the displacements they cause, the theorem reads

$$\int_S t_i u'_i \, dS + \int_V F_i u'_i \, dV = \int_S t'_i u_i \, dS + \int_V F'_i u_i \, dV. \tag{9.1}$$

Now let the unprimed quantities refer to the problem of a dislocation in an elastic half-space and the primed quantities refer to the deformation caused by a point force F'_i at the point x'_i in the elastic half-space. The surface integral on the left-hand side of the reciprocal theorem vanishes, because t_i is zero on the free surface of the half-space and because it changes sign across the dislocation while u'_i is continuous there. The volume integral on the left-hand side of the reciprocal theorem vanishes because there are no body forces in the dislocation problem. To evaluate the surface integral on the right-hand side of the reciprocal theorem, let ν_i be the unit outward normal vector from the side of the dislocation with displacement u_i^-. The traction on this side of the dislocation is then $t'_i = \tau_{ij}\nu_j$. On the other side of the dislocation, the normal vector is reversed and the displacement is u_i^+. The traction on that side is then $t'_i = -\tau_{ij}\nu_j$. If the dislocation is $\Delta u_i = u_i^+ - u_i^-$, the contribution to the surface integral on the right-hand side of the reciprocal theorem from the dislocation is $-\int_S t'_i \Delta u_i \, dS$. Since t'_i vanishes on the free surface of the half-space there is no contribution to the surface integral from there. If the point force F'_i is of unit strength in the k-direction, generating a displacement field $u_i'^k$, the right-hand side of the reciprocal theorem gives

$$u_k(x'_i) = \int_S t'_i \Delta u_i \, dS, \tag{9.2}$$

where the integral is over the surface of dislocation. By Hooke's law,

$$t'_i = \tau'_{ij}\nu_j = \nu_j \left[\lambda \frac{\partial u_\ell'^k}{\partial x'_\ell} \delta^i_j + \mu \left(\frac{\partial u_i'^k}{\partial x'_j} + \frac{\partial u_j'^k}{\partial x'_i} \right) \right]. \tag{9.3}$$

We are thus led to Volterra's formula,

$$u_k(x'_i) = \int_S \Delta u_i \left[\lambda \frac{\partial u_\ell'^k}{\partial x'_\ell} \delta^i_j + \mu \left(\frac{\partial u_i'^k}{\partial x'_j} + \frac{\partial u_j'^k}{\partial x'_i} \right) \right] \nu_j \, dS. \tag{9.4}$$

In this form of Volterra's formula, the displacement field is found by integrating derivatives of the point force solution over the fault surface, with the point force located at the field point. If we denote the radius to the field point as $\boldsymbol{r}$, and the radius to a point on the dislocation surface by $\boldsymbol{r}'$, then a point force in the k-direction at $\boldsymbol{r}$ generates a displacement field $u_i'^k(\boldsymbol{r}', \boldsymbol{r})$ observed at $\boldsymbol{r}'$. With these conventions, Volterra's formula takes the form

$$u_k(\boldsymbol{r}) = \int_{S'} \Delta u_i(\boldsymbol{r}') \left[\lambda \frac{\partial u_\ell'^k(\boldsymbol{r}', \boldsymbol{r})}{\partial x'_\ell} \delta^i_j + \mu \left(\frac{\partial u_i'^k(\boldsymbol{r}', \boldsymbol{r})}{\partial x'_j} + \frac{\partial u_j'^k(\boldsymbol{r}', \boldsymbol{r})}{\partial x'_i} \right) \right] \nu_j \, dS'. \tag{9.5}$$

Now introduce a second displacement field $u_k''^i(\boldsymbol{r}, \boldsymbol{r}')$, which by our conventions is the kth component of displacement produced by a unit point force in the i-direction at $\boldsymbol{r}'$, observed at $\boldsymbol{r}$. Applying Betti's reciprocal theorem to the two point-force systems of displacement fields, the surface integrals vanish, since the surface of the half-space is force free, and the volume integrals produce the identity

$$u_i'^k(\boldsymbol{r}', \boldsymbol{r}) = u_k'^i(\boldsymbol{r}, \boldsymbol{r}'). \tag{9.6}$$

With this identity, Volterra's formula can be modified to give the displacement at the field point as an integral over the dislocation surface of derivatives of the displacements at the point forces:

$$u_k(\boldsymbol{r}) = \int_{S'} \Delta u_i(\boldsymbol{r}') \left[\lambda \frac{\partial u_k'^\ell(\boldsymbol{r}, \boldsymbol{r}')}{\partial x_\ell'} \delta^i_j + \mu \left(\frac{\partial u_k'^i(\boldsymbol{r}, \boldsymbol{r}')}{\partial x_j'} + \frac{\partial u_k'^j(\boldsymbol{r}, \boldsymbol{r}')}{\partial x_i'} \right) \right] \nu_j \, dS'. \tag{9.7}$$

With this formulation, the point forces are on the surface of dislocation and the partial derivatives may be interpreted as opposing double forces. The term proportional to the Kronecker delta δ^i_j arises when the discontinuity over the surface of dislocation is normal to the surface. The opposing double forces then represent points of dilatation. When the discontinuity over the surface of dislocation is parallel to the surface, paired double forces without moment occur. These double force combinations are often called *nuclei of strain*.

9.1.1 The displacement fields of inclined faults

We begin by adopting a right-hand system of Cartesian co-ordinates (x_1, x_2, x_3), with (x_1, x_2) being the co-ordinates on the surface of the force-free elastic half-space, with x_3 measured vertically downward into the half-space. u_i^j then represents the ith component of displacement observed at (x_1, x_2, x_3) due to a point force of unit magnitude acting in the j-direction at (ξ_1, ξ_2, ξ_3). The dislocation surface is then taken to be the rectangular fault surface shown in Figure 9.1. In this configuration, with a fault inclined at angle θ to the horizontal, and with ξ the down-dip co-ordinate, the fault surface extends over $-L \le \xi_1 \le L$ and $d \le \xi \le D$, with the total fault length given by $2L$.

In applications of dislocation theory to earthquake displacement fields, the discontinuity over the surface of dislocation is parallel to that surface. The faults are then described as *slip faults*. The term proportional to δ^i_j is then missing in Volterra's formula (9.7). The outward unit normal vector to the fault surface is $\hat{\nu} = (0, \sin\theta, -\cos\theta)$. If the slip is down the fault surface it is called *dip-slip*. For a dip-slip of magnitude U, the slip vector is $\Delta \boldsymbol{u} = (0, U\cos\theta, U\sin\theta)$. Carrying out the implied summations over i and j, Volterra's formula for a dip-slip fault gives

$$u_k = \mu U \int_{S'} \left[\left(\frac{\partial u_k'^2}{\partial \xi_2} - \frac{\partial u_k'^3}{\partial \xi_3} \right) \sin 2\theta - \left(\frac{\partial u_k'^2}{\partial \xi_3} + \frac{\partial u_k'^3}{\partial \xi_2} \right) \cos 2\theta \right] dS'. \tag{9.8}$$

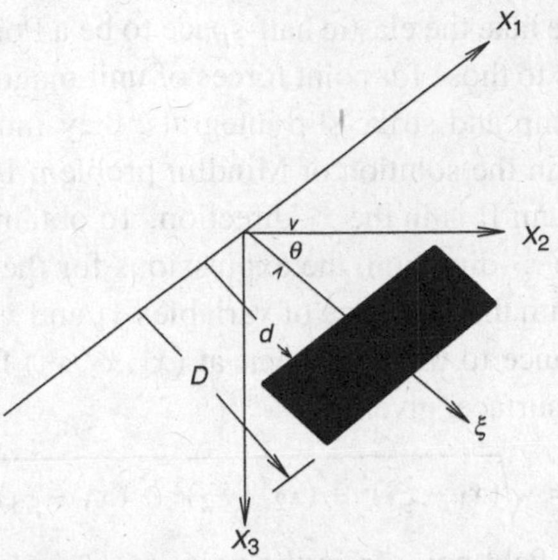

Figure 9.1 Inclined fault geometry and co-ordinates.

If the slip is along the strike of the fault, it is called a *strike-slip*. For a strike-slip U_1 in the direction of co-ordinate x_1, the slip vector is $\Delta \boldsymbol{u} = (U_1, 0, 0)$. Once again, carrying out the implied summations over i and j, Volterra's formula for a strike-slip fault gives

$$u_k = \mu U_1 \int_{S'} \left[\left(\frac{\partial u_k'^1}{\partial \xi_2} + \frac{\partial u_k'^2}{\partial \xi_1} \right) \sin\theta - \left(\frac{\partial u_k'^1}{\partial \xi_3} + \frac{\partial u_k'^3}{\partial \xi_1} \right) \cos\theta \right] dS'. \tag{9.9}$$

By the chain rule for partial derivatives, with $u(\xi_2, \xi_3)$ a function of the co-ordinates ξ_2 and ξ_3,

$$\frac{\partial u}{\partial \xi} = \cos\theta \frac{\partial u}{\partial \xi_2} + \sin\theta \frac{\partial u}{\partial \xi_3}. \tag{9.10}$$

Using this relation, the expression for the displacement field of the dip-slip fault may be transformed to

$$u_k = \mu U \int_d^D \int_{-L}^{L} \left[2\left(\sin\theta \frac{\partial u_k'^2}{\partial \xi} - \cos\theta \frac{\partial u_k'^3}{\partial \xi} \right) + \left(\frac{\partial u_k'^3}{\partial \xi_2} - \frac{\partial u_k'^2}{\partial \xi_3} \right) \right] d\xi_1 d\xi, \tag{9.11}$$

while the expression for the displacement field of the strike-slip fault remains as

$$u_k = \mu U_1 \int_d^D \int_{-L}^{L} \left[\left(\frac{\partial u_k'^1}{\partial \xi_2} + \frac{\partial u_k'^2}{\partial \xi_1} \right) \sin\theta - \left(\frac{\partial u_k'^1}{\partial \xi_3} + \frac{\partial u_k'^3}{\partial \xi_1} \right) \cos\theta \right] d\xi_1 d\xi. \tag{9.12}$$

The point force displacement fields required for the evaluation of the dip-slip and strike-slip integrals are provided by the solutions of Mindlin problems I (1.527) and II (1.529). The Mindlin problems are solutions for point forces of magnitude

$8\pi\mu(\lambda+2\mu)/(\lambda+\mu)$. We take the elastic half-space to be a Poisson solid with $\lambda = \mu$. To reduce the solutions to those for point forces of unit magnitude, required for the evaluation of the dip-slip and strike-slip integrals, they must then be divided by $12\pi\mu$. The point force in the solution of Mindlin problem I is in the x_3-direction. That for Mindlin problem II is in the x_1-direction. To obtain the required solution for a point force in the x_2-direction, the expressions for the first two components are exchanged along with the exchange of variables x_1 and x_2, as well as ξ_1 and ξ_2. In addition to the distance to the field point at (x_1, x_2, x_3) from a source point at (ξ_1, ξ_2, ξ_3) on the fault surface, given by

$$R = \sqrt{(x_1 - \xi_1)^2 + (x_2 - \xi_2)^2 + (x_3 - \xi_3)^2}, \tag{9.13}$$

and the distance to the field point from the point $(\xi_1, \xi_2, -\xi_3)$ on the image of the fault surface in the upper half-space, given by

$$Q = \sqrt{(x_1 - \xi_1)^2 + (x_2 - \xi_2)^2 + (x_3 + \xi_3)^2}, \tag{9.14}$$

it is convenient to introduce the abbreviations

$$r_2 = x_2 \sin\theta - x_3 \cos\theta, \qquad q_2 = x_2 \sin\theta + x_3 \cos\theta, \tag{9.15}$$

$$r_3 = x_2 \cos\theta + x_3 \sin\theta, \qquad q_3 = -x_2 \cos\theta + x_3 \sin\theta. \tag{9.16}$$

r_2 and r_3 are the distances to the field point, measured normal and down-dip to the fault, while q_2 and q_3 are the distances to the field point, measured normal and up-dip to the image fault. With these abbreviations, we have

$$R^2 = (x_1 - \xi_1)^2 + r_2^2 + (r_3 - \xi)^2, \tag{9.17}$$

$$Q^2 = (x_1 - \xi_1)^2 + q_2^2 + (q_3 + \xi)^2 \tag{9.18}$$

$$= (x_1 - \xi_1)^2 + h^2 = k^2 + (q_3 + \xi)^2, \tag{9.19}$$

where h is the projection of Q on the plane $x_1 = 0$ and k is its projection on the plane $q_3 = 0$.

Expressed in indefinite integral form, the Volterra integrals for the dip-slip dislocation (9.11) give

$$\begin{aligned}
12\pi\frac{u_1}{U} = {} & (x_2 - \xi_2)\sin\theta\left[\frac{2}{R} + \frac{4}{Q} - 4\frac{\xi_3 x_3}{Q^3} - \frac{3}{Q + x_3 + \xi_3}\right] \\
& - \cos\theta\left[3\ln(Q + x_3 + \xi_3) + 2\frac{x_3 - \xi_3}{R} + 4\frac{x_3 - \xi_3}{Q} + 4\frac{\xi_3 x_3\,(x_3 + \xi_3)}{Q^3}\right] \\
& + \frac{3}{\cos\theta}\left[\ln(Q + x_3 + \xi_3) - \sin\theta\ln(Q + q_3 + \xi)\right] \\
& + 6x_3\left[\frac{\cos\theta}{Q} - \frac{q_2 \sin\theta}{Q\,(Q + q_3 + \xi)}\right],
\end{aligned} \tag{9.20}$$

$$
\begin{aligned}
12\pi\frac{u_2}{U} = {} & \sin\theta\Bigg[-\ln(R+x_1-\xi_1)+\ln(Q+x_1-\xi_1)+4\frac{\xi_3 x_3}{Q(Q+x_1-\xi_1)} \\
& +3\frac{x_1-\xi_1}{Q+x_3+\xi_3}+(x_2-\xi_2)^2\left\{\frac{2}{R(R+x_1-\xi_1)}+\frac{4}{Q(Q+x_1-\xi_1)}\right. \\
& \left.-\,4\xi_3 x_3\frac{2Q+x_1-\xi_1}{Q^3(Q+x_1-\xi_1)^2}\right\}\Bigg] \\
& -\cos\theta\Bigg[(x_2-\xi_2)\left\{2\frac{x_3-\xi_3}{R(R+x_1-\xi_1)}+4\frac{x_3-\xi_3}{Q(Q+x_1-\xi_1)}\right. \\
& \left.+\,4\xi_3 x_3(x_3+\xi_3)\frac{2Q+x_1-\xi_1}{Q^3(Q+x_1-\xi_1)^2}\right\}+6\tan^{-1}\left\{\frac{(x_1-\xi_1)(x_2-\xi_2)}{(h+x_3+\xi_3)(Q+h)}\right\} \\
& -\,3\tan^{-1}\left\{\frac{(x_1-\xi_1)(r_3-\xi)}{r_2 R}\right\}+6\tan^{-1}\left\{\frac{(x_1-\xi_1)(q_3+\xi)}{q_2 Q}\right\}\Bigg] \\
& +6\Bigg[\frac{1}{\cos\theta}\tan^{-1}\left\{\frac{(k-q_2\cos\theta)(Q-k)+(q_3+\xi)k\sin\theta}{(x_1-\xi_1)(q_3+\xi)\cos\theta}\right\} \\
& +\,x_3\left\{\frac{(\sin^2\theta-\cos^2\theta)(q_3+\xi)+2q_2\cos\theta\sin\theta}{Q(Q+x_1-\xi_1)}+\frac{(x_1-\xi_1)\sin^2\theta}{Q(Q+q_3+\xi)}\right\}\Bigg],
\end{aligned}
\tag{9.21}
$$

$$
\begin{aligned}
12\pi\frac{u_3}{U} = {} & \sin\theta\Bigg[(x_2-\xi_2)\left\{2\frac{x_3-\xi_3}{R(R+x_1-\xi_1)}+4\frac{x_3-\xi_3}{Q(Q+x_1-\xi_1)}\right. \\
& \left.-\,4\xi_3 x_3(x_3+\xi_3)\frac{2Q+x_1-\xi_1}{Q^3(Q+x_1-\xi_1)^2}\right\}-6\tan^{-1}\left\{\frac{(x_1-\xi_1)(x_2-\xi_2)}{(h+x_3+\xi_3)(Q+h)}\right\} \\
& +\,3\tan^{-1}\left\{\frac{(x_1-\xi_1)(r_3-\xi)}{r_2 R}\right\}-6\tan^{-1}\left\{\frac{(x_1-\xi_1)(q_3+\xi)}{q_2 Q}\right\}\Bigg] \\
& +\cos\theta\Bigg[\ln(R+x_1-\xi_1)-\ln(Q+x_1-\xi_1)-2\frac{(x_3-\xi_3)^2}{R(R+x_1-\xi_1)} \\
& -\,4\frac{(x_3+\xi_3)^2-\xi_3 x_3}{Q(Q+x_1-\xi_1)}-4\xi_3 x_3(x_3+\xi_3)^2\frac{2Q+x_1-\xi_1}{Q^3(Q+x_1-\xi_1)^2}\Bigg] \\
& +6x_3\Bigg[\cos\theta\sin\theta\left\{\frac{2(q_3+\xi)}{Q(Q+x_1-\xi_1)}+\frac{x_1-\xi_1}{Q(Q+q_3+\xi)}\right\} \\
& -\,q_2\frac{\sin^2\theta-\cos^2\theta}{Q(Q+x_1-\xi_1)}\Bigg].
\end{aligned}
\tag{9.22}
$$

Again expressed in indefinite integral form, the Volterra integrals for the strike-slip dislocation (9.12) give

$$
\begin{aligned}
12\pi\frac{u_1}{U_1} &= (x_1-\xi_1)\left[\frac{2r_2}{R(R+r_3-\xi)} - \frac{4q_2-2x_3\cos\theta}{Q(Q+q_3+\xi)} - \frac{3\tan\theta}{Q+x_3+\xi_3}\right.\\
&\quad\left. + \frac{4q_2x_3\sin\theta}{Q^3} - 4q_2q_3x_3\sin\theta\frac{2Q+q_3+\xi}{Q^3(Q+q_3+\xi)^2}\right]\\
&\quad - 6\tan^2\theta\tan^{-1}\left[\frac{(k-q_2\cos\theta)(Q-k)+(q_3+\xi)k\sin\theta}{(x_1-\xi_1)(q_3+\xi)\cos\theta}\right]\\
&\quad + 3\tan^{-1}\left[\frac{(x_1-\xi_1)(r_3-\xi)}{r_2R}\right] - 3\tan^{-1}\left[\frac{(x_1-\xi_1)(q_3+\xi)}{q_2Q}\right],
\end{aligned}
\tag{9.23}
$$

$$
\begin{aligned}
12\pi\frac{u_2}{U_1} &= \sin\theta\left[3\tan\theta\sec\theta\ln(Q+x_3+\xi_3) - \ln(R+r_3-\xi)\right.\\
&\quad\left. - (1+3\tan^2\theta)\ln(Q+q_3+\xi)\right] + \frac{2r_2^2\sin\theta}{R(R+r_3-\xi)} + \frac{2r_2\cos\theta}{R}\\
&\quad - 2\sin\theta\frac{2x_3(q_2\cos\theta-q_3\sin\theta)+q_2(q_2+x_2\sin\theta)}{Q(Q+q_3+\xi)}\\
&\quad - 3\tan\theta\frac{x_2-\xi_2}{Q+x_3+\xi_3} + 2\frac{q_2\cos\theta-q_2\sin\theta-x_3\sin^2\theta}{Q}\\
&\quad + 4q_2x_3\sin\theta\frac{x_2-\xi_2+q_3\cos\theta}{Q^3} - 4q_2^2q_3x_3\sin^2\theta\frac{2Q+q_3+\xi}{Q^3(Q+q_3+\xi)^2},
\end{aligned}
\tag{9.24}
$$

$$
\begin{aligned}
12\pi\frac{u_3}{U_1} &= \cos\theta\left[\ln(R+r_3-\xi) + (1+3\tan^2\theta)\ln(Q+q_3+\xi)\right.\\
&\quad\left. - 3\tan\theta\sec\theta\ln(Q+x_3+\xi_3)\right] + \frac{2r_2\sin\theta}{R} + 2\sin\theta\frac{q_2+x_2\sin\theta}{Q}\\
&\quad - \frac{2r_2^2\cos\theta}{R(R+r_3-\xi)} + \frac{4q_2x_3\sin^2\theta-2(q_2+x_2\sin\theta)(x_3+q_3\sin\theta)}{Q(Q+q_3+\xi)}\\
&\quad + 4q_2x_3\sin\theta\frac{x_3+\xi_3-q_3\sin\theta}{Q^3}\\
&\quad - 4q_2^2q_3x_3\cos\theta\sin\theta\frac{2Q+q_3+\xi}{Q^3(Q+q_3+\xi)^2}.
\end{aligned}
\tag{9.25}
$$

An actual slip dislocation may combine both dip-slip and strike-slip components, as shown in Figure 9.2. The relative magnitudes of the two components are given by the *rake*, or pitch angle, measured positive from the down-dip direction to the total slip vector.

The programme DISPF.FOR calculates the displacement fields of inclined faults for both dip-slip and strike-slip faults, or combinations thereof, as indicated by

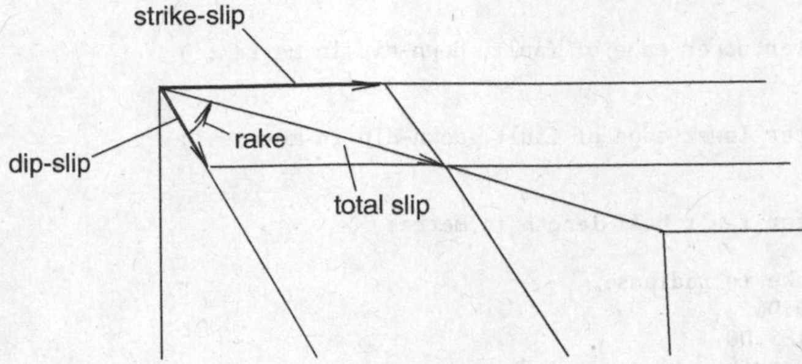

Figure 9.2 The total slip on a slip dislocation may have both dip-slip and strike-slip components, as measured by the rake angle.

the rake angle. At the request of the programme, fault parameters DIP (dip angle), SLIP (slip), RAKE (rake angle), DU (upper edge of fault plane measured down-dip), DL (lower edge of fault plane measured down-dip) and L (fault half-length), are entered. Subroutine DIPS calculates the three components of displacement resulting from the dip-slip, while subroutine STRIKES calculates the three components of displacement resulting from the strike-slip. Subroutine DIPS in turn calls the subroutines DIP1, DIP2 and DIP3 to evaluate the double indefinite integrals expressing the three displacement components in terms of fault limits. Similarly, subroutine STRIKES in turn calls the subroutines STRIKE1, STRIKE2 and STRIKE3 to evaluate the double indefinite integrals expressing the three displacement components in terms of fault limits.

```
C                    PROGRAMME DISPF.FOR
C
C This programme calculates the three components of the displacement
C fields of dip-slip and strike-slip faults and combinations thereof.
C Surface values of the three components of displacement are placed
C in the output file plot.dat, as functions of surface co-ordinates
C along the strike and orthogonal to the strike, in preparation for
C contour plotting.
C
      IMPLICIT DOUBLE PRECISION(A-H,O-Z)
      DIMENSION U(3),UD(3),US(3),X(3)
      DOUBLE PRECISION L
      OPEN(UNIT=1,FILE='plot.dat',STATUS='UNKNOWN')
C Define pi.
      PI=3.141592653589793D0
C Enter fault parameters.
      WRITE(6,10)
 10   FORMAT(1X,'Enter dip angle in degrees.')
      READ(5,*)DIP
      WRITE(6,11)
 11   FORMAT(1X,'Enter slip in metres.')
      READ(5,*)SLIP
      WRITE(6,12)
 12   FORMAT(1X,'Enter rake angle in degrees.')
      READ(5,*)RAKE
```

```
      WRITE(6,13)
 13   FORMAT(1X,'Enter upper edge of fault, down-dip in metres.')
      READ(5,*)DU
      WRITE(6,14)
 14   FORMAT(1X,'Enter lower edge of fault, down-dip in metres.')
      READ(5,*)DL
      WRITE(6,15)
 15   FORMAT(1X,'Enter fault half-length in metres.')
      READ(5,*)L
C Convert dip and rake to radians.
      DIP=DIP*PI/180.D0
      RAKE=RAKE*PI/180.D0
C Find dip-slip and strike-slip components.
      SLIPD=SLIP*DCOS(RAKE)
      SLIPS=SLIP*DSIN(RAKE)
C Generate surface plot file, plot.dat, for ten fault lengths
C by ten fault lengths.
      X(3)=0.D0
      DO 16 I=0,50
        AI=DFLOAT(I)
        X(1)=0.4D0*AI*L
        DO 17 J=0,50
          AJ=DFLOAT(J)
          X(2)=0.4D0*AJ*L
          CALL DIPS(UD,X,DIP,SLIPD,DU,DL,L)
          CALL STRIKES(US,X,DIP,SLIPS,DU,DL,L)
          DO 18 K=1,3
            U(K)=UD(K)+US(K)
 18       CONTINUE
          WRITE(1,19)X(1),X(2),U(1),U(2),U(3)
 19       FORMAT(1X,5D14.6)
 17     CONTINUE
 16   CONTINUE
      END
C
      SUBROUTINE DIPS(U,X,DIP,SLIP,DU,DL,L)
C
C This subroutine calculates the displacement vector U(3),
C at the field point X(3), for a dip-slip fault, dipping at angle DIP,
C with slip magnitude SLIP, upper boundary measured down-dip at DU,
C lower boundary down-dip at DL, fault extending along the strike
C from -L to L.
C
      IMPLICIT DOUBLE PRECISION(A-H,O-Z)
      DIMENSION U(3),X(3)
      DOUBLE PRECISION L
C Set value of pi.
      PI=3.141592653589793D0
C Find parameters of field point distances normal and parallel to
C the fault and its image.
      R2=X(2)*DSIN(DIP)-X(3)*DCOS(DIP)
      R3=X(2)*DCOS(DIP)+X(3)*DSIN(DIP)
      Q2=X(2)*DSIN(DIP)+X(3)*DCOS(DIP)
      Q3=-X(2)*DCOS(DIP)+X(3)*DSIN(DIP)
C Find U(1).
      CALL DIP1(U1,L,DL,X,DIP,R2,R3,Q2,Q3)
      U(1)=U1
      CALL DIP1(U1,-L,DU,X,DIP,R2,R3,Q2,Q3)
      U(1)=U(1)+U1
      CALL DIP1(U1,-L,DL,X,DIP,R2,R3,Q2,Q3)
      U(1)=U(1)-U1
```

```
      CALL DIP1(U1,L,DU,X,DIP,R2,R3,Q2,Q3)
      U(1)=U(1)-U1
C Find U(2).
      CALL DIP2(U2,L,DL,X,DIP,R2,R3,Q2,Q3)
      U(2)=U2
      CALL DIP2(U2,-L,DU,X,DIP,R2,R3,Q2,Q3)
      U(2)=U(2)+U2
      CALL DIP2(U2,-L,DL,X,DIP,R2,R3,Q2,Q3)
      U(2)=U(2)-U2
      CALL DIP2(U2,L,DU,X,DIP,R2,R3,Q2,Q3)
      U(2)=U(2)-U2
C Find U(3).
      CALL DIP3(U3,L,DL,X,DIP,R2,R3,Q2,Q3)
      U(3)=U3
      CALL DIP3(U3,-L,DU,X,DIP,R2,R3,Q2,Q3)
      U(3)=U(3)+U3
      CALL DIP3(U3,-L,DL,X,DIP,R2,R3,Q2,Q3)
      U(3)=U(3)-U3
      CALL DIP3(U3,L,DU,X,DIP,R2,R3,Q2,Q3)
      U(3)=U(3)-U3
C Scale displacements.
      U(1)=SLIP*U(1)/(12.D0*PI)
      U(2)=SLIP*U(2)/(12.D0*PI)
      U(3)=SLIP*U(3)/(12.D0*PI)
      RETURN
      END
C
      SUBROUTINE DIP1(U1,XI1,XI,X,DIP,R2,R3,Q2,Q3)
C
C This subroutine evaluates the one-component, U1,
C of the double integral for specified limits XI1 and XI.
C
      IMPLICIT DOUBLE PRECISION (A-H,O-Z)
      DIMENSION X(3)
C Find distance R from fault to field point and
C distance Q from image fault to field point.
      R=(X(1)-XI1)*(X(1)-XI1)+R2*R2+(R3-XI)*(R3-XI)
      R=DSQRT(R)
      Q=(X(1)-XI1)*(X(1)-XI1)+Q2*Q2+(Q3+XI)*(Q3+XI)
      Q=DSQRT(Q)
C Find fault plane co-ordinates XI2 and XI3.
      XI2=XI*DCOS(DIP)
      XI3=XI*DSIN(DIP)
C Evaluate U1.
      U1=(X(2)-XI2)*DSIN(DIP)*(2.D0/R+4.D0/Q-4.D0*XI3*X(3)/Q**3
     1 -3.D0/(Q+X(3)+XI3))-DCOS(DIP)*(3.D0*DLOG(Q+X(3)+XI3)
     2 +2.D0*(X(3)-XI3)/R+4.D0*(X(3)-XI3)/Q
     3 +4.D0*XI3*X(3)*(X(3)+XI3)/Q**3)
     4 +3.D0*(DLOG(Q+X(3)+XI3)-DSIN(DIP)*DLOG(Q+Q3+XI))/DCOS(DIP)
     5 +6.D0*X(3)*(DCOS(DIP)/Q-Q2*DSIN(DIP)/(Q*(Q+Q3+XI)))
      RETURN
      END
C
      SUBROUTINE DIP2(U2,XI1,XI,X,DIP,R2,R3,Q2,Q3)
C
C This subroutine evaluates the two-component, U2,
C of the double integral for specified limits XI1 and XI.
C
      IMPLICIT DOUBLE PRECISION(A-H,O-Z)
      DIMENSION X(3)
      DOUBLE PRECISION K
```

```
C Find distance R from fault to field point and
C distance Q from image fault to field point.
      R=(X(1)-XI1)*(X(1)-XI1)+R2*R2+(R3-XI)*(R3-XI)
      R=DSQRT(R)
      Q=(X(1)-XI1)*(X(1)-XI1)+Q2*Q2+(Q3+XI)*(Q3+XI)
      Q=DSQRT(Q)
C Find projections of Q on plane X(1)=0 and on plane Q3=0.
      H=Q2*Q2+(Q3+XI)*(Q3+XI)
      H=DSQRT(H)
      K=(X(1)-XI1)*(X(1)-XI1)+Q2*Q2
      K=DSQRT(K)
C Find fault plane co-ordinates XI2 and XI3.
      XI2=XI*DCOS(DIP)
      XI3=XI*DSIN(DIP)
C Evaluate U2.
      U2=DSIN(DIP)*(-DLOG(R+X(1)-XI1)+DLOG(Q+X(1)-XI1)
     1 +4.D0*XI3*X(3)/(Q*(Q+X(1)-XI1))+3.D0*(X(1)-XI1)/(Q+X(3)+XI3)
     2 +(X(2)-XI2)**2*(2.D0/(R*(R+X(1)-XI1))+4.D0/(Q*(Q+X(1)-XI1))
     3 -4.D0*XI3*X(3)*(2.D0*Q+X(1)-XI1)/(Q**3*(Q+X(1)-XI1)**2)))
     4 -DCOS(DIP)*((X(2)-XI2)*(2.D0*(X(3)-XI3)/(R*(R+X(1)-XI1))
     5 +4.D0*(X(3)-XI3)/(Q*(Q+X(1)-XI1))+4.D0*XI3*X(3)*
     6 (X(3)+XI3)*(2.D0*Q+X(1)-XI1)/(Q**3*(Q+X(1)-XI1)**2))
     7 +6.D0*DATAN((X(1)-XI1)*(X(2)-XI2)/((H+X(3)+XI3)*(Q+H)))
     8 -3.D0*DATAN((X(1)-XI1)*(R3-XI)/(R2*R))
     9 +6.D0*DATAN((X(1)-XI1)*(Q3+XI)/(Q2*Q)))
      U2=U2+6.D0*DATAN(((K-Q2*DCOS(DIP))*(Q-K)+(Q3+XI)*K*DSIN(DIP))/
     1 ((X(1)-XI1)*(Q3+XI)*DCOS(DIP)))/DCOS(DIP)
     2 +6.D0*X(3)*(((DSIN(DIP)**2-DCOS(DIP)**2)*(Q3+XI)
     3 +2.D0*Q2*DCOS(DIP)*DSIN(DIP))/(Q*(Q+X(1)-XI1))
     4 +(X(1)-XI1)*DSIN(DIP)**2/(Q*(Q+Q3+XI)))
      RETURN
      END
C
      SUBROUTINE DIP3(U3,XI1,XI,X,DIP,R2,R3,Q2,Q3)
C
C This subroutine evaluates the three-component, U3,
C of the double integral for specified limits XI1 and XI.
C
      IMPLICIT DOUBLE PRECISION(A-H,O-Z)
      DIMENSION X(3)
C Find distance R from fault to field point and
C distance Q from image fault to field point.
      R=(X(1)-XI1)*(X(1)-XI1)+R2*R2+(R3-XI)*(R3-XI)
      R=DSQRT(R)
      Q=(X(1)-XI1)*(X(1)-XI1)+Q2*Q2+(Q3+XI)*(Q3+XI)
      Q=DSQRT(Q)
C Find projection of Q on plane X(1)=0.
      H=Q2*Q2+(Q3+XI)*(Q3+XI)
      H=DSQRT(H)
C Find fault plane co-ordinates XI2 and XI3.
      XI2=XI*DCOS(DIP)
      XI3=XI*DSIN(DIP)
C Evaluate U3.
      U3=DSIN(DIP)*((X(2)-XI2)*(2.D0*(X(3)-XI3)/(R*(R+X(1)-XI3))
     1 +4.D0*(X(3)-XI3)/(Q*(Q+X(1)-XI1))-4.D0*XI3*X(3)*(X(3)+XI3)*
     2 (2.D0*Q+X(1)-XI1)/(Q**3*(Q+X(1)-XI1)**2))
     3 -6.D0*DATAN((X(1)-XI1)*(X(2)-XI2)/((H+X(3)+XI3)*(Q+H)))
     4 +3.D0*DATAN((X(1)-XI1)*(R3-XI)/(R2*R))
     5 -6.D0*DATAN((X(1)-XI1)*(Q3+XI)/(Q2*Q)))
     6 +DCOS(DIP)*(DLOG(R+X(1)-XI1)-DLOG(Q+X(1)-XI1)
     7 -2.D0*(X(3)-XI3)**2/(R*(R+X(1)-XI1))
```

```
     8 -4.D0*((X(3)+XI3)**2-XI3*X(3))/(Q*(Q+X(1)-XI1))-4.D0*XI3*X(3)*
     9 (X(3)+XI3)**2*(2.D0*Q+X(1)-XI1)/(Q**3*(Q+X(1)-XI1)**2))
      U3=U3+6.D0*X(3)*(DCOS(DIP)*DSIN(DIP)*
     1 (2.D0*(Q3+XI)/(Q*(Q+X(1)-XI1))+(X(1)-XI1)/(Q*(Q+Q3+XI)))
     2 -Q2*(DSIN(DIP)**2-DCOS(DIP)**2)/(Q*(Q+X(1)-XI1)))
      RETURN
      END
C
      SUBROUTINE STRIKES(U,X,DIP,SLIP,DU,DL,L)
C
C This subroutine calculates the displacement vector U(3),
C at the field point X(3), for a strike-slip fault, dipping at angle DIP,
C with slip magnitude SLIP, upper boundary measured down-dip at DU,
C lower boundary down-dip at DL, fault extending along the strike
C from -L to L.
C
      IMPLICIT DOUBLE PRECISION(A-H,O-Z)
      DIMENSION U(3),X(3)
      DOUBLE PRECISION L
C Set value of pi.
      PI=3.141592653589793D0
C Find parameters of field point distances normal and parallel to
C the fault and its image.
      R2=X(2)*DSIN(DIP)-X(3)*DCOS(DIP)
      R3=X(2)*DCOS(DIP)+X(3)*DSIN(DIP)
      Q2=X(2)*DSIN(DIP)+X(3)*DCOS(DIP)
      Q3=-X(2)*DCOS(DIP)+X(3)*DSIN(DIP)
C Find U(1).
      CALL STRIKE1(U1,L,DL,X,DIP,R2,R3,Q2,Q3)
      U(1)=U1
      CALL STRIKE1(U1,-L,DU,X,DIP,R2,R3,Q2,Q3)
      U(1)=U(1)+U1
      CALL STRIKE1(U1,-L,DL,X,DIP,R2,R3,Q2,Q3)
      U(1)=U(1)-U1
      CALL STRIKE1(U1,L,DU,X,DIP,R2,R3,Q2,Q3)
      U(1)=U(1)-U1
C Find U(2).
      CALL STRIKE2(U2,L,DL,X,DIP,R2,R3,Q2,Q3)
      U(2)=U2
      CALL STRIKE2(U2,-L,DU,X,DIP,R2,R3,Q2,Q3)
      U(2)=U(2)+U2
      CALL STRIKE2(U2,-L,DL,X,DIP,R2,R3,Q2,Q3)
      U(2)=U(2)-U2
      CALL STRIKE2(U2,L,DU,X,DIP,R2,R3,Q2,Q3)
      U(2)=U(2)-U2
C Find U(3).
      CALL STRIKE3(U3,L,DL,X,DIP,R2,R3,Q2,Q3)
      U(3)=U3
      CALL STRIKE3(U3,-L,DU,X,DIP,R2,R3,Q2,Q3)
      U(3)=U(3)+U3
      CALL STRIKE3(U3,-L,DL,X,DIP,R2,R3,Q2,Q3)
      U(3)=U(3)-U3
      CALL STRIKE3(U3,L,DU,X,DIP,R2,R3,Q2,Q3)
      U(3)=U(3)-U3
C Scale displacements.
      U(1)=SLIP*U(1)/(12.D0*PI)
      U(2)=SLIP*U(2)/(12.D0*PI)
      U(3)=SLIP*U(3)/(12.D0*PI)
      RETURN
      END
C
```

```
      SUBROUTINE STRIKE1(U1,XI1,XI,X,DIP,R2,R3,Q2,Q3)
C
C This subroutine evaluates the one-component, U1,
C of the double integral for specified limits XI1 and XI.
C
      IMPLICIT DOUBLE PRECISION (A-H,O-Z)
      DIMENSION X(3)
      DOUBLE PRECISION K
C Find distance R from fault to field point and
C distance Q from image fault to field point.
      R=(X(1)-XI1)*(X(1)-XI1)+R2*R2+(R3-XI)*(R3-XI)
      R=DSQRT(R)
      Q=(X(1)-XI1)*(X(1)-XI1)+Q2*Q2+(Q3+XI)*(Q3+XI)
      Q=DSQRT(Q)
C Find projection of Q on plane Q3=0.
      K=(X(1)-XI1)*(X(1)-XI1)+Q2*Q2
      K=DSQRT(K)
C Find fault plane co-ordinates XI2 and XI3.
      XI2=XI*DCOS(DIP)
      XI3=XI*DSIN(DIP)
C Evaluate U1.
      U1=(X(1)-XI1)*(2.D0*R2/(R*(R+R3-XI))
     1 -(4.D0*Q2-2.D0*X(3)*DCOS(DIP))/(Q*(Q+Q3+XI))
     2 -3.D0*DTAN(DIP)/(Q+X(3)+XI3)+4.D0*Q2*X(3)*DSIN(DIP)/Q**3
     3 -4.D0*Q2*Q3*X(3)*DSIN(DIP)*(2.D0*Q+Q3+XI)/(Q**3*(Q+Q3+XI)**2))
     4 -6.D0*DTAN(DIP)**2*DATAN(((K-Q2*DCOS(DIP))*(Q-K)
     5 +(Q3+XI)*K*DSIN(DIP))/((X(1)-XI1)*(Q3+XI)*DCOS(DIP)))
     6 +3.D0*DATAN((X(1)-XI1)*(R3-XI)/(R2*R))
     7 -3.D0*DATAN((X(1)-XI1)*(Q3+XI)/(Q2*Q))
      RETURN
      END
C
      SUBROUTINE STRIKE2(U2,XI1,XI,X,DIP,R2,R3,Q2,Q3)
C
C This subroutine evaluates the two-component, U2,
C of the double integral for specified limits XI1 and XI.
C
      IMPLICIT DOUBLE PRECISION(A-H,O-Z)
      DIMENSION X(3)
C Find fault to field point distance R, and image fault to field point
C distance Q.
      R=(X(1)-XI1)*(X(1)-XI1)+R2*R2+(R3-XI)*(R3-XI)
      R=DSQRT(R)
      Q=(X(1)-XI1)*(X(1)-XI1)+Q2*Q2+(Q3+XI)*(Q3+XI)
      Q=DSQRT(Q)
C Find fault plane co-ordinates XI2 and XI3.
      XI2=XI*DCOS(DIP)
      XI3=XI*DSIN(DIP)
C Evaluate U2.
      U2=DSIN(DIP)*(3.D0*DTAN(DIP)*DLOG(Q+X(3)+XI3)/(DCOS(DIP))
     1 -DLOG(R+R3-XI)-(1.D0+3.D0*DTAN(DIP)**2)*DLOG(Q+Q3+XI))
     2 +2.D0*R2*R2*DSIN(DIP)/(R*(R+R3-XI))+2.D0*R2*DCOS(DIP)/R
     3 -2.D0*DSIN(DIP)*(2.D0*X(3)*(Q2*DCOS(DIP)-Q3*DSIN(DIP))
     4 +Q2*(Q2+X(2)*DSIN(DIP)))/(Q*(Q+Q3+XI))
     5 -3.D0*DTAN(DIP)*(X(2)-XI3)/(Q+X(3)+XI3)
     6 +2.D0*(Q2*DCOS(DIP)-Q3*DSIN(DIP)-X(3)*DSIN(DIP)**2)/Q
     7 +4.D0*Q2*X(3)*DSIN(DIP)*(X(2)-XI2+Q3*DCOS(DIP))/Q**3
     8 -4.D0*Q2*Q2*Q3*X(3)*DSIN(DIP)**2*(2.D0*Q+Q3+XI)/
     9 (Q**3*(Q+Q3+XI)**2)
      RETURN
      END
```

```
C
      SUBROUTINE STRIKE3(U3,XI1,XI,X,DIP,R2,R3,Q2,Q3)
C
C This subroutine evaluates the three-component, U3,
C of the double integral for specified limits XI1 and XI.
C
      IMPLICIT DOUBLE PRECISION(A-H,O-Z)
      DIMENSION X(3)
C Find distance R from fault to field point and
C distance Q from image fault to field point.
      R=(X(1)-XI1)*(X(1)-XI1)+R2*R2+(R3-XI)*(R3-XI)
      R=DSQRT(R)
      Q=(X(1)-XI1)*(X(1)-XI1)+Q2*Q2+(Q3+XI)*(Q3+XI)
      Q=DSQRT(Q)
C Find fault plane co-ordinates XI2 and XI3.
      XI2=XI*DCOS(DIP)
      XI3=XI*DSIN(DIP)
C Evaluate U3.
      U3=DCOS(DIP)*(DLOG(R+R3-XI)
     1 +(1.D0+3.D0*DTAN(DIP)**2)*DLOG(Q+Q3+XI)
     2 -3.D0*DTAN(DIP)*DLOG(Q+X(3)+XI3)/DCOS(DIP))
     3 +2.D0*R2*DSIN(DIP)/R+2.D0*DSIN(DIP)*(Q2+X(2)*DSIN(DIP))/Q
     4 -2.D0*R2*R2*DCOS(DIP)/(R*(R+R3-XI))
     5 +(4.D0*Q2*X(3)*DSIN(DIP)**2-2.D0*(Q2+X(2)*DSIN(DIP))*
     6 (X(3)+Q3*DSIN(DIP)))/(Q*(Q+Q3+XI))
     7 +4.D0*Q2*X(3)*DSIN(DIP)*(X(3)+XI3-Q3*DSIN(DIP))/Q**3
     8 -4.D0*Q2*Q2*Q3*X(3)*DCOS(DIP)*DSIN(DIP)*(2.D0*Q+Q3+XI)/
     9 (Q**3*(Q+Q3+XI)**2)
      RETURN
      END
```

As an illustration of results from the programme DISPF.FOR, we calculate the surface displacement components for the 11 March 2011 Japanese earthquake. It is described by Hayes (2011) and by the article "Preliminary Geodetic Slip Model of the 2011 M9.0 Tohoku-chiho Taiheiyo-oki Earthquake" on the United States Geological Survey website (http://earthquake.usgs.gov/earthquakes/world/japan). The Hayes paper identifies the source as a nearly pure thrust fault with dip 14°, rake (measured positive from the down-dip direction) 177°, and seismic moment 3.85×10^{22} N m. The USGS article gives the total fault length as 700 km, with the upper and lower edges of the fault at 3 km and 57 km, respectively. With a dip of 14°, the top edge is measured 12.4 km down-dip and the bottom edge is 235.6 km down-dip. This gives a fault area of 223.2 km $\times$ 700 km $= 1.56 \times 10^5$ km$^2 = 1.56 \times 10^{11}$ m^2. For a near-surface shear modulus of 1.143×10^{10} N/m^2 (Cal8), a slip of 21.6 m yields the seismic moment of 3.85×10^{22} N m.

Contours of the three components of displacement, computed for the 11 March 2011 Japanese earthquake, are shown in Figures 9.3, 9.4 and 9.5. Contour levels are given in millimetres for ten fault lengths in the direction of the strike and orthogonal to the strike.

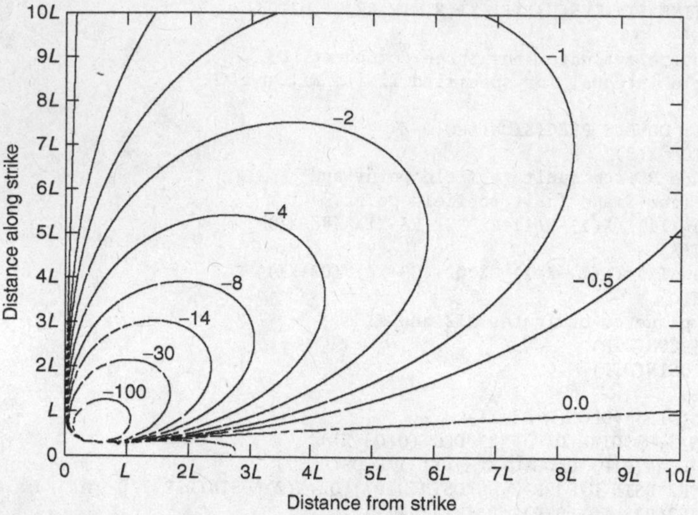

Figure 9.3 Contours of the displacement field component along the strike direction in millimetres. The field is shown for 10 fault lengths along the strike ($L = 700$ km), from the midpoint, and 10 fault lengths orthogonal to the strike, from the midpoint.

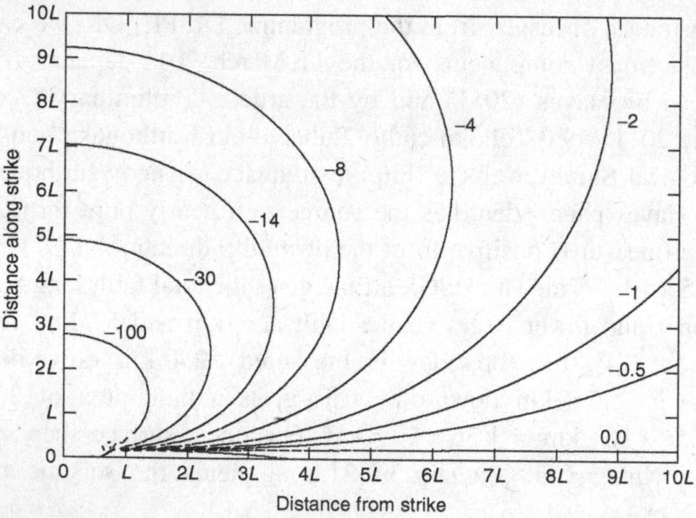

Figure 9.4 Contours of the displacement field component orthogonal to the strike direction in millimetres. The field is shown for 10 fault lengths along the strike ($L = 700$ km), from the midpoint, and 10 fault lengths orthogonal to the strike, from the midpoint.

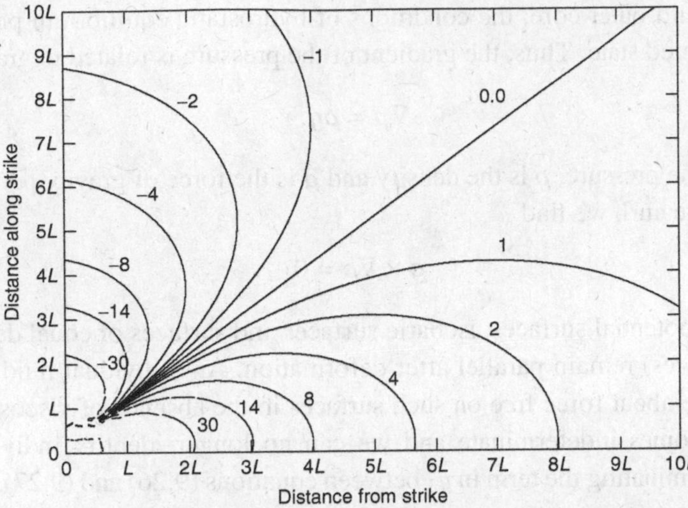

Figure 9.5 Contours of the displacement field component vertically downward in millimetres. The field is shown for 10 fault lengths along the strike ($L = 700$ km), from the midpoint, and 10 fault lengths orthogonal to the strike, from the midpoint. Uplift as shown is negative.

9.2 The theory for realistic Earth models

The elasticity theory of dislocations presented in the previous section was for a uniform, not self-gravitating, elastic half-space, taken to be a Poisson solid ($\lambda = \mu$). We now extend the theory to more realistic, rotating, self-gravitating, radially inhomogeneous Earth models with a liquid outer core and solid inner core.

Deformations in such realistic Earth models have been considered in Chapter 3. We first consider the equations (3.63) through (3.67) governing spheroidal deformations in the liquid outer core for degree $n \geq 1$. In the static case, the dynamical body force terms vanish and the governing system becomes

$$\frac{dy_1}{dr} = -\frac{2}{r}y_1 + \frac{1}{\lambda}y_2 + \frac{n(n+1)}{r}y_3, \tag{9.26}$$

$$\frac{dy_2}{dr} = -\frac{2\rho_0}{r}\left(2g_0 + r\Omega^2\right)y_1 + \frac{n(n+1)}{r}\rho_0 g_0 y_3 - \rho_0 y_6, \tag{9.27}$$

$$0 = \frac{\rho_0 g_0}{r}y_1 - \frac{1}{r}y_2 - \frac{\rho_0}{r}y_5, \tag{9.28}$$

$$\frac{dy_5}{dr} = 4\pi G\rho_0 y_1 + y_6, \tag{9.29}$$

$$\frac{dy_6}{dr} = -4\pi G\rho_0\frac{n(n+1)}{r}y_3 + \frac{n(n+1)}{r^2}y_5 - \frac{2}{r}y_6. \tag{9.30}$$

In the liquid outer core, the conditions of hydrostatic equilibrium prevail, even in the deformed state. Thus, the gradient of the pressure is related to gravity by

$$\nabla p = \rho \boldsymbol{g}, \tag{9.31}$$

where p is the pressure, ρ is the density and $\boldsymbol{g}$ is the force of gravity per unit mass. On taking the curl, we find

$$\boldsymbol{g} \times \nabla \rho = 0. \tag{9.32}$$

Hence, equipotential surfaces, isobaric surfaces and surfaces of equal density (isopycnic surfaces) remain parallel after deformation. An individual fluid particle is able to move about force free on such surfaces in the absence of viscosity. Therefore, y_3 becomes indeterminate and we can no longer identify individual fluid particles. Eliminating the term in y_3 between equations (9.26) and (9.27), we obtain

$$\frac{dy_2}{dr} = \rho_0 g_0 \frac{dy_1}{dr} - \frac{2\rho_0}{r}\left(g_0 + r\Omega^2\right) y_1 - \frac{\rho_0 g_0}{\lambda} y_2 - \rho_0 y_6. \tag{9.33}$$

Substituting for y_6 from equation (9.29) and using relation (3.47), we find that

$$\frac{dy_2}{dr} = \rho_0 \frac{d}{dr}(g_0 y_1) - \frac{\rho_0 g_0}{\lambda} y_2 - \rho_0 \frac{dy_5}{dr}. \tag{9.34}$$

Multiplying equation (9.28) by r/ρ_0 and differentiating gives

$$\rho_0 \frac{d}{dr}\left(g_0 y_1\right) = \rho_0 \frac{d}{dr}\left(\frac{y_2}{\rho_0}\right) + \rho_0 \frac{dy_5}{dr}, \tag{9.35}$$

which, on substitution in (9.34), finally yields

$$\left[\frac{d}{dr}\left(\ln \rho_0\right) + \frac{\rho_0 g_0}{\lambda}\right] y_2 = 0. \tag{9.36}$$

Thus, $y_2 = 0$, unless

$$\frac{d}{dr}\left(\ln \rho_0\right) = -\frac{\rho_0 g_0}{\lambda}, \tag{9.37}$$

which is known as the Adams–Williamson condition and becomes

$$1 + \frac{\lambda}{\rho_0^2 g_0} \frac{d\rho_0}{dr} = 1 + \frac{\alpha^2}{\rho_0 g_0} \frac{d\rho_0}{dr} = 0. \tag{9.38}$$

This is just the condition for the density profile to be perfectly adiabatic (6.166) or neutrally stratified, a condition unlikely to be met everywhere in the liquid outer core. We conclude that $y_2 = 0$. From equation (9.28), this implies that $y_1 = y_5/g_0$ and that the left side of equation (9.27) vanishes, allowing substitution for y_3 in equation (9.30). For $n \geq 1$, the equations governing deformation of the liquid outer core degenerate to

$$y_1 = \frac{1}{g_0} y_5, \tag{9.39}$$

$$y_2 = 0, \tag{9.40}$$

$$y_4 = 0, \tag{9.41}$$

$$\frac{dy_5}{dr} = \frac{4\pi G \rho_0}{g_0} y_5 + y_6, \tag{9.42}$$

$$\frac{dy_6}{dr} = \left[-\frac{16\pi G \rho_0}{g_0 r} \left(1 + \frac{r\Omega^2}{2g_0} \right) + \frac{n(n+1)}{r^2} \right] y_5 - \left[\frac{4\pi G \rho_0}{g_0} + \frac{2}{r} \right] y_6. \tag{9.43}$$

Traditionally, it had been assumed that the variable y_1 was continuous at the boundaries of the liquid core, as is the case for dynamical problems. This led Jeffreys and Vincente (1966) to conclude that a solution may be impossible without the assumptions that the Adams–Williamson condition holds and that the density profile in the liquid outer core is perfectly adiabatic. The solution to the dilemma of Jeffreys and Vincente was given by Smylie and Mansinha (1971), who pointed out that in the static case the solid boundaries of the liquid core penetrate the equipotential, isobaric and equal-density surfaces (just as a loaded ship does, with the degree of penetration being an indication of load). Since the equipotential and its normal derivative are continuous, as shown by Israel *et al.* (1973), equation (9.29) requires a compensating discontinuity in y_6 to make dy_5/dr continuous. Of course, an extra hydrostatic pressure is exerted on the penetrating solid boundaries of the liquid core. Taking the radius of the core–mantle boundary to be $r = b$, with b^+ the radius just outside the core–mantle boundary and b^- the radius just inside, we have for spheroidal deformations of degree $n \geq 1$,

$$y_1(b^-) = \frac{1}{g_0} y_5, \tag{9.44}$$

$$y_2(b^-) = 0, \tag{9.45}$$

$$y_2(b^+) = -\rho_0(b^-)\, g_0 \left(\frac{y_5}{g_0} - y_1(b^+) \right), \tag{9.46}$$

$$y_6(b^+) = y_6(b^-) + 4\pi G \rho_0(b^-) \Delta y_1, \tag{9.47}$$

where

$$\Delta y_1 = y_1(b^-) - y_1(b^+) \tag{9.48}$$

is the discontinuity of y_1. Similar conditions prevail at the inner core.

The generalisation of the reciprocal theorem of Betti to elasto-gravitational systems whose elastic properties are spatially varying, and which are subject to hydrostatic pre-stress in the equilibrium reference state, yields (3.31)

$$\begin{aligned}\int_{\mathcal{S}} (t_i - \rho_0 g_0 u_s \nu_i)\, u'_i d\mathcal{S} + \int_{\mathcal{V}} F_i u'_i d\mathcal{V} \\ = \int_{\mathcal{S}} \left(t'_i - \rho_0 g_0 u'_s \nu_i\right) u_i d\mathcal{S} + \int_{\mathcal{V}} F'_i u_i d\mathcal{V} \\ + \frac{1}{4\pi G} \int_{\mathcal{S}} \left[V_1 \left(\frac{\partial V'_1}{\partial x_i} - 4\pi G \rho_0 u'_i \right) - V'_1 \left(\frac{\partial V_1}{\partial x_i} - 4\pi G \rho_0 u_i \right) \right] \nu_i d\mathcal{S}, \end{aligned} \tag{9.49}$$

for two systems of surface tractions, body forces and decreases in gravitational potential (primed and unprimed), t_i, F_i, V_1 and t'_i, F'_i, V'_1, acting on material contained in a volume $\mathcal{V}$ by a surface $\mathcal{S}$, producing displacement fields u_i and u'_i, respectively. u_s and u'_s are the respective components of displacement in the orthometric direction (opposite to gravity). While this form of the theorem applies to realistic Earth models, it requires generalisation to the case where the variable y_6 is discontinuous across the boundaries of the liquid outer core. Considering first the core–mantle boundary, the generalisation requires only that we demonstrate that the surface integrals in (9.49) evaluated over the top surface of the deformed outer core cancel those evaluated over the bottom surface of the mantle. A similar cancellation is required at the inner core boundary. y_6, as defined by equation (3.45), is the radial coefficient of the outward normal component of the gravitational flux vector $\nabla V_1 - 4\pi G \rho_0 \boldsymbol{u}$. Writing

$$f_s = \left(\frac{\partial V_1}{\partial x_i} - 4\pi G \rho_0 u_i \right) \nu_i \tag{9.50}$$

as a shorthand for the outward normal component of the gravitational flux vector, the physical equivalent of equation (9.47) becomes

$$f_s(b^+) = f_s(b^-) + 4\pi G \rho_0(b^-) \Delta u_s, \tag{9.51}$$

where

$$\Delta u_s = u_s(b^-) - u_s(b^+). \tag{9.52}$$

Similarly, writing t_s for the normal traction, the physical equivalent of equation (9.46) becomes

$$t_s(b^+) = -\rho_0(b^-) V_1 + \rho_0(b^-) g_0 u_s(b^+). \tag{9.53}$$

Of course, $t_s(b^-) = 0$ since $y_2(b^-) = 0$. The sum of the surface integrals in (9.49) at the core–mantle boundary is then found to be

$$\int_{\mathcal{S}} (t'_s u_s - t_s u'_s)_{r=b^+}\, d\mathcal{S} + \int_{\mathcal{S}} \rho_0(b^-) \left(V_1 \Delta u'_s - V'_1 \Delta u_s \right) d\mathcal{S}. \tag{9.54}$$

Substituting for the normal tractions from equation (9.53), the first integral in (9.54) is transformed to

$$- \int_S \rho_0(b^-)\left(V_1' u_s - V_1 u_s'\right)_{r=b^+} dS + \int_S g_0\rho_0(b^-)\left(u_s' u_s - u_s u_s'\right)_{r=b^+} dS$$

$$= - \int_S \rho_0(b^-)\left(V_1' u_s - V_1 u_s'\right)_{r=b^+} dS. \tag{9.55}$$

The physical equivalent of equation (9.44) is

$$u_s(b^-) = \frac{1}{g_0} V_1. \tag{9.56}$$

This allows the second integral in (9.54) to be written as

$$\int_S \rho_0(b^-)\left(V_1 \frac{1}{g_0} V_1' - V_1' \frac{1}{g_0} V_1\right) dS - \int_S \rho_0(b^-)\left(V_1 u_s' - V_1' u_s\right)_{r=b^+} dS$$

$$= - \int_S \rho_0(b^-)\left(V_1 u_s' - V_1' u_s\right)_{r=b^+} dS. \tag{9.57}$$

Adding expressions (9.55) and (9.57), we find that the extra surface integrals over the core–mantle boundary (9.54) cancel each other. Similar arguments may be constructed for the inner core boundary. Thus, the form (9.49) of Betti's reciprocal theorem is unaltered for discontinuities in the variable y_6 given by equation (9.47).

Applied to the mantle and crust (the shell), the surface integrals in Betti's reciprocal theorem remain to be considered for the surface of the Earth and the faces of the dislocation. By arguments analogous to those used at the core–mantle and inner core boundaries, the surface integrals can be shown to vanish over the surface of the Earth. The surface integrals

$$\frac{1}{4\pi G} \int_S \left[V_1 \left(\frac{\partial V_1'}{\partial x_i} - 4\pi G \rho_0 u_i' \right) - V_1' \left(\frac{\partial V_1}{\partial x_i} - 4\pi G \rho_0 u_i \right) \right] \nu_i dS \tag{9.58}$$

over the faces of the dislocation cancel each other for slip faults, since for slip faults there is no displacement component normal to the faces of the dislocation and the normal vectors on the two sides of the dislocation are in opposite directions. The surface integrals

$$\int_S (t_i - \rho_0 g_0 u_s \nu_i)\, u_i' dS \quad \text{and} \quad \int_S \left(t_i' - \rho_0 g_0 u_s' \nu_i\right) u_i dS \tag{9.59}$$

over the faces of the dislocation cancel each other since, for slip faults, there is no displacement component normal to the faces of the dislocation, and the tractions required to maintain the dislocation are equal and opposite on the two faces. The remaining steps in the derivation of Volterra's formula using Betti's reciprocal theorem follow those in Section 9.1.1. The radius vector to the field

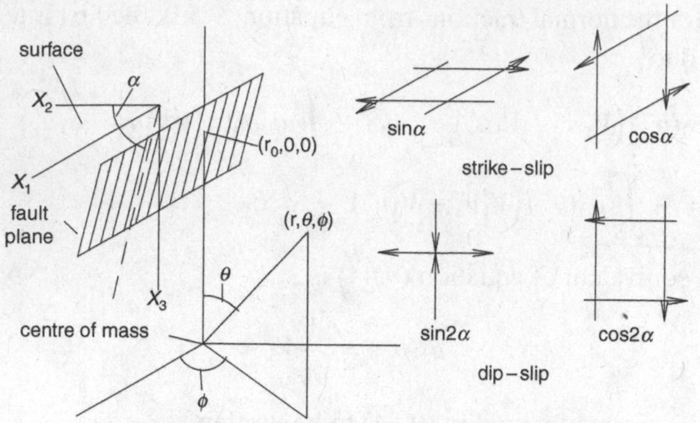

Figure 9.6 Fault geometry and focal force systems.

point is $\boldsymbol{r} = (x_1, x_2, x_3)$, while the radius vector to a point on the fault surface is $\boldsymbol{r}' = (x_1', x_2', x_3')$ with the fault dip at an angle α to the horizontal. Volterra's formula for a dip-slip fault gives the displacement vector at the field point as

$$u_i(\boldsymbol{r}) = \int_{S'} \mu\Delta u \left[\left(\frac{\partial u_i^2(\boldsymbol{r}, \boldsymbol{r}')}{\partial x_2'} - \frac{\partial u_i^3(\boldsymbol{r}, \boldsymbol{r}')}{\partial x_3'} \right) \sin 2\alpha - \left(\frac{\partial u_i'^2(\boldsymbol{r}, \boldsymbol{r}')}{\partial x_3'} + \frac{\partial u_i^3(\boldsymbol{r}, \boldsymbol{r}')}{\partial x_2'} \right) \cos 2\alpha \right] dS', \tag{9.60}$$

with the slip, Δu, having components $\Delta u_2 = \Delta u \cos\alpha$ and $\Delta u_3 = \Delta u \sin\alpha$. Volterra's formula for a strike-slip fault gives the displacement vector at the field point as

$$u_i(\boldsymbol{r}) = \int_{S'} \mu\Delta u_1 \left[\left(\frac{\partial u_i^1(\boldsymbol{r}, \boldsymbol{r}')}{\partial x_2'} + \frac{\partial u_i^2(\boldsymbol{r}, \boldsymbol{r}')}{\partial x_1'} \right) \sin\alpha - \left(\frac{\partial u_i^1(\boldsymbol{r}, \boldsymbol{r}')}{\partial x_3'} + \frac{\partial u_i^3(\boldsymbol{r}, \boldsymbol{r}')}{\partial x_1'} \right) \cos\alpha \right] dS', \tag{9.61}$$

with the slip having the single component Δu_1. These forms of Volterra's formula allow the displacement field to be interpreted as being due to the superposition of the displacement fields of a continuous distribution of dipole forces over the fault surface. We take the systems of dipole force distributions shown in Figure 9.6 to be located at $\boldsymbol{r}_0$ with spherical polar co-ordinates $(r_0, \theta_0, 0)$ on the fault surface.

The unit forces entering Volterra's formula are products of unit vectors with the scalar densities

$$\delta = \frac{\delta(r - r_0)\,\delta(\theta - \theta_0)\,\delta(\phi)}{r^2 \sin\theta}, \tag{9.62}$$

since the volume element, $r^2 \sin\theta\, dr\, d\theta\, d\phi$, is a scalar capacity. Taking the local Cartesian focal co-ordinates to be oriented so that x_1', x_2' and x_3' are in the directions of increasing θ, decreasing ϕ and decreasing r, respectively, for spherical polar co-ordinates (r, θ, ϕ) in this *epicentral* system of the field point, we have the dipole force densities for the dip-slip system as

$$\frac{1}{r^2 \sin\theta}\left[\hat{\boldsymbol{r}}\left\{\frac{1}{r\sin\theta}\delta(r-r_0)\,\delta(\theta-\theta_0)\,\delta'(\phi)\cos 2\alpha \right.\right.$$
$$\left. + \delta'(r-r_0)\,\delta(\theta-\theta_0)\,\delta(\phi)\sin 2\alpha\right\}$$
$$+ \hat{\boldsymbol{\phi}}\left\{-\frac{1}{r\sin\theta}\delta(r-r_0)\,\delta(\theta-\theta_0)\,\delta'(\phi)\sin 2\alpha\right.$$
$$\left.\left. + \delta'(r-r_0)\,\delta(\theta-\theta_0)\,\delta(\phi)\cos 2\alpha\right\}\right], \tag{9.63}$$

and the dipole force densities for the strike-slip system as

$$\frac{1}{r^2 \sin\theta}\left[-\frac{\hat{\boldsymbol{r}}}{r}\delta(r-r_0)\,\delta'(\theta-\theta_0)\,\delta(\phi)\cos\alpha\right.$$
$$+ \hat{\boldsymbol{\theta}}\left\{\frac{1}{r\sin\theta}\delta(r-r_0)\,\delta(\theta-\theta_0)\,\delta'(\phi)\sin\alpha\right.$$
$$\left. - \delta'(r-r_0)\,\delta(\theta-\theta_0)\,\delta(\phi)\cos\alpha\right\}$$
$$\left. + \frac{\hat{\boldsymbol{\phi}}}{r}\delta(r-r_0)\,\delta'(\theta-\theta_0)\,\delta(\phi)\sin\alpha\right]. \tag{9.64}$$

While expressions (9.63) and (9.64) take account of the direct effects of the transformation to spherical polar co-ordinates, there are additional terms arising from co-ordinate curvature.

The spherical polar unit vectors $(\hat{\boldsymbol{r}}, \hat{\boldsymbol{\theta}}, \hat{\boldsymbol{\phi}})$ in the epicentral system are related to the geocentric Cartesian unit vectors $(\hat{\mathbf{e}}_1, \hat{\mathbf{e}}_2, \hat{\mathbf{e}}_3)$ in that system by

$$\hat{\boldsymbol{r}} = \hat{\mathbf{e}}_1 \sin\theta\cos\phi - \hat{\mathbf{e}}_2 \sin\theta\sin\phi - \hat{\mathbf{e}}_3\cos\theta, \tag{9.65}$$

$$\hat{\boldsymbol{\theta}} = \hat{\mathbf{e}}_1 \cos\theta\cos\phi - \hat{\mathbf{e}}_2 \cos\theta\sin\phi + \hat{\mathbf{e}}_3\sin\theta, \tag{9.66}$$

$$\hat{\boldsymbol{\phi}} = -\hat{\mathbf{e}}_1 \sin\phi - \hat{\mathbf{e}}_2\cos\phi. \tag{9.67}$$

Now consider the double-couple force system arising from the strike-slip expression and illustrated in the upper left of Figure 9.7. Here, unit vectors $\pm\hat{\boldsymbol{\theta}}$ have been transported in the ϕ-direction and unit vectors $\pm\hat{\boldsymbol{\phi}}$ have been transported in the θ-direction. From (9.66) and (9.67), we find

$$\frac{\partial\hat{\boldsymbol{\theta}}}{\partial\phi} = \hat{\boldsymbol{\phi}}\cos\theta \quad \text{and} \quad \frac{\partial\hat{\boldsymbol{\phi}}}{\partial\theta} = 0. \tag{9.68}$$

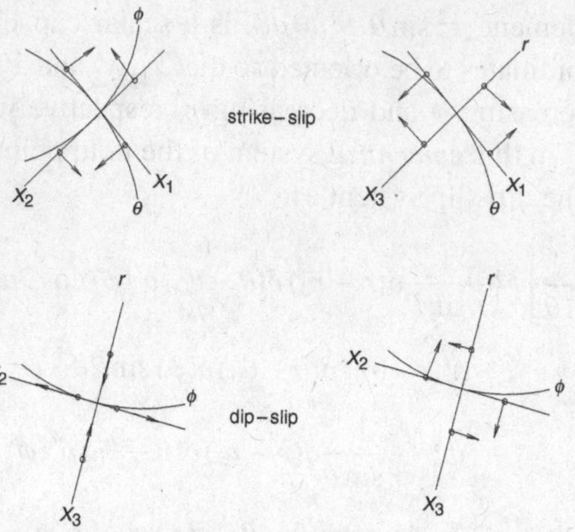

Figure 9.7 Double force and moment transport.

Thus, only transport in the ϕ-direction gives a non-null result. In Cartesian geometry the double couple involved is

$$\hat{e}_1' \sin\alpha\, \delta(x_1')\, \delta(x_3') \lim_{\delta x_2' \to 0} \frac{\delta(x_2' + \delta x_2'/2) - \delta(x_2' - \delta x_2'/2)}{\delta x_2'}. \tag{9.69}$$

Each of the two forces gain small extra components

$$\frac{1}{2}\delta\phi \frac{\partial \hat{\theta}}{\partial \phi} \delta(\phi \pm \delta\phi/2)\, \delta(r - r_0)\, \delta(\theta - \theta_0) \sin\alpha \tag{9.70}$$

before division by $\delta x_2'$ and parallel transport into the limit as $\delta x_2' \to 0$. Since $\delta x_2' = r_0 \sin\theta_0\, \delta\phi$, this produces the extra term

$$\frac{1}{r_0 \sin\theta_0\, \delta\phi} \frac{1}{r_0^2 \sin\theta_0} \delta\phi\, \hat{\phi} \cos\theta_0\, \delta(\phi)\, \delta(r - r_0)\, \delta(\theta - \theta_0) \sin\alpha, \tag{9.71}$$

expressed per unit volume. Hence, co-ordinate curvature contributes the term in double-force density

$$\hat{\phi} \frac{\cot\theta_0}{r_0} \delta \sin\alpha, \tag{9.72}$$

arising from the double-couple term for strike-slip faults and illustrated in the top left of Figure 9.7.

In the case of the double-couple force system arising from the strike-slip expression and illustrated in the upper right of Figure 9.7, the unit vectors $\pm\hat{r}$ have been transported in the θ-direction and the unit vectors $\pm\hat{\boldsymbol{\theta}}$ have been transported in the r-direction. From (9.65) and (9.66), we find that

$$\frac{\partial \hat{\mathbf{r}}}{\partial \theta} = \hat{\boldsymbol{\theta}} \quad \text{and} \quad \frac{\partial \hat{\boldsymbol{\theta}}}{\partial r} = 0. \tag{9.73}$$

Thus, only transport in the θ-direction gives a non-null result. With arguments that are similar to those made before, we find co-ordinate curvature contributes the term in double-force density,

$$-\hat{\boldsymbol{\theta}}\frac{\delta}{r_0}\cos\alpha, \tag{9.74}$$

arising from the double-couple term for strike-slip faults and illustrated in the top right of Figure 9.7.

For the double-force system arising from the dip-slip expression and illustrated in the lower left of Figure 9.7, the unit vectors $\pm\hat{\mathbf{r}}$ have been transported in the r-direction and the unit vectors $\pm\hat{\boldsymbol{\phi}}$ have been transported in the ϕ-direction. From (9.65) and (9.67), we find

$$\frac{\partial \hat{\mathbf{r}}}{\partial r} = 0 \quad \text{and} \quad \frac{\partial \hat{\boldsymbol{\phi}}}{\partial \phi} = -\hat{\mathbf{r}}\sin\theta - \hat{\boldsymbol{\theta}}\cos\theta. \tag{9.75}$$

Thus, only transport in the ϕ-direction gives a non-null result. With arguments similar to those made before, we find co-ordinate curvature contributes the term in double-force density,

$$\hat{\mathbf{r}}\frac{\delta}{r_0}\sin 2\alpha + \hat{\boldsymbol{\theta}}\frac{\cot\theta}{r_0}\delta\sin 2\alpha, \tag{9.76}$$

arising from the double-force term for dip-slip faults and illustrated in the lower left of Figure 9.7.

For the double-couple force system arising from the dip-slip expression and illustrated in the lower right of Figure 9.7, the unit vectors $\pm\hat{\mathbf{r}}$ have been transported in the ϕ-direction and the unit vectors $\pm\hat{\boldsymbol{\phi}}$ have been transported in the r-direction. From (9.65), (9.66) and (9.67), we find that

$$\frac{\partial \hat{\mathbf{r}}}{\partial \phi} = \hat{\boldsymbol{\phi}}\sin\theta \quad \text{and} \quad \frac{\partial \hat{\boldsymbol{\phi}}}{\partial r} = 0. \tag{9.77}$$

Thus, only transport in the ϕ-direction gives a non-null result. With arguments similar to those made before, we find co-ordinate curvature contributes the term in double-force density,

$$\hat{\boldsymbol{\phi}}\frac{\delta}{r_0}\cos 2\alpha, \tag{9.78}$$

arising from the double-couple term for dip-slip faults and illustrated in the lower right of Figure 9.7.

In summary, co-ordinate curvature contributes the extra double-force density terms for the dip-slip system,

$$\hat{r}\frac{\delta}{r_0}\sin 2\alpha + \hat{\theta}\frac{\cot\theta_0}{r_0}\delta\sin 2\alpha + \hat{\phi}\frac{\delta}{r_0}\cos 2\alpha, \tag{9.79}$$

and the extra double-force density terms for the strike-slip system,

$$-\hat{\theta}\frac{\delta}{r_0}\cos\alpha + \hat{\phi}\frac{\cot\theta_0}{r_0}\delta\sin\alpha. \tag{9.80}$$

Using the orthogonality relations (1.205) and (1.209), as outlined, the radial spheroidal, transverse spheroidal and torsional radial coefficients can be extracted from the dipole force densities for the dip-slip system (9.63). In the limit as θ_0 approaches zero, we find

$$\begin{aligned}
u_n^0(r) &= \frac{2n+1}{4\pi r^2}\delta'(r-r_0)\sin 2\alpha,\\
u_n^{-1}(r) &= -\frac{i(2n+1)}{8\pi r^3}n(n+1)\,\delta(r-r_0)\cos 2\alpha,\\
u_n^1(r) &= -\frac{i(2n+1)}{8\pi r^3}\delta(r-r_0)\cos 2\alpha,\\
v_n^{-1}(r) &= \frac{i(2n+1)}{8\pi r^2}\delta'(r-r_0)\cos 2\alpha,\\
v_n^1(r) &= \frac{i(2n+1)}{8\pi r^2 n(n+1)}\delta'(r-r_0)\cos 2\alpha,\\
v_n^{-2}(r) &= -\frac{2n+1}{8\pi r^3}(n-1)(n+2)\,\delta(r-r_0)\sin 2\alpha,\\
v_n^2(r) &= -\frac{2n+1}{8\pi r^3 n(n+1)}\delta(r-r_0)\sin 2\alpha,\\
t_n^{-1}(r) &= -\frac{2n+1}{8\pi r^2}\delta'(r-r_0)\cos 2\alpha,\\
t_n^1(r) &= \frac{2n+1}{8\pi r^2 n(n+1)}\delta'(r-r_0)\cos 2\alpha,\\
t_n^{-2}(r) &= -\frac{i(2n+1)}{8\pi r^3}(n-1)(n+2)\,\delta(r-r_0)\sin 2\alpha,\\
t_n^2(r) &= \frac{i(2n+1)}{8\pi r^3 n(n+1)}\delta(r-r_0)\sin 2\alpha.
\end{aligned} \tag{9.81}$$

Similarly, for the strike-slip system (9.64) we find

$$\begin{aligned}
u_n^{-1}(r) &= \frac{2n+1}{8\pi r^3} n\,(n+1)\,\delta(r-r_0)\cos\alpha, \\
u_n^{1}(r) &= -\frac{2n+1}{8\pi r^3}\delta(r-r_0)\cos\alpha, \\
v_n^{-1}(r) &= -\frac{2n+1}{8\pi r^2}\delta'(r-r_0)\cos\alpha, \\
v_n^{1}(r) &= \frac{2n+1}{8\pi r^2 n\,(n+1)}\delta'(r-r_0)\cos\alpha, \\
v_n^{-2}(r) &= -\frac{3i\,(2n+1)}{16\pi r^3}(n-1)\,(n+2)\,\delta(r-r_0)\sin\alpha, \\
v_n^{2}(r) &= \frac{3i\,(2n+1)}{16\pi r^3 n\,(n+1)}\delta(r-r_0)\sin\alpha, \\
t_n^{0}(r) &= -\frac{2n+1}{8\pi r^3}\delta(r-r_0)\sin\alpha, \\
t_n^{-1}(r) &= -\frac{i\,(2n+1)}{8\pi r^2}\delta'(r-r_0)\cos\alpha, \\
t_n^{1}(r) &= -\frac{i\,(2n+1)}{8\pi r^2 n\,(n+1)}\delta'(r-r_0)\cos\alpha, \\
t_n^{-2}(r) &= \frac{3\,(2n+1)}{16\pi r^3}(n-1)\,(n+2)\,\delta(r-r_0)\sin\alpha, \\
t_n^{2}(r) &= \frac{3\,(2n+1)}{16\pi r^3 n\,(n+1)}\delta(r-r_0)\sin\alpha.
\end{aligned} \tag{9.82}$$

Once again, extracting the radial coefficients of the radial spheroidal, transverse spheroidal and torsional parts of the extra double-force densities (9.79), arising from co-ordinate curvature for the dip-slip system, and taking the limit as θ_0 goes to zero, we find

$$\begin{aligned}
u_n^{0}(r) &= \frac{2n+1}{4\pi r^3}\delta(r-r_0)\sin 2\alpha, \\
v_n^{0}(r) &= -\frac{2n+1}{8\pi r^3}\delta(r-r_0)\sin 2\alpha, \\
v_n^{-1}(r) &= \frac{i\,(2n+1)}{8\pi r^3}\delta(r-r_0)\cos 2\alpha, \\
v_n^{1}(r) &= \frac{i\,(2n+1)}{8\pi n\,(n+1)\,r^3}\delta(r-r_0)\cos 2\alpha,
\end{aligned}$$

$$v_n^{-2}(r) = \frac{(2n+1)(n-1)(n+2)}{16\pi r^3}\delta(r-r_0)\sin 2\alpha, \tag{9.83}$$

$$v_n^{2}(r) = \frac{2n+1}{16\pi n(n+1)r^3}\delta(r-r_0)\sin 2\alpha,$$

$$t_n^{-1}(r) = -\frac{2n+1}{8\pi r^3}\delta(r-r_0)\cos 2\alpha,$$

$$t_n^{1}(r) = \frac{2n+1}{8\pi n(n+1)r^3}\delta(r-r_0)\cos 2\alpha,$$

$$t_n^{-2}(r) = \frac{i(2n+1)(n-1)(n-2)}{16\pi r^3}\delta(r-r_0)\sin 2\alpha,$$

$$t_n^{2}(r) = -\frac{i(2n+1)}{16\pi n(n+1)r^3}\delta(r-r_0)\sin 2\alpha.$$

Similarly, for the extra double-force densities (9.80) arising from co-ordinate curvature for the strike-slip system, we find

$$v_n^{-1}(r) = -\frac{2n+1}{8\pi r^3}\delta(r-r_0)\cos\alpha,$$

$$v_n^{1}(r) = \frac{2n+1}{8\pi n(n+1)r^3}\delta(r-r_0)\cos\alpha,$$

$$v_n^{-2}(r) = \frac{i(2n+1)(n-1)(n+2)}{16\pi r^3}\delta(r-r_0)\sin\alpha,$$

$$v_n^{2}(r) = -\frac{i(2n+1)}{16\pi n(n+1)r^3}\delta(r-r_0)\sin\alpha,$$

$$t_n^{0}(r) = \frac{2n+1}{8\pi r^3}\delta(r-r_0)\sin\alpha, \tag{9.84}$$

$$t_n^{-1}(r) = -\frac{i(2n+1)}{8\pi r^3}\delta(r-r_0)\cos\alpha,$$

$$t_n^{1}(r) = -\frac{i(2n+1)}{8\pi n(n+1)r^3}\delta(r-r_0)\cos\alpha,$$

$$t_n^{-2}(r) = -\frac{(2n+1)(n-1)(n+2)}{16\pi r^3}\delta(r-r_0)\sin\alpha,$$

$$t_n^{2}(r) = -\frac{2n+1}{16\pi n(n+1)r^3}\delta(r-r_0)\sin\alpha.$$

9.3 Changes in the inertia tensor and the secular polar shift

The effect the displacement fields in realistic Earth models, just described, have on the rotation of the Earth can be found from the changes they produce in the components of the inertia tensor (Smylie and Mansinha, 1967; Smylie and Zuberi, 2009). The changes in the components of the inertia tensor, ΔI_{ij}, to first order in the small quantities u_r/r, u_θ/r, and $u_\phi/(r\sin\theta)$ in the epicentral system are

$$\begin{aligned}
\Delta I_{11} &= 2\int r\left[u_r\left(1-\sin^2\theta\cos^2\phi\right) - u_\theta\cos\theta\sin\theta\cos^2\phi + u_\phi\sin\theta\cos\phi\sin\phi\right]dm,\\
\Delta I_{22} &= 2\int r\left[u_r\left(1-\sin^2\theta\sin^2\phi\right) - u_\theta\cos\theta\sin\theta\sin^2\phi - u_\phi\sin\theta\cos\phi\sin\phi\right]dm,\\
\Delta I_{33} &= 2\int r\left[u_r\sin^2\theta + u_\theta\cos\theta\sin\theta\right]dm,\\
\Delta I_{12} &= -\int r\left[2u_r\sin^2\theta\cos\phi\sin\phi + 2u_\theta\cos\theta\sin\theta\cos\phi\sin\phi\right.\\
&\qquad\left. + u_\phi\sin\theta\left(\cos^2\phi-\sin^2\phi\right)\right]dm,\\
\Delta I_{13} &= -\int r\left[2u_r\cos\theta\sin\theta\cos\phi + u_\theta\left(\cos^2\theta-\sin^2\theta\right)\cos\phi - u_\phi\cos\theta\sin\phi\right]dm,\\
\Delta I_{23} &= -\int r\left[2u_r\cos\theta\sin\theta\sin\phi + u_\theta\left(\cos^2\theta-\sin^2\theta\right)\sin\phi + u_\phi\cos\theta\cos\phi\right]dm.
\end{aligned} \tag{9.85}$$

The integrands in (9.85) can all be cast in the form of scalar products of the displacement field with particular spheroidal vectors of degrees zero and two. For example, we may write

$$\Delta I_{11} = 2\int r\boldsymbol{u}\cdot\boldsymbol{S}'dm, \tag{9.86}$$

where $\boldsymbol{S}'$ is the spheroidal vector with radial coefficients

$$u_0^{0\prime} = \frac{2}{3}, \tag{9.87}$$

$$u_2^{0\prime} = \frac{1}{3}, \qquad v_2^{0\prime} = \frac{1}{6}, \tag{9.88}$$

$$u_2^{2\prime} = -\frac{1}{12}, \qquad v_2^{2\prime} = -\frac{1}{24}, \tag{9.89}$$

$$u_2^{-2\prime} = -2, \qquad v_2^{-2\prime} = -1. \tag{9.90}$$

The orthogonality property (1.205) then allows the expressions (9.85) to be reduced to

$$\begin{aligned}
\Delta I_{11} &= 4\pi \int_0^d r^3\rho_0(r)\left[\frac{4}{3}u_0^0(r) + \frac{2}{15}u_2^0(r) - \frac{4}{5}u_2^2(r) - \frac{1}{30}u_2^{-2}(r)\right.\\
&\quad \left. + \frac{2}{5}v_2^0(r) - \frac{12}{5}v_2^2(r) - \frac{1}{10}v_2^{-2}(r)\right] dr,\\
\Delta I_{22} &= 4\pi \int_0^d r^3\rho_0(r)\left[\frac{4}{3}u_0^0(r) + \frac{2}{15}u_2^0(r) + \frac{4}{5}u_2^2(r) + \frac{1}{30}u_2^{-2}(r)\right.\\
&\quad \left. + \frac{2}{5}v_2^0(r) + \frac{12}{3}v_2^2(r) + \frac{1}{10}v_2^{-2}(r)\right] dr,\\
\Delta I_{33} &= 4\pi \int_0^d r^3\rho_0(r)\left[\frac{4}{3}u_0^0(r) - \frac{4}{15}u_2^0(r) - \frac{4}{5}v_2^0(r)\right] dr, \qquad (9.91)\\
\Delta I_{12} &= 4\pi i \int_0^d r^3\rho_0(r)\left[-\frac{4}{5}u_2^2(r) + \frac{1}{30}u_2^{-2}(r) - \frac{12}{5}v_2^2(r) + \frac{1}{10}v_2^{-2}(r)\right] dr,\\
\Delta I_{13} &= 4\pi \int_0^d r^3\rho_0(r)\left[\frac{2}{5}u_2^1(r) - \frac{1}{15}u_2^{-1}(r) + \frac{6}{5}v_2^1(r) - \frac{1}{5}v_2^{-1}(r)\right] dr,\\
\Delta I_{23} &= 4\pi i \int_0^d r^3\rho_0(r)\left[\frac{2}{5}u_2^1(r) + \frac{1}{15}u_2^{-1}(r) + \frac{6}{5}v_2^1(r) + \frac{1}{5}v_2^{-1}(r)\right] dr,
\end{aligned}$$

with d the radius of the Earth.

As shown in Section 9.2, the solid boundaries of the outer core can penetrate the gravitational equipotentials coincident with the boundaries of the outer core in the undeformed state. This gives further contributions to the changes in the inertia tensor, which for the core–mantle boundary amount to

$$\begin{aligned}
\Delta J_{11} &= 2\pi b^4\rho_0(b^-)\left[\frac{4}{3}\Delta u_0^0 + \frac{2}{15}\Delta u_2^0 - \frac{4}{5}\Delta u_2^2 - \frac{1}{30}\Delta u_2^{-2}\right],\\
\Delta J_{22} &= 2\pi b^4\rho_0(b^-)\left[\frac{4}{3}\Delta u_0^0 + \frac{2}{15}\Delta u_2^0 + \frac{4}{5}\Delta u_2^2 + \frac{1}{30}\Delta u_2^{-2}\right],\\
\Delta J_{33} &= 2\pi b^4\rho_0(b^-)\left[\frac{4}{3}\Delta u_0^0 - \frac{4}{15}\Delta u_2^0\right], \qquad (9.92)\\
\Delta J_{12} &= 2\pi b^4\rho_0(b^-)\left[-\frac{4}{5}\Delta u_2^2 + \frac{1}{30}\Delta u_2^{-2}\right],\\
\Delta J_{13} &= 2\pi b^4\rho_0(b^-)\left[\frac{2}{5}\Delta u_2^1 - \frac{1}{15}\Delta u_2^{-1}\right],\\
\Delta J_{23} &= 2\pi b^4\rho_0(b^-)\left[\frac{2}{5}\Delta u_2^1 + \frac{1}{15}\Delta u_2^{-1}\right],
\end{aligned}$$

where Δu_n^m is the excess of the radial displacement of the base of the mantle over that of the gravitational equipotential coincident with the core–mantle boundary in the undeformed state. Similar results are found for the inner core boundary.

A sudden redistribution of mass within the Earth, such as that accompanying a major earthquake, produces a secular polar shift (Smylie and Mansinha, 1967; Smylie and Zuberi, 2009),

$$\frac{\Omega \Delta c}{\sigma_0 A}, \tag{9.93}$$

where Ω is the angular velocity of Earth's rotation, $\Delta c = \Delta c_{13} + i\Delta c_{23}$ is the change in the off-diagonal components of the inertia tensor in the geographical system of reference, σ_0 is the angular frequency of the Chandler wobble and A is the equatorial moment of inertia of the Earth. The real part of the secular polar shift is measured towards Greenwich and the imaginary part is measured in the direction of 90° E. The changes in the off-diagonal components Δc_{13} and Δc_{23} of the inertia tensor in the geographical system of reference are found by similarity transformation of the changes $\Delta I_{ij} + \Delta J_{ij}$ in the inertia tensor in the epicentral system. A closely equal and opposite contribution to the Chandler wobble occurs, so that only a small instantaneous shift in pole position is realised, with the pole path embarking on a new circular arc about a new centre of rotation.

From expressions (9.91) and (9.92), the changes in the inertia tensor in the epicentral system of reference depend on spheroidal displacement fields of zeroth and second degree. The zeroth-degree displacement field makes an equal contribution only to the diagonal components of the inertia tensor. This contribution will, therefore, be the same in all centre of mass co-ordinate systems. Since the excitation of wobble and secular polar shift depend only on two of the off-diagonal components of the inertia tensor, the degree-zero spheroidal displacements can be ignored. In the mantle and crust, the second-degree spheroidal displacement fields obey the system of ordinary differential equations (3.102) through (3.107). Four separate solutions of the non-homogeneous equations are involved with singular sources proportional to

$$u_{F2}^m = \frac{5}{8\pi r^3}\mu\,\delta(r - r_0), \tag{9.94}$$

$$u_{F2}^m = \frac{5}{8\pi r^2}\mu\,\delta'(r - r_0), \tag{9.95}$$

$$v_{F2}^m = \frac{5}{8\pi r^3}\mu\,\delta(r - r_0), \tag{9.96}$$

$$v_{F2}^m = \frac{5}{8\pi r^2}\mu\,\delta'(r - r_0). \tag{9.97}$$

Solutions of the homogeneous sixth-order spheroidal system can be represented by the propagator matrix $\boldsymbol{Y}(r,\rho)$ (3.299), which is the solution of the homogeneous differential system

$$\frac{d\boldsymbol{Y}(r,\rho)}{dr} = \boldsymbol{A}(r)\cdot\boldsymbol{Y}(r,\rho), \tag{9.98}$$

with the initial condition

$$\boldsymbol{Y}(\rho,\rho) = \boldsymbol{I}, \tag{9.99}$$

where $\boldsymbol{I}$ is the unit matrix that propagates the fundamental solutions from radius ρ to radius r. Each column vector of $\boldsymbol{Y}$ is a linearly independent solution of

$$\frac{d\boldsymbol{y}(r)}{dr} = \boldsymbol{A}(r)\cdot\boldsymbol{y}(r). \tag{9.100}$$

The solution of the non-homogeneous differential system

$$\frac{d\boldsymbol{y}(r)}{dr} = \boldsymbol{A}(r)\cdot\boldsymbol{y}(r) + \boldsymbol{g}(r) \tag{9.101}$$

is then given by

$$\boldsymbol{y}(r) = \int_b^r \boldsymbol{Y}(r,\rho)\,\boldsymbol{g}(\rho)\,d\rho + \boldsymbol{Y}(r,b)\,\boldsymbol{y}(b). \tag{9.102}$$

If $\boldsymbol{g}(r) = \boldsymbol{G}\delta(r-r_0)$, where $\boldsymbol{G}$ is a constant vector, the solution is

$$\boldsymbol{y}(r) = \begin{cases} \boldsymbol{Y}(r,r_0)\,\boldsymbol{G} + \boldsymbol{Y}(r,b)\,\boldsymbol{y}(b), & r > r_0 \\ \boldsymbol{Y}(r,b)\,\boldsymbol{y}(b), & r < r_0. \end{cases} \tag{9.103}$$

If $\boldsymbol{g}(r) = \boldsymbol{G}\delta'(r-r_0)$, the solution is

$$\boldsymbol{y}(r) = \begin{cases} -\dfrac{d\boldsymbol{Y}(r,r_0)}{dr_0}\boldsymbol{G} + \boldsymbol{y}(r,b)\,\boldsymbol{y}(b), & r > r_0 \\ \boldsymbol{Y}(r,b)\,\boldsymbol{y}(b), & r < r_0. \end{cases} \tag{9.104}$$

Integration from r_0 to r and then back to r_0 gives

$$\boldsymbol{Y}(r,r_0)\,\boldsymbol{Y}(r_0,r) = \boldsymbol{I}, \tag{9.105}$$

while integration from r to r_0 and then back to r gives

$$\boldsymbol{Y}(r_0,r)\,\boldsymbol{Y}(r,r_0) = \boldsymbol{I}. \tag{9.106}$$

Differentiating (9.105) with respect to r_0 yields

$$\frac{d\boldsymbol{Y}(r,r_0)}{dr_0}\boldsymbol{Y}(r_0,r) = -\boldsymbol{Y}(r,r_0)\frac{d\boldsymbol{Y}(r_0,r)}{dr_0}. \tag{9.107}$$

Substitution from (9.98), with r_0 replacing r and r replacing ρ, transforms the right side, and we find that

$$\frac{d\boldsymbol{Y}(r, r_0)}{dr_0}\boldsymbol{Y}(r_0, r) = -\boldsymbol{Y}(r, r_0) \cdot \boldsymbol{A}(r_0) \cdot \boldsymbol{Y}(r_0, r). \tag{9.108}$$

Right multiplying both sides by $\boldsymbol{Y}(r, r_0)$ and using (9.106) reduces this relation to

$$\frac{d\boldsymbol{Y}(r, r_0)}{dr_0} = -\boldsymbol{Y}(r, r_0) \cdot \boldsymbol{A}(r_0). \tag{9.109}$$

This then allows (9.104) to be expressed as

$$\boldsymbol{y}(r) = \begin{cases} \boldsymbol{Y}(r, r_0) \cdot \boldsymbol{A} \cdot \boldsymbol{G} + \boldsymbol{Y}(r, b)\, \boldsymbol{y}(b), & r > r_0 \\ \boldsymbol{Y}(r, b)\, \boldsymbol{y}(b), & r < r_0. \end{cases} \tag{9.110}$$

Programme SHIFT.FOR computes the four required solutions of the non-homogeneous equations and evaluates the changes in the inertia tensor in the epicentral co-ordinate system. It then transforms the changes in the inertia tensor to geocentric co-ordinates and finds the secular polar shift for specified fault parameters.

Appendix A
Elementary results from vector analysis

In this appendix we tabulate some of the elementary results, assumed as background, from vector analysis.

A.1 Vector identities

For arbitrary vectors $\boldsymbol{a}$, $\boldsymbol{b}$, $\boldsymbol{c}$, $\boldsymbol{d}$, we have a number of identities frequently used in elementary analysis. The *scalar triple product* obeys the cyclic identity

$$\boldsymbol{a} \cdot (\boldsymbol{b} \times \boldsymbol{c}) = \boldsymbol{b} \cdot (\boldsymbol{c} \times \boldsymbol{a}) = \boldsymbol{c} \cdot (\boldsymbol{a} \times \boldsymbol{b}) . \tag{A.1}$$

The *vector triple product* can be expanded as

$$\boldsymbol{a} \times (\boldsymbol{b} \times \boldsymbol{c}) = (\boldsymbol{a} \cdot \boldsymbol{c})\, \boldsymbol{b} - (\boldsymbol{a} \cdot \boldsymbol{b})\, \boldsymbol{c}. \tag{A.2}$$

The *scalar quadruple product* can be successively developed as

$$\begin{aligned} (\boldsymbol{a} \times \boldsymbol{b}) \cdot (\boldsymbol{c} \times \boldsymbol{d}) &= \boldsymbol{a} \cdot [\boldsymbol{b} \times (\boldsymbol{c} \times \boldsymbol{d})] \\ &= \boldsymbol{a} \cdot [(\boldsymbol{b} \cdot \boldsymbol{d})\, \boldsymbol{c} - (\boldsymbol{b} \cdot \boldsymbol{c})\, \boldsymbol{d}] \\ &= (\boldsymbol{a} \cdot \boldsymbol{c})\, (\boldsymbol{b} \cdot \boldsymbol{d}) - (\boldsymbol{a} \cdot \boldsymbol{d})\, (\boldsymbol{b} \cdot \boldsymbol{c}) . \end{aligned} \tag{A.3}$$

Finally, the *vector quadruple product* obeys the identity

$$(\boldsymbol{a} \times \boldsymbol{b}) \times (\boldsymbol{c} \times \boldsymbol{d}) = [(\boldsymbol{a} \times \boldsymbol{b}) \cdot \boldsymbol{d}]\boldsymbol{c} - [(\boldsymbol{a} \times \boldsymbol{b}) \cdot \boldsymbol{c}]\boldsymbol{d}. \tag{A.4}$$

A.2 Vector calculus identities

There are a variety of identities from vector calculus involving the vector operator *del* or *nabla*, written ∇. These involve the arbitrary scalars ϕ and ψ as well as the arbitrary vectors $\boldsymbol{a}$ and $\boldsymbol{b}$. We first have the gradient of the sum of scalars,

$$\nabla (\phi + \psi) = \nabla \phi + \nabla \psi, \tag{A.5}$$

and then the gradient of the product of scalars,

$$\nabla(\phi\psi) = \phi\nabla\psi + \psi\nabla\phi. \tag{A.6}$$

The divergence of the sum of two vectors becomes

$$\nabla\cdot(\boldsymbol{a}+\boldsymbol{b}) = \nabla\cdot\boldsymbol{a} + \nabla\cdot\boldsymbol{b}. \tag{A.7}$$

The curl of the sum of two vectors is

$$\nabla\times(\boldsymbol{a}+\boldsymbol{b}) = \nabla\times\boldsymbol{a} + \nabla\times\boldsymbol{b}. \tag{A.8}$$

The divergence of the product of a scalar and a vector yields

$$\nabla\cdot(\phi\boldsymbol{a}) = \boldsymbol{a}\cdot\nabla\phi + \phi\nabla\cdot\boldsymbol{a}, \tag{A.9}$$

while the curl of the product of a scalar and a vector gives

$$\nabla\times(\phi\boldsymbol{a}) = \nabla\phi\times\boldsymbol{a} + \phi\nabla\times\boldsymbol{a}. \tag{A.10}$$

The gradient of the scalar product of two vectors develops as

$$\nabla(\boldsymbol{a}\cdot\boldsymbol{b}) = (\boldsymbol{a}\cdot\nabla)\,\boldsymbol{b} + (\boldsymbol{b}\cdot\nabla)\,\boldsymbol{a} + \boldsymbol{a}\times(\nabla\times\boldsymbol{b}) + \boldsymbol{b}\times(\nabla\times\boldsymbol{a})\,. \tag{A.11}$$

The divergence of the cross product of two vectors can be expanded as

$$\nabla\cdot(\boldsymbol{a}\times\boldsymbol{b}) = \boldsymbol{b}\cdot\nabla\times\boldsymbol{a} - \boldsymbol{a}\cdot\nabla\times\boldsymbol{b}, \tag{A.12}$$

while the curl of the cross product of two vectors becomes

$$\nabla\times(\boldsymbol{a}\times\boldsymbol{b}) = \boldsymbol{a}\nabla\cdot\boldsymbol{b} - \boldsymbol{b}\nabla\cdot\boldsymbol{a} + (\boldsymbol{b}\cdot\nabla)\,\boldsymbol{a} - (\boldsymbol{a}\cdot\nabla)\,\boldsymbol{b}. \tag{A.13}$$

In Cartesian co-ordinates, the curl of a vector, taken twice, can be expressed as

$$\nabla\times(\nabla\times\boldsymbol{a}) = \nabla(\nabla\cdot\boldsymbol{a}) - \nabla^2\boldsymbol{a}. \tag{A.14}$$

The curl of the gradient of a scalar vanishes identically,

$$\nabla\times\nabla\phi \equiv 0, \tag{A.15}$$

and the divergence of the curl of a vector also vanishes identically,

$$\nabla\cdot(\nabla\times\boldsymbol{a}) \equiv 0. \tag{A.16}$$

A.3 Integral theorems

For a volume $\mathcal{V}$ enclosed by a surface $\mathcal{S}$, the *divergence theorem of Gauss* reads

$$\iiint_{\mathcal{V}} \nabla \cdot \boldsymbol{a}\, d\mathcal{V} = \iint_{\mathcal{S}} \boldsymbol{a} \cdot \hat{\boldsymbol{\nu}} d\mathcal{S}, \tag{A.17}$$

where $\hat{\boldsymbol{\nu}}$ is the unit outward vector normal to the surface. Two analogous integral theorems are

$$\iiint_{\mathcal{V}} \nabla\phi\, d\mathcal{V} = \iint_{\mathcal{S}} \phi\hat{\boldsymbol{\nu}}\, d\mathcal{S} \tag{A.18}$$

and

$$\iiint_{\mathcal{V}} \nabla \times \boldsymbol{a}\, d\mathcal{V} = \iint_{\mathcal{S}} \hat{\boldsymbol{\nu}} \times \boldsymbol{a}\, d\mathcal{S}. \tag{A.19}$$

If the vector $\boldsymbol{a}$ in the divergence theorem of Gauss (A.17) is $\psi\nabla\phi$, where ψ and ϕ are scalar functions of position, we have

$$\iiint_{\mathcal{V}} \nabla \cdot (\psi\nabla\phi)\, d\mathcal{V} = \iint_{\mathcal{S}} (\psi\nabla\phi) \cdot \hat{\boldsymbol{\nu}} d\mathcal{S}. \tag{A.20}$$

Expanding the divergence

$$\nabla \cdot (\psi\nabla\phi) = \nabla\psi \cdot \nabla\phi + \psi\nabla^2\phi, \tag{A.21}$$

we find *Green's first identity*,

$$\iiint_{\mathcal{V}} \nabla\psi \cdot \nabla\phi\, d\mathcal{V} + \iiint_{\mathcal{V}} \psi\nabla^2\phi\, d\mathcal{V} = \iint_{\mathcal{S}} (\psi\nabla\phi) \cdot \hat{\boldsymbol{\nu}} d\mathcal{S}. \tag{A.22}$$

On interchanging ψ and ϕ, we get

$$\iiint_{\mathcal{V}} \nabla\phi \cdot \nabla\psi\, d\mathcal{V} + \iiint_{\mathcal{V}} \phi\nabla^2\psi\, d\mathcal{V} = \iint_{\mathcal{S}} (\phi\nabla\psi) \cdot \hat{\boldsymbol{\nu}} d\mathcal{S}. \tag{A.23}$$

Subtracting (A.23) from (A.22), we obtain *Green's second identity* or *Green's theorem*,

$$\iiint_{\mathcal{V}} \left(\psi\nabla^2\phi - \phi\nabla^2\psi\right) d\mathcal{V} = \iint_{\mathcal{S}} (\psi\nabla\phi - \phi\nabla\psi) \cdot \hat{\boldsymbol{\nu}} d\mathcal{S}. \tag{A.24}$$

For a surface $\mathcal{S}$ bounded by a closed curve C, with $d\boldsymbol{s}$ an infinitesimal vector displacement along the curve in a direction such that the surface is on the left, we have the *theorem of Stokes*,

$$\iint_{\mathcal{S}} (\nabla \times \boldsymbol{a}) \cdot \hat{\boldsymbol{\nu}} d\mathcal{S} = \oint_C \boldsymbol{a} \cdot d\boldsymbol{s}. \tag{A.25}$$

A related integral theorem is

$$\iint_{\mathcal{S}} \hat{\boldsymbol{\nu}} \times \nabla\phi\, d\mathcal{S} = \oint_C \phi\, d\boldsymbol{s}. \tag{A.26}$$

Appendix B

Properties of Legendre functions

Legendre functions arise from the separation of variables for Laplace's equation in spherical polar co-ordinates (r, θ, ϕ). The separated functional dependence on r for integer n can be either r^n for internal sources or $1/r^{n+1}$ for external sources, while that for ϕ is $e^{im\phi}$ for integer m. The separated functional dependence on θ, say $w(\theta)$, obeys Legendre's equation,

$$\frac{1}{\sin\theta}\frac{d}{d\theta}\left(\sin\theta\frac{dw}{d\theta}\right)+\left\{n\,(n+1)-\frac{m^2}{\sin^2\theta}\right\}w=0, \qquad \text{(B.1)}$$

where n is the integer degree and m is the integer order, with $|m| \le n$. Legendre's equation has regular singularities at infinity and at ± 1 on the real axis. Legendre functions of the first kind are regular in the finite plane and are denoted by $P_n^m(\cos\theta)$. For $m = 0$, they are polynomials of degree n in $z = \cos\theta$ and can be generated by Rodrigues' formula,

$$P_n(z)=\frac{1}{2^n n!}\frac{d^n}{dz^n}\left(z^2-1\right)^n, \qquad \text{(B.2)}$$

giving $P_0(z) = 1$, $P_1(z) = z$, etc. For $m \neq 0$, they are called associated Legendre functions of the first kind, and for $0 \le m \le n$ are generated, in turn, by

$$P_n^m(z)=(-1)^m\frac{\left(1-z^2\right)^{m/2}}{2^n n!}\frac{d^{n+m}}{dz^{n+m}}\left(z^2-1\right)^n \qquad \text{(B.3)}$$

$$=(-1)^m\left(1-z^2\right)^{m/2}\frac{d^m P_n(z)}{dz^m}. \qquad \text{(B.4)}$$

For $m \le 0$,

$$P_n^{-m}=(-1)^m\frac{(n-m)!}{(n+m)!}P_n^m \qquad \text{(B.5)}$$

with $|m| \le n$. Spherical harmonic functions have the form

$$P_n^m(\cos\theta)\,e^{im\phi}. \qquad \text{(B.6)}$$

They are orthogonal under integration over a sphere, or

$$\int_0^{2\pi}\int_0^{\pi} P_n^m(\cos\theta)P_l^k(\cos\theta)\sin\theta\, e^{i(m+k)\phi}\,d\theta\,d\phi = (-1)^m\frac{4\pi}{2n+1}\delta_l^n\delta_{-k}^m. \tag{B.7}$$

For two points with spherical polar co-ordinates (θ, ϕ) and (θ', ϕ') on a sphere of arbitrary radius, by the law of cosines for spherical triangles,

$$\cos\Theta = \cos\theta\cos\theta' + \sin\theta\sin\theta'\cos(\phi - \phi'), \tag{B.8}$$

with Θ the angle between the radius vectors through the two points. The addition theorem for spherical harmonics (Sommerfeld, 1964, pp. 133–144) gives

$$P_n(\cos\Theta) = \sum_{m=-n}^{n}(-1)^m P_n^m(\cos\theta)P_n^{-m}(\cos\theta')e^{im(\phi-\phi')}. \tag{B.9}$$

Associated Legendre functions of the first kind have the closed-form expression,

$$P_n^m(z) = (-1)^m\frac{(2n)!}{2^n n!\,(n-m)!}\left(1-z^2\right)^{m/2}\left\{z^{n-m} + c_2 z^{n-m-2} + \cdots\right\}, \tag{B.10}$$

where $\left\{z^{n-m} + c_2 z^{n-m-2} + \cdots\right\}$ is an even polynomial in z for $n-m$ even, an odd polynomial in z for $n-m$ odd. Associated Legendre functions of the second kind have logarithmic singularities at ± 1 on the real axis and are denoted by $Q_n^m(\cos\theta)$.

The associated Legendre functions of the first kind have properties that we will use in many contexts, while the associated Legendre functions of the second kind will find only rare application in this work.

B.1 Recurrence relations

Among the most important properties of the Legendre functions of the first and second kinds are the recurrence relations that both obey. We record them here for functions of $\cos\theta = z$ of the first kind, with degree n and azimuthal number m, denoted by $P_n^m(\cos\theta)$.

The recurrence relations for fixed azimuthal number but variable degree are

$$\cos\theta\, P_n^m = \frac{1}{2n+1}\left[(n-m+1)\,P_{n+1}^m + (n+m)\,P_{n-1}^m\right], \tag{B.11}$$

$$\sin\theta\frac{dP_n^m}{d\theta} = \frac{1}{2n+1}\left[n\,(n-m+1)\,P_{n+1}^m - (n+1)\,(n+m)\,P_{n-1}^m\right]. \tag{B.12}$$

The recurrence relations for fixed degree but variable azimuthal number are

$$\frac{dP_n^m}{d\theta} = \frac{1}{2}\left[P_n^{m+1} - (n-m+1)\,(n+m)\,P_n^{m-1}\right], \tag{B.13}$$

$$m\cos\theta\, P_n^m = -\frac{\sin\theta}{2}\left[P_n^{m+1} + (n+m)\,(n-m+1)\,P_n^{m-1}\right]. \tag{B.14}$$

Those for which both degree and azimuthal number are variable are

$$\sin\theta\, P_n^{m+1} = \frac{1}{2n+1}\left[(n-m)\,(n-m+1)\,P_{n+1}^m - (n+m)\,(n+m+1)\,P_{n-1}^m\right], \tag{B.15}$$

$$\sin\theta\, P_n^m = \frac{1}{2n+1}\left[P_{n-1}^{m+1} - P_{n+1}^{m+1}\right], \tag{B.16}$$

$$P_{n-1}^m = \cos\theta\, P_n^m + (n-m+1)\sin\theta\, P_n^{m-1}, \tag{B.17}$$

$$2mP_{n-1}^m = -\sin\theta\left[P_n^{m+1} + (n-m)\,(n-m+1)\,P_n^{m-1}\right]. \tag{B.18}$$

B.2 Evaluation of Legendre functions

We examine the evaluation of Legendre functions of the first kind only. For this task (Press *et al.*, 1992, p. 247), we start with the closed expression (B.10) for the sectoral harmonic $P_m^m(z)$,

$$P_m^m(z) = (-1)^m \frac{(2m)!}{2^m m!}\left(1-z^2\right)^{m/2}. \tag{B.19}$$

The factor $(2m)!/2^m m!$ can be rearranged to give

$$\begin{aligned}\frac{(2m)!}{2^m m!} &= \frac{(2m-1)\,(2m-3)\cdots(1)\cdot(2m)\,(2m-2)\cdots(2)}{2^m m!}\\ &= (2m-1)\,(2m-3)\cdots(1)\\ &= (2m-1)!!,\end{aligned} \tag{B.20}$$

a double factorial. It is thus the product of all positive odd integers less than $2m$. By the recurrence relation (B.11),

$$(n-m+1)\,P_{n+1}^m = (2n+1)\,zP_n^m - (n+m)\,P_{n-1}^m. \tag{B.21}$$

This, then, relates the current Legendre function to the two Legendre functions of the next lower degrees. For $m = n$,

$$P_{m+1}^m = (2m+1)\,zP_m^m, \tag{B.22}$$

since $P_n^m = 0$ for $n < m$. Thus, starting with P_m^m as given by (B.19), we can generate P_{m+1}^m. Using (B.21), we can then generate P_{m+2}^m, and so on, to the required P_n^m.

The function subprogramme PMN(Z,M,N) calculates $P_n^m(z)$ for given values of z and the integers m and n. The programme LEGENDRE.FOR is an interactive programme that calls PMN(Z,M,N) to give the value of the Legendre function of the first kind for azimuthal number m, degree n and argument z.

```
C          PROGRAMME LEGENDRE.FOR
C
C Programme to evaluate Legendre functions of the first kind.
C
      IMPLICIT DOUBLE PRECISION(A-H,O-Z)
 40   WRITE(6,10)
 10   FORMAT(1X,'Enter m, n, z')
      READ(5,*)M,N,Z
      F=PMN(Z,M,N)
      WRITE(6,20)M,N,Z,F
 20   FORMAT(1X,'Order=',I2,5X,'Degree=',I2,5X,'Argument=',
     1 F7.4,5X,'Function=',F20.15)
      WRITE(6,30)
 30   FORMAT(1X,'Enter 0 to end, 1 to calculate another value')
      READ(5,*)IBR
      IF(IBR.EQ.1) GO TO 40
      END
C
      DOUBLE PRECISION FUNCTION PMN(Z,M,N)
C
C PMN computes the value of the Legendre function of degree N, azimuthal
C number M, for argument Z. The procedure begins with the formula for
C the sectoral function for N=M, and then uses the recurrence relation
C across degree to arrive at the value for degree N.
C
      IMPLICIT DOUBLE PRECISION(A-H,O-Z)
C Test if input arguments are out of range.
      IF(M.LT.0.OR.DABS(Z).GT.1.D0) WRITE(6,10)
 10   FORMAT(1X,'ONE OR MORE INPUT PARAMETERS OUT OF RANGE')
C Set Legendre function to zero if M>N and return.
      PMN=0.D0
      IF(M.GT.N) GO TO 11
C Begin calculation of sectoral value PMM.
C Set initial value of PMM.
      PMM=1.D0
C Test if Legendre polynomial.
      IF(M.EQ.0) GO TO 12
C Find square root of 1-Z*Z.
      SRUMZ2=DSQRT((1.D0-Z)*(1.D0+Z))
C Complete calculation of sectoral value PMM.
      FACT=1.D0
      DO 13 I=1,M
        PMM=-PMM*FACT*SRUMZ2
        FACT=FACT+2.D0
 13   CONTINUE
C Test if sectoral Legendre function.
 12   IF(M.EQ.N) GO TO 14
C Convert M, N to floating point variables.
      AM=DFLOAT(M)
      AN=DFLOAT(N)
C Find next higher degree Legendre function.
      PMMP1=(2.D0*AM+1.D0)*Z*PMM
C Test if M+1=N.
      IF(M+1.EQ.N) GO TO 15
C Prepare to complete recursion.
      MP2=M+2
      P1=PMM
      P2=PMMP1
C Complete recursion.
      DO 16 I=MP2,N
        AI=DFLOAT(I)
```

Table B.1 *Sample values of Legendre functions of the first kind calculated with the programme LEGENDRE.FOR.*

z	$P_1^1(z)$	$P_2^0(z)$	$P_2^1(z)$	$P_2^2(z)$	$P_3^0(z)$	$P_3^1(z)$	$P_3^2(z)$	$P_3^3(z)$
1.0	0.0000	1.0000	0.0000	0.0000	1.0000	0.0000	0.0000	0.0000
0.9	−0.4359	0.7150	−1.1769	0.5700	0.4725	−1.9942	2.5650	−1.2423
0.8	−0.6000	0.4600	−1.4400	1.0800	0.0800	−1.9800	4.3200	−3.2400
0.7	−0.7141	0.2350	−1.4997	1.5300	−0.1925	−1.5533	5.3550	−5.4632
0.6	−0.8000	0.0400	−1.4400	1.9200	−0.3600	−0.9600	5.7600	−7.6800
0.5	−0.8660	−0.1250	−1.2990	2.2500	−0.4375	−0.3248	5.6250	−9.7428
0.4	−0.9165	−0.2600	−1.0998	2.5200	−0.4400	0.2750	5.0400	−11.5481
0.3	−0.9539	−0.3650	−0.8585	2.7300	−0.3825	0.7870	4.0950	−13.0213
0.2	−0.9798	−0.4400	−0.5879	2.8800	−0.2800	1.1758	2.8800	−14.1091
0.1	−0.9950	−0.4850	−0.2985	2.9700	−0.1475	1.4179	1.4850	−14.7756
0.0	−1.0000	−0.5000	0.0000	3.0000	0.0000	1.5000	0.0000	−15.0000

```
      P3=((2.D0*AI-1.D0)*Z*P2-(AI+AM-1.D0)*P1)/(AI-AM)
      P1=P2
      P2=P3
16    CONTINUE
      PMN=P3
      GO TO 11
14    PMN=PMM
      GO TO 11
15    PMN=PMMP1
11    CONTINUE
      RETURN
      END
```

Sample values of $P_n^m(z)$ as output of these codes are given in Table B.1.

The first few Legendre functions in explicit form, beginning with the Legendre polynomials, are as follows:

$$P_0(z) = 1, \quad P_2(z) = \tfrac{1}{2}\left(3z^2 - 1\right), \quad P_3(z) = \tfrac{1}{2}\left(5z^3 - 3z\right),$$

$$P_1(z) = z, \quad P_2^1(z) = -3\left(1 - z^2\right)^{1/2} z, \quad P_3^1(z) = -\tfrac{3}{2}\left(1 - z^2\right)^{1/2}\left(5z^2 - 1\right),$$

$$P_1^1(z) = -\left(1 - z\right)^{1/2}, \quad P_2^2(z) = 3\left(1 - z^2\right), \quad P_3^2(z) = 15\left(1 - z^2\right)z,$$

$$P_3^3(z) = -15\left(1 - z^2\right)^{3/2}.$$

Appendix C
Numerical Earth models

The inversion of seismological observations leads to the construction of models of the physical properties of Earth. The basic models take Earth to be spherically symmetric with properties, such as density, the Lamé coefficients of elasticity, and gravity, listed as functions of radius alone. These models form the numerical basis for the calculation of Earth's dynamics, including its short-period and long-period free oscillations, its response to tidal forcing and the dynamics of its rotation.

C.1 The Earth models

We list here the files cal8.dat, 1066a.dat, prem.dat and core11.dat for four well-known Earth models, Cal8 (Bullen and Bolt, 1985, pp. 471–473), 1066A (Gilbert and Dziewonski, 1975), PREM (Dziewonski and Anderson, 1981) and Core11 (Widmer *et al.*, 1988), respectively. The first line in each case gives the name of the Earth model in the format 10A8. The second line gives the number of Love numbers to be calculated in the format I10. The third line gives the number of model points and the initial number of integration steps for the inner core, outer core, mantle and crust in the format 8I10. Although a variable stepsize Runge–Kutta method is used in calculations, initial stepsizes are specified. The columns tabulate the radius, density, the Lamé coefficients of elasticity, and gravity in the format 1X,F10.1,F10.2,F10.1,F10.1,F10.1. Following tradition, radius is expressed in kilometres, density in grams per cubic centimetre, the Lamé coefficients in kilobars and gravity in centimetres per second per second. After they are read in, these Earth properties are scaled to SI units before proceeding to calculations.

BOLT AND UHRHAMMER MODEL CAL8

24

r (km)	ρ_0 $(\mathrm{g\,cm^{-3}})$	λ (kbar)	μ (kbar)	g_0 $(\mathrm{cm\,s^{-2}})$			
6	512	10	10	22	704	5	64
0.0	13.58	13912.7	1760.0	0.0			
171.0	13.59	13923.0	1761.0	76.0			
771.0	13.55	13645.0	1737.0	294.0			
971.0	13.49	13310.7	1700.0	368.0			
1171.0	13.38	12735.3	1639.0	442.0			
1216.0	13.34	12570.7	1625.0	459.0			
1216.0	12.17	12637.0	0.0	459.0			
1371.0	12.14	12581.0	0.0	502.0			
1571.0	12.03	12320.0	0.0	560.0			
1821.0	11.84	11840.0	0.0	634.0			
2171.0	11.52	10929.0	0.0	736.0			
2571.0	11.11	9671.0	0.0	848.0			
2971.0	10.62	8299.0	0.0	953.0			
3171.0	10.33	7569.0	0.0	1001.0			
3371.0	10.01	6846.0	0.0	1046.0			
3486.0	9.82	6427.0	0.0	1071.0			
3486.0	5.92	4847.0	2868.0	1071.0			
3521.0	5.74	4789.0	2862.0	1066.0			
3591.0	5.52	4639.0	2838.0	1058.0			
3671.0	5.43	4443.7	2807.0	1048.0			
3971.0	5.23	3917.7	2615.0	1020.0			
4371.0	5.06	3388.0	2451.0	999.0			
4771.0	4.92	2984.3	2242.0	992.0			
4971.0	4.83	2766.0	2130.0	992.0			
5171.0	4.74	2522.7	2027.0	994.0			
5371.0	4.61	2251.3	1912.0	997.0			
5571.0	4.43	1986.7	1736.0	1001.0			
5701.0	4.22	1925.7	1442.0	1002.0			
5731.0	4.09	1829.7	1286.0	1002.0			
5771.0	3.98	1617.7	1169.0	1002.0			
5871.0	3.84	1503.0	1062.0	1000.0			
5971.0	3.61	1310.2	888.5	998.0			
6071.0	3.47	993.3	760.0	994.0			
6151.0	3.42	943.7	686.0	991.0			
6221.0	3.40	926.0	645.0	998.0			
6291.0	3.36	873.0	639.0	986.0			
6331.0	3.35	801.2	663.4	985.0			
6351.0	3.34	751.3	682.0	984.0			
6351.0	3.26	628.2	602.8	984.0			
6356.0	3.26	628.2	602.8	983.0			
6361.0	2.16	170.9	114.3	983.0			
6366.0	2.16	170.9	114.3	982.0			
6371.0	2.16	170.9	114.3	982.0			

GILBERT AND DZIEWONSKI EARTH MODEL 1066A

24

33	512	33	10	89	704	5	64
0.0	13.42	13717.0	1768.0	0.0			
38.4	13.42	13711.0	1767.0	14.4			
76.8	13.41	13696.0	1766.0	28.8			

115.2	13.41	13678.0	1765.0	43.2
153.6	13.40	13653.0	1763.0	57.5
192.0	13.39	13624.0	1761.0	71.9
230.4	13.38	13588.0	1759.0	86.2
268.8	13.38	13547.0	1757.0	100.6
307.2	13.37	13502.0	1754.0	114.9
345.6	13.36	13452.0	1751.0	129.2
384.0	13.35	13398.0	1747.0	143.5
422.4	13.33	13340.0	1743.0	157.7
460.8	13.32	13280.0	1738.0	171.9
499.2	13.30	13218.0	1734.0	186.1
537.6	13.28	13158.0	1728.0	200.3
576.0	13.27	13098.0	1723.0	214.4
614.4	13.25	13041.0	1717.0	228.5
652.9	13.23	12987.0	1711.0	242.6
691.3	13.22	12935.0	1705.0	256.6
729.7	13.21	12887.0	1699.0	270.7
768.1	13.19	12842.0	1693.0	284.7
806.5	13.18	12801.0	1686.0	298.6
844.9	13.16	12761.0	1679.0	312.6
883.3	13.14	12724.0	1672.0	326.5
921.7	13.13	12689.0	1665.0	340.4
960.1	13.12	12656.0	1657.0	354.3
998.5	13.10	12624.0	1650.0	368.2
1036.9	13.09	12594.0	1642.0	382.0
1075.3	13.07	12563.0	1635.0	395.8
1113.7	13.06	12533.0	1627.0	409.6
1152.1	13.04	12503.0	1619.0	423.4
1190.5	13.03	12473.0	1612.0	437.2
1229.5	13.02	12455.0	1605.0	451.1
1229.5	12.15	13180.0	0.0	451.1
1299.4	12.11	12976.0	0.0	471.2
1369.8	12.07	12783.0	0.0	491.8
1440.3	12.03	12597.0	0.0	512.5
1510.7	11.98	12421.0	0.0	533.5
1581.2	11.94	12255.0	0.0	554.6
1651.7	11.90	12091.0	0.0	575.7
1722.1	11.86	11930.0	0.0	596.9
1792.6	11.82	11768.0	0.0	618.1
1863.0	11.77	11606.0	0.0	639.3
1933.5	11.72	11444.0	0.0	660.4
2003.9	11.66	11279.0	0.0	681.4
2074.4	11.60	11102.0	0.0	702.3
2144.9	11.54	10905.0	0.0	723.0
2215.3	11.47	10690.0	0.0	743.6
2285.8	11.41	10468.0	0.0	763.9
2356.2	11.34	10254.0	0.0	784.1
2426.7	11.27	10052.0	0.0	804.0
2497.2	11.19	9853.0	0.0	823.8
2567.6	11.12	9648.0	0.0	843.2
2638.1	11.03	9428.0	0.0	862.5
2708.5	10.95	9195.0	0.0	881.4
2779.0	10.86	8958.0	0.0	900.0
2849.5	10.77	8722.0	0.0	918.4
2919.9	10.69	8487.0	0.0	939.4
2990.4	10.60	8248.0	0.0	954.2
3060.8	10.52	8000.0	0.0	971.7
3131.3	10.43	7742.0	0.0	989.0
3201.8	10.33	7471.0	0.0	1005.9
3272.2	10.23	7188.0	0.0	1022.5
3342.7	10.13	6906.0	0.0	1038.7

3413.1	10.03	6631.0	0.0	1054.6
3484.3	9.91	6362.0	0.0	1070.2
3484.3	5.53	4590.0	2905.0	1070.2
3518.2	5.52	4595.0	2890.0	1065.2
3552.9	5.51	4600.0	2873.0	1060.4
3587.5	5.50	4571.0	2856.0	1055.8
3622.1	5.48	4539.0	2839.0	1051.5
3656.7	5.47	4511.0	2823.0	1047.4
3691.4	5.46	4472.0	2806.0	1043.6
3726.0	5.45	4430.0	2790.0	1040.0
3760.6	5.43	4380.0	2773.0	1036.5
3795.3	5.42	4324.0	2757.0	1033.3
3829.9	5.41	4262.0	2740.0	1030.3
3864.5	5.39	4195.0	2723.0	1027.4
3899.2	5.37	4126.0	2705.0	1024.8
3933.8	5.36	4054.0	2687.0	1022.2
3968.4	5.34	3982.0	2669.0	1019.9
4003.1	5.32	3912.0	2650.0	1017.6
4037.7	5.30	3845.0	2630.0	1015.5
4072.3	5.28	3781.0	2611.0	1013.5
4107.0	5.25	3721.0	2591.0	1011.7
4141.6	5.23	3668.0	2570.0	1009.9
4176.2	5.21	3619.0	2549.0	1008.3
4210.8	5.19	3575.0	2528.0	1006.7
4245.5	5.17	3537.0	2507.0	1005.3
4280.1	5.15	3504.0	2485.0	1003.9
4314.7	5.13	3473.0	2464.0	1002.7
4349.4	5.11	3446.0	2442.0	1001.5
4384.0	5.09	3421.0	2420.0	1000.4
4418.6	5.07	3395.0	2399.0	999.4
4453.3	5.05	3369.0	2378.0	998.5
4487.9	5.03	3341.0	2358.0	997.6
4522.5	5.01	3311.0	2338.0	996.9
4557.2	4.99	3278.0	2318.0	996.2
4591.8	4.97	3243.0	2299.0	995.6
4626.4	4.95	3204.0	2281.0	995.0
4661.1	4.93	3165.0	2262.0	994.5
4695.7	4.91	3122.0	2243.0	994.1
4730.3	4.89	3079.0	2224.0	993.7
4764.9	4.87	3035.0	2204.0	993.4
4799.6	4.86	2991.0	2184.0	993.1
4834.2	4.84	2944.0	2165.0	992.9
4868.8	4.83	2899.0	2145.0	992.7
4903.5	4.81	2852.0	2126.0	992.7
4938.1	4.80	2807.0	2107.0	992.6
4972.7	4.78	2761.0	2089.0	992.7
5007.4	4.77	2714.0	2072.0	992.7
5042.0	4.76	2667.0	2054.0	992.9
5076.6	4.74	2617.0	2036.0	993.1
5111.3	4.72	2564.0	2017.0	993.3
5145.9	4.71	2509.0	1996.0	993.6
5180.5	4.68	2451.0	1971.0	993.9
5215.2	4.66	2390.0	1944.0	994.2
5249.8	4.63	2331.0	1915.0	994.5
5284.4	4.61	2278.0	1886.0	994.8
5319.1	4.59	2233.0	1860.0	995.2
5353.7	4.57	2199.0	1837.0	995.6
5388.3	4.55	2180.0	1820.0	996.0
5422.9	4.54	2175.0	1809.0	996.4
5457.6	4.54	2178.0	1802.0	996.9
5492.2	4.54	2191.0	1803.0	997.4

5526.8	4.54	2200.0	1801.0	998.1
5561.5	4.51	2183.0	1770.0	998.7
5596.1	4.47	2138.0	1714.0	999.3
5630.7	4.40	2066.0	1633.0	999.9
5665.4	4.32	1970.0	1531.0	1000.3
5700.0	4.21	1844.0	1403.0	1000.5
5731.2	4.11	1731.0	1291.0	1000.4
5762.5	4.03	1644.0	1206.0	1000.2
5793.8	3.96	1574.0	1140.0	999.8
5825.0	3.90	1513.0	1082.0	999.4
5856.2	3.85	1461.0	1031.0	998.8
5887.5	3.81	1419.0	989.0	998.2
5918.8	3.76	1384.0	953.0	997.6
5950.0	3.71	1343.0	909.0	996.9
5975.6	3.66	1301.0	866.0	996.2
6001.3	3.60	1259.0	822.0	995.5
6026.9	3.55	1218.0	784.0	994.7
6052.5	3.51	1180.0	749.0	993.8
6078.1	3.47	1143.0	718.0	992.9
6103.8	3.44	1101.0	693.0	992.0
6129.4	3.42	1055.0	674.0	991.1
6155.0	3.40	1006.0	661.0	990.1
6180.6	3.39	951.0	653.0	989.2
6206.3	3.39	897.0	649.0	988.3
6231.9	3.38	841.0	651.0	987.4
6257.5	3.37	764.0	662.0	986.6
6283.1	3.37	693.0	676.0	985.7
6308.8	3.36	627.0	691.0	984.9
6334.4	3.35	572.0	710.0	984.2
6360.0	3.34	539.0	722.0	983.4
6360.0	2.18	191.0	145.0	983.4
6362.0	2.18	191.0	145.0	983.1
6365.5	2.18	191.0	145.0	982.7
6369.0	2.18	191.0	145.0	982.3
6371.0	2.18	190.0	145.0	982.0

EARTH MODEL PREM ISOTROPIC VERSION

24

14 512 24 10 43 704 5 64

0.0	13.09	13079.5	1760.8	0.0
100.0	13.09	13074.9	1759.4	36.6
200.0	13.08	13060.8	1755.4	73.1
300.0	13.07	13037.3	1748.7	109.6
400.0	13.05	13004.5	1739.3	146.0
500.0	13.03	12962.5	1727.3	182.4
600.0	13.01	12911.1	1712.8	218.6
700.0	12.98	12850.6	1695.7	254.7
800.0	12.95	12780.9	1676.0	290.7
900.0	12.91	12702.1	1654.0	326.5
1000.0	12.87	12614.4	1629.6	362.0
1100.0	12.83	12517.8	1602.8	397.4
1200.0	12.77	12412.4	1573.9	432.5
1221.5	12.76	12388.6	1567.4	440.0
1221.5	12.17	13047.3	0.0	440.0
1300.0	12.12	12887.8	0.0	463.7
1400.0	12.07	12679.3	0.0	494.1
1500.0	12.01	12464.4	0.0	524.8
1600.0	11.95	12242.4	0.0	555.5
1700.0	11.88	12012.7	0.0	586.1
1800.0	11.81	11775.0	0.0	616.7
1900.0	11.73	11528.6	0.0	647.0
2000.0	11.65	11273.3	0.0	677.2

2100.0	11.57	11008.8	0.0	707.0
2200.0	11.48	10734.8	0.0	736.5
2300.0	11.39	10451.2	0.0	765.6
2400.0	11.29	10157.9	0.0	794.3
2500.0	11.19	9854.9	0.0	822.5
2600.0	11.08	9542.3	0.0	850.2
2700.0	10.97	9220.4	0.0	877.5
2800.0	10.85	8889.4	0.0	904.1
2900.0	10.73	8549.8	0.0	930.2
3000.0	10.60	8201.9	0.0	955.7
3100.0	10.47	7846.5	0.0	980.5
3200.0	10.33	7484.3	0.0	1004.6
3300.0	10.18	7116.1	0.0	1028.0
3400.0	10.03	6742.8	0.0	1050.7
3480.0	9.90	6441.3	0.0	1068.2
3480.0	5.57	4597.6	2937.7	1068.2
3500.0	5.56	4581.6	2932.6	1065.3
3600.0	5.51	4502.5	2906.9	1052.0
3630.0	5.49	4479.2	2899.2	1048.4
3700.0	5.46	4375.6	2855.5	1040.7
3800.0	5.41	4232.6	2794.2	1031.0
3900.0	5.36	4094.5	2734.2	1022.7
4000.0	5.31	3960.7	2675.2	1015.8
4100.0	5.26	3830.5	2616.9	1010.1
4200.0	5.21	3703.2	2559.2	1005.4
4300.0	5.16	3578.3	2501.8	1001.6
4400.0	5.11	3455.1	2444.7	998.6
4500.0	5.05	3333.3	2387.6	996.4
4600.0	5.00	3212.2	2330.5	994.7
4700.0	4.95	3091.6	2273.1	993.7
4800.0	4.90	2971.0	2215.4	993.1
4900.0	4.84	2850.2	2157.2	993.0
5000.0	4.79	2728.8	2098.4	993.3
5100.0	4.73	2606.6	2039.0	993.8
5200.0	4.68	2483.5	1978.9	994.7
5300.0	4.62	2359.3	1918.0	995.7
5400.0	4.56	2234.0	1856.3	997.0
5500.0	4.50	2107.5	1793.7	998.4
5600.0	4.44	1979.9	1730.3	999.9
5650.0	4.41	1974.6	1638.8	1000.6
5701.0	4.19	1848.6	1393.5	1001.4
5736.0	3.98	1706.4	1224.1	1000.9
5771.0	3.98	1682.9	1209.7	1000.4
5821.0	3.91	1579.6	1128.4	999.7
5871.0	3.85	1480.6	1050.7	998.8
5921.0	3.79	1385.6	976.6	997.9
5971.0	3.63	1246.2	856.1	996.9
6016.0	3.52	1155.6	789.5	995.2
6061.0	3.49	1114.5	773.1	993.6
6106.0	3.46	1074.3	756.9	992.0
6151.0	3.40	933.8	698.5	990.5
6186.0	3.36	837.4	660.6	989.1
6221.0	3.37	842.4	665.2	987.8
6256.0	3.37	847.3	669.8	986.6
6291.0	3.37	852.3	674.5	985.5
6311.0	3.38	855.1	677.1	984.9
6331.0	3.38	858.0	679.8	984.4
6346.6	3.38	860.2	681.9	983.9
6346.6	2.90	458.8	441.1	983.9
6356.0	2.75	400.5	353.7	983.3
6368.0	1.81	181.8	133.1	982.2

6369.5	1.02	21.4	0.1	981.9
6371.0	1.02	21.4	0.1	981.6

WIDMER EARTH MODEL CORE11

24

33	512	33	10	84	704	5	64

0.0	13.02	13078.0	1709.3	0.0
38.2	13.02	13077.0	1709.0	13.9
76.3	13.02	13075.0	1708.4	27.8
114.5	13.02	13072.0	1707.4	41.6
152.7	13.01	13067.0	1706.0	55.5
190.9	13.01	13061.0	1704.2	69.4
229.0	13.01	13053.0	1702.1	83.2
267.2	13.00	13044.0	1699.5	97.1
305.4	13.00	13034.0	1696.5	110.9
343.6	12.99	13023.0	1693.2	124.7
381.7	12.99	13010.0	1689.4	138.6
419.9	12.98	12995.0	1685.3	152.4
458.1	12.97	12979.0	1680.8	166.2
496.2	12.96	12962.0	1675.9	179.9
534.4	12.96	12944.0	1670.7	193.7
572.6	12.95	12924.0	1665.1	207.5
610.8	12.94	12903.0	1659.1	221.2
648.9	12.93	12881.0	1652.7	234.9
687.1	12.91	12857.0	1646.0	248.6
725.3	12.90	12832.0	1638.8	262.3
763.4	12.89	12805.0	1631.4	275.9
801.6	12.88	12778.0	1623.6	289.5
839.8	12.86	12749.0	1615.4	303.1
878.0	12.85	12718.0	1606.9	316.7
916.1	12.84	12686.0	1598.0	330.2
954.3	12.82	12653.0	1588.8	343.7
992.5	12.80	12619.0	1579.2	357.2
1030.6	12.79	12584.0	1569.3	370.7
1068.8	12.77	12547.0	1559.1	384.1
1107.0	12.75	12508.0	1548.5	397.4
1145.2	12.73	12469.0	1537.6	410.8
1183.3	12.71	12428.0	1526.4	424.1
1218.8	12.69	12386.0	1514.9	436.4
1218.8	12.22	13036.0	0.0	436.4
1292.1	12.18	12893.0	0.0	459.0
1362.7	12.14	12747.0	0.0	480.9
1433.2	12.09	12597.0	0.0	502.9
1503.8	12.05	12445.0	0.0	524.9
1574.4	12.00	12288.0	0.0	546.8
1645.0	11.95	12128.0	0.0	568.7
1715.5	11.90	11964.0	0.0	590.5
1786.1	11.85	11797.0	0.0	612.2
1856.7	11.79	11624.0	0.0	633.8
1927.3	11.74	11448.0	0.0	655.3
1997.9	11.68	11267.0	0.0	676.6
2068.4	11.62	11081.0	0.0	697.8
2139.0	11.55	10890.0	0.0	718.7
2209.6	11.49	10695.0	0.0	739.5
2280.2	11.42	10495.0	0.0	760.0
2350.8	11.35	10290.0	0.0	780.4
2421.3	11.27	10081.0	0.0	800.5
2491.9	11.20	9866.4	0.0	820.4
2562.5	11.12	9647.1	0.0	840.0
2633.1	11.04	9423.2	0.0	859.3
2703.6	10.96	9194.7	0.0	878.4
2774.2	10.87	8961.6	0.0	897.1

2844.8	10.78	8724.1	0.0	915.6
2915.4	10.69	8482.5	0.0	933.8
2986.0	10.60	8236.8	0.0	951.6
3056.5	10.50	7987.3	0.0	969.1
3127.1	10.40	7734.2	0.0	986.2
3197.7	10.30	7477.9	0.0	1003.0
3268.3	10.20	7218.5	0.0	1019.4
3338.8	10.09	6956.5	0.0	1035.5
3409.4	9.98	6692.2	0.0	1051.1
3482.4	9.86	6425.8	0.0	1066.9
3482.4	5.65	4589.1	2954.2	1066.9
3517.5	5.63	4559.1	2944.6	1062.1
3555.0	5.62	4529.4	2935.0	1057.3
3592.5	5.60	4499.8	2925.3	1052.7
3630.0	5.54	4482.0	2908.3	1048.5
3666.5	5.49	4439.1	2877.7	1044.3
3703.0	5.47	4385.7	2854.9	1040.4
3739.4	5.46	4333.2	2832.2	1036.7
3775.9	5.44	4281.3	2809.8	1033.3
3812.4	5.42	4230.1	2787.5	1030.0
3848.9	5.40	4179.6	2765.3	1026.9
3885.4	5.38	4129.7	2743.3	1024.0
3921.9	5.36	4080.5	2721.4	1021.3
3958.3	5.34	4031.6	2699.7	1018.7
3994.8	5.32	3983.4	2678.0	1016.3
4031.3	5.31	3935.7	2656.5	1014.1
4067.8	5.29	3888.4	2635.0	1012.0
4104.3	5.27	3841.5	2613.7	1010.1
4140.7	5.25	3795.0	2592.4	1008.2
4177.2	5.23	3748.9	2571.2	1006.6
4213.7	5.21	3703.0	2550.0	1005.0
4250.2	5.19	3657.5	2528.9	1003.6
4286.7	5.17	3612.3	2507.8	1002.2
4323.1	5.15	3567.3	2486.7	1001.0
4359.6	5.14	3522.5	2465.7	999.9
4396.1	5.12	3478.0	2444.7	998.9
4432.6	5.10	3433.6	2423.7	998.0
4469.1	5.08	3389.3	2402.7	997.1
4505.6	5.06	3345.2	2381.7	996.4
4542.0	5.04	3301.2	2360.6	995.7
4578.5	5.02	3257.3	2339.6	995.2
4615.0	5.00	3213.4	2318.5	994.7
4651.5	4.98	3169.6	2297.4	994.2
4688.0	4.96	3125.8	2276.2	993.9
4724.4	4.94	3082.1	2255.0	993.6
4760.9	4.92	3038.3	2233.8	993.3
4797.4	4.90	2994.5	2212.5	993.1
4833.9	4.88	2950.7	2191.1	993.0
4870.4	4.86	2906.8	2169.7	993.0
4906.9	4.84	2862.8	2148.2	992.9
4943.3	4.82	2818.8	2126.6	993.0
4979.8	4.80	2774.7	2104.9	993.0
5016.3	4.78	2730.5	2083.2	993.2
5052.8	4.76	2686.2	2061.4	993.3
5089.3	4.74	2641.8	2039.4	993.5
5125.7	4.72	2597.2	2017.4	993.7
5162.2	4.70	2552.5	1995.3	994.0
5198.7	4.67	2507.7	1973.1	994.3
5235.2	4.65	2462.8	1950.8	994.6
5271.7	4.63	2417.7	1928.3	995.0
5308.1	4.61	2372.4	1905.8	995.3

5344.6	4.59	2327.0	1883.2	995.7
5381.1	4.57	2281.4	1860.4	996.1
5417.6	4.54	2235.7	1837.5	996.6
5454.1	4.52	2189.9	1814.6	997.0
5490.6	4.50	2143.9	1791.5	997.5
5527.0	4.48	2097.7	1768.3	998.0
5563.5	4.46	2051.4	1745.0	998.5
5600.0	4.43	1984.6	1729.3	999.0
5633.7	4.40	1960.9	1675.0	999.4
5667.3	4.38	1956.7	1614.3	999.9
5701.0	4.17	1834.4	1399.8	1000.3
5771.0	3.99	1677.4	1209.8	999.2
5804.3	3.96	1615.4	1149.2	998.8
5837.7	3.92	1547.9	1096.2	998.3
5871.0	3.88	1482.3	1044.9	997.8
5904.3	3.84	1418.6	995.1	997.3
5937.7	3.79	1356.7	946.8	996.7
5971.0	3.65	1246.4	852.7	996.0
5993.5	3.53	1175.1	796.9	995.2
6016.0	3.52	1154.2	788.6	994.3
6038.5	3.50	1133.6	780.4	993.5
6061.0	3.49	1113.2	772.2	992.7
6083.5	3.47	1092.9	764.1	991.9
6106.0	3.46	1072.9	756.0	991.1
6128.5	3.45	1053.1	748.0	990.3
6151.0	3.38	928.0	696.0	989.5
6179.0	3.33	826.8	655.4	988.3
6207.0	3.33	831.1	659.0	987.2
6235.0	3.33	835.2	662.6	986.1
6263.0	3.34	839.2	666.3	985.0
6291.0	3.37	855.7	665.8	984.1
6318.8	3.40	872.0	665.2	983.3
6346.6	3.40	875.4	669.1	982.5
6346.6	2.90	458.8	441.1	982.5
6356.0	2.75	400.5	353.7	981.9
6368.0	1.81	181.8	133.1	980.8
6369.5	1.02	21.4	0.1	980.5
6371.0	1.02	21.4	0.1	980.1

References

Abramowitz, M., and Stegun, I. A. 1964. *Handbook of Mathematical Functions, Applied Mathematics Series, vol. 55*. National Bureau of Standards, Washington.

Acheson, D. J. 1975. On hydromagnetic oscillations within the Earth and core–mantle coupling. *Geophys. J. R. Astr. Soc.*, **43**, 253–268.

Aldridge, K. D., and Toomre, A. 1969. Axisymmetric inertial oscillations of a fluid in a rotating spherical container. *J. Fluid Mech.*, **37(2)**, 307–323.

Alterman, Z., Jarosch, H., and Pekeris, C. L. 1959. Oscillations of the Earth. *Proc. Roy. Soc. A*, **252**, 80–95.

Backus, G. 1958. A class of self-sustaining dissipative spherical dynamos. *Ann. Phys. (N. Y.)*, **4**, 372–447.

Bluestein, L. I. 1970. A linear filtering approach to the computation of the discrete Fourier transform. *IEEE Trans. Audio Electroacoust.*, **AU-18**, 451–455.

Box, G. E. P., Jenkins, G. M., and Reinsel, G. C. 1994. *Time Series Analysis, Forecasting and Control, 3rd ed.* Prentice-Hall, Englewood Cliffs, NJ.

Brazhkin, V. V. 1998. Investigation of the crystallization of liquid iron under pressure: extrapolation of the melt viscosity into the megabar range. *JETP Lett.*, **68**, 502–508.

Brazhkin, V. V., and Lyapin, A. G. 2000. Universal viscosity growth in metallic melts at megabar pressures: the vitreous state of the Earth's inner core. *Physics-Uspekhi*, **43(5)**, 493–508.

Bryan, G. H. 1889. The waves on a rotating liquid spheroid of finite ellipticity. *Phil. Trans. R. Soc. Lond. A*, **180**, 187–219.

Bullen, K. E. 1975. *The Earth's Density*. Chapman and Hall, London.

Bullen, K. E., and Bolt, B. A. 1985. *An Introduction to the Theory of Seismology, 4th ed.* Cambridge University Press.

Burden, R. L., and Faires, J. D. 1988. *Numerical Analysis, 4th ed.* PWS-Kent Publishing Co., Boston.

Burg, J. P. 1967. Maximum entropy spectral analysis. In: *Proc. 37th Annual Meeting Soc. Explor. Geophys.* Oklahoma City, OK.

Burg, J. P. 1968. A new analysis technique for time series data. In: *Proc. NATO Advan. Study Inst. Signal Processing*. Enschede, Netherlands.

Carnahan, B., Luther, H. A., and Wilkes, J. O. 1969. *Applied Numerical Methods*. John Wiley & Sons, New York.

Chandrasekhar, S. 1969. *Ellipsoidal Figures of Equilibrium.* Yale University Press.

Claerbout, J. F. 1985. *Fundamentals of Geophysical Data Processing*. Blackwell Scientific Publications, Palo Alto, CA.

Cole, R. H. 1968. *Theory of Ordinary Differential Equations*. Appleton-Century-Crofts, New York.

Cooley, J. W., and Tukey, J. W. 1965. An algorithm for the machine calculation of complex Fourier series. *Math. Computat.*, **19**, 297–301.

Copson, E. T. 1955. *Theory of Functions of a Complex Variable*. Oxford University Press.

Courtier, N., Ducarme, B., Goodkind, J., *et al.* 2000. Global superconducting gravimeter observations and the search for the translational modes of the inner core. *Phys. Earth Planet. Inter.*, **117**, 3–20.

Crossley, D. J., and Smylie, D. E. 1975. Electromagnetic and viscous damping of core oscillations. *Geophys. J. R. Astr. Soc.*, **42**, 1011–1033.

Dahlen, F. A., and Tromp, J. 1998. *Theoretical Global Seismology*. Princeton University Press.

Darwin, G. H. 1899. The theory of the figure of the Earth carried to the second order of small quantities. *Mon. Not. R. Astr. Soc.*, **60**, 82–124.

de Sitter, W. 1924. On the flattening and the constitution of the Earth. *Bull. Astr. Inst. Neth.*, **55**, 97–108.

Dziewonski, A. M., and Anderson, D. L. 1981. Preliminary reference Earth model. *Phys. Earth Planet. Inter.*, **25**, 297–356.

Erdélyi, A., Magnus, W., Oberhettinger, F., and Tricomi, F. G. 1953. *Higher Transcendental Functions, vol. II, Bateman MS Project*. McGraw-Hill, New York.

Farrell, W. E. 1972. Deformation of the Earth by surface loads. *Reviews of Geophysics and Space Physics*, **10**, 761–797.

Feissel, M., and Mignard, F. 1998. Letter to the Editor: The adoption of ICRS on 1 January 1998: Meaning and consequences. *Astronomy and Astrophysics*, **331**, L33–L36.

Fischer, I. 1975. The figure of the Earth — changes in concepts. *Geophys. Surveys*, **2**, 3–54.

Fricke, W., Schwan, H., and Lederle, T. 1988. *Fifth Fundamental Catalogue, part 1*. Veroff. Astron. Rechen Inst., Heidelberg.

Friedlander, S. 1985. Internal oscillations in the Earth's fluid core. *Geophys. J. R. Astr. Soc.*, **80**, 345–361.

Gilbert, F., and Backus, G. 1966. Propagator matrices in elastic wave and vibration problems. *Geophysics*, **31**, 326.

Gilbert, F., and Dziewonski, A. M. 1975. An application of normal mode theory to the retrieval of structural parameters and source mechanisms from seismic spectra. *Phil. Trans. R. Soc. Lond. A*, **278**, 187–269.

Golub, G., and Kahan, W. 1965. Calculating the singular values and pseudo-inverse of a matrix. *J. SIAM Numer. Anal., Ser. B*, **2(2)**, 205–224.

Greenspan, H. P. 1969. *The Theory of Rotating Fluids*. Cambridge University Press.

Guinot, B., and Feissel, M. 1969. *Annual Report for 1968*. Bureau International de l'Heure, 61 avenue de l'Observatoire, Paris-14e.

Hayes, G. H. 2011. Rapid source characterization of the 2011 M 9.0 off the Pacific coast of Tohoku earthquake. *Earth Planets Space*, **63**, 529–534.

Heideman, M. T., Johnson, D. H., and Burris, C. S. 1995. Gauss and the history of the FFT. *IEEE Acoustics, Speech, and Signal Processing Magazine*, **1**, 14–21.

Hida, Y., Li, X., and Bailey, D. 2000. *Quad Double Arithmetic: Algorithms, Implementation, and Application*. Tech. rept. LBNL-46996. Lawrence Berkeley National Laboratory, Berkeley, CA.

Hofmann-Wellenhof, B., and Moritz, H. 2006. *Physical Geodesy*. Springer-Verlag, Vienna.

IAG. 1971. *Geodetic Reference System 1967*. Bureau Central de l'Association Internationale de Géodésie, Paris.

Israel, M., Ben-Menahem, A., and Singh, J. 1973. Residual deformation of real Earth models with application to the Chandler wobble. *Geophys. J. R. Astr. Soc.*, **32**, 219–247.

Jeffreys, H., and Vincente, R. O. 1966. Comparison of forms of the elastic equations for the Earth. *Mem. Acad. R. Belgique*, **37**, 5–31.

Jenkins, G. M., and Watts, D. G. 1968. *Spectral Analysis and Its Applications*. Holden-Day, San Francisco.

Jiang, X. 1993. *Wobble-Nutation Modes of the Earth*. Ph.D. thesis, York University, Toronto.

Johnson, I. M., and Smylie, D. E. 1977. A variational approach to whole-Earth dynamics. *Geophys. J. R. Astr. Soc.*, **50**, 35–54.

Kreyszig, E. 1967. *Advanced Engineering Mathematics*. John Wiley & Sons, New York, NY.

Kutta, W. 1901. Beitrag zur naherungsweisen integration totaler differentialgleichungen. *Z. angew. Math. u. Phys.*, **46**, 435–453.

Lacoss, R. T. 1971. Data adaptive spectral analysis methods. *Geophysics*, **36**, 661–675.

Lamb, H. 1881. On the oscillations of a viscous spheroid. *Proc. Lond. Math. Soc.*, **13**, 51–56.

Lambeck, K. 1980. *The Earth's Variable Rotation*. Cambridge University Press.

Lambeck, K. 1988. *Geophysical Geodesy*. Oxford University Press.

Lancaster, P. 1966. *Lambda-Matrices and Vibrating Systems*. Pergamon Press, Oxford.

Lanczos, C. 1964. A precision approximation of the gamma function. *J. SIAM Numer. Anal., Ser. B*, **1**, 86–96.

Liapounov, A. 1930. Sur la figure des corpes célestes. *Bull. Acad. Sci. USSR*, **7**.

Love, A. E. H. 1927. *A Treatise on the Mathematical Theory of Elasticity, 4th ed.* Cambridge University Press.

MacCullagh, J. 1845. On the rotation of a solid body. *Proc. Royal Irish Academy*, **3**, 370–371.

Mansinha, L., and Smylie, D. E. 1967. Effect of earthquakes on the Chandler wobble and the secular polar shift. *J. Geophys. Res.*, **72**, 4731–4743.

Maron, M. J., and Lopez, R. J. 1991. *Numerical Analysis: A Practical Approach*. Wadsworth Publishing Co., Belmont, CA.

Mathews, P. M., Herring, T. A., and Buffet, B. A. 2002. Modeling of nutation-precession: new nutation series for nonrigid Earth and insights into the Earth's interior. *J. Geophys. Res.*, **107**, ETG3–1 to ETG3–30.

Middleton, D. 1960. *An Introduction to Statistical Communication Theory*. McGraw-Hill, New York.

Mindlin, R. D. 1936. Force at a point in the interior of a semi-infinite solid. *Physics*, **7**, 195–202.

Mindlin, R. D., and Cheng, D. H. 1950. Nuclei of strain in the semi-infinite solid. *Journal of Applied Physics*, **21**, 926–930.

Mohr, P. J., Taylor, B. N., and Newell, D. B. 2008. CODATA Recommended values of the fundamental physical constants: 2006. *Reviews of Modern Physics*, **80**, 633–691.

Mueller, I. I. 1969. *Spherical and Practical Astronomy as Applied to Geodesy*. Frederick Ungar Publishing, New York.

Munk, W. H., and MacDonald, G. J. F. 1960. *The Rotation of the Earth*. Cambridge University Press.

Neuber, H. 1934. Ein neurer Censatz zur losing raumlicher Probleme der Elastezetatstheorie. *Z. angew. Math. Mech.*, **14**, 203.

O'Keefe, J. A., Hertz, H. G., and Marchant, M. 1958. Oblateness of the Earth by artificial satellite. *Harvard College Observatory Announcement Card, June 24*, **1408**, 1.

Palmer, A. 2005. *Free Core Nutations of the Earth*. Ph.D. thesis, York University, Toronto.

Palmer, A., and Smylie, D. E. 2005. VLBI Observations of free core nutations and viscosity at the top of the core. *Phys. Earth Planet. Inter.*, **148**, 285–301.

Papkovich, P. F. 1932. An expression for a general integral of the equations of the theory of elasticity in terms of harmonic functions. *Izvestiya Akademii Nauk SSSR, Physics-Mathematics Series*, **10**, 1425–1435.

Pekeris, C. L., and Accad, Y. 1972. Dynamics of the liquid core of the Earth. *Phil. Trans. R. Soc. Lond. A*, **273**, 237–260.

Poincaré, H. 1885. Sur l'équilibre d'une masse fluide animée d'un mouvement de rotation. *Acta Math.*, **7**, 259–380.

Poincaré, H. 1910. Sur la précession des corps déformables. *Bulletin Astronomique*, **27**, 321–356.

Poirier, J. P. 1988. Transport properties of liquid metals and viscosity of Earth's core. *Geophys. J. Int.*, **92**, 99–105.

Press, F. 1965. Displacements, strains, and tilts at teleseismic distances. *J. Geophys. Res.*, **70**, 2395–2412.

Press, W. H., Teukolsky, S. A., Vettering, W. T., and Flannery, B. P. 1992. *Numerical Recipes in Fortran, 2nd ed.* Cambridge University Press.

Ralston, A., and Rabinowitz, P. 1978. *A First Course in Numerical Analysis, 2nd ed.* McGraw-Hill Book Co.

Rapp, R. H. 1974. Current estimates of mean Earth ellipsoid parameters. *Geophys. Res. Lett.*, **1**, 35–38.

Rochester, M. G. 1956. *The Application of Dislocation Theory to Fracture of the Earth's Crust*. M.Phil. thesis, University of Toronto.

Rochester, M. G., and Smylie, D. E. 1974. On changes in the trace of the Earth's inertia tensor. *J. Geophys. Res.*, **79**, 4948–4951.

Rochester, M. G., Jensen, O. G., and Smylie, D. E. 1974. A search for the Earth's 'nearly diurnal free wobble'. *Geophys. J. R. Astr. Soc.*, **38**, 349–363.

Runge, C. 1895. Uber die numerische Auflosung von Differentialgleichungen. *Math. Ann.*, **46**, 167–178.

Shannon, C. E. 1948. A mathematical theory of communication. *Bell Sys. Tech. J.*, **27**, 379–423, 623–656.

Smylie, D. E. 1999. Viscosity near Earth's solid inner core. *Science*, **284**, 461–463.

Smylie, D. E., and Jiang, X. 1993. Core oscillations and their detection in superconducting gravimeter records. *Journal of Geomagnetism and Geoelectricity*, **45**, 1347–1369.

Smylie, D. E., and Mansinha, L. 1967. Effect of earthquakes on the Chandler wobble and the secular polar shift. *J. Geophys. Res.*, **72**, 4731–4743.

Smylie, D. E., and Mansinha, L. 1971. The elasticity theory of dislocations in real Earth models and changes in the rotation of the Earth. *Geophys. J. R. Astr. Soc.*, **23**, 329–354.

Smylie, D. E., and McMillan, D. G. 2000. The inner core as a dynamic viscometer. *Phys. Earth Planet. Inter.*, **117**, 71–79.

Smylie, D. E., and Palmer, A. 2007. Viscosity of Earth's outer core. *Cornell University Library, arXiv:0709.3333v1 [physics.geo-ph]*, 1–30.

Smylie, D. E., and Rochester, M. G. 1981. Compressibility, core dynamics and the subseismic wave equation. *Phys. Earth Planet. Inter.*, **24**, 308–319.

Smylie, D. E., and Rochester, M. G. 1986. Long period core dynamics. In: Cazenave, A. (ed), *Earth Rotation: Solved and Unsolved Problems*. Reidel, Amsterdam.

Smylie, D. E., and Zuberi, M. 2009. Free and forced polar motion and modern observations of the Chandler wobble. *Journal of Geodynamics*, **48**, 226–229.

Smylie, D. E., Clarke, G. K. C., and Ulrych, T. J. 1973. Analysis of irregularities in the Earth's rotation. *Methods in Computational Physics*, **13**, 391–430.

Smylie, D. E., Szeto, A. M. K., and Rochester, M. G. 1984. The dynamics of the Earth's inner and outer cores. *Rep. Prog. Phys.*, **47**, 855–906.

Smylie, D. E., Jiang, X., Brennan, B. J., and Sato, K. 1992. Numerical calculation of modes of oscillation of the Earth's core. *Geophys. J. Int.*, **108**, 465–490.

Smylie, D. E., Hinderer, J., Richter, B., and Ducarme, B. 1993. The product spectra of gravity and barometric pressure in Europe. *Phys. Earth Planet. Inter.*, **80**, 135–157.

Smylie, D. E., Brazhkin, V. V., and Palmer, A. 2009. Direct observations of the viscosity of Earth's outer core and extrapolation of measurements of the viscosity of liquid iron. *Physics-Uspekhi*, **52**, 79–92.

Sokolnikoff, I. S. 1956. *Mathematical Theory of Elasticity, 2nd ed.* McGraw-Hill, New York.

Sommerfeld, A. 1964. *Partial Differential Equations in Physics*. Academic Press, New York.

Sorenson, H. V., Burris, C. S., and Heidman, M. T. 1995. *Fast Fourier Transform Database*. PWS Publishing, Boston. (Update of Technical Report TR-8402).

Stacey, F. D. 1992. *Physics of the Earth, 3rd ed.* Brookfield Press, Brisbane.

Steketee, J. A. 1958a. On Volterra's dislocations in a semi-infinite elastic medium. *Canadian Journal of Physics*, **36**, 192–205.

Steketee, J. A. 1958b. Some geophysical applications of the elasticity theory of dislocations. *Canadian Journal of Physics*, **36**, 1168–1198.

Stewart, G. W. 2001. *Matrix Algorithms II: Eigensystems*. Society for Industrial and Applied Mathematics, Philadelphia.

Szegö, G. 1920. Beiträge zur theorie der toeplitzschen formen. *Math. Zeitschr.*, **6**, 167–190.

Szeto, A. M. K., and Smylie, D. E. 1984a. Coupled motions of the inner core and possible geomagnetic implications. *Phys. Earth Planet. Inter.*, **36**, 27–42.

Szeto, A. M. K., and Smylie, D. E. 1984b. The rotation of the Earth's inner core. *Phil. Trans. R. Soc. Lond. A*, **313**, 171–184.

Volterra, V. 1903. Sur la stratification d'une masse fluide en équilibre. *Acta Math.*, **27**, 105–124.

Von Zeipel, H. 1924. The radiative equilibrium of a rotating system of gaseous masses. *Mon. Not. R. Astr. Soc.*, **84**, 665, 684, 702.

Wahr, J. M. 1981. The forced nutations of an elliptical, rotating, elastic, and oceanless Earth. *Geophys. J. R. Astr. Soc.*, **64**, 705–727.

Wavre, R. 1932. *Figures Planétaires et Géodésie*. Gauthier-Villars, Paris.

Welch, P. D. 1967. The use of the fast Fourier transform for estimation of spectra: a method based on time averaging over short, modified periodograms. *IEEE Trans. Audio Electroacoust.*, **AU-15**, 70–74.

Widmer, R., Masters, G., and Gilbert, F. 1988 (June). The spherical Earth revisited. In: *17th Inter. Conf. on Math. Geophys.* IUGG, Blanes, Spain.

Widom, H. 1965. Toeplitz matrices. In: Hirschman, I. I. Jr. (ed), *Studies in Real and Complex Analysis*. Mathematical Association of America.

Wilkinson, J. H. 1965. *The Algebraic Eigenvalue Problem*. Oxford University Press.

Wilkinson, J. H. 1968. Global convergence of tridiagonal QR algorithm with origin shifts. *Linear Algebra and Its Applications*, **1**, 409–420.

Wold, H. O. 1938. *A Study in the Analysis of Stationary Time Series, 2nd ed.* Almqvist & Wiksell, Uppsala.

Wooding, R. A. 1956. The multivariate distribution of complex normal variables. *Biometrika*, **43**, 212–215.

Fortran index

Subject index

Printed in the United States
By Bookmasters